W9-AQB-564

Plant Response to Air Pollution

Dedicated to

Professor A.K.M. Ghouse
Our Beloved Teacher

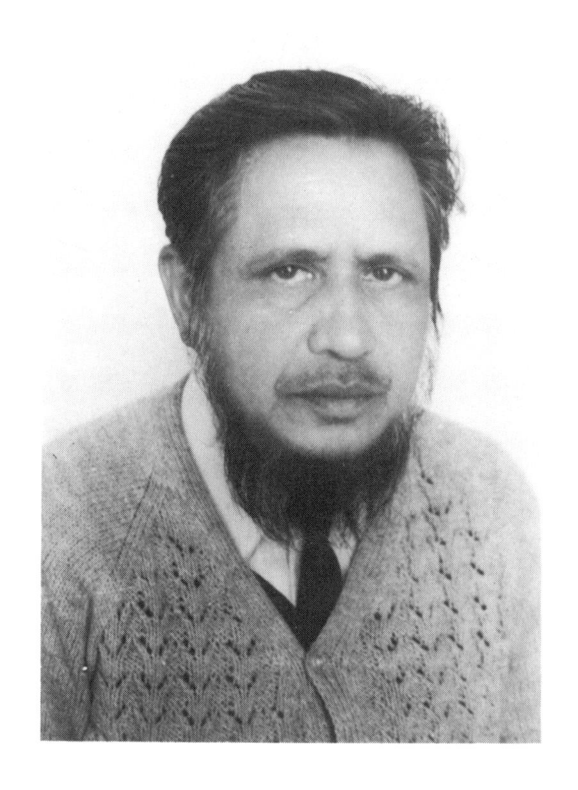

Plant Response to Air Pollution

Edited by

MOHAMMAD YUNUS

National Botanical Research Institute, Lucknow, India

MUHAMMAD IQBAL

Hamdard University, New Delhi, India

JOHN WILEY & SONS

Chichester • New York • Brisbane • Toronto • Singapore

Baffins Lane, Chichester,
West Sussex PO19 1UD, England

National 01243 779777
International (+44) 1243 779777

Reprinted June 1997 and December 1997

Other Wiley Editorial Offices

John Wiley & Sons, Inc., 605 Third Avenue,
New York, NY 10158-0012, USA

Jacaranda Wiley Ltd, 33 Park Road, Milton,
Queensland 4064, Australia

John Wiley & Sons (Canada) Ltd, 22 Worcester Road,
Rexdale, Ontario M9W 1L1, Canada

John Wiley & Sons (Asia) Pte Ltd, 2 Clementi Loop #02-01,
Jin Xing Distripark, Singapore 129809

Library of Congress Cataloging-in-Publication Data

Plant response to air pollution / edited by Mohammad Yunus and Mohammad Iqbal.
p. cm.
Includes bibliographical references and index.
ISBN 0-471-96061-6
1. Plants, Effect of air pollution on. 2. Crops — Effect of air pollution on. 3. Air — Pollution — Environmental aspects. I. Yunus, Mohammad. II. Iqbal, Mohammad.
QK751.P57 1996
581.5′222 — dc20 95-42780
CIP

British Library Cataloguing in Publication Data

A catalogue record for this book is available from the British Library

ISBN 0-471-96061-6

Typeset in 10/12pt Times by Laser Words, Madras, India
Printed and bound in Great Britain by Bookcraft (Bath) Ltd
This book is printed on acid-free paper responsibly manufactured from sustainable forestation, for which at least two trees are planted for each one used for paper production.

Contents

Contributors

Abdin, M.Z., Department of Botany, Faculty of Science, Hamdard University, New Delhi 110 062, India

Agrawal, M., Ecology Research Laboratory, Department of Botany, Banaras Hindu University, Varanasi 221 005, India

Atkinson, C.J., Horticulture Research International, East Malling, West Malling, Kent ME19 6BJ, UK

Barnes, J.D., Department of Agricultural and Environmental Sciences, Ridley Building, The University, Newcastle upon Tyne, NE1 7RU, UK

Davison, A.W., Department of Agricultural and Environmental Sciences, Ridley Building, The University, Newcastle upon Tyne, NE1 7RU, UK

Ferris, R., School of Biological Sciences, University of Sussex, Falmer, Brighton, Sussex BN1 9QG, UK

Freer-Smith, P.H., The Forestry Authority, Research Division, Alice Holt Lodge, Wrecclesham, Farnham, Surrey GU10 4LH, UK

Heath, R.L., Department of Botany and Plant Sciences, University of California, Riverside, CA 92521-0124, USA

Herschbach, C., Albert-Ludwigs Universität Freiburg, Institut für Forstbotanik & Baumphysiologie, AM Flughafen 17, D-79085 Freiburg i.Br., Germany

Hull, M.R., Department of Agricultural and Environmental Sciences, Ridley Building, The University, Newcastle upon Tyne, NE1 7RU, UK

Iqbal, M., Department of Botany, Faculty of Science, Hamdard University, New Delhi 110 062, India

Kerstiens, G., Institute of Environmental and Biological Sciences, Lancaster University, Lancaster, LA1 4YQ, UK

Krupa, S.V., Department of Plant Pathology, 495 Borlaugh Hall, 1991 Buford Circle, University of Minnesota, St Paul, MN 55108, USA

Kucera, L.J., D-WAHO Holzwissenschaften, ETH-Zentrum, CH-8092 Zurich, Switzerland

Lea, P.J., Institute of Environmental and Biological Sciences, Lancaster University, Lancaster, LA1 4YQ, UK

Mahmooduzzafar, Department of Botany, Faculty of Science, Hamdard University, New Delhi 110 062, India

Mansfield, T.A., Institute of Environmental and Biological Sciences, Lancaster University, Lancaster, LA1 4YQ, UK

Mudd, J.B., Statewide Air Pollution Research Centre & Department of Botany and Plant Sciences, University of California, Riverside, CA 92521 USA

Oren, R., School of the Environment, Duke University, Box 90328, Durham, NC 27708-0328, USA

Ormrod, D.P., Department of Horticultural Science, Faculty of Graduate Studies, University of Guelph, Guelph, Ontario, Canada N1G 2WI

Pearson, M., Institute of Environmental and Biological Sciences, Lancaster University, Lancaster LA1 4YQ, UK

Požgaj, A., Faculty of Wood Technology, Technical University of Zvolen, 96053 Zvolen, Slovak Republic

Rennenberg, H., Albert-Ludwigs Universität Freiburg, Institut für Forstbotanik & Baumphysiologie, AM Flughafen 17, D-79085 Freiburg i.Br., Germany

Rowland-Bamford, A.J., Institute of Environmental and Biological Sciences, Lancaster University, Lancaster, LA1 4YQ, UK

Sane, P.V., Plant Molecular Biology Lab., National Botanical Research Institute, Lucknow 226 001, India

Saxe, H., Institute of Botany, Dendrology and Forest Genetics, The Royal Veterinary and Agricultural University, Kirkegårdsvej 3A, DK-2970, Hørsholm, Denmark

Singh, N., Environmental Research Cell, National Botanical Research Institute, Lucknow 226 001, India

Taylor, G., School of Biological Sciences, University of Sussex, Falmer, Brighton, Sussex BN1 9QG, UK

Tripathi, R.D., Environmental Research Cell, National Botanical Research Institute, Lucknow 226 001, India

Wolfenden, J., Institute of Environmental and Biological Sciences, Lancaster University, Lancaster, LA1 4YQ, UK

Yunus, M., Environmental Research Cell, National Botanical Research Institute, Lucknow 226 001, India

Zobel, A.M., Department of Chemistry, Trent University, Peterborough, Ontario, Canada K9J 7B8

Foreword

Air pollution poses a serious threat to human health and the environment worldwide. It contributes significantly to regional and global atmospheric issues such as global warming, acidification and depletion of the ozone layer. It affects all living things, including all kinds of vegetation on which we depend for our survival.

This book is a laudable effort to bring together in one volume the enormous body of knowledge we have regarding the impact of air pollution on plants. I am conscious of the fact that by the time this volume is published new knowledge will have qualified many of the interpretations contained in this volume. But the central thesis of this volume will stand. It is that we must judge pollution in its context. It is also important to determine at frequent intervals whether a pollutant released in an environment could have far-ranging effects in the biosphere. Perhaps the most important advice is that we must probe all the interconnections in order to expose any weak link in the ecological chain.

I am confident that this volume will advance scientific research, promote global awareness on the severity of the problem and also stimulate corrective action. This book should be of interest to the scientific community as well as all those who have a stake in conserving our environment.

Elizabeth Dowdeswell
United Nations Under Secretary-General
and Executive Director of UNEP

Preface

The importance of the study of environmental hazards and their impact on living beings needs no emphasis. The effect of environmental pollution on vegetation is astounding. As a sequel to the current popular trend of plant biology research synchronized with studies of phenomena in relation to environmental stress, an enormous body of knowledge about the *plant–pollution relationship* has been generated. This knowledge has been manifested throughout the scientific literature — in seminars and symposia, public press and the specialized books.

Although several excellent works have appeared on air pollution biology, fewer are available to provide the broader background that encompasses the whole gamut of plant responses to atmospheric insult. Most of the woks hitherto produced have brought out the effect of certain individual pollutants or pollutant mixtures on plant survival and performance. This multi-author work, *Plant Response to Air Pollution,* aims at integrating the varied plant growth responses and knitting a more meaningful story by elucidating the structural, functional and biochemical responses in addition to the whole-plant or plant community response to the pollution stress; the focus of attention is the plant rather than the pollutant. This, we believe, portrays a clearer picture of plant performance versus air pollution, and helps develop a better insight into the pollution-based disturbances that the plant experiences at different stages of the life cycle. The book contains 20 chapters authored by a select group of the eminent scientists who are known globally for their contributions. The chapters have been arranged so as first to portraya broad picture of atmospheric pollution and the factors affecting it, then to address the uptake and metabolism of the pollutants, and later the cellular resistance, visible injury/damage and the whole-plant or community decline. Effort has been made not only to present the state-of-the-art on the subject in question, but also to identify frontline areas for future research. The book should interest not only the students and researchers of environmental botany, ecology and forestry, but all those who love plants and have any interest in global vegetation and environmental health.

We are indebted to our contributors who despite their various commitments accepted readily the additional burden of writing the chapters. Besides, we thank M.S. Alam and M. Athar (Hamdard University, New Delhi, India), V.J. Black (Loughborough University of Technology, UK), A.D. Bradshaw (University of Liverpool, UK), J.V.H. Constable (University of Nevada, Reno, USA), P.J. Lee, T.A. Mansfield, N. Paul, A.R. Wellburn and J. Wolfenden (Lancaster University, UK) and P.V. Sane (National Botanical Research Institute, Lucknow, India) who helped in vetting of some of the chapters. We also appreciate Mahmooduzzafar (Hamdard University, New Delhi) and Nandita Singh (NBRI, Lucknow) who placed at our disposal their alert and versatile services for developing this book.

Professor Allauddin Ahmad, Vice-Chancellor, Hamdard University, New

Delhi, and Dr P.V. Sane, Director, National Botanical Research Institute, Lucknow, encouraged publication of this book by moral as well as material support. The electronic version of the text was developed at Hamdard University by Sabah Ashraf, Mueed Ahmad and M. Ziauddin, while Indu Mehra helped with illustrations. The publication plan finally came to fruition through the efforts of Amanda Hewes, Nicky Christopher and Isabelle Strafford of John Wiley & Sons, UK. We sincerely thank all these people.

We take pride in dedicating this work to commemorate the scientific contributions of Professor Abdul Kasim Muhammad Ghouse, the former Chairman of the Botany Department at the Aligarh Muslim University, Aligarh, India, who initiated us into botanical research.

Mohammad Yunus
Muhammad Iqbal

1 Global Status of Air Pollution: An Overview

M. YUNUS, N. SINGH & M. IQBAL

INTRODUCTION

Consideration of air pollution problems on a larger scale requires revision in thinking — from a narrow local context to a broad global environment. The "global" does not necessarily imply that the pollutant in question encompasses the entire world. Rather, similar air pollution problems occur at various locations around the globe, and are not unique to individual locations. The world occurrence of air pollutants has made the problem universal. The approach of thinking globally about air quality can be emphasized by the following interlinks between air pollutants and their host medium, the atmosphere:

- Air pollutants are inseparably associated with changes in the earth's atmosphere.
- The various air pollution problems occurring at a local scale involve many of the pollutants that also cause problems globally and the problems are interconnected.
- Secondary pollutant formation of various species occurs under short and long-range transport processes.

SOURCES OF POLLUTION

Air pollution has been exacerbated by four specific events that typically occur as countries industrialize: expansion of cities, increase in traffic, rapid economic development and higher levels of energy consumption. The number of cities with more than 7 million inhabitants grew from 13 to 35 between 1950 and 1985 (UN, 1989). It is projected that by the year 2000 the population in more than 60 cities will be above 10 million. The majority of these cities will be in developing countries, which are undergoing the most rapid urbanization (Figure 1.1a, b). According to a WHO/UNEP report (1992), if the current trend continues, Mexico City, São Paulo and Tokyo will be the three largest cities in the world, having over 50 million inhabitants each, by the year 2000.

In many developing cities, the growth of both industrial and residential areas is unplanned, unstructured and unzoned, thus exacerbating the air pollution problems. In urban areas the main sources of pollution are power plants, industries, motor vehicles and domestic sources. Combustion of fossil fuel in

Plant Response to Air Pollution. Edited by Mohammad Yunus and Muhammad Iqbal

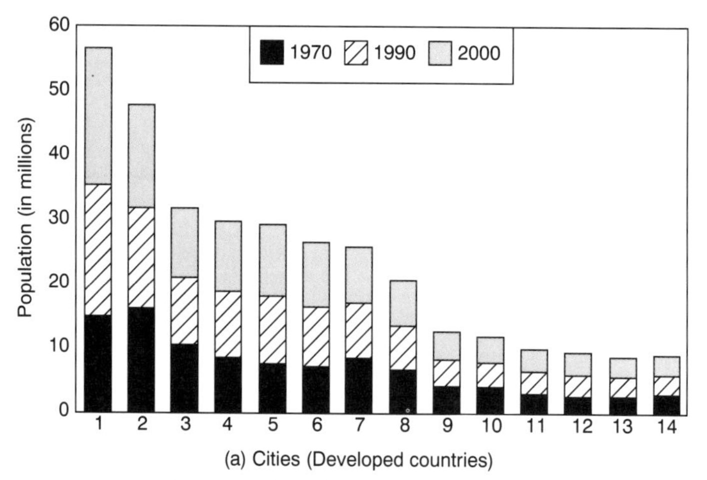

1-Tokyo, 2-New York, 3-London, 4-Los Angeles, 5-Osaka, 6-Moscow, 7-Paris, 8-Chicago, 9-Philadelphia, 10-Detroit, 11-San Francisco, 12-Toronto, 13-Washington DC, 14-Birmingham

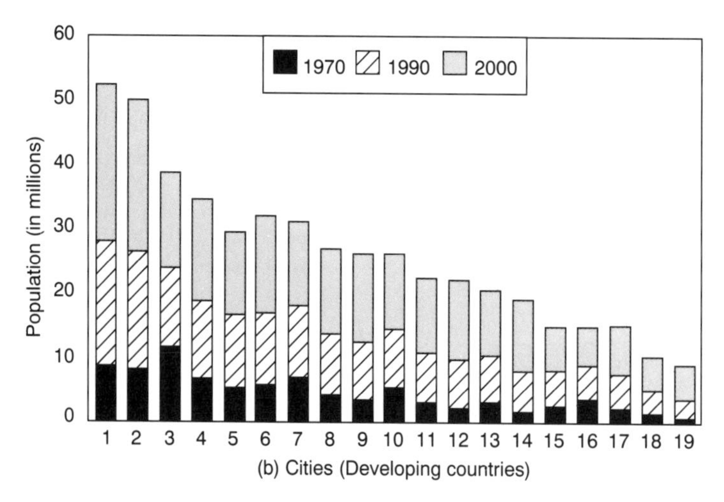

1-Mexico City, 2-São Paulo, 3-Shanghai, 4-Calcutta, 5-Seoul, 6-Bombay, 7-Rio de Janeiro, 8-Jakarta, 9-Tehran, 10-Cairo, 11-Karachi, 12-Lagos, 13-Bangkok, 14-Dhaka, 15-Bogota, 16-Hong Kong, 17-Baghdad, 18-Ankara, 19-Medan

Figure 1.1. Human population in some cities of the developed countries (a), and the developing countries (b)

stationary and mobile sources leads to the production of sulphur dioxide, nitrogen oxides and particulates in the form of fly-ash and soot, and secondary particulates like sulphate (SO_4^{2-}) and nitrate (NO_3^-) aerosols, etc.

Use of motor vehicles is growing fast in the developing countries (Figure 1.2). Since 1979, the greatest increases have been in Asia. Motor vehicle contributions are less important in cities with lower levels of motorization and in cities located where strong control measures have been established, e.g., in temperate regions.

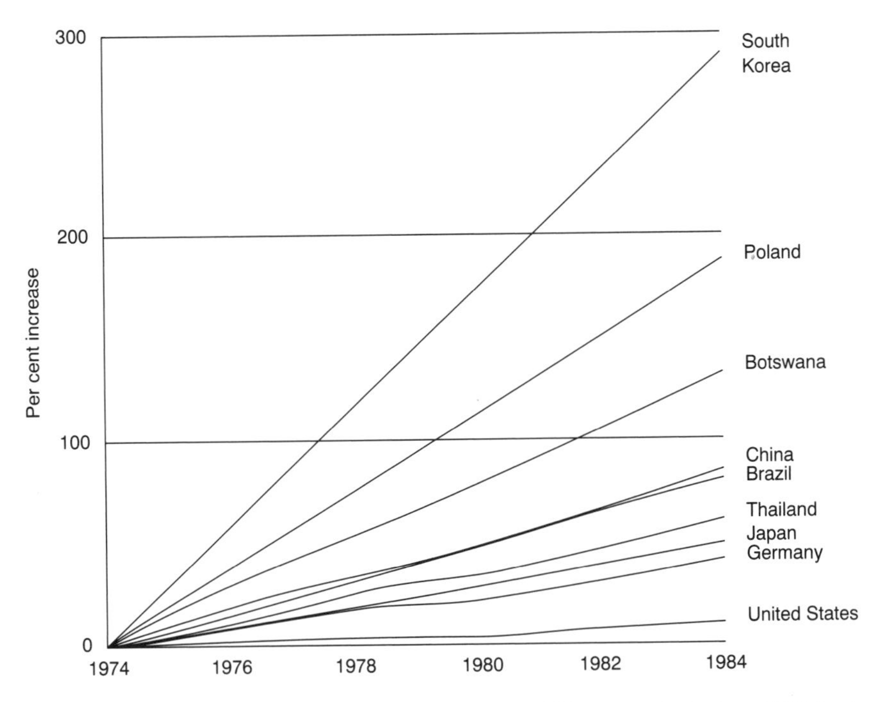

Figure 1.2. Per cent increase in vehicle numbers (per 1000 population) since 1974 (redrawn after UNEP, 1987)

In many developing countries, vehicle fleets tend to be older and poorly maintained, thus increasing the significance of motor vehicles as a pollutant source. The number of vehicles is expected to grow faster in the developing countries than in the developed countries. The share of motor vehicles in the pollution load is thus set to increase in the developing countries, and, in the absence of the introduction of stringent control measures for traffic-related pollutants, air quality will deteriorate further. Petrol-fuelled motor vehicles are the principal source of nitrogen oxides.

There are three major methods by which acid gases such as sulphur dioxide and nitrogen dioxide are eventually transformed into dilute acid concentrations in rain-water. The first, homogeneous gas-phase reactions occur in dry atmospheres associated with photolytic oxidation processes. The second, homogeneous aqueous-phase reactions occur between individual species in a liquid medium such as a cloud or raindrop. The third, heterogeneous aqueous-phase reactions, occur during adsorption on solid surfaces and are extremely complex. The heterogeneous reactions probably assist in creating rain-water acidity but are not considered to be as important as the other two. The relative importance of any process operating in the atmosphere depends strongly on the meteorological conditions, such as the presence of clouds, relative humidity, intensity of solar radiation, temperature, etc.

Suspended particulate matter (SPM) consists of solid and liquid particles emitted from numerous natural and man-made sources. SPM is a complex and variable mixture of different-sized particles with many chemical components. Larger particles may come from wind-blown (industrial) dust, volcanic eruptions, pollen/spores and algal filaments; finer particles tend to be formed by combustion and gas to particle conversions (chiefly from SO_2). The constituents can vary, although in urban areas they typically include carbon particles and polynuclear aromatic hydrocarbons (PAHs) produced by incomplete fuel combustion. The modern-day SPM problems are aggravated by the lack of emission regulations on the increasing number of diesel-powered vehicles being used for freight transport. Exhausts from diesel-powered vehicles are a major source of SPM.

Nature keeps carbon dioxide (CO_2) in balance. Human activities are upsetting this balance by (a) releasing extra CO_2 by burning fossil fuel, and (b) destroying the forest and other vegetation, thus reducing the CO_2 sink.

Nitrous oxide emissions result naturally from microbial processes in soil and water. Human activities add to them through burning of biomass and fossil fuels. The annual total emissions of nitrous oxides have been estimated at about 30 million tonnes. Fertilizer application enhances the flux of nitrous oxide to the atmosphere. A study (SCOPE, 1986) estimates the emissions of nitrous oxide from the use of fertilizers at 600–2300 tonnes a year.

Methane is produced by methanogenic bacteria found in anaerobic conditions in natural wetland ecosystems, in waterlogged paddy fields, in the anoxic rumen of cattle, and in the guts of termites and other wood-consuming insects. Methane emissions from wetlands vary depending on soil and air temperature, soil moisture, the amount and composition of organic substrate, and the vegetation.

Chlorofluorocarbons, especially CFC-11 and CFC-12, have been (and still are) emitted from industrial sources to the atmosphere for the past half a century. They are used as propellants in spray bottles, refrigerators, aerosol cans, air-conditioners and the like. On molecule to molecule basis CFCs are 20 000 times more efficient trappers of heat than CO_2. And while CO_2 has its sinks in forests and oceans, CFCs have no natural sinks. Several other halocarbons are also potentially significant both in terms of radiative activity and ability to deplete stratospheric ozone: e.g., CFC-113 ($CFCl_2CF_2Cl$), used as a cleaning solvent in the electronics industry; HCFC-22 (CHF_2Cl), used as a refrigerant in home air-conditioners, and methyl chloroform (CH_3CCl_3), used as an industrial solvent. At the present time, however, the impact of these gases is thought to be less important than that of CFC-11 and CFC-12.

STATUS IN THE DEVELOPED COUNTRIES

The concern about air pollution has triggered national and international action. Programmes have been drawn up to monitor and assess air quality and observe trends. In many developed countries early legislation and monitoring efforts were focused on sulphur dioxide and suspended particulate matter alone. However, since the late 1970s, the networks have typically been expanded to incorporate the routine monitoring of pollutants such as carbon monoxide, nitrogen oxides and lead. During the 1980s, urban air quality monitoring for the traditional air pollutants has also been established in the less developed countries, especially in Asia and South America.

Sulphur dioxide

The annual global emissions of sulphur dioxide (SO_2) currently stand at about 294 million tonnes, of which 160 million tonnes are anthropogenic (UNEP/GEMS, 1991). Total figures are calculated mainly from estimates of the production or consumption of fossil fuels, the sulphur content of fuels, and mean emission factors.

About 90% of these emissions come from the northern hemisphere (the United States and the Soviet Union being the biggest sources). The dispersal area, the amount of acid produced and the polluting effects of SO_2 depend on factors such as weather conditions (wind, cloud cover, humidity and sunlight), the presence of other pollutants, the height at which SO_2 is released and the length of time it stays in the atmosphere.

SO_2 emissions have fallen in most industrialized countries as a result of air pollution control measures (Figure 1.3). In London, total emissions of SO_2 fell from 179 000 tonnes in 1975 to 49 000 tonnes in 1983. Fuel oil is the dominant source of SO_2 followed by coal and gas oil (Munday *et al.*, 1989). There has been

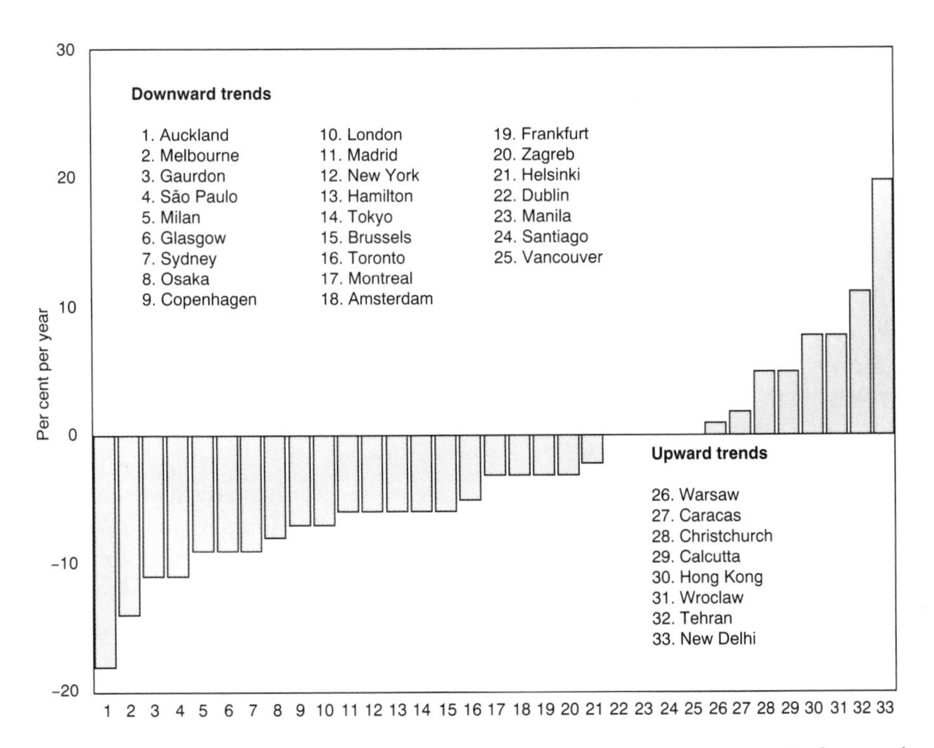

Figure 1.3. Trends in the annual average SO_2 concentration in some cities (redrawn after GEMS/WHO, 1988)

a marked shift from fuel oil and coal, to natural gas. The shut-down of electricity generating stations in the UK has also helped to accelerate reduction in SO_2 levels. In 1975, there were 15 major power-stations operative in London, but by 1988 only three remained. Specific legislation restricting the sulphur content of fuel to below 1% has been in force since 1972. Annual mean concentrations lay around 300–400 $\mu g\ m^{-3}$ until the mid-1960s, when a steady decline began. Currently levels are around 20–30 $\mu g\ m^{-3}$. A similar trend is observed throughout the UK (Laxen & Thompson, 1987; LSS, 1990a).

In Los Angeles, total anthropogenic emissions of oxides of sulphur (SO_x) were approximately 50 000 tonnes in 1987, with mobile sources accounting for 62% (SCAG, 1991). SO_2 emissions from refineries and power-plants have been drastically curtailed by emission regulations and switching over to natural gas.

Sulphur dioxide levels now meet the National Ambient Air Quality Standard (NAAQS) and WHO Annual Mean Guidelines. In 1990, the annual arithmetic mean concentrations ranged from 0.2 to 10 $\mu g\ m^{-3}$, well below the WHO Annual Mean Guidelines. However, the maximum hourly SO_2 as measured at Howthorne was 886 $\mu g\ m^{-3}$, which exceeds the WHO Hourly Guidelines of 350 $\mu g\ m^{-3}$. This was caused by a breakdown at a nearby oil refinery (SCAG, 1991).

In Moscow, total SO_2 emissions were approximately 130 000 tonnes in 1990. An overall decrease was observed between 1987 and 1990 (Rovinsky *et al.*, 1991). The most significant reason for this was the reduction in the use of coal and heavy fuel oil in power generation. The use of coal was reduced by over 70% and of heavy fuel oil by approximately one-third during the period 1987–1990. SO_2 levels linked with the increase in fuel consumption, increase in winter, and the highest SO_2 concentrations are observed in the industrial zones. The mean levels of SO_2 do not generally exceed the WHO guidelines (40–60 $\mu g\ m^{-3}$) or the national standard (50 $\mu g\ m^{-3}$) (Rovinsky *et al.*, 1991).

In New York, SO_2 emissions have decreased significantly over the past 20 years as a result of lowering of the sulphur content of bituminous coal and residual fuel oil (USEPA, 1991), followed by a shift to natural gas and light oils. The annual mean SO_2 concentrations have been within the WHO guideline range of 40–60 $\mu g\ m^{-3}$ since 1985. A rare high pollution episode due to stagnation in January 1989 caused daily SO_2 levels to reach 230 $\mu g\ m^{-3}$ at Green Point Avenue (a maximum of 269 $\mu g\ m^{-3}$ was recorded at the Mabel Dean High School) (NYDEC, 1990). The highest annual arithmetic mean reported by USEPA in 1989 was 60 $\mu g\ m^{-3}$ (USEPA, 1991).

In Tokyo, the SO_2 emissions have also been reduced drastically since 1970. This is because of fuel desulphurization, and the development and wide use of emission gas treatment technologies for mobile and stationary sources. The total SO_2 emission was 835 000 tonnes in 1986. This equals an emission of 6.9 kg SO_2 per person per annum which is one of the lowest per capita rates in the industrialized countries (Komeiji *et al.*, 1990; OECD, 1991).

In Germany (erstwhile FRG), the annual SO_2 emissions were 3.2 million tonnes in 1980. New legislation (Ordinance on Large Combustion Installation), aimed at halving SO_2 emissions by 1993, was passed in 1983. Formerly, East Germany was the world's biggest producer of brown coal, producing more than a quarter of the world's total. In 1983 it used 280 million tonnes of coal. With a water content of just over 50% and carbon content of just under 25%, brown coal has a low thermal and energy value. Larger quantities have to be burnt to generate the same amount of energy. The annual SO_2 emissions are thought to be about 4 million tonnes, of which about 2.5 million tonnes are deposited within the country, and the rest exported to neighbouring countries, mainly Poland, Czechoslovakia and Sweden (UNEP/GEMS, 1991).

Suspended particulate matter

Data for the period 1982–1984 show emissions of about 27 million tonnes of suspended particulate matter (SPM) per year, but the current total is probably closer to 135 million tonnes (UNEP/GEMS, 1991). SPM emissions has fallen in most industrialized countries between 1970 and 1990 — particularly in recent years, due to massive and costly air pollution reduction programmes (Figure 1.4).

In London, enforcement of the Clean Air Acts (1956, 1964 and 1968) has resulted in dramatic reduction in SPM. The reduction in solid fuel use, together with the introduction of smokeless fuels, caused a steady decline in emissions throughout the 1970s and 1980s. In 1983 total smoke emissions were 11 000 tonnes, compared with 14 000 tonnes in 1978 (WHO/UNEP, 1992). Transport accounts for 76% of the total emission. Smoke emissions from diesel-engined vehicles were estimated to be 66% of the total smoke emissions in 1984. The

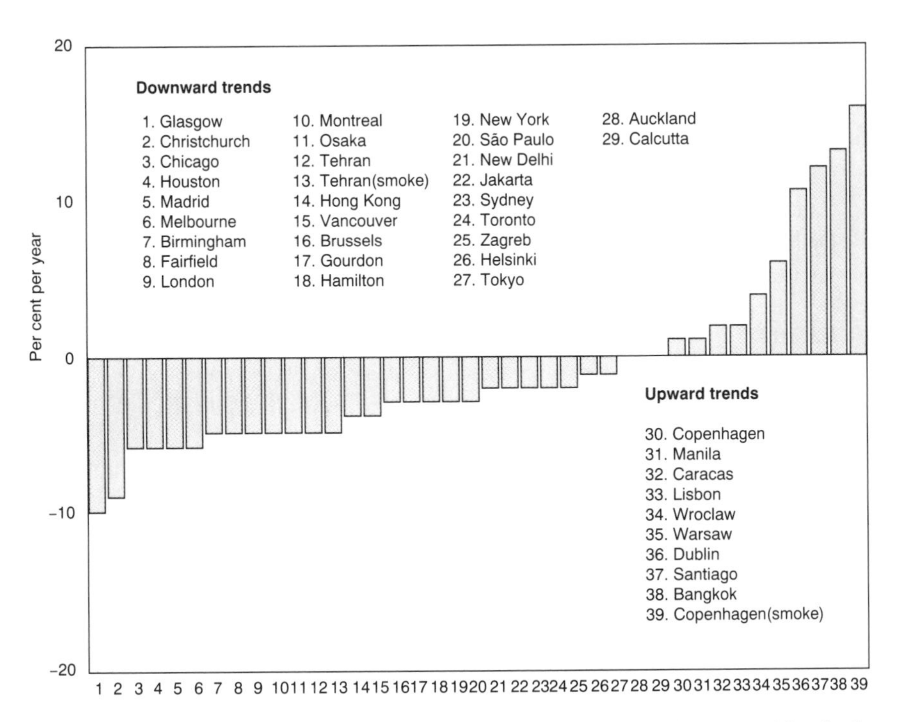

Figure 1.4. Trends in the annual average particulate concentration in some cities (redrawn after GEMS/WHO, 1988)

annual mean SPM concentrations have now come below 30 μg m^{-3}. In 1988, London Scientific Services (LSS) presented results of a baseline metal-in-dust sampling survey in Greater London. Industrial areas (e.g. Lea Valley) and areas with high traffic densities (e.g. Central London) had the highest concentrations of cadmium, copper, iron, lead and zinc (Schwar *et al.*, 1988).

In Los Angeles, total suspended particulates (TSP) continue to exceed WHO guidelines, the greatest proportion coming from anthropogenic sources. Total SPM emissions in 1987 laid around 400 000 tonnes. Of this, 5% was from vehicular traffic, 3% from industry and 91% from other miscellaneous sources mostly related to commercial/domestic activities (e.g., paved road dust stirred by vehicle movement). Los Angeles has the highest annual mean in the US (94 μg m^{-3}). The annual arithmetic mean TSP concentrations in 1990 ranged from 46 μg m^{-3} at Crestline to 115 μg m^{-3} at Fontana; the daily maximum (1770 μg m^{-3}) also occurred at Fontana (SCAG, 1991).

Table 1.1. Lead content of petrol in different cities

City	Lead content of petrol (g l^{-1})	Remarks
Bangkok	0.84 (1980)	
	0.45 (1984)	
	0.40 (1989)	
	0.15 (1993)	
Beijing	0.4–0.8	80% unleaded petrol
Bombay	0.80 (1986)	Premium petrol
	0.56 (1986)	Regular petrol
	0.15 (1991)	
Buenos Aires	0.6–1.0	
Cairo	0.8	
Calcutta	0.1	
Delhi	1.8	
Jakarta	0.60 (1987)	Premium petrol
	0.73 (1987)	Super petrol
Karachi	1.5–2.0	
London	0.15	>33% unleaded petrol
Los Angeles	0.026	>95% unleaded petrol
Manila	1.16	
Mexico City	0.54	
Moscow	0.0	No leaded fuel sales in city
New York	0.026	>95% unleaded petrol
Rio de Janeiro	0.45	Ethanol and gasohol used
São Paulo	0.45	40% ethanol, 60% petrol/gasohol
Seoul	0.15	
Shanghai	0.40	
Tokyo	0.15	>95% unleaded petrol

In Moscow, concentrations of SPM for the period 1980–1988 were rather stable with mean annual concentrations of 150–250 μg m^{-3} (MCGNPS, 1991). A considerable decrease was noted between 1988 and 1990 (Rovinsky *et al.*, 1991). Reductions in the use of liquid and solid fuel by power-stations probably account for the decrease in SPM in recent years. Seasonal variation of SPM concentration is not explicit, but a slight increase is observed in April–July and October–December (MCGNPS, 1991).

In New York, total estimated emissions of particulate matter <10 μm (PM_{10}) were approximately 55 000–290 000 tonnes in 1985 (USEPA, 1991). Over a 25-year period (1965–1990) a striking reduction in annual SPM occurred via enforcement of Federal and New York State particulate emission regulations. These include laws concerning incinerators, industrial processes, fossil fuel combustion and diesel-fuelled motor vehicles. The minimization of allowable fuel sulphur content also caused a shift to cleaner, less particulate-emitting fuels such as natural gas. SPM decreased slightly from 1975 to 1989, all well below the upper WHO guideline of 90 μg m^{-3}.

In Tokyo, SPM levels decreased steadily from 1975 to 1989. Concentrations that were about 55–75 μg m^{-3} between 1975 and 1977 came below 60 μg m^{-3} at all monitoring sites in 1989. High short-term pollution levels varied from 131 to 148 μg m^{-3} in the various stations. This is well below the WHO daily guideline of 230 μg m^{-3} (Government of Japan, 1990).

Lead

Airborne lead (Pb) is closely associated with the density of motor-vehicle traffic using leaded fuel and the concentration of Pb additives in the fuel (Table 1.1). Concentrations range from 0 to 2 g l^{-1}. In the developed countries, which are moving towards totally lead-free petrol, Pb levels are low and well within the WHO guidelines (Figure 1.5).

Airborne Pb has been a major political and scientific issue in the UK for over 20 years. Reduction in permissible Pb levels in petrol from 0.4 to 0.15 g l^{-1} came into force in 1986. It is estimated that Pb emissions are of the order of 350–700 tonnes per annum. Continuous monitoring between 1986 and 1989 showed that ambient Pb levels have decreased due to increased availability and use of unleaded petrol (LSS, 1990a). Annual concentrations are now well below WHO guidelines at background sites. Airborne Pb levels are likely to fall further as the proportion of vehicles using unleaded petrol increases.

In Los Angeles, petrol is primarily Pb-free, and the little leaded petrol sold contains only 0.026 g l^{-1} Pb. The percentage of days exceeding levels of standard was 100 during 1975–1977, but this fell to 50% in 1979–1981, and to zero in 1984–1986 (SCAG, 1991). However, several violations of the National Ambient Air Quality Standards occurred in 1991, in the vicinity of secondary Pb smelters, leading to a call for a new Pb control plan.

In Moscow, prohibition on sale of leaded petrol and a reduction in motor-vehicle traffic has helped to reduce Pb levels since the late 1980s (Rovinsky

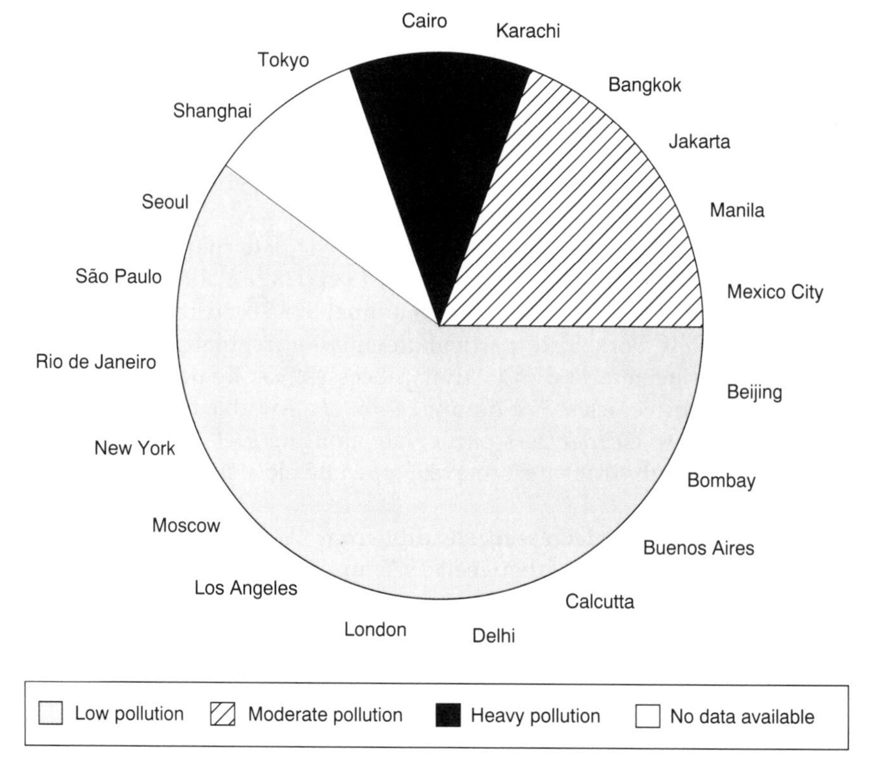

Figure 1.5. Level of lead in some cities. Low pollution = WHO guidelines are normally met; Moderate pollution = WHO guidelines exceeded by up to a factor of two; Heavy pollution = WHO guidelines exceeded by over a factor of two (redrawn after WHO/UNEP, 1992)

et al., 1991). The Pb content in air during the period 1985–1990 varied from 0.01 to 0.04 μg m^{-3}. The levels are well below the WHO guidelines (1 μg m^{-3}).

In Japan, the maximum level of Pb additive in petrol is 0.15 g l^{-1}. Pb emissions were reduced when unleaded petrol was introduced.

Carbon monoxide

Road traffic is by far the largest source of man-made carbon monoxide (CO) emissions, accounting for virtually all the emissions in some urban areas. Daily CO concentrations in urban area rise and fall with traffic density and changes in weather conditions, e.g., levels are maximal during morning and evening rush-hours, the highest concentrations often occur in confined spaces such as tunnels, garages, loading bays and underpasses. Temperature inversions may also cause a local build-up of high concentrations at very low altitudes.

In the United States, the number of vehicles increased four-fold between 1940 and 1976, while annual CO emissions rose from 73 to over 100 million tonnes (UNEP/GEMS, 1991). Between 1977 and 1985, the number of vehicles grew by 25%, but CO emissions fell from 80 million to 60 million tonnes annually due to control measures. The United States and Japan pioneered the first effective control strategy in the late 1960s by improving engine design and reducing exhaust emissions by developing a method of converting most of the CO content of exhaust into carbon dioxide. More stringent controls have since been introduced in Canada and several European countries including Austria, Sweden and Switzerland. In the United States, emissions from vehicles fell by 33% between 1980 and 1989. In the Netherlands and West Germany, emissions fell by 32 and 37%, respectively, between 1975 and 1985. In Britain, Ireland and several East European countries, however, emissions grew over the same period by 5.2% because of increased traffic congestion (UNEP/GEMS, 1991).

Carbon monoxide concentrations touch peaks in winter during high pollution episodes. These episodes tend to correspond with ground-based temperature inversions and low wind speed which trap vehicle emissions close to the ground.

Acid deposition

Because fossil fuel emissions of one country are often converted into acid deposition in another, acid pollution has become a major international issue. Scandinavian governments have been critical of Britain's unwillingness to curb its pollutive emissions. In March 1984, Canadian government officials said that US inaction on acid pollution was one of the biggest irritants in US-Canadian relations. However some international cooperation in 1984 the Ottawa conference on acidification resulted in the formation of the "30% Club". The member countries pledged to reduce their national emissions of acid-forming SO_2 by at least 30% in 10 years.

The northeast USA, in and downwind of the industrial heartland of the Ohio River basin, suffers the worst damage from acid rain. Ten states, viz. Ohio, Pennsylvania, Indiana, Illinois, Missouri, Wisconsin, Kentucky, Florida, West Virginia and Tennessee, have rainfall with a pH of 4.0 or 4.5 and individual rainstorms with pHs below 3.0. The National Academy of Sciences estimates that damage worth 5 billion US$ or more is inflicted every year in the Eastern United States alone, where thin or sandy soils and limited buffering capacity make the soils susceptible to damage. In Pennsylvania, 36 out of 40 high mountain streams surveyed have lost much of their buffering capacity since 1962.

By 1986, Canada had a century of acid pollution. Canada today is the major source and a major victim of acid rain. As reported by the Ontario environment department, 140 acidified lakes in the province were devoid of fish. The current estimates suggest that acid pollution causes damage worth $180 to $250 million every year, and puts at risk agriculture, fishing and tourism over large areas of Canada.

Japan, by reducing SO_2 and NO_x emissions, has almost eliminated the damage due to acid pollution. However, China is now exporting acid emission to Japan in high quantities. On the whole, since the developed countries are cutting back sulphur and nitrogen oxide emissions to acceptable levels, acid deposition is consequently reduced to a great extent.

Greenhouse gases

Of all the global air pollution controversies that have captured the imagination of the public and the media, none has received as much attention as the global "greenhouse" and the potential world-wide warming associated with anthropogenic emissions of greenhouse gases. The concept that human beings could inadvertently alter global climate through waste emissions from the burning of fossil fuels has been acknowledged by atmospheric scientists for several decades (Budyko, 1977; Ramanathan, 1988).

Sunlight heats up sea, land and vegetation. The warmed surface of the earth then radiates heat, back towards space. But this outward flow of radiation is at much longer wavelengths than the incident sunlight, in the infrared part of the spectrum. On its way out, some of this infrared radiation is absorbed by trace gases.

The natural greenhouse effect of CO_2 on the climate has been known for more than a century. Human activities are artificially increasing the amount of CO_2 and other greenhouse gases in the atmosphere, disturbing their natural geochemical cycles. It is now generally accepted that increasing greenhouse gases in the atmosphere will make the earth warmer. The rate of this global warming, its regional distribution and its impacts, however, remain a subject of debate.

Carbon dioxide

Nature maintains the carbon dioxide (CO_2) balance. Natural processes produce vast amounts of the gas; terrestrial life emits, simply by breathing, about 100 000 million tonnes of CO_2 yearly while decomposing vegetation releases another 2000–5000 million tonnes. But photosynthesis takes up almost the same amount as these processes release. Human activities are upsetting the balance in two ways. They release extra CO_2 by burning fossil fuel on the one hand and by destroying forests and other vegetation on the other. Natural processes maintain the carbon balance through the atmosphere, oceans and biosphere in what is known as the geochemical carbon cycle, whereas human activities disturb the cycle by injecting CO_2 into the atmosphere.

Level of CO_2 in the atmosphere is thought to have been about 270 parts per million (ppm) during the first half of the 19th century, i.e., before industrialization. Scientists are able to determine past CO_2 levels by analysing the air trapped within ice in glaciers, some of which is very old and can be accurately dated (Boden *et al.*, 1994).

The Mauna Loa, Hawaii, USA, atmospheric CO_2 measurements constitute the longest continuous record of atmospheric CO_2 concentrations available in the world. The Mauna Loa site is considered one of the most favourable locations for measuring undisturbed air because possible local influences of vegetation or human activities on atmospheric CO_2 concentrations are minimal and any influences from volcanic events may be excluded from the records (Keeling & Whorf, 1994). The Mauna Loa record shows a 12% increase in the mean annual concentrations in 32 years, from 315.83 ppm by volume of dry air (ppmv) in 1959 to 353.95 ppmv in 1990. The data from Alert, Northwest Territories, Canada, show an increase in the annual value from 348.4 ppmv in 1986 to 355.5 ppmv in 1990.

Data on atmospheric CO_2 concentrations from Amsterdam Island in the Indian Ocean have shown an increase in the annual value from 341.1 ppmv in 1983 to 350.5 ppmv in 1989 and revealed a seasonal pattern, with the annual drawdown typically occurring in December–January and the annual build-up occurring in

May–June (see Conway *et al.*, 1994). The average peak-to-peak amplitude for Amsterdam Island was 0.84 ppmv (Conway *et al.*, 1988).

From 1985 to 1992, the annual average concentrations of CO_2 at Cape Grim, Tasmania, Australia, calculated on the basis of undried air samples, have risen from 343.5 ppmv to 353.5 ppmv. This represents an annual increase of 1.43 ppmv (Conway *et al.*, 1994).

The annual atmospheric CO_2 concentration at Ragged Point, Barbados, has risen from 351.5 ppmv in 1988 to 356.0 ppmv in 1992. The record exhibits a seasonal cycle with the yearly minimum occurring in September/October and the yearly maximum in May (Conway *et al.*, 1994). In Kotelny Island, the northernmost part of the former USSR, the yearly mean concentration of atmospheric CO_2 has risen from 352.03 ppmv in 1987 to 358.8 ppmv in 1993 (Brounshtein *et al.*, 1994). The CO_2 value has risen from 334.0 ppmv in 1982 to 349.4 ppmv in 1992 at Zugspitze, Germany (Sladkovic *et al.*, 1994).

Oxides of nitrogen

Natural emissions of nitrogen oxides (NO_x) are caused by lightning, forest fires and the microbial activity in soils. Being globally distributed, they create only low background values. A 1980 estimate puts the total natural and man-made emissions of NO_x at 150 million tonnes per year, over half of it from natural sources (UNEP/GEMS, 1991). In the industrial regions of Europe and North America, however, the man-made NO_x outweighs the natural NO_x by 5–10 times.

In Western Europe, about 30–50% of man-made emissions come from motor vehicles, and another 30–40% from power-plants, mainly those fired by coal. In 1989, 40% of the NO_x emissions in the United States came from road transport. In 1985, 64% of the Canadian NO_x emissions were attributable to road transport. In Sweden, about 30–40% of man-made nitrogen compounds come from agriculture and forestry (UNEP/GEMS, 1991).

Despite increases in energy consumption, particularly by vehicles, NO_x emissions in many developed countries have remained steady (Figure 1.6). Japan achieved a 21% decline in emissions between 1974 and 1983; it was the first country to set emission standards for NO_x from stationary sources. By 1986, Japan had installed more than 320 flue gas denitrification units in power-plants and other industrial units (UNEP/GEMS, 1991).

Catalytic converters for road vehicles were first introduced in the late 1970s and are now compulsory on new cars in Australia, Canada, Japan and the United States. As a result, emissions of NO_x from new cars in Japan were cut by 92% between 1972 and 1978. During this period, the USA reduced emissions by 75%. However, these reductions in emissions have been largely offset by increases in the volume of the road traffic. Several European cities, including Amsterdam, Frankfurt and London, have seen increases in NO_x, mainly from vehicles. Few European countries have stringent emissions standards.

Transport, particularly motor vehicles, is the major anthropogenic source of NO_x in London. It accounted for about 75% of total NO_x emission in 1984 in comparison to only 57% in 1975. The industrial/institutional emissions fell from 30 500 tonnes (34%) in 1975 to 12 200 tonnes (15%) in 1983 (Munday *et al.*,

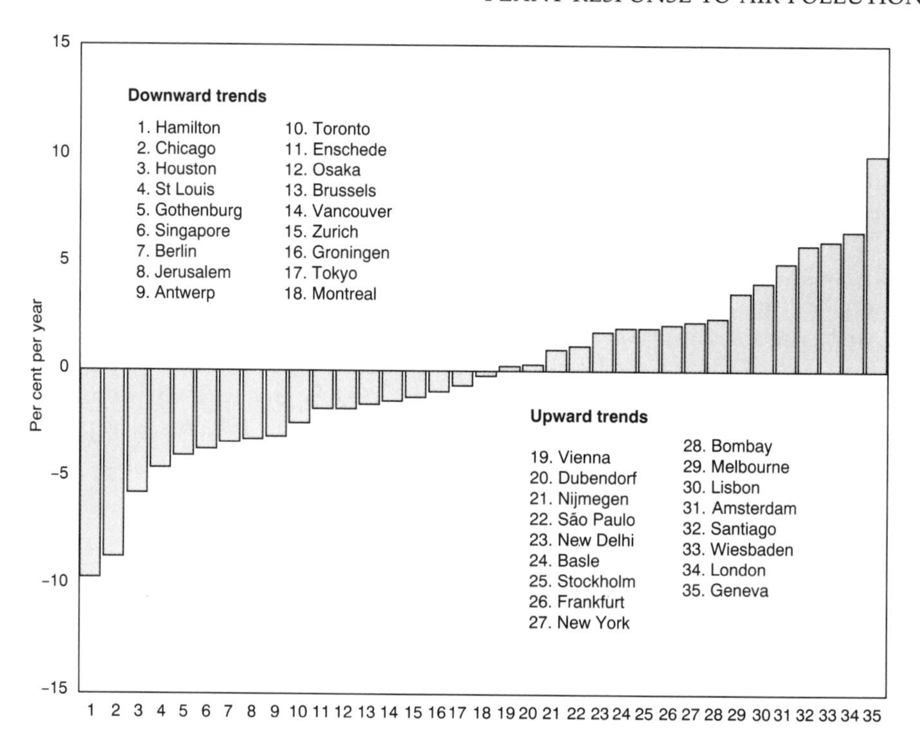

Figure 1.6. Trends in the annual average NO_x concentration in some cities (redrawn after GEMS/WHO, 1988)

1989). London experienced the highest NO_2 concentrations in December 1991, with a maximum hourly average of 867 μg m^{-3} (423 ppmv). During the episode (12–15 Dec.), NO_2 levels were in excess of 205 μg m^{-3} (100 ppmv) for an average of 72 hours and over 600 μg m^{-3} (300 ppmv) for 8 hours. The episode is believed to have been caused by motor-vehicle emissions trapped during a period of cold calm weather (WSL, 1992).

In Los Angeles, total NO_x emissions in 1987 were over 440 000 tonnes per annum, 76% of which was attributable to mobile sources. According to USEPA (1991), Los Angeles is the only city in the United States that has failed to attain the Federal NO_2 NAAQS. The WHO one-hour guideline of 400 μg m^{-3}, corresponding to 0.22 ppm NO_2, was exceeded in 1990 at eight of the 24 stations reporting NO_2 data. The highest reported hourly value was 526 μg m^{-3} (0.28 ppm) (SCAG, 1991).

In Moscow, the major source of oxides of nitrogen is electricity generation which accounts for 70% of the total anthropogenic NO_x emissions of 210 000 tonnes per annum. The contribution of motor vehicles to total anthropogenic NO_x did not exceed 19% in 1990. Daily mean NO_2 concentrations reached 100–150 μg m^{-3}, which is comparable to the air quality guideline recommended by WHO (150 μg m^{-3}) (CVGMO, 1986; MCGNPS, 1990, 1991).

In New York, man-made NO_x emissions in 1985 totalled over 120 000 tonnes per annum in New York City (NYC) and over 513 000 tonnes in the New

York City Metropolitan Area (NYCMA). NO_2 concentrations in the NYCMA have decreased through motor-vehicle emission controls. In 1990, NYC met the US NAAQS of 94 $\mu g\ m^{-3}$ annual arithmetic mean with a value of 87 $\mu g\ m^{-3}$ (WHO/UNEP, 1992).

In Tokyo, the man-made NO_x comes from high-temperature combustion processes and the chemical industry (nitric acid and fertilizer production). Emissions were estimated to be 52 700 tonnes per annum in 1985. Most of these emissions (35 400 tonnes per annum or 67%) come from motor-vehicle exhausts. Of the total NO_x emissions, 18% come from industrial sources, 9.5% from domestic sources and 5% from ships and aircraft (TMG, 1989a). With respect to the NO_x emission from motor-vehicle sources, diesel-powered vehicles (especially cargo vehicles) are of particular importance: although they contribute only about 20% of the total mileage travelled, they contribute more than 50% of all NO_x emissions. Conversely, petrol-powered vehicles that are equipped with exhaust-pipe emission control devices contribute more than 50% of the total mileage travelled, but emit only about 21% of NO_x (TMG, 1989a).

For the whole of Japan, the 0.06 ppm standard is exceeded at about 25% of the automobile exhaust monitoring stations, which are located primarily in the large metropolitan areas such as Kanagawa, Osaka and Tokyo. In Tokyo, during 1988, the 0.04–0.06 ppm daily average national standard was exceeded on several occasions (TMG, 1989b).

Germany has been more active recently than any other European nation, except Switzerland, in promoting regulations on car exhaust gases. NO_x is central to the problem of forest damage and forest death in Germany, so any strategy to solve the problem must include a substantial cut in NO_x. Introduction of catalytic converters has cut the emissions. A reduction of car speed limits from the present "free speed" to a maximum 100 kph (62 mph), would, according to the calculations made by the Berlin Environment Agency (Umweltbundesamt), reduce Germany's NO_x emissions by 185 000 tonnes per year (i.e., 6%). "Transportation optimizing", i.e., the encouragement of rail and water transport, public transport in the large cities, and the use of more vehicles powered by natural gas, is regarded as the key in minimizing NO_x emission from road vehicles.

Methane

Methane (CH_4) concentrations were not measured until the late 1960s. The measurements made during the 1980s show a strong upward trend of about 1.1% annually. Ice cores containing trapped air have been used to analyse long-term trends. Data reveal that methane levels have been rising more or less parallel with the growth of human population. This makes sense, because methane is generated through a number of human activities, e.g., rice cultivation, coal mining and fossil fuel burning, and also from the stomachs of ruminating cattle. How much methane is produced individually by each of these activities is not certain. About 425 million tonnes of methane are released in the atmosphere each year. However, methane does not have a long lifetime in the atmosphere as it is oxidized to other chemicals fairly rapidly.

Natural wetland ecosystems represent a large source of gas. Estimated annual fluxes range from 100 to 150 million tonnes a year. Methane emissions from wetlands vary depending on soil and air temperature, soil moisture, the amount and composition of organic substrate and the vegetation. The extensive, organic-rich arctic and boreal wetlands are especially important sources of the gas, and account for about half of the total world-wide emissions from natural wetlands. Rice paddies also produce the gas; estimates range from 35 to 110 million tonnes a year. Altering the productivity of rice paddies and/or extending their cropping area increases methane emission. Seiler & Conrad (1987) indicate that methane emissions may have risen from 75 million tonnes a year in 1950 to 115 million tonnes in 1980 because of an increase in the extent of paddy cultivation. Domestic animals are estimated to produce 74 million tonnes a year and termites between 15 and 150 million tonnes.

Some data collected from the monitoring programme of the National Oceanic and Atmospheric Administrations and the Climate Monitoring and Diagnostic Laboratory (NOAA/CMDL) reveal that:

- The annual CH_4 concentrations at Mould Bay (Northwest Territories, Canada) have risen from 1713.7 ppbv in 1984 to 1779.8 ppbv in 1989. The average annual growth rate was 12.2 ppbv and the degree of scatter for the CH_4 data from Mould Bay was much less than that from the other northernmost sites (Steele *et al.*, 1987). The reduced scatter could be a reflection of greater distances between the Mould Bay site and significant sources of CH_4.
- The average annual CH_4 concentrations at Mauna Loa, Hawaii, USA, increased from 1648.5 ppbv in 1985 to 1724.0 ppbv in 1992. For the period 1983–1992 the globally averaged background concentration of CH_4 in the marine boundary layer increased by an average of 11.4 ppbv per year, or 0.78% per year when referenced to the globally averaged concentration at the midpoint of that period (Dlugokencky *et al.*, 1994).
- The annual atmospheric CH_4 concentration at Alert, North West Territories, Canada, rose from 1742.2 ppbv in 1986 to 1802.1 ppbv in 1992 (Dlugokencky *et al.*, 1994).
- The annual concentration of CH_4 at Amsterdam Island in the Indian Ocean, rose from 1561.1 ppbv in 1983 to 1635.4 ppbv in 1989. The NOAA/CMDL data for the period May 1983 to April 1987 show an average global growth rate of 12.9 ppbv per year (Ehhalt *et al.*, 1990).
- Fraser *et al.* (1986) reported an average growth rate of 18.2 ppbv per year at Cape Grim (Tasmania, Australia) for the 7-year period 1978–1984. The NOAA/CMDL data show that the average annual concentration at Cape Grim rose from 1588.0 ppbv in 1984 to 1666.2 ppbv in 1992 (Dlugokencky *et al.*, 1994).

Chlorofluorocarbons

Chlorofluorocarbons, especially CFC-11 and CFC-12, have been emitting from industrial sources to the atmosphere for the past half a century. The annual emission of each of the two main CFCs is estimated to be about 0.4 tonnes (Rasmussen & Khalil, 1986). Photochemical destruction, mainly in the stratosphere, and very slow uptake by the oceans are the only known significant sinks for chlorofluorocarbons (see Boden *et al.*, 1994).

Significant production of CFC-11 began in the late 1930s for use in refrigeration. The production increased very slowly until after World War II, when CFCs began to be used as propellants for aerosol sprays. By the 1950s, the use of CFCs in refrigeration as the blowing agent in open cell (from rubber) and closed cell (rigid polyurethane) foams constituted an additional significant source of release. During the 1960s and early 1970s, production and release of CFCs grew very rapidly. Over the 10-year period 1964–1974, for instance, the total CFC-11 release grew exponentially, at the rate of 12.3% per year, primarily due to the rapid world-wide expansion of CFC use in aerosols, whereas from 1952 to 1974, CFC-12 release grew at a constant rate of 11.9% per year. From 1975, however, fluorocarbon production and release actually declined, following the US ban on the use of CFCs in aerosols. By the early 1980s, however, this decline reversed, partly due to continued growth in CFC production for aerosols by countries not participating in the earlier ban. In addition, the use of CFC-11 as the blowing agent in rigid polyurethane foams grew unabatedly throughout the 1970s and 1980s and CFC-12 remained the preferred cooling agent in home refrigerators and automobile air-conditioners. For the period 1974–1989, CFCs release from this use grew exponentially, at the rate of 9.6% per year. The release of CFCs from all sources (except refrigeration) fell precipitously in 1989, largely in response to efforts to limit emissions of fully halogenated CFCs as prescribed by the 1987 Montreal Protocol.

Ozone

Ozone (O_3) is a natural constituent of the atmosphere, and life on earth depends on its presence. Ozone concentrations vary widely with height, latitude, season and time of day which makes long-term detection of changing ozone concentrations very difficult. However, ozone concentrations appear to be increasing in the lower atmosphere. This is because of increased concentrations of other molecules, such as CO, NO_x, hydrocarbons, etc., that catalyse the reaction in which O_3 is formed from oxygen.

Higher up, say above 25 km (stratosphere), there is some evidence that O_3 levels begin to show a small decline. This may be as CFC levels rise, release free chlorine, and hence accelerate the breakdown of ozone. At these heights, O_3 plays a quite different role in altering climate. The ozone molecule absorbs solar radiation. Hence if ozone levels were to fall high up in the atmosphere, less solar radiation would be absorbed there, and more would be available to travel down through the atmosphere and strike the earth's surface. The problem of predicting future O_3 levels is quite complex as this depends almost entirely on the future concentrations of other chemicals which may themselves be highly interactive, and the information on the chemistry of what is happening in the upper atmosphere is very inadequate and complicated (UNEP, 1992).

Surveys have revealed that precursor emissions generated in London can increase downwind O_3 concentrations between 38 and 154 $\mu g\ m^{-3}$ (18–72 ppb) within a few hours (up to 10 hours) (Ball, 1987; Varley *et al.*, 1988). The number of days when hourly O_3 concentrations exceeded 171 $\mu g\ m^{-3}$ (80 ppb) at the central London monitoring site were at a maximum of 25 days in 1976 to about 15 days in 1975 and 1989, and 10 days in 1983 (LSS, 1990b). Peaks in O_3 levels correspond with exceptional meteorological conditions; the summers of 1975, 1976 and 1989 were among the sunniest since O_3 records began.

Los Angeles has the most serious O_3 problem in the USA (USEPA, 1991). In 1990, the maximum hourly O_3 concentration was 660 $\mu g\ m^{-3}$ (0.33 ppm) at the Crestline site; here the US NAAQS was exceeded on 103 days and the State Standard on 144 days (SCAG, 1991). The State Standard was exceeded at all stations in 1990.

Ozone is not considered a problem in Moscow. Occasional studies show that, even when suitable meteorological conditions occur, levels rarely exceed the guidelines.

In New York, annual emissions of reactive organic compounds (ROCs) in 1985 totalled 216 000 tonnes in New York City, and over 870 000 tonnes in NYCMA. Areas of O_3 maxima in NYCMA are "downwind" of the city proper depending on wind direction, e.g., in New Jersey or Connecticut. The O_3 situation is improving as the number of violations of the NAAQS (0.12 ppm O_3 one-hour mean, the expected value not to be exceeded more than once per year) has decreased from almost 120 to less than 10 (USEPA, 1991). The fall in O_3 maxima is primarily a result of automobile emission controls on NO_x and ROCs, plus stricter inspection, maintenance of motor vehicles and application of "Reasonably Available Control Technology" (RACT) to be applied to stationary hydrocarbon emission sources (NYDEC, 1984).

In Tokyo, the annual mean concentrations of photochemical oxidants have decreased since the late 1960s (TMG, 1989b). While the annual mean concentrations of oxidants were as high as 0.035–0.045 ppm (70–90 $\mu g\ m^{-3}$, expressed as O_3) in 1968–1970, the levels dropped to about 0.02 ppm (40 $\mu g\ m^{-3}$) in 1978. Since then, levels have been stable. On days when any one-hour O_3 concentration exceeds 0.12 ppm (240 $\mu g\ m^{-3}$), an "oxidant warning" is given. At this level adverse health effects may occur in a significant proportion of the exposed population. In the whole of Japan, the number of "oxidant warning days" per year varied between 59 and 171.

STATUS IN THE DEVELOPING COUNTRIES

In almost all developing countries, air quality is worsening. Rapidly growing cities, more traffic on roads, use of dirtier fuels, reliance on outdated industrial processes, growing energy consumption, and lack of industrial zoning and environmental regulations are all contributing to the reduced urban air quality and deteriorating public health. To improve economic and social well-being, many developing countries give priority to rapid industrial development. This must lead to progress and improve the material quality of life, but it can also result in serious environmental deterioration if not carefully controlled. In many developing cities, the growth of both industrial and residential areas is unplanned, unstructured and unzoned.

Sulphur dioxide

Man-made SO_2 emissions have been rising about 4% annually — a rate equivalent to the rise in global energy consumption. About 90% of these emissions come from the northern hemisphere (the USA and the Soviet Union), but the developing countries will contribute more as they develop their industrial base. SO_2 pollution is becoming particularly evident in countries such as China, Mexico and India (Figure 1.7).

Based on scientific findings and epidemiological studies, WHO has established two guidelines for SO_2 concentrations which should apply simultaneously. One relates to safe long-term exposure by recommending an average annual level of SO_2, the other to safe levels of daily exposure which are higher as it is assumed that such exposure will not be prolonged. The recommended annual mean level is 40–60 $\mu g\ m^{-3}$. The 100–150 μg^{-3} limit for 98% of daily averages is designed to ensure that even particularly sensitive members of the population be protected from short-term adverse effects. Many cities continue to exceed the guidelines, in some cases by large margins (WHO/UNEP/GEMS, 1988).

In Bangkok, the major sources of SO_2 are light and heavy industries using fuel oil and domestic lignite, the latter having a very high sulphur content (about 2.8%) (WHO/UNEP, 1992). Concentrations of SO_2 in the ambient air are very low. This

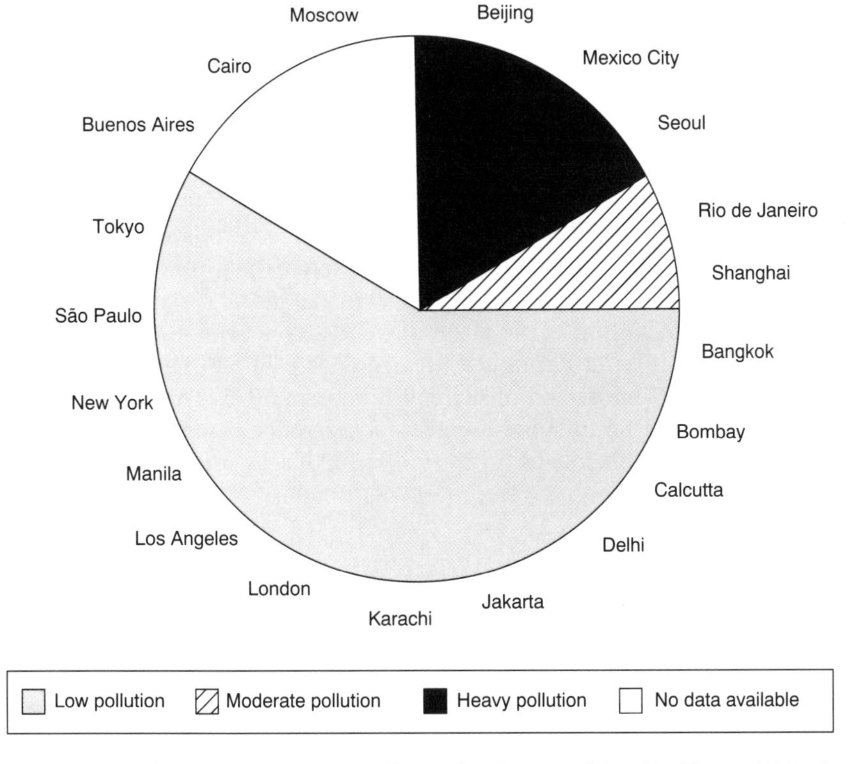

Figure 1.7. Level of SO_2 in some cities. Pollution levels as explained in Figure 1.5 (redrawn after WHO/UNEP, 1992)

is due to the meteorological conditions favouring the dispersion of industrial emissions outside metropolitan Bangkok. The WHO 98 percentile, short-term guideline value of 150 $\mu g\ m^{-3}$, is not generally exceeded on more than 7 days per annum. The SO_2 annual maxima decreased from 50–80 $\mu g\ m^{-3}$ in the period 1981–1983 to about 20 $\mu g\ m^{-3}$ in 1986–1989. These maxima will probably decrease further during the 1990s when sulphur content in diesel fuel will be lowered to 0.5%.

In Beijing the dominant sources of SO_2 are heavy industries and power-plants. Industrial sources account for 87% of the SO_2 emissions (Krupnick & Sebastian, 1990). According to estimates, SO_2 emissions were about 526 000 tonnes per annum in 1985. The annual mean SO_2 concentrations have remained relatively constant during the past few years. The city centre had annual means over 100 $\mu g\ m^{-3}$ until 1989 and thus exceeded the WHO Annual Mean Guideline (40–60 $\mu g\ m^{-3}$). The days with highest SO_2 pollution (up to 250 $\mu g\ m^{-3}$) are between November and March (the summer period).

In Bombay, industrial sources account for nearly all SO_2 emissions. Following a 66% increase in SO_2 emissions between 1970 and 1980, the rate of increase slowed significantly over the following 10 years. In 1990, total SO_2 emissions were around 157 000 tonnes. This levelling-off in the emissions is largely due to the introduction of natural gas as a major fuel source, from the newly opened gas fields located on the West Coast (NEERI, 1991b).

In Delhi, industrial sources, and power-stations in particular, are responsible for most of the SO_2 emissions. The total anthropogenic SO_2 emission is approximately 59 000 tonnes per annum. SO_2 emissions from transport sources have increased and will continue to increase due to the increasing motor vehicle population (16 658 in 1971 to 75 709 in 1987) (NEERI, 1991a).

In Karachi, total SO_2 emissions amount to about 77 000 tonnes per annum, about 85% of which is emitted by fuel-oil-fired power-stations (Ghauri *et al.*, 1988; Beg, 1990). Traffic SO_2 emissions are thought to be relatively small. Motor vehicles emit 2800, ships 1200 and aircraft 200 tonnes per annum.

In Mexico City, mean levels range from 80 to 200 $\mu g\ m^{-3}$, daily maxima being between 200 and 550 $\mu g\ m^{-3}$. No clearcut trends have been noted between 1986 and 1991 (Gobierno de la Republica Mexico, 1990). In São Paulo, Nefussi *et al.* (1977) reported SO_2 emissions of 600 000 tonnes in 1976. This had dropped to 280 000 tonnes by 1981 and subsequently stabilized at around 110 000 tonnes per annum by 1988/89. This means an overall reduction by a factor of over five.

Suspended particulate matter

Particulate emission is increasing in some developing countries, mainly because coal washing is less common and abatement equipment is poorly maintained (or absent altogether). WHO has recommended guidelines for annual and daily levels of SPM (annual mean of 60–90 $\mu g\ m^{-3}$ and a daily average of 150–230 $\mu g\ m^{-3}$). A number of cities in developing countries exceed the guidelines on particulates (Figure 1.8).

In Bangkok, anthropogenic SPM emissions were estimated to be about 40 000 tonnes in 1980. Since then, emissions have increased by a factor of two

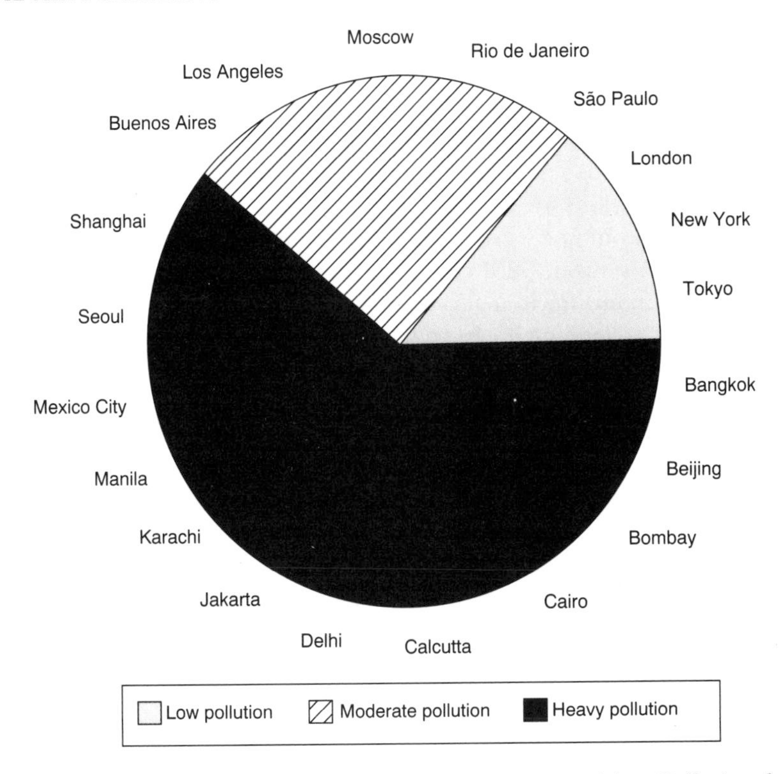

Figure 1.8. Level of suspended particulate matter in some cities. Pollution levels as explained in Figure 1.5 (redrawn after WHO/UNEP, 1992)

(Faiz *et al.*, 1990). Even in suburban areas, SPM concentrations exceed WHO guidelines. Particle size analyses of SPM samples have shown that more than 60% of particulates are smaller than 10 μm, and particles in the ranges of 0.6–1 and 5–7 μm are especially common. This indicates the seriousness of the SPM problem in Bangkok as the particles fall into the inhalable size range (ONEB, 1989).

In Beijing, anthropogenic SPM emissions were estimated by the World Bank to be about 116 000 tonnes in 1985; industry accounts for 59% (68 000 tonnes per annum) of it (Krupnick & Sebastian, 1990). Annual mean SPM levels ranged between 200 and 500 $\mu g\ m^{-3}$ from 1981 to 1990, which exceeds the WHO guidelines (60–90 $\mu g\ m^{-3}$).

In Bombay, SPM emissions have increased significantly in recent years and are projected to continue rising into the next century. In 1973, 67% of the total industrial SPM emissions were caused by power-plants (NEERI, 1991b). SPM emissions attributable to transport have increased greatly. The transport sector doubled its share in the total estimated SPM emissions between 1970 (2%) and 1990 (4%). The proportion is likely to increase further with increasing motor-vehicle traffic.

Calcutta has a very severe SPM problem. Industrial sources account for 98% of the total (NEERI, 1991c). The high emissions and ambient concentrations of SPM result from the excessive use of coal by Calcutta's industry. Estimates indicate a

66% increase in SPM emissions between 1970 and 1980. The annual arithmetic mean concentrations in 1990s were lower than in 1987 (268–453 $\mu g\ m^{-3}$) but still well above the WHO guidelines.

In Delhi, total SPM emissions were around 115 700 tonnes per annum in 1990 and are likely to increase to 122 600 tonnes per annum by 2000 (NEERI, 1991a). Anthropogenic emissions are not the only source of SPM in Delhi. Natural dust is also blown in from the surrounding arid areas, such as the Great Indian Desert. The monthly mean SPM concentration in June is over 600 $\mu g\ m^{-3}$. This natural dustfall remains in circulation for long periods. Wind speeds also increase during the summer leading to further resuspension. Some estimates attribute one-third of all anthropogenic SPM in Delhi's atmosphere to road transport. Transport SPM is probably of greatest concern in terms of health effects due to the carcinogenic nature of certain diesel-smoke constituents (e.g., polynuclear aromatic hydrocarbons — PAHs).

Annual mean ambient SPM concentrations show no significant trend. However, 98 percentile concentrations at all stations have increased and at the residential and commercial sites concentrations more than doubled between 1982 and 1990 (NEERI, 1991a). These concentrations are far in excess of the recognized WHO guidelines and also exceed the Indian Air Quality Standards (360 $\mu g\ m^{-3}$).

In Karachi, there are natural sources of particulate matter (such as wind-blown dusts from surrounding desert areas) as well as anthropogenic sources (e.g., motor-vehicle exhausts, industrial activities like iron and steel production, asphalt mixing plants, cement kilns, etc.). The total anthropogenic SPM emissions in Karachi amount to 106 000 tonnes per annum, with about half of it coming from industrial activities. Traffic, mostly motor vehicles, contributes about 7000 tonnes SPM per annum. Annual mean concentrations calculated from the monthly mean data show a clearly increasing trend of SPM. In the SPARCENT (Space Science Division) site, the annual means were 239 $\mu g\ m^{-3}$ in 1985, 265 $\mu g\ m^{-3}$ in 1986 and 275 $\mu g\ m^{-3}$ in 1987 (Ghauri *et al.*, 1988).

In São Paulo, the emissions totalled 68 000 tonnes in 1988. Although there has been a net decrease in the yearly mean and in the extreme daily levels, it is clear that the net result is less successful than for SO_2 (CETESB, 1990).

Lead

Recognizing the health hazards of lead, almost all industrialized countries have progressively reduced lead content of petrol over the past decade. Emissions of lead from stationary sources (such as non-ferrous metal-smelting and battery plants) have also been reduced. Little has changed with regard to reduced lead levels in petrol in developing countries (Table 1.1). With more vehicles on the roads, lead reduction in petrol has become a major priority. In developing countries, control on stationary sources may be more important than that on vehicular emissions. Severe lead pollution from a single point source can expose large numbers of people living in the area to excessive lead levels.

In Bangkok, lead in commercial petrol was reduced from 0.84 $g\ l^{-1}$ to 0.45 $g\ l^{-1}$ in 1984 and to 0.40 $g\ l^{-1}$ in 1989. Data show that the annual arithmetic mean airborne Pb concentrations are below the WHO long-term guideline of 1 $\mu g\ m^{-3}$. However, roadside monitoring results revealed much higher values; a maximum of 2.5 $\mu g\ m^{-3}$ during the period 1985–1987 (ONEB, 1989).

In Beijing, there is a 0.8 $g\ l^{-1}$ limit for Pb content in petrol, and 80% of the petrol consumed is Pb-free. Furthermore, traffic in Beijing is still limited, so emissions of Pb are small and well below 100 tonnes per annum (Vahter & Slorach, 1990).

In Bombay, petrol-driven vehicles are the main source of ambient Pb. Their number has risen from approximately 125 000 in 1971, to 468 000 in 1987 and to 588 000 in 1989. In 1986, the Pb content of petrol in Bombay was 0.8 $g\ l^{-1}$ for premium and 0.56 $g\ l^{-1}$ for regular. The current mean Pb content of petrol from Bombay refineries is 0.155 $g\ l^{-1}$ (NEERI, 1991b). Monitoring indicates that the annual airborne Pb levels have fallen significantly since the 1970s to between 0.25 and 0.33 $\mu g\ m^{-3}$, well below the WHO guideline of 1 $\mu g\ m^{-3}$.

In Cairo, the maximum Pb content of petrol is 0.8 $g\ l^{-1}$, which is relatively high. It may be estimated that a motor vehicle population of 900 000 could produce Pb emissions of the order of 1000–2000 tonnes per annum. Pb levels exceed the WHO Annual Mean Guideline almost all over Cairo (Ali *et al.*, 1986). Pb concentrations were greatest in the summer, e.g., 6.4 $\mu g\ m^{-3}$ in June 1984.

Lead content of petrol from the Haldia refinery, which supplies Calcutta, is 0.1 $g\ l^{-1}$. Despite the relatively low lead content of petrol, the annual airborne lead levels are the highest in India (NEERI, 1991d). The annual concentrations are highest at the residential and commercial sites (0.73 $\mu g\ m^{-3}$), but below 1 $\mu g\ m^{-3}$.

The Pb content of petrol distributed from the Mathura refinery which serves Delhi is relatively high (1.8 $g\ l^{-1}$). Given this Pb content and the number of petrol-driven motor vehicles registered in Delhi, it is estimated that Pb emissions from this source are of the order of 600 tonnes per annum (NEERI, 1991d). While monitoring the air quality along road transactions of Lucknow city, Singh *et al.* (1995) observed high concentrations of lead (the highest being 2.96 $\mu g\ m^{-3}$).

Lead concentrations in petrol in Pakistan are between 1.5 and 2.0 $g\ l^{-1}$, which is relatively high on a world scale. Concentrations of inorganic Pb are very high in Karachi. Mean concentrations in urban areas were about 1–3 $\mu g\ m^{-3}$ with the maximum value of 7–9 $\mu g\ m^{-3}$ in areas of heavy traffic; far above the WHO range (0.5–1.0 $\mu g\ m^{-3}$) (Beg, 1990).

In Rio de Janeiro (FEEMA, 1989) and São Paulo (CETESB, 1989), the Pb levels are low because of reliance on alcohol-powered vehicles coupled with reduced Pb levels in petrol.

Carbon monoxide

Dealing with CO emissions in the developing countries will be more complex. New cars from the North are now increasingly fitted with NO_x and CO controls, demanding the use of unleaded petrol, but this is not widely available in the developing world. CO emissions in newly industrialized countries are expected to rise over the next few years (Figure 1.9).

CO emission in Bangkok was 120 000–160 000 tonnes per annum in 1980 and could go up to 420 000 tonnes per annum in 2000 (Faiz *et al.*, 1990). In the

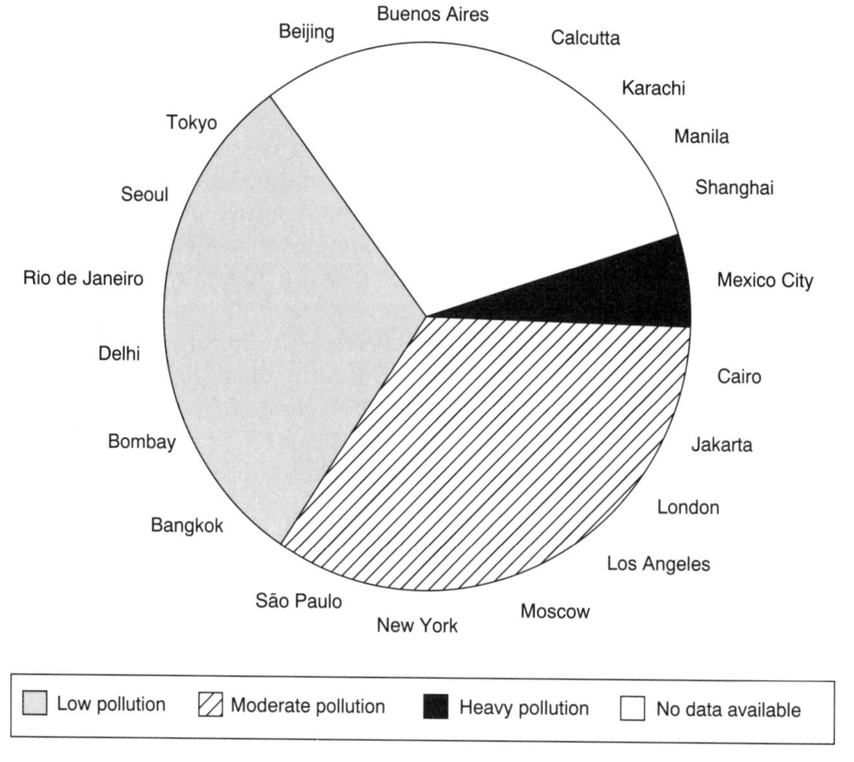

Figure 1.9. Level of CO in some cities. Pollution levels as explained in Figure 1.5 (redrawn after WHO/UNEP, 1992)

absence of domestic heating nearly all CO emissions result from incomplete combustion of fuel in motors. Thus, CO emissions will increase with growing motor-vehicle traffic unless effective emission controls are implemented. For example, a 49 mg m^{-3}, 1-hour mean and a 30 mg m^{-3}, 8-hour mean, were recorded, which were much above WHO guidelines.

In Bombay, estimated emissions increased from 69 000 tonnes per annum in 1970 to 188 000 tonnes per annum in 1990–1991, and will probably be 255 500 tonnes per annum by 2000. Most of this increase is attributed to motor-vehicle transport which was estimated to be responsible for 89% of the total CO emissions in 1990 (NEERI, 1991b).

In Delhi, CO emissions increased from 140 000 tonnes per annum in 1980 to 265 000 tonnes per annum in 1990 (NEERI, 1991a). With the increase in motor vehicles, CO emissions are projected to exceed 400 000 tonnes per annum by the year 2000.

In Karachi, total CO emissions were estimated to be 271 000 tonnes in 1989. More than 45% of this (123 000 tonnes per annum) arises from open burning of 2.9 million tonnes of waste. During 1969 to 1987, CO concentrations varied from 10 to 30 mg m^{-3} at the several monitoring sites throughout the city (Beg, 1990).

Acid deposition

Acid pollution was, until recently, thought of as a North American and Northern European problem, and is still commonly seen as a developed country's problem. However, evidence of acid pollution — or conditions that could cause acid pollution — has already been found in South and Southeast Asia, the Middle East, Southern Africa and Latin America.

As the world's third largest consumer of primary commercial energy and emitter of SO_2, and the second largest producer of hard coal, China faces greater risks of acid pollution than any other Third World country. Ninety per cent of the locations with pH less than 5.6 occur south of the Yangtze River, southwest China being the most seriously affected region. Sulphate concentration in rain in Guiyang (281 $\mu eq\ l^{-1}$) is about six times higher than in New York City (Galloway *et al.*, 1987). It is likely that rain-water acidity in China is due more to local influences than long-range transport (Zhao *et al.*, 1988), suggesting that below-cloud scavenging of SO_4^{-2} aerosols and SO_2 is the primary mechanism creating acid rain. In Beijing, SO_4^{-2} concentrations are still high but pHs are close to neutral. In this area, acidity in rain-water is neutralized mainly by Ca^{2+} from calcareous soil aerosols created by agricultural activities, from calcareous material used for building, and from alkaline emissions that arise from many small to medium furnaces.

Southeast Asia is rapidly industrializing and its energy demands are rising. Coal-fired power-plants in the region are expected to expand, which would raise coal consumption. A more immediate problem is the rapid increase in the volume of road traffic. Traffic accounts for 92% of the air pollution in Kuala Lumpur and 75% in Metro Manila. Cities in the region share several characteristics that suggest that air pollution will become worse, courting the risk of dry acid deposition. Monthly samples of rainfall in Petaling Jaya (outside Kuala Lumpur) and Tanah Rata (in Perak) returned pH readings of 4.4–4.8 and 4.9–5.5, respectively.

Brazil's most heavily industrialized areas are in the southeastern states of São Paulo, Rio de Janiero, Minas Gerais, Parana and Rio Grande do Sul. Soil samples from the eastern parts of São Paulo state have shown extremely low pH values (3.7–4.7).

India, being a major producer and consumer of coal, emits huge amounts of SO_2 and NO_x and this emission is likely to rise with the growth in road traffic and the operations of fertilizer plants, refineries and petrochemical and other industries. These pollutants along with favourable climatic conditions are prone to acid deposition. Rain of pH 4.8 has been reported in Bombay (NEERI, 1991d).

With few concentrations of heavy industry or heavily populated urban conurbations, and deriving most of its energy needs from non-commercial fuels such as firewood, acid pollution is generally a distant concern in most of Africa. Evidence of acid pollution has, nevertheless, been found near concentrations of industry in South Africa and Zambia.

In Criciuma, the heart of Chile's coal-producing area, two-thirds of the region's rivers, lakes and reservoirs are reported to be "highly acidified".

Greenhouse gases

The fate of the earth's climate depends on how much concentrations of carbon dioxide and other trace gases are likely to increase in the future. The developing countries have become significant emitters of greenhouse gases only in the past two decades, due to large populations and rising material ambitions (Boden *et al.*, 1994).

Carbon dioxide

Burning of fossil fuels is the main source of man-made carbon dioxide emissions; it is estimated to release about 5000 million tonnes of carbon a year. Deforestation is the next most important factor (Figure 1.10). A study (Ditwiler & Hall, 1988) has shown that deforestation in the tropics may be responsible for the release of 310 to 1300 million tonnes of carbon a year, while conversion of forest soil to other uses causes another 110 to 250 million tonnes to be given off. In all, man-made changes in the earth's biota, including forest destruction and the destruction of grassland and other vegetation cover through desertification, result in a net annual release of 1600 million tonnes of CO_2. Deforestation in Brazil, Colombia,

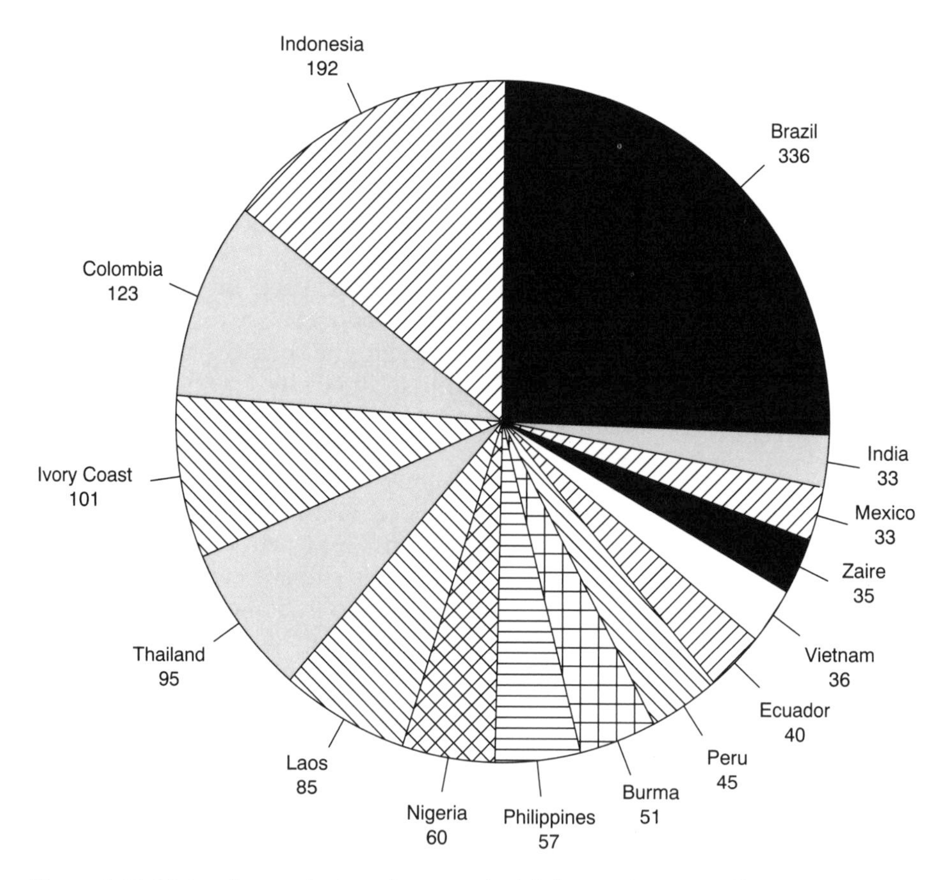

Figure 1.10. Net carbon emissions from tropical deforestation (in million tonnes, 1980)

Ecuador, Indonesia, Mexico, Peru, the Philippines, Thailand and Zaire accounts for three-quarters of these CO_2 releases.

In India, CO_2 emission from fossil fuels alone was 32.1 million tonnes in 1960 which rose to 192.0 million tonnes in 1991 (Marland *et al.*, 1994). In the Middle East it has risen from 25.0 to 358.0 million tonnes per annum within that time. China, emitting 213.5 million tonnes per annum of CO_2 in the 1960s, had raised its emissions to 555.2 million tonnes per annum by the end of the 1980s. In Taiwan emissions were only 3.1 million tonnes in 1960 but had gone up to 23.6 million tonnes in 1990. CO_2 emissions in Brazil had risen from 12.0 million tonnes in 1960 to 58.8 million tonnes in 1991 (Marland *et al.*, 1994).

Oxides of nitrogen

Data for developing countries are scarce, but the urban NO_x levels are probably rising in rapidly industrializing countries (Figure 1.11), such as Brazil, Chile, Hong Kong and India (albeit decreasing in Singapore). The emissions are probably high in these countries because of the many old and poorly maintained vehicles on road. The ambient levels can be affected by several local characteristics such as the "Canyon effect" of tall buildings, meteorological conditions (temperature inversions can increase levels), and photochemical conversion.

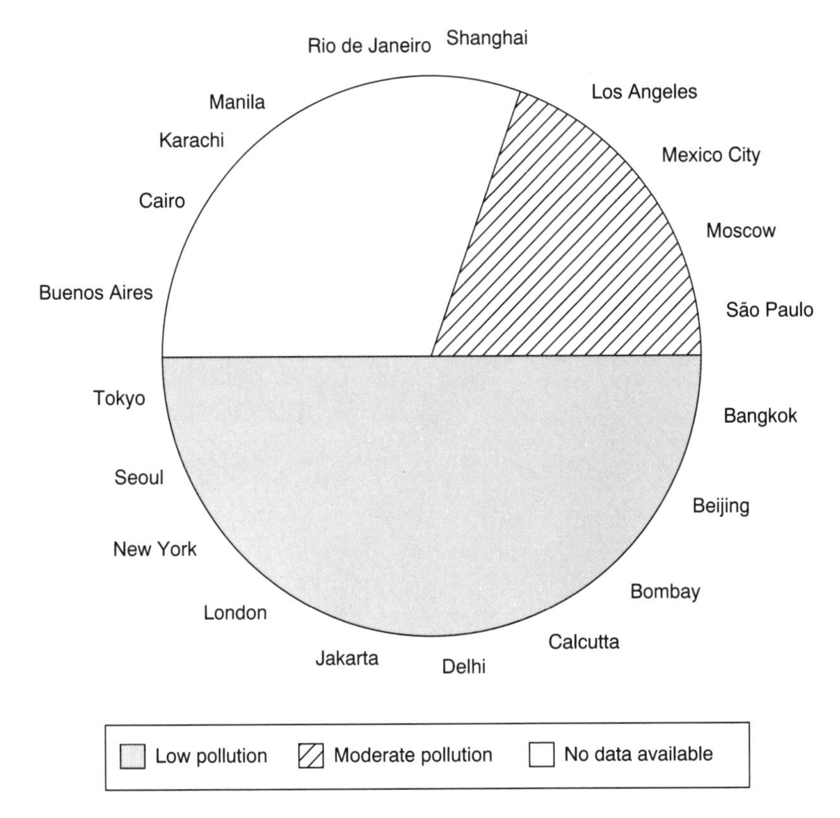

Figure 1.11. Level of NO_x in some cities. Pollution levels as explained in Figure 1.5 (redrawn after WHO/UNEP, 1992)

In Bombay, transport was estimated to account for 52% of NO_x emissions in 1990 and will probably increase by approximately 14 600 tonnes per annum by 2000 (NEERI, 1991b). Detailed vehicle-emission inventories produced by the Department of the Environment indicate that diesel vehicles (predominantly trucks) are the dominant source of motor-vehicle-derived NO_x in Bombay.

In Cairo, NO_x emissions from motor vehicles were estimated to be about 7300 tonnes per annum in 1990, with two-thirds of it coming from cars and one-third from buses (WHO/UNEP, 1992).

In 1970 industry was the major source of NO_x (69%) in Calcutta, but now the dominant source is transport. Emissions from industrial plant had increased to over 11 000 tonnes per annum by 1980, which has stabilized since then. Transport emissions have risen from an estimated 1825 tonnes per annum in 1970 to 25 550 tonnes per annum in 1990 (NEERI, 1991c).

In Seoul, the emissions of NO_x increased markedly from 1984 to 1989. An increase in motor-vehicle traffic has probably been the main factor for this change. The transport sector accounts for 78% of the NO_x emission. The total man-made NO_x emissions in 1990 were estimated to be 270 000 tonnes per annum in the Greater Seoul Area and 130 000 tonnes per annum in Seoul (Rhee, 1991).

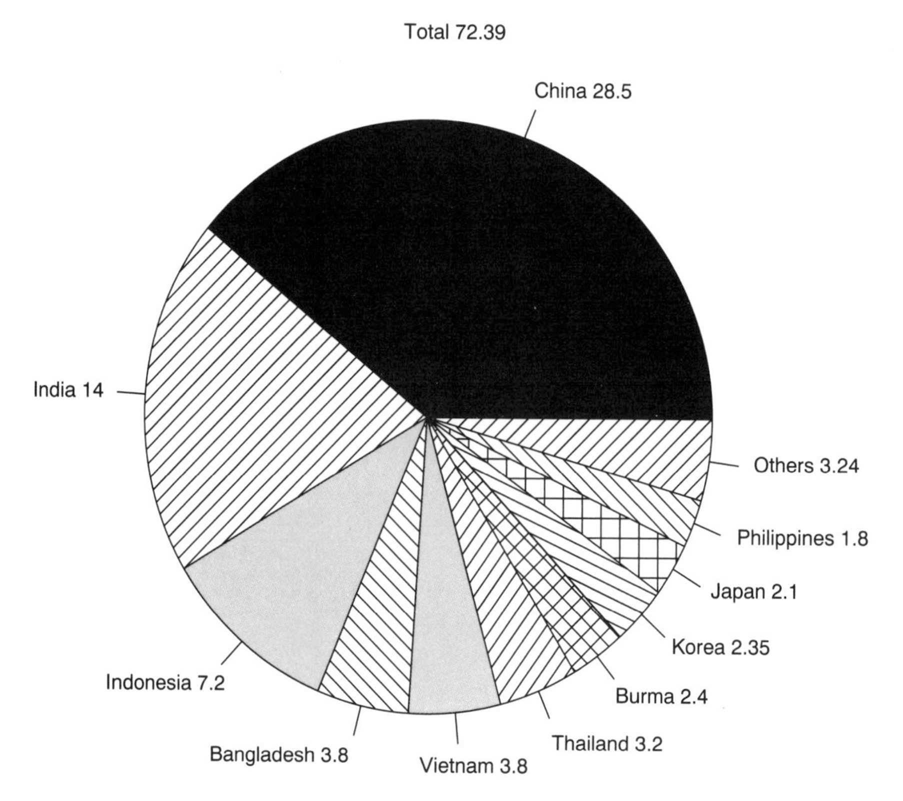

Figure 1.12. Annual emission of methane from biomass approach (paddy fields) for the Asian region (in g $\times 10^{11}$)

Methane

The major sources of methane are enteric fermentation in cattle and insects, biomass burning and garbage landfills, coal mines and natural gas leaks, rice paddies and swamps. Rice paddy fields are the main emanator of methane. The developed countries do not cultivate rice, neither do their people eat it much. But in the developing world including Bangladesh, Burma, China, India and Thailand and other countries of Asia rice is the staple diet for a large number of people and rice cultivation is widespread (Parashar *et al.*, 1994). Emission of methane from biomass approach has been estimated for the Asian region (see Figure 1.12) (Mitra, 1991).

Chlorofluorocarbons

CFCs are the genies in spray bottles, refrigerators, aerosol cans, air-conditioners and other such systems. The developing countries are minor producers of CFCs in comparison to the developed countries. By 1986 about 2.5 million tonnes of CFC compounds were being produced in the world — 35% of them in the US alone, and another 35% in Western Europe. India releases less than 2% of the world CFC production.

Ozone

Tropospheric ozone is a secondary pollutant originating from hydroxy radical, nitrogen oxides, methane, carbon monoxide, etc. Ozone concentration varies with season, time of day and other meteorological conditions. The developing world may be a minor emitter of ozone precursors but the climatic conditions are favourable for its production in many countries (Figure 1.13). Concentration of O_3 is higher in the warmer months due to increased insolation, higher temperatures and greater stagnation. The increase in fossil fuel use in industries and automobiles will definitely increase O_3 emissions in the developing countries.

In Bangkok, O_3 is not a problem because of the year-round monsoon which prevents build-up of pollutants. The highest concentrations were found in the March–May season, when solar radiation is strongest. The maximum 1-hour mean O_3 measured was around 100 $\mu g\ m^{-3}$, which is below the WHO guideline value of 200 $\mu g\ m^{-3}$ (WHO/UNEP, 1992).

In Beijing, O_3 was 158–262 $\mu g\ m^{-3}$ (hourly) in June 1986 and 133–320 $\mu g\ m^{-3}$ in July 1987, which was maximum (Tang *et al.*, 1988). These high O_3 concentrations are due to an increase in NO_x emissions (because of a greater abundance of motor vehicles without control measures). O_3 is not routinely measured in Bombay, Cairo, Calcutta or Delhi, but increasing NO_x and hydrocarbon emissions and relatively high insolation imply that if O_3 is not already a problem, very soon it is likely to be (Yunus *et al.*, unpublished data). In Delhi measurements taken during 1989/90 revealed that ozone varied from 9.4 to 128.31 ppbv, exhibiting a wide temporal and seasonal variation. On many occasions 1-hour ozone concentration was more than 113 ppbv (Varshney & Aggarwal, 1992).

In Karachi, the O_3 concentration was maximum in the afternoon, between 12.00 and 15.00 hours, irrespective of climate conditions and site location. The concentration also varied upwind and downwind, being as low as 20–50 $\mu g\ m^{-3}$

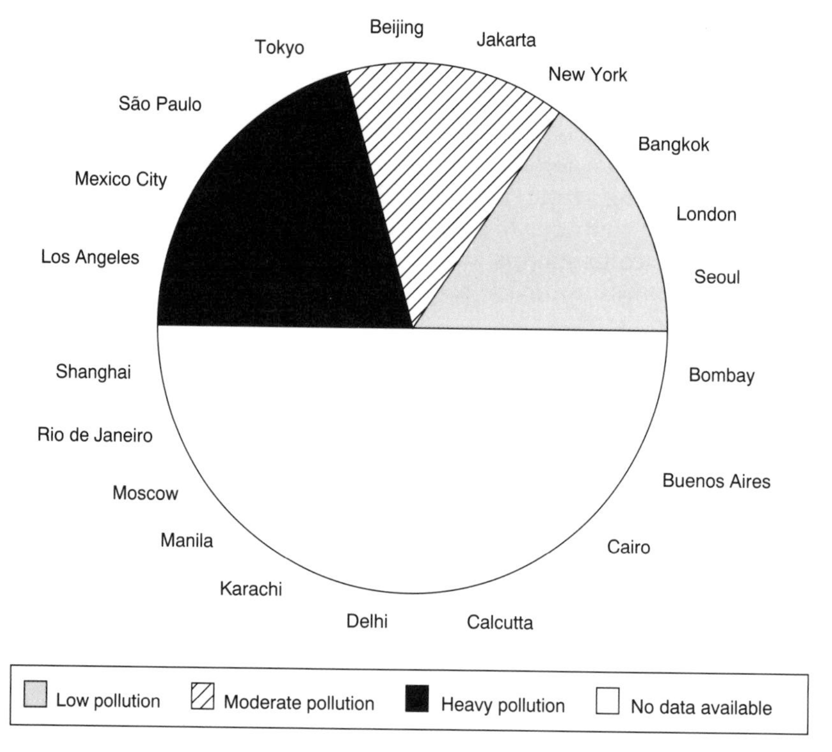

Figure 1.13. Level of O_3 in some cities. Pollution levels as explained in Figure 1.5 (redrawn after WHO/UNEP, 1992)

at the upwind sites, and as high as 80–100 μg m^{-3} at the downwind sites (Ghauri, *et al.*, 1992).

Ozone levels in Mexico City are exceptionally high (DDF, 1989; Gobierno de la Republica Mexico, 1990). The annual mean O_3 fluctuated around 200 μg m^{-3} with lows around 100–150 μg m^{-3} and highs between 300 and 400 μg m^{-3}. Hourly O_3 levels often reach 600 μg m^{-3} there, with extreme values up to 850–900 μg m^{-3}. The number of hours during which the national air quality standard of 110 ppb is exceeded is consequently high; 80 to 100 hours per month is not unusual.

AIR POLLUTION IMPACTS

At present, assessment of air quality for public health purposes consists essentially of examining ambient air quality against the established guidelines. The WHO has recommended air quality guidelines for a wide range of pollutants; these guidelines indicate the level and exposure time at which no adverse effects on human health are expected to occur. In addition to human health impacts, the pollutants also have adverse impacts on aquatic and terrestrial ecosystems. SO_2, NO_x, acidic aerosols and O_3 are phytotoxic.

SO_2 has a number of adverse health effects, and is linked to bronchitis, tracheitis and other respiratory problems. Environmental effects of SO_2 include acidification of soils, lakes and rivers and damage to plants and crops. SO_2 is also responsible for corroding buildings, monuments and works of art.

SPM can cause or aggravate respiratory diseases, especially asthma and pulmonary emphysema, and may damage lungs. Many of the organic compounds present in SPM are known to be carcinogenic. SPM can also damage the vital photosynthetic systems of plants by covering leaves, plugging stomata and reducing absorption of CO_2 and sunlight. Particles also stain fabrics, painted surfaces and buildings, reducing the life of materials and surface finishes.

Lead can adversely affect blood and the human nervous system. Anaemia is a common early toxic effect. Higher levels of lead in blood can lead to brain dysfunctions, acute or chronic encephalopathy and kidney damage.

Carbon monoxide has no known effects on vegetation or materials. Its effect mainly relate to human health. Absorbed through the lungs, CO reacts with haemoproteins, particularly the haemoglobin in blood forming carboxyhaemoglobin, restricting oxygen binding and the transport of oxygen in the blood.

Acid deposition may release aluminium in soils, or it may reduce the availability of other chemicals, such as magnesium and calcium. The aluminium released by acidification can enter rivers and lakes, where it poisons fish and can ultimately be ingested by humans.

Greenhouse gases are predicted to bring about a climate change. For example, if the earth's surface warms up, more snow and ice will melt, less solar radiation will be reflected from the earth's surface and temperatures will tend to increase even more. When climate changes, society suffers. Essentially this is because society is as well adapted to existing climates — with their inherent variability — as it can be.

REFERENCES

Ali, E.A., Nasralla, M.M. & Shakour, A.A. 1986. Spatial and seasonal variation of lead in Cairo atmosphere. *Environ. Pollut.* **11B**: 205–210.

Ball, D.J. 1987. A history of secondary pollutant measurements in the London region 1971–1986. *Sci. Total Environ.* **59**: 181–206.

Beg, M.A.A. 1990. *Report on Status of Air Pollution in Karachi: Past, Present and Future.* Pakistan Council of Scientific and Industrial Research, Karachi.

Bekker, A.A. & Boikova, R.A. 1990. *Basic Problems of Improving Air and Aquatic Basins of Moscow and Moscow Region.* All-Union Institute of Scientific and Technical Information (VINITI) 9, Moscow.

Boden, T.A., Kaiser, D.P., Sepanski, R.J. & Stoss, F.W. (eds) 1994. *Trends '93: A Compendium of Data on Global Change.* ORNL/CDIAC-65. Carbon Dioxide Information Analysis Center, Oak Ridge National Laboratory, Oak Ridge, TN.

Brounshtein, A.M., Shashkov, A.A., Paranonova, N.N., Privalov, V.I. & Starodubtsev, Y.A. 1994. Atmospheric CO_2 records from sites in the main Geophysical Observatory air sampling network. In Boden, T.A., Kaiser, D.P., Sepanski, R.J. & Stoss, F.W. (eds) *Trends '93: A Compendium of Data on Global Change*: 180–192. ORNL/CDIAC-65. Carbon Dioxide Information Analysis Center, Oak Ridge National Laboratory, Oak Ridge, TN.

Budyko, M.I. 1977. *Climatic Changes.* American Geophysical Union, Washington, DC, 261 pp.

CETESB, 1989. *Relatório de Qualidade do ar no Estado de São Paulo 1988.* Série Relatórios, Companhia de Tecnologia de Saneamento Ambiental, São Paulo, Brazil.

CETESB, 1990. *Relatório de Qualidade do ar no Estado de São Paulo 1989*. Série Relatórios, Companhia de Tecnologia de Saneamento Ambiental, São Paulo, Brazil.

Conway, T.J., Tans, P., Waterman, L.S., Thonina, K.W., Masarie, K.A. & Gammon, R.H. 1988. Atmospheric carbon dioxide measurements in the remote global troposphere, 1981-1984. *Tellus* **40**(b): 81-115.

Conway, T.J., Tans, P. & Waterman, L.S. 1994. Atmospheric CO_2 records from sites in the NOAA/CMDL air sampling network. In Boden, T.A., Kaiser, D.P., Sepanski, R.J. & Stoss, F.W. (eds) *Trends '93: A Compendium of Data on Global Change*: 41-119. ORNL/CDIAC-65. Carbon Dioxide Information Analysis Center, Oak Ridge National Laboratory, Oak Ridge, TN.

CVGMO, 1986. *Review of the Atmospheric Pollution and Harmful Substance Emissions in Moscow Region for 1985*. Moscow Centre for Hydrometeorology and Environmental Observations, Moscow.

DDF, 1989. *Programa Integral Contra la Contaminacion Atmosferica*. Programa de Emergenicia, Proyectos, Departmento del Distrito Federal, Mexico DF.

Ditwiler, R.P. & Hall, C.A.S. 1988. Tropical forests and the global carbon cycle. *Science* **239**: 42.

Dlugokenky, E.J., Lang, P.M., Masarie, K.A. & Steele, L.P. 1994. Atmospheric CH_4 records from sites in the NOAA/CMDL air sampling network. In Boden, T.A., Kaiser, D.P., Sepanski, R.J. & Stoss, F.W. (eds) *Trends '93: A Compendium of Data on Global Change*: 274-350. ORNL/CDIAC-65. Carbon Dioxide Information Analysis Center, Oak Ridge National Laboratory, Oak Ridge, TN.

Ehhalt, D.H., Fraser, P.J., Albritton, D., Cicerone, R.J., Khalil, M.A.K., Legrand, M., Makide, V., Rowland, F.S., Steele, L.P. & Zander, R. 1990. Trends in source gases. In *Report of the International Ozone Trends Panel — 1988*: 543-569. World Meteorological Organization, Global Ozone Research and Monitoring Project — Report No. 18, Geneva.

Faiz, A., Sinha, K., Walsh, M. & Varma, A. (eds) 1990. *Automotive Air Pollution: Issues and Options for Developing Countries*. World Bank Policy and Research Working Paper WPS 492, World Bank, Washington, DC.

FEEMA, 1989. *Qualidade do ar na Regiao Metropolitana do Rio de Janeiro 1984/1987*. Findacao Estadual de Engenharia dó Meio Ambiente, Rio de Janeiro, Brazil.

Fraser, P.J., Khalil, M.A.K., Rasmussen, R.A. & Steele, L.P. 1986. Tropospheric methane in the mid-latitudes of the Southern Hemisphere. *J. Atmos. Chem.* **1**: 125-135.

Galloway, J.N., Zhao, D., Xiong, J. & Likens, G.E. 1987. Acid rain: China, United States and a remote area. *Science* **236**: 1559-1562.

GEMS/WHO, 1988. *Assessment of Urban Air Quality Worldwide*. WHO, Geneva, Switzerland.

Ghauri, B.M.K., Salam, M. & Mirza, M.I. 1988. *A Report on Assessment of Air Pollution in the Metropolitan Karachi*. Space Science Division, Pakistan Space Upper Atmosphere Research Commission (SUPARCO), Karachi.

Ghauri, B.M.K., Salam, M. & Mirza, M.I. 1992. Surface ozone in Karachi. In Ilyas, M. (ed.) *Ozone Depletion Implications for the Tropics*: 169-177. University of Science Malaysia/United Nations Environment Programme, Nairobi.

Gobierno de la Republica Mexico, 1990. *Programa Integral Contra la Contaminacion Atmosferica*. Gabierno de la Republica, Mexico DF.

Government of Japan, 1990. *Japanese Performance of Energy Conservation and Air Pollution Control*. Environment Agency, Government of Japan, Tokyo.

Keeling, C.D. & Whorf, T.P. 1994. Atmospheric CO_2 records from sites in the SIO air sampling network. In Boden, T.A., Kaiser, D.P., Sepanski, R.J. & Stoss, F.W. (eds) *Trends '93: A Compendium of Data on Global Change*: 16-26. ORNL/CDIAC-65. Carbon Dioxide Information Analysis Center, Oak Ridge National Laboratory, Oak Ridge, TN.

Komeiji, T., Akoki, K., Koyama, I. & Okita, T. 1990. Trends of air quality and atmospheric deposition in Tokyo. *Atmos. Environ.* **24A**: 2099-2103.

Krupnick, A. & Sebastian, I. 1990. *Issues in Urban Air Pollution: Review of the Beijing Case.* Environment Working Paper No. 31, World Bank Washington, DC.

Laxen, D.P.H. & Thompson, M.A. 1987. Sulphur dioxide in Greater London, 1931-1985. *Environ. Pollut.* **43**: 103-114.

LSS, 1990a. *London Air Pollution Monitoring Network — Fourth Report 1989.* LSS/LWMP/120, London Scientific Services, London.

LSS, 1990b. *London Wide Ozone Monitoring Programme: Summary of Results for 1989.* LSS/LWMP/109, London Scientific Services, London.

Marland, G., Andres, R.J. & Boden, T.A. 1994. Global, regional and national CO_2 emissions. In Boden, T.A., Kaiser, D.P., Sepanski, R.J. & Stoss, F.W. (eds) *Trends '93: A compendum of Data on Global change*: 505-584. ORNL/CDIAC-65. Carbon Dioxide Information Analysis Center, Oak Ridge, TN.

MCGNPS, 1990. *Review of the Atmospheric Pollution in Moscow City and Towns and Settlements of the Moscow Region for 1989.* Moscow Centre for Hydrometeorology and Environmental Observations, Moscow.

MCGNPS, 1991. *Yearbook of the Atmospheric Pollution in Moscow City and Towns and Settlements of the Moscow Region for 1990.* Moscow Centre for Hydrometeorology and Environmental Observations, Moscow.

Mitra, A.P. 1991. *Global Change — Greenhouse Gas Emissions in India: A Preliminary Report.* Scientific Report No. 1. CSIR, Publication & Information Directorate, New Delhi.

Munday, P.K., Timmis, R.J. & Walker, C.A. 1989. *A Dispersion Modelling Study of Present Air Quality and Future Oxides of Nitrogen Concentrations in Greater London.* Warren Spring Laboratory, Stevenage, UK.

NEERI, 1991a. *Air Pollution Aspects of Three Indian Megacities. I: Delhi.* National Environmental Engineering Research Institute, Nagpur, India.

NEERI, 1991b. *Air Pollution Aspects of Three Indian Megacities. II: Bombay.* National Environmental Engineering Research Institute, Nagpur, India.

NEERI, 1991c. *Air Pollution Aspects of Three Indian Megacities. III: Calcutta.* National Environmental Engineering Research Institute, Nagpur, India.

NEERI, 1991d. *Air Quality Status: Toxic Metals, Polycyclic Hydrocarbons, Anionic Composition and Rain Water Characteristics (Delhi, Bombay and Calcutta).* National Environmental Engineering Research Institute, Nagpur, India.

Nefussi, N., Guimaraes, F.A. & Oliveira, G. 1977. Air pollution control programmes in the State of São Paulo, Brazil. In *Proceedings of the 4th International Clean Air Congress, Tokyo*: 942-946.

NYDEC, 1984. *Air Quality Implementation Plan, Control of Carbon Monoxide and Hydrocarbons in New York City Metropolitan Area.* AIR P-174 (8/83), New York Department of Environmental Conservation, New York.

NYDEC, 1990. *New York State Air Quality Report.* Ambient Air Monitoring System Annual 1989, US Environmental Protection Agency, Research Triangle Park, NC.

OECD, 1991. *OECD Environmental Data Compendium 1991.* Organization for Economic Cooperation and Development, Paris.

ONEB, 1989. *Air and Noise Pollution in Thailand 1989.* No. 07-03-33. Office of the National Environment Board, Bangkok.

Parashar, D.C., Mitra, A.P., Gupta, P.K., Rai, J., Sharma, R.C., Singh, N., Kaul, S., Lal, G., Chaudhary, A., Ray, H.S., Das, S.N., Parida, K.M., Rao, S.B., Kanungo, S.P., Ramasami, T., Nair, B.U., Swamy, M., Singh, G., Gupta, S.K., Singh, A.R., Saikia, B.K., Barua, A.K.S., Pathak, M.G., Iyer, C.S.P., Gopalakrishnan, M., Sane, P.V., Singh, S.N., Banerjee, R., Sethunathan, N., Adhya, T.K., Rao, V.R., Palit, P., Saha, A.K., Purkait, N.N., Chaturvedi, G.S., Sen, S.P., Sen, M., Sarkar, B., Banik, A., Subbaraya, B.H., Lal, S., Venkatramani, S. & Sinha, S.K. 1994. Methane budget from paddy field in India. *Cur. Sci.* **66**: 938-940.

Ramanathan, V. 1988. The greenhouse theory of climatic change: A test by an inadvertent global experiment. *Science* **240**: 293-299.

Rasmussen, R.A. & Khalil, M.A. 1986. The behaviour of trace gases in the troposphere. *Sci. Total Environ.* **48**: 169.

Rhee, D.G. 1991. *Air Pollution in the Megacity of Seoul.* Ministry of Environment, Seoul, South Korea (unpublished).

Rovinsky, F.V., Egorov, V.I. & Starodubsky, I.A. 1991. *Urban Air Quality in Moscow.* Institute of Global Climate and Ecology, Moscow (unpublished report).

SCAG, 1991. *Air Quality Management Plan — South Coast Air Basin.* South Coast Air Quality Management District, Southern California Association of Governments, Los Angeles.

Schwar, M.J.E., Moorcraft, J.S., Laxen, D.P.H., Thompson, M. & Armorgie, C. 1988. Baseline metal-in-dust concentrations in greater London. *Sci. Total Environ.* **68**: 25-43.

SCOPE, 1986. The greenhouse effect, climate change and ecosystems. In Bolin, B., Doos, B.R., Jager, J. & Warrik, R.A. (eds) *SCOPE Report 29.* International Council of Scientific Unions, Wiley, Chichester, 541 pp.

Seiler, W. & Conrad, R. 1987. *The Geophysiology of Amazonia.* Wiley, New York.

Singh, N., Yunus, M., Srivastava, K., Singh, S.N., Pandey, V., Misra, J. & Ahmad, K.J. 1995. Monitoring of auto exhaust pollution by roadside plants. *Env. Mon. Assess.* **34**: 13-26.

Sladkovic, R., Scheel, H.E. & Seiler, W. 1994. Atmospheric CO_2 records from sites operated by the Fraunhofer Institute for Atmospheric Environmental Research. In Boden, T.A., Kaiser, D.P., Sepanski, R.J. & Stoss, F.W. (eds) *Trends' 93: A Compendium of Data on Global Change*: 148-156. ORNL/CDIAC-65. Carbon Dioxide Information Analysis Center, Oak Ridge National Laboratory, Oak Ridge, TN.

Steele, L.P., Fraser, P.J., Rasmussen, R.A., Khalil, M.A.K., Conway, T.J., Crawford, M.J., Gammon, R.H., Masarie, K.A. & Thoning, K.W. 1987. The global distribution of methane in the troposphere. *J. Atmos. Chem.* **5**: 125-171.

Tang, X., Li, I., Chen, D., Bai, Y., Li, X., Wu, X. & Chen, J. 1988. The study of photochemical smog pollution in China. In *Proceedings Third Joint Conference of Air Pollution Studies in Asia*: 238-251. Japan Society of Air Pollution, Tokyo.

TMG, 1989a. *Automobile Pollution Control Plan: Towards a Better Living Environment in Tokyo.* Tokyo Metropolitan Government, Tokyo.

TMG, 1989b. *Summary of the Results of Regular Monitoring for Air Pollution in the FY 1989.* Air Monitoring Division, Tokyo.

UN, 1989. *Prospects of World Urbanization, 1988.* Population Studies No. 112, United Nations, New York.

UNEP, 1987. *Environmental Data Report.* Blackwell, Oxford.

UNEP, 1992. *The Impact of Ozone-Layer Depletion.* United Nations Environment Programme, Nairobi.

UNEP/GEMS, 1991. *Urban Air Pollution.* United Nations Environment Programme, Oxford.

USEPA, 1991. *National Air Quality and Emissions Trends, Report 1989.* US Environmental Protection Agency, Research Triangle Park, NC.

Vahter, M. & Slorach, S. 1990. *Exposure Monitoring of Lead and Cadmium: An International Pilot Study within the WHO/UNEP Human Exposure Assessment Location (HEAL) Programme.* United Nations Environment Programme, Nairobi.

Varey, R.H., Ball, D.J., Crane, A.J., Laxen, D.P.H. & Sandalls, F.J. 1988. Ozone formation in the London plume. *Atmos. Environ.* **22**: 1335-1346.

Varshney, C.K. & Aggarwal, M. 1992. Ozone pollution in the urban atmosphere of Delhi. *Atmos. Environ.* **26B**: 291-294.

WHO/UNEP, 1992. *Urban Air Pollution in Megacities of the world.* World Health Organization/United Nations Environment Programme. Blackwell, Oxford.

WHO/UNEP/GEMS, 1988. *Assessment of Urban Air Quality.* Geneva.

WSL, 1992. *Initial Analysis of NO_2 Pollution Episode, December 1991.* Warren Spring Laboratory, Stevenage, UK.

Zhao, D., Xiong, J., Xu, Y. & Chan, W.H. 1988. Acid rain in southwestern China. *Atmos. Environ.* **22**: 349-358.

2 The Role of Atmospheric Chemistry in the Assessment of Crop Growth and Productivity

S.V. KRUPA

INTRODUCTION

The chemical climatology of the earth is governed by natural processes and by human activities. It is the latter that is of much concern at local, national and international levels. Without significant alterations by people, the atmosphere of the earth can remain within the limits of its homeostasis or equilibrium (Lovelock, 1987). However, depending on their physical and chemical properties, additions of man-made air pollutants, when present at sufficient concentrations and frequencies, can cause adverse effects on appropriate environmental receptors such as crops (Krupa *et al.*, 1994). Furthermore, the original air pollutants emitted can be transformed in the atmosphere to form secondary pollutants and can also be involved in regional, continental or global-scale phenomena such as "acidic precipitation" and "climate change."

Ambient air is always composed of pollutant mixtures, with the concentrations of individual pollutants varying in time and with location. Based on their physical and chemical properties, different air pollutants exhibit varying levels of toxicity to different environmental receptors. Although present in a mixture, depending on the nature of the exposure, a particular air pollutant can be dominant in exerting adverse effects on a receptor sensitive to that pollutant. For example, sulphur dioxide (SO_2) can cause adverse effects on vegetation in the vicinity of its sources. In comparison, ozone (O_3) can be a regional or even a continental-scale problem. These two pollutants are produced by different mechanisms and also differ in their atmospheric fate (Table 2.1). Because of this type of diversity, in the following sections of this chapter an effort is made to provide a general synthesis of atmospheric processes including the mechanisms of transfer of pollutants from the atmosphere to the biosphere, a prelude to environmental impacts.

For definitions of the technical terms used in this chapter, the reader should consult Table 2.2.

The natural sources and/or sinks of air pollutants consist of volcanoes, the ocean floor, soils, vegetation and animals. In comparison, man-made sources consist of: (a) single event, point; (b) continuous, point; (c) line; (d) area; and (e) regional emitters. Given this variety of sources, air pollutants occur as gases,

Plant Response to Air Pollution. Edited by Mohammad Yunus and Muhammad Iqbal

Table 2.1. Comparison of the nature and phytotoxic properties of sulphurous pollutants and photochemical smog[a]

Characteristic	Sulphurous compounds	Photochemical smog
1. First recognized	Centuries ago	Mid-1940s
2. Place of origin of modern-day concern	London	Los Angeles
3. Primary pollutants	SO_2, sooty particles	Organic compounds, NO_x ($NO + NO_2$)
4. Secondary pollutants	H_2SO_4, aerosol sulphates, sulphonic acids, etc.	O_3, PAN, HNO_3, aldehydes, particulate nitrates, sulphates, etc.
5. Temperature	Cool ($\lesssim$35°F)	Hot ($\gtrsim$75°F)
6. Relative humidity	High, usually foggy	Low, usually hot and dry
7. Type of inversion	Radiation (ground)	Subsidence (overhead)
8. Time of air pollution peak concentrations	Early morning	Noon-evening
9. Major phytotoxic pollutant	SO_2	O_3
10. Geographic magnitude of impact	SO_2 : immediate vicinity of source regions	O_3 : regional, interregional and even continental
11. Nature of major environmental concern	SO_2 : fine particle SO_4 aerosols, acidic precipitation	Global climate change
12. Background concentration	SO_2 : $\lesssim$1 ppb	O_3 : $\lesssim$40 ppb
13. Concentration in heavily polluted areas	SO_2 : 0.2–2.0 ppm	O_3 : 0.2–0.5 ppm
14. Nature of deposition	SO_2 : dry, closer to sources. SO_4 : wet, further away from sources	O_3 : dry only
15. Velocity of dry deposition (V_d)	SO_2 : 0.1–4.5 cm s^{-1}	O_3 : 0.1–2.1 cm s^{-1}
16. Examples of sensitive plant species	SO_2 : alfalfa, barley, red clover, oats, aster, aspen, birch, jack pine	Alfalfa, barley, bean, red and white clover, grape, oats, potato, radish, soybean, tobacco, wheat, ash, aspen, black cherry, tulip poplar, white pine
17. Typical symptoms on broadleaved[b] plants	Interveinal chlorosis, necrosis	Upper surface chlorosis, bleaching, bronzing, flecking, stippling, unifacial and bifacial necrosis
18. Foliar accumulation of the pollutant	SO_2 : yes (sulphur)	O_3 : no
19. Preliminary mechanism of foliar injury	SO_2 : sulphite (SO_3^-)	O_3 : free radicals
20. Chronic effects on crops and forests	SO_2 : known (e.g., Europe)	O_3 : known (e.g., USA)

[a]Modified from Finlayson-Pitts & Pitts (1986).

[b]Symptoms on conifers similar in both cases: tip chlorosis or necrosis spreading downward. Mottling and banding can be observed with O_3.

SO_2 = sulphur dioxide; O_3 = ozone; SO_4 = sulphate.

Table 2.2. Definitions of technical terms used in this chapter

Term	Definition
Aerosol	A solid or liquid particle in a gas phase. Chemical aerosols are generally less than 2 μm in size.
Air pollutant	A gas, vapour or particle (solid or liquid) added to the atmosphere by human activities resulting in an increase in their concentration above a range, considered to represent the background.
Albedo	The ratio of the amount of electromagnetic radiation reflected by a body to the amount incident upon it, commonly expressed as a percentage. The albedo should be distinguished from the reflectivity, which refers to one specific wavelength (monochromatic radiation).
Area source	An urban complex or a city.
Brownian motion	The constant zig-zag movement of colloidal dispersions in a liquid or gas phase caused by collision between molecules (named after Robert Brown, 1773–1858).
Cloud-radiative forcing	A measure of cloud–climate interaction, indicated by the modulation of the short and long wavelength fluxes by clouds.
Coarse particle	Particles with a size of >2.0 μm.
Continuous, point source	A continuous stationary source of pollution, e.g., a coal-burning electricity generating facility.
Diffusion	A gradual mixing of the molecules of two or more substances, as a result of random thermal motion.
Dry deposition	Deposition of air pollutants on to surfaces by diffusion, Brownian motion, sedimentation, impaction or interception.
Ekman layer	The layer of transition between the surface boundary, where the shearing stress is constant, and the free atmosphere where the atmosphere is treated as an ideal fluid in approximate geostrophic (pressure gradient) equilibrium.
Fine particle	Particles with a size of <2.0 μm.
Geostrophic wind	Horizontal wind velocity for which the Coriolis acceleration exactly balances the horizontal pressure force. The geostrophic wind is thus directed along the contour lines on a constant pressure surface (or along the isobars in a geopotential surface) with low elevations (or low pressure) to the left in the northern hemisphere and to the right in the southern hemisphere.
Greenhouse effect	Inappropriate term used to define the heating of the ambient, free atmosphere due to changing tropospheric processes (atmospheric reradiative effect).
Heterogeneous oxidation	Oxidation due to gas–particle (solid or liquid) collision.
Homogeneous oxidation	Oxidation due to gas–gas collision.
Inversion	Increase, rather than a decrease in air temperature with altitude. Can start at the surface or at some height above the surface.

continued overleaf

Table 2.2 (*continued*)

Term	Definition
Line source	A source of pollution that has the configuration of a line, e.g., a highway.
Photochemical oxidants	Those substances which oxidize I^- in the KI measurement method ($2H^+ + 2I^- + \text{oxidant} \rightarrow I_2 + O_2 + H_2O$).
Photochemical smog	A combination of smoke and fog consisting of pollutants produced through chemical reactions driven by radiant energy (sunlight).
Planetary boundary layer (PBL)	Also known as atmospheric boundary layer. That layer of the atmosphere from the earth's surface to the geostrophic wind level including, therefore, the surface boundary layer and the Ekman layer. Above this layer lies the free atmosphere.
Primary air pollutant	An air pollutant that is directly emitted by a source.
Rainout	Origin of raindrops from within the cloud leading to the removal of pollutants from within and from the atmosphere between the clouds and the surface.
Regional source	A large geographic region consisting of two or more area sources.
Secondary air pollutants	Air pollutants derived from the primary pollutants, secondarily as a result of reactions in the atmosphere, e.g., ozone.
Sedimentation	Deposition of particles greater than 1 μm in size through the influence of gravitational (centrifugal) forces.
Single-event, point source	A random, discontinuous source of air pollution, e.g., a chemical spill or an accident.
Stratosphere	Earth's atmosphere between altitudes of 10 and 50 km where temperature increases with altitude.
Subsidence	A descending motion of the air in the atmosphere, usually with the implication that the condition exists over a broad area. This involves a temperature inversion produced by the adiabatic warming of a layer of subsiding air.
Surface boundary layer	That thin layer of air adjacent to the earth's surface extending up to the so-called anemometer level (the base of the Ekman layer). Within this layer the wind distribution is determined largely by the vertical temperature gradient and the nature and contours of the underlying surface.
Troposphere	Earth's atmosphere for approximately the first 10 km where temperature decreases with altitude (ignoring localized radiation or subsidence inversions).
UV-B	Ultraviolet radiation in the wavelength band of 280–320 nm.
Washout	Origin of rain from an air layer below the cloud layer. Therefore, pollutants within the cloud are not removed.
Wet deposition	Deposition process of air pollutants from the atmosphere to surfaces by rain, snow, sleet, fog, etc.

Table 2.3. Selected air pollutants and their sources of origin[a]

Air pollutant	Some sources of origin
Carbon dioxide (CO_2)	Fossil fuel combustion; natural processes; land-use conversion
Ozone (O_3)	Photochemical reactions in the atmosphere; storm centres; natural occurrence in the upper atmosphere
Sulphur dioxide (SO_2)	Combustion of fossil fuels; petroleum and natural gas industries; metal smelting and refining processes
Hydrogen fluoride (HF)	Aluminium industries; steel manufacturing industries; phosphate fertilizer plants; brick plants
Peroxyacetyl nitrate (PAN)	Photochemical reactions in the atmosphere
Oxides of nitrogen (NO_x)	Cars and trucks; combustion of natural gas, fuel oil and coal; refining of petroleum; compressors of natural gas; incineration of organic waste; wood burning
Particulate matter (PM):	
Fine (<2.0 μm)	Predominant portion from homogeneous and heterogeneous reactions in the atmosphere; coal and fossil fuel combustion; metal smelting and refining processes; a variety of other industries including waste incineration
Coarse (>2.0 μm)	Cement mills; lime kilns; incinerators; combustion of coal, petrol and fuel oil; a variety of industries
Ethylene (C_2H_4) and related olefins	Motor vehicles; refuse burning; combustion of coal and oil; leaky natural gas heaters; natural occurrence
Ammonia (NH_3)	Leaks or breakdown in industrial operations; spillage of anhydrous ammonia; feedlots and stockyards
Chlorine (Cl) and hydrogen chloride (HCl)	Petroleum refineries; glass industries; plastic incineration; scrap burning; accidental spills
Metals	Same as sources of particulate matter (with the exception of reactions in the atmosphere); industry involving some form of combustion

[a]With the exception of PAN, ethylene and related olefins, other organic pollutants are not considered.

as vapours and as liquid or solid particles (Table 2.3). The particulate matter in the atmosphere can be classified as fine (<2.0 μm) and coarse (>2.0 μm) fractions. While coarse particles are mechanically (physically) generated, a predominant portion of the fine particles is produced by homogeneous and heterogeneous reactions in the atmosphere.

The photo-oxidation cycle (see Figure 2.1) serves as the key scheme for gas-phase homogeneous reaction chemistry. The sources of the precursor pollutants represented in Figure 2.1 are listed in Table 2.3. Demerjian (1986), Finlayson-Pitts & Pitts (1986) and Wayne (1987) have summarized information relevant to the chemistry of the clean troposphere. Alterations introduced as a result of human activity on the photo-oxidation cycle within the atmosphere are predominantly due to two classes of compounds: volatile organic compounds (VOCs) and oxides

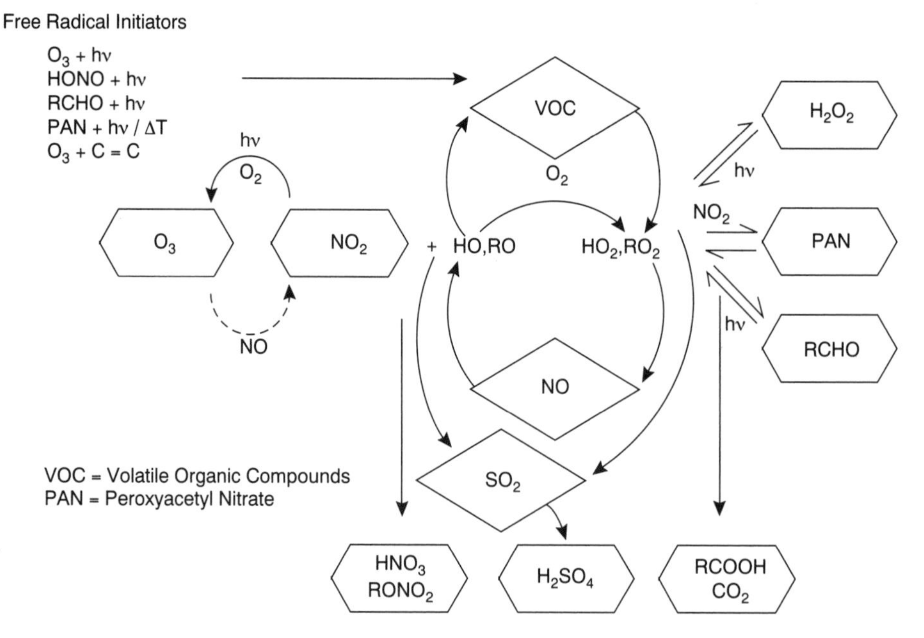

Figure 2.1. Schematic representation of the polluted atmosphere photo-oxidation cycle. Pollutants emitted from anthropogenic sources (primary pollutants) are shown in diamond-shaped boxes; pollutants formed in the atmosphere (secondary pollutants) are shown in hexagonal boxes. The reaction scheme represents gas-phase chemistry only. (Source: courtesy of Demerjian, 1986, and John Wiley & Sons, New York)

of nitrogen (NO_x) (see Figure 2.1). Free radical reaction on VOCs is initiated by a select group of compounds which, for the most part, are activated by sunlight. Formaldehyde (HCHO) and nitrous acid (HONO or HNO_2), in particular, have been shown to have high potential as free radical initiators during the early morning sunrise period. After the initial free radical attack, the VOCs decompose through paths resulting in the production of peroxy radicals (HO_2, RO_2, $^1RO_2$, etc.), and partially oxidized products which in themselves may be photoactive radical producing compounds. The peroxy radicals react with nitric oxide (NO) converting it to nitrogen dioxide (NO_2) and in the process produce hydroxy and alkoxy radicals (OH, RO, R^1O, etc.). The alkoxy radicals can be further oxidized, forming additional peroxy radicals and partially oxidized products, thereby completing the inner loop reaction chain shown in Figure 2.1, or they may attack (as would be the major path for the hydroxyl radical) the VOC pool present in the polluted atmosphere, thereby completing the outer loop of the reaction chain. In addition to the gas-phase chemistry described in Figure 2.1, SO_2 can also be oxidized by heterogeneous reactions (Finlayson-Pitts & Pitts, 1986; Hidy & Mueller, 1986).

In power-plant plumes, SO_2 oxidation rates of up to 4% h^{-1} have been reported (Husar *et al.*, 1978). Often such rates are much higher if the plume passes through clouds or fog banks (Eatough *et al.*, 1984). Similarly, the rates of production of SO_4^{2-} from SO_2 are much higher during the summer compared with the winter

(Richards *et al.*, 1981). According to Gillani (1978) and Forrest *et al.* (1981), noon-time SO_2 conversion rates in a power-plant plume were 1–4% h^{-1} compared with the night-time rates of <0.5% h^{-1}. However, significant SO_4^{2-} production (4.5–10.8% h^{-1}) can occur at night-time if clouds were a contributing factor in the SO_2 conversion (Cass & Shair, 1984). For details of the suggested mechanisms for the aqueous oxidation of SO_2, the reader is referred to Graedel & Goldberg (1983), Bielski *et al.* (1985), and Graedel *et al.* (1986).

Compared to SO_2, the oxidation of NO_x to nitrate (NO_3^-) in power-plant plumes and in ambient air is less understood. Homogeneous oxidation of NO to HNO_3 is through its interaction with O_3 and $HO_2(RO_2)$ and OH radicals. NO_2 can be oxidized to nitrate (NO_3^-) through gas-phase followed by aqueous-phase reactions (Schwartz, 1989). In power-plant plumes, rates of conversion of NO_x from roughly 0.2% up to 12% h^{-1} have been observed, with rates being much greater during the middle of the day than at night (Hegg & Hobbs, 1979; Richards *et al.*, 1981). The products of NO_x oxidation appear to be PAN (peroxy-acetyl nitrate) and nitric acid (HNO_3). In urban plumes, rates of conversion of NO_x of <5% h^{-1} up to 24% h^{-1} have been reported (Chang *et al.*, 1979; Spicer, 1982a, b).

Two of the products shown in Figure 2.1, the H_2SO_4 (sulphuric acid) and HNO_3 (nitric acid) differ in their physical and chemical behavior in the atmosphere. HNO_3 is more volatile and thus significant concentrations of that substance can exist in the gas phase. In comparison, H_2SO_4 has a low vapour pressure under ambient conditions and thus exists in the fine-particle phase (Whitby, 1978; Roedel, 1979). These particles are hygroscopic and, in addition to causing visibility degradation, act as cloud-condensation nuclei (Husar *et al.*, 1978). Both H_2SO_4 and HNO_3 react with ammonia (NH_3) and other cations (e.g., Ca^{2+}, Mg^{2+}) to form salts, for example, ammonium bisulphate (NH_4HSO_4), letovicite ($(NH_4)_3H(SO_4)_2$) and ammonium sulphate ($(NH_4)_2SO_4$). Because of the characteristics of the equilibrium between HNO_3, NH_3 and NH_4NO_3 (ammonium nitrate),

Table 2.4. The distribution of some chemical species of relevance to dry deposition, among trace gas, fine particle and coarse particle fractions

	Trace gases	Fine particles	Coarse particles
Sulphur species	SO_2	Most SO_4^{2-}	Some SO_4^{2-}
Nitrogen species	HNO_3 NH_3 NO_x	Most NH_4^+ Most NO_3^-	Some NH_4^+ Some NO_3^-
Oxidants	O_3 H_2O_2*		
Carbon species	VOCs*	Organic acids	Graphitic compounds
Trace metals		Dominant	Some
Crustal materials		Some	Dominant
Oceanic materials		Some NaCl	Most NaCl

*H_2O_2 = hydrogen peroxide; VOCs = volatile organic compounds.
Source: US NAPAP (1987).

HNO_3 can volatalize relatively easily even after forming NH_4NO_3 (Finlayson-Pitts & Pitts, 1986). No analogous physical and chemical changes exist for H_2SO_4. Nevertheless, H_2SO_4 and other fine particles containing a variety of anions and cations are produced by gas-phase reactions, by the reactions of gases on the surfaces of pre-existing solid and liquid particles, and by reactions within the particle volume. It is very important to note that these fine particles are the predominant regional or continental atmospheric source of sulphur and nitrogen species and trace metals. In comparison, oceanic and crustal elements predominate in the coarse fraction (Table 2.4).

The preceding discussion of sulphur and nitrogen chemistry is closely linked to the phenomenon of acidic precipitation. This subject is discussed in greater detail in a later section of this chapter.

ATMOSPHERIC TRACE GASES AND GLOBAL CHANGE

There is an increasingly growing international concern regarding human influence on global climate (IPCC, 1990). The issues of concern are: (1) depletion of beneficial, naturally occurring stratospheric O_3 and a consequent increase in tropospheric ultraviolet (UV)-B radiation; (2) the increases in the surface emissions of atmospheric trace gases resulting in the modified "greenhouse effect" and global warming, and changes in precipitation and wind patterns; and thus, (3) possible drastic alterations in terrestrial and aquatic ecosystems.

The "greenhouse effect" and climate modification are governed by the interactions between tropospheric and stratospheric processes (Wuebbles *et al.*, 1989). A key atmospheric constituent participating in these interactions is O_3. Ozone concentrations vary with altitude above the earth's surface roughly as follows: 0–10 km (troposphere), 10% by volume; 10–35 km, 80%; and above 35 km, 10% (Cicerone, 1987). In the troposphere O_3 concentrations also vary with the latitude (Pruchniewicz, 1973).

In the stratosphere, a series of photochemical reactions involving O_3 and molecular O_2 occur. Ozone strongly absorbs solar radiation in the region from $\approx$210 to 290 nm, whereas O_2 absorbs at $< \approx 200$ nm. The absorption of light primarily by O_3 is a major factor causing the increase in temperature with altitude in the stratosphere. Excited O_2 and O_3 photodissociate, initiating a series of reactions in which O_3 is both formed and destroyed leading to a steady-state concentration of O_3 (Finlayson-Pitts & Pitts, 1986).

In comparison, in the troposphere NO_2 exposed to radiation photodissociates at $\lesssim$420 nm.

$$NO_2 + h\nu \rightarrow NO + O$$

The quantum yield for O atom production has been studied extensively because of its role as the only significant anthropogenic source of O_3 in the troposphere.

$$O_2 + O \overset{M}{\rightarrow} O_3$$

The threshold wavelength for the production of ground-state NO and O atoms is 397.8 nm. This is a very important point, since some effects scientists seem to

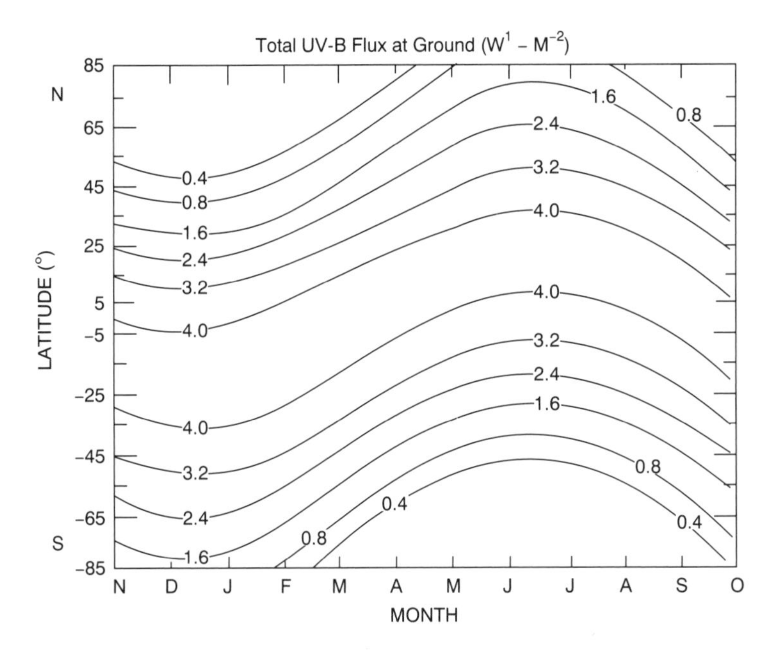

Figure 2.2. The latitudinal and monthly distribution of UV-B radiation at the ground computed for clear sky conditions and a local time of 10:00 am. Values include all wavelengths between 280 and 320 nm. (Source: Frederick, 1986; US EPA and UNEP)

think that increased surface UV-B (280–320 nm) will increase tropospheric O_3 production. Figure 2.2 shows the latitudinal and monthly distribution of UV-B radiation at the ground level computed for clear sky conditions.

Stratospheric O_3 serves as a shield against biologically harmful solar UV radiation, initiates key stratospheric chemical reactions, and transforms solar radiation into heat and the mechanical energy of winds. Also, the downward intrusions of stratospheric air supply the troposphere with the O_3 necessary to initiate photochemical processes in the lower atmosphere (see Figure 2.1). The flux of photochemically active UV-B photons (wavelength 280–320 nm) into the troposphere is limited by the amount of stratospheric O_3 (Cicerone, 1987). In addition to this protective effect of stratospheric O_3 against UV, clouds reflect a large part of the incoming solar radiation, causing the albedo of the entire earth to be about twice what it would be in the absence of clouds (Cess, 1976). Clouds cover about one-half of the earth's surface, doubling the proportion of sunlight reflected back into space to 30% (see Figure 2.3).

Ever since the publications of Johnston (1971) and Molina and Rowland (1974) human activities have been projected to substantially deplete the stratospheric O_3 through anthropogenic increases in the global concentrations of key atmospheric trace chemicals (methane, CH_4; nitrous oxide, N_2O; synthetic chlorofluorocarbons, CFCs; and organobromines, OBs). Cicerone (1987) has provided an excellent treatment of this problem. Figure 2.4 shows examples of large-scale processes that produce and transfer source gases, which undergo irreversible photo-oxidation to yield important gaseous radicals to the stratosphere.

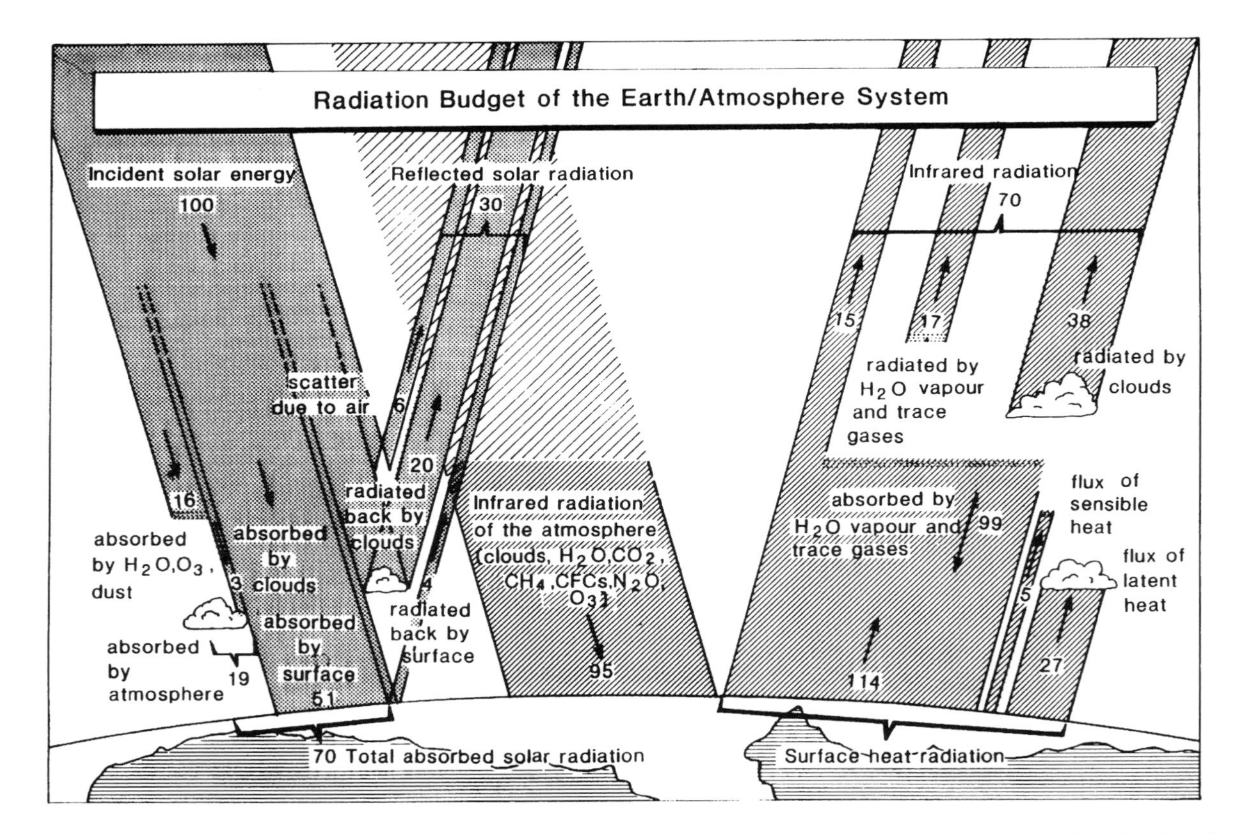

Figure 2.3. Radiation budget of the earth/atmosphere system. Distribution of incident solar radiation at surface level and in the atmosphere. Totals are shown at the boundaries of the atmosphere and the earth's surface. Short-wave radiation fluxes are depicted on the left-hand side, while long-wave radiation fluxes of the earth and the atmosphere are shown on the right-hand side. In both cases, the incident solar radiation at the outer boundary of the atmosphere is used as a reference, i.e., 100. Changes in greenhouse gas concentrations provoke changes in temperature. (Source: German Bundestag, 1991)

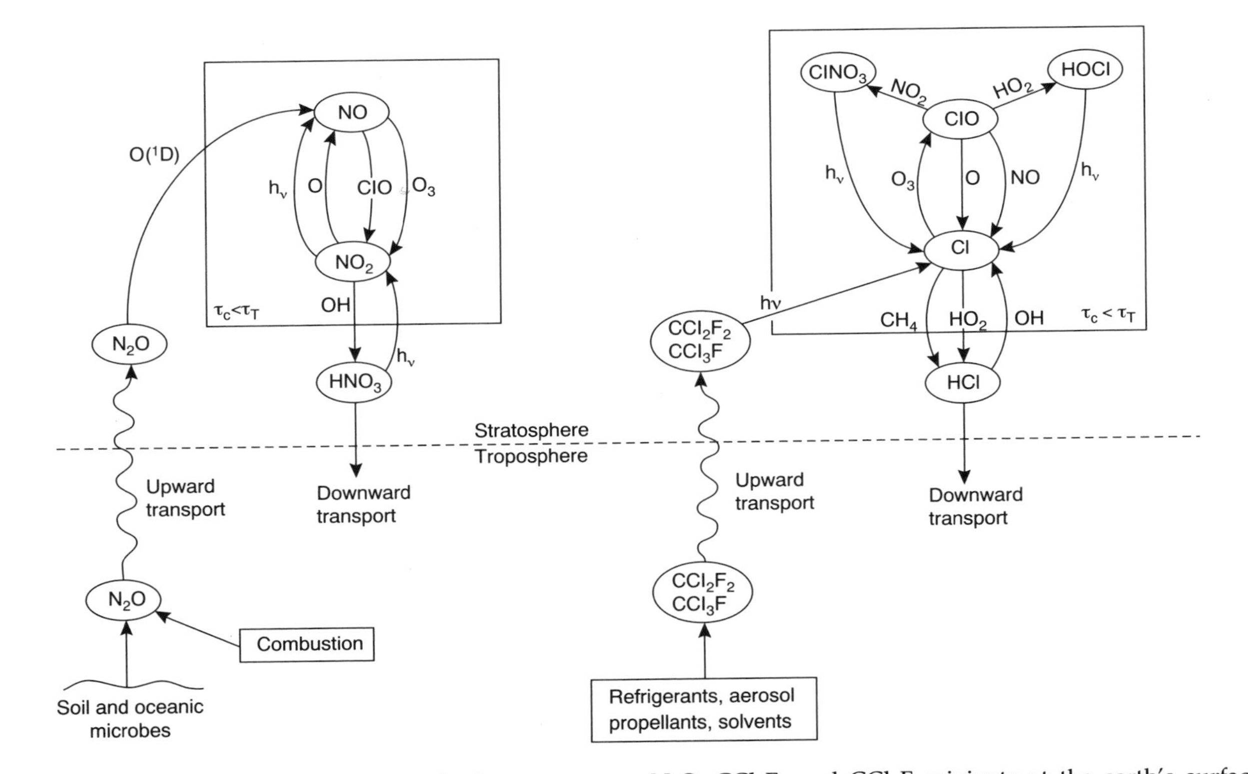

Figure 2.4. A schematic depiction of how stratospheric source gases N_2O, CCl_2F_2 and CCl_3F originate at the earth's surface and are transported upward into the stratosphere, where they are irreversibly photo-oxidized to yield key gas-phase radicals. Reactants shown inside the boxes undergo reactions with time constants τ_c that are less than τ_T (the time required for vertical transport). Similarly, some CH_4 reaches the stratosphere, where it gives rise to H_2O, H_2 and HO_x. (Source: Cicerone, 1987. Copyright 1987 by the AAAS)

Attempts to predict the future effects of continued increases in stratospheric source gases (e.g. CFCs) have given rise to various mathematical models. Simulated CFC releases lead to decreases in the O_3 column at all latitudes (Isaksen & Stordal, 1986). Larger decreases in the O_3 column were calculated for high latitudes (>40°) than for low latitudes.

Although UV-B constitutes only 1.5% of the total solar radiation reaching the surface prior to attenuation by the earth's atmosphere, reduced amounts of stratospheric O_3 will permit disproportionately large amounts of specific wavelengths of UV-B radiation to penetrate through the atmosphere (Cutchis, 1974).

According to Frederick *et al.* (1991), the long-term data on atmospheric O_3 combined with a set of radiative transfer calculations provides estimates of the variability in UV that should have occurred over the period 1957–1988 under clear, pollution-free skies. In this analysis no statistically significant trends were observed for the entire 32-year time period for the latitudes 30 to 64°N. However, over a shorter period (1970–1988), the annually integrated erythemal irradiance showed an upward trend of $+2.1 \pm 1.2\%$ per decade based on all O_3 data at latitudes from 40 to 52°N. No trends existed in lower (30–39°N) and higher (53–64°N) latitude bands. The authors caution that their trend line provides a very simple index of the variability in UV and their results should not necessarily be extrapolated into the future.

Since the studies of Frederick *et al.* (1991), evidence has been provided for the first time to show a clear increase in surface-level UV-B radiation at a North American site. Spectral measurements of UV-B radiation at Toronto (44°N) since 1989 indicate that the intensity of light at wavelengths near 300 nm has increased by 35% per year in winter and 7% per year in summer. The wavelength dependence of these trends indicates that the increase is caused by the downward trend in total O_3 that was measured at Toronto during the same period. The trend at wavelengths between 320 and 325 nm is essentially zero (Kerr & McElroy, 1993). However, these types of measurements need to be extended across all continents before broad generalizations can be made regarding surface-level UV-B radiation.

With or without the predicted and/or observed changes, the incoming solar radiation to the surface of the earth is in the ultraviolet to near infrared wavelength bands. After some absorption, surfaces reradiate heat energy back to the atmosphere at long-wavelength infrared. This energy is trapped by certain atmospheric chemical constituents and by clouds, leading to a warming of the atmosphere above the surface, the "greenhouse effect" (see Figure 2.3). The critical concern at this time is whether human influence has increased and accelerated this "greenhouse effect" towards progressive global warming leading to disastrous ecological consequences (Houghton & Woodwell, 1989).

Table 2.5 provides a summary of the information concerning atmospheric trace gases of significance in global change. Surface concentrations of radiatively active trace gases are increasing. Using general circulation models (GCMs) global-surface temperatures have been predicted to increase by 1.5 to 4.5°C from the 19th to the 21st century (Wuebbles *et al.*, 1989). This effect is attributed predominantly to increasing CO_2 concentrations (Table 2.5).

Table 2.5. Atmospheric trace gases that are radiatively active and of significance to global change

	Carbon dioxide (CO_2)	Methane (CH_4)	Nitrous oxide (N_2O)	Chlorofluoro-carbons (CFCs)	Tropospheric ozone (O_3)	Carbon monoxide (CO)	Water vapour (H_2O)
Greenhouse role	Heating	Heating	Heating	Heating	Heating	None	Heats in air; cools in clouds
Effect on stratospheric ozone layer	Can increase or decrease	Can increase or decrease	Can increase or decrease	Decrease	None	None	Decrease
Principal anthropogenic sources	Fossil fuels; deforestation	Rice culture; cattle; fossil fuels; biomass burning	Fertilizer; land-use conversion	Refrigerants; aerosols; industrial processes	Hydrocarbons (with NO_x); biomass burning	Fossil fuels; biomass burning	Land-use conversion; irrigation
Principal natural sources	Balanced in nature	Wetlands	Soils; tropical forests	None	Hydrocarbons	Hydrocarbon oxidation	Evapotran-spiration
Atmospheric lifetime	50–200 years.	10 years	150 years	60–100 years	Weeks to months	Months	Days
Present atmospheric concentration in parts per billion by volume (ppbv) at surface	353 000	1720	310	CFC-11: 0.28 CFC-12: 0.48	25–45*	100*	3000–6000 in stratosphere

continued overleaf

Table 2.5 (*continued*)

	Carbon dioxide (CO_2)	Methane (CH_4)	Nitrous oxide (N_2O)	Chlorofluoro-carbons (CFCs)	Tropospheric ozone (O_3)	Carbon monoxide (CO)	Water vapour (H_2O)
Preindustrial concentration (1750–1800) at surface (ppbv)	280 000	790	288	0	10	40–80	Unknown
Present annual rate of increase (%)	0.5	0.9	0.3	4	0.5–2.0*	0.7–1.0*	Unknown
Relative contribution to the anthropogenic greenhouse effect (%)	60	15	5	12	8	None	Unknown

*Northern hemisphere data.
Source: Modified from EarthQuest (1990).

Many predictions of global climate change have been through the use of one or two, rather than three-dimensional circulation models. Certainly the application of three-dimensional models is limited by the present-day availability of computer power. Many of the models have been unable to fully include the role of cloud radiative forcing. Studies on cloud radiative forcing based on the Earth Radiation Budget Experiment (ERB) show atmospheric cooling over North America (Ramanathan *et al.*, 1989). Clouds appear to have a net cooling effect globally of about four times as much energy as would be trapped by doubling CO_2 levels. However, while in mid and high latitudes the net cooling from clouds is large, over the tropics, their cooling is nearly cancelled by heating. Even within these general tendencies, a great deal of geographic patchiness and variability must be expected. Long-term average values describing the global change cannot be as valuable as defining the standard deviations of the mean values at a local level. Given the geographic patchiness, locations may vary in the future in greater or fewer number of days of high or low temperatures and wet or dry conditions that will be ecologically significant. This is an area that requires major research emphasis in the future.

ATMOSPHERIC DEPOSITION

Depending on their chemical reactivity or conversely their inertness in the troposphere, while certain molecules such as the CFCs are transported upward into the stratosphere, others such as SO_2 are converted to secondary pollutants. Nevertheless, in the troposphere deposition processes limit the lifetime of most pollutants in the atmosphere, control the distance travelled before deposition, and limit their atmospheric concentrations.

Dry deposition leads to the direct collection of gases, vapours and particles on land and water surfaces. Dry deposition begins with turbulent transport and/or sedimentation to the near-surface layer which is then penetrated by convection, diffusion or inertial processes, and ends with physical or chemical capture of the pollutants by the surface (Voldner *et al.*, 1986). The rate of pollutant transfer by dry deposition is controlled by a wide range of physical, chemical and biological factors which vary in their relative importance according to the nature and state of the surface, the characteristics of the pollutant and the state of the atmosphere (US National Research Council, 1983). The complexity of the individual processes involved and the variety of possible interactions among them combine to prohibit easy generalization; nevertheless, a "deposition velocity" (V_d), analogous to a gravitational falling speed, is of considerable use. In practice, knowledge of V_d enables fluxes (F) to be estimated from air concentrations (C) as the simple product, $V_d \cdot C$ (US National Research Council, 1983). Table 2.6 provides a summary of some of the results obtained in studies on the dry deposition of air pollutants on a variety of surfaces. For more information on this subject, the reader is referred to Garland (1978), Sehmel (1980), International Electric Research Exchange (1981), Hosker & Lindberg (1982), US National Research Council (1983), Hicks (1984), Voldner *et al.* (1986), VDI (1987), and Davidson & Wu (1989). However, it is important to note that dry deposition is the dominant removal process for gases and coarse particles closer to their sources of origin.

Table 2.6. Typical ranges of deposition velocities (V_d) reported for some pollutants and surfaces

Species	Surface	Deposition velocity[a] V_d (cm s^{-1})
O_3	Soil, short grass	0.10–2.10
	Grass, soil, water	0.47–0.55
	Maize, soybean field	0.20–0.84
NO	Soil, cement	0.10–0.20
NO_2	Soil, cement	0.30–0.80
	Alfalfa	1.90
PAN	Grass, soil	0.14–0.30
	Alfalfa	0.63
HNO_3	Grassy field[b]	1.0–4.7
SO_2	Grass[c]	0.1–4.5
	Pine forest[d]	0.1–1.0
Particulate sulphur (includes H_2SO_4)	Frozen bare soil[e]	[g]
	Deciduous forest, wintertime[e]	[g]
	Grassy pasture, drought[e]	0.17–0.24
	Pine forest[e]	0.48–0.90
	Grass[e]	0.02–0.2
	Grassy field[f]	0.1–1.2

[a]Except where stated elsewhere, from McRae & Russell (1984) and references therein.
[b]Huebert & Robert (1985).
[c]Davis & Wright (1985).
[d]Garland (1978) and references therein.
[e]Wesely *et al.* (1985).
[f]Feely *et al.* (1985).
[g]Indistinguishable from zero.
Source: Finlayson-Pitts & Pitts (1986). Refer to this source for specific citations given here.

As opposed to the discussion of dry deposition, both SO_2 and sulphate (SO_4^{2-}) contribute significantly to the dissolved sulphur in rain. The contribution of SO_4^{2-} appears inevitable, since SO_4^{2-} particles serve as cloud condensation nuclei. However, the incorporation of SO_2 may be suppressed if the condensation nuclei are initially acidic. Garland (1978) has summarized the information on mechanisms contributing to sulphur in rain-water. Overall, wet deposition constitutes the major removal process for SO_4^{2-} and NO_3^- further away from their sources of origin.

In addition to the rainout of condensation nuclei, several other physical processes may contribute to SO_4^{2-} in rain. Diffusophoresis and Brownian diffusion may result in collection of small particles to the cloud droplets and raindrops may further collect particles by impaction, interception or diffusion. However, according to Garland (1978), only the rainout of condensation nuclei appears capable of explaining the concentrations of several milligrams per litre of SO_4^{2-} observed in practice.

The washout of large particles by raindrops may make a significant contribution, but this fraction of the aerosol will be exhausted by the first few millimetres of rain and may, therefore, account for the enhancement in SO_4^{2-} concentration observed at the beginning of some periods of rain (Garland, 1978; Pratt *et al.*, 1983a).

Diffusion and interception may be of greater significance in snow because of the larger surface area of the precipitation elements. However, the concentration of condensation nuclei collected in precipitation may be greatly reduced if distillation from liquid to solid phase dominates the aggregation of cloud droplets in the growth of snowflakes.

In summary, the probable contribution of dissolved SO_2 is smaller than the contribution due to the rainout of SO_4^{2-}. However, as previously stated, oxidation of SO_2 in clouds can make a substantial contribution to the SO_4^{2-} in rain.

In contrast to SO_4^{2-}, less information has been published regarding the removal mechanisms of nitrogen species by precipitation. Both theory and experimental evidence suggest that HNO_3 can be formed rapidly from a combined gas-phase/liquid-phase process (Schwartz, 1989). Evidence favours the probable importance of the formation of dinitrogen pentoxide (N_2O_5) from NO_2 followed by its reaction with water droplets to form HNO_3 which can be effectively scavenged by precipitation.

Because of the significant concern arising from the occurrence of "acidic precipitation", numerous investigators have examined the qualitative and quantitative aspects of precipitation chemistry in the last 20 years (Chamberlain *et al.*, 1981; US National Research Council, 1983; Teasley, 1984; US National Academy Press, 1986; VDI, 1987; Knapp *et al.*, 1988; Longhurst, 1989).

The pH of natural precipitation is often assumed to be regulated by the dissociation of dissolved carbon dioxide (CO_2), thus having a value of 5.6, and precipitation pH values below 5.6 are due to the addition of acidic components (primarily related to SO_4^{2-} and NO_3^-) by human activity (Garrels & MacKenzie, 1971; Likens & Bormann, 1974; Galloway *et al.*, 1976). Table 2.7 provides as an example, a summary of the molar ratios of SO_4^{2-} to NO_3^- for several sites in the northeastern United States during 1977-1981.

If SO_4^{2-} and NO_3^- are responsible for most of the free acidity (pH) in precipitation, a strong statistical relationship should be observed between H^+ and SO_4^{2-} and/or NO_3^-. This degree of association is site specific, and proximity

Table 2.7. Molar ratios of sulphate to nitrate in precipitation in the United States, 1977-1981

Sampling location	April-September*	October-March*	Annual average
Whiteface Mt, NY	1.4 (28)	0.65 (26)	1.08
Ithaca, NY	1.4 (25)	0.62 (21)	1.07
University Park, PA	1.3 (27)	0.74 (29)	1.04
Charlottesville, VA	1.3 (27)	0.81 (21)	1.16
Urbana, IL	1.4 (15)	0.94 (13)	1.25
Brookhaven, NY	1.12 (17)	1.0 (17)	1.10
Lewes, DE	1.4 (18)	0.93 (18)	1.16
Oxford, OH	1.5 (16)	1.2 (19)	1.37
Average	1.4 ± 0.1	0.86 ± 0.19	

*Numbers in parenthesis are the number of months of data in each sample.
Source: Adapted from MAP3S/RAINE Research Community (1982).

to sources of SO_2 and NO_x as well as sources of other substances may influence the chemical properties of the acidic substances in the atmosphere (Lefohn & Krupa, 1988a). At a given site, differences in the meteorology between precipitation events may result in wide variations in the measured acidity, concentrations of SO_4^{2-} and NO_3^- and relationships of these anions with H^+ and other cations (Pratt & Krupa, 1985).

The pH of atmospheric precipitation at a given location depends on the chemical nature and relative proportions of acids and bases in the solution. Sequeira (1982) concluded that a pH of 5.6 may not be a reasonable reference value for unpolluted precipitation. Charlson and Rodhe (1982) also questioned the validity of using pH 5.6 as the background reference point, citing naturally occurring acids as possibly responsible for low pH values of rain. They stated that consideration of the natural cycling of water and sulphate through the atmosphere, precipitation rates, and experimentally determined rates of SO_4^{2-} scavenging, indicates that average pH values of approximately 5.0 would be expected in pristine locations in the absence of basic materials. This value will vary considerably due to the variability in scavenging efficiencies as well as geographic patchiness in the sulphur and hydrological cycles. Thus, precipitation pH values might range from 4.5 to 5.6 due to these variabilities alone (Charlson & Rodhe, 1982). Lefohn & Krupa (1988a) found that the pH of precipitation with minimum concentrations of SO_4^{2-} and NO_3^- (relatively clean) was in the range of 4.6 to 5.5 for the northeastern United States.

The preceding discussion illustrates the complexity of precipitation chemistry and the difficulty in deriving generalized, uniform temporal and spatial bases for the acidity of rain. Nevertheless, Table 2.8 provides a listing of inorganic ions important in precipitation chemistry. Wet deposition of these ions over a unit area is frequently comparable or greater (particularly further away from source regions) than the dry deposition of these and related chemical species. However, the relative contribution of the two processes is known to exhibit significant temporal and spatial variability and, thus, their individual significance should not be generalized in the context of effects research.

Table 2.8. Some inorganic ions important in precipitation chemistry*

Cations	Anions
H^+	Cl^-
NH_4^+	NO_3^-
Na^+	SO_3^{2-}
K^+	SO_4^{2-}
Ca^{2+}	PO_4^{3-}
Mg^{2+}	CO_3^{2-}

*All ions are presented here in their completely dissociated states. The reader should note, however, that various states of partial dissociation are possible as well (e.g., HSO_3^-, HCO_3^-).
Source: US National Research Council (1983).

GEOGRAPHIC TRANSPORT AND PATTERNS OF DISTRIBUTION

Tables 2.9 and 2.10 provide a summary of the relative concentrations of various chemical constituents in clean and polluted atmospheres. As stated previously, concentrations of individual pollutants vary in time and with location. Locally,

Table 2.9. Concentrations of some trace gases in the clean troposphere and in typical urban polluted atmospheres

Species	Clean troposphere (ppb)	Polluted atmosphere (ppb)	References
O_3	30	100-200	National Research Council (1976); Singh *et al.* (1978)
HO	1.0×10^{-5}-2.0×10^{-4}	4.1×10^{-4}-2.4×10^{-3}	Wang *et al.* (1975); Davis *et al.* (1976)
HO_2	10^{-4}-10^{-2}	0.1-0.2	Calvert & McQuigg (1975); Cox *et al.* (1976)
N_2O	330	—	Cicerone *et al.* (1978)
NO	0.01-0.05	60-740	Drummond (1977); Ritter *et al.* (1978); Air Quality Criteria for Oxides of Nitrogen (1980)
NO_2	0.1-0.5	40-220	Noxon *et al.* (1978); Ritter (1978); Air Quality Criteria for Oxides of Nitrogen (1980)
HONO	10^{-3}-10^{-1}	4-21	Nash (1974)
$HONO_2$	0.02-0.3	6-20	Huebert & Lazrus (1978); Doyle *et al.* (1979)
PAN[a]	1	10-65	Lonneman *et al.* (1976)
NH_3	0.1-1	20-80	Dawson (1977); Doyle *et al.* (1979)
NH_4NO_3	0.03-0.5	8-30	Doyle *et al.* (1979)
H_2	500	—	Schmidt (1974)
H_2O_2	0.1-1	5-40	Bufalini *et al.* (1972); Kok *et al.* (1978)
CH_4	1.6-1.7×10^3	2-3×10^3	Fink *et al.* (1964); Altshuller *et al.* (1973)
NMHC[b]	5-10	10^2-10^3	Robinson *et al.* (1973); Leonard *et al.* (1976)
HCHO	0.1-1	10-40	Altshuller & McPherson (1963)
CO	50-200	10^2-10^4	Seiler (1974); EPA National Air Quality and Emissions Trends Report (1977)
CO_2	3.3×10^5	—	Lowe *et al.* (1979)

[a]PAN = peroxyacetyl nitrate.
[b]NMHC = non-methane hydrocarbons.
Source: Demerjian (1986), refer to this paper for full references cited in the table. (Reprinted with permission, John Wiley & Sons, Inc. Copyright 1986.)

Table 2.10. Typical peak concentrations (as 1-hour averages) of gas-phase criteria pollutants observed in the troposphere over the continents

	Type of atmosphere			
Pollutant	Remote	Rural	Moderately polluted	Heavily polluted
CO	$\lesssim$0.2 ppm[a]	0.2–1 ppm[b]	~1–10 ppm[c]	10–50 ppm
NO_2	$\lesssim$1 ppb[d,a,e]	1–20 ppb[f,g]	0.02–0.2 ppm[c]	0.2–0.5 ppm
O_3	$\lesssim$0.05 ppm[d,a]	0.02–0.08 ppm	0.1–0.2 ppm	0.2–0.5 ppm
SO_2	$\lesssim$1 ppb[h]	~1–30 ppb[g]	0.03–0.2 ppm	0.2–2 ppm
NMHC	$\lesssim$65 ppb[a]	100–500 ppb[b]	300–1500 ppb[i]	$\gtrsim$1.5 ppm

NMHC = Non-methane hydrocarbons.
[a]Kelly *et al.* (1982); Hoell *et al.* (1984).
[b]Seila (1979).
[c]Ferman *et al.* (1981).
[d]Kelly *et al.* (1980).
[e]Johnston & McKenzie (1984).
[f]Spicer *et al.* (1982); Pratt *et al.* (1983).
[g]Martin & Barber (1981).
[h]Ludwick *et al.* (1980); Maroulis *et al.* (1980).
[i]Sexton *et al.* (1982).
Source: Finlayson-Pitts & Pitts (1986). Refer to that book for full references cited here.

ground-level concentrations of air pollutants will be high during radiation and subsidence inversions. These types of situations can lead to acute foliar injury on vegetation. At some distance from a source, the plume will effectively fill a layer or dispersing volume of the atmosphere. It will then disperse over large distance with apparently much less dilution. This property relates to long-distance transport. Long-distance transport is favoured by large, slow-moving high-pressure systems, frequently characterized by combinations of stability, wind and wind-shear, such that the persistence of pollutants within the air mass is favoured. High-pressure systems are also characterized by nearly cloud-free conditions. This leads to strong sunlight and conditions for the production of photochemical pollutants such as O_3 and fine particle SO_4^{2-}. Such conditions are also not conductive for rainfall, thus pollutants may persist in the atmosphere for a considerable amount of time and be carried over hundreds to thousands of kilometres.

Wind-shear or the change in wind direction with height becomes important in long-range transport, since the shear provides for a mechanism of dilution within the polluted air mass. However, even with the wind-shear, other properties of the air mass may be favourable for the persistence of relatively high concentrations of the pollutants.

One of the most commonly discussed consequences of long-range transport is the phenomenon of acidic precipitation. Table 2.11 and Figure 2.5 provide an example of the relationships between major source regions, modelled path of transport and precipitation chemistry in one part of north central United States Similar examples are available for other parts of North America, Europe and other parts of the world (Husar *et al.*, 1978; Pack *et al.*, 1978; Smith & Hart, 1978; Ottar *et al.*, 1984; US National Academy Press, 1986; VDI, 1987; Schwartz, 1989).

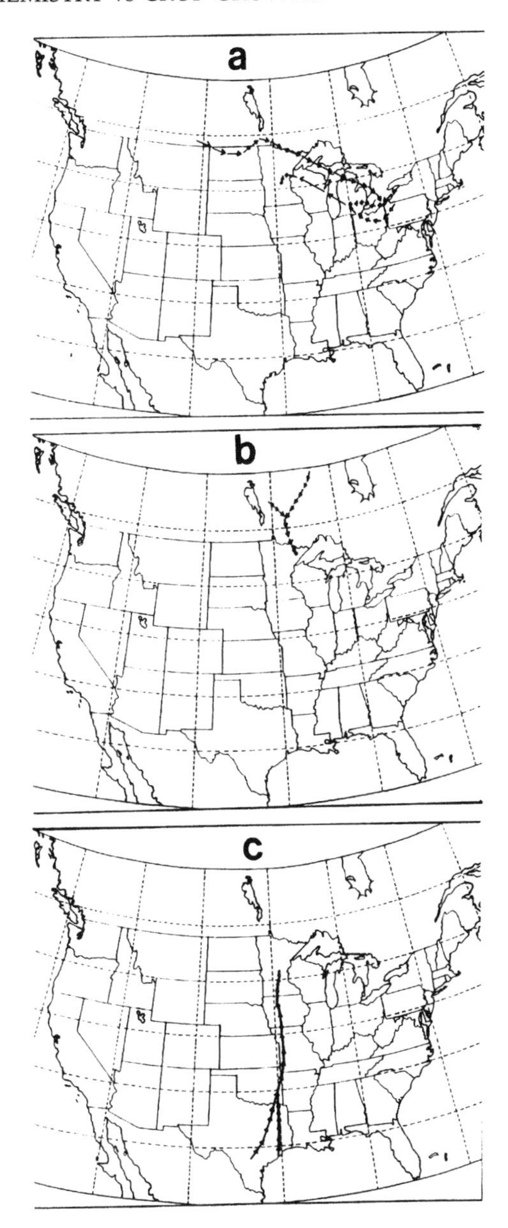

Figure 2.5. Branching air trajectories for the time period preceding three individual types of rain chemistry in Minnesota, US (refer to Table 2.11). The individual arrows on the trajectories represent 3-h travel time for the air mass. (a) Air trajectory for rain with relatively *high* $SO_4^{2-} + NO_3^-$, and H^+ concentrations (Type I). The trajectory goes through geographic regions of high SO_2 and NO_x emissions. (b) Air trajectory for rain with relatively *lower* $SO_4^{2-} + NO_3^-$, and H^+ concentrations (Type II). The trajectory goes through geographic regions with relatively low SO_2 and NO_x emissions. (c) Air trajectory for rain with relatively *high* $SO_4^{2-} + NO_3^-$, but *very* low H^+ concentrations (Type III). This situation represents aged air masses where much of the SO_4^{2-} and NO_3^- appear to be neutralized. (Source: Krupa *et al.*, 1987; by permission of Selper Limited)

Table 2.11. Classification of daily rain samples at Minnesota, USA, by their relative chemistry[a]

Characteristic	Type I	Type II	Type III
$\bar{x}$ pH	4.20	4.44	5.80
$\bar{x}$ H^+ (μeq l^{-1})	63	36	2
$\bar{x}$ SO_4^{2-} (μeq l^{-1})	125	39	100
$\bar{x}$ NO_3^- (μeq l^{-1})	54	21	64
$\bar{x}$ $NH_4^+ : H^+$ (μeq l^{-1})[b]	1.44	0.76	168

[a]Type I, II and III correspond to (a), (b) and (c) in Figure 2.5.
[b]$NH_4^+ : H^+$ (μeq l^{-1}) can be used as an indicator of the neutralization of the acid SO_4^{2-} aerosol by NH_3.
The reaction products are NH_4HSO_4, $(NH_4)_3H(SO_4)_2$ and $(NH_4)_2SO_4$. Thus, the higher the $NH_4^+ : H^+$ ratio, the greater is the concentration of $(NH_4)_2SO_4$, up to 100%.
Source: Krupa *et al.* (1987); by permission of Selper Limited.

TYPES OF CROP RESPONSES TO AMBIENT GASEOUS EXPOSURES

Traditionally, based on their responses to ambient gaseous air-pollutant exposures, crop effects are classified as acute and chronic. Acute responses involve rapid changes in the physiological and biochemical processes of the plant (Schulte-Hostede *et al.*, 1988). Under appropriate conditions, such changes frequently lead to the production of the symptoms of foliar injury (Krupa & Manning, 1988). Acute responses are known to be induced by relatively high hourly pollutant concentrations from a few to several hours on a given day or on recurring days and the injury symptoms manifest within a few to several days after such exposures. Most recent evidence suggests that acute and chronic responses to O_3 appear to be due to moderate or intermediate, but not the highest hourly pollutant concentrations (Tonneijck & Bugter, 1991; Grünhage *et al.*, 1993; Krupa *et al.*, 1993, 1995). This phenomenon may be due to lower atmospheric gaseous pollutant (O_3) flux or transfer *per se* and/or stomatal closure by the plant during periods of high pollutant (O_3) concentrations (US EPA, 1986) as a protective reaction to stress, since the entry of gases into the leaves is predominantly through the stomata.

In contrast to the acute responses, chronic effects are due to the exposures to gaseous pollutants for weeks, months, whole growth seasons and the entire life cycles. Chronic exposures involve relatively low hourly pollutant concentrations, with periodic, intermittent, highly variable episodes (acute exposures) or relative high concentrations. While acute responses may or may not lead to crop yield reductions, chronic responses can lead to foliar chlorosis, retarded growth, foliar abscission, retarded flowering, flower abscission, reduced biomass and economic yield and altered nutritional quality of the consumable plant product (Kickert & Krupa, 1990). The chronic effects are of much concern in the context of agriculture, particularly since growth and biomass effects can occur without attendant visible foliar symptoms.

IMPORTANT CHARACTERISTICS OF AMBIENT AIR POLLUTANT EXPOSURE PROFILES

Historically, biologists interested in air pollution research have viewed single air pollutants as causal agents of adverse effects on appropriate biological receptors (Mudd & Kozlowski, 1975). However, as stated previously, ambient air is always composed of mixtures of pollutants of differing physical and chemical properties (Finlayson-Pitts & Pitts, 1986). The joint effects of these pollutants in the mixture can result in more than additive, additive or less than additive effects, particularly in chronic exposures (Torn *et al.*, 1987). However, this does not mean that at a given time a particular pollutant cannot be present in high enough concentrations to lead to an acute effect characteristic of that particular pollutant. Because of these considerations, a number of investigators have examined the exposure characteristics of ambient air pollutants and a brief general summary from these studies is provided here.

Certain primary pollutants such as sulphur dioxide (SO_2) *per se* are considered to be important at a relatively local scale in relation to the sources of their origin (Table 2.1). In comparison, secondary pollutants such as ozone (O_3) and fine particle sulphate (SO_4^{2-}) are important at regional and even continental scales. Still others, such as elevated levels of carbon dioxide (CO_2), are of concern at the global scale.

Air pollutants (both dry and wet) exhibit dynamic variability in space and time. These features are governed by the variability in: (a) emissions of primary pollutants; (b) meteorology governing pollutant transport and transformation; and (c) patterns of deposition (dry and wet). Thus, ambient air pollutant profiles exhibit latitudinal, altitudinal, inter and intra annual (seasonal), intra seasonal and even daily variations (Pruchniewicz, 1973; Husar *et al.*, 1978; Singh *et al.*, 1978; US National Research Council, 1983; Logan, 1985; Lee *et al.*, 1986; Knapp *et al.*, 1988; Lefohn *et al.*, 1990; Legge & Krupa, 1990; Lovett & Kinsman, 1990). In developing field research on biological responses, site or area-specific characterization of the ambient exposure profiles of pollutants of interest must be performed *a priori* to have experimental relevancy to real-world situations.

Again, most biological systems respond rapidly to acute pollutant exposures. Sensitive receptors or processes may respond after one or a series of such exposures. Thus, the frequency of pollutant episodes is a major consideration in biological-effects research. High ground-level air concentrations of pollutants occur during radiation and subsidence inversions and, therefore, are governed by local or regional meteorology (Robinson, 1984). There are a number of meteorological models that can predict reasonably well the potential or possible frequency of acute exposures of primary air pollutants at specific geographic areas in the vicinity of point sources (Lyons & Scott, 1990; McVehil & Nosal, 1990). There are also models for predicting pollutant inputs at the regional and even continental scales (Husar *et al.*, 1978; Eliassen, 1980; Venkatram, 1986; US NAPAP, 1989; Chang & Dennis, 1990; Dennis *et al.*, 1990). The uncertainties attached to many such models increase with geographic scaling and with the increasing

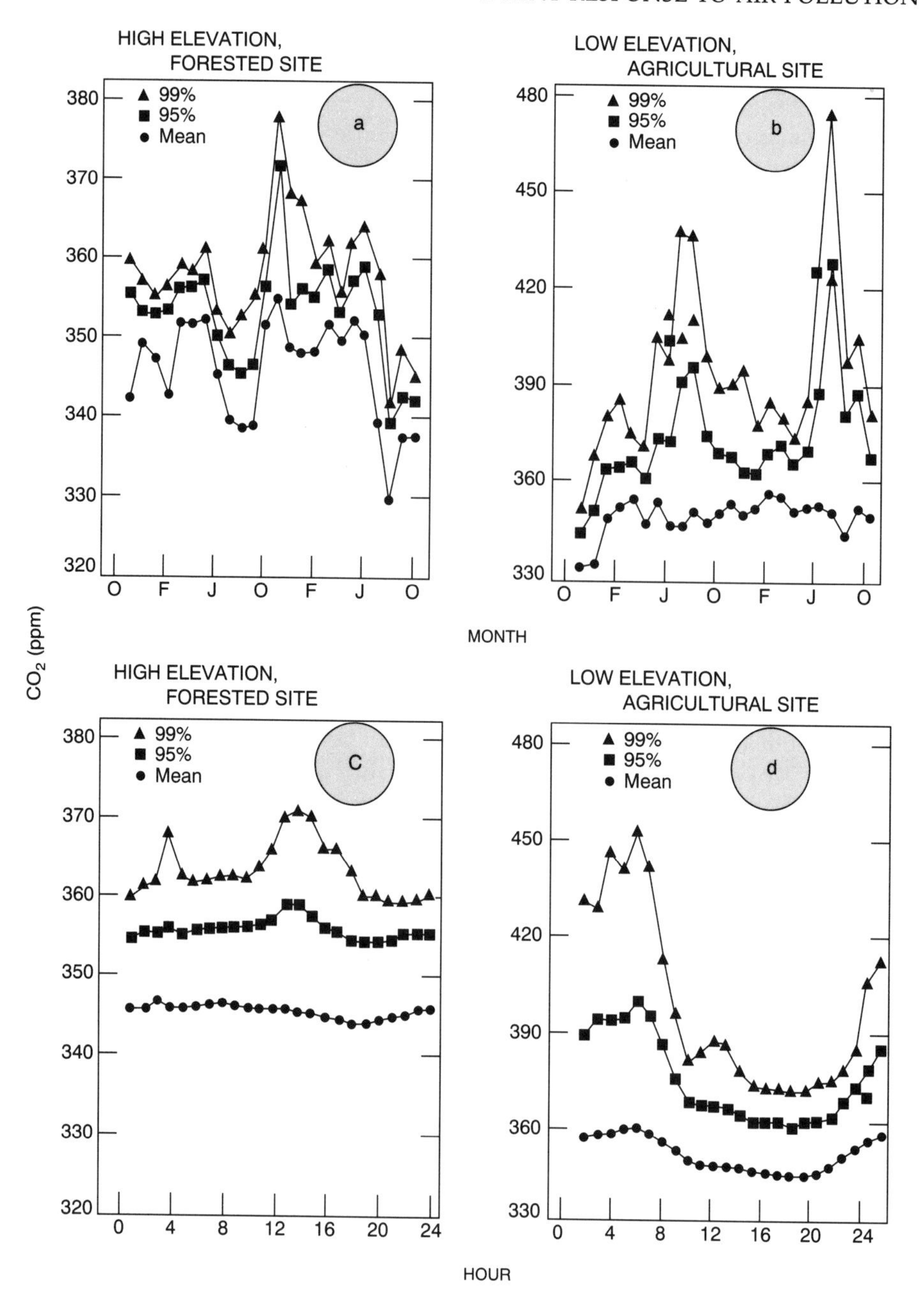

Figure 2.6. Profiles of hourly mean, 95th and 99th percentile concentrations of CO_2 by the month (a) at Fortress Mountain, Alberta, Canada (2103 m MSL) and (b) at Crossfield East, Alberta, Canada (1098 m MSL), and by the hour (c) at Fortress Mountain, Alberta, Canada (2103 m MSL) and (d) at Crossfield East, Alberta, Canada (1098 m MSL). Note the difference in the scales of the vertical axes. All data are from October 1985 through October 1987. (Source: Legge & Krupa 1988, 1990; Lewis Publishers, Inc.)

complexity of stochastic atmospheric processes included in the model. It is critical that any model developed for the first time or any subsequent modification of the original model for a new application, be validated prior to its application.

As previously stated there are seasonal differences in the profiles of ambient air pollutants. For example, while at relatively clean sites peak monthly O_3 concentrations occur during spring, at other sites (e.g., those influenced by urban plumes) such peak concentrations may be extended into the autumn (Singh *et al.*, 1978; Logan, 1985; Lefohn *et al.*, 1990).

Seasonal and hourly variations in ambient CO_2 profiles at a relatively high and a relatively low elevation site are presented in Figure 2.6(a–d). The observed differences in the CO_2 profiles between the high and the low-elevation site may be due to ecosystem types and land-use patterns. The low monthly CO_2 concentrations during the middle to late summer at the high elevation, forested site (see Figure 2.6a) may be due to rapid growth of vegetation and higher carbon assimilation (CO_2 sink). Conversely the high ambient CO_2 concentrations during the autumn at the low-elevation, agricultural site (see Figure 2.6b) may be due

Table 2.12. Concentrations and exposure characteristics of major acidic and acidifying pollutants and ozone in crop growing regions of the United States

Pollutant	Average growing season concentration	Exposure characteristics
Acidic rain	Highest weighted average annual acidity, pH 4.1, in OH-PA-NY area, decreasing south and west to the Plains States (to pH 5.1); 2-4 mg l^{-1} SO_4^{2-} and 1-3 mg l^{-1} NO_3^- nationwide; SO_4^{2-} : NO_3^--ratio ~2 in the east (NADP, 1983, 1984, 1985).	Fluctuations of H^+, SO_4^{2-}, and NO_3^- within and between events. Regional patterns in H^+ and SO_4^{2-} : NO_3^- ratio. Total deposition related to amount of rainfall. Concentrations of H^+ and SO_4^{2-} highest during growing season. Maximum recorded event acidity occurred in eastern US rain (~pH 3.0) and Los Angeles fogs (~pH 1.7).
SO_2	5-10 μg m^{-3} (2-4 ppb); monthly average as high as 25-50 μg m^{-3} (10-20 ppb) (US EPA, 1982). Hourly peaks of 100-3000 ppb (250-7500 μg m^{-3}) major sources.	In rural areas generally ≤20 μg m^{-3}. Peaks localized near sources and greatly affected by meteorological conditions; exposures often last only a few hours separated by days between fumigations. No strong diurnal variation on a regional scale.
NO_2	3-25 μg m^{-3} (2-13 ppb) in eastern US <2 μg m^{-3} in west; maximum hourly concentrations up to 150 μg m^{-3} (80 ppb) (US EPA, 1984).	Short-duration peak exposures separated by days between fumigations. May accompany SO_2 peaks but co-occurrence is <50 h during the growing season. Decreases with increasing O_3 diurnally.
O_3	7-hour (0900-1600) average 88 μg m^{-3} (45 ppb). Hourly peaks of 235 μg m^{-3} (120 ppb).	Strong diurnal variation (highest in early afternoon, lowest in early morning). Peaks related to stagnant urban plumes.

Modified from: US NAPAP (1987).

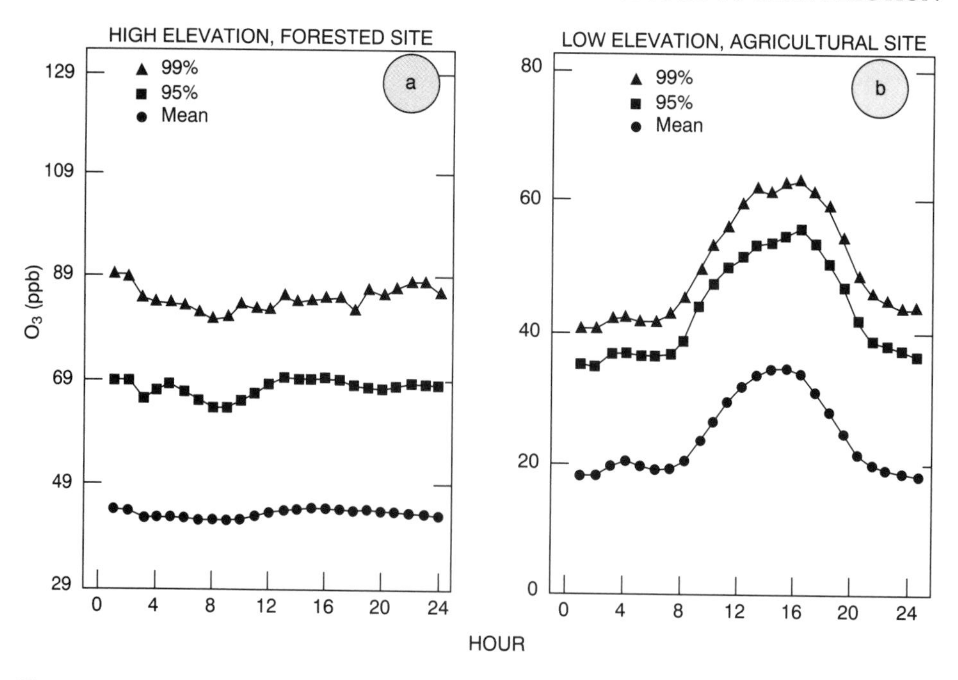

Figure 2.7. Profiles of hourly mean, 95th and 99th percentile concentrations of O_3 by the hour (a) at Fortress Mountain, Alberta, Canada (2103 m MSL) and (b) at Crossfield East, Alberta, Canada (1098 m MSL). Note the difference in the scales of the vertical axes between Figures 2.7(a) and (b). All data are from October 1985 through October 1987. (Source: Legge & Krupa 1988, 1990; Lewis Publishers, Inc.)

to the crop harvest (absence of a major CO_2 sink), soil turnover and autumn fertilization.

Similarly, small seasonal differences in emissions of SO_2, the rate at which these emissions are dispersed by atmospheric transport and mixing processes, and the rate at which SO_2 is converted by homogeneous and heterogeneous reactions to SO_4^{2-} give rise to a seasonal variability in the concentrations of SO_2 at ground level. These processes lead to seasonality of H^+ and SO_4^{2-} in precipitation (Table 2.12).

Because the free troposphere (the atmosphere above the mixed layer) serves as a reservoir for pollutants such as CO_2 and O_3, at many high-elevation monitoring sites, the mean of daily exposure profiles of these pollutants is essentially flat (see Figures 2.6c and 2.7a) (Legge & Krupa, 1988; Lefohn *et al.*, 1990). At these locations source-sink relationships may also be in rough equilibrium. While this discussion pertains to mean profiles, the standard deviations of those mean

---

Figure 2.8. Examples of different forms of co-occurrences: (a) simultaneous-only co-occurrence of SO_2/O_3 at concentrations $\geq$0.05 ppm for a set of hourly data; (b) sequential-only co-occurrence of SO_2/NO_2 at concentrations $\geq$0.05 ppm; (c) complex-sequential co-occurrence of O_3/NO_2 at concentrations $\geq$0.05 ppm. The horizontal dashed line designates the minimum concentrations (0.05 ppm). (Source: Lefohn *et al.*, 1987; by permission of Elsevier Science Limited)

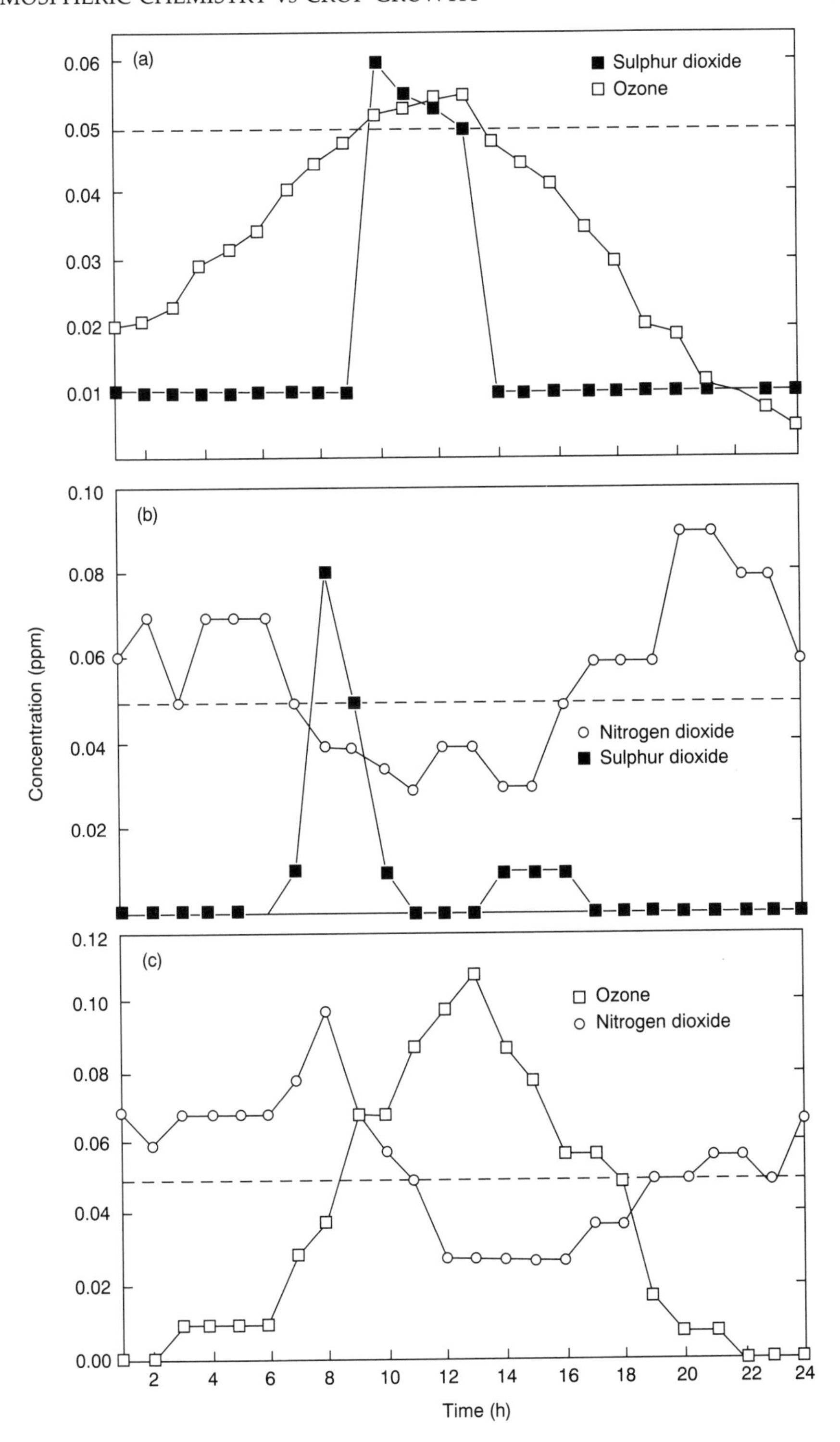
(a)
Sulphur dioxide
Ozone
(b)
Nitrogen dioxide
Sulphur dioxide
(c)
Ozone
Nitrogen dioxide
Concentration (ppm)
Time (h)
0.06
0.05
0.04
0.03
0.02
0.01
0.10
0.08
0.06
0.04
0.02
0.12
0.10
0.08
0.06
0.04
0.02
0.00
2
4
6
8
10
12
14
16
18
20
22
24

values are extremely important in the context of pollutant-exposure dynamics and biological responses (Legge *et al.*, 1991).

In contrast to the previous discussion, at low-elevation sites, secondary pollutants such as O_3 exhibit a distinct daily exposure profile (see Figure 2.7b) (Wolff *et al.*, 1987; Lefohn *et al.*, 1990; Legge & Krupa, 1990). Since tropospheric O_3 production is primarily through photochemistry (see Figure 2.1), high surface O_3 concentrations are observed during daily periods of high radiation (Finlayson-Pitts & Pitts, 1986; Legge & Krupa, 1990).

Coarse particulate air pollutants (>2.0 μm) are generated by mechanical processes, constitute the dominant component for crustal materials and sea salt (Table 2.4) and, in general, are rapidly removed from the atmosphere by sedimentation (Whitby, 1978). In contrast, fine particles (<2.0 μm) are mainly produced by atmospheric processes. Because of their low deposition velocity, they tend to accumulate in the atmosphere (Lee *et al.*, 1986). Accumulation of fine particle acid aerosols in the atmosphere can increase the atmospheric residence times of O_3 and the potential for adverse environmental effects (Chevone *et al.*, 1986). Photochemically generated fine particles accumulate in the atmosphere at night-time (Stevens *et al.*, 1978), thus occurring at high concentrations in a sequential order to the mid-afternoon highs of O_3. There are few studies, if any, on the effects of fine particle aerosols on vegetation (Herzfeld, 1982; Gmur *et al.*, 1983).

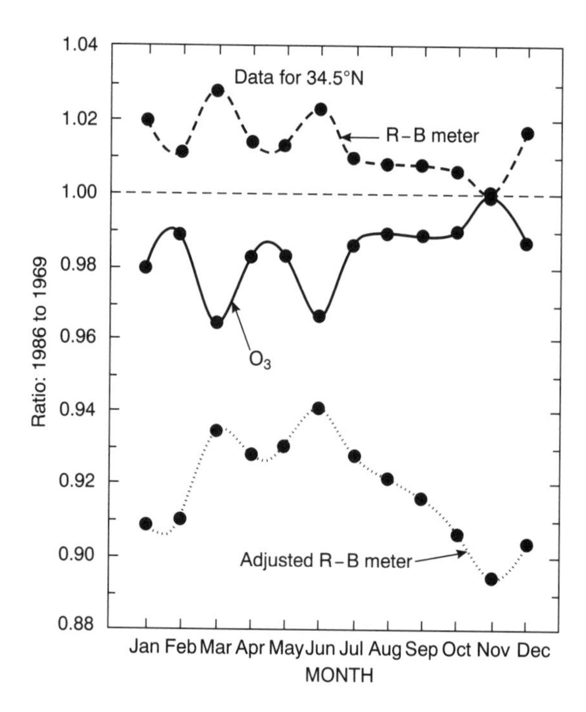

Figure 2.9. Ratios of column ozone and surface UV-B (erythemal flux units) between 1986 and 1969. R-B meter: Robertson-Berger meter data; Adjusted R-B meter: see text. Adapted from Figure 10 (upper curves) and Figure 11 (lowest curve) of Frederick *et al.* (1989)

Because of the nature of atmospheric processes in a 24-hour period two or more pollutants can exhibit increases or decreases in their concentrations simultaneously, sequentially or in some variable or poorly defined pattern relative to each other (Legge & Krupa, 1990). Lefohn *et al.* (1987) examined the occurrence patterns of ambient SO_2, NO_2, NO and O_3 at several rural sites in the United States and provided some examples of exposure profiles (see Figure 2.8). These and other more extensive types of site-specific ambient air quality characterization must be performed in designing future field experiments.

Inasmuch as air concentrations of many pollutants exhibit dynamic variability in time and space, so does the composition of precipitation. In addition to the

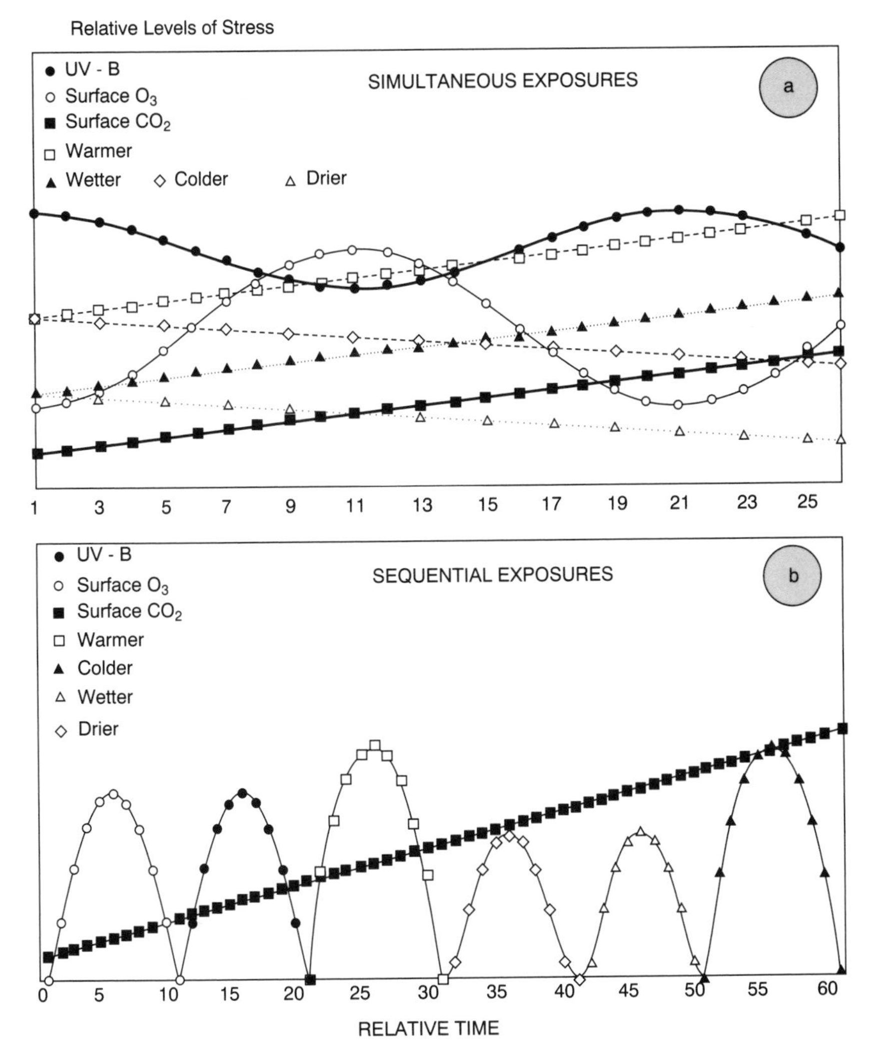

Figure 2.10. A hypothetical scenario of (a) simultaneous versus (b) sequential exposures to potential changes in multiple climatic variables. (Source: Krupa & Kickert, 1993; by permission of Kluwer Academic Publishers)

seasonal variations in precipitation chemistry (Pratt & Krupa, 1983) and the amounts of ions deposited (Knapp *et al.*, 1988), there are qualitative and quantitative differences in the composition of precipitation between and within individual events at any given geographic location (Pratt *et al.*, 1983b, 1984). The properties of precipitation are stochastic in nature. Because of this complexity, all of the studies on the effects of acidic precipitation on crops have used "simulated" precipitation (US NAPAP, 1987). While these types of studies may be valuable in examining worst-case scenarios, they are not very relevant to the present-day ambient conditions (Legge & Krupa, 1990).

In the context of global climate change, one environmental variable of concern is the possible increases in the surface UV-B radiation due to losses in the stratospheric O_3 column (Krupa & Kickert, 1989). Inasmuch as stratospheric O_3 has a protective effect against the penetration of UV-B to the surface, such a protective effect is also provided by tropospheric O_3 (see Figure 2.9). Therefore, high levels of tropospheric O_3 and UV-B probably will not occur simultaneously, but in a sequential order. This is a principle that has been largely ignored in effects research.

Another aspect of global climate change relates to alterations in temperature and precipitation patterns. Specific geographic regions may become warmer, colder, wetter, drier or remain unchanged (Kickert & Krupa, 1990). Nevertheless, Krupa and Kickert (1993) provided graphic representations of hypothetical scenarios of simultaneous and sequential exposure profiles of several variables in the context of global climate changes (see Figure 2.10a, b).

MODELLING CAUSE-EFFECT RELATIONSHIPS

In the field of air pollution research, much of what we know on the direct effects on plants relates to gaseous air pollutants. Therefore, the following narrative is primarily directed to this subject. A number of investigators have used statistical or process models to describe the relationships between gaseous air pollutant exposure and plant response.

It is beyond the scope of this chapter to provide detailed information on all possible approaches to cause-effects modelling. Among other publications, the reader should consult Enoch and Kimball (1986), Krupa and Kickert (1987), *Environmental Pollution* (1988), Heck *et al.* (1988), Dixon *et al.* (1990), Kickert and Krupa (1991) and Geijn *et al.* (1993).

As stated by Kickert & Krupa (1991), one of the fundamental requirements of any cause-effects modelling is to provide a balanced definition and description of both the cause and the effect that makes sense in the context of known atmospheric and biological processes. Many publications on cause-effects modelling (e.g., Dixon *et al.*, 1990) emphasize biological response analysis at the expense of an inadequate treatment of the atmospheric variables. This is an area of concern. Frequently in modelling cause-effects relationships, the randomly varying atmospheric parameters are reduced to stationary functions (average seasonal or annual values) and correspondingly the dynamics of plant growth and response is reduced to a single measurement (season-end harvest, for example). A chronic pollutant exposure should be viewed as the random occurrence of one or more hours or days of pollutant episodes (relatively high concentrations) with intermittent one or more non-episodal (relatively low concentrations) periods. While the occurrence of episodes should be viewed as periods of stress, depending on the tolerance capacity of the plant and the nature of the other prevalent environmental conditions, non-episodal periods should be viewed as repair, compensation, or respite time for the plants. If pollutant episode or stress results in the impairment of normal plant health, repair or compensation can adjust plant health to varying degrees of the stress effect. If these processes are viewed as a recurring time series in the life cycle of a plant, then the final observed product is due to: [stress effect] – [repair]. If stress exceeds repair, then there will be an adverse effect. To the contrary, if repair exceeds stress, there will be no adverse effects. It is important to note that the repair costs are far and above the normal maintenance costs of the plant. Nevertheless, in many cases, energy assimilation and allocation can offset maintenance and repair costs, with the net result being no adverse effects.

In summary, the previous narrative emphasizes the importance of the episodicity of ambient pollutant (gas) concentrations, the frequency of occurrence of such episodes, and the length of the repair periods between such episodes in defining pollutant-exposure profile and plant response (Krupa & Kickert, 1987; Lefohn & Runeckles, 1987; Lefohn *et al.*, 1989).

Epidemiologists modelling the effects of pathogens and pests on plant productivity have shown the significance of the relationship between stress occurrence and plant phenology in regulating the final yield (Griffiths *et al.*, 1973; Calpouzos *et al.*, 1976; Lockwood *et al.*, 1977; Lynch, 1980; Gottwald & Bertrand, 1983; Agrios *et al.*, 1985). van Haut and Stratmann (1970) and more recently Krupa and Nosal (1989b) recognized this concept in their studies on the effects of SO_2 on plants. Blum and Heck (1980) examined the effects of acute O_3 exposures on snap bean at various stages of its life cycle under controlled conditions. Similarly, in controlled experiments, Kasana (1991) found differing O_3 sensitivity in three tropical legume species with changing growth stages. There has been only one attempt so far, however, to examine the relationship between chronic ambient O_3 exposure and plant growth stages (Krupa & Nosal, 1989c). It is important to note that this concept of defining plant phenological stages and their differing sensitivity to ambient pollutant exposures has largely been ignored in our understanding of cause–effect relationships. Luxmoore (1988), in his treatment of crop mechanistic models, defines those process models which consider phenology, for example, as "phenomenological models." The occurrence of pollutant episodes must coincide with sufficient frequency with the sensitive growth stages of the plant to result in adverse effects on plant growth and productivity. As previously stated, most empirical models do not consider this aspect, since they are single-point models.

Based on the previous discussion, emphasis in future cause–effect modelling must be directed to the use of multipoint models (Krupa & Kickert, 1987; Kickert & Krupa, 1991). Experimental designs will have to include either destructive or non-destructive measurements of biological responses at various time intervals during the experiment to account for the effects of pollutant-exposure dynamics on the variable plant growth and response properties leading to the final biomass harvested.

Many of us tend to view plant health in the ambient environment as a univariate system, as we have, for example, in many of our studies on the potential adverse effects of acidic precipitation on terrestrial vegetation (Lefohn & Krupa, 1988b). In many of our field experiments, what is expected to serve as a control for a given independent variable, most likely does not serve as a realistic control for another independent variable of concern. Given what we know, greater emphasis should be directed to experiments conducted in open, ambient environments so as to increase our confidence in the results we obtain. In conducting these types of studies, monitoring of all potential stress (abiotic and biotic) variables as well as the conventional atmospheric and edaphic growth-regulating parameters will have to be performed. Mathematical models will have to be developed to integrate these data with the corresponding biological-response measurements in establishing cause–effect relationships. First-order numerical time series models are presently available (Krupa & Nosal, 1989a, b). Although, such models will have to be improved further and integrated with theories of plant physiology and disease epidemiology, an example from Krupa and Nosal (1989b) is provided below.

Relationships between alfalfa leaf dry weight per 100 stems and ambient exposures to sulphur dioxide (in the vicinity of a coal-fired power-plant) and other environmental variables during the second crop growth season at seven study sites in east central Minnesota were modelled as:

$$y = \sum_i \sum_n \sum_t a_i^{(n)}(t) x_i^n(t)$$

where y is the yield (e.g., alfalfa leaf dry weight per 100 stems); $i (= 1, 2, 3, \ldots)$ is the exposure parameter type (e.g., number of SO_2 exposures, peak concentration, cumulative integral of exposure, weekly maximum temperature, weekly minimum temperature, weekly total precipitation); $n (= 1, 2, 3, \ldots)$ is the power of the non-linear term; $t (= 1, 2, 3, \ldots)$ is the time series of alfalfa height growth; $a_i^{(n)}(t)$ is the regression coefficient; and $x_i^{(n)}(t)$ is the exposure quantification term. The regression coefficient is estimated by the least squares method. The results showed that:

$$\begin{aligned}
&\text{Alfalfa leaf dry weight per 100 stems} \\
&= -6.03 \times Epis(S_1) + 1.26 \times Peak(S_1) + -2.6 \times T_{max}(S_1) + -0.97 \times T_{min}(S_1) \\
&\quad + -0.14 \times Prec(S_1) + 2.84 \times T_{min}(S_2) + -1.99 \times T_{max}(S_3) \\
&\quad + 0.27 \times \{Epis(S_1)\}^2 + -0.02 \times \{Peak(S_1)\}^2 + -0.0004 \times \{Int(S_1)\}^2 \\
&\quad + 0.001 \times \{Int(S_2)\}^2.
\end{aligned}$$

Multiple correlation = 0.99; significance (tail probability) = 0.0000; Mallows $C_p = 9.64$; $R^2 = 0.97$; intercept = 111.37.

Where, $Epis$ = number of SO_2 episodes; $Peak$ = peak SO_2 concentration; Int = cumulative integral of SO_2 exposure; T_{max} and T_{min} = maximum and minimum weekly temperature; $Prec$ = total weekly precipitation; and S_1, S_2 and S_3 = first, second and third portion of a three-part (three growth stages) growth curve leading to the harvest.

As opposed to the direct effects of air pollutants on plants, for literature regarding the models for potential effects of acidic precipitation on soils, the reader should consult Reuss & Johnson (1986), Turchenek *et al.* (1987), US NAPAP (1987) and Longhurst (1989). Similarly Kabata-Pendias & Pendias (1984), Adriano (1986), Kramer & Allen (1988), Alloway (1990) and Shaw (1990) describe approaches to studying the fate of trace metal inputs on to soils.

In the final analysis, it is critical to note that the time resolution required for the definition of the environmental variables in response modelling will vary with the type of receptor under consideration. For example, in direct vegetation response research, time resolution of environmental variables can vary from less than an hour (SO_2), to an hour (O_3) to an event (precipitation chemistry), to a day (air temperature) and to a week (precipitation depth). Additional intensive efforts are needed to develop and validate stochastic, time series multivariate models to establish cause–effect relationships in this case. To the contrary, in soil-response research seasonal or annual values may be sufficient. These types of considerations must be placed in a proper perspective in the collection and application of data in cause–effect modelling.

A PERSPECTIVE OF THE PRESENT AND THE FUTURE

It can be concluded from what has been presented in the previous sections of this chapter that the quantitative interpretations presented to describe the dynamics of the atmosphere–biosphere (crop canopy) interface are not completely satisfactory at the present time. The limitations of our current understanding are due to: (a) our continued view of air pollutant effects on crop growth and productivity under ambient conditions as univariate, single-point (season-end yield) systems; (b) our hesitancy to accept the rule that under ambient conditions control for one independent variable generally does not serve as a control for another variable; and (c) our continued reliance on retrospective numerical analyses of previously collected field data. This includes the collapse of the importance of the dynamics and stochasticity of cause–effect relationships, because of the use of various time-dependent averaging or summation techniques to express pollutant exposures. Future research will have to include more open-field pollutant exposure–crop response studies, rather than continued reliance on various types of field exposure chambers. In such studies the main growth-regulating variables (including stress factors, both abiotic and biotic) will have to be monitored and their individual contribution to the overall crop growth and productivity apportioned within an integrated system. Scientists created the concept of Occam's Razor, which is the principle that forces many of them to look for the simplest explanation, but it does not mean that nature necessarily must oblige.

REFERENCES

Adriano, D.C. 1986. *Trace Elements in the Terrestrial Environment*. Springer-Verlag, New York, 533 pp.

Agrios, G.N., Walker, M.E. & Ferro, D.N. 1985. Effects of cucumber mosaic virus inoculation at successive weekly intervals on growth and yield of pepper (*Capsicum annum*) plants. *Plant Dis.* **69**: 524-555.

Alloway, B.J. (ed.) 1990. *Heavy Metals in Soils*. Wiley, New York, 339 pp.

Bielski, B.H.J., Cabelli, D.E., Arudi, R.L. & Ross, A.B. 1985. Reactivity of H_2O_2/O_2-radicals in aqueous solutions. *J. Phys. Chem. Ref. Data* **14**: 1041-1100.

Blum, U.T. & Heck, W.W. 1980. Effects of acute ozone exposures on snap-bean at various stages of its life-cycle. *Environ. Exp. Bot.* **20**: 73-85.

Calpouzos, L., Roelfs, A.P., Madson, M.E., Martin, F.B., Welsh, J.R. & Wilcoxson, D. 1976. *A New Model to Measure Yield Losses Caused by Stem Rust in Spring Wheat*. Tech. Bull. 307, Agricultural Experiment Station, University of Minnesota, St Paul, MN, 23 pp.

Cass, G.R. & Shair, F.H. 1984. Sulfate accumulation in a sea breeze/land breeze circulation system. *J. Geophys. Res.* **89**(D1): 1429-1438.

Cess, R.D. 1976. Climate change-Appraisal of atmospheric feedback mechanisms employing zonal climatology. *J. Atmos. Sci.* **33**: 1831-1843.

Chamberlain, J., Foley, H., Hammer, D., MacDonald, G., Rothaus, D. & Ruderman, M. 1981. *The Physics and Chemistry of Acid Precipitation*. Technical Report JSR-81-25. SRI International, Palo Alto, CA.

Chang, J. & Dennis, R. 1990. *SOS/T 4: The Regional Acid Deposition Model and Engineering Model*. US NAPAP, Washington, DC.

Chang, T.Y., Norbeck, J.M. & Weinstock, B. 1979. An estimate of the NO_x removal rate in an urban atmosphere. *Environ. Sci. Technol.* **13**: 1534-1537.

Charlson, R.J. & Rodhe, H. 1982. Factors controlling the acidity of natural rainwater. *Nature* **295**: 683-685.

Chevone, B.I., Herzfeld, D.E., Krupa, S.V. & Chappelka, A.H. 1986. Direct effects of atmospheric sulfate deposition on vegetation. *JAPCA* **36**: 813-816.

Cicerone, R.J. 1987. Changes in stratospheric ozone. *Science* **237**: 35-42.

Cutchis, P. 1974. Stratospheric ozone depletion and solar ultraviolet radiation on earth. *Science* **184**: 13-19.

Davidson, C.I. & Wu, Y.L. 1989. Dry deposition of particles and vapors. In Adriano, D.C. (ed.) *Acid Precipitation Volume 2. Sources, Emissions, and Mitigation*: 103-216. Advances in Environmental Sciences Series, Springer-Verlag, New York.

Demerjian, K.L. 1986. Atmospheric chemistry of ozone and nitrogen oxides. In Legge, A.H. & Krupa, S.V. (eds) *Air Pollutants and their Effects on the Terrestrial Ecosystem*: 105-127. Wiley, New York.

Dennis, R., Clark, T., Barchet, R., Roth, P., Reynolds, S., Seilkop, S., Laulainen, N., Clarke, J., Chang, J. & Samson, P. 1990. *SOS/T 5: Evaluation of Regional Acidic Deposition Models*. US NAPAP, Washington, DC.

Dixon, R.K., Meldahl, R.S., Ruark, G.A. & Warren, W.G. 1990. *Process Modeling of Forest Growth Responses to Environmental Stress*. Timber Press, Portland, OR, 441 pp.

EarthQuest, 1990. University Corporation for Atmospheric Research, Boulder, CO.

Eatough, D.J., Arthur, R.J., Eatough, N.L., Hill, M.W., Mangelson, N.F., Richter, B.E., Hansen, L.D. & Cooper, J.A. 1984. Rapid conversion of $SO_{2(g)}$ to sulfate in a fog bank. *Environ. Sci. Technol.* **18**: 855-859.

Eliassen, A. 1980. A review of long-range transport modeling. *J. Appl. Meteorol.* **19**: 231-240.

Enoch, H.Z. & Kimball, B.A. (eds) 1986. *Carbon Dioxide Enrichment of Greenhouse Crops. II. Physiology, Yield, and Economics*. CRC Press, Boca Raton, FL.

Environmental Pollution, 1988. Response of crops to air pollutants. *Environ. Pollut.* **53**: 1-478.

Finlayson-Pitts, B.J. & Pitts, J.N. Jr 1986. *Atmospheric Chemistry: Fundamentals and Experimental Techniques*. Wiley, New York, 1098 pp.

Forrest, J., Garber, R.W. & Newman, L. 1981. Conversion rates in power plant plumes based on filter pack data: The coal-fired Cumberland plume. *Atmos. Environ.* **15**: 2273-2282.
Frederick, J.E. 1986. The ultraviolet radiation environment of the biosphere. In *Effects of Changes in Stratospheric Ozone and Global Climate*: 121-128. US EPA and UNEP Publ. 1.
Frederick, J.E., Snell, H.E. & Haywood, E.K. 1989. Solar ultraviolet radiation at the earth's surface. *Photochem. Photobiol.* **50**: 443-450.
Frederick, J.E., Weatherhead, E.C. & Haywood, E.K. 1991. Long-term variations in ultraviolet sunlight reaching the biosphere-Calculations for the past 3 decades. *Photochem. Photobiol.* **54**: 781-788.
Galloway, J.N., Likens, G.E. & Edgerton, E.S. 1976. Acid precipitation in northeastern United States: pH and acidity. *Science* **194**: 722-724.
Garland, J.A. 1978. Dry and wet removal of sulphur from the atmosphere. *Atmos. Environ.* **12**: 349-362.
Garrels, R.M. & MacKenzie, F.T. 1971. *Evolution of Sedimentary Rocks*. Norton, New York, 397 pp.
Geijn, S.C. van de, Goudriaan, J. & Berendse, F. (eds) 1993. *Climate Change; Crops and Terrestrial Ecosystems*. Agrobiologische Thema's 9 (Agrobiological Themes 9), DLO Centre for Agrobiological Research (CABO-DLO), Wageningen, The Netherlands, 144 pp.
German Bundestag, 1991. Ozone depletion, changes in UV-B radiation and their effects. In *Protecting the Earth-A Status Report with Recommendations for a New Energy Policy*: 571-616. Deutscher Bundestag, Referat Öffentlichkeitsarbeit, Bonn.
Gillani, N.V. 1978. Project MISTT: Meso scale plume modelling of the dispersion, transformation, and ground removal of SO_2. *Atmos. Environ.* **12**: 569-588.
Gmur, N.F., Evans, L.S. & Cunningham, E.A. 1983. Effects of ammonium sulfate aerosols on vegetation. II. Mode of entry and responses of vegetation. *Atmos. Environ.* **17**: 715-721.
Gottwald, T.R. & Bertrand, P.F. 1983. Effect of time of inoculation with *Cladosporium caryigenum* on pecan scab development and nut quality. *Phytopath.* **73**: 714-718.
Graedel, T.E. & Goldberg, K.I. 1983. Kinetic studies of raindrop chemistry. I. Inorganic and organic processes. *J. Geophys. Res.* **88**: 865-882.
Graedel, T.E., Mandich, M.L. & Weschler, C.J. 1986. Kinetic model studies of atmospheric droplet chemistry. II. Homogeneous transition metal chemistry in raindrops. *J. Geophys. Res.* **91**(D4): 5021-5025.
Griffiths, E., Jones, D.G. & Valentine, M. 1973. Effects of powdery mildew at different growth stages on grain yield of barley. *Ann. Appl. Biol.* **80**: 343-349.
Grünhage, L., Dämmgen, U., Haenel, H.-D. Jäger, H.-J., Holl, A., Schmitt, J. & Hanewald, K. 1993. A new potential air quality criterion derived from vertical flux densities of ozone and from plant response. *Angew. Bot.* **67**: 9-13.
Heck, W.W., Tingey, D.T. & Taylor, O.C. (eds) 1988. *Assessment of Crop Loss from Air Pollutants*. Elsevier Applied Science, London, 552 pp.
Hegg, D.A. & Hobbs, P.V. 1979. Some observations of particulate nitrate concentrations in coal-fired power plant plumes. *Atmos. Environ.* **13**: 1715-1716.
Herzfeld, D.E. 1982. Interactive effects of sub-micron sulphuric acid aerosols and ozone on soybean and pinto bean. MS Thesis, University of Minnesota, St Paul, MN, 105 pp.
Hicks, B.B. 1984. The acidic deposition phenomenon and its effect. A-7 Dry deposition processes. *Critical Assessment Review Papers*, Vol. 1, Atmospheric Sciences, EPA-600/8-83-016AF, pp. 7-1 to 7-70.
Hidy, G.M. & Mueller, P.K. 1986. The sulphur oxide-particulate matter complex. In Legge, A.H. & Krupa, S.V. (eds) *Air Pollutants and their Effects on the Terrestrial Ecosystem*: 51-104. Wiley, New York.
Hosker, R.P. & Lindberg, S.E. 1982. Review: Atmospheric deposition and plant assimilation of gases and particles. *Atmos. Environ.* **16**: 889-910.
Houghton, R.A. & Woodwell, G.M. 1989. Global climatic change. *Sci. Amer.* **4**: 36-44.

Husar, R.B., Lodge, J.P. Jr. & Moore, D.J. (eds) 1978. *Sulphur in the Atmosphere*. Pergamon Press, New York, 816 pp.

International Electric Research Exchange, 1981. *Effects of SO_2 and its Derivatives on Health and Ecology. 2. Natural Ecosystems, Agriculture, Forestry and Fisheries*. Central Electricity Generating Board, Leatherhead, UK, pp. 1-1 to 6-13.

IPCC (Intergovernmental Panel on Climate Change), 1990. *Climate Change: The IPCC Scientific Assessment* (Houghton, J.T., Jenkins, G.J. & Ephraums, J.J. (eds)). Cambridge University Press, Cambridge, 364 pp.

Isaksen, I.S.A. & Stordal, F. 1986. Ozone perturbations by enhanced levels of chlorofluorocarbons, nitrous oxide (N_2O) and methane: A two-dimensional diabetic circulation study including uncertainty estimates. *J. Geophys. Res.* **91**: 5249-5263.

Johnston, H.S. 1971. Reduction of stratospheric ozone by nitrogen oxide catalysts from supersonic transport exhaust. *Science* **173**: 517-522.

Kabata-Pendias, A. & Pendias, H. 1984. *Trace Elements in Soils and Plants*. CRC Press, Boca Raton, FL, 315 pp.

Kasana, M.S. 1991. Sensitivity of three leguminous crops to O_3 as influenced by different stages of growth and development. *Environ. Pollut.* **69**: 131-150.

Kerr, J.B. & McElroy, C.T. 1993. Evidence for large upward trends of ultraviolet-B radiation linked to ozone depletion. *Science* **262**: 1032-1034.

Kickert, R.N. & Krupa, S.V. 1990. Forest responses to tropospheric ozone and global climate change: An analysis. *Environ. Pollut.* **68**: 29-65.

Kickert, R.N. & Krupa, S.V. 1991. Modeling plant response to tropospheric ozone: A critical review. *Environ. Pollut.* **70**: 271-383.

Knapp, W.W., Bowersox, V.C., Chevone, B.I., Krupa, S.V., Lynch, J.A. & McFee, W.W. 1988. *Precipitation Chemistry in the United States: Summary of Ion Concentration Variability 1978-1984*. Technical Bulletin, Water Resources Research Institute, Cornell University, Ithaca, New York, pp. 1-1 to D-1.

Kramer, J.R. & Allen, H.E. (eds) 1988. *Metal Speciation: Theory, Analysis and Application*. Lewis Publishers Inc., Chelsea, MI, 357 pp.

Krupa, S.V. & Kickert, R.N. 1987. An analysis of numerical models of air pollutant exposure and vegetation response. *Environ. Pollut.* **44**: 127-158.

Krupa, S.V. & Kickert, R.N. 1989. The greenhouse effect: Impacts of ultraviolet-B (UV-B) radiation, carbon dioxide (CO_2), and ozone (O_3) on vegetation. *Environ. Pollut.* **61**: 263-393.

Krupa, S.V. & Kickert, R.N. 1993. The greenhouse effect: The impacts of carbon dioxide (CO_2), ultraviolet-B (UV-B) radiation and ozone (O_3) on vegetation (crops). *Vegetatio* **104/105**: 321-328.

Krupa, S.V. & Manning, W.J. 1988. Atmospheric ozone: Formation and effects on vegetation. *Environ. Pollut.* **50**: 101-137.

Krupa, S.V. & Nosal, M. 1989a. Effects of ozone on agricultural crops. In Schneider, T., Lee, S.D., Wolters, G.J.R. & Grant, L.D. (eds) *Atmospheric Ozone Research and Its Policy Implications*: 229-238. Elsevier Science Publ. B.V., Amsterdam.

Krupa, S.V. & Nosal, M. 1989b. A multivariate, time series model to relate alfalfa responses to chronic, ambient sulphur dioxide exposures. *Environ. Pollut.* **61**: 3-10.

Krupa, S.V. & Nosal, M. 1989c. Application of spectral coherence analysis to describe the relationships between ozone exposure and crop growth. *Environ. Pollut.* **60**: 319-330.

Krupa, S.V., Lodge, J.P. Jr, Nosal, M. & McVehil, G.E. 1987. Characteristics of aerosol and rain chemistry in north central USA. In Perry, R., Harrison, R.M., Bell, J.N.B. & Lester, J.N. (eds) *Acidic Rain: Scientific and Technical Advances*: 121-128. Selper Ltd, London.

Krupa, S.V., Manning, W.J. & Nosal, M. 1993. Use of tobacco cultivars as biological indicators of ambient ozone pollution: An analysis of exposure-response relationships. *Environ. Pollut.* **81**: 137-146.

Krupa, S.V., Nosal, M. & Legge, A.H. 1994. Ambient ozone and crop loss: Establishing a cause-effect relationship. *Environ. Pollut.* **83**: 269-276.

Krupa, S.V., Grünhage, L., Jäger, H.-J., Nosal, M., Manning, W.J., Legge, A.H. & Hanewald, K. 1995. Ambient ozone (O_3) and adverse crop response: A unified view of cause and effect. *Environ. Pollut.* **87**: 119-126.

Lee, S.D., Schneider, T., Grant, L.D. & Verkerk, P.J. 1986. *Aerosols*. Lewis Publishers, Inc., Chelsea, MI, 1221 pp.

Lefohn, A.S. & Krupa, S.V. 1988a. The relationship between hydrogen and sulfate ions in precipitation-A numerical analysis of rain and snowfall chemistry. *Environ. Pollut.* **49**: 289-311.

Lefohn, A.S. & Krupa, S.V. (eds) 1988b. *Acidic Precipitation: A Technical Amplification of NAPAP's Findings*. Air Pollution Control Association, Pittsburgh, PA, 239 pp.

Lefohn, A.S. & Runeckles, V.C. 1987. Establishing a standard to protect vegetation — ozone exposure/dose considerations. *Atmos. Environ.* **21**: 561-568.

Lefohn, A.S., Davis, C.E., Jones, C.K., Tingey, D.T. & Hogsett, W.E. 1987. Co-occurrence patterns of gaseous air pollutant pairs at different minimum concentrations in the United States. *Atmos. Environ.* **21**: 2435-2444.

Lefohn, A.S., Runeckles, V.C., Krupa, S.V. & Shadwick, D.S. 1989. Important considerations for establishing a secondary ozone standard to protect vegetation. *JAPCA* **39**: 1039-1045.

Lefohn, A.S., Krupa, S.V. & Winstanley, D. 1990. Surface ozone exposures measured at clean locations around the world. *Environ. Pollut.* **63**: 189-224.

Legge, A.H. & Krupa, S.V. 1988. *The Present and Potential Effects of Acidic and Acidifying Air Pollutants on Alberta's Environment: Final Report*. Alberta Government-Industry Acidic Deposition Programme, 664 pp.

Legge, A.H. & Krupa, S.V. (eds) 1990. *Acidic Deposition: Sulphur and Nitrogen Oxides*. Lewis Publishers, Inc., Chelsea, MI, 659 pp.

Legge, A.H., Nosal, M., McVehil, G.E. & Krupa, S.V. 1991. Ozone and the clean troposphere: Ecological implications. *Environ. Pollut.* **70**: 157-175.

Likens, G.E. & Bormann, F.H. 1974. Acid rain: A serious environmental problem. *Science* **184**: 1176-1179.

Lockwood, J.L., Percich, J.A. & Maduewisi, J.N.C. 1977. Effect of leaf removal simulating pathogen-induced defoliation on soybean yields. *Plant Dis.* **61**: 458-462.

Logan, J.A. 1985. Tropospheric ozone: Seasonal behavior, trends, and anthropogenic influence. *J. Geophys. Res.* **90**: 10463-10482.

Longhurst, J.N.S. (ed.) 1989. *Acid Deposition: Sources, Effects and Controls*. Technical Communications, British Library, London, 348 pp.

Lovelock, J.E. 1987. *Gaia: A New Look at Life on Earth*. Oxford University Press, Oxford, 157 pp.

Lovett, G.M. & Kinsman, J.D. 1990. Atmospheric pollutant deposition to high elevation ecosystems. *Atmos. Environ.* **24A**: 2767-2786.

Luxmoore, R.J. 1988. Assessing the mechanisms of crop loss from air pollutants with process models. In Heck, W.W., Taylor, O.C. & Tingey, D.T. (eds) *Assessment of Crop Loss from Air Pollutants*: 417-443. Elsevier Applied Science, London.

Lynch, R.E. 1980. European corn borer yield losses in relation to hybrid and stages of corn development. *J. Econ. Ent.* **73**: 159-164.

Lyons, T.J. & Scott, W.D. (eds) 1990. *Principles of Air Pollution Meteorology*. CRC Press, Boca Raton, FL, 224 pp.

MAP3S/RAINE Research Community, 1982. The MAP3S/RAINE precipitation chemistry network: Statistical overview for the period 1976-1980. *Atmos. Environ.* **16**: 1603-1631.

McVehil, G.E. & Nosal, M. 1990. Model-generated air quality statistics for application in vegetation response models in Alberta. In Legge, A.H. & Krupa, S.V. (eds) *Acidic Deposition: Sulphur and Nitrogen Oxides*: 481-498. Lewis Publishers, Chelsea, MI.

Molina, M.J. & Rowland, F.S. 1974. Stratospheric sink for chlorofluoromethanes: Chlorine atom catalyzed destruction of ozone. *Nature* **249**: 810-812.

Mudd, J.B. & Kozlowski, T.T. 1975. *Responses of Plants to Air Pollution*. Academic Press, New York, 383 pp.

NADP, 1983, 1984, 1985. *NADP Annual Data Summary. Precipitation Chemistry in the United States* (Gibson, J.H., Programme Coordinator), Colorado State University, Fort Collins, CO.

Ottar, B., Dovland, H. & Semb, A. 1984. Long-range transport of air pollutants and acid precipitation. In Treshow, M. (ed.) *Air Pollution and Plant Life*: 39-71. Wiley, New York.

Pack, D.H., Ferber, G.J., Heffter, J.L., Telegadas, K., Angell, J.K., Hoecker, W.H. & Machta, L. 1978. Meteorology of long-range transport. *Atmos. Environ.* **12**: 425-444.

Pratt, G.C. & Krupa, S.V. 1983. Seasonal trends in precipitation chemistry. *Atmos. Environ.* **17**: 1845-1847.

Pratt, G.C. & Krupa, S.V. 1985. Aerosol chemistry in Minnesota and Wisconsin and its relation to rainfall chemistry. *Atmos. Environ.* **19**: 961-971.

Pratt, G.C., Coscio, M.R., Gardner, D.W., Chevone, B.I. & Krupa, S.V. 1983a. An analysis of the chemical properties of rain in Minnesota. *Atmos. Environ.* **17**: 347-355.

Pratt, G.C., Hendrickson, R.C., Chevone, B.I., Christopherson, D.A., O'Brien, M. & Krupa, S.V. 1983b. Ozone and oxides of nitrogen in the rural upper midwestern USA. *Atmos. Environ.* **10**: 2013-2023.

Pratt, G.C., Coscio, M.R. & Krupa, S.V. 1984. Regional rainfall chemistry in Minnesota and West Central Wisconsin. *Atmos. Environ.* **18**: 173-182.

Pruchniewicz, P.G. 1973. The average tropospheric ozone content and its variation with season and latitude as a result of the global ozone circulation. *Pure. Appl. Geophys.* **106**: 1058-1073.

Ramanathan, V., Cess, R.D., Harrison, E.F., Minnis, P., Barkstrom, B.R., Ahmad, E. & Hartmann, D. 1989. Cloud-radiative forcing and climate: Results from the earth radiation budget experiment. *Science* **243**: 57-63.

Reuss, J.O. & Johnson, D.W. 1986. *Acid Deposition and the Acidification of Soils and Waters.* Springer-Verlag, New York, 119 pp.

Richards, L.W., Anderson, J.A., Blumenthal, D.L., Brandt, A.A., MacDonald, J.A., Watus, N., Macias, E.S. & Bhardwaja, P.S. 1981. The chemistry, aerosol physics, and optical properties of a western coal-fired power plant plume. *Atmos. Environ.* **15**: 2111-2134.

Robinson, E. 1984. Dispersion and fate of atmospheric pollutants. In Treshow, M. (ed.) *Air Pollution and Plant Life:* 15-37. Wiley, New York.

Roedel, W. 1979. Measurements of sulfuric acid saturation vapor pressure: Implications for aerosol formation by heteromolecular nucleation. *J. Aerosol Sci.* **10**: 375-386.

Schulte-Hostede, S., Darrall, N.M., Blank, L.W. & Wellburn, A.R. (eds) 1988. *Air Pollution and Plant Metabolism.* Elsevier Applied Science, London, 379 pp.

Schwartz, S.E. 1989. Acid deposition: Unraveling a regional phenomenon. *Science* **243**: 753-762.

Sehmel, G.A. 1980. Particle and gas dry deposition: A review. *Atmos. Environ.* **14**: 983-1011.

Sequeira, R. 1982. Acid rain: An assessment based on acid-base considerations. *JAPCA* **32**: 241-245.

Shaw, A.J. 1990. *Heavy Metal Tolerance in Plants: Evolutionary Aspects.* CRC Press, Boca Raton, FL, 355 pp.

Singh, H.B., Ludwig, F.L. & Johnson, W.B. 1978. Tropospheric ozone: Concentrations and variabilities in clean, remote atmospheres. *Atmos. Environ.* **12**: 2185-2196.

Smith, F.B. & Hart, R.D. 1978. Meteorological aspects of the transport of pollution over long distances. *Atmos. Environ.* **12**: 461-478.

Spicer, C.W. 1982a. The distribution of oxidized nitrogen in urban air. *Sci. Total Environ.* **24**: 183-192.

Spicer, C.W. 1982b. Nitrogen oxide reactions in the urban plume of Boston. *Science* **215**: 1095-1097.

Stevens, R.K., Dzubay, T.G., Russwurm, G. & Rickel, D. 1978. Sampling and analysis of atmospheric sulfates and related species. *Atmos. Environ.* **12**: 55-68.

Teasley, J.I. (ed.) 1984. *Acid Precipitation Series. Vol. 1-9.* Butterworth Publishers, Boston, MA.

Tonneijck, A.E.G. & Bugter, R.J.F. 1991. Biological monitoring of ozone effects on indicator plants in the Netherlands: Initial research on exposure-response functions. *VDI-Berichte* **901**: 613-624.

Torn, M.S., Degrange, J.E. & Shinn, J.H. 1987. *The Effects of Acidic Deposition on Alberta Agriculture.* Alberta Acid Deposition Research Programme, ADRP-B-08/87, The University of Calgary, Calgary, Alberta, 160 pp.

Turchenek, L.W., Abboud, S.A., Tomas, C.J., Fessenden, R.J. & Holowaychuk, N. 1987. *Effects of Acid Deposition on Soils in Alberta.* Prepared for the Alberta Government-Industry Acid Deposition Research Programme, ADRP-B-05-87, Alberta Research Council, Edmonton, 202 pp.

US EPA, 1982. *Air Quality Criteria for Particulate Matter and Sulfur Oxides.* EPA-600/8-82-029b, Research Triangle Park, NC.

US EPA, 1984. *The Acid Deposition Phenomenon and Its Effects.* Critical Assessment Review Papers. Vol. 1. EPA-600/8-83-016AF, Office of Research and Development, Washington, DC.

US EPA, 1986. Effects of ozone and other photochemical oxidants on vegetation. In *Air Quality Criteria for Ozone and Other Photochemical Oxidants,* Vol. 3, pp 6-1-6-298; EPA-600/8-84-020cF, Environmental Criteria and Assessment Office, Research Triangle Park, NC.

US NAPAP, 1987. *The National Acid Precipitation Assessment Program.* Washington, DC. Refer to Volumes 1 through 4, Interim Assessment Report.

US NAPAP, 1989. *Models Planned for Use in the NAPAP Integrated Assessment.* Washington, DC.

US National Academy Press, 1986. *Acid Deposition. Long Term Trends.* Washington, DC, 506 pp.

US National Research Council, 1983. *Acidic Deposition. Atmospheric Processes in Eastern North America.* National Academy Press, Washington, DC, 375 pp.

van Haut, H. & Stratmann, H. 1970. *Farbtafelatlas über Schwefeldioxidwirkungen und Pflanzen.* W. Girardet, Essen, West Germany, 206 pp.

VDI (Verein Deutscher Ingenieure), 1987. *Acidic Precipitation.* Dusseldorf, Germany, 281 pp.

Venkatram, A. 1986. Statistical long-range transport models. *Atmos. Environ.* **20**: 1317-1324.

Voldner, E.C., Barrie, L.A. & Sirois, A. 1986. A literature review of dry deposition of oxides of sulphur and nitrogen with emphasis on long-range transport modelling in North America. *Atmos. Environ.* **20**: 2101-2123.

Wayne, R.P. 1987. The photochemistry of ozone. *Atmos. Environ.* **21**: 1683-1694.

Whitby, K.T. 1978. The physical characteristics of sulfur aerosols. *Atmos. Environ.* **12**: 135-159.

Wolff, G.T., Lioy, P.J. & Taylor, R.S. 1987. The diurnal variations of ozone at different altitudes on a rural mountain in the eastern United States. *JAPCA* **37**: 45-48.

Wuebbles, D.J., Grant, K.E., Connell, P.S. & Penner, J.E. 1989. The role of atmospheric chemistry in climate change. *JAPCA* **39**: 22-28.

3 Nutritional Disharmony in Plants: Soil and Weather Effects on Source–Sink Interactions

R. OREN

INTRODUCTION

In this chapter, a reduction in the long-term growth rate of forest stands is viewed in relation to a continuously declining availability of various nutrient elements in the soil. Trees adjust their demand for an increasingly limiting element by reducing growth, but at a rate lower than the reduction in the element's supply. Under such conditions, trees function continuously at a level very near nutritional limitations. As a result, a sudden and relatively small increase in the ratio of demand-to-uptake of elements can cause a nutritional imbalance as reflected in typical symptoms of forest decline, including chlorosis and unusually low canopy leaf-area. In crowns of evergreen conifer species, these symptoms can result from short-term increases in the growth rates of new crown components which, under limited nutrient supply, must be supported by retranslocation of elements from mature needles (Oren & Schulze, 1989). A decrease in the ratio of demand to uptake, due to a growth reduction, can restore a nutritional balance in which the symptoms of nutrient shortage are alleviated.

Temporary shifts in demand relative to uptake are induced mostly by variation in weather conditions (e.g., in moisture availability and temperature) which impel changes in growth. Thus, decline symptoms are most clearly reflected in stands growing on soils poor in certain nutrients, during seasons and years when conditions are very favourable for growth. Stands will be particularly prone to nutritional limitations during phases in their development when stand growth is high and a large proportion of the site's nutrients is already incorporated into the standing biomass. These imbalances in demand/uptake ratios can be described as nutritional disharmony.

The concept of nutritional disharmony will be described herein. Each component of the concept will be supported by both literature and an individual study — the investigation of forest decline in the Fichtelgebirge, northeastern Bavaria, FRG. A model based on the nutritional disharmony concept was used to evaluate decline phenomena in the Appalachian Mountains of the southeastern USA. In the conclusion of the chapter, the output of several model runs will

Plant Response to Air Pollution. Edited by Mohammad Yunus and Muhammad Iqbal

be used as a demonstration of the potential effects of soil quality and weather conditions on canopy production and nutritional status.

NUTRITIONAL DISHARMONY

The nutritional disharmony concept was developed based on information from many studies on various aspects of plant nutrition (Linder & Ingestad, 1977; Brix, 1981; Ingestad, 1982; Chapin & Kedrowski, 1983; Linder & Rook, 1984; Timmer & Morrow, 1984; Nambiar & Fife, 1987), and is discussed in detail elsewhere (Oren *et al.*, 1988a; Oren & Schulze, 1989). The results from a comprehensive study performed in the Fichtelgebirge were used to first evaluate the concept. The study centred on a detailed comparison of two *Picea abies* (L.) Karst. (Norway spruce) stands at different stages of decline (Oren *et al.*, 1988b; Schulze *et al.*, 1989a). One of the stands appeared healthy (hereafter referred to as "healthy"), while the other displayed clear signs of decline ("declining"). The ostensibly declining stand exhibited needle chlorosis and various degrees of needle loss of non-current-age classes. Individual trees or small groups of trees exhibiting these decline symptoms were interspersed among apparently healthy ones.

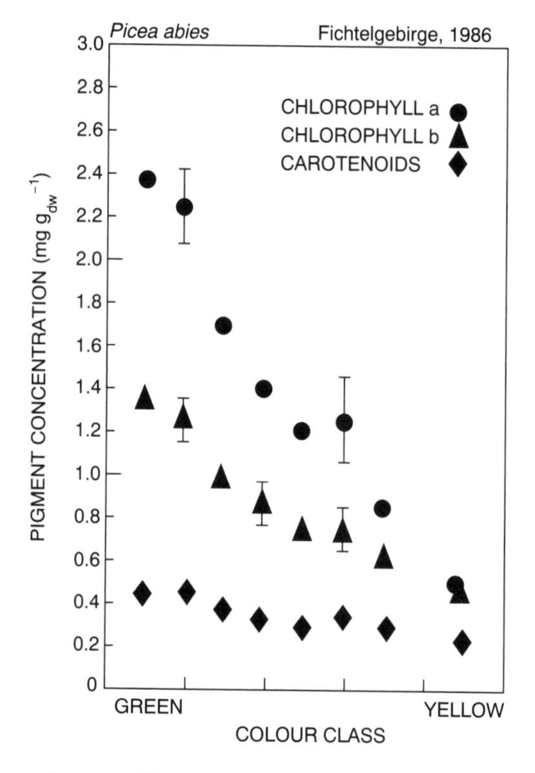

Figure 3.1. Chlorophyll, as well as carotenoids in *Picea abies* needles decreased with increasing chlorosis in needles of various age classes, from 30-year-old trees at a declining site (after Oren *et al.*, 1993)

A possible cause for the foliage damage at the declining site at the Fichtelgebirge is air pollution because of the site's elevation and position downwind from sources of pollution. Acidic precipitation at the declining site resulted in a proton input which was 60% higher than in the healthy site, mostly due to greater sulphate and nitrate deposition (Hantschel *et al.*, 1988). Potential causes of forest decline in polluted areas include direct damage to foliage (Guderian 1977; Winner *et al.*, 1985; Lange *et al.*, 1989). Thus, an obvious explanation for the symptoms could have been direct damage to needles, as well as reduction in needle performance due to the greater dose of air pollutants in the declining site.

Evidence of direct damage to needles at the declining site were not uncovered despite exhaustive research. Needle colour in the declining stand was directly related to chlorophyll concentration (Figure 3.1; Zimmermann *et al.*, 1988, Oren *et al.*, 1993) and, along a range in needle colour from green to yellow, chlorophyll concentration was directly related to needle magnesium concentration (Figure 3.2; Köstner *et al.*, 1990; Oren *et al.*, 1993). During the growing period, strong chlorosis in *P. abies* needles was apparent at a magnesium concentration below 9 μmol g_{dw}^{-1}, as was shown in studies on seedlings (Ingestad, 1959), associated with a chlorophyll concentration of *ca.* 1.8 mg g_{dw}^{-1}. However, measurements of all gas-exchange characteristics of green needles at the two sites, and both green and chlorotic needles at the declining site, failed to reveal any differences until the magnesium concentrations were below 6 μmol g_{dw}^{-1} (Figure 3.3; Zimmermann *et al.*, 1988; Lange *et al.*, 1989), at a chlorophyll concentration of *ca.* 1.5 mg g_{dw}^{-1}, just above the lowest concentration of magnesium found in intact needles (4 μmol g_{dw}^{-1}; Figure 3.2). All physiological and biochemical studies of trees at the declining site failed to detect evidence of direct detrimental impact of atmospheric pollutants (Lange *et al.*, 1989). Indeed, as has been suggested in earlier studies (Zech & Popp, 1983; Küppers *et al.*, 1985), the difference between magnesium nutrition of trees at the two sites was the most likely cause of forest decline in this area. Magnesium nutrition may differ among stands due to lower magnesium availability and lower capability of roots to access and absorb nutrients (Meyer *et al.*, 1988).

Although many decline processes have no early visual expression, decline in stands is typically identified by visual or structural changes above-ground (Innes, 1988; Vales & Bunnel, 1988). Decline is indicated by chlorosis, thinning of crowns due to premature needle senescence, reduction in stemwood production, and a rate of tree mortality which cannot be explained based on density-related self-thinning. Declining stands, for example, produce less stemwood than is expected based on site characteristics. However, a reduction in stemwood production, perhaps the earliest of all the decline symptoms, is not a response unique to air pollution. Stemwood production can differ among stands of similar age, and can change with age and between years for reasons unrelated to air pollution. Thus, before nutrition can be used to explain forest decline, other potential causes must be excluded (Oren *et al.*, 1989).

Low stemwood production in a stand may reflect a host of stand and site factors other than the adverse effects of air pollution (Oren *et al.*, 1989). Stand

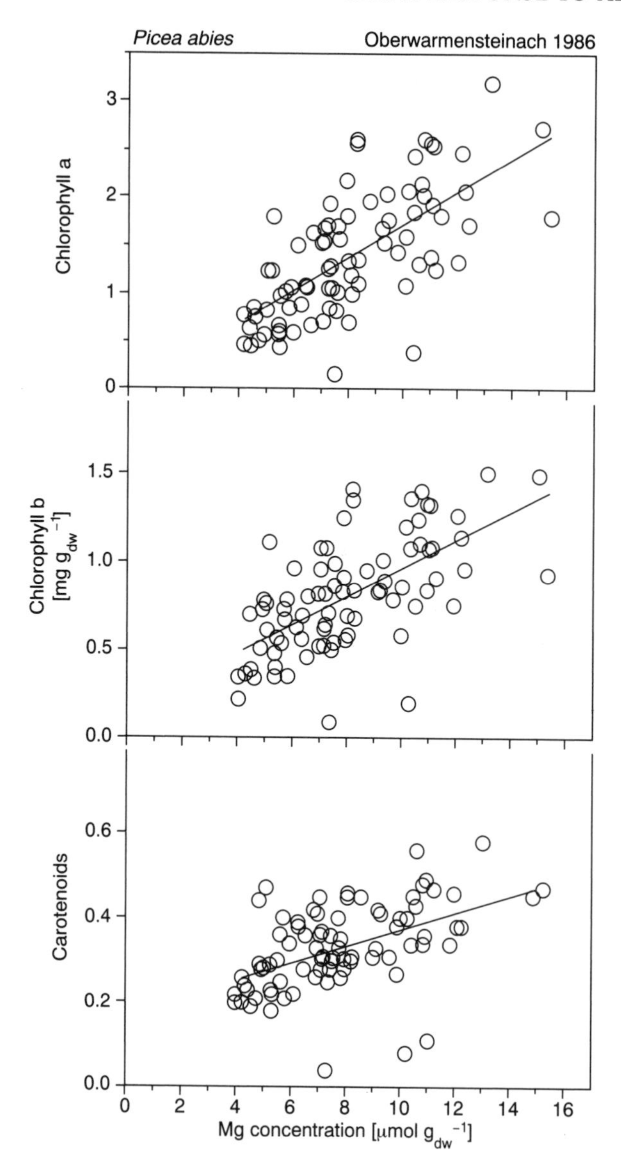

Figure 3.2. The concentration of chlorophyll a and b and of carotenoids increased with magnesium concentration in needles of 30-year-old declining *Picea abies* stand in the Fichtelgebirge where magnesium availability is thought to disturb plant function and growth (from Oren *et al.*, 1993; by permission of National Research Council of Canada)

factors associated with lower stemwood growth include the degree of site occupancy; this can be evaluated based on leaf-area index, basal area or number of trees per hectare, or a combination of both tree volume and density in relation to density-related mortality. Stands also grow at a lower rate when their leaf-area ratio, the ratio of photosynthesizing area to respiring support biomass, declines with mean tree size. Site factors which affect the length of the growing season

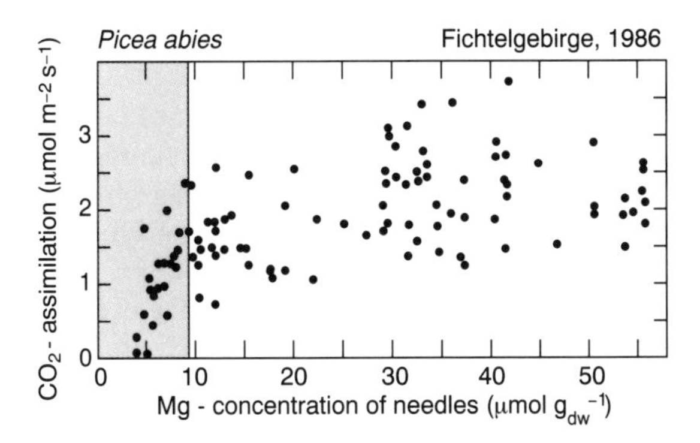

Figure 3.3. Magnesium concentration in needles of various ages from healthy and declining stands affected net CO_2 assimilation rate only at levels below that in which chlorosis begins (*ca.* 6 μmol g_{dw}^{-1}). CO_2 assimilation values are on an all-sided needle-area basis. Measurements were made at light saturation with leaf temperature of 19°C and various levels of intercellular CO_2 concentrations (c_i). Mg concentration below 6 μmol g_{dw}^{-1} affected net CO_2-assimilation rate of needles of all ages in both healthy and declining stands, while chlorosis appeared below 9 μmol g_{dw}^{-1} (shaded portion) (after Zimmermann *et al.*, 1988)

(e.g., frost-free days) or site quality indicators (e.g., water and nutrient availability) will result in predictable differences in growth rate among stands. To account for these potential causes of variation in stemwood production among stands, sites in the Fichtelgebirge study of similar site occupancy and with a similar growing season were chosen. The visibly declining site had a more favourable leaf-area ratio, and similar or better supply of water and availability of most nutrients. These conditions allowed exclusion of many potential causes other than pollution. However, the declining site did have lower magnesium and calcium availability.

The effect on forests of high nitrogen deposition in stands poorly supplied with nutrient cations was postulated by the "ammonium hypothesis" (Nihlgård 1985). The "ammonium hypothesis" implicated high nitrogen deposition, particularly in ammonium form, in causing a nutritional imbalance in trees, by both increasing nitrogen uptake relative to other nutrients and reducing the availability of cations by a chain of processes, beginning with acidic deposition on cation-poor soils and culminating with the leaching of cations out of the rooting zone (Johnson *et al.*, 1982; Mohren *et al.*, 1986; Kelly & Strickland, 1987; Kaupenjohann, 1989). However, this hypothesis is too coarse to explain some of the phenomena associated with forest decline, including (1) the long period of reduced stemwood production preceding chlorosis and, once chlorosis develops, (2) the occasional recovery of chlorotic trees, and (3) stand foliage production being unaffected even when some trees become chlorotic. The conceptual model of "nutritional disharmony", or temporally changing nutritional status, was conceived as a hypothetical explanation of the complex symptoms of forest decline (Oren *et al.*, 1988a) that could be used to explain the various decline phenomena observed in forests, as well as the variation in decline among sites and stands.

FACTORS AFFECTING EXTERNAL NUTRIENT SOURCE ACTIVITY

Soil characteristics

In addition to within-stand variation in the appearance of trees, there are large variations in the severity of forest decline among stands receiving the same amount of acidic deposition (Figure 3.4; Schulze & Freer-Smith, 1991). These variations may be caused by several factors. The critical load for nitrogen deposition, i.e., the nitrogen load at which there is no net change in the soil's base saturation, decreases with sulphur deposition (Figure 3.5; Schulze *et al.*, 1989b). Moreover, the critical nitrogen load is lower in soils developed from parent material of lower cation content, or which releases cations more slowly due to lower weathering rates, and is affected by nitrogen accumulation in the biomass, long-term nitrogen accumulation in soil, and background nitrogen leaching.

Mineral horizons of soils originating from phyllitic or granitic parent material have a higher acid neutralizing capacity than the overlying organic layer but, nevertheless, may have little exchangeable calcium and magnesium (Kaupenjohann, 1989). Therefore, the aluminium released in proton buffering remains in the soil solution and competes with calcium and magnesium for exchange sites, allowing more calcium and magnesium leaching as soil acidification proceeds. This reduces the storage of these elements in the mineral horizons and, at advanced podzolization stages, leaves the organic layer as the main source of nutrient cations.

Soil structure appreciably affects the nutrient supply to trees. In particular, gravel content affects the weathering rate, and aggregation characteristics influence the flux of elements in the soil (Horn, 1987). The main proton buffering reactions occur at the surface of aggregates. In the Fichtelgebirge study, the soils of the two sites developed to somewhat different stages from the phyllitic parent material (Kaupenjohann, 1989). The calcium and magnesium mass flow out of aggregates in the soil at the declining site was about half that found at the healthy site. Because of a smaller concentration gradient at the declining site between the aggregate surface and the soil solution, an equilibrium was approached fairly quickly and the ion flux from the interior of the aggregate to its surface was reduced.

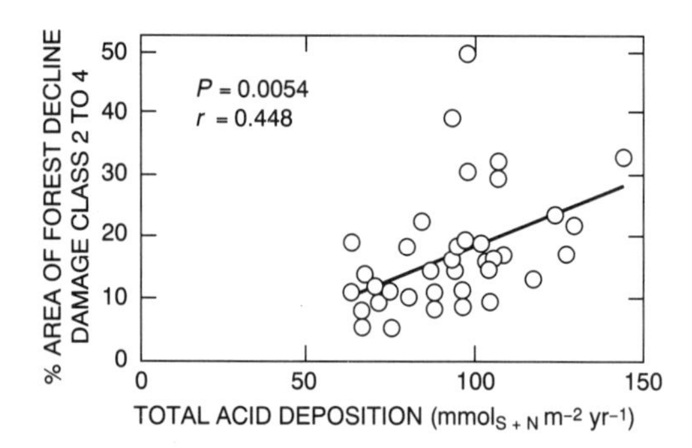

Figure 3.4. Areas of moderate to severe decline of *Picea abies* in Germany during 1986 as related to wet deposition of sulphate and nitrate (from Schulze & Freer-Smith, 1991; by permission of the Royal Society of Edinburgh)

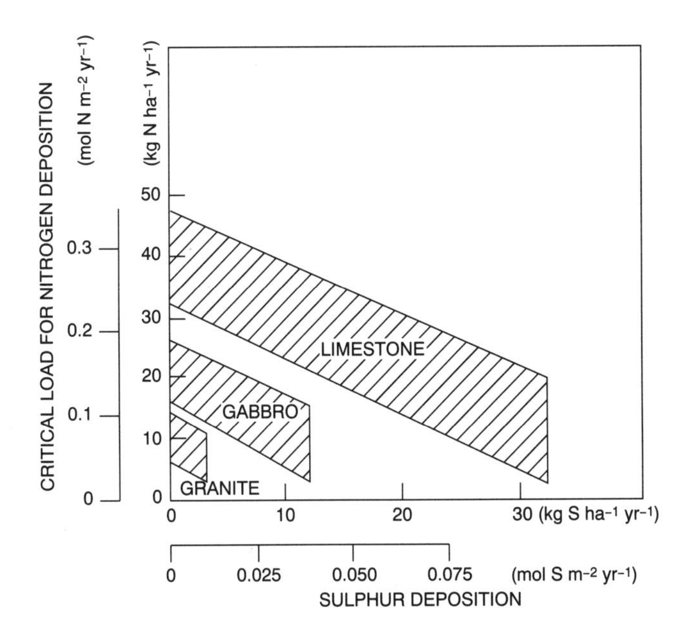

Figure 3.5. Critical loads for nitrogen as dependent on sulphur deposition and parent rock material. The critical loads are estimated under conditions of zero net change in soil base saturation. The range for each exemplified parent rock material represents certain site-dependent processes (see text). Weathering rates: limestone >0.2 mol m^{-2} yr^{-1}; gabbro 0.1-0.2 mol m^{-2} yr^{-1}; granite 0.02-0.05 mol m^{-2} yr^{-1} (from Schulze *et al.*, 1989b; by permission of Kluwer Academic Publishers)

The balance between ammonium and nitrate ions in the soil solution, and the preferred form of nitrogen (nitrate or ammonium) for uptake by roots affect soil chemistry and acidity. Thus, if roots take up more ammonium than nitrate, due to either preference for, or prevalence of this nitrogen form, soil acidification may be enhanced (Arnold & van Diest, 1991). However, many processes are involved in the sequence of nitrogen transformation in forest ecosystems, and the effect of one process on soil acidity should be evaluated together with the effects of others.

Roots and mycorrhiza

Because roots are unable to penetrate into aggregates, their main source of elements is the surface of aggregates. However, as a result of the leaching of the aggregate's surface by acidic deposition and, in advanced stages of acidification, by the low ion flux from inside, this source may be limited (Horn, 1987). Moreover, root penetration into mineral horizons is impaired by the conditions created by the acidification processes. Proton buffering in soils at advanced stages of acidification is due to aluminium, while at earlier stages other cations participate more in proton buffering. High aluminium-to-calcium ratios have been shown to reduce root development (Rost-Siebert, 1983, 1985).

In the Fichtelgebirge, roots at both sites began the season by growing new tips in the organic horizons (Meyer *et al.*, 1988). By the middle of the growing season, most formation of new root tips at the healthy site occurred in the mineral soil. In the declining site such a shift was not clearly apparent and root tip growth was restricted mostly to the organic layer. Moreover, mycorrhiza, which can penetrate aggregates, were significantly lower at the declining site, probably reducing nutrient uptake from the interior of aggregates. It is difficult to isolate a single variable at the declining site as the cause of reduced root growth in the mineral horizon, the reduction in the number of root tips and the decrease in mycorrhizal frequency (Meyer *et al.*, 1988; Schneider *et al.*, 1989). However, the impaired root system at the declining site, combined with lower availability of calcium and magnesium at that site, resulted in greatly reduced uptake rates of these elements in comparison to the healthy site.

Analysing the relationship between roots, soil and foliar nutrient concentrations (Meyer *et al.*, 1988; Schneider *et al.*, 1989) showed that: (1) the number of root tips in the mineral soil increased with the Ca:Al ratio (Figure 3.6), and (2) with increasing number of root tips per unit of leaf area, calcium and magnesium concentrations increased in the foliage. The second association was interpreted to mean that either (1) plant nutrition improved with increasing ratio of absorbing to transpiring surfaces, or (2) in comparison to highly podzolized soils, less podzolized soils both permit a better development of the root system and have a higher availability and consequently uptake of magnesium per unit of root surface, without affecting root-to-leaf ratio. Regardless of the exact mechanism, trees in the healthy stand reached bud break containing more calcium and magnesium in all of their organs than trees in the declining stand (Oren *et al.*, 1988a).

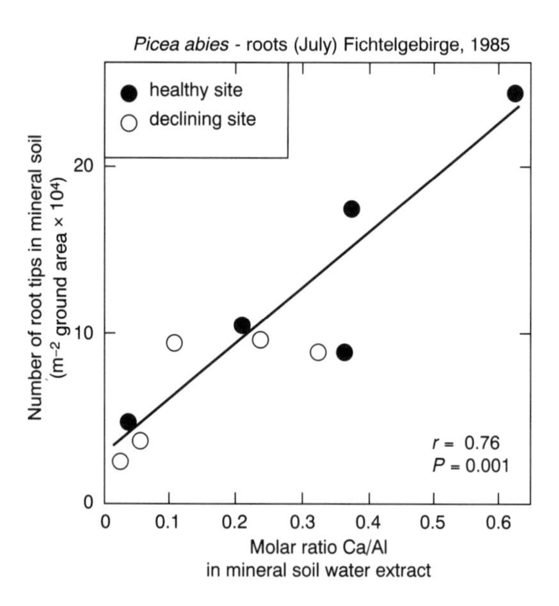

Figure 3.6. Relation between the total number of root tips per square metre ground area in the first 20 cm of mineral soil and the molar ratio of Ca:Al in mineral soil water extracts, in healthy and declining stands (from Meyer *et al.*, 1988; by permission of Springer-Verlag)

Stand development and species composition

As managed forest stands develop after a complete harvest, an increasing proportion of the site's nutrient capital is retained above ground, reducing the availability of nutrients for uptake (see Pritchett & Fisher, 1987). The rate at which this process proceeds depends on the initial availability of a given element, the site quality and the species composition. The rate of this process for a specific nutrient increases in situations where soils are poor in that nutrient, where sites are conducive for high growth rate and where the species grows rapidly.

The rate at which stands produce biomass above ground is highest when trees reach maturity in fully-stocked stands. In *P. abies* stands of central Europe this occurs at an age of *ca.* 30 years (Assmann, 1970), probably at the time when maximum canopy leaf-area is reached (Oren *et al.*, 1988b). Sites and species which reach their maximum canopy leaf-area soon after the stand is perturbed [e.g., *Pseudotsuga menziesii* (Mirb.) Franco after thinning on a site of high quality, Waring *et al.*, 1981] increase their demand for nutrients at a high rate, while those which develop canopy leaf-area slowly (e.g., *Pinus ponderosa* Dougl. ex Laws on dry sites, Oren *et al.*, 1987), increase their nutrient-uptake rate at a lower rate. Annual stemwood production declines from the maximum as more carbon is allocated to support the respiration requirements of increasing sapwood (Waring *et al.*, 1986), and as stands become over-mature (Assmann, 1970). Uptake of nitrogen, for example, was estimated and modelled to depend on water availability, but reached higher rates for species which can maintain higher canopy leaf-area values (Aber *et al.*, 1991).

Based on the above information, the period in which nutritional disharmony may develop most readily should be shortly after maximum canopy leaf-area is reached, when a large proportion of a generally scarce element has already accumulated in biomass above ground. In the Fichtelgebirge, both 30-year-old stands had just reached maximum canopy leaf-area, although stemwood production at the healthy stand was greater (Oren *et al.*, 1988b). However, much smaller magnesium capital in the soil of the declining site resulted in a larger proportion of this element being retained above ground, and a smaller amount being available for uptake (Horn *et al.*, 1989), preconditioning this stand to nutritional disharmony.

Not all stands at the declining site showed decline symptoms. An immediately adjacent 100-year-old stand of *P. abies* had a higher magnesium concentration in the foliage than the declining 30-year-old stand and did not display chlorosis (Kaupenjohann *et al.*, 1989). A second, adjacent 10-year-old stand, developing after a complete harvest, also had a higher foliar magnesium concentration (Figure 3.7; Buchmann *et al.*, 1994). In this case, in one treatment, magnesium concentration in needles increased as a result of weekly addition of magnesium during three growing seasons and, in another treatment, with 30% leaf-area reduction at the beginning of the study. The 30 kg nitrogen ha^{-1} deposited annually in the site did not prevent a decline in nitrogen concentration in the foliage; even nitrogen addition to over three times the natural level in one treatment did not alter this pattern of decline.

Ingestad (1987) has demonstrated that his concepts in seedling nutrition are applicable to forest trees and stands. Perhaps, as he demonstrated with seedlings,

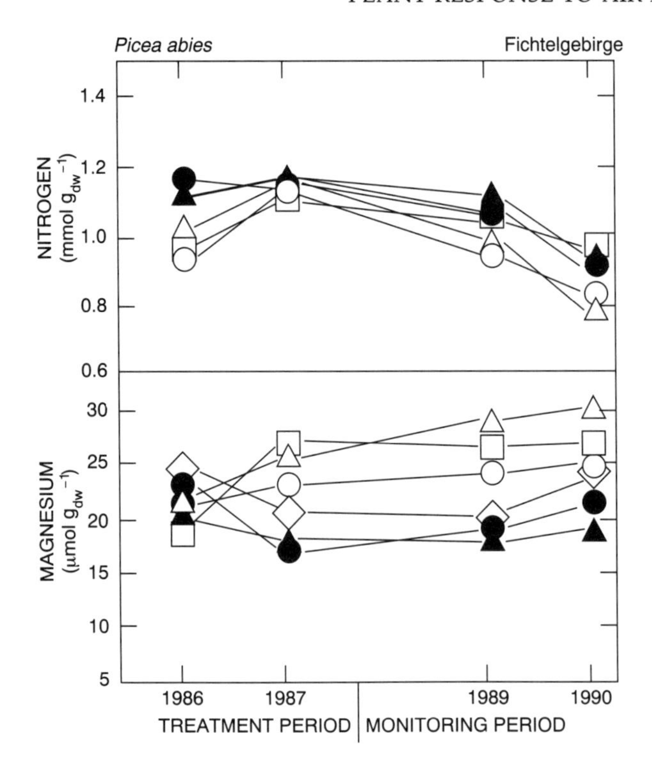

Figure 3.7. The concentrations of nitrogen and magnesium in 1-year-old needles of 10-year-old *Picea abies* saplings in a stand adjacent to the declining 30-year-old stand, in October — after growth ceased, during two treatment years and, a year later, two additional years. Treatments were: *squares* — nitrogen, *open triangles* — magnesium, *diamonds* — sugar, *open circles* — watered control, *filled-in circles* — control, *filled-in triangles* — reduced leaf-area. See text for details. Nitrogen addition (*ca.* 70 kg ha^{-1} yr^{-1}) did not alter the nitrogen concentration; magnesium addition increased needle magnesium concentration without affecting growth or physiology (from Buchmann *et al.*, 1994; by permission of Elsevier Science Limited)

when stand production rate is limited by factors other than nitrogen supply, nitrogen uptake rate and foliar concentration eventually decline. A reduction in nitrogen uptake rate under limited water supply was demonstrated in both broadleaf and conifer stands (Aber *et al.*, 1991). The uptake of phosphorus by *Pinus taeda* L. was directly related to stand growth, which, in turn, was promoted by nitrogen supply; however, due to the resulting increase in biomass, phosphorus concentration in foliage was relatively unaffected by both nitrogen and phosphorus supply (Vose & Allan, 1988).

In young, open stands, a large proportion of deposited nitrogen is incorporated into the litter layer and taken up by vegetation in the understorey (Buchmann *et al.*, 1992). This probably occurs also in over-mature, open stands. Thus, it appears that in open stands, growth may be more limited by canopy leaf-area than by either nitrogen or magnesium. In this situation, added nitrogen and magnesium may simply increase the leaching rates of these elements (Buchmann *et al.*, 1994), but the effect on growth, and thus nutrient sink activity, should be negligible. These open stands, then, are not conducive to the development of nutritional disharmony, unless the leaching rate of a scarce element is high.

FACTORS AFFECTING SINK ACTIVITY

In conifers, growth is more often limited by carbon utilization (sink activity) than carbon gain (source activity) because most environmental stress factors tend to limit growth processes before they affect photosynthesis (Luxmoore *et al.*, 1995). Thus, if the source of carbohydrates (i.e., current photosynthesis and storage) after bud break is sufficiently strong to support a high rate of needle production, and in the absence of limitations other than nitrogen, needle production will increase with increasing supply of nitrogen. As a result, the sink for all other nutrient elements also increases with nitrogen supply (Oren & Schulze, 1989). In growing needles, a dilution in the concentration of these elements occurs if they are supplied at a rate lower than the growth rate (Figure 3.8; Timmer & Morrow, 1984). Growth ceases and nitrogen accumulates in needles when another element is diluted to a physiologically limiting level. However, growth will correspond to the nitrogen supplied to the needles as long as other elements are present in sufficient quantities for proper physiological performance of the growing needle (Oren *et al.*, 1988a).

Mobile elements used in the formation of new tissues are supplied from mature tissues and from current uptake from the soil. The relative contribution of each source varies depending on several factors affecting the rate at which

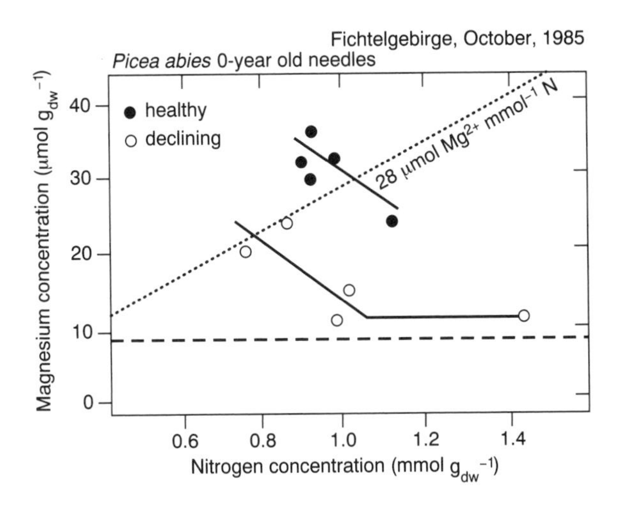

Figure 3.8. Concentrations of magnesium and calcium in new needles in October declined with increasing nitrogen concentration from an increase in needle weight in both healthy and declining sites. Levels did not reach those at which chlorosis begins (*horizontal dashed line*, after Ingestad, 1959). The *diagonal dotted line* represents the magnesium to nitrogen ratio above which production is within 95% of maximum (Ingestad, 1959) (from Oren *et al.*, 1988a; by permission of Springer-Verlag)

plants extract elements from the soil relative to the rate at which mature tissues retranslocate elements to support the growth of new tissues (Chapin & Kedrowski, 1983; Nambiar & Fife, 1987). There are several comprehensive reviews of factors which affect the rate of element uptake from the soil (see Oren & Sheriff, 1995). The rate at which elements retranslocate from mature tissues depends on the potential growth rate, as determined by the growth environment, and by the relative growth rate, which decreases with plant size (Evens, 1972; Waring *et al.*, 1986).

Sink activity in seedlings

The relative growth rate of seedlings is high, which means that a large amount of new tissue is produced relative to the quantity of existing mature tissue. Thus, the amounts of different elements which can be retranslocated from mature tissues to support the growth of new tissues are small relative to the total amounts necessary. Because the nutrient storage capacity, or capacitance, is relatively small in seedlings, the buffer between demand and supply of nutrients is reduced; this makes seedlings useful for evaluating nutrient source–sink interactions.

Ingestad has conducted numerous studies on seedling nutrition (see Ingestad, 1982, 1987). His data on *Betula verrucosa* Ehrh. (Ingestad, 1979, 1981; Ingestad & Lund, 1979) and *P. abies* (Ingestad, 1959) permit evaluation of the effects on

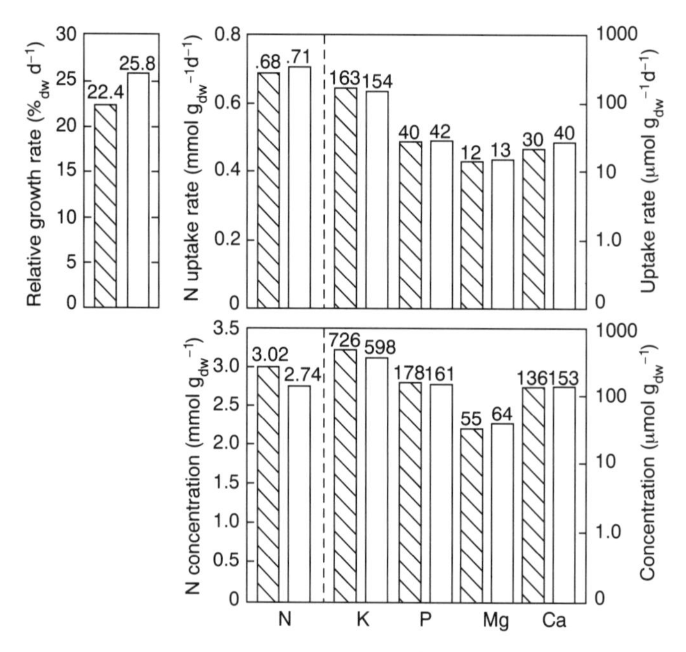

Figure 3.9. The relative growth rates, uptake rates and concentrations of five elements (*log scale* for all except nitrogen) in *Betula verrucosa* seedlings grown at 18-hour days in high nutrient solution (*hatched bars*) or 24-hour days in medium strength nutrient solution (*open bars*). Despite the lower nutrient supply, faster-growing plants took up similar or higher amounts of nutrients (after Ingestad, 1981)

nutrient source–sink interactions of increasing growth rate associated with experimental enhancement of, respectively, non-nutritional and nutritional resources. The results of the studies on *B. verrucosa* showed that increasing day-length from 18 to 24 hours increased the daily relative growth rate by 3% (Figure 3.9). Although nutrient supply under the 24-hour light day was lower, it did not limit growth. The high growth rate was associated with an increased uptake rate of all elements except potassium, while the tissue concentration of elements responded variably — increasing (calcium and magnesium), decreasing (nitrogen and potassium), or remaining unchanged (phosphorus). The decrease in the potassium uptake, associated with a drastic reduction in its tissue concentration when seedlings were grown at the lower nutrient supply, may indicate a luxury consumption of potassium when supplied at a higher rate. A smaller decline in tissue phosphorus concentration was not associated with a reduction in uptake of phosphorus. Luxmoore *et al.* (1986) have demonstrated an increase in uptake of all nutrients except potassium as growth increased 2.5-fold with three-fold increase in atmospheric CO_2 concentration. The concentration of all elements in that study must have decreased because the maximum increase in uptake among the macro-nutrients was only 30%. Thus, it appears that when growth rate is enhanced by a non-nutrient resource, uptake of nutrients is enhanced although their tissue concentration may decline. Growth rate will probably remain high until tissue concentration of one or more elements becomes so low as to affect the growth processes.

The results of another study (Ingestad, 1959) can be used to evaluate the effects of increasing nitrogen supply on growth, nutrient uptake and tissue concentration under conditions of constant supplies of calcium (non-retranslocatable element) and magnesium (retranslocatable element). Increasing nitrogen supply resulted in an increasing growth rate within the suboptimal supply range; the growth rate decreased at higher nitrogen supply rates (Figure 3.10). Nitrogen uptake and tissue concentrations increased with increasing growth. Calcium and magnesium uptake also generally increased with growth; however, their concentration declined. Thus, when growth is enhanced by both nutritional and non-nutritional factors, nutrient uptake is also enhanced, although often at a relatively lower rate. The resulting reduction in the concentrations of elements should not cause chlorosis or impair physiological function in growing needles. More mature needles may experience such damage as they retranslocate elements to growing needles.

Foliage production in trees

More so than seedlings, trees can rely on resources stored in mature tissues to supply different sinks (Oren & Sheriff, 1995). Mature tissues can supply large proportions of needed water, carbohydrates and retranslocatable nutrients over short periods of great demand. Thus, water supplied from storage in the sapwood of some species can meet 30–50% of daily transpiration (Waring & Running, 1978; Waring *et al.*, 1979), and carbohydrates and certain nutrients stored in the crown provide a large proportion of these resources used in the period of intensive extension growth after bud break (Oren *et al.*, 1988a, b; Horn *et al.*, 1989).

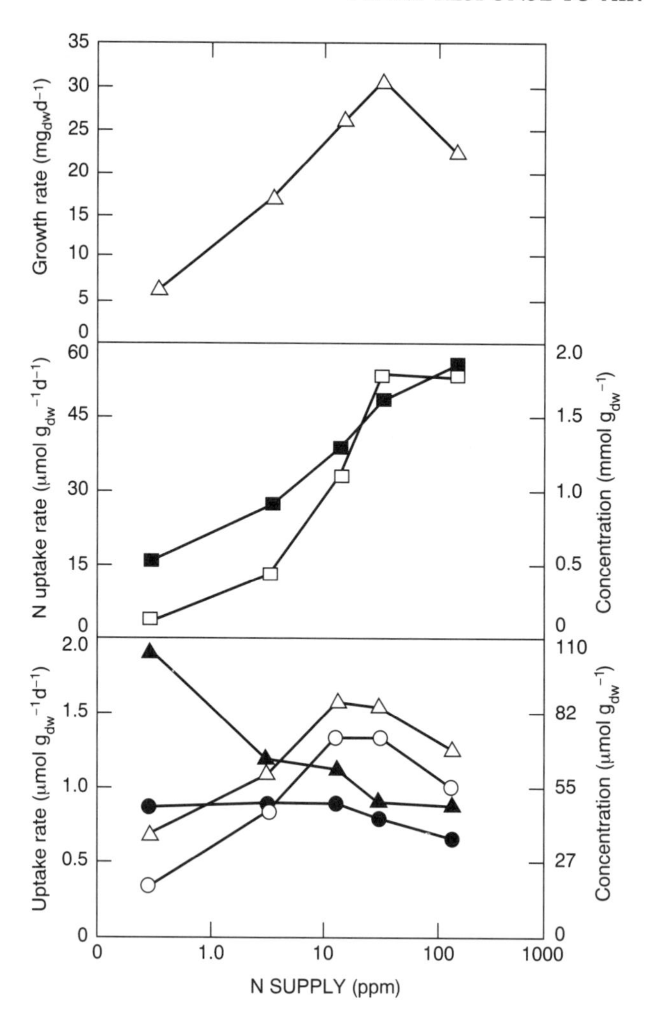

Figure 3.10. Growth rate of *Picea abies* seedlings increased with nitrogen supply in the suboptimal range and decreased thereafter; the supply of all other elements was kept constant. Both nitrogen uptake rate (*open squares*) and concentration (*filled-in squares*) increased with nitrogen supply. The uptake rate (*bottom, open symbols*) of calcium (*triangles*) and magnesium (*circles*) followed the pattern of the growth rate; however, their concentrations (*filled-in symbols*) generally declined (after Ingestad, 1959)

Most of the nitrogen supplied to support the growth of needles is provided by retranslocation from mature needles. Once another resource, e.g., magnesium, becomes limiting to proper physiological activity of the growing needle, growth ceases. This occurs at a concentration just above that of chlorosis (Figure 3.8).

Older needles are less fortunate under conditions of limited uptake of retranslocatable element. If their store of such an element is high prior to bud break, contributing a large amount to support the growth of needles may not reduce the concentration to a level where it can either cause chlorosis or affect physiological performance (Figure 3.11; Oren *et al.*, 1988a; Oren & Schulze, 1989). However, if

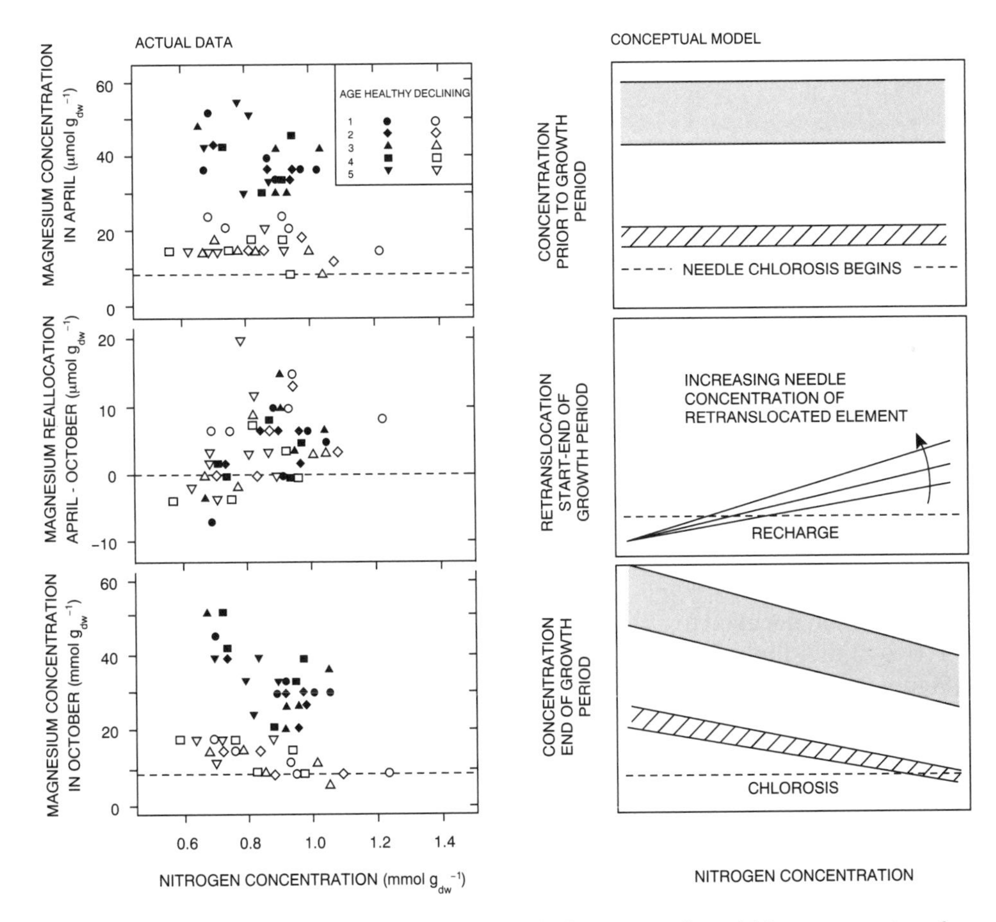

Figure 3.11. Actual data and conceptual model illustrating the within-season retranslocation of magnesium within needles of healthy (*filled-in symbols* and *shaded area*) and declining (*open symbols* and *hatched area*) stands of *Picea abies* (from Oren & Schulze, 1989; by permission of Springer-Verlag)

the stores are poorly stocked, such contribution is likely to reduce physiological vigour, cause chlorosis, and, eventually, premature shedding.

In the Fichtelgebirge, chlorophyll content of mature needles in both sites increased before bud break (Figure 3.12), as magnesium concentration increased with uptake before growth began (i.e., during the last part of the *recharge phase*, before the *canopy growth phase* begins; Oren *et al.*, 1988a; Oren *et al.*, 1993), and probably as conditions became more favourable for photosynthesis, thus reducing chlorophyll degradation. However, after bud break, only needles at the healthy site were sufficiently supplied with magnesium such that, although some of the magnesium was retranslocated, chlorophyll concentration remained high, and the needles green. In contrast, mature needles at the declining site, representing an apparently healthy tree in each plot, were barely able to maintain magnesium concentration above the level where chlorophyll

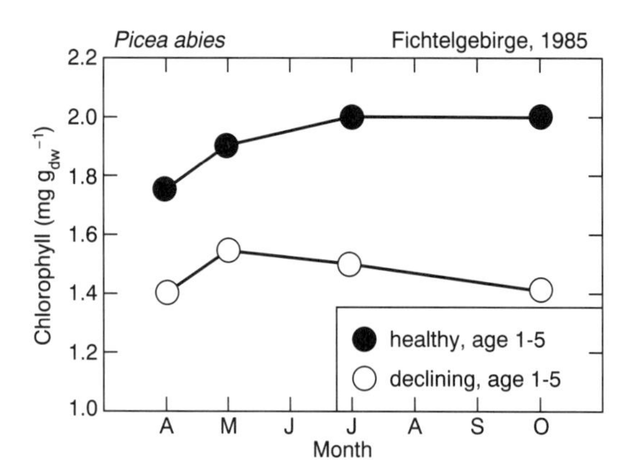

Figure 3.12. Seasonal trends in total chlorophyll concentration in needles of healthy and declining *Picea abies* stands, show a general decline in concentration after bud break in needles of the declining site due to magnesium retranslocation out of mature needles to support the growth of new needles (from Oren *et al.*, 1993; by permission of National Research Council of Canada)

concentration declined to a chlorotic level. Indeed, several classes of older needles of two individuals became chlorotic before the end of the growing season (Figure 3.10).

Although needles poorly stocked with nutrient elements attempt to retain more of their nutrients, releasing less to developing needles, foliage production appears to be affected only after mature needles of a tree become chlorotic (Oren & Schulze, 1989). Thus, foliage production is affected by a limited supply of certain elements only after the function of mature foliage is already impaired and its longevity reduced. For this reason, the production rate of canopy components may remain unchanged for some time while mature needles are becoming chlorotic and canopy leaf-area declines. However, the growth decline observed in many stands may not be attributable to the chlorosis and needle loss because a large proportion of foliage in forests persists in deep shade contributing little to the stand carbon budgets. In stands of *P. abies*, the needles in the lower 40% of the canopy accounted for 4% of the carbon uptake (Schulze *et al.*, 1977). Thus, in stands of shade-tolerant species, a relatively large reduction of leaf-area index should occur before their carbon budget is strongly affected (Waring, 1991). In contrast, a relatively small reduction in leaf-area index in stands of shade-intolerant species may result in an appreciable reduction in the carbon budget and, in turn, a rapid decline in the production of new canopy components.

Mature needles can replenish their stores when weather conditions reduce growth proportionally more than they reduce uptake of nutrients. Preventing growth by removing buds, thus decreasing the rate of growth relative to nutrient uptake, resulted in re-greening of magnesium-deficient *P. abies* needles (Lange *et al.*, 1987). A moderate soil drought (Berger, 1990), and cool but not too cold weather may also produce similar results. Thus, under certain weather conditions, chlorotic trees may re-green, a phenomenon which has been observed in central Europe.

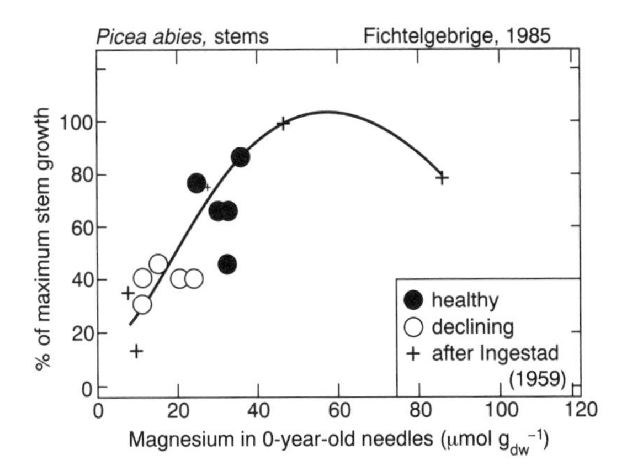

Figure 3.13. The relationship between magnesium concentrations in new needles and percentage of maximum growth of stem recalculated from a *Picea abies* seedling experiment (+; Ingestad, 1959). A line was fitted by eye to the seedling data. Data from the plot with the highest growth of stemwood per unit of leaf area were drawn on the line according to the October concentrations of magnesium in new foliage at the plot. Growth of all other plots at the healthy and declining sites was expressed as a percentage of growth in that plot and drawn according to the foliar concentration (from Oren & Schulze, 1989; by permission of Springer-Verlag)

Stemwood production in trees

Stemwood growth, which occurs in *P. abies* after foliage production is nearly complete (i.e., stem growth phase, Oren *et al.*, 1988a), depends to a large extent on the uptake of nutrients from the soil because the stores within trees are mostly used at that time (Oren *et al.*, 1988a; Horn *et al.*, 1989). Thus, once the maximum canopy leaf-area is reached, variation in stemwood growth should be better correlated with variation in the supply of a limiting nutrient than with variation in crown growth. Indeed, fertilizer applications which increase stemwood production but do not necessarily induce changes in canopy leaf-area (Binkley, 1986) support this contention.

In the Fichtelgebirge, production of new canopy components, which depended on retranslocation of magnesium, was similar in plots of the healthy and declining sites (Oren *et al.*, 1988b). However, stemwood growth in the plots was highly correlated with the foliar magnesium concentrations (Figure 3.13, Oren & Schulze, 1989) which, in turn, were highly correlated with the magnesium supply (Schulze *et al.*, 1989a). The difference in the source of magnesium, storage *vs.* uptake, used in the production of these two components, canopy and stems, respectively, explains situations in which foliage production is unaffected although stemwood production is reduced.

PREDICTING FOREST DECLINE PHENOMENA WITH THE NUTRITIONAL DISHARMONY MODEL

Huang (1995) developed a model designed to simulate the effects of weather conditions (e.g., temperature and precipitation), nitrogen availability, and the availability of retranslocatable elements on growth and nutrient concentrations in different organs of *Picea rubens*. This was done in order to evaluate whether the effects of weather and soil conditions on trees of different ages may cause seasonal nutrition patterns and decline symptoms similar to those observed in *P. rubens* stands of the southern Appalachian Mountains, North Carolina.

The decline in the spruce–fir forest of the southern Appalachians, like the decline in large parts of Europe, has been attributed to soil acidification by precipitation (Johnson & Reuss, 1984; Bormann, 1985), although this has not yet been conclusively demonstrated (Johnson *et al.*, 1991). Soils acidified by precipitation tend to contain fewer exchangeable cations and more nitrogen and sulphur than less acidified soils which have developed from a similar parent rock material (Johnson, 1990; Johnson *et al.*, 1991). The model output (Figure 3.14) represents the effect on magnesium concentration in mature (2-year-old) needles of four combinations of nitrogen and magnesium availability. Magnesium was considered representative of retranslocatable elements which tend to leach out of forest ecosystems under high input of acidic precipitation.

The four combinations represent: (1) high availability of magnesium relative to nitrogen, a condition typical of many temperate coniferous forests, (2) low

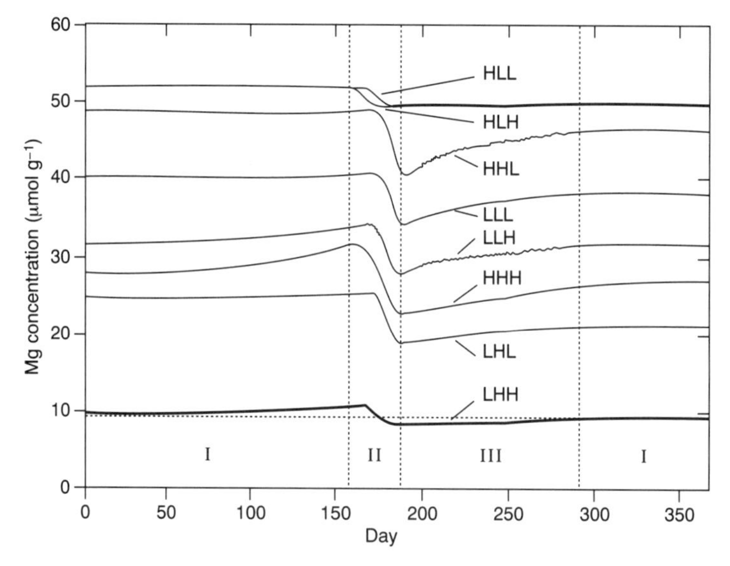

Figure 3.14. Model predictions of seasonal trends in magnesium concentrations in 2-year-old needles of *Picea rubens* under various combinations of magnesium (first letter in the three-letter combination) and nitrogen (second letter), and growth temperature (third letter). The levels were set to favourable (H) or unfavourable (L). The season is partitioned to three phases: recharge phase (I), canopy growth phase (II), and stem growth phase (III), indicating the major process in each part of the season. The model was initialized using typical data for 30-year-old trees, and the 10th season is shown. Based on the model, only under conditions where magnesium supply is low, and growth is promoted by favourable nitrogen supply and temperature (LHH combination), chlorosis may result and persist (indicated by needle magnesium concentrations below the horizontal dotted line) (from Huang, 1995)

availabilities of both magnesium and nitrogen, representing temperate coniferous forests on soils which have developed from magnesium-poor or slowly weathering parent rock material, (3) high availability of both nitrogen and magnesium, typical of forest ecosystems similar to the first condition during the early phase of pollution, and (4) high availability of nitrogen relative to magnesium, representing ecosystems similar to the first condition at an advanced stage of acidification, and the second condition at all stages of increased nitrogen load. In addition to affecting the sink strength for magnesium through the effect of nitrogen on growth, sink strength was also manipulated in each of the four combinations by simulating growth during the growing season in optimal and low air temperatures.

The simulated pattern of growth and magnesium concentration in mature foliage closely matched the observed pattern in both *P. abies* (Oren *et al.*, 1988a) and *P. rubens* (Huang, 1995), including such changes in phenology as earlier bud break under more favourable temperatures and greater nitrogen availability. The model simulation predicts that under favourable magnesium supply, magnesium concentration in mature needles will not decrease to chlorosis-inducing levels, regardless of how favourable other conditions are for growth. Moreover, even when magnesium uptake is low, high nitrogen deposition alone may not be sufficient to promote growth, thus increasing the sink for magnesium, and causing mature needles to lose magnesium and become chlorotic. Chlorosis of mature needles due to magnesium retranslocation may occur only after several years of otherwise optimal conditions for growth (e.g., temperature and moisture) when trees are well supplied with nitrogen.

Thus, in condition (4) — ecosystems poor in magnesium and rich in nitrogen — trees may alternate in nutrient status and colour depending on weather conditions: yellow when temperature and moisture conditions are optimal for growth, green when conditions are suboptimal. Thus, based on the model simulation, needle shedding occurs only after needles had been chlorotic for several seasons, during a long period of optimal conditions for growth.

Huang's model is based on a conceptual model of nutritional disharmony conceived as a result of a study in central Europe (Oren & Schulze, 1989); however, this incorporates many more growth and nutrition processes than implied in the conceptual model. The predictions produced by the model are in good agreement with observations in the high-elevation spruce–fir forests of the southern Appalachians. Not only are the seasonal and age-related trends in nutrient concentrations predicted correctly, but predicted effects of temperature and nitrogen supply on phenology are in agreement with observed phenomena as well. For example, both high temperature and high nitrogen supply hasten bud break and delay the end of the growing season. Indeed, frost damage was implicated in causing decline in forest trees which were not ready for low temperature due to unseasonably high activity resulting from high tissue nitrogen (Roelofs *et al.*, 1988). Although these are compelling simulation outcomes, the model still requires validation. A long-term monitoring of forest conditions, i.e., stemwood growth, needle colour, and canopy leaf-area, and weather-related growth conditions, (temperature and moisture regimes), would be helpful in evaluating nutritional disharmony as a concept for explaining several of the forest decline phenomena.

ACKNOWLEDGEMENT

I thank Sharon Billings for her thorough review of an earlier version of this chapter.

REFERENCES

Aber, J.D., Melillo, J.M., Nadelhoffer, K.N., Pastor, J. & Boone, R.D. 1991. Factors controlling nitrogen cycling and nitrogen saturation in northern temperate forest ecosystems. *Ecol. Appl.* **1**: 303-315.

Arnold, G. & van Diest, A. 1991. Nitrogen supply, tree growth and soil acidification. *Fert. Res.* **27**: 29-38.

Assmann, E. 1970. *The Principles of Forest Yield Study*. Pergamon Press, Oxford.

Berger, A. 1990. Nährstofftransport in Xylemsaft der Fichte (*Picea abies* (L.) Karst) in Beziehung zur Tageszeit, Transpiration, zum Bodenwassergehalt und zur Nährstoffkonzentration in der Bodenlösung. Unpublished work for Diploma, Universität Bayreuth, Bayreuth, Germany.

Binkley, D. 1986. *Forest Nutrition Management*. Wiley, New York.

Bormann, F.H. 1985. Air pollution and forests: an ecosystem perspective. *Bioscience* **35**: 434-441.

Brix, H. 1981. Effects of nitrogen fertilizer source and application rates on foliar nitrogen concentration, photosynthesis, and growth of Douglas-fir. *Can. J. For. Res.* **11**: 775-780.

Buchmann, N., Oren. R., Gebauer, G., Dietrich, P. & Schulze, E.-D. 1992. The use of stable isotopes in ecosystem research. First results of a field study with ^{15}N. *Isotopenpraxis* **28**: 51-71.

Buchmann, N., Oren, R., Schulze, E.-D. & Zimmermann, R. 1994. Response of magnesium-deficient saplings in a young, open stand of *Picea abies* (L.) Karst. to elevated soil magnesium, nitrogen and carbon. *Environ. Pollut.* **87**: 31-43.

Chapin, F.S. III & Kedrowski, R.A. 1983. Seasonal changes in nitrogen and phosphorus fractions and autumn retranslocation in evergreen and deciduous taiga trees. *Ecology* **64**: 376-391.

Evens, G.C. 1972. *The Quantitative Analysis of Plant Growth*. Studies in Ecology, Vol. 1, University of California Press, Berkeley, Los Angeles.

Guderian, R. 1977. *Air Pollution*. Ecological Studies 22. Springer-Verlag, Berlin.

Hantschel, R., Kaupenjohann, M., Horn, R. & Zech, W. 1988. Acid rain studies in the Fichtelgebirge (NE-Bavaria). In Mathy, P. (ed.) *Air Pollution and Ecosystems, Proc. Comm. Eur. Communities Symp., Grenoble, France*: 881-886. Reidel, Dordrecht.

Horn, R. 1987. The role of structure for nutrient sorptivity of soils. *Z. Pflanzenernähr Bodenkd* **150**: 13-16.

Horn, R., Schulze, E.-D. & Hantschel, R. 1989. Nutrient balance and element cycling in healthy and declining Norway spruce stands. In Schulze, E.-D., Lange, O.L. & Oren, R. (eds) *Forest Decline and Air Pollution: A Study of Spruce on Acid Soils*. Ecological Studies 77: 444-455. Springer-Verlag, Berlin.

Huang, C. 1995. Modeling physiological and ecological effects of nutritional disharmony on spruce-fir forests. Unpublished PhD Thesis, Duke University, Durham, NC.

Ingestad, T. 1959. Studies on the nutrition of forest tree seedlings. II. Mineral nutrition of spruce. *Physiol. Plant.* **12**: 568-593.

Ingestad, T. 1979. Nitrogen stress in birch seedlings. II. N, K, P, Ca, and Mg nutrition. *Physiol. Plant.* **45**: 149-157.

Ingestad, T. 1981. Nutrition and growth of birch and gray alder seedlings in low conductivity solutions and at varied relative rates of nutrient addition. *Physiol. Plant.* **52**: 454-466.

Ingestad, T. 1982. Relative addition rate and external concentration: Driving variables used in plant nutrition research. *Plant Cell Environ.* **5**: 443-453.

Ingestad, T. 1987. New concepts on soil fertility and plant nutrition as illustrated by research on forest trees and stands. *Geoderma* **40**: 237-252.

Ingestad, T. & Lund, A.-B. 1979. Nitrogen stress in birch seedlings. I. Growth technique and growth. *Physiol. Plant.* **45**: 137-148.

Innes, J.L. 1988. Forest health surveys: problems in assessing observer objectivity. *Can. J. For. Res.* **16**: 560-565.

Johnson, D.W. 1990. Nitrogen retention in forest soils. *J. Environ. Qual.* **21**: 1-12.

Johnson, D.W. & Reuss, J.O. 1984. Soil mediated effects of atmospherically deposited sulphur and nitrogen. *Phil. Trans. R. Soc. Lond. B. Biol. Sci.* **305**: 383-392.

Johnson, D.W., Turner, J. & Kelly, J.M. 1982. The effect of acid rain on soil nutrient status. *Water Resour. Res.* **18**: 449-461.

Johnson, D.W., Van Miegroet, H., Lindberg, S.E., Todd, D.E. & Harrison, R.B. 1991. Nutrient cycling in red spruce forests of the Great Smoky Mountains. *Can. J. For. Res.* **21**: 769-787.

Kaupenjohann, M. 1989. Effects of acid rain on soil chemistry and nutrient availability in the soil. In Schulze, E.-D., Lange, O.L. & Oren, R. (eds) *Forest Decline and Air Pollution: A Study of Spruce on Acid Soils.* Ecological Studies **77**: 297-340. Springer-Verlag, Berlin.

Kaupenjohann, M., Zech, W., Hantschel, R., Horn, R. & Schneider, B.U. 1989. Mineral nutrition of forest trees: A regional survey. In Schulze, E.-D., Lange, O.L. & Oren, R. (eds) *Forest Decline and Air Pollution: A Study of Spruce on Acid Soils.* Ecological Studies **77**: 282-296. Springer-Verlag, Berlin.

Kelly, J.M. & Strickland, R.C. 1987. Soil nutrient leaching in response to simulated acid rain treatment. *Water Air Soil Pollut.* **34**: 167-181.

Köstner, B., Czygan, F.-C. & Lange, O.L. 1990. An analysis of needle yellowing in healthy and chlorotic Norway spruce (*Picea abies*) in a forest decline area of the Fichtelgebirge (N.E. Bavaria). Annual time-course changes in chloroplast pigments for five different needle age classes. *Trees* **4**: 55-67.

Küppers, M., Zech, W., Schulze, E.-D. & Beck, E. 1985. CO_2-Assimilation, Transpiration und Wachtum von *Pinus sylvestris* L. bei unterschiedlicher Magnesiumversorgung. *Forstw. Cbl.* **104**: 23-36.

Lange, O.L., Zellner, H., Gebel, J., Schramel, P., Köstner, B. & Czygan, F.C. 1987. Photosynthetic capacity, chloroplast pigments, and mineral content of previous year's spruce needles with and without the new flush: analysis of the forest-decline phenomenon of needle bleaching. *Oecologia (Berl.)* **73**: 351-357.

Lange, O.L., Heber, U., Schulze, E.-D. & Ziegler, H. 1989. Atmospheric pollutants and plant metabolism. In Schulze, E.-D., Lange, O.L. & Oren, R. (eds) *Forest Decline and Air Pollution: A Study of Spruce on Acid Soils.* Ecological Studies **77**: 238-273. Springer-Verlag, Berlin.

Linder, S. & Ingestad, T. 1977. Ecophysiological experiments under limiting and non-limiting conditions of mineral nutrition in field and laboratory. In *Bicentenary Celebration of C.P. Thunberg Visit to Japan*: 69-76. R. Swed. Embassy and the Bot. Soc. Japan.

Linder, S. & Rook, D.A. 1984. Effects of mineral nutrition on carbon dioxide exchange and partitioning of carbon in trees. In Bowen, G.D. & Nambiar, E.K.S. (eds) *Nutrition of Plantation Forests*: 211-236. Academic Press, London.

Luxmoore, R.J., O'Neill, E.G., Ells, J.M. & Rogers, H.H. 1986. Nutrient uptake and growth responses of Virginia pine to elevated atmospheric carbon dioxide. *J. Environ. Qual.* **15**: 244-251.

Luxmoore, R.J., Oren, R., Sheriff, D.W. & Thomas, R.B. 1995. Source-sink-storage relationships of conifers. In Smith, W.K. & Hinckly, T.M. (eds) *Resource Physiology of Conifers.* Physiol. Ecol. Ser.: 179-216. Academic Press, London.

Meyer, J., Schneider, B.U., Werk, K.S., Oren, R. & Schulze, E.-D. 1988. Performance of two *Picea abies* (L.) Karst. stands at different stages of decline. V. Root tips and mycorrhiza development and their relation to aboveground and soil nutrients. *Oecologia (Berl.)* **77**: 7-13.

Mohren, G.M.J., Van Den Burg, J. & Burger, F.W. 1986. Phosphorus deficiency induced by nitrogen input in Douglas fir in the Netherlands. *Plant Soil.* **95**: 191-200.

Nambiar, E.K.S. & Fife, D.N. 1987. Growth and nutrient retranslocation in needles of radiata pine in relation to nitrogen supply. *Ann. Bot.* **60**: 147-156.

Nihlgård, B. 1985. The ammonium hypothesis — an additional explanation to the forest dieback in Europe. *Ambio* **14**: 2-8.

Oren, R. & Schulze, E.-D. 1989. Nutritional disharmony and forest decline: A conceptual model. In Schulze, E.-D., Lange, O.L. & Oren, R. (eds) *Forest Decline and Air Pollution: A Study of Spruce on Acid Soils.* Ecological Studies **77**: 425-443. Springer-Verlag, Berlin.

Oren, R. & Sheriff, D.W. 1995. Water and nutrient acquisition by roots and canopies. In Smith, W.K. & Hinckly, T.M. (eds) *Resource Physiology of Conifers.* Physiol. Ecol. Ser.: 39-74. Academic Press, London.

Oren, R., Waring, R.H., Stafford, S.G. & Barrett, J.W. 1987. Twenty-four years of ponderosa pine growth in relation to canopy leaf area and understorey competition. *For. Sci.* **33**: 538-547.

Oren, R., Schulze, E.-D., Werk, K.S. & Meyer, J. 1988a. Performance of two *Picea abies* (L.) Karst. stands at different stages of decline. VII. Nutrient relations and growth. *Oecologia (Berl.)* **77**: 163-173.

Oren, R., Schulze, E.-D., Werk, K.S., Meyer, J., Schneider, B.U. & Heilmeier, H. 1988b. Performance of two *Picea abies* (L.) Karst. stands at different stages of decline. I. Carbon relations and stand growth. *Oecologia (Berl.)* **75**: 25-37.

Oren, R., Werk, K.S., Meyer, J. & Schulze, E.D. 1989. Potential and limitations of field studies on forest decline associated with anthropogenic pollution. In Schulze, E.-D., Lange, O.L. & Oren, R. (eds) *Forest Decline and Air Pollution: A Study of Spruce on Acid Soils.* Ecological Studies **77**: 23-36. Springer-Verlag, Berlin.

Oren, R., Werk, K.S., Buchmann, N. & Zimmermann, R. 1993. Chlorophyll — nutrient relationships identify nutritionally-caused decline in *Picea abies* (L.) Karst. stands. *Can. J. For. Res.* **23**: 1187-1195.

Pritchett, W.L. & Fisher, R.F. 1987. *Properties and Management of Forest Soils.* Wiley, New York.

Roelofs, J.G.M., Boxman, A.W. & van Dijk, H.F.G. 1988. Effects of airborne ammonium on natural vegetation and forests. In Mathy, P. (ed.) *Air Pollution and Ecosystems.* CEC Air Pollution Report Series **7**: 876-880. Reidel, Dordrecht.

Rost-Siebert, K. 1983. Aluminum-Toxizität und - Toleranz an Keimpflanzen von Fichte (*Picea abies* (L.) Karst.) und Buche (*Fagus sylvatica* L.). *AFZ* **38**: 686-689.

Rost-Siebert, K. 1985. Untersuchungen zur H-und Al-Ionentoxizität an Keimpflanzen von Fichte (*Picea abies* (L.) Karst.) und Buche (*Fagus sylvatica* L.) in Lösungskultur. *Ber Forschungszentrums Waldökosysteme/Waldsterben*, Bd 12, Universität Göttingen, p. 219.

Schneider, B.U., Meyer, J., Schulze, E.-D. & Zech, W. 1989. Root and mycorrhizal development in healthy and declining Norway spruce stands. In Schulze, E.-D., Lange, O.L. & Oren, R. (eds) *Forest Decline and Air Pollution: A Study of Spruce on Acid Soils.* Ecological Studies **77**: 370-391. Springer-Verlag, Berlin.

Schulze, E.-D. & Freer-Smith, P.H. 1991. An evaluation of forest decline based on field observations focused on Norway spruce, *Picea abies.* In Last, F.T. & Whatling, R. (eds) *Acid Deposition, Its Nature and Impacts*: 155-168. The Royal Society of Edinburgh, Edinburgh.

Schulze, E.-D., Fuchs, M. & Fuchs, M.I. 1977. Spatial distribution of photosynthetic capacity and performance in a mountain spruce forest of Northern Germany. III. The significance of the evergreen habit. *Oecologia (Berl.)* **30**: 239-248.

Schulze, E.-D., Oren, R. & Lange, O.L. 1989a. Processes leading to forest decline: A synthesis. In Schulze, E.-D., Lange, O.L. & Oren, R. (eds) *Forest Decline and Air Pollution: A Study of Spruce on Acid Soils.* Ecological Studies **77**: 459-468. Springer-Verlag, Berlin.

Schulze, E.-D., De Vries, W., Hauhs, M., Rosén, K., Rasmussen, L., Tamm, C.-O. & Nilsson, J. 1989b. Critical loads for nitrogen deposition on forest ecosystems. *Water, Air Soil Pollut.* **48**: 451-456.

Timmer, V.R. & Morrow, L.D. 1984. Predicting fertilizer growth response and nutrient status of jack pine by foliar diagnosis. In Stone, E.L. (ed.) *Forest Soils and Treatment Impacts*: 335-351. University of Tennessee, Knoxville, TN.

Vales, D.J. & Bunnel, F.L. 1988. Comparison of methods for estimating forest overstorey cover. I. Observer. Effects. *Can. J. For. Res.* **18**: 606-609.

Vose, J.M. & Allan, H.L. 1988. Leaf area, stemwood growth, and nutrition relationships in loblolly pine. *For. Sci.* **34**: 547-563.

Waring, R.H. 1991. Responses of evergreen trees to multiple stresses. In Mooney, H. (ed.) *Response of Plants to Multiple Stresses*: 371-390. Academic Press, London.

Waring, R.H. & Running, S.W. 1978. Sapwood water storage: Its contribution to transpiration and effects upon water conductance through the stems of old-growth Douglas-fir. *Plant Cell Environ.* **1**: 131-140.

Waring, R.H., Whitehead, D. & Jarvis, P.G. 1979. The contribution of stored water to transpiration in Scots pine. *Plant Cell Environ.* **2**: 309-317.

Waring, R.H., Newman, K. & Bell, J. 1981. Efficiency of tree crowns and stemwood production at different canopy leaf densities. *Forestry* **54**: 129-137.

Waring, R.H., Aber, J.D., Melillo, J.M. & Moore, B. III 1986. Precursors of change in terrestrial ecosystems. *Bioscience* **36**: 433-438.

Winner, W., Mooney, H.A. & Goldstein, R.A. 1985. *Sulphur Dioxide and Vegetation*. Stanford University Press, Stanford.

Zech, W. & Popp, E. 1983. Magnesiummangel, einer der Gründe für das Fichten- und Tannensterben in NO-Bayern. *Forstw. Cbl.* **102**: 50-55.

Zimmermann, R., Oren, R., Schulze, E.-D. & Werk, K.S. 1988. Performance of two *Picea abies* (L.) Karst. stands at different stages of decline. II. Photosynthesis and leaf conductance. *Oecologia (Berl.)* **76**: 513-518.

4 Global Changes in Atmospheric Carbon Dioxide: The Influence on Terrestrial Vegetation

C.J. ATKINSON

BACKGROUND TO THE "GREENHOUSE EFFECT"

Carbon dioxide does not significantly absorb short-wave solar radiation entering the earth's atmosphere. The re-emission of this energy back into space as long-wave infrared radiation occurs because it is able to escape from the earth's atmosphere by remaining unabsorbed. This "atmospheric window" (8 to 12 μm) in the absorption of long-wave infrared radiation, has the effect of allowing the earth to cool (Mitchell, 1989). It is by coincidence that molecular carbon dioxide [CO_2] strongly absorbs infrared radiation within the wave bands of this "window", and by so doing perturbs the earth's energy balance (see Schneider, 1989; Figure 4.1). In part, the extent to which the capacity of the atmosphere to cool becomes impaired depends on the concentration of carbon dioxide [CO_2].

Within the atmospheric "window" CO_2 is not the only gas to possess this absorptive capacity, the others include water vapour, methane, nitrous oxide, chlorofluorocarbons (CFCs) and ozone. Because of the analogy of this effect with the process which causes greenhouses to heat up, these gases are frequently described as "greenhouse gases". The other "greenhouse gases", with the exception of water vapour, occur on a global scale at much lower atmospheric concentrations than CO_2, and may be considered as less important contributors to potential greenhouse warming (climate change). However, their effectiveness as infrared absorbers on a molecule for molecule basis is much greater than that of CO_2. For example, methane has been calculated to be at least one order of magnitude more effective at absorbing infrared radiation than CO_2 (Lashof & Ahuja, 1990).

The concentrations of all of the recognized "greenhouse gases" has been shown to be on the increase (e.g., methane by approximately 1% per annum) and they are therefore likely to become considerably more important in determining the planet's energy balance. With its present rate of increase, methane in 50 years time will make a contribution to global warming amounting to about 30% of the existing CO_2 contribution (Mansfield, 1991), while at its 1990 concentration CO_2 contributes approximately half of the calculated greenhouse heating effect (McElroy, 1988).

Plant Response to Air Pollution. Edited by Mohammad Yunus and Muhammad Iqbal

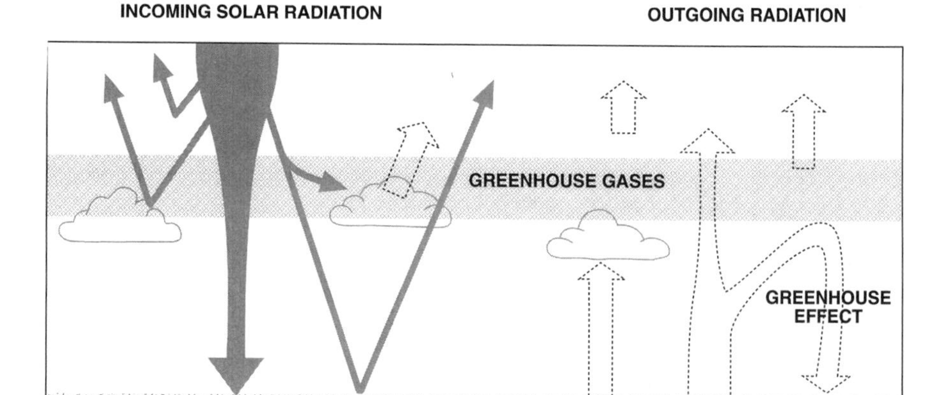

Figure 4.1. The heat-trapping ability of earth's atmosphere as a feature of the planet's energy balance. Incoming solar radiation and its deflection is depicted on the left (short wavelength), while the re-emitted infrared radiation (long wavelength) is shown on the right. The numbers attached to each process refer to their percentage contribution in a total energy balance. Greenhouse gases and clouds can be seen to trap a high proportion of the surface radiation and cause the greenhouse heating effect. Data are from Schneider (1989)

The retention of solar energy after conversion into long-wave radiation has provided an environment in which it was possible for life to evolve and richly diversify, which on a planet without an atmosphere that absorbs thermal radiation, could be some 33°C colder (Mitchell, 1989). It is only comparatively recently, in the evolutionary history of the planet, that the importance of "greenhouse gases" other than water vapour and CO_2 has been understood. It is now established that the [CO_2] has been in a steady-state flux at around 270 vpm (volume parts per million or μl CO_2 l^{-1} or cm^3 m^{-3}) for probably 2000 years, and possibly for as long as 10 million years, during the evolution of terrestrial vegetation. Marked seasonal changes in CO_2 are apparent, however, and relate to seasonal changes in global photosynthesis.

The earth's atmospheric [CO_2] has been monitored at a number of locations from the early 1960s onwards. From one such remote site at Hawaii's Mauna Loa Observatory, the [CO_2] has been shown to increase from 315 vpm in 1958 to 350 vpm in 1988 (Keeling *et al.*, 1976). These data have now been supplemented with measurements made from gas trapped within ice core bubbles and $^{13}C/^{12}C$ ratios in tree rings (Stuiver, 1982). From ice core samples extracted from depths down to 2 km, extrapolations have been made of the atmospheric [CO_2] many thousands of years before the beginning of the industrial revolution (Neftel *et al.*, 1982; Friedli *et al.*, 1986; Barnola *et al.*, 1987). From the pre-industrial atmospheric [CO_2] (275 vpm), to the present-day level at 350 vpm, we have already seen a rise in [CO_2] of approximately 25%.

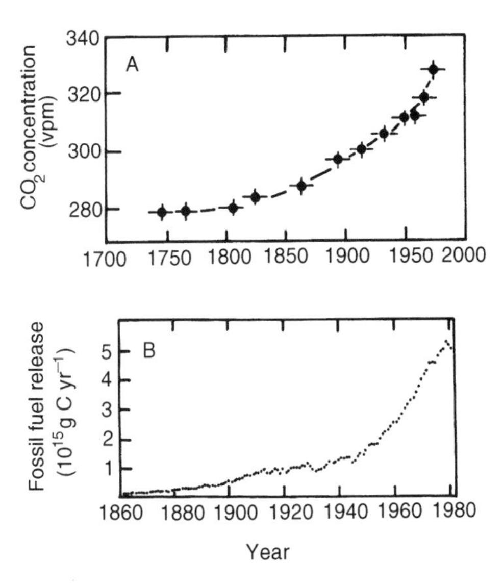

Figure 4.2. The atmospheric concentration of CO_2 measured in ice from a glacier formed over the last 250 years (A), and the fossil fuel CO_2 emissions from 1860 to 1982 (B). Data are from Neftel *et al.* (1985) and Marland & Rotty (1983), respectively (reprinted with permission from *Nature*, Macmillan Magazines Ltd and the American Geophysical Union)

The ice core data also tell us that the observed increase in atmospheric [CO_2] is a feature of the industrial era (Figure 4.2a). This increase can be strongly correlated with the global increase in the consumption of fossil fuels (*ca.* 5.3×10^{12} kg carbon yr^{-1}: Figure 4.2b) and therefore it is expected to continue to increase into the next century (Rotty & Marland, 1986). There are also other important factors in the global carbon budget equation for consideration, particularly those associated with forest clearances and the fluxes of uptake and release by these sinks. The debate is particularly focused on the role of tropical forests.

Using average global temperatures it has already been calculated that the observed increase in global [CO_2] has been responsible for a detectable temperature rise of between 0.5 and 0.7°C over the last 140 years (NASA, 1988). At present it is believed that atmospheric [CO_2] is increasing at a rate of 1.5 vpm (0.48%) per annum and this will yield a concentration of 420 vpm by the year 2035, or a doubling towards the end of the next century (reviewed by Crane, 1985). This projected rise in [CO_2] has been calculated using general (atmospheric) circulation models (GCMs). Depending on which model is used a wide range of possible effects on temperature and other climatological variables have been predicted (see Mitchell, 1983; Crane, 1985; Roeckner, 1992; Cess *et al.*, 1993). For example, Mitchell (1989) suggests a further warming of 1.5°C is likely over the next 40 years. The greatest extent of this warming would appear to be most likely to occur towards the polar regions, while minimal changes are expected in equatorial regions.

Due to a complex web of feedback interactions within the atmosphere (Crane, 1985), it is clearly difficult to predict how increasing concentrations of "greenhouse gases" will modify the terrestrial environment of plants. As yet even the highly complex GCMs are unable to resolve accurately the important environmental details necessary to predict whole plant or community responses to climate change. This inability has much to do with the uncertainties associated with assessing the effects of elevated CO_2 and temperature changes on the hydrological cycle (Chahine, 1992). However, this review focuses primarily on the effects that changes in atmospheric [CO_2] have on plant functioning. Other chapters within this book will discuss the implications of some of the aspects of other "greenhouse gases".

PHYSIOLOGICAL EFFECTS OF CARBON DIOXIDE

Photosynthesis

At its present atmospheric concentration CO_2 limits the ability of C_3 species to fix carbon, so any increase in CO_2 tends to enhance the rate of assimilation (see reviews by Warrick *et al.*, 1986; Lawlor & Michell, 1991). This is usually, but not invariably associated with an increase in plant growth. The reason why net photosynthesis is enhanced can be related to a number of factors which are connected to the characteristics of the primary carboxylating enzyme (ribulose bisphosphate carboxylase-oxygenase [Rubisco, E.C. 4.1.1.39]), (reviewed by Lawlor & Michell, 1991).

1. This enzyme occupies a unique position of importance, as it is the primary rate-limiting step in C_3 photosynthesis. As with all enzymes, its rate is enhanced by an increase in the concentration of a rate-limiting substrate.
2. A typical substrate concentration of CO_2, in the chloroplast stroma (4–8 mmol m^{-3}) is considerably less than the K_m value (36 mmol m^{-3}) for Rubisco carboxylation (reviewed by Bowes, 1991), but CO_2 has the ability to activate Rubisco directly (Pearcy & Bjorkman, 1983).
3. Rubisco can act as either a carboxylase or oxygenase, depending on the partial pressures of CO_2 and O_2 at the active site (reviewed by Bowes, 1991). It follows that increasing the CO_2 concentration can increase the photosynthetic quantum efficiency of C_3 plants under low levels of illumination, due to the suppression of photorespiration (Ehleringer & Bjorkman, 1977).

Carbon dioxide activates Rubisco through binding at a site discrete from that involved in the enzymic carboxylation reaction (Edwards & Walker, 1983). This binding of CO_2 to the specific lysine group produces a carbamate residue which reacts rapidly and reversibly with a magnesium ion to form the active complex (Andrews & Lorimer, 1987). The preferential binding to the inactive enzyme, by RuBP will block carbamylation, which must be removed by the ATP-requiring Rubisco activase (Salvucci, 1989). As the activity of this enzyme is irradiance dependent, the activation state of Rubisco does not solely reflect the CO_2 supply (reviewed by Bowes, 1991). The metabolic advantages gained from modulating the activity of Rubisco in relation to photosynthesis have been discussed (see review by Woodrow, 1994). Changes in the activation state will alter both V_{max} and the K_m of Rubisco (Edwards & Walker, 1983). From A/c_i analysis it is frequently inferred that herbs and trees grown at elevated CO_2 can show a decline in V_{max}. This suggests a decrease in the amount, activity or kinetic properties of Rubisco. But this is not the case with all species (Sage, 1994).

Those species which fix CO_2 by using phosphoenolpyruvate carboxylase (PEP) into C_4 dicarboxylic acids (C_4 metabolism) show very little response of photosynthesis or quantum efficiency to an increase in [CO_2] (Pearcy & Bjorkman, 1983). This absence of a CO_2 response can be explained by the fact that at the C_4 photosynthetic carboxylation site the [CO_2] is close to its saturation concentration for PEP carboxylase. This occurs in C_4 species because of their ability to raise the local [CO_2] internally by means of "CO_2 pumping" and Kranz anatomy. The competitive carboxylase/oxygenase reaction is also absent with PEP carboxylase (reviewed by Monson, 1989). Those carbon losses associated with the photorespiration in C_3 species are therefore absent (Bjorkman, 1966).

Photorespiratory losses of carbon come about when O_2 is utilized as the substrate by RuBP carboxylase–oxygenase, and the products, phosphoglycolate and 3-phospho-glycerate, enter the photorespiratory carbon oxidation cycle (PCO) from which CO_2 is released. This release of previously fixed carbon is apparent as a reduction in net fixed carbon and can in fact be considerable; at 340 vpm CO_2 it may be as high as 40%. Any increase in CO_2 will reduce the competitive influence of O_2 and photorespiratory carbon loss (Pearcy & Bjorkman, 1983). Despite these drawbacks, C_3 photosynthesis has been retained by virtually all autotrophs for 3500 million years, and only recently (the last 150 million years) has a significant alternative, C_4 metabolism appeared (Andrews & Lorimer, 1987).

It has been shown that the photosynthetic rate of C_3 plants when grown at elevated [CO_2] increases on average by 52% with a doubling of the [CO_2] (Cure & Acock, 1986), while an averaged 44% stimulation was apparent for trees (Gunderson & Wullschleger, 1994). This increase in assimilation rate may not, however, be sustained and may fall on average to a value of around 29% of the original enhancement (Cure & Acock, 1986; Sage *et al.*, 1989; reviewed by Eamus & Jarvis, 1989). This fall in assimilation is not evident in all cases (Campbell *et al.*, 1988, 1990) and in some incidences assimilation may even increase (Wong, 1979; Hicklenton & Jolliffe, 1980). In different studies even the same species may show both increases and decreases.

When assimilation rate does fall it is often associated with a decline in Rubisco activity and/or an increase in the accumulation of non-structural carbohydrate (Wong, 1979; Besford & Hand, 1989; Yelle *et al.*, 1989b; reviewed by Bowes, 1991). The most likely causes for this decline are either a loss of Rubisco protein, or a change in enzyme-specific activity (Sage *et al.*, 1989; Besford, 1990; Besford *et al.*, 1990; reviewed by Bowes, 1991; Van Oosten *et al.*, 1992). For example, with tomato an accelerated loss in Rubisco protein at high CO_2, during leaf development, accounted for the reduced photosynthetic performance (see Figure 4.3: Besford *et al.*, 1984; 1990). The timing of the decrease or acclimation of the photosynthetic rate to an increase in [CO_2] would appear to vary considerably (reviewed by

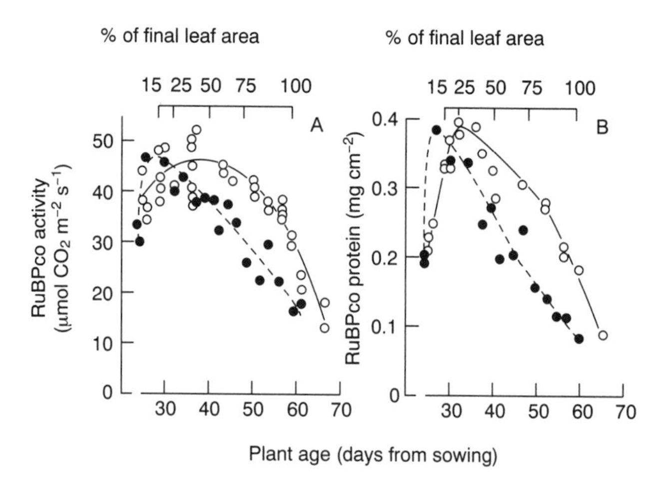

Figure 4.3. The activity of *in vitro* RuBP- and CO_2-saturated RuBPco activity (A), and the RuBPco protein (B) extracted from the 5th leaf of *Lycopersicon esculentum* (Mill.) cv. Findon Cross, at various stages of development. Plants were grown in either 340 vpm CO_2 (○) or 1000 vpm CO_2 (●) (from Besford *et al.*, 1990; by permission of Oxford University Press)

Mott, 1990). As yet there is no complete consensus regarding the mechanism(s), or the strategic importance of changes in the photosynthetic rate (but see Van Oosten & Besford, 1994).

The optimization of the allocation of growth-limiting resources, such as nitrogen, may take place in order to avoid a single process limiting a metabolic pathway (Cowan, 1986; Woodrow & Berry, 1988; Tissue *et al.*, 1993; Woodrow, 1994). Under normal light and CO_2 conditions the rate of photosynthesis could be expected to be regulated by RuBP regeneration, RuBP carboxylation capacity and sucrose synthesis capacity (von Caemmerer & Farquhar, 1984). It appears that when the $[CO_2]$ increases so does the capacity of Rubisco, and in order for an increase in assimilation to be maintained, the balance would have to shift towards the now limiting RuBP regeneration and sucrose synthesis capacities.

The actual number of active Rubisco sites may even decline, removing any excess capacity in response to high CO_2 (Sage *et al.*, 1988). Theoretically, the acclimation to increases in $[CO_2]$ should come about in response to the reallocation of protein nitrogen away from the non-limiting component Rubisco, to light harvesting, RuBP regeneration and carbohydrate synthesis (Sage *et al.*, 1989; Tissue *et al.*, 1993). Despite the fact that a number of studies have demonstrated a decrease in Rubisco activity on acclimation to elevated CO_2 (see above), the evidence for a reallocation of nitrogen away from Rubisco to the light harvesting and carbohydrate synthesizing enzymes is limited (see Mott, 1990). Recently, Woodrow (1994) has attempted to model the acclimation of the C_3 photosynthesis system under elevated CO_2.

Besford (1990) suggests that prior to CO_2 enrichment the Rubisco level is in excess of the light harvesting and electron transport capacities. Acclimation occurs because carboxylation capacity falls relative to that of light harvesting and electron transport, which would clearly have to be maintained to achieve sustained CO_2 fixation rates. It is also possible that plants acclimating to elevated CO_2 may have greater photosynthetic capacity than that of the synthesis and utilization of sucrose and this itself may cause feedback inhibition of photosynthesis and/or chloroplast disruption (reviewed by Bowes, 1991; Arp, 1991; Van Oosten & Besford, 1994). This is an area hotly debated, in part because we do not fully understand the mechanisms controlling the interactions between carbon acquisition and allocation. Recent evidence has shown that the accumulation of sucrose and glucose specifically represses the nuclear transcription of genes which encode for part of the Rubisco molecule (Van Oosten & Besford, 1994).

The accumulation of non-structural carbohydrate has been well documented for plants grown under high CO_2 (Figure 4.4: and see below). This accumulation can be reduced by transferring plants back to ambient CO_2 (Sasek *et al.*, 1985). If the sink utilization rate of sucrose is lower than that of synthesis, the accumulation results in feedback on the enzymes of sucrose synthesis (see review by Woodrow, 1994). This leads to a lower concentration of inorganic orthophosphate (Pi), which causes an increased rate of starch synthesis (Herold, 1980). This may only occur when sinks for carbohydrate storage or growth are unavailable or inactive. For example, limiting the number of soybean pods on a plant reduced its photosynthetic rate and appeared to exacerbate the problem of high CO_2 acclimation (i.e., a decline in assimilation rate) (Clough *et al.*, 1981). Assimilation rate was maintained only in rapidly growing and fruiting citrus trees (Koch *et al.*, 1986). Fruit thinning (sink removal) also has an inhibitory influence on photosynthesis (see Flore & Lakso, 1989). Simulated grazing of white clover (*Trifolium repens*) at elevated CO_2 maintained the increase in photosynthesis (Ryle & Powell, 1992). Perennial plants with large storage organs such as rhizomes and tubers are also suggested to be unlikely to experience sink limitations (Arp & Drake, 1991). This may also be the case with the grasses, where tiller production provides the rapid development of a new sink (Baxter *et al.*, 1994). Experiments where the root growth of cotton plants grown at elevated CO_2 was restricted showed that a reduction in photosynthesis was not associated with a fall in stomatal conductance. It was only carboxylation efficiency that declined when

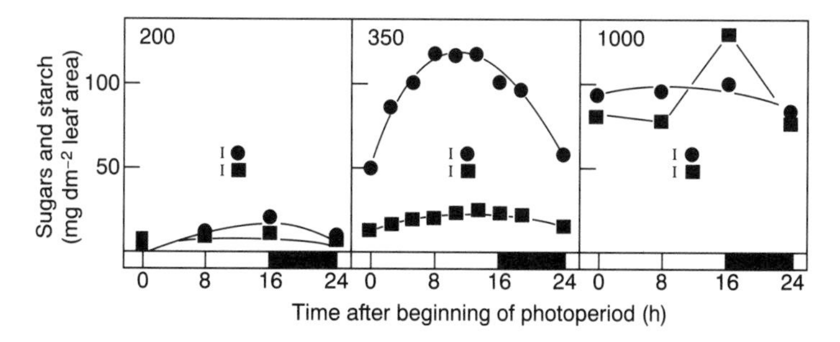

Figure 4.4. The diurnal variation of starch (●) and sugar content (■) of leaf laminae of *Trifolium repens* L. at three different CO_2 concentrations (μl l^{-1}). The unshaded and shaded areas on the figure represent the light and dark periods (from Scheidegger & Nosberger, 1984; by permission of Academic Press)

root growth was restricted (Arp, 1991; Thomas & Strain, 1991). These and other experiments support the notion that a photosynthetic product causes sink-limited feedback inhibition of soluble protein content, Rubisco activity and quantity through changes in gene expression (Sasek *et al.*, 1985; Sheen, 1990; Van Oosten & Besford, 1994). Not all the evidence is supporting, as when two tomato cultivars, which are known to accumulate carbohydrates to different levels, were grown under elevated CO_2 their photosynthetic rates were not different (Yelle *et al.*, 1989a, b).

Stomatal responses to CO_2

It can generally be concluded that the open stomata of most species will close to varying degrees in response to an increase in atmospheric [CO_2]. Assuming a doubling in atmospheric [CO_2] it has been calculated that on average the stomatal conductance of a C_3 species will decline by between 30 and 40% (see reviews by Cure & Acock, 1986; Morison, 1987; Eamus & Jarvis, 1989). This sensitivity of stomata to [CO_2] has been well documented (Linsbauer, 1916; Meidner & Mansfield, 1968; Morison & Gifford, 1983). Some exceptions to this general rule apply as there are a number of coniferous species from the northern temperate forests which show stomatal insensitivity to increasing CO_2 (Beadle *et al.*, 1979; Higginbotham *et al.*, 1985; reviewed by Eamus & Jarvis, 1989). Stomatal conductance increases in response to increased CO_2 in *Liriodendron tulipifera* also (Norby & O'Neill, 1991).

Both the intensity and direction of the stomatal response to CO_2, may change due to environmental influences (Beadle *et al.*, 1979; Morison & Gifford, 1983; Tolley & Strain, 1985; Conroy *et al.*, 1986; Hollinger, 1987; Mansfield & Atkinson, 1990). The stomatal insensitivity of some tree species may enable photosynthesis to respond positively to further increases in [CO_2] whereas sensitive species (herbaceous species) would be stomatally limited.

A decline in stomatal conductance will reduce the rate at which water vapour is lost through transpiration. This occurs because stomata create a greater resistance to the loss of water vapour than they do for the diffusive entry of CO_2 into the intercellular spaces. The greater resistance, and therefore the rate-limiting step, in the CO_2 uptake pathway actually exists within the cells of the mesophyll. This effect combined with any stimulatory influence of CO_2 on photosynthesis will produce a potentially beneficial increase in water-use efficiency (WUE). These benefits should equate with changes in [CO_2] (Polley *et al.*, 1993).

Despite our predictions about changes in WUE our knowledge of the mechanistic responses of stomata to CO_2 is still very limited. As yet we do not even have a widely accepted understanding of the basic action of CO_2 on stomatal guard cells (reviewed by Mansfield *et al.*, 1990). We do know, however, from epidermal strip studies with isolated guard cells that they are able to respond directly to CO_2 (Squire & Mansfield, 1972). The mechanisms of stomatal opening and closing involve changes in guard cell turgor brought about by the movement of K^+ and H^+ ions. The interaction of CO_2 with these processes seems not to be due to either pH changes or the photosynthetic carbon-reduction pathway in guard cells (Outlaw, 1989). It has been suggested that CO_2 may inhibit an electrogenic pump in the plasmalemma which extrudes H^+ ions from guard cells (Edwards & Bowling, 1985). Other proposals that have been put forward involve CO_2-enhanced malate formation which can act as a counter ion for K^+ uptake. These and other possible stomatal mechanisms have recently been reviewed (Mansfield *et al.*, 1990).

In many cases where the regulation of stomatal conductance by CO_2 has been explored by indirectly calculating the internal [CO_2] (c_i), it is suggested that values are maintained around 100 and 230 μmol mol^{-1} CO_2 for C_4 and C_3 species, respectively. It is important to have knowledge of the [CO_2] within intercellular spaces as this value is similar to the [CO_2] at the chloroplast (Woodrow & Berry, 1988). It would also appear that this is the location at which CO_2 influences stomatal aperture (Mott, 1988), as the CO_2-impermeable cuticle on the external surfaces makes direct sensing highly unlikely. The relationship between c_i and ambient [CO_2] (c_a) is complicated because not only do stomata respond to it, but so does photosynthesis. It has been suggested that stomatal conductance and photosynthesis respond to changes in c_a in a manner in which the ratio of c_i to c_a remains fairly constant, though no mechanistic basis has been proposed. The effect of being able to balance this ratio would enable adjustments in the diffusional limitations to assimilation in response to biochemical changes (Woodrow *et al.*, 1990).

The ability of stomatal guard cells to respond to CO_2 at c_i has, therefore, been viewed as a means by which stomatal conductance reflects changes in assimilation rate and enable the optimization of stomatal aperture and therefore transpiration. Ideally, the response of stomata to CO_2 at c_i should only prevail when this concentration is at a subsaturation level for photosynthesis (Farquhar & Sharkey, 1982). Sensitivity to CO_2 above the saturation point would enable stomatal closure to limit transpiration and not assimilation. Such a situation would be more relevant for C_4 plants but, due to a lack of a saturation point for C_3 species, partial stomatal closure will always have some restriction on assimilation.

A similar argument has been put forward for changes in c_i sensitivity which depends on the water balance of the plant (Davies & Mansfield, 1987). Limited stomatal sensitivity to CO_2 when water is plentiful, would enable plants to maximize assimilation without stomatal restrictions. When water becomes limiting, the priorities change from maximizing assimilation to conserving the transpirational losses. Some evidence does suggest that when plants are grown without exposure to a water deficit, their stomata have a very limited response to CO_2 (Raschke, 1975; reviewed by Mansfield & Atkinson, 1990). This lack of sensitivity, or an increase in insensitivity, can in fact be removed by either the application of the water stress hormone abscisic acid (ABA) or by the induction of its endogenous supply (Heath & Mansfield, 1962; Raschke *et al.*, 1976; Dubbe *et al.*, 1978; Radin *et al.*, 1988; reviewed by Mansfield & Atkinson, 1990). The hormonal influence on stomatal sensitivity appears not to be simple. Experiments with isolated epidermal tissue suggest that a complex balance between auxins, cytokinins and ABA may prevail (Snaith & Mansfield, 1982).

The long-term exposure of plants to moderately elevated CO_2 concentrations would not appear to alter their stomatal sensitivity to CO_2. Recent observations of leaf epidermis collected from stored historical herbarium specimens dating back to 1750, suggest that the number of stomata per unit area may in fact have declined with the increase in global [CO_2] in the range 225–700 μmol mol^{-1} (Woodward, 1987; Penuelas & Matamala, 1990). These data showed that over a 200-year period the decline in the stomatal density varied between 20 and 40% for a range of trees, shrubs and herbs. Examination of fossil leaves of *Salix* extending over at least 16 000 years BP and perhaps much further, also shows a decline in stomatal density associated with apparent increases in atmospheric CO_2 (Beerling *et al.*, 1993). More direct studies with *Nardus stricta*, collected at various altitudes, indicated that plants from higher elevations had a greater decrease in stomatal density when exposed to an increasing [CO_2] (Woodward & Bazzaz, 1988). It does, however, remain unclear as to whether this was a direct response to CO_2 or a response to a change in plant water use or WUE.

Further studies with two tropical tree species, *Plentaclethra macroloba* and *Ochroma lagopus* (Oberbauer *et al.*, 1985), and loblolly pine, sweet gum, maize and soybean (Thomas & Harvey, 1983) as well as perennial ryegrass (Ryle & Stanley, 1992) were unable to show convincing stomatal density responses to CO_2. Despite this species variability, these observations are particularly exciting as they highlight the potentially important role that phenotypic plasticity may play, through adjustments in stomatal density in the acclimation of some species to changes in [CO_2]. This will enable increases in WUE to be achieved. Radoglou & Jarvis (1990a) have attempted to study the influence of increasing CO_2 on the epidermal cell development of a woody species (various clones of *Populus*) but they were unable to demonstrate any direct influence of CO_2 during ontogenesis of stomata or later during epidermal cell expansion. The decline in stomatal conductance in response to elevated CO_2 was, therefore, not due to changes either in stomatal numbers or size.

The evolutionary implications of stomatal regulation have recently been reviewed on a floristic basis and highlight the likelihood that a single pattern of stomatal regulation cannot be assumed (Robinson, 1994).

Response to water deficits

As already mentioned, a reduction in stomatal conductance under elevated levels of CO_2 will probably influence the ratio of assimilation rate relative to that of transpiration. The implications for an increase in WUE are of great importance (see review by Chaves & Pereira, 1992). Evidence does exist which indicates that the long-term water consumption of a species can be lowered when grown at elevated CO_2 (Carlson & Bazzaz, 1980; Polley *et al.*, 1993). Theoretically, a CO_2-caused reduction in transpiration should increase a plant's water potential (i.e., make it less negative), postpone or delay the rate at which a plant dehydrates, and by so doing increase its tolerance to water deficits (Nijs *et al.*, 1989; Sasek & Strain, 1989; Bhattacharya *et al.*, 1990; Sasek & Strain, 1990; Townend, 1993).

A decrease in stomatal density in response to a long-term increase in [CO_2] may suggest that the evolutionary importance that lower [CO_2] had as a limiting factor, may be declining. A more important limiting factor may well now be water availability, which would clearly be enhanced by a decline in stomatal density improving WUE.

In a comparative experiment with *Liquidambar styraciflua* and *Pinus taeda*, during exposure to a water deficit, it was with sweet gum that the water potentials declined more slowly when grown under elevated compared to ambient CO_2 (Tolley & Strain, 1984). A similar situation was found with sweet potato (Bhattacharya *et al.*, 1990). The stomatal response to changes in the ΔW (water vapour pressure gradient between the leaf and the air) is also an important mechanism in restricting the transpirational water loss. The degree to which stomatal closure occurs in response to an increase in ΔW varies between species and with plant water status. But growth at elevated CO_2 can influence the stomatal response to ΔW. When changes in ΔW were made for *Pinus menziessii* and *P. radiata*, the stomata of plants grown at an elevated [CO_2] were less sensitive to drying air; the same was not true for *Nothofagus fusca* (Hollinger, 1987).

The ability of plants to withstand periods of soil water deficit is also an important determinant of growth and survival. One means by which growth rates can be maintained as soil water potential falls is for the cell's turgor pressure to be increased (Sasek & Strain, 1988, 1989; Bhattacharya *et al.*, 1990). This can be achieved by an increase in the cellular solute content (the osmotic potential becomes more negative), for example, by the accumulation of simple sugars (solute or osmotic adjustment). Plants which have been grown under elevated CO_2 may typically have an excess accumulation of photosynthetic products which can increase the cells' osmotic pressure (Idso, 1988; Sasek & Strain, 1988, 1989).

Increases in the atmospheric [CO_2] can also bring about changes in gross morphology, but between species these changes can often be inconsistent. There is frequently evidence to indicate a decrease in leaf-area ratio. Plants which have been acclimated to elevated CO_2 may also show a shift in biomass allocation which could also be beneficial to the conservation or acquisition of resources. The shift in allocation pattern away from leaves into root growth may be a case where nutrient and water uptake can be increased (reviewed by Farrar & Williams, 1991). For the plant this is a trade off between its ability to acquire carbon versus uptake of water and nutrients (Norby *et al.*, 1992). Such a shift in the ratio of leaf to root mass would have the potential advantage of increasing water uptake while simultaneously reducing its loss, and perhaps improving the plant's effective drought response.

Experiments with wheat have indeed shown that plants grown at elevated [CO_2] were in fact able to compensate for restrictions in growth caused by water deficits. In these experiments an increase in root biomass was observed at a depth of 0–10 cm (Chaudhuri *et al.*, 1990). This strategy is also apparent for a number of other species during CO_2 enrichment (see, e.g. Figure 4.5; Hollinger, 1987; Conroy *et al.*, 1988; Norby & O'Neill, 1991; reviewed by Eamus & Jarvis, 1989). A shift away from shoot growth may not, however, be advantageous when plants are in competition in a water-limited habitat (DeLucia & Heckathorn, 1989). A CO_2-induced increase in leaf area may result in greater water use per plant. This may negate any advantage, on an individual leaf-area basis, of a CO_2-related increase in WUE. In fact, transpiration rate per unit ground area may not decrease (reviewed by Warrick *et al.*, 1986).

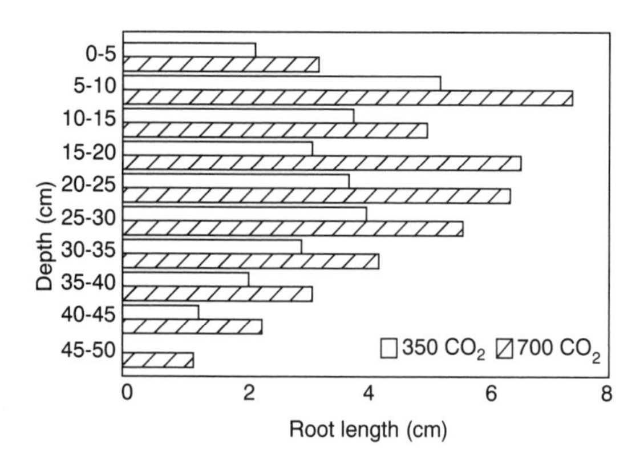

Figure 4.5. The length of first-order lateral roots with depth in *Glycine max* L. Merr. 'Lee' non-nodulating plants, grown in either 350 (□) or 700 (▨) μl l^{-1} CO_2 (from Rogers *et al.*, 1992; by permission of Blackwell Science)

Respiration, carbon allocation and growth

As CO_2 is the carbon substrate for growth, it is not surprising that changes in atmospheric [CO_2] will have a profound influence on plant metabolism, and, in particular, that of respiration. While respiration associated with growth would be substrate-concentration-limited (CO_2 responsive), the same would not be true for respiration associated with tissue maintenance, which is regulated by ATP use (Amthor, 1991). It is from respiration that both growth and the maintenance of existing tissue is supplied with energy, with growth being stoichiometrically linked to respiration rate (reviewed by Farrar & Williams, 1991). A considerable proportion (>50%) of daily fixed carbon can be shown to be rapidly (within 24 hours) lost by respiration (Atkinson & Farrar, 1983).

It is highly likely that an increase in [CO_2] through enhanced assimilation will bring about a rise in the mass of the non-structural carbohydrate fraction in leaf (Tolbert & Zelitch, 1983; reviewed by Farrar & Williams, 1991; Ryle *et al.*, 1992; Wullschleger *et al.*, 1992). This accumulation may not always be associated with source leaves, as shown, for example, in studies with sweet potato tubers (Bhattacharya *et al.*, 1990). An increase in carbohydrate can in itself lead to a rise in dark respiration rate (Hrubec *et al.*, 1984, 1985; Farrar, 1985; Bunce, 1990; Van Oosten *et al.*, 1992), but this may not always be the case (Gifford *et al.*, 1985; Amthor, 1991). However, when time-dependent changes in the CO_2 stimulation of growth are viewed with those of the dynamic nature of the respiratory capacity (reviewed by Farrar & Williams, 1991), dark respiration rates correlate with size of non-structural carbohydrate fraction in source leaves. This correlation was not apparent for plants in the absence of CO_2 enrichment (Hrubec *et al.*, 1985).

Direct effects of CO_2 on the respiration of whole plant parts, with exception of leaves, is unlikely to occur, because the envisaged changes in atmospheric CO_2 are small enough to have little impact on the cellular [CO_2] (Amthor, 1991). Some observations have, however, been made of the direct effect of CO_2 on the respiratory enzymes, particularly on fruit and seeds, and indeed there are changes in the activity of a number of enzymes (see Amthor, 1991). These observations were made at very high CO_2 levels which do not represent the potential range of expected changes in atmospheric [CO_2], and, therefore, any inferences which can be drawn from them are likely to be limited.

The reasons for, and the implications of, accumulations in non-structural carbohydrate are important. Increases in photosynthetic products such as starch may be due to an imbalance between the activities of source and sink (see Figure 4.4: reviews by Arp, 1991; Farrar & Williams, 1991; Stitt, 1991; Baxter *et al.*, 1994). Clearly, then, the ability of source leaves to export carbohydrate and the activity of storage and/or growing tissues to utilize it will be particularly important. It may well be that photosynthetic acclimation takes place because of direct or indirect changes in these processes. Why then should carbohydrate accumulate in plants grown under elevated [CO_2]? Evidence can be presented which shows that the export of assimilate from source leaves increases in response to elevations in CO_2 (Ho, 1978; Cure *et al.*, 1987). During a detailed examination of assimilate partitioning in soybean plants grown under ambient CO_2, there was little export from source leaves during the light period (Cure *et al.*, 1991). However, 6 to 7 days after exposure to elevated CO_2, the proportion of assimilate exported in the light began to increase. It is well known that the capacity for export to take place in the light does not increase in synchrony with changes in photosynthetic capacity (Ryle & Powell, 1976; Ho, 1977; Cure *et al.*, 1991).

Also, the enzymes of sucrose synthesis and assimilate export can change in response to assimilate demand irrespective of photosynthetic rate (Rufty & Huber, 1983; Stitt & Quick, 1989). It was also apparent with the soybean experiment that, as the fraction of export in the light increased, so did the amount of dry matter (Cure *et al.*, 1991). There would appear to be a strong correlation between growth and export, which suggests a feedback control. Why export during the light period should be stimulated remains somewhat unclear. The timing of export may influence the location to which carbohydrates are distributed (Cure *et al.*, 1991). It is assumed that during the light period assimilate goes to root growth rather that of leaves (Huber, 1983); a CO_2-induced increase in export in the light would, therefore, promote water and nutrient uptake.

Little is known about the potential influence of CO_2 on the enzymes involved with the partitioning of assimilate, but it appears that the activity of the rate-controlling enzyme for sucrose synthesis (sucrose phosphate synthase, SPS) can be negatively correlated with starch accumulation (Huber *et al.*, 1984). This negative correlation is evident with both soybeans and *Cucumis sativus* (Huber *et al.*, 1984; Peet *et al.*, 1986). Knowledge about the influence of CO_2 on assimilate transport is equally limited and also contradictory (reviewed by Farrar & Williams, 1991).

Effects of elevated CO_2 on the allocation of carbon to secondary metabolites are also important. Many plants, and in particular woody perennials, release considerable quantities of carbon into the atmosphere. Hydrocarbons such as isoprene are released by leaves during photosynthesis and when combined with the oxides of nitrogen they are able to form atmospheric ozone. Recent experimental evidence suggests that the capacity for isoprene production in *Quercus rubra* could double in response to a 26% increase in [CO_2] (Sharkey *et al.*, 1991), while with *Eucalyptus globulus, Populus tremuloides, Quercus alba* and *Mucuna pruriens* emission rates at the higher CO_2 level were actually reduced (Monson *et al.*, 1991; Sharkey *et al.*, 1991).

Plant growth is controlled by a number of interactive processes and, when plants are grown under elevated CO_2, changes in growth are likely to come about for a number of reasons. An increase in [CO_2], for example, will lead to an enhancement in carbon supply; plant water and osmotic potentials will increase and it is likely that the ability of tissue to osmoregulate will also be increased (Sionit *et al.*, 1980). All of these factors can have a role in determining the potential of cells and tissues to grow.

Despite the fact that photosynthesis determines available substrate, it is carbon allocation that determines both the fate and efficiency with which it is used. In terms of growth, what is important is the fraction of this accumulation that is made available for sink growth. So, not only do we have to consider rates of assimilation, but also the direction and efficiency with which carbon is allocated. It has been experimentally calculated for a range of species that with a doubling of the [CO_2] there was, on average, an increase in biomass accumulation of 28% (Cure & Acock, 1986). Studies have examined the quantities and activities of the enzymes associated with assimilation and growth. For example, Peet (1986) showed that when an increase in cucumber assimilation rate failed to translate into an increase in biomass, it was due to a loss in the activity of Rubisco and carbonic anhydrase (Peet *et al.*, 1986). There is some evidence from a few species that there may be changes in the proportional allocation of carbon to various sinks (Sasek & Strain, 1988, 1989).

Allocation to reproduction was enhanced with white clover (*Trifolium repens*), but not in perennial ryegrass (*Lolium perenne*) (Overdieck, 1986). In many agronomic studies, where yield increases occur, they are relative to increases in total biomass (Chaudhuri *et al.*, 1990; Mott, 1990). The elevated CO_2 increased the biomass in both leaves and tubers of sweet potato, but not in a strict proportion, favouring the roots (Bhattacharya *et al.*, 1990; reviewed by Farrar & Williams, 1991). Also, when growth increases occur, they often do so early in the life of the plant. Subsequently, when assimilation may decline along with relative growth rates (RGR), plants growing under elevated CO_2 may continue to grow faster than those at ambient [CO_2] due to an initial increase in leaf area.

Increases in leaf thickness may be associated with an increase in cell size (Conroy *et al.*, 1986) and in the number of cell layers (Ho, 1977; Thomas & Harvey, 1983). Such acclimation may be particularly important if excessive carbohydrate accumulation causes chloroplast disruption (Cave *et al.*, 1981; reviewed by Bowes, 1991). Soybean plants may in fact be able to maintain high rates of assimilation in response to increasing CO_2 because of an ability to provide a sufficient quantity of sites for starch storage (Vu *et al.*, 1989). The same argument has been put forward as an explanation of why, frequently, there is no accumulation of reserves in plants with large active storage sinks. It may also be equally appropriate for field plants and natural vegetation, in which the growth potential of such organs, particularly below ground, is not restricted as with experimental containment (see Figure 4.6: Curtis *et al.*, 1989a). This may also explain why acclimation to CO_2 occurs with pot plants (Sasek *et al.*, 1985), but not

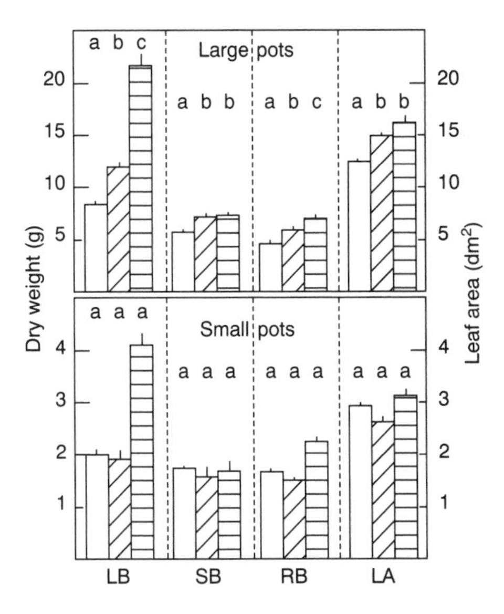

Figure 4.6. The effect of growth for 28 days in 270 (□), 350 (▨) and 650 (▤) μl l^{-1} CO_2 on the biomass accumulation of *Gossypium hirsutum* L. cv. Coker 315, plants grown in either large or small pots. Within pot size treatments, bars which are designated by the same letter are not different at the 0.05 level of significance. Symbols: LB, leaf biomass; SB, stem biomass; RB, root biomass; LA, leaf area (from Thomas & Strain, 1991; by permission of American Society of Plant Physiologists)

in field-grown plants, where extensive new root growth can occur (Radin *et al.*, 1987; Curtis *et al.*, 1989b). Some support for this suggestion is apparent with species with indeterminate as opposed to determinate growth forms (Chaves & Pereira, 1992). In some cases the former have shown larger increases in dry weight than determinant species, and again this has been correlated with the sink strength (size) and numbers (Mauney *et al.*, 1979; Paez *et al.*, 1980, 1984; Bhattacharya *et al.*, 1985b). With the limited number of appropriate experimental species comparisons available, there would appear to be no evidence to indicate growth rate *per se* as a factor influencing sink strength and CO_2 responsiveness (see review by Gunderson & Wullschleger, 1994).

Nutrient acquisition

The extent to which plants are able to respond to changes in [CO_2] will probably be reflected in their ability to capture and utilize mineral resources. When nutrients and other resources limit growth there are examples where, despite enrichment of atmospheric CO_2, differences in growth between individuals were not apparent (Patterson & Flint, 1982; Goudriaan & de Ruiter, 1983; Brown & Higginbotham, 1986). The situation may, however, be different for woody perennials where it is suggested that CO_2 responses persist despite low nutrient supply (see Jarvis, 1989). The results of such experiments imply an inconsistency with "Liebeg's law" of a single limiting factor determining plant growth rate (Sinclair, 1992). In this analysis Sinclair considers resources, whether minerals or carbon, to be simultaneously limiting growth. It is also interesting to note that in most if not all cases when tree species have been exposed to elevated CO_2, nutrient concentrations appear to fall (Mousseau & Saugier, 1992).

A substantial proportion of the plant's nitrogen content is partitioned into protein, of which 30 to 50% in C_3 plants can be Rubisco. Inevitably this means that there is a very close relationship between carbon fixation and that of the acquisition of minerals, particularly nitrogen. Changes in the carbon:nitrogen (C/N) ratio are, therefore, likely to be important. When a plant's tissue nitrogen concentration increases so does the amount of Rubisco protein (Makino *et al.*, 1984) and its activity (Medina, 1971; Thomas & Thorne, 1975). Increases in tissue nitrogen concentration in plants grown at elevated CO_2 can prevent starch accumulation (Thomas *et al.*, 1975). Declines in tissue nitrogen concentration have also been observed (Curtis *et al.*, 1989a, b); this may in part provide an explanation of CO_2-related falls in assimilation rate (Sage *et al.*, 1989). Less nitrogen may, in fact, be invested in Rubisco and more in the light harvesting and electron transport components, i.e., towards other limiting processes of photosynthesis (Tissue *et al.*, 1993).

It may be expected that when mineral acquisition and carbon fixation are closely related, as is the case for symbiotic nitrogen fixation, an increase in CO_2 concentration should enable nitrogen fixation to increase; in some cases this apparently does not happen (see review by Sinclair, 1992). Changes in the C/N ratio may also have important consequences for herbivore selection (Lincoln *et al.*, 1986) and microbial decomposition (Melillo *et al.*, 1982), as well as changes in plant-to-plant interactions (Garbutt *et al.*, 1990). Here, for example, CO_2-induced increases in seed-nitrogen concentration affect early events in a plant's competitive performance.

Imbalance between sources and sinks is also important. The limited availability of nitrogen in tundra habitats has been used to explain why *Eriophorum vaginatum* showed an initial positive growth response to increasing CO_2, but it only lasted for a single growing season. This can be contrasted with the sustained response of *Scripus olneyi* from a nutrient-rich estuarine habitat (Tissue & Oechel, 1987; Curtis *et al.*, 1989a). Differences of this type may in part also be explained by consideration not only of mineral concentrations but also the total amount of nutrient available to a plant (Sinclair, 1992).

The fate of acquired carbon may also be influenced by a plant's nutrient status. Growth-limiting nutrients may promote the allocation of carbon to root growth (Luxmore *et al.*, 1986; Norby *et al.*, 1987).

IMPACT ON ECOSYSTEMS

Changes in the composition of plant communities

The geographical distribution of numerous species can be closely correlated with local and regional environmental factors (e.g., see the UK flora, Grace, 1987). The potential for changes in temperature to influence processes such as flowering, seed germination, whole plant responses to water deficits and chilling are likely to be more obvious and perhaps more extensive than those of direct effects from increases in CO_2 (Cannell, 1990). The potential impact of increases in CO_2 on vegetation diversity is being studied through modelling (Rochefort & Woodward, 1992; Woodward, 1992). The inclusion of CO_2-related changes in vegetation transpiration within such models suggests that dry regions are most likely to experience the greatest increases in diversity. Measured changes in WUE have been used to calculate the potential for the extension of a species' geographic range based on precipitation information (Polley *et al.*, 1993).

A particularly important question to answer must be to what extent are the native plants of a region likely to have the capacity to acclimate and/or adapt to future increases in [CO_2] (Strain & Cure, 1985). From an evolutionary standpoint species tend to be rather conservative, i.e., they respond to environmental perturbations not by evolving but by migration (Bradshaw & McNeilly, 1991). As Bradshaw & McNeilly point out, the mechanism of the Darwinian process is constrained by the quantity of variability within a population. Despite the fact that changes in the environment can generate intense selection pressures, the pace at which natural selection can work is ultimately limited by the amount of variability. Clearly, this limitation combined with geographical constraints on species migration would suggest that many species must inevitably become extinct.

Competitive interactions will likely be altered by changes in CO_2, as is the case for other environmental factors (Garbutt *et al.*, 1990; Ferris & Taylor, 1994). It is therefore likely that the early events of vegetation responses to CO_2 may well be detected as perturbations in competitive hierarchies and changes in the abundances of species (see Strain & Cure, 1985; Bazzaz & Garbutt, 1988; Bazzaz *et al.*, 1989, 1990; Bazzaz & Williams, 1991). Those species which have dynamic, short, life cycles are most likely to be able to adapt to environmental changes whereas, perennial, slow-maturing, long-lived species may have limited capacity to evolve. This does not preclude the ability of perennials to migrate. Tree-line vegetation, for example, can and has in the past shown dramatic change, at least to climate warming, over what is a very short time period (150 years) (MacDonald *et al.*, 1993). If migration is not a practical option then much will also depend on the degree of phenotypic plasticity a species possesses.

It has already been suggested that there are likely to be major differences in the potential of C_3 and C_4 species to respond photosynthetically to increases in [CO_2]. Many of the published field studies have attempted to determine if these predicted differences are functionally important in natural ecosystems. Some of these studies have not only looked at monocultures of C_3 and C_4 species but have also included habitats in which they coexist. There is some justification for focusing on the comparative differences between C_3 and C_4 species. Natural tropical grasslands in which both C_3/C_4 species coexist are important elements of the global carbon cycle (Hall & Scurlock, 1991). Their importance will probably increase with their abundance, as they replace tropical forest after clearance.

Field experiments where plants have been exposed to CO_2 enrichment *in situ* have already answered some important questions. As with "pot experiments" (see Figure 4.5), it has often been shown that photosynthetic capacity can be enhanced, but with field plants such increases are not so frequently lost by acclimation. These increases in assimilation can in fact be maintained for considerable periods of time, in some cases years (Ziska *et al.*, 1990; Arp & Drake, 1991). Predictions regarding the greater potential of C_3 photosynthesis to respond to elevated [CO_2] have been substantiated and do in fact translate into increases in biomass production (Bazzaz & Carlson, 1984; Zangerl & Bazzaz, 1984; Curtis, *et al.*, 1989a, b; Bazzaz *et al.*, 1989; Bazzaz, 1990; Ziska *et al.*, 1990).

The outcome of competitive interactions for light, particularly with shoot densities (Gifford, 1977; Tissue & Oechel, 1987; Curtis *et al.*, 1989a) is also influenced by CO_2-induced changes in canopy architecture (Figure 4.7: Reekie & Bazzaz, 1989). In a comparative study with two co-occurring honeysuckle species (*Lonicera*), one native and the other an introduced alien, it was shown that the alien species was more responsive to increases in CO_2 (Sasek & Strain, 1991). This came about due to change in carbon allocation, which was evident as an increase in leaf area and stem growth.

Similar results were obtained in a study with another weedy vine species introduced into the United States (Sasek & Strain, 1990). When kudzu (*Pueraria lobata*), a leguminous vine native to eastern Asia, was exposed to elevated CO_2, considerable changes in the ratio of carbon allocated to stem height relative to that allocated to stem thickness were apparent (Sasek & Strain, 1988). Such events are also likely to have considerable influence on the reproductive capacity of species and their potential to exploit and migrate into new habitats (Figure 4.8: Sasek & Strain, 1988). Overdieck (1986) has shown that seed weight in white clover increased with elevated CO_2. It was also suggested that CO_2 induced an increase in clover flower size which may promote attractiveness to pollinators, while observations of flower phenology from a group of coexisting species suggest that

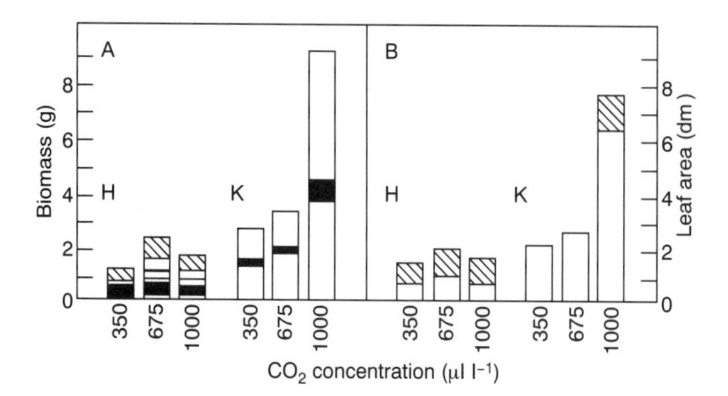

Figure 4.7. The biomass (A) and leaf area (B) production of *Lonicera japonica* Thumb. (Japanese honeysuckle, H) and *Pueraria lobata* Ohwi (Kudzu, K) after 60 days of growth at 18/12°C day/night and at 350, 675 or 1000 $\mu l\ l^{-1}$ CO_2. Each honeysuckle plant was separated into five parts: roots (□, bottom), main stem (■), main stem leaves (▤), branch stems (□, top) and branch leaves (▧), while Kudzu consisted of roots (□, bottom), main stem (■) and main stem leaves (□, top). For both species main stem (□) and branch leaf (▧) areas are shown (from Sasek & Strain, 1990; by permission of Kluwer Academic Publishers)

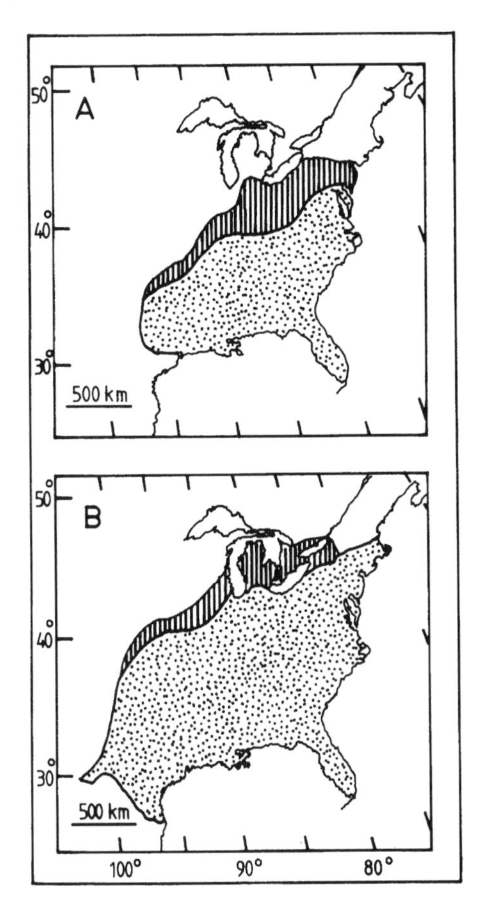

Figure 4.8. The current range (stippled) in North America and potential range (slashed) of (A) *Pueraria lobata* Ohwi (Kudzu) and (B) *Lonicera japonica* Thumb. (Japanese honeysuckle) as a result of a doubling in CO_2 concentration and global warming (from Sasek & Strain, 1990; by permission of Kluwer Academic Publishers)

reproduction can be altered by CO_2, either being advanced (Garbutt & Bazzaz, 1984; Garbutt *et al.*, 1990) or delayed with different species (Carter & Peterson, 1983). This may have the effect of phenologically uncoupling insect pollinator and flowering, which will be particularly important and have considerable influence on community structure through the capacity for regeneration.

The timing of seedling emergence is also important in determining the outcome of competitive interactions, as of course will be differential growth rates in early life. There is some evidence to indicate that small seeds actually germinate quicker when exposed to enhanced [CO_2] (Heichel & Jaynes, 1974).

It would be difficult, retrospectively, to determine whether past increases in atmospheric CO_2 were responsible for species migration, but despite the obvious and complex nature of this subject some attempts have been made to model the potential for changes in species distributions. In one such study detailed changes in the composition and productivity of North American forests were implicated in response to increasing CO_2. It has been possible to identify species with different sensitivities to CO_2 (Pastor & Post, 1988). CO_2 response studies with the American beech suggest that, at least in relative terms, this species is likely to be able to respond positively to changes in CO_2 (Bazzaz *et al.*, 1990). However, other work shows that this species may be particularly susceptible to rapid increases in temperature and, because of its large seed size, will have difficulty in migrating to cooler regions (Davis *et al.*, 1986).

The complexity of the fate of the American beech highlights an extremely important area of limited information. The observed and the proposed changes in [CO_2] have occurred, and will occur in a very short space of time, at least in evolutionary terms. The question has to be asked, how will a future rapid increase in CO_2 constrain the ability of plants to either adapt or migrate? The data on *Fagus grandifolia* and *Tsuga canadensis* suggest that they can move northward but only at a rate of 20 to 25 km per 100 years. In order to track the expected rate of change in temperature, they will have to move at a rate of 100 km per 100 years (Davis, 1989). One should clearly not overlook the major contribution that man has had in influencing plant migration (Woodward, 1992).

Recent reports have highlighted the potential value that natural CO_2 springs may have in examining the long-term ecological and genetical development of vegetation under elevated CO_2. Research presently under way may well prove very interesting (see Miglieta *et al.*, 1993).

Consequences for agriculture

Despite the intention of this review to focus directly on the response of plants to increases in [CO_2] it can often be difficult to separate out the effects of closely associated environmental factors. This is particularly so for the CO_2-induced changes in temperature. It is, therefore, also worth including some mention of one of the most important influences that temperature can have on agriculture.

As was true for native plants, temperature can have a significant influence on the geographical distribution of major crops. This subject was reviewed by Parry *et al.* (1989; see also Carter *et al.*, 1992) in which, for example, they suggest that a 3°C increase in UK temperature could potentially extend the geographical range of grain maize (*Zea mays*) as far north as Inverness. Their calculations were based on an 850 degree-day requirement above 10°C. Even an increase of only 1.5°C would enable maize to be grown over most of England. However, the situation is by no means as simple as to allow the importation of farming practices and cultivars from more environmentally appropriate regions (i.e., in this case the warmer regions of France) (Mansfield, 1991). The cereals and grasses grown in northwest Europe require critical day-lengths for flower induction by short days or low temperatures. The idea of shifts in crop-production boundaries as interpreted by both changes in temperature, precipitation and evapotranspiration has, due to its importance, been extensively modelled (reviewed by Warrick *et al.*, 1986).

The influence of temperature on flowering date is particularly important, and frequently elevated temperatures are used to induce earlier cropping. Horticulture has for many years been interested in establishing models which allow predictions about flowering dates from temperature records (Browning & Miller, 1992; Kronenberg, 1994). Experimentally, elevating temperature during critical periods of flower development may, with some crops, have significant effects on cropping as well as flowering date (Atkinson & Taylor, 1994). Changes in flowering date, induced by winter warming, may have repercussions on the likelihood of perennial crops being exposed to damaging low temperatures (Atkinson & Taylor, 1994; Kramer, 1994).

It comes as no surprise that the socioeconomic importance of the world's agricultural crops has stimulated extensive research into the effects of elevated CO_2 on major crop yields (reviewed by Krupa & Kickert, 1989). Much effort has been applied to plant biochemical and physiological studies, combined with detailed modelling studies of climate interactions. It does, however, have to be said, that despite even this effort many of the conclusions that are drawn from these models are not necessarily in close agreement (Adams *et al.*, 1990). This failure inevitably reflects many weaknesses in our understanding of a whole range of processes. There are obvious doubts regarding the extent of climate change, its influence on weather patterns and the potential for elevation in [CO_2] to compensate (Chahine, 1992; Cess *et al.*, 1993).

Effective answers must also include assessments of many other aspects, e.g., the hydrological balances (irrigation, etc.), potential for regional and local climate changes, changes in nutrient supply and demand, and appropriate cost-benefit analysis of economic returns. The apparent inconsistencies in our interpretation of this type of analysis must, at least in part, be reflections of the nature of the research undertaken. Much of our knowledge has been acquired from pot experiments where restricted carbon sinks may be a considerable problem; this will probably have the potential to influence photosynthetic ability (see above and Figure 4.6).

It has already been suggested that C_4 species will not respond photosynthetically to increases in [CO_2]. But this does not mean that in all cases growth responses to elevated CO_2 are absent; this may in part be due to an interaction with the availability of water to the plant (Warrick *et al.*, 1986). When wheat was subject to water deficits in combination with elevated CO_2, yields were in fact similar to those obtained from plants growing without water deficits at ambient CO_2 (Sionit *et al.*, 1980).

The dominant position of C_3 species in many agricultural ecosystems is incentive enough to provide a complete understanding of the important processes which apparently cause their photosynthetic rates to acclimate (decline) to CO_2 (see above). Clearly, the importance of sink size in determining this down regulation of assimilation must be fully explored by field experimentation. It is also equally important to understand the extent and implications of CO_2-related changes in leaf senescence rates (Bhattacharya *et al.*, 1985a, b).

The importance of sink size may explain why immature wheat often shows little response to increased CO_2 (Neales & Nicholls, 1978), but when tillering and grain filling have begun a much greater response is apparent (Gifford, 1977). This same reasoning has been applied to the increased yields of soybeans at elevated [CO_2] (see Campbell *et al.*, 1990 and references within). This was due to a number of factors, a season-long increase in canopy-assimilation rates, the rapid development of the leaf-area index and no down regulation of photosynthetic capacity.

Consequences for forests

Woody plants are a very important part of the global carbon cycle; they occupy about a third of the land surface, 70% of the terrestrial biomass, and carry out two-thirds of the total photosynthesis. These factors, combined with their long life spans, provide them with the potential to be considerable sinks for the storage of carbon. The gas exchange and carbon storage capacity of forest trees, therefore, plays a central role in the regulation of global carbon flux and carbon sink sizes.

As yet, it is still difficult to predict the role that woody perennials, and forest trees in general, may play as dynamic sinks for CO_2 storage. It is equally difficult to interpret how short-term exposures to increased [CO_2] will translate in the long term to changes in growth (see Gunderson & Wullschleger, 1994). Such questions remain unanswered primarily because of the nature of the experiments which have been performed. Typically, experiments are carried out on juvenile material and for only a short period of time, but this is not always the case (Gunderson *et al.*, 1993). It is well known that many features of the physiology and functioning of woody perennials dramatically change with age, particularly their environmental responses (Higginbotham *et al.*, 1985). It is no surprise that the response of trees to short-term CO_2 enrichment can be particularly transient (see the photosynthesis section, Van Oosten *et al.*, 1992). As Jarvis (1989) points out, important differences in short-term physiological acclimation must be considered along with the longer-term selection of genotypes over generations of exposure to elevated levels of CO_2. Ideally, experiments should last for a significant period of the life of a woody perennial, without the below-ground biomass being restricted by a limited rooting volume. The extent to which the leaf and its atmosphere are coupling can also be particularly important with respect to potential changes in water-use efficiency (Jarvis & McNaughton, 1986). Only after such efforts could we expect, for example, structural changes in xylem differentiation to perhaps impose important functional changes in hydraulic conductivity which are related to whole plant water-use efficiency (Atkinson, unpublished).

From those experiments where woody plants have been exposed to elevated [CO_2] for varying periods of time, it would appear that, as for herbaceous species, photosynthesis, growth rate and biomass of C_3 species can be increased, e.g. *Liriodendron tulipifera* (Norby & O'Neill, 1991; Gunderson *et al.*, 1993), *Nothofagus fusca* (Hollinger, 1987), *Picea sitchensis* (Canham & McCavish, 1981), *Pinus contorta* (Higginbotham *et al.*, 1985), *Pinus ponderosa* (Surano *et al.*, 1986), *Pinus radiata* (Hollinger, 1987; Conroy *et al.*, 1988) *Pinus taeda* (Rogers *et al.*, 1983), *Populus grandidentata* (Jurik *et al.*, 1984), *Quercus alba* (Norby *et al.*, 1986a; Gunderson *et al.*, 1993) and six tropical C_3 species (Ziska *et al.*, 1991) (see also Kramer, 1981; Gunderson & Wullschleger, 1994). There are some exceptions to this general rule, as in the experiments carried out with *Castanea sativa* (Mousseau & Enoch, 1989).

Physiological changes in response to elevated CO_2 may also be accompanied by morphological changes. In some instances leaves appear to become thicker at elevated [CO_2] (Mousseau & Enoch, 1989; Radoglou & Jarvis, 1990a; Norby &

O'Neill, 1991). Leaf growth, with respect to thickening, is stimulated by a greater expansion of mesophyll cells, rather than any increase in cell number (Radoglou & Jarvis, 1990b). Delays in senescence have been reported for *Quercus alba* (Norby *et al.*, 1986a, b), while enhancements have been found for both *Castanea sativa* and *Populus tremuloides* (Mousseau & Enoch, 1989; Brown, 1991).

There is clearly a need to determine whether changes in assimilation will translate into differences in xylem growth. In a few experiments with increased CO_2, both quantitative and qualitative changes have been noted with the differentiation of xylem (Atkinson, unpublished). But these are not always described as being obviously beneficial to the plant (Norby *et al.*, 1992). An increase in annual-ring width may be primarily due to a more sensitive response of earlywood (Telewski & Strain, 1987). Tracheid wall thickening also increases and by so doing enhances wood density (Conroy *et al.*, 1990). As already suggested, the relevance of these observations is clearly limited by the fact that they refer to short-term changes in juvenile wood. From the point of view of commercial forestry it is essential to understand the mechanisms which control carbon allocation. For there to be commercial benefits in increased carbon fixation, additional photosynthate has to be allocated to xylem production in a manner which equates the desirable qualitative aspects of the wood which match its intended use. There may well be conflicts between such commercially desirable traits and those essential for maintaining physiological functioning. This conflict of interest may already be apparent, as evidence suggests that elevated CO_2 may increase the proportion of carbon allocated to root biomass (Higginbotham *et al.*, 1985; Norby & O'Neill, 1991; Norby *et al.*, 1992). This situation would provide an effective store for carbon but, outside a coppicing system, would be of limited value to the forester.

With the possibility for an increase in the amount of carbon allocated to below-ground organs, particularly roots, it is likely that changes in the soil microorganisms will occur. The growth of root mycorrhizal systems, for example, appears to increase for plants grown at elevated CO_2. This is due to an increase in the rate of flux of exudates from the host roots (Norby *et al.*, 1987). As yet we have little insight into the importance of such a response for nutrient recycling or any cost–benefit analysis to the host as opposed to the mycorrhiza.

Attempts have also been made to draw conclusions about the effects of global increases in CO_2 on trees growing either within natural habitats or forest plantations. Because of the long-lived nature of much of this material it is possible to select trees which began growing prior to the industrially induced increase in [CO_2]. It is known that the dominant effect on growth-ring thickness (wood production), particularly at tree-lines, is one of temperature (MacDonald *et al.*, 1993), whereas the influence of water availability is likely to increase in environments which experience frequent summer water deficits. Species which are extremely long-lived are particularly suitable for this analysis, as they provide a chronology of annual-ring widths which may extend back thousands of years. This analysis has not been restricted to the use of living material, and chronologies have been overlapped and extended further back into the past by use of preserved material. The results from this approach are, however, complex and frequently difficult to interpret.

There would appear to be considerable differences in the potential for increases in atmospheric [CO_2] to translate into a positive increase in wood production. Also, not only may such increases be difficult to detect, but their detectibility may vary with species. The physiological age of a tree will also determine the vigour of its growth and its carbon-allocation priorities. There may also be complications with seasonal/diurnal variations in [CO_2] within the canopy which are different to those of the bulk air (see review by Jarvis, 1989). This type of variation in the canopy means that as a tree develops and matures it may well experience a very different [CO_2]. Clearly, these factors are not independent of the tree's climatic environment, but would have considerable influence on ring width. Even the position within the tree from which samples of wood are taken can be very important. This is evident from differences in the sensitivity of the whole-tree hierarchical carbon-allocation pattern to environmental change (Larson, 1963). It therefore comes as no surprise that the experimental evidence is conflicting.

An examination of growth trends in widely distributed sites within the northern hemisphere from 1950 onwards, using tree-ring analysis, was unable to attribute the observed increases in ring width solely to enhanced atmospheric [CO_2] (Kienast & Luxmore, 1988). The conclusions of this research are supported by another study with *Pinus balfouriana, P. murayana* and *Juniperus occidentalis,* collected in the Sierra Nevada (Graumlich, 1991). On the other hand, a study of ring widths from the extremely long-lived bristlecone pines (*Pinus longaeva, P. aristata*) has been used to suggest such an increase is present (LaMarche *et al.*, 1984). The same may also be true for other high-elevation trees (*Pinus flexilis*) in the southwestern United States and in the arctic (Jacoby, 1986; Graybill, 1987; Hari *et al.*, 1984). These data are the subject of much discussion regarding the validity of the techniques adopted and the analyses used (Cooper, 1986; Gale, 1986). Some of this discrepancy may be due to the confounding effects of changes in other biological or environmental factors of which we have limited or no present knowledge. We may also expect different habitats to respond in rather different ways. Increases in CO_2 may enhance photosynthesis with a concomitant increase in WUE and such a fertilization effect may, therefore, be more evident in arid regions. Clearly, if our understanding and interpretation of the climate change scenario is correct, we should also find concurrent increase in ring width in response to rising temperature (Hansen *et al.*, 1981; Wigley *et al.*, 1985). Any tree-ring analysis would firstly have to be able to predict accurately this response in terms of its affect on growth before it could be removed from any direct CO_2-response analysis. The very nature and complexity of this temperature response alone may make it difficult to expect any enlightening results from annual-ring-width analysis. Even geographical regional differences in temperature change may alter the rate at which any warming trends are reflected in tree-ring growth (Briffa *et al.*, 1990).

One of the key questions being asked about forest ecosystems focuses on their potential to be sinks for the removal of atmospheric CO_2. As already noted, woody perennials have the potential to store considerable quantities of carbon in their tissues, but the questions now being addressed go far beyond that (Woodwell *et al.*, 1983; Tans *et al.*, 1990). Obviously, further reductions in the size of forests will exacerbate the rise in atmospheric CO_2 if further CO_2 is released by combustion, and by reducing the potential size of the CO_2 sink. Despite there being increases in the area of northern and boreal forests, these increases in no way balance the considerably more important clearances taking place in the tropics. It is this situation that prevails when assessments are made of the potential for reforestation to solve the problems of increasing atmospheric [CO_2]. Jarvis (1989) suggests that a young actively growing forest, about twice the size of Europe, could potentially assimilate all the CO_2 produced at the present rate from combustion and oxidation. This of course would only be achieved during the early life of a forest, while it was an active sink for carbon.

It would appear that despite many difficulties, reforestation may at least in part provide a solution to the problem of an increasing [CO_2] (Schroeder & Ladd, 1991). The use of woody perennials in the production of biomass as a replacement for fossil-fuel-derived energy has been suggested as better solution for reducing the rate of increases in atmospheric CO_2 (Vitousek, 1991). It is clearly important that the ways in which trees and wood products are utilized to provide a carbon store must be effectively assessed (see Jones, 1992).

Even attempts to provide simple solutions, e.g., by planting to re-establish forest growth on old sites, may be fraught with problems. Particularly apparent in the Amazon region after tree removal are dramatic changes in both the soil and the above-ground microclimate, which now prevent the establishment of primary rain forest species (Shukla *et al.*, 1990).

ACKNOWLEDGEMENTS

This chapter is dedicated to the memory of my first botany teacher, G.M. Bird. I thank Drs J.D. Quinlan and R.T. Besford and Professor H.G. Jones for their helpful comments on an earlier version of this chapter.

REFERENCES

Adams, R.M., Rosenzweig, C., Peart, R.M., Ritchie, J.T., McCarl, B.A., Glyer, J.D., Curry, R.B., Jones, J.W., Boote, K.J. & Allen L.H. Jr 1990. Global climate change and US agriculture. *Nature* **345**: 219-224.

Amthor, J.S. 1991. Respiration in a future, higher-CO_2 world. *Plant, Cell Environ.* **14**: 13-20.

Andrews, T.J. & Lorimer, G.H. 1987. Rubisco: structure, mechanisms, and prospects for improvements. In Hatch, M.D. & Boardman, N.K. (eds) *The Biochemistry of Plants, Volume 10, Photosynthesis*: 131-218. Academic Press, New York.

Arp, W.J. 1991. Effects of source-sink relations on photosynthetic acclimation to elevated CO_2. *Plant, Cell Environ.* **14**: 869-875.

Arp, W.J. & Drake, B.G. 1991. Increased photosynthetic capacity of *Scirpus olneyi* after 4 years of exposure to elevated CO_2. *Plant, Cell Environ.* **14**: 1003-1006.

Atkinson, C.J. & Farrar, J.F. 1983. Allocation of photosynthetically-fixed carbon in *Festuca ovina* L. and *Nardus stricta* L. *New Phytol.* **95**: 519-531.

Atkinson, C.J. & Taylor, L. 1994. The influence of autumn temperature on flowering time and cropping of *Pyrus communis* cv. Conference. *J. Hort. Sci.* **69**: 1067-1075.

Barnola, J.M., Raynard, D., Korotkevich, Y.S. & Lorius, C. 1987. Vostok ice core provides 160,000-year record of atmospheric carbon dioxide. *Nature* **329**: 408-414.

Baxter, R., Ashenden, T.W., Sparks, T.H. & Farrar, J.F. 1994. Effects of elevated carbon dioxide on three montane grass species. I. Growth and dry matter partitioning. *J. Exp. Bot.* **45**: 305-315.

Bazzaz, F.A. 1990. The response of natural ecosystems to the rising global CO_2 levels. *Annu. Rev. Ecol. Syst.* **21**: 167-196.

Bazzaz, F.A. & Carlson, R.W. 1984. The response of plants to elevated CO_2. I. Competition among an assemblage of annuals at two levels of soil moisture. *Oecologia* **62**: 196-198.

Bazzaz, F.A. & Garbutt, K. 1988. The response of annuals in competitive neighborhoods: effects of elevated CO_2. *Ecology* **69**: 937-946.

Bazzaz, F.A. & Williams, W.E. 1991. Atmospheric CO_2 concentrations within a mixed forest: implications for seedling growth. *Ecology* **72**: 12-16.

Bazzaz, F.A., Garbutt, K., Reekie, E.G. & Williams, W.E. 1989. Using growth analysis to interpret competition between a C_3 and a C_4 annual under ambient and elevated CO_2. *Oecologia* **79**: 223-235.

Bazzaz, F.A., Coleman, J.S. & Morse, S.R. 1990. Growth responses of seven major co-occurring tree species of the northeastern United States to elevated CO_2. *Can. J. For. Res.* **20**: 1479-1484.

Beadle, C.L., Jarvis, P.G. & Neilson, R.E. 1979. Leaf conductance as related to xylem water potential and carbon dioxide concentration in Sitka spruce. *Physiol. Plant.* **45**: 158-166.

Beerling, D.J., Chaloner, W.G., Huntley, B., Pearson, J.A. & Tooley, M.J. 1993. Stomatal density responds to the glacial cycle of environmental change. *Proc. R. Soc. London B* **251**: 133-138.

Besford, R.T. 1990. The greenhouse effect: acclimation of tomato plants growing in high CO_2, relative changes in Calvin cycle enzymes. *J. Plant Physiol.* **136**: 458-463.

Besford, R.T. & Hand, D.W. 1989. The effects of CO_2 enrichment and nitrogen oxides on some Calvin cycle enzymes and nitrate reductase in glasshouse lettuce. *J. Exp. Bot.* **40**: 329-336.

Besford, R.T., Withers, A.C. & Ludwig, L.J. 1984. Ribulose bisphosphate carboxylase and net CO_2 fixation in tomato leaves. In Sybesma, C. (ed.) *Advances in Photosynthetic Research III*: 779-782. Martinus Nijhoff/Dr. W. Junk, The Hague.

Besford, R.T., Ludwig, L.J. & Withers, A.C. 1990. The greenhouse effect: acclimation of tomato plants growing in high CO_2, photosynthesis and ribulose-1,5-bisphosphate carboxylase protein. *J. Exp. Bot.* **41**: 925-931.

Bhattacharya, N.C., Biswas, P.K., Bhattacharya, S., Sionit, N. & Strain, B.R. 1985a. Growth and yield response of sweet potato to atmospheric CO_2 environment. *Crop Science* **25**: 975-981.

Bhattacharya, S., Bhattacharya, N.C., Biswas, P.K. & Strain, B.R. 1985b. Response of cowpea (*Vigna unguiculata* L.) to CO_2 enrichment environment on growth, dry-matter production and yield components at different stages of vegetative and reproductive growth. *J. Agric. Sci. (Cambridge)* **105**: 527-534.

Bhattacharya, N.C., Hileman, D.R., Ghosh, P.P., Musser, R.L., Bhattacharya, S. & Biswas, P.K. 1990. Interaction of enriched CO_2 and water stress on the physiology of and biomass production in sweet potato grown in open-top chambers. *Plant, Cell Environ.* **13**: 933-940.

Bjorkman, O. 1966. The effect of oxygen concentration on photosynthesis in higher plants. *Physiol. Plant.* **19**: 618-633.

Bowes, G. 1991. Growth at elevated CO_2: photosynthetic responses mediated through Rubisco. *Plant, Cell Environ.* **14**: 795-806.

Bradshaw, A.D. & McNeilly, T. 1991. Evolutionary response to global climate change. *Ann. Bot.* (suppl.) **67**: 5-14.

Briffa, K.R., Bartholin, T.S., Eckstein, D., Jones, P.D., Karlen, W., Schweingruber, F.H. & Zetterberg, P. 1990. A 1,400-year tree-ring record of summer temperatures in Fennoscandia. *Nature* **346**: 434-439.

Brown, G.D. & Higginbotham, K.O. 1986. Effects of carbon dioxide enrichment and nitrogen supply on growth of boreal tree seedlings. *Tree Physiol.* **2**: 223-232.

Brown, K.R. 1991. Carbon dioxide enrichment accelerates the decline in nutrient status and relative growth rate of *Populus tremuloides* Michx. seedlings. *Tree Physiol.* **8**: 161-173.
Browning, G. & Miller, J.M. 1992. The association of year-to-year variation in average yield of pear cv. Conference in England with weather variables. *J. Hort. Sci.* **67**: 593-599.
Bunce, J.A. 1990. Short- and long-term inhibition of respiratory carbon dioxide efflux by elevated carbon dioxide. *Ann. Bot.* **65**: 637-642.
Campbell, W.J., Allen, L.H. Jr & Bowes, G. 1988. Effects of CO_2 concentration on Rubisco activity, amount, and photosynthesis in soybean leaves. *Plant Physiol.* **88**: 1310-1316.
Campbell, W.J., Allen, L.H. Jr & Bowes, G. 1990. Response of soybean canopy photosynthesis to CO_2 concentration, light, and temperature. *J. Exp. Bot.* **41**: 427-433.
Canham, A.E. & McCavish, W.J. 1981. Some effects of CO_2, daylength and nutrition on the growth of young forest tree plants. I. In the seedling stage. *Forestry* **54**: 169-182.
Cannell, M.G.R. 1990. Carbon dioxide and the global carbon cycle. In Cannell, M.G.R. & Hooper, M.D. (eds) *The Greenhouse Effect and Terrestrial Ecosystems of the UK*: 6-9. Institute of Terrestrial Ecology Research Publication No. 4, HMSO, London.
Carlson, R.W. & Bazzaz, F.A. 1980. The effects of elevated CO_2 concentration on growth, photosynthesis, transpiration, and water-use efficiency of plants. In Singh J.J. & Deepak A. (eds) *Environmental and Climatic Impact of Coal Utilization*: 177-222. Academic Press, New York.
Carter, R.W. & Peterson, 1983. Effects of a CO_2-enriched atmosphere on the growth and competitive interaction of a C_3 and a C_4 grass. *Oecologia* **58**: 188-193.
Carter, T.R., Porter, J.H. & Parry, M.L. 1992. Some implications of climatic change for agriculture in Europe. *J. Exp. Bot.* **43**: 1159-1167.
Cave, G., Tolley, L.C. & Strain, B.R. 1981. Effect of carbon dioxide enrichment on chlorophyll content, starch content and starch grain structure in *Trifolium subterraneum* leaves. *Physiol. Plant.* **51**: 171-174.
Cess, R.D., Zhang, M.-H., Potter, G.L., Barker, H.W., Colman, R.A., Dazlich, D.A., Del Genio, A.D., Esch, M., Fraser, J.R., Galin, V., Gates, W.L., Hack, J.J., Ingram, W.J., Kichl, J.T., Lacis, A.A., Le Treut, H., Li, Z.-X., Liang, X.-Z., Mahfouf, J.-F., McAvaney, B.J., Meleshko, V.P., Morcrette, J.-J., Randall, D.A., Roeckner, E., Royer, J.-F., Sokolov, A.P., Sporyshev, P.V., Taylor, K.E., Wang, W.-C. & Wetherald, R.T. 1993. Uncertainties in carbon dioxide radiative forcing in atmospheric general circulation models. *Science* **262**: 1252-1255.
Chahine, M.T. 1992. The hydrological cycle and its influence on climate. *Nature* **359**: 373-380.
Chaudhuri, U.N., Kirkham, M.B. & Kanesmasu, E.T. 1990. Root growth of winter wheat under elevated carbon dioxide and drought. *Crop Sci.* **30**: 853-857.
Chaves, M.M. & Pereira, J.S. 1992. Water stress, CO_2 and climate change. *J. Exp. Bot.* **43**: 1131-1139.
Clough, J.M., Peet, M.M. & Kramer, P.J. 1981. Effects of high atmospheric CO_2 and sink size on rates of photosynthesis of a soybean cultivar. *Plant Physiol.* **67**: 1007-1010.
Conroy, J.P, Barlow, E.W.R. & Bevege, D.I. 1986. Responses of *Pinus radiata* seedlings to carbon dioxide enrichment at different levels of water and phosphorus: growth, morphology and anatomy. *Ann. Bot.* **57**: 165-177.
Conroy, J.P., Küppers, M., Küppers, B., Virgona, J. & Barlow, E.W.R. 1988. The influence of CO_2 enrichment, phosphorus deficiency and water stress on the growth, conductance and water use of *Pinus radiata* D. Don. *Plant, Cell Environ.* **11**: 91-98.
Conroy, J.P., Milham, P.J., Mazur, M. & Barlow, E.W.R. 1990. Growth, dry weight partitioning and wood properties of *Pinus radiata* D. Don after 2 years of CO_2 enrichment. *Plant, Cell Environ.* **13**: 329-337.
Cooper, C.F. 1986. Carbon dioxide enhancement of tree growth at high elevations. *Science* **231**: 860.
Cowan, I.R. 1986. Economics of carbon fixation in higher plants. In Givnish, T.J. (ed.) *On the Economy of Plant Form and Function*: 133-170. Cambridge University Press, Cambridge.

Crane, A.J. 1985. Possible effects of rising CO_2 on climate. *Plant, Cell Environ.* **8**: 371-379.
Cure, J.D. & Acock, B. 1986. Crop response to carbon dioxide doubling: a literature survey. *Agric. Forest Meteorol.* **38**: 127-145.
Cure, J.D., Rufty T.W. Jr & Israel, D.W. 1987. Assimilate utilization in the leaf canopy and whole-plant growth of soybean during acclimation to elevated CO_2. *Bot. Gaz.* **148**: 67-72.
Cure, J.D., Rufty T.W. Jr & Israel, D.W. 1991. Assimilate relations in source and sink leaves during acclimation to a CO_2-enriched atmosphere. *Physiol. Plant.* **83**: 687-695.
Curtis, P.S., Drake, B.G., Leadly, P.W., Arp, W.J. & Whigham, D.F. 1989a. Growth and senescence in plant communities exposed to elevated CO_2 concentrations on an estuarine marsh. *Oecologia* **78**: 20-26.
Curtis, P.S., Drake, B.G., & Whigham, D.F. 1989b. Nitrogen and carbon dynamics in C_3 and C_4 estuarine marsh plants grown under CO_2 in situ. *Oecologia* **78**: 297-301.
Davies, W.J. & Mansfield, T.A. 1987. Auxins and stomata. In Zeiger, E., Farquhar, G.D. & Cowan, I.R. (eds) *Stomatal Function*: 293-309. Stanford University Press, Stanford, CA.
Davis, M.B. 1989. Lags in vegetation response to greenhouse warming. *Climate Change* **15**: 75-82.
Davis, M.B., Woods, K.D., Webb, S.L. & Futyma, R.P. 1986. Dispersal versus climate: expansion of *Fagus* and *Tsuga* into the upper Great Lakes region. *Vegetatio* **67**: 93-103.
DeLucia, E.H. & Heckathorn, S.A. 1989. The effect of soil drought on water-use efficiency in contrasting Great Basin desert and Sierran montane species. *Plant, Cell Environ.* **12**: 935-940.
Dubbe, D.R., Farquhar, G.D. & Raschke, K. 1978. Effects of abscisic acid on the gain of the feedback loop involving carbon dioxide and stomata. *Plant Physiol.* **62**: 413-417.
Eamus, D. & Jarvis, P.G. 1989. The direct effects of increase in the global atmospheric CO_2 concentration on natural and commercial temperate trees and forests. *Adv. Ecol. Res.* **16**: 1-55.
Edwards, A. & Bowling, D.J.F. 1985. Evidence for a CO_2 inhibited proton extrusion pump in the stomatal cells of *Tradescantia virginiana. J. Exp. Bot.* **36**: 91-98.
Edwards, G. & Walker, G. 1983. *C_3 : C_4 Mechanisms, and Cellular and Environmental Regulation of Photosynthesis.* Blackwell, Oxford.
Ehleringer, J. & Bjorkman, O. 1977. Quantum yields for CO_2 uptake in C_3 and C_4 plants: dependence on temperature, CO_2 and O_2 uptake. *Plant Physiol.* **59**: 86-90.
Farquhar, G.D. & Sharkey, T.D. 1982. Stomatal conductance and photosynthesis. *Ann. Rev. Plant Physiol.* **33**: 317-345.
Farrar, J.F. 1985. The respiratory source of CO_2. *Plant, Cell Environ.* **8**: 427-438.
Farrar, J.F. & Williams, M.L. 1991. The effects of increased atmospheric carbon dioxide and temperature on carbon partitioning, source-sink relations and respiration. *Plant, Cell Environ.* **14**: 819-830.
Ferris, R. & Taylor, G. 1994. Stomatal characteristics of four native herbs following exposure to elevated CO_2. *Ann. Bot.* **73**: 447-453.
Flore, J.A. & Lakso, A.N. 1989. Environmental and physiological regulation of photosynthesis in fruit crops. *Hort. Rev.* **11**: 111-157.
Friedli, H., Lotscher, H., Oeschger, H., Siegenthaler, U. & Stauffer, B. 1986. Ice core record of the $^{13}C/^{12}C$ ratio of atmospheric CO_2 in the past two centuries. *Nature* **324**: 237-238.
Gale, J. 1986. Carbon dioxide enhancement of tree growth at high elevations. *Science* **231**: 859-860.
Garbutt, K. & Bazzaz, F.A. 1984. The effects of elevated CO_2 on plants. III. Flower, fruit and seed production and abortion. *New Phytol.* **98**: 433-446.
Garbutt, K., Williams, W.E. & Bazzaz, F.A. 1990. Analysis of the differential responses of five annuals to elevated CO_2 during growth. *Ecology* **71**: 1185-1194.
Gifford, R.M. 1977. Growth pattern, CO_2 exchange and dry weight distribution in wheat growing under differing photosynthetic environments. *Aust. J. Plant Physiol.* **4**: 99-110.
Gifford, R.M., Lambers, H. & Morison, J.I.L. 1985. Respiration of crop species under CO_2 enrichment. *Physiol. Plant.* **63**: 351-356.

Goudriaan, J. & de Ruiter, H.E. 1983. Plant growth in response to CO_2 enrichment, at two levels of nitrogen and phosphorus supply. 1. Dry matter, leaf area and development. *Netherl. J. Agric. Sci.* **31**: 157-169.

Grace, J. 1987. Climate tolerance and the distribution of plants. *New Phytol.* (suppl.) **106**: 113-130.

Graumlich, L.J. 1991. Subalpine tree growth, climate, and increasing CO_2: An assessment of recent growth trends. *Ecology* **72**: 1-11.

Graybill, D.A. 1987. A network of high elevation conifers in the western US for detection of tree-ring growth response to increasing atmospheric carbon dioxide. In Jacoby, G.C. Jr & Hornbeck, J.W. (eds) *Proceedings International Symposium on Ecological Aspects of Tree-Ring Analysis*: 463-474. US Department of Energy, New York.

Gunderson, C.A. & Wullschleger, S.D. 1994. Photosynthetic acclimation in trees to rising atmospheric CO_2: A broader perspective. *Photosyn. Res.* **39**: 369-388.

Gunderson, C.A., Norby, R.J. & Wullschleger, S.D. 1993. Foliar gas exchange responses of two deciduous hardwoods during 3 years of growth in elevated CO_2: No loss of photosynthetic enhancement. *Plant, Cell & Environ.* **16**: 797-807.

Hall, D.O. & Scurlock, J.M.O. 1991. Climate change and productivity of natural grasslands. *Ann. Bot.* (suppl.) **67**: 49-55.

Hansen, J., Johnson, D., Lacis, A., Lebedeff, S., Lee, P., Rind, D. & Russell, G. 1981. Climatic impact of increasing atmospheric carbon dioxide. *Science* **213**: 957-966.

Hari, P. Arovaara, H., Raunemaa, Y. & Hautojarvi, A. 1984. Forest growth and the effects of energy production: A method for detecting trends in the growth potential of trees. *Can. J. For. Res.* **14**: 437-440.

Heath, O.V.S. & Mansfield, T.A. 1962. A recording porometer with detachable cups operating on four different leaves. *Proc. Royal Soc. London* **156B**: 1-13.

Heichel, G.H. & Jaynes, R.A. 1974. Stimulating emergence and growth of *Kalmia* genotypes with carbon dioxide. *Hortsci.* **9**: 60-62.

Herold, A. 1980. Regulation of photosynthesis by sink activity — the missing link. *New Phytol.* **86**: 131-144.

Hicklenton, P.R. & Jolliffe, P.A. 1980. Alterations in the physiology of CO_2 exchange in tomato plants grown in CO_2-enriched atmospheres. *Can. J. Bot.* **58**: 2181-2189.

Higginbotham, K.O., Mayo, J.M., L'Hirondelle, S. & Krystofiak, D.K. 1985. Physiological ecology of Lodgepole pine (*Pinus contorta*) in an enriched CO_2 environment. *Can. J. For. Res.* **15**: 417-421.

Ho, L.C. 1977. Effects of CO_2 enrichment on the rates of photosynthesis and translocation of tomato leaves. *Ann. Applied Biol.* **87**: 191-200.

Ho, L.C. 1978. The regulation of carbon transport and the carbon balance of mature tomato leaves. *Ann. Bot.* **42**: 155-164.

Hollinger, D.Y. 1987. Gas exchange and dry matter allocation responses to elevation of atmospheric CO_2 concentration in seedlings of three tree species. *Tree Physiol.* **3**: 193-202.

Hrubec, T.C., Robinson, J.M. & Donaldson, R.P. 1984. Effect of CO_2 enrichment on soybean leaf and mitochondrial respiration. *Plant Physiol.* (suppl.) **75**: 158.

Hrubec, T.C., Robinson, J.M. & Donaldson, R.P. 1985. Effect of CO_2 enrichment and carbohydrate content on the dark respiration of soybean. *Plant Physiol.* **79**: 684-689.

Huber, S.C. 1983. Relation between photosynthetic starch formation and dry-weight partitioning between the shoot and root. *Can. J. Bot.* **61**: 2709-2716.

Huber, S.C., Rogers, H. & Israel, D.W. 1984. Effects of CO_2 enrichment on photosynthesis and photosynthate partitioning in soybean (*Glycine max*) leaves. *Physiol. Plant.* **62**: 95-101.

Idso, S.B. 1988. Three phases of plant response to atmospheric CO_2 enrichment. *Plant Physiol.* **87**: 5-7.

Jacoby, G.C. 1986. Long-term temperature trends and a positive departure from the climate-growth response since the 1950s in high elevation lodgepole pine from California. In Rosenzweig, C. & Dickinson, R. (eds) *Proceedings of the NASA Conference on*

Climate-Vegetation Interactions: 81-83. Office for Interdisciplinary Earth Studies (OIES), University Corporation for Atmospheric Research (UCAR), Colorado.
Jarvis, P.G. 1989. Atmospheric carbon dioxide and forests. *Phil. Trans. Royal Soc. London B* **324**: 369-392.
Jarvis, P.G. & McNaughton, K.G. 1986. Stomatal control of transpiration. *Adv. Ecol. Res.* **15**: 1-29.
Jones, H.G. 1992. *Plants and Microclimate: A Quantitative Approach to Environmental Plant Physiology*, 2nd edition. Cambridge University Press, Cambridge.
Jurik, T.W., Weber, J.A. & Gates, D.M. 1984. Short-term effects of CO_2 on gas exchange of leaves of bigtooth aspen (*Populus grandidentata*) in the field. *Plant Physiol.* **75**: 1022-1026.
Keeling, C.D., Bacastow, R.B., Bainbridge, A.E., Ekdahl, C.A., Guenther, P.R., Waterman, L.S. & Chin, J.F.S. 1976. Atmospheric carbon dioxide variations at Mauna Loa Observatory, Hawaii. *Tellus* **28**: 538-551.
Kienast, F. & Luxmore, R.J. 1988. Tree-ring analysis and conifer growth responses to increased atmospheric CO_2 levels. *Oecologia* **76**: 487-495.
Koch, K.E., Jones, P.H., Avigne, W.T. & Allen, L.H. 1986. Growth, dry matter partitioning, and diurnal activities of RuBP carboxylase in citrus seedlings maintained at two levels of CO_2. *Physiol. Plant.* **67**: 477-484.
Kramer, K., 1994. A modelling analysis of the effects of climatic warming on the probability of spring frost damage to tree species in the Netherlands and Germany. *Plant, Cell Environ.* **17**: 367-377.
Kramer, P.J. 1981. Carbon dioxide concentration, photosynthesis, and dry matter production. *Bioscience* **31**: 29-33.
Kronenberg, H.G. 1994. Temperature influences on the flowering dates of *Syringa vulgaris* L. and *Sorbus aucuparia* L. *Scientia Horticulturae* **57**: 59-71.
Krupa, S.V. & Kickert, R.N. 1989. The greenhouse effect: Impacts of ultraviolet-B (UV-B) radiation, carbon dioxide (CO_2), and ozone (O_3) on vegetation. *Environ. Pollut.* **61**: 263-393.
LaMarche, V.C., Graybill, D.A., Fritts, H.C. & Rose, M.R. 1984. Increasing atmospheric carbon dioxide: tree-ring evidence for growth enhancement in natural vegetation. *Science* **225**: 1019-1021.
Larson, P.R. 1963. Stem form development in forest trees. *For. Sci. Monogr.* **5**: 1-42.
Lashof, D.A. & Ahuja, D.A. 1990. Relative contributions of greenhouse gas emissions to global warming. *Nature* **344**: 529-531.
Lawlor, D.W. & Michell, R.A.C. 1991. The effects of increasing CO_2 on crop photosynthesis and productivity: a review of field studies. *Plant, Cell Environ.* **14**: 807-818.
Lincoln, D.E., Couvet, D. & Sionit, N. 1986. Response of an insect herbivore to host plants grown in enriched carbon dioxide atmospheres. *Oecologia* **69**: 556-560.
Linsbauer, K. 1916. Beitrage zur Kenntnis der Spaltoffnungs-bewegung. *Flora* **9**: 100-143.
Luxmore, R.J., O'Neill, E.G., Ellis, J.M. & Rogers, H.H. 1986. Nutrient uptake and growth responses of Virginia pine to elevated atmospheric carbon dioxide. *J. Environ. Qual.* **15**: 244-251.
MacDonald, G.M., Edwards, T.W.D., Moser, K.A., Pienitz, R. & Smol, J.P. 1993. Rapid response of tree line vegetation and lakes to past climate warming. *Nature* **361**: 243-246.
Makino, A., Mae, T. & Ohira, K. 1984. Relation between nitrogen and ribulose-1,5-*bis*phosphate-carboxylase in rice leaves from emergence through senescence. *Plant Cell Physiol.* **25**: 429-437.
Mansfield, T.A. 1991. Global warming and the greenhouse effect. *Prof. Hort.* **5**: 3-9.
Mansfield, T.A. & Atkinson, C.J. 1990. Stomatal behaviour in water stressed plants. In Alscher, R. (ed.) *Stress Responses in Plants: Adaptation and Acclimation Mechanisms*: 241-264. Wiley-Liss Inc., New York.
Mansfield, T.A., Hetherington, A.M. & Atkinson, C.J. 1990. Some current aspects of stomatal physiology. *Annu. Rev. Plant Physiol. Plant Mol. Biol.* **41**: 55-75.
Marland, G. & Rotty, R.M. 1983. *Carbon Dioxide Emissions from Fossil Fuel CO_2 Emissions.* Report, US Department of Energy, Washington, DC.

Mauney, J.R., Guinn, G., Fry, K.E. & Hesketh, J.D. 1979. Correlation of photosynthetic carbon dioxide uptake and carbohydrate accumulation in cotton, soybean, sunflower and sorghum. *Photosynthetica* **13**: 260-266.

McElroy, M. 1988. The challenge of global change. *New Scientist* **119**: 34-36.

Medina, E. 1971. Relationship between nitrogen level, photosynthetic capacity, and carboxydismutase activity in *Atriplex patula* leaves. *Annu. Rev. Carnegie Inst. US*, 1969-1970: 655-662. Stanford, CA.

Meidner, H. & Mansfield, T.A. 1968. *Physiology of Stomata*. McGraw-Hill, London.

Melillo, J.M., Aber, J.D. & Muratore, J.F. 1982. Nitrogen and lignin control of hardwood leaf litter decomposition dynamics. *Ecology* **63**: 621-626.

Miglietta, F., Raschi, A., Bettarini, I., Resti, R. & Selvi, F. 1993. Natural CO_2 springs in Italy: A resource for examining long-term response of vegetation to rising atmospheric CO_2 concentrations. *Plant, Cell Environ.* **16**: 873-878.

Mitchell, J.F.B. 1983. The seasonal response of a general circulation model to changes in CO_2 and sea temperatures. *Quart. J. Royal Meteorol. Soc.* **109**: 113-152.

Mitchell, J.F.B. 1989. The "greenhouse effect" and climate change. *Rev. Geophys.* **27**: 115-139.

Monson, R.K. 1989. On the evolutionary pathways resulting in C_4 photosynthesis and crassulacean acid metabolism (CAM). *Adv. Ecol. Res.* **19**: 57-110.

Monson, R.F., Hills, A.J., Zimmerman, P.R. & Fall, R.R. 1991. Studies of the relationship between isoprene emission rate and CO_2 or photon-flux density using a real-time isoprene analyser. *Plant, Cell Environ.* **14**: 517-523.

Morison, J.I.L. 1987. Intercellular CO_2 concentration and stomatal response to CO_2. In Zeiger, E., Farquhar, G.D. & Cowan, I.R. (eds) *Stomatal Function*: 229-251. Stanford University Press, CA.

Morison, J.I.L. & Gifford, R.M. 1983. Stomatal sensitivity to CO_2 and humidity. *Plant Physiol.* **71**: 789-796.

Mott, K.A. 1988. Do stomata respond to CO_2 concentrations other than intercellular? *Plant Physiol.* **86**: 200-203.

Mott, K.A. 1990. Sensing the atmospheric CO_2 by plants. *Plant, Cell Environ.* **13**: 731-737.

Mousseau, M. & Enoch, H.Z. 1989. Carbon dioxide enrichment reduces shoot growth in sweet chestnut seedlings (*Castanea sativa* Mill). *Plant, Cell Environ.* **12**: 927-934.

Mousseau, M. & Saugier, B. 1992. The direct effect of increased CO_2 on gas exchange and growth of forest tree species. *J. Exp. Bot.* **43**: 1121-1130.

NASA (National Aeronautics and Space Administration) 1988. *Earth System Science, A Closer View. Report of the Earth System Sciences Committee.* NASA Advisory Council, NASA, Washington, DC.

Neales, T.F. & Nicholls, A.O. 1978. Growth responses of young wheat plants to a range of ambient CO_2 levels. *Aust. J. Plant Physiol.* **5**: 45-59.

Neftel, A., Oeschger, H., Schwander, J., Stauffer, B. & Zumbrunn, R. 1982. Ice sample measurements give atmospheric CO_2 content during the past 40,000 years. *Nature* **295**: 220-223.

Neftel, A., Moor, E., Oeschger, H. & Stauffer, B. 1985. Evidence from polar ice cores for the increase in atmospheric CO_2 in the past two centuries. *Nature* **315**: 45-47.

Nijs, I., Impens, I. & Behaeghe, T. 1989. Effects of long-term elevated atmospheric CO_2 concentration on *Lolium repens* canopies in the course of a terminal drought stress period. *Can. J. Bot.* **67**: 2720-2725.

Norby, R.J. & O'Neill, E.G. 1991. Leaf area compensation and nutrient interactions in CO_2-enriched seedlings of yellow-poplar (*Liriodendron tulipifera* L.). *New Phytol.* **117**: 515-528.

Norby, R.J., O'Neill, E.G. & Luxmore, R.J. 1986a. Effects of atmospheric CO_2 enrichment on the growth and mineral nutrition of *Quercus alba* seedlings in nutrient-poor soil. *Plant Physiol.* **82**: 83-89.

Norby, R.J., Pastor, J. & Melillo, J. 1986b. Carbon-nitrogen interactions in CO_2-enriched white oak: physiological and long-term perspectives. *Tree Physiol.* **2**: 233-241.

Norby, R.J., O'Neill, E.G., Hood, W.G. & Luxmore, R.J. 1987. Carbon allocation, root exudation and mycorrhizal colonisation of *Pinus echinata* seedlings grown under CO_2 enrichment. *Tree Physiol.* **3**: 203-210.

Norby, R.J., Gunderson, C.A., Wullschleger, S.D., O'Neill, E.G. & McCracken, M.K. 1992. Productivity and compensatory responses of yellow-poplar trees in elevated CO_2. *Nature* **357**: 322-324.

Oberbauer, S., Strain, B. & Fetcher, N. 1985. Effect of CO_2 enrichment on seedling physiology and growth of two tropical tree species. *Physiol. Plant.* **65**: 352-356.

Outlaw, W.H. 1989. Critical examination of the quantitative evidence for and against photosynthetic CO_2 fixation by guard cells. *Physiol. Plant.* **77**: 275-281.

Overdieck, D. 1986. Long-term effects of an increased CO_2 concentration on terrestrial plants in model ecosystems. Morphology and reproduction of *Trifolium repens* L. and *Lolium perenne* L. *Int. J. Biometeorol.* **30**: 323-332.

Paez, A., Hellmers, H. & Strain, B.R. 1980. CO_2 effects on apical dominance in *Pisum sativum*. *Physiol. Plant.* **50**: 43-46.

Paez, A., Hellmers, H. & Strain, B.R. 1984. Carbon dioxide enrichment and water stress interactions on growth of two tomato cultivars. *J. Agric. Sci. (Cambridge)* **102**: 687-693.

Parry, M.L., Carter, T.R. & Porter, J.H. 1989. The greenhouse effect and the future of UK agriculture. *J. Royal Agric. Soc.* **150**: 120-131.

Pastor, J. & Post, W.M. 1988. Responses of northern forests to CO_2 induced climate change. *Nature* **334**: 55-58.

Patterson, D.T. & Flint, E.P. 1982. Interacting effects of CO_2 and nutrient concentration. *Weed Science* **30**: 389-394.

Pearcy, R.W. & Bjorkman, O. 1983. Physiological effects. In Lemon, E.R. (ed.) *CO_2 and Plants: The Response of Plants to Rising Levels of Atmospheric Carbon Dioxide*: 65-106. Westview Press, Boulder, CO.

Peet, M.M. 1986. Acclimation to high CO_2 in monoecious cucumbers. 1. Vegetative and reproductive growth. *Plant Physiol.* **80**: 59-62.

Peet, M.M., Huber, S.C. & Patterson, D.T. 1986. Acclimation to high CO_2 in monoecious cucumbers. II Alterations in gas exchange rates, enzyme activities and starch and nutrient concentration. *Plant Physiol.* **80**: 63-67.

Penuelas, J. & Matamala, R. 1990. Changes in N and S leaf content, stomatal density and specific leaf area of 14 plant species during the last three centuries of CO_2 increase. *J. Exp. Bot.* **41**: 1119-1124.

Polley, H.W., Johnson, H.B., Marino, B.D. & Mayeux, H.S. 1993. Increase in C_3 plant water-use efficiency and biomass over glacial to present CO_2 concentrations. *Nature* **361**: 61-64.

Radin, J.W., Kimball, B.A., Hendrix, D.L. & Mauney, J.R. 1987. Photosynthesis of cotton plants exposed to elevated levels of carbon dioxide in the field. *Photosynt. Res.* **12**: 191-203.

Radin, J.W., Hartung, W., Kimball, B.A. & Mauney, J.R. 1988. Correlation of stomatal conductance with photosynthetic capacity of cotton only in a CO_2-enriched atmosphere: mediation by abscisic acid. *Plant Physiol.* **88**: 1058-1062.

Radoglou, K.M. & Jarvis, P.G. 1990a. Effects of CO_2 enrichment on four poplar clones. I. Growth and leaf anatomy. *Ann. Bot.* **65**: 616-626.

Radoglou, K.M. & Jarvis, P.G. 1990b. Effects of CO_2 enrichment on four poplar clones. II. Leaf surface properties. *Ann. Bot.* **65**: 627-632.

Raschke, K. 1975. Stomatal action. *Ann. Rev. Plant Physiol.* **26**: 309-340.

Raschke, K., Pierce, M. & Popiela, C.C. 1976. Abscisic acid content and stomatal sensitivity to CO_2 in leaves of *Xanthium strumarium* L. after pretreatments in warm and cold growth chambers. *Plant Physiol.* **57**: 115-121.

Reekie, E.G. & Bazzaz, F.A. 1989. Competition and patterns of resource use among seedlings of five tropical trees grown at ambient and elevated CO_2. *Oecologia* **79**: 212-222.

Robinson, J.M. 1994. Speculations on carbon dioxide starvation, late Tertiary evolution of stomatal regulation and floristic modernization. *Plant, Cell Environ.* **17**: 348-354.
Rochefort, L. & Woodward, F.I. 1992. Effects of climate change and a doubling of CO_2 on vegetation diversity. *J. Exp. Bot.* **43**: 1169-1180.
Roeckner, E. 1992. Past, present and future levels of greenhouse gases in the atmosphere and model projections of related climatic changes. *J. Exp. Bot.* **43**: 1097-1109.
Rogers, H.H., Bingham, G.E., Cure, J.D., Smith, J.M. & Surano, K.A. 1983. Responses of selected plant species to elevated carbon dioxide in the field. *J. Environ. Qual.* **12**: 569-574.
Rogers, H.H., Peterson C.M., McCrimmon, J.N., Cure, J.D. 1992. Response of plant roots to elevated atmospheric carbon dioxide. *Plant, Cell Environ.* **15**: 749-752.
Rotty, R.M. & Marland, G. 1986. Fossil fuel consumption: Recent amounts, patterns and trends of CO_2. In Trabalka, J.R. & Reichie, D.E. (eds) *A Global Analysis*: 474-490. Springer-Verlag, New York.
Rufty, T.W. Jr & Huber, S.C. 1983. Changes in starch formation and the activities of sucrose phosphate synthase and cytoplasmic fructose 1,6 bisphosphate in response to source sink alterations. *Plant Physiol.* **72**: 474-480.
Ryle, G.J.A. & Powell, C.E. 1976. Effects of rate of photosynthesis on the pattern of assimilate distribution in the graminaceous plant. *J. Exp. Bot.* **27**: 189-199.
Ryle, G.J.A. & Powell, C.E. 1992. The influence of elevated CO_2 and temperature on biomass production of continuously defoliated white clover. *Plant, Cell Environ.* **15**: 593-599.
Ryle, G.J.A. & Stanley, J. 1992. Effect of elevated CO_2 on stomatal size and distribution in perennial ryegrass. *Ann. Bot.* **69**: 563-565.
Ryle, G.J.A., Powell, C.E. & Davidson, I.A. 1992. Growth of white clover dependent on N_2 fixation, in elevated CO_2 and temperature. *Ann. Bot.* **70**: 221-228.
Sage, R.F. 1994. Acclimation of photosynthesis to increasing atmospheric CO_2: The gas exchange perspective. *Photosyn. Res.* **39**: 351-368.
Sage, R.F., Sharkey, T.D. & Seemann, J.R. 1988. The *in vivo* response of the ribulose-1,5-bisphosphate carboxylase activation state and the pool sizes of photosynthetic metabolites to elevated CO_2 in *Phaseolus vulgaris* L. *Planta* **174**: 407-416.
Sage, R.F., Sharkey, T.D. & Seemann, J.R. 1989. Acclimation of photosynthesis to elevated CO_2 in five C_3 species. *Plant Physiol.* **89**: 590-596.
Salvucci, M.E. 1989. Regulation of Rubisco activity *in vivo*. *Physiol. Plant.* **77**: 164-171.
Sasek, T.W. & Strain, B.R. 1988. Effects of carbon dioxide enrichment on the growth and morphology of kudzu (*Pueraria lobata*). *Weed Sci.* **36**: 28-36.
Sasek, T.W. & Strain, B.R. 1989. Effects of carbon dioxide enrichment on the expansion and size of kudzu (*Pueraria lobata*) leaves. *Weed Sci.* **37**: 23-28.
Sasek, T.W. & Strain, B.R. 1990. Implications of atmospheric CO_2 enrichment and climatic change for the geographical distribution of two introduced vines in the USA. *Climate Change* **16**: 31-51.
Sasek, T.W. & Strain, B.R. 1991. Effects of CO_2 enrichment on the growth morphology of a native and an introduced honeysuckle vine. *Am. J. Bot.* **78**: 69-75.
Sasek, T.W., DeLucia, E.H. & Strain, B.R. 1985. Reversibility of photosynthetic inhibition in cotton after long-term exposure to elevated CO_2 concentrations. *Plant Physiol.* **78**: 619-622.
Scheidegger, U.C. & Nosberger, J. 1984. Influence of carbon dioxide concentration on growth, carbohydrate content, translocation and photosynthesis of white clover. *Ann. Bot.* **54**: 735-742.
Schneider, S.H. 1989. The changing climate. *Sci. Amer.* **261**: 38-47.
Schroeder, P. & Ladd, L. 1991. Slowing the increase in atmospheric carbon dioxide: a biological approach. *Climate Change* **19**: 283-290.
Sharkey, T.D., Loreto, F. & Delwiche, C.F. 1991. High carbon dioxide and sun/shade effects on isoprene emissions from Oak and Aspen tree leaves. *Plant, Cell Environ.* **14**: 333-338.
Sheen, J. 1990. Metabolic repression of transcription in higher plants. *Plant Cell* **2**: 1027-1038.

Shukla, J., Nobre, C. & Sellers, P. 1990. Amazon deforestation and climate change. *Science* **247**: 1322-1325.

Sinclair, T.R. 1992. Mineral nutrition and plant growth response to climate change. *J. Exp. Bot.* **43**: 1141-1146.

Sionit, T.W., Hellmers, H. & Strain, B.R. 1980. Growth and yield of wheat under CO_2 enrichment and water stress. *Crop Sci.* **20**: 687-690.

Snaith, P.J. & Mansfield, T.A. 1982. Control of the CO_2 responses of stomata by indol-3-ylacetic acid and abscisic acid. *J. Exp. Bot.* **33**: 360-365.

Squire, G.R. & Mansfield, T.A. 1972. A simple method for isolating stomata on detached epidermis by low pH-treatment: Observations of the importance of the subsidiary cells. *New Phytol.* **71**: 1033-1043.

Stitt, M. 1991. Rising CO_2 levels and their potential significance for carbon flow in photosynthetic cells. *Plant, Cell Environ.* **14**: 741-762.

Stitt, M. & Quick, W.P. 1989. Photosynthetic carbon partitioning: its regulation and possibilities for manipulation. *Physiol. Plant.* **77**: 633-641.

Strain, B.R. & Cure, J.D. 1985. *Direct Effects of Increasing Carbon Dioxide on Vegetation.* DOE/ER-0238, US Department of Energy, Springfield, VA.

Stuiver, M. 1982. The history of the atmosphere as recorded by carbon isotopes. In Goldburg, E.D. (ed.) *Atmospheric Chemistry*: 159-179. Springer-Verlag, Berlin.

Surano, K.A., Daley, P.F., Houpis, J.L.J., Shinn, J.H., Helms, J.A., Palassou, R.J. & Castella, M.P. 1986. Growth and physiological responses of *Pinus ponderosa* Dougl. ex P. Laws to long-term elevated CO_2 concentrations. *Tree Physiol.* **2**: 243-259.

Tans, P.P., Fung, I.Y. & Takahashi, T. 1990. Observational constraints on the global atmospheric CO_2 budget. *Science* **247**: 1431-1438.

Telewski, F.W. & Strain, B.R. 1987. Densitometric and ring width analysis of 3-year-old *Pinus taeda* L. and *Liquidambar styraciflua* L. grown under three levels of CO_2 and two water regimes. In Jacoby, G.C. Jr & Hornbeck, J.W. (eds) *Proceedings International Symposium on Ecological Aspects of Tree-Ring Analysis*: 494-500. US Department of Energy, New York.

Thomas, J.F. & Harvey, C.N. 1983. Leaf anatomy of four species grown under continuous long-term CO_2 enrichment. *Bot. Gaz.* **144**: 303-309.

Thomas, J.F., Raper, C.D., Anderson, C.E. & Downs, R.J. 1975. Growth of young tobacco plants as affected by carbon dioxide and nutrient variables. *Agron.* J. **67**: 685-689.

Thomas, R.B. & Strain, B.R. 1991. Root restriction as a factor in photosynthetic acclimation of cotton seedlings grown in elevated carbon dioxide. *Plant Physiol.* **96**: 627-634.

Thomas, S.M. & Thorne, G.N. 1975. Effects of nitrogen fertilizer on photosynthesis and ribulose 1,5-diphosphate carboxylase activity in spring wheat in the field. *J. Exp. Bot.* **26**: 43-51.

Tissue, D.T. & Oechel, W.C. 1987. Responses of *Eriophorum vaginatum* to elevated CO_2 and temperature in the Alaskan tussock tundra. *Ecology* **68**: 401-410.

Tissue, D.T., Thomas, R.B. & Strain, B.R. 1993. Long-term effects of elevated CO_2 and nutrients on photosynthesis and Rubisco in loblolly pine seedlings. *Plant, Cell Environ.* **16**: 859-865.

Tolbert, N.E. & Zelitch, I. 1983. Carbon metabolism. In Lemon, E.R. (ed.) *CO_2 and Plants: The Response of Plants to Rising Levels of Atmospheric Carbon Dioxide*: 21-64. Westview Press, Boulder, CO.

Tolley, L.C. & Strain, B.R. 1984. Effects of CO_2 enrichment on growth of *Liquidambar styraciflua* and *Pinus taeda* tree seedlings under different irradiance levels. *Can. J. For. Res.* **14**: 343-350.

Tolley, L.C. & Strain, B.R. 1985. Effects of CO_2 enrichment and water stress on gas exchange of *Liquidambar styraciflua* and *Pinus taeda* seedlings grown under different irradiance levels. *Oecologia* **65**: 166-172.

Townend, J. 1993. Effects of elevated carbon dioxide and drought on the growth and physiology of clonal sitka spruce [*Picea sitchensis* (Bong.) Car.]. *Tree Physiol.* **13**: 389-399.

Van Oosten, J.-J. & Besford, R.T. 1994. Sugar feeding mimics effect of acclimation to high CO_2-rapid down regulation of Rubisco small subunit transcripts but not of the large subunit transcripts. *J. Plant Physiol.* **143**: 306-312.

Van Oosten, J.-J., Afif, D. & Dizengremel 1992. Long-term effects of a CO_2 enriched atmosphere on enzymes of the primary carbon metabolism of spruce trees. *Plant Physiol. Biochem.* **30**: 541-547.

Vitousek, P.M. 1991. Can planted forests counteract increasing atmospheric carbon dioxide? *J. Environ. Qual.* **20**: 348-354.

von Caemmerer, S. & Farquhar, G.D. 1984. Effects of partial defoliation, changes in irradiance during growth, short-term water stress and growth at enhanced $p(CO_2)$ on the photosynthetic capacity of leaves of *Phaseolus vulgaris* L. *Planta* **160**: 320-329.

Vu, J.C.V., Allen, L.H. & Bowes, G. 1989. Leaf ultrastructure, carbohydrates and protein of soybeans grown under CO_2 enrichment. *Environ. Exp. Bot.* **29**: 141-147.

Warrick, R.A., Gifford, R.M. & Parry, M.L. 1986. CO_2, climate change and agriculture. Assessing the response of food crops to the direct effects of increased CO_2 and climatic change. In Bolin, B., Doos, B.R., Jager, J. & Warrick, R.A. (eds) *The Greenhouse Effect, Climatic Change, and Ecosystems*, Scope **29**: 393-473. Wiley, Chichester.

Wigley, T.M.L., Angell, J.K. & Jones, P.D. 1985. Analysis of the temperature record. In MacGracken, M.C. & Luther, F.M. (eds) *Detecting the Climatic Effects of Increasing Carbon Dioxide*: 55-90. US Department of Energy, Washington DC.

Wong, S.C. 1979. Elevated atmospheric partial pressure of CO_2 and plant growth. *Oecologia* **44**: 68-74.

Woodrow, I.E. 1994. Optimal acclimation of the C_3 photosynthetic system under enhanced CO_2. *Photosyn. Res.* **39**: 401-412.

Woodrow, I.E. & Berry, J.A. 1988. Enzymatic regulation of photosynthetic CO_2 fixation in C_3 plants. *Annu. Rev. Plant Physiol. Plant Mol. Biol.* **39**: 533-594.

Woodrow, I.E., Ball, J.T. & Berry, J.A. 1990. Control of photosynthetic carbon dioxide fixation by the boundary layer, stomata and ribulose bisphosphate carboxylase/oxygenase. *Plant, Cell Environ.* **13**: 339-347.

Woodward, F.I. 1987. Stomatal numbers are sensitive to increases in CO_2 from pre-industrial levels. *Nature* **327**: 617-618.

Woodward, F.I. 1992. Predicting plant responses to global environmental change. *New Phytol.* **122**: 239-251.

Woodward, F.I. & Bazzaz, F.A. 1988. The response of stomatal density to CO_2 partial pressure. *J. Exp. Bot.* **39**: 1771-1781.

Woodwell, G.M., Hobbie, J.E., Houghton, R.A., Melillo, J.M., Moore, B., Peterson, B.J. & Shaver, G.R. 1983. Global deforestation: contribution to atmospheric carbon dioxide. *Science* **222**: 1081-1086.

Wullschleger, S.D., Norby, R.J. & Hendrix, D.L. 1992. Carbon exchange rates, chlorophyll content, and carbohydrate status of two forest tree species exposed to carbon dioxide enrichment. *Tree Physiol.* **10**: 21-31.

Yelle, S., Beeson, R.C., Trudel, M.J. & Gosselin, A. 1989a. Acclimation of two tomato species to high atmospheric CO_2. I. Sugar and starch concentrations. *Plant Physiol.* **90**: 1464-1472.

Yelle, S., Beeson, R.C., Trudel, M.J. & Gosselin, A. 1989b. Acclimation of two tomato species to high atmospheric CO_2. II. Ribulose-1,5-bisphosphate carboxylase/oxygenase and phosphoenolpyruvate carboxylase. *Plant Physiol.* **90**: 1473-1477.

Zangerl, A.R. & Bazzaz, F.A. 1984. The response of plants to elevated CO_2. II. Competition interactions between annual plants under varying light and nutrients. *Oecologia* **62**: 412-417.

Ziska, L.H., Drake, B.G. & Chamberlain, S. 1990. Long-term photosynthetic response in single leaves of a C_3 and C_4 salt marsh species grown at elevated atmospheric CO_2. *Oecologia* **83**: 469-472.

Ziska, L.H., Hogan, K.P., Smith, A.P. & Drake, B.G. 1991. Growth and photosynthesis response of nine tropical species with long-term exposure to elevated carbon dioxide. *Oecologia* **86**: 383-389.

5 Impacts of Air Pollutants and Elevated Carbon Dioxide on Plants in Wintertime

J.D. BARNES, M.R. HULL & A.W. DAVISON

INTRODUCTION

Plant responses to air pollution vary with season, with relatively short-term exposures to phytotoxic concentrations resulting in damage during the winter months when metabolically regulated detoxification processes are slow or lacking. Furthermore, sulphur-containing air pollutants, NH_3, and to a lesser extent O_3, can predispose plants to greater risk of winter injury. Our understanding of these subtle and complex interactions has improved dramatically over the past decade and it is now recognized that they may be of considerable ecological significance. However, there is still little known about the biochemical mechanisms that underlie effects. In this chapter we review what is known, and question the theory that increasing levels of atmospheric N deposition will increase the risk of winter damage. The same combustion sources that produce traditional air pollutants also inject CO_2 into the atmosphere, so the impact of rising atmospheric concentrations of CO_2 on plants at low temperatures is also discussed.

The generation of energy by burning fossil fuels in power-stations and engines, all manner of industrial processes, the biodegradation of wastes, and some farming and forestry operations leads to the release of thousands of different chemicals into the atmosphere. Most have little or no discernible effect on the environment because the concentrations are very low or because they are not toxic to biological systems; some, however, are known to have effects on human health, and the plants and animals of natural and managed ecosystems. Current concerns focus on the increasing introduction of carbon dioxide (CO_2), phytotoxic air pollutants (O_3, NO_x, NH_x, SO_2, NMHCs) and other trace gases (CH_4, CFCs) into the atmosphere. There is irrefutable evidence that the atmospheric concentration of many of these gases is increasing, and it is expected that this trend will continue into the 21st century. This has prompted widespread interest in the direct effects of these gases on plant growth, productivity and competition, and indirect effects via possible changes in climate. Because of the uncertainties regarding the extent of global and regional climate change (Idso, 1990), and the fact that there are likely to be direct effects of changing atmospheric composition on the plants and animals of natural and managed ecosystems before discernible

Plant Response to Air Pollution. Edited by Mohammad Yunus and Muhammad Iqbal

climate change is evident, this article will concentrate on what is known of the direct effects of rising atmospheric gas concentrations on plant growth and cold hardiness, and vice versa.

There is a wealth of literature documenting the effects of air pollutants and elevated CO_2 on the growth and productivity of crops and trees, and competition in natural and managed ecosystems. However, a vast majority of experiments have been conducted under controlled, favourable conditions, with the result that we have a poor understanding of the way in which combinations of factors influence plant growth. In reality, plants will be exposed to rising CO_2 against a perpetually changing background of air pollution, and to other stresses such as frost, drought or attack by pests and pathogens that may occur concurrently or be separated in time. Over the past decade it has become increasingly clear that a number of these stresses may modify plant response to air pollutants, CO_2, and vice versa. Indeed, in the case of a number of toxic air pollutants there are indications that such interactions may be just as, if not more, important than the direct effects of the pollutants themselves (see Bell, 1993). Although both CO_2 and air pollutants originate in a large part from the combustion of fossil fuels, differences at the temporal and spatial scale, and in the mode of action of the individual gases have resulted in a dissociation of research on the effects of rising CO_2 from that of the impact of air pollutants. Consequently, despite repeated research recommendations to the contrary, there have been few definitive experiments investigating interactions between CO_2 and air pollutants, and none of these have involved exposure under winter conditions. This article, therefore, has no alternative but to consider the impacts of air pollutants and rising CO_2 separately.

NATURE OF WINTER INJURY

Chilling injury

One of the major factors determining the geographical distribution and agricultural productivity of plants is the ability to withstand low-temperature stress. Many crop plants, particularly those from warmer climates, are sensitive to temperatures between 5 and 12°C. These chill-sensitive species exhibit restricted germination and growth, reduced yield, and even death following exposure to temperatures below a critical threshold for a matter of hours (Lyons, 1973). The consensus view is that chilling injury arises as a result of inactivation of membrane-bound enzymes caused by (a) membrane lipid phase transition from a normal liquid-crystalline form to a solid gel state; (b) a decrease in enzyme-complex free energy below that required for catalytic activation; and (c) protein denaturation. Of these three factors, lipid-phase transitions are considered to be the most important. The critical temperature at which the molecular reordering of the lipids takes place is believed to be governed by the degree of unsaturation of certain membrane lipid fatty acids (Wilson & Crawford, 1974). However, an alternative hypothesis for the basis of chilling sensitivity has been proposed by Wilson (1976), based upon his own work with beans (*Phaseolus vulgaris*). His work suggested that chilling injury was primarily due to excessive water loss as a result of chill-induced prevention of stomatal closure, combined with a decreased permeability of the roots to water. However, despite more than a decade of research the mechanisms underlying chilling injury are still not completely understood. Although chilling injury is of considerable economic importance, we are unaware of any studies that have specifically investigated whether exposure to air pollutants can modify the risk of chilling injury in plants growing close to their critical temperature. Many crop plants grown in Europe, the United States and Asia are at risk from chilling injury, but it is the effects of frost that have captured the attention of most researchers.

Freezing injury

Freezing injury occurs principally as a consequence of extracellular ice formation in plant tissues. This results in freeze-induced dehydration of cells, a loss of turgor, concentration of cell solutes and denaturation of proteins. Continued dehydration can result in freeze-induced cell plasmolysis which is associated with specific "lesions" (expansion-induced lysis, lamellar-to-hexagonal$_{II}$ phase transitions, fracture-jump lesions) in the plasma membrane which vary depending on the stage of acclimation and the temperature to which cells are cooled (Steponkus *et al.*, 1992). Prolonged periods at subzero temperatures can also result in depletion of stored cellular metabolites, whilst ice encasement can lead to damage resulting from anoxia (Levitt, 1980). Plants that overwinter in temperate regions undergo an orchestrated series of metabolic and physiological changes in response to environmental stimuli (low temperatures, short days) that minimize the effects of extracellular ice formation. Although the exact nature of this "hardening" process remains to be established, it is known to involve changes in membrane lipid composition, changes in the content and composition of cytoplasmic lipids, accumulation of a number of low molecular-weight metabolites (e.g., raffinose, stachyose, sucrose, amino acids, organic acids) and an increase in the amount of bound-hydrate water on the surface of cytoplasmic protein molecules (see Wellburn *et al.*, 1996). These metabolic changes ultimately contribute to increased cryostability of cellular membranes, destabilization of which is the primary cause of freezing injury. The rate at which hardening occurs and the timing of spring dehardening vary with species. These differences are important because most often injury occurs as a result of early or late frosts, rather than as a result of lower temperatures in midwinter. For example, evergreen trees that are highly resistant to freezing when dormant in the winter can be seriously damaged or killed by early (autumn) and late (spring) frosts. In some sensitive species, freezing and, therefore, injury is facilitated by certain strains of bacteria that produce ice-nucleating proteins. Although these bacteria play a vital role in determining freezing resistance in these species (Lindow *et al.*, 1978, 1982), little is known of their significance to a wider range of plants. The possibility that phytotoxic air pollutants or rising CO_2 may influence such bacteria has not been examined.

Winter desiccation

Winter desiccation is another important facet of winter injury, and in the field symptoms are very similar to those resulting from frost damage *per se*. In evergreens, the relative importance of freezing injury and desiccation is still a subject of controversy. In many instances, damage assumed to be the direct effect of low temperature in experimental studies may have been confused with the effects of desiccation (Lucas & Penuelas, 1990). All plants exposed to winter conditions are potentially at risk from desiccation, but effects are most often observed in evergreens, and, in extreme cases, deciduous species. Injury occurs when water loss exceeds the supply. Freezing of the soil and/or water in the conducting vessels interrupts the supply of water to the aerial parts of the plant and consequently water lost to the atmosphere cannot be replenished. In evergreens, resistance to winter desiccation is effected through stomatal closure which accompanies the onset of dormancy. However, even in midwinter, rapid changes in air temperature and cold drying winds can induce stomatal re-opening (Tranquillini, 1982). Consequently, although the cuticle forms the main barrier to winter water loss (see Davison & Barnes, 1986) and the rate of winter transpiration is generally very low, there are certain conditions when it may exceed the rate at which water can be replaced. When the critical minimum water content is reached, further desiccation results in irreversible injury. Severe damage of this type results in necrosis that is not apparent until the spring when the temperature rises.

Photodynamic damage

When temperatures are low during winter, the rate of net photosynthesis approaches zero and thus the primary sink for electrons from photosystem I is removed. Consequently, periods of high irradiance during winter can result in extensive photodynamic damage to the chloroplast if energy from absorbed photons is not dissipated (Gillies & Vidaver, 1990). Cold-acclimated species have two strategies for overcoming this problem: (a) decreased efficiency of photosystem II (photoinhibition) which provides a mechanism for the thermal and radiative dissipation of excess excitation energy from the photosynthetic apparatus (Gillies & Vidaver, 1990); (b) increased activity of endogenous scavengers of activated oxygen and hydrogen peroxide; cold acclimation has been associated with increased levels of ascorbate, ascorbate peroxidase, dehydroascorbate reductase, glutathione, glutathione reductase and superoxide dismutase (see Jahnke *et al.*, 1991; Barber & Andersson, 1992). Despite the fact that gaseous air pollutants are known to stimulate the production of free radical species in plant tissues, and trigger the plant's natural defences against oxidative stress, we are unaware of any studies that have specifically investigated the effects of air pollutants, or elevated carbon dioxide on the sensitivity of overwintering plants to photodynamic damage.

INTERACTIONS BETWEEN AIR POLLUTANTS AND WINTER STRESS

Airborne sulphur

Sulphur dioxide is undoubtedly the most important and widespread sulphur-containing air pollutant. However, there are others (in particular hydrogen sulphide and dimethylsulphide), the importance and terrestrial effects of which are largely unknown. These pollutants are deposited on terrestrial ecosystems via three mechanisms, namely dry deposition, wet deposition and occult deposition. Dry deposition is the process by which pollutants are transferred from the air by impaction and uptake. Wet deposition is the process by which pollutants are removed by precipitation and transferred to the ground via gravitational processes (rain, snow). The third route, occult deposition, includes pollutant transfer in mist, fog and cloud water. In Britain, and other parts of Europe, there has been a decrease in sulphur emissions over the past decade, as a direct consequence of legislation, changes in fuel and warmer weather (Davison & Barnes, 1992). However, global SO_2 emissions from fossil fuel combustion continue to increase, possibly doubling between 1960 and 1980 (Brunold, 1990). This trend is most evident in industrially developing countries, where sulphur still presents a considerable problem.

The concentration of SO_2 varies both in time and space, reflecting variations in energy demand with season, dispersion and the rate of deposition and chemical transformation (UK-TERG, 1988). During the wintertime when demand for energy is high, atmospheric concentrations of SO_2 are typically in the order of 50% higher than the annual mean (Figure 5.1). Superimposed upon these seasonal variations are diurnal variations in the concentration in the order of ±10–20% of the daily mean, reflecting variations in energy demand and meteorological conditions. Consequently, SO_2 concentrations are at their highest when plants are undergoing metabolic changes that result in increased cold hardiness. During this period, there is also the greatest likelihood of freezing temperatures.

Many of the older field observations that relate to the effects of pollutants during winter are discussed elsewhere (Davison & Barnes, 1986; Davison *et al.*,

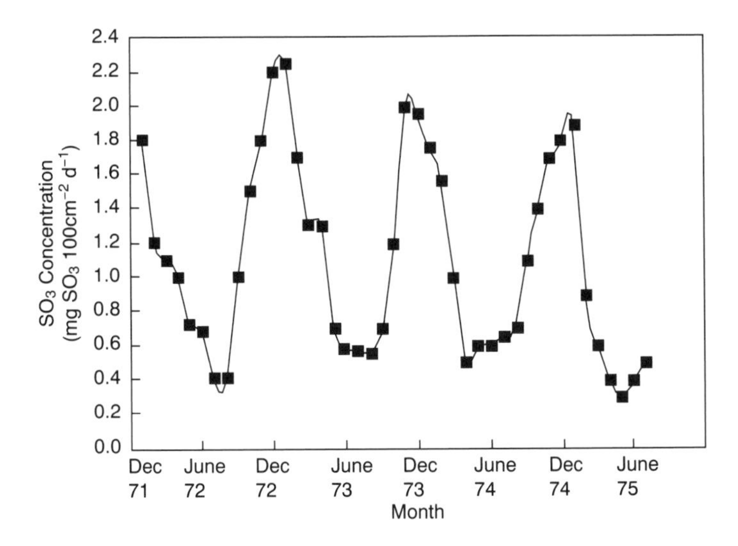

Figure 5.1. Graph showing seasonal trends in monthly mean SO_2 levels at an urban site in northeast England for the period December 1971 to December 1975. A smoothing procedure using 3-month moving averages has been applied to demonstrate seasonality (redrawn from Renner, 1990)

1987). Although difficult to evaluate because higher SO_2 and NO_x levels during the winter undoubtedly contributed to reports of greater pollution damage at this time of the year, taken as a whole they do provide circumstantial, although anecdotal, evidence that the effects of SO_2 are greatest during the winter months. Whitmore & Mansfield (1983) were among the first to demonstrate the phenomenon experimentally (Figure 5.2). When these researchers grew *Poa pratensis* throughout the year in replicated glasshouses receiving 68 nmol mol^{-1} SO_2, and 68 nmol mol^{-1} each of $SO_2 + NO_2$, there were clearly greater effects of the pollutants during the winter months. Laboratory-based studies have demonstrated that this effect is due to the low temperatures and low irradiance to which plants are exposed in wintertime (Davies, 1980; Jones & Mansfield, 1982). The low rate of assimilation that occurs under these conditions reduces the supply of carbon skeletons, reducing power and ATP, decreasing the capacity of the plant to detoxify the pollutant and/or repair damage. In support of this theory, significant amounts of sulphite and nitrite have been shown to accumulate in the extracellular fluid of plants exposed to $SO_2 + NO_2$ under winter conditions (Renner, 1990; Wolfenden *et al.*, 1991).

Early field observations established that there may be sufficient uptake of sulphur from the atmosphere during winter to produce effects at the time or later in the year (Materna, 1974; Keller, 1978, 1981) and there is now a wealth of evidence showing that SO_2 predisposes plants to frost damage (see Davison & Barnes, 1986; Davison *et al.*, 1987). Among the first to report a controlled experiment in which plants showed an effect of SO_2 on frost resistance were Davison & Bailey (1982). They exposed perennial ryegrass (*Lolium perenne* L.) to 100 nmol mol^{-1} SO_2 under controlled conditions and found that treated plants suffered greater frost damage than those maintained in clean air. The effect

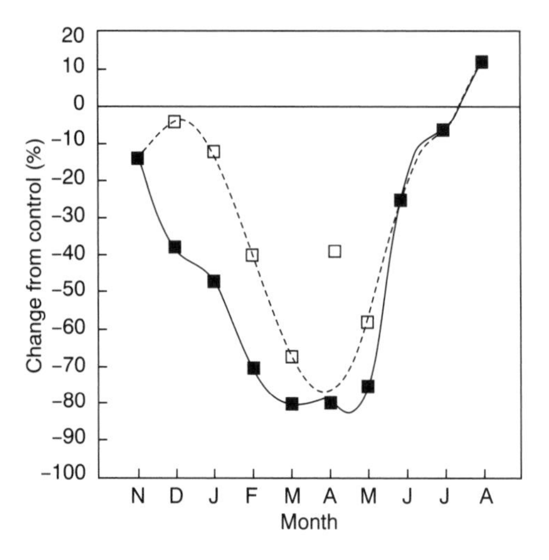

Figure 5.2. Seasonal differences in the effects of SO_2 (and $SO_2 + NO_2$) on the growth of *Poa pratensis*: □ 68 nmol mol^{-1} SO_2; ■ 68 nmol mol^{-1} SO_2 + 68 nmol mol^{-1} NO_2 (redrawn from Whitmore & Mansfield, 1983; by permission of Elsevier Science Limited)

depended on mineral nutrition; when soil levels of nitrogen and sulphur were limiting, there was no effect of SO_2 on frost hardiness. The same authors (reported in Davison & Barnes, 1986) showed that exposure of seedlings to 100 nmol mol^{-1} SO_2 for only two hours under hardening conditions significantly reduced post-freezing survival. There was also a significant interaction between SO_2 and the effect of low temperature on post-freezing growth of the root, but not shoot. Renner (1990) continued this work and Figure 5.3 shows some of her data relating to the post-freezing survival of ryegrass grown in sand culture and exposed to 100 nmol mol^{-1} SO_2 under hardening conditions. Controlled studies on conifers have shown that SO_2 can predispose buds, as well as needles, to frost damage (Keller, 1978; Freer-Smith & Mansfield, 1987). These reports of enhanced frost damage during controlled environment studies have been confirmed by observations made during winter exposure to SO_2 using field-release systems (see Davison *et al.*, 1987).

Other forms of sulphur deposition have also been shown to predispose plants to freezing injury. For example, exposure to 250 nmol mol^{-1} H_2S under hardening conditions has been shown to predispose seedlings of winter wheat (*Triticum aestivum* L.) to freezing injury (Stuiver *et al.*, 1992). As in the case with SO_2, this effect was accompanied by reduced shoot growth in plants exposed to H_2S at 3°C but not at 20°C. In addition, reduced frost hardiness in red spruce [*Picea rubens* Sarg. syn. *P. rubra* (Du Roi) Link] exposed to acidic mist seems to be related to the S:N content of the applied solution (Fowler *et al.*, 1989), with effects attributable to the SO_4 and NH_4 in the mist, but not the NO_3 (Cape *et al.*, 1991).

Despite the fact that it has been recognized for over a decade that SO_2 can predispose plants to freezing injury, the physiological and/or biochemical

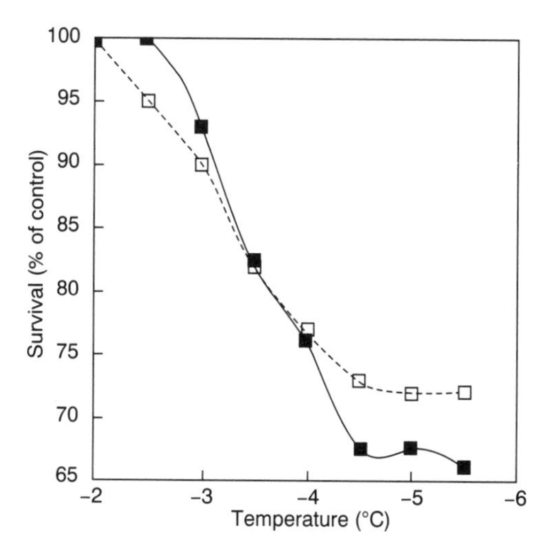

Figure 5.3. The effects of fumigation with 100 nmol mol^{-1} SO_2 on the post-freezing survival of *Lolium perenne*. Control plants were maintained in clean, charcoal-scrubbed air. The graphs show seedling survival as a percentage of control. Lines were calculated by probit analysis of the raw data. Data for two cultivars are presented: □ Peramo; ■ Augusta (redrawn from Renner, 1990)

changes that underlie this effect remain to be elucidated. One reason for this is the lack of fundamental information on the mechanisms underlying winter injury. However, several effects have been observed during overwinter exposures to sulphur-containing air pollutants that would not be commensurate with increased frost hardiness: (a) changes in the levels of cellular constituents (e.g., amino acids, soluble sugars) which are characteristic of an increase in frost tolerance are much less pronounced in plants exposed to SO_2, SO_2/NO_2 and H_2S under hardening conditions (Davison *et al.*, 1987; Renner, 1990; Stuiver *et al.*, 1992); (b) changes in the degree of unsaturation of chloroplast membrane lipid fatty acids, which are characteristic of an increase in the cryostability of cell membranes, are less pronounced in plants exposed to SO_2, and $SO_2 + NO_2$ (Wolfenden *et al.*, 1991; Antonnen, 1992); (c) energy storage compounds (e.g., triacylglycerol) which are used to fuel recovery processes after winter stress, are reduced in plants exposed to SO_2 (Malhotra & Khan, 1978; Antonnen, 1992); (d) free radicals produced as a result of the metabolism of sulphur-containing pollutants are known to result in oxidation of membrane-bound proteins (Heath, 1980); and (e) changes in the turgor relations of individual cells leading to the production of larger cells with stiffer cell walls and lower solute concentrations (Mansfield *et al.*, 1986; Freer-Smith & Mansfield, 1987). It is likely that some, if not all these changes contribute to the enhanced frost sensitivity of SO_2-treated plants. However, the relative importance of these effects remains to be established.

Sulphur dioxide can also predispose plants to winter desiccation, through effects on stomatal performance and changes in the proportional allocation of assimilate. Field evidence demonstrating that conifers growing in areas heavily polluted with SO_2 experience greater winter water deficits is reviewed elsewhere (Davison & Barnes, 1986; Davison *et al.*, 1987). More recently, controlled studies have indicated that exposure to SO_2 may seriously impair the ability of polluted leaves to effectively regulate their water loss. In some species exposure to low levels of SO_2 and NO_2 has been shown to increase stomatal conductance over the day, with the result that polluted leaves loose more water than their counterparts raised in clean air (see Mansfield *et al.*, 1990). SO_2 and NO_2-treated stomates show less response to the ABA formed within leaves when they are suddenly deprived of water (Neighbour *et al.*, 1988), or to ABA transported in to leaves via the xylem (Atkinson *et al.*, 1991). In addition, SO_2 and SO_2/NO_2 mixtures have a major effect on the relative distribution of growth above and below ground. In general, root growth is more severely affected than shoot growth (Wolfenden & Mansfield, 1990). The net result of these effects is that plants exposed to SO_2 and NO_2 deplete their soil-water reserves faster than plants maintained in clean air. This phenomenon persists under conditions of soil-water deficit with the result that plants exposed to SO_2 and NO_2 are more sensitive to drought and probably winter desiccation.

Ozone

Ozone (O_3) is present both in the troposphere and the stratosphere. In the troposphere it is a secondary pollutant, produced via a complex series of photochemical reactions involving reactive hydrocarbons and NO_x emitted into the atmosphere by natural processes and to a much larger extent by anthropogenic activities. As a result of increasing emissions of the precursors for O_3 production, there has been a marked increase in the background concentration of O_3 in the troposphere since the industrial revolution (Ashmore & Bell, 1991). There has also been a concurrent increase in the frequency and duration of photochemical episodes, and the concentration of O_3 in the troposphere continues to rise at a rate of $\approx$1–2% per annum (Hough & Derwent, 1990); a trend expected to continue in to the 21st century. This has caused concern because O_3 can adversely affect human health, and damage crops, trees and natural vegetation. Ambient levels of O_3 over widespread areas of the industrialized world are already known to be high enough to result in extensive visible damage to sensitive species, and to have adverse effects on the growth of plants in natural and managed plant communities (Krupa & Manning, 1988; UK-TERG, 1988; Davison & Barnes, 1992, 1993). Furthermore, O_3 is implicated as a factor contributing to the widespread decline of high-elevation forests in central Europe and eastern North America (Blank, 1985; McLaughlin, 1985; Barnes & Davison, 1988; Wellburn *et al.*, 1995).

For many years it was believed that O_3 in the troposphere was more or less inert. However, relatively recent advances in measurement techniques have demonstrated that there are pronounced spatial and temporal variations in tropospheric O_3 concentrations (Hov, 1984). This variability is partly due to natural processes, but anthropogenic sources of other pollutants also affect the distribution and abundance of O_3. Production in the troposphere is favoured by high temperatures and high irradiance, and at low altitudes there are marked diel and seasonal variations in O_3 concentrations. However, phytotoxic levels in Europe and the United States are generally restricted to the summer months (April–September). If O_3 is to affect winter hardiness under these circumstances, the effects of the pollutant must be persistent; effects of summer exposure to O_3 must carry over into the following winter and/or spring. There is growing evidence that this is indeed the case. Recent studies have shown that O_3-induced physiological and biochemical disturbances may persist for many months after exposure to the pollutant has terminated and possibly even persist from one year's summer exposure to the next (see Barnes *et al.*, 1990b; Eamus *et al.* 1990; Davison & Barnes, 1993).

At upland sites, which are better coupled to the free troposphere, the concentration of O_3 remains relatively high throughout the day and seasonal variations in O_3 concentration are less pronounced. Consequently, there is a greater likelihood of potentially phytotoxic concentrations occurring during the period that plants are hardening or dehardening, and less temporal separation between O_3 exposure and winter stress. In many areas of central and southern Europe, meteorological conditions and anthropogenic activities can result in O_3 generation throughout the year. In Athens, for example, where there is a severe photochemical oxidant problem (Velissariou *et al.*, 1992), potentially phytotoxic O_3 episodes can occur all year round (Velissariou, Barnes & Davison, unpublished). Under such conditions it is possible that plants are exposed, particularly during spring and autumn, to relatively high concentrations of O_3 during the day, and low (even freezing) temperatures at night. In these situations, O_3 damage may be exacerbated by low temperatures prior to, or following, exposure to the pollutant (Davis & Wood 1973; Pearson, Davison & Barnes, unpublished).

Although effects are generally not as dramatic as those observed with SO_2, there is growing evidence that winter injury is increased in plants exposed to O_3. The first report of an interaction between ozone and frost was made by scientists working at the laboratories of National Power in the UK, when a few trees of clonal Norway spruce (*Picea abies* [L.] Karst) that were left over after a long-term O_3 fumigation were inadvertantly exposed to an early out-of-season frost. The ensuing damage was restricted to the previous year's needles of certain clones that had previously been exposed to O_3 (Brown *et al.*, 1987). Since this observation exposure to O_3 under controlled conditions has been shown to enhance the frost sensitivity of conifers (Fincher *et al.*, 1989; Chappelka *et al.*, 1990; Edwards *et al.*, 1990; Fincher & Alscher, 1992), broadleaved trees (Eissenstat *et al.*, 1991) and herbaceous species (Barnes *et al.*, 1988b). Although some of this work is difficult to evaluate because the O_3 used may have been contaminated by oxides of nitrogen (Brown & Roberts, 1988), there is accumulating evidence that O_3 *per se* was responsible for the observed effects. Proof that O_3 affects the timing, rather than the ultimate extent of frost hardiness, was provided by Lucas *et al.* (1988). These authors observed a dose-related increase in frost damage in Sitka spruce (*Picea sitchensis*) in autumn following exposure to O_3 the

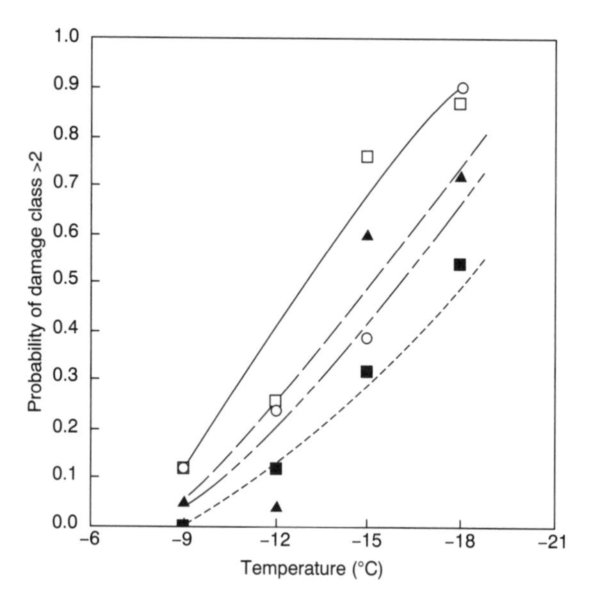

Figure 5.4. Relationship between needle damage and freezing temperatures for shoots of *Picea sitchensis*. An ordinal regression model was fitted to the data and observed data with temperature treated as a covariate against the probability of a damage score greater than two. The raw probability scale is presented. Ozone treatments are: ■ <5 nmol mol^{-1} O_3; ○ 70 nmol mol^{-1} O_3; ▲ 120 nmol mol^{-1} O_3; □ 170 nmol mol^{-1} O_3 (redrawn from Lucas *et al.*, 1988; courtesy of Cambridge University Press)

previous summer (Figure 5.4). This effect occurred in the absence of any effects on growth or visible symptoms during the fumigation. Although the mechanisms underlying the impact of O_3 on frost hardiness are not yet fully understood (see Wellburn *et al.*, 1995), a number of physiological and biochemical disturbances have been found that may contribute to the observed effects: (a) the increase in unsaturation of specific membrane lipids that is considered to be of central importance in the cold-hardening process is less pronounced in plants that have previously been exposed to O_3 (Wolfenden & Wellburn, 1991); (b) increases in low molecular weight sugars (i.e., ethanol-soluble sugars), and specific sugars (i.e., raffinose, stacchyose) that accrue in parallel with increasing cold hardiness are delayed in plants previously exposed to O_3 (Alscher *et al.*, 1989; Barnes *et al.*, 1990a); (c) increases in antioxidant compounds in winter in plants previously exposed to O_3 indicate greater oxidative stress (Hausladen *et al.*, 1990); (d) there are long-term effects of O_3 on pigment composition (Robinson & Wellburn, 1991).

As well as effects of O_3 on frost hardiness, there is circumstantial evidence to suggest that O_3 may increase the risk of winter desiccation. Ozone is known to have a major influence on the relative partitioning of assimilates between root and shoot, with the shoot generally gaining at the expense of the root (Cooley & Manning, 1987). This is reflected in a marked decrease in K (allometric root:shoot coefficient) in plants exposed to O_3. This effect may be exacerbated by impaired stomatal functioning and pollutant-induced changes in the epicuticular waxes that result in a reduced efficiency of O_3-treated leaves to regulate water loss. In conifers, O_3 can promote stomatal opening in current-year needles, while inducing simultaneous stomatal closure in previous-year needles (Barnes *et al.*, 1990c; Pfirrmann, 1992). Exposure to O_3 can also modify diel stomatal functioning, with the result that at night the stomata of polluted plants fail to close to the same extent as plants maintained in clean air (Skärby *et al.*, 1987; Barnes *et al.*, 1990a, b). Moreover, exposure to O_3 modifies stomatal response to water stress; when leaves are exposed to severe water deficits the stomata on leaves from O_3-treated plants do not close as rapidly as those on leaves from plants maintained in clean air (Barnes *et al.*, 1990c, Mansfield *et al.*, 1992; Barnes & Bahnemann, unpublished). The possible importance of these effects is underlined by an experiment conducted by Barnes & Davison (1988), where eight clones of Norway spruce were exposed to 120 nmol mol^{-1} O_3 for 70 days and then excised shoots frozen to temperatures between $-6°C$ and $-19°C$. In four clones severe post-freezing necrosis developed on the previous year's needles of plants previously exposed to O_3. At the time this damage was suggested to provide evidence that O_3 predisposed plants to freezing; however, the damage observed was different from frost damage observed in the field and the experiment was conducted late in the year when plants would have been expected to have already acquired considerable frost resistance. It therefore seems much more likely that the damage observed was in fact a result of O_3-promoted desiccation. The relative contribution of changes in the structure and chemical composition of the epicuticular waxes (Barnes *et al.*, 1988a, 1990c; Barnes & Brown, 1990; Percy *et al.*, 1992) to these effects remains to be established, but studies using isolated cuticles indicate that realistic concentrations of O_3 have little effect on the permeability of the cuticle to water (Kersteins & Lendzian, 1989). Furthermore, changes in the turgor relations of individual cells may, at least to some extent, compensate for an increased risk of cell dehydration in O_3-exposed plants during winter (Barnes *et al.*, 1990b).

Nitrogenous air pollutants

Over the past few decades, there has been a well-documented increase in the atmospheric concentration of a number of nitrogen (N)-containing compounds as a result of the combustion of fossil fuels and a shift toward intensive farming methods. Compounds include oxidized N-containing trace gases — principally nitric oxide (NO), nitrogen dioxide (NO_2), higher oxides of N (e.g., N_2O_5), nitric acid vapour (HNO_3); and reduced compounds — principally, nitrate aerosol (NA), gaseous ammonia (NH_3) and nitrate (NO_3^-) and ammonium (NH_4^+) in precipitation. Increasing atmospheric concentrations of these compounds have led to enhanced rates of N deposition and there is growing concern throughout Europe that ecosystems formerly limited by the availability of N are being affected by N deposited from the atmosphere. In the Netherlands, excess N deposition is believed to be responsible for a specific kind of regional forest decline as well as playing a central role in the transition of large areas of *Calluna*-dominated heathland into grassland dominated by nitrophilous species. One of the major effects of increasing atmospheric deposition of N is believed to be an increase in the sensitivity of plants to winter stress. Indeed, reduced frost hardiness induced by increased ecosystem loading with inorganic N has been accepted as a viable explanation for the decline of red spruce at high elevations in eastern North America (Friedland *et al.*, 1985) and may be a factor contributing to the widespread decline of coniferous forests in central Europe (Nihlgard, 1985). But what proof is there for endorsement of the concept that increasing atmospheric N deposition will predispose plants to greater risk of winter injury? To date, few experiments have specifically investigated the effects of N loading from the atmosphere on terrestrial ecosystems, so a major source of information has been the myriad of experiments that have investigated the effects of different rates of fertilizer N on plant growth and cold hardiness. Data depicting the effects of fertilizer N on plants' cold hardiness are summarized in Table 5.1. However, there are many problems in extrapolating from these studies to the effects of long-term low-level inputs of atmospheric N on cold tolerance: (a) it is often impossible to differentiate between different forms of N (i.e., nitrate vs. ammonium) because of the nature of the applied fertilizer; (b) many early reports were based solely on field observations of overwinter survival, using crops or trees with past histories of high fertilizer applications preceding severe winters; (c) many early studies did not characterize the nutrient status of plant tissue; (d) effects of N on frost hardiness may be obscured or misinterpreted as a result of deficiencies in other elements, e.g., boron concentrations are negatively related to N concentration and boron deficiency is known to reduce frost hardiness (Cooling, 1967); (e) cause and effect relationships are often inconclusive, e.g., in experiments where conifers have been shown to be more frost sensitive following N fertilization there is no clear relationship between frost hardiness and the

Table 5.1. Influence of nitrogen fertilization on the ability of plants to tolerate cold temperatures (modified from Pellett & Carter, 1981)

Species	Reference	Cold tolerance reaction
Gymnosperms		
Abies balsamea	†Salonius (1981)	O
Abies grandis	*Benzian *et al.* (1974)	O
Juniperus chinensis	*Pellet & White (1969)	O
Picea abies	Puempel *et al.* (1975)	—
	*Aldhous (1972)	+O
	†Tamm (1968)	O
Picea rubens	DeHayes *et al.* (1989)	++
	†Safford *et al.* (1977)	O
	†Sheppard & Shottafer (1979)	O
	Hirondelle *et al.* (1992)	++
Picea sitchensis	*Benzian (1967)	+
	*Benzian & Freeman (1967)	+
	*Benzian *et al.* (1974)	—
Pinus sylvestris	†Radwan & deBell (1980)	O
	*Hellergren (1981)	— O
Pinus taeda	†Penning de Vries *et al.* (1975)	O
	†Miller (1986)	O
Pseudotsuga menziesii	†Gessel & Walker (1956)	O
Tsuga heterophylla	*Benzian (1967)	+
	*Benzian *et al.* (1974)	O
	Aronsson (1980)	—
Angiosperms-Monocots		
Agrostis alba	*Carroll (1943)	O
Agrostis canina	*Carroll (1943)	—
Agrostis tenuis	*Carroll (1943)	—
Anthoxanthum odoratum	*Carroll (1943)	—
Avena sativa	*Dexter (1935)	—
Cynodon dactylon	*Adams & Twersky (1960)	—
	*Alexander & Gilbert (1963)	—
	*Gilbert & Davis (1971)	—
Cynosurus cristatus	*Carroll (1943)	—
Dactylis glomerata	*Howell & Jung (1965)	—
Festuca rubra	*Carroll (1943)	—
Hordeum vulgare	*Dexter (1935)	—
Lolium multiflorum	*Carroll (1943)	—
L. multiflorum × *L. perenne*	*Baker & Davis	—
Lolium perenne	*Carroll (1943)	—
Poa annua	*Carroll (1943)	—
Poa compressa	*Carroll (1943)	—
Poa nemoralis	*Carroll (1943)	—
Poa pratensis	*Carroll (1943)	—
	*Carroll & Welton (1939)	—
	*Ferguson (1966)	+
Poa trivialis	*Carroll (1943)	—

continued overleaf

Table 5.1. (*continued*)

Species	Reference	Cold tolerance reaction
Secale cereale	*Dexter (1935)	–
Triticum aestivum	*Dexter (1935)	–
	*Kuksa (1939)	–
	*Slavnyi & Musienko (1972)	–
Dicots – Temperate woody		
Cornus sericea	*Pellet (1973)	O
Forsythia × intermedia	*Pellet (1973)	– O
Ilex crenata	*Havis *et al.* (1972)	O
Malus domestica	*Rawlings & Potter (1937)	–
	*Tingley *et al.* (1938)	– O
	*Sudds & Marsh (1943)	–
	*Way (1954)	–
	*Edgerton (1957)	–
Prunus cerasus	*Kennard (1949)	O
Prunus persica	*Chandler (1913)	+
	*Crane (1924)	–
	*Cullinan (1931)	+
	*McCunn & Dorsey (1935)	O
	*Higgins *et al.* (1943)	+O
	*Edgerton & Harris (1950)	O
	*Proebsting (1961)	+
Pycanthra coccinea	*Kelley (1972)	–
Rhododendron sp.	*Joiner & Ellis (1964)	O
Ribes uva-crispa	*Faustov (1965)	–
Rosa sp.	*Carrier (1953)	–
Vitis labrusca	*Gladwin (1917)	O
Dicots – Subtropical plants		
Aleurites fordii	*Brown & Potter (1949)	+
	*Sitton (1954)	+ –
Citrus limon	*Bunina (1957)	+
Citrus reticulata	*Bedrikovskaja (1952)	+
Citrus sinensis	*Smith & Rasmussen (1958)	+
Litchi chinensis	*Young & Noonan (1959)	O
Sansevieria trifas	*Marlatt (1974)	–
	*Conover & Poole (1977)	–
Solanum tuberosum	*Drozdov & Sycheva (1965)	+
Dicots – Temperate herbaceous plants		
Brassica oleracea	*Boswell (1925)	–
	*Kimball (1927)	–
	*Dexter (1935)	–
	*Ragan & Nylund (1977)	–
Fragaria	*Zurawicz & Stushnoff (1977)	–

\+ = Greater cold tolerance from increasing levels of N.
— = Less cold tolerance with increasing levels of N.
O = No influence on cold tolerance from increasing levels of N.
Note: More than one symbol indicates more than two concentrations of N and indicates that results differed between concentrations, or that more than one experiment was reported with differing results.
Note: See Klein *et al.* (1989) for references marked by a dagger (†), and Pellett & Carter (1981) for those marked by an asterisk (*).

N content of needles (Aronsson, 1980); (f) N has often been added to conifers after the cessation of growth in autumn. We are aware of only a few studies that have specifically investigated the effects of simulated long-term inputs of atmospheric N (by repeated application of ammonium sulphate) on plant cold hardiness. Without exception these studies have been conducted with *Calluna vulgaris*, presumably reflecting current interest in the factors responsible for the dieback of this species in the Netherlands. Uren and Ashmore (pers. comm.) applied ammonium sulphate (10 kg N ha^{-1} yr^{-1}) to fully randomized blocks in a *Calluna*-dominated heath in southern England, but found no significant effects on frost hardiness in the autumn or the following spring despite increasing application rates to 15 and 20 kg N ha^{-1} yr^{-1} over the winter period. Yet, there was a stimulation of growth and more flowers were produced in the higher N treatments. In contrast, higher rates of N (ammonium sulphate; 50, 100 and 150 kg N ha^{-1} yr^{-1}) applied to *Calluna* at an upland site in North Wales have been shown to have no effect on frost hardiness in the autumn, but earlier growth was stimulated and this new season's growth was especially frost sensitive (Caporn *et al.*, 1990). Similar effects of simulated N deposition on the frost hardiness of the new season's growth in *Calluna* have been reported in controlled studies conducted in the Netherlands (IPO, 1990).

A critical assessment of the available literature indicates that the impacts of foliar-derived N on plant cold hardiness are equally as controversial as the effects of N taken up from soil. Although plants may be more sensitive to nitrogenous air pollutants when they are growing slowly in wintertime (see the previous section), consistent decreases in frost hardiness have only been observed in plants exposed to NH_3. Increased foliar N as a result of the uptake of NO_2 has been shown to have little impact on cold hardiness. A recent detailed study, in which several cultivars of *Lolium perenne* were exposed to NO_2 (and/or SO_2) under simulated autumn conditions has demonstrated no effect of NO_2 on frost hardiness (even at high concentrations), despite some stimulation of growth (Renner, 1990). Similarly, Freer-Smith & Mansfield (1987) reported only marginal effects of 30 ppb SO_2 and/or 30 ppb NO_2 on the frost hardiness of *Picea sitchensis* following prolonged exposure to the pollutants, although budbreak in trees exposed to NO_2 occurred earlier which could enhance their sensitivity to spring frosts. As NO_2 generally co-exists with SO_2 in polluted atmospheres, many studies have investigated the combined effects of the two pollutants, which means that it is often impossible to establish cause–effect relationships for the individual gases. However, it is interesting to note that interactions between sulphur deposition and frost hardiness have been shown to be dependent upon soil N status (Davison & Bailey, 1982; L'Hirondelle *et al.*, 1992). There may also be important complex effects of combinations of pollutants on frost hardiness. For instance, Neighbour *et al.* (1990) reported that frost damage in *Picea rubens* exposed to O_3 in glasshouses occurred only in trees grown in charcoal-filtered air (contaminated with NO) but not in purafil-filtered air (reduced NO).

In contrast to the effects of NO_2, there is accumulating evidence that exposure of plants to elevated levels of NH_3 can predispose plants to winter stress. In *Calluna*, it has been shown that overwinter survival is significantly lower in plants exposed to long-term mean concentrations of 75 and 150 nmol mol^{-1} NH_3 (Dueck, 1990) and recent experiments have shown that NH_3 affects both the rate of hardening and dehardening in this species (IPO, 1990; Uren & Ashmore, pers.comm.). However, the underlying mechanisms remain to be elucidated. Increased frost sensitivity in these experiments was associated with a significant increase in the N content of plant tissues, but there were also increases in the N/K, N/Mg and N/Ca ratios which could be of importance. Ammonia has also been shown to predispose needles of Scots pine (*Pinus sylvestris* L.) to frost injury (IPO, 1990). When frost hardiness was assessed in the spring after a 5-month exposure to NH_3, electrolyte leakage was found to increase significantly at −10°C, but not at −4 or −7°C (Figure 5.5).

In addition to the uptake of gaseous nitrogen and sulphur, into foliage via stomata, these ions may also be taken up from the soil or from the thin film of water covering leaf surfaces following precipitation events. Ion exchange processes on leaf surfaces may be influenced by the effects of dry and wet deposited pollutants on epicuticular wax structure (Barnes & Brown, 1990; Percy *et al.*, 1992). Evidence that wet-deposited N may influence frost hardiness is equivocal. Several experimental studies have demonstrated that long-term high inputs of N in acidic mist may have beneficial effects on the frost hardiness of red spruce (DeHayes *et al.*, 1989; L'Hirondelle *et al.*, 1992), whilst others have found no effects (Klein *et al.*, 1989). In contrast, reduced frost hardiness has been found in studies that have exposed red spruce seedlings to mist containing $(NH_4)_2SO_4$ and HNO_3 (Fowler *et al.*, 1989; DeHayes *et al.*, 1991). But the effect may be due to

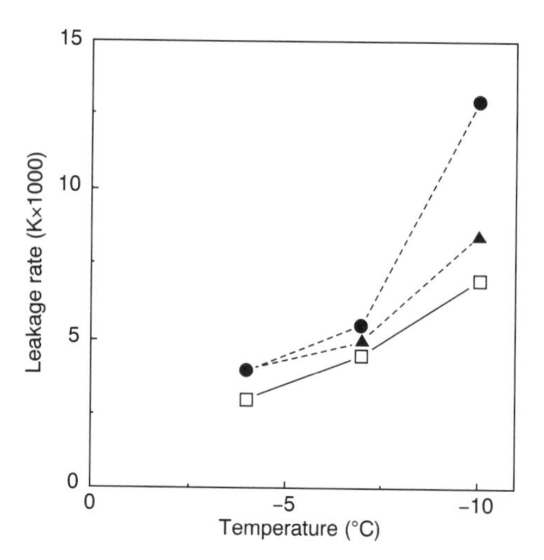

Figure 5.5. Rate of electrolyte leaching from current-year needles of Scots pine (*Pinus sylvestris* L.) in April 1988 following fumigation with ambient air or ambient air supplemented with NH_3, and exposure to freezing temperatures (−4, −7 and −10°C). ▲ 75 nmol mol^{-1} NH_3, ● 150 nmol mol^{-1} NH_3, and □ ambient, non-filtered air (redrawn from IPO, 1990)

the sulphate and ammonium content of the applied solutions, rather than nitrate (Cape *et al.*, 1991). Indeed, the nitrate content of mist may even mitigate the detrimental effects of sulphate and ammonium on frost hardiness (Cape *et al.*, 1991). The differential effects of ammonium and nitrate on frost hardiness observed in these experiments could reflect differences in the rate of uptake of the different ions from solution; the rate of foliar uptake of $^{15}NH_4^+$ may be as much as five times that of $^{15}NO_3^-$ (Bowden *et al.*, 1989).

In contrast to the effects of N on plant cold hardiness, there is evidence, albeit largely circumstantial, that fertilization may increase sensitivity to winter desiccation. Increased N supply, whether through foliar or root uptake, results in a reallocation of available photosynthate with the shoot favoured at the expense of root. Furthermore, there is growing evidence that combinations of $SO_2 + NO_2$, gaseous NH_3 and acid mist (containing nitrate and/or ammonium) can impair stomatal performance, such that the ability of exposed shoots to conserve water is seriously reduced (Neighbour *et al.*, 1988; Barnes *et al.*, 1990b, c; IPO, 1990). This results in a more rapid decline in shoot water potential in treated plants under conditions of soil-moisture deficit (IPO, 1990; Uren & Ashmore, pers. comm.). In addition, exposure to NO_x is reported to increase cuticle permeability by a factor of five (Lendzian, 1984). Imbalanced growth of roots and shoots, combined with reductions in the ability of treated shoots to conserve water would be expected to increase plant sensitivity to soil-water deficit and possibly desiccation during the winter months. On the other hand, changes in cellular water relations that would favour survival under conditions of cellular dehydration have also been observed in plants exposed to dry and wet deposition of N-containing compounds. Trees exposed to NO_2 or mist containing sulphate and/or nitrate have been shown to display a decrease in maximum turgor ($\psi_{p\,max}$) at full hydration and an increase in cell-wall elasticity (Freer-Smith & Mansfield, 1987; Eamus *et al.*, 1989; Barnes *et al.*, 1990c). This would be expected to allow cells to shrink more during drying, resulting in smaller decreases in turgor for a given loss of water compared with more rigid tissues–an effect that would be expected to favour survival in times of severe cell dehydration.

IMPACT OF RISING CO_2 ON PLANTS IN WINTER

The concentration of carbon dioxide in the atmosphere $[CO_2]_a$ has been rising at a steadily increasing rate since the industrial revolution. There is a consenus of opinion that the pre-industrial atmospheric CO_2 concentration was $\approx 275\ \mu mol\ mol^{-1}$ dry air, and this has already increased by $\approx 25\%$ to the present-day level of $354\ \mu mol\ mol^{-1}$ (Keeling, 1986). The increase continues unabated at a rate of $1.5\ \mu mol\ mol^{-1}$ per annum, and $[CO_2]_a$ is expected to attain a concentration of $530\ \mu mol\ mol^{-1}$ by 2050 AD and possibly $700\ \mu mol\ mol^{-1}$ by the end of the 21st century (Conway *et al.*, 1988). Although the global carbon cycle is complex and involves large exchanges between the atmosphere, the oceans and the biosphere, most of this rise is attributable to anthropogenic activities

(Rotty & Marland, 1986). As it is widely accepted that this change in atmospheric composition has profound implications for plant growth and productivity, plant responses to elevated CO_2 have attracted considerable research interest (Lemon, 1983; Bazzaz, 1990; Kimball *et al.*, 1990). It is not, however, the purpose of this article to provide a detailed account of the impact of rising CO_2 on C_3 species, as it is considered in detail within this book and elsewhere (Lemon, 1983; Strain & Cure, 1985; Kimball, 1986; Bazzaz, 1990; Kimball *et al.*, 1990; Anonymous, 1991; Eamus, 1991). Discussion in this section will concentrate on the interactive effects of elevated CO_2 and low temperature on plants.

Different physiological and biochemical responses to elevated CO_2 have been observed at the leaf, plant, community and ecosystem levels of organization (Long, 1991; Long & Hutchin, 1991). This knowledge has led to a more mechanistic approach in modelling the response to elevated $[CO_2]_a$ on a global scale, as opposed to previous extrapolations of primary production from measured standing crops (Long & Hutchin, 1991). However, despite this, the vast majority of experiments on the effects of elevated CO_2 have been conducted on plants raised under optimal experimental conditions with little attention paid to interactions with other important environmental variables (e.g., temperature, light, UV-B radiation, water and nutrient availability and gaseous pollutants). Consequently, present model predictions almost certainly overestimate the impact of increasing concentrations of CO_2, and it will not be possible to refine current models until the response of plants to combinations of environmental variables are understood (Barnes & Pfirrmann, 1992). Of particular concern is that increased concentrations of CO_2 together with a number of trace gases will result in an increase in mean global temperature (Schneider, 1989). However, the extent of this increase has been widely debated (Idso, 1990). In addition, some climate models predict increased variation in temperatures on a regional scale, such that differences between maximum and minimum yearly temperatures may increase sharply as well as the variation in temperature from day to day (Wigley, 1985; Balling & Idso, 1990). One possible scenario for a high CO_2 world is an increased frequency and intensity of extreme climatic events, including early and late frosts (Eamus, 1991). These fluctuations in temperature patterns, together with a higher atmospheric CO_2 concentration, are likely to have important implications for the future productivity of natural and managed plant communities (Bazzaz, 1990).

As a result of the effects of temperature on photosynthetic rates, rising $[CO_2]_a$ will cause greater stimulations in photosynthesis and growth at higher temperatures (Cure, 1985; Kimball, 1986; Allen *et al.*, 1990; Long, 1991). In C_3 plants, both temperature and CO_2 are expected to influence the rate of CO_2 assimilation (A) via effects on the enzyme ribulose bisphosphate carboxylase/oxygenase (Rubisco; EC 4.1.1.39) (Sage *et al.*, 1990), the capacity to regenerate ribulose-1,5-bisphosphate (RuBP) (Farquhar & Von Caemmerer, 1982) and the capacity to synthesize starch and sucrose (Sharkey & Vandeveer, 1989). Rubisco catalyses two reactions. One involves carboxylation of RuBP and produces two molecules of glycerate-3-P, whereas the other involves oxygenation of RuBP and produces one molecule of glycerate-3-P and one molecule of glycollate-2-P (photorespiration). Increased temperature favours oxygenation by decreasing, relative to O_2, both the solubility of CO_2 and the specificity of Rubisco for CO_2 (Jordan & Ogren, 1984). For any given CO_2 concentration, increase in temperature will increase the rate of oxygenation relative to carboxylation, so that the proportion of potential photosynthesis lost to photorespiration will increase with temperature (Long, 1991). Consequently, as well as the direct effects of temperature on leaf photosynthetic rates, there will be interactive effects of rising CO_2 and temperature at the level of primary carboxylation which will result in an increase in the temperature optimum of photosynthesis by as much as 5°C (Long, 1991).

At the leaf level, species grown long term in elevated CO_2 have often shown a decrease in the *in vivo* carboxylation efficiency, correlated with a decrease in the extractable activity of Rubisco (Sage *et al.*, 1989; Rowland-Bamford *et al.*, 1991). Long (1991) examined the significance of these decreases by modelling the decrease in V_{cmax} (the maximum RuBP saturated rate of carboxylation) and V_{omax} (the maximum RuBP saturated rate of oxygenation) over a range of temperatures. It was found that when leaf temperatures exceeded 22.5°C, even with a decrease in total activity by 40% of controls, A_{sat} would still be greater in leaves

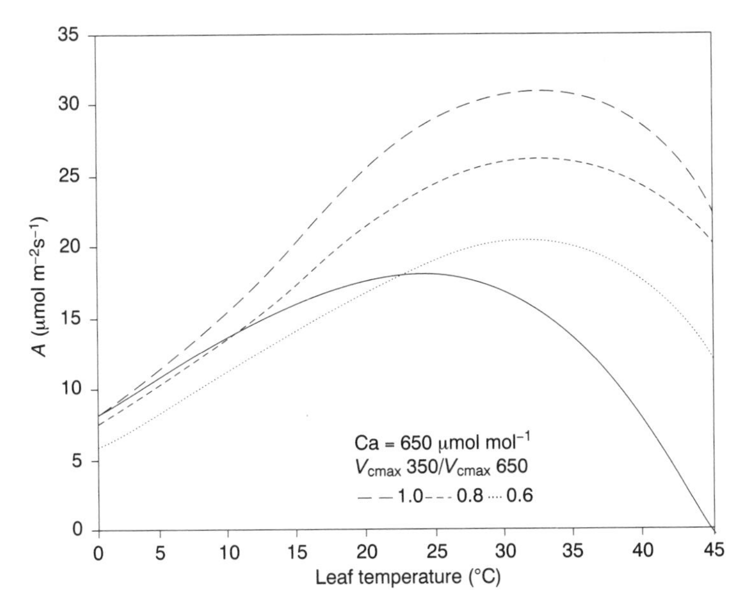

Figure 5.6. Predicted light saturated rates of leaf CO_2 uptake (A_{sat}) at the current atmospheric CO_2 concentration $[CO_2]_a$ of 350 μmol mol^{-1} and with an elevated $[CO_2]_a$ of 650 μmol mol^{-1}. The effects of decreased Rubisco activity with acclimation to the elevated $[CO_2]_a$ are simulated by decreasing the maximum RuBP saturated velocities of carboxylation and oxygenation to 0.8 (a 20% decrease) and 0.6 (a 40% decrease) of the values assumed for plants grown at $[CO_2]_a$ of 350 μmol mol^{-1} ($V_{cmax,650}/V_{cmax,350}$) (redrawn from Long, 1991; courtesy of Blackwell Scientific Publications)

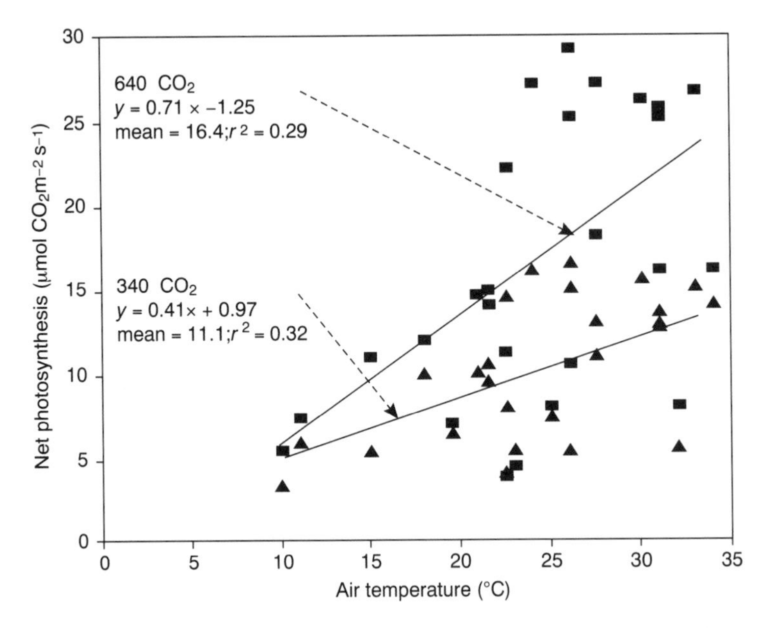

Figure 5.7. Relationship between net photosynthesis and temperature for water-lilies grown at 340 (▲) and 640 (■) μmol CO_2 mol^{-1} air. Correlation coefficients for both lines are significant at the 0.01 level. Slopes of the two fitted lines are significantly different ($P < 0.01$; 25 d.f.) (redrawn from Allen *et al.*, 1990; courtesy of Elsevier Science Limited)

at 650 ppm CO_2 than at 350 ppm (Figure 5.6). However, where temperature was lower than 22.5°C, the depression of carboxylation velocity (V_c) due to decreased Rubisco activity, exceeded the increase attributable to inhibition of oxygenation by elevated CO_2 (Long, 1991). Consequently, if acclimation to elevated CO_2 does result in reduced activity of Rubisco in the field, a matter currently of some debate (Arp, 1991), then at low temperatures rising CO_2 could result in decreased rates of leaf photosynthesis (Long, 1991). In support of these model predictions experimental work has shown the stimulatory effect of CO_2 on photosynthesis and growth to be strongly temperature dependent (see Figure 5.7) with the benefits of CO_2 enhancement increasing linearly with increasing temperature (Idso *et al.*, 1987; Allen *et al.*, 1988, 1990; Idso & Kimball, 1989).

Furthermore, data consistent with the modelled effects of elevated CO_2 on photosynthesis at low temperatures have been obtained in experiments. For example, Idso *et al.* (1987) studying the effects of elevated CO_2 over a range of temperatures in five species (radish, carrot, cotton, water hyacinth and water fern) showed there were minimal effects of CO_2 on growth at lower temperatures, and in some species there was evidence of a detrimental effect on growth at temperatures below 18.5°C. More recent studies on the temperature response of two temperate species (radish and carrot) have indicated similar interactive effects of CO_2 and temperature on growth, but the temperature below which CO_2 enhancement became detrimental for growth was lower; 11–12°C (Idso &

Kimball, 1989). However, one of the major drawbacks with these experimental studies is that they have largely been confined to summer annuals, or species of subtropical origin, and therefore the results may not be representative of the responses shown by species that are active at low temperatures, and which have a much lower temperature response for photosynthesis. In support of criticism, two C_3 species (*Brassica juncea* and *Chenopodium album*) and two C_4 plants (*Amaranthus retroflexus* and *Echinochloa crus-galli*) have been shown to respond positively to elevated CO_2 (675 μmol mol^{-1}) at low temperature (12°C day/8°C night), with average growth modification factors (675/350 μmol mol^{-1}) of 1.60 at 22/18°C and 1.51 at 12/8°C (Potvin, 1994). While our own-controlled environment studies on winter wheat (*Triticum aestivum* L. cv. Hereward) have shown a strong positive response to CO_2 enrichment (700 μmol mol^{-1}) at temperatures as low as 8°C (Figure 5.8). Ziska & Bunce (1994) report similar findings for the effects of elevated CO_2 at low temperature on the growth of two herbaceous perennials, alfalfa (*Medicago sativa* L.) and orchard grass (*Dactylis glomerata* L.). Findings are consistent with field data for *M. sativa* where the greatest growth enhancement in response to elevated CO_2 was found in the spring and autumn during periods of low temperature (J.A. Bunce reported in Ziska & Bunce, 1994). These recent data draw into serious question the validity of predictions of plant productivity with elevated CO_2 which are simply based on models of the temperature response of photosynthesis. Temperature response curves for photosynthesis at various CO_2 concentrations are highly consistent between species, so the validity of modelling approaches like that undertaken by Long (1991) cannot be questioned. Future efforts must therefore be directed at verifying the positive responses of

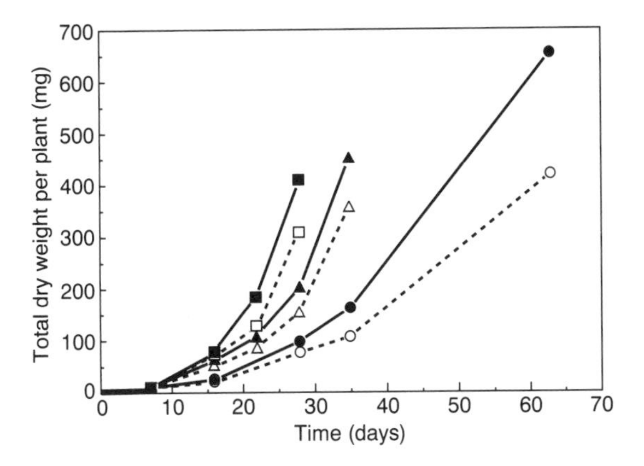

Figure 5.8. Interactive effects of elevated CO_2 temperature on the growth of winter wheat (*Triticum aestivum* L. cv. Hereward). Plants were raised at two CO_2 concentrations (355 μmol mol^{-1} [open symbols] and 700 μmol mol^{-1} [closed symbols]). Seeds were germinated at 16°C to provide a uniform start, then as soon as the coleoptile emerged three temperature treatments were imposed: 17°C [□, ■], 13°C [△, ▲] and 8°C [○, ●]. Harvests were made at intervals during the exposure, but because plants in different treatments developed at different rates, to avoid pot limitation the final harvest was made when plants had reached the three-leaf, 5-7 tiller stage

cool climate species to elevated CO_2 at low temperatures and to elucidating the mechanisms which may underlie this effect. In this context it is of interest to note that Ziska & Bunce (1993, 1994) have found that CO_2-induced growth enhancement at lower temperatures in alfalfa and orchard grass is associated with a greater reduction in respiration under CO_2-enriched conditions, but adjustments in growth pattern and changes in source:sink relationships may also contribute (Potvin, 1994).

At the canopy level, mathematical models based on current understanding of the effects of temperature and CO_2 on the subprocesses of photosynthesis have been used to simulate diurnal courses of canopy photosynthesis for July days at three latitudes, assuming the same canopy characteristics and sky conditions for each site (Long, 1991). This has shown that stimulation of net canopy rates of CO_2 uptake per unit ground area (A_c) were much lower at high-latitude sites (mean temperature 5°C) than at low-latitude sites (mean temperature 30°C). These findings are supported by the results of two long-term field studies in which tundra vegetation has been exposed to elevated CO_2. At a high-latitude site, CO_2 enrichment produced no significant long-term increase in canopy carbon exchange (Oechel & Strain, 1985; Oechel & Reichers, 1987), whereas marked increases in canopy carbon exchange were found at a low-latitude site (Drake & Leadley, 1991). However, Grulke *et al.* (1990) working on arctic tundra vegetation reported that field exposure to elevated CO_2 stimulated the rate of canopy photosynthesis even though there was complete homeostatic adjustment of assimilation at the leaf level in the dominant species, *Eriophorum vaginatum*, after 3 weeks of exposure (Tissue & Oechel, 1987). Two factors were considered to contribute to this phenomenon: (a) *E. vaginatum* showed considerable tillering in response to elevated CO_2 and (b) there were a number of species other than *E. vaginatum* that may not have acclimated. This study underlines the need for long-term field-based experimental investigations at the community level.

In addition to recent evidence that elevated CO_2 may have a positive effect on the growth of cool-climate species at low temperatures, there is growing evidence that rising levels of CO_2 may protect against chilling injury in species of subtropical and tropical origin. In okra (*Abelmoschus esculentus*), a chilling-sensitive species, exposure to elevated CO_2 has been shown to have beneficial effects on growth at low temperatures (Sionit *et al.*, 1981). At ambient CO_2, plants grew best at the highest temperatures (32°C day/26°C night), but exposure to elevated CO_2 resulted in increased survival, growth and reproduction under suboptimal conditions (20°C day/14°C night). Similarly, Potvin (1985) showed that photosynthetic rates in the grass *Echinochloa crus-galli* subjected to one night of chilling at 7°C were depressed the next morning from 58 to 27% in plants raised at ambient CO_2 (350 μmol mol^{-1}) compared with 5 and 0% for plants raised at elevated CO_2 (675 μmol mol^{-1}). If these effects on chilling injury are a general phenomenon, it may mean that rising CO_2 will enable chilling-sensitive species to survive under temperature regimes where they cannot at present. Consequently, chilling-sensitive species may shift or expand their distribution to higher elevations to a greater extent than is currently predicted from models based on expected temperature changes alone (Hogan *et al.*, 1991).

Considering the fact that climate-change scenarios include the possibility of an increase in the frequency and intensity of early and late frosts, little attention has been paid to interactions between rising CO_2, atmospheric warming and plant cold hardiness. Preliminary data for two cultivars of winter wheat (*Triticum aestivum* cvs. Frederick and Norstar) indicate that exposure to elevated CO_2 (895 μmol mol^{-1}) under simulated autumn conditions may increase frost damage (Andrews, unpublished; Table 5.2). Similarly, relatively brief exposure of containerized seedlings of black spruce (*Picea mariana* Mill.) to 1000 μmol mol^{-1} CO_2 has been shown to predispose both needles and buds to frost damage (Margolis & Vezina, 1990). It was proposed that CO_2-induced foliar nitrogen deficiency contributed to the reduced capacity of seedlings to develop frost resistance. However, the relatively short-term exposure in this experiment allowed insufficient time for long-term physiological adjustments to the elevated CO_2 atmosphere.

It has been predicted that atmospheric warming will be especially pronounced during autumn, winter and spring at high latitudes (Kettunen *et al.*, 1987). This could have considerable implications for the timing of budburst in temperate and boreal trees, which in turn may have various implications for the growth and survival of tree species in these regions. In Finland, the effects of atmospheric warming resulting from a doubling of $[CO_2]_a$ on the timing of budburst and the risk of post-dormant frost damage on trees have been assessed using computer models (Hanninen, 1991). Simulations indicate that atmospheric warming will result in hastened budburst, due to accelerated development during mild spells in winter, and increase the risk of subsequent frost damage. These findings suggest that the survival and growth of trees in central Finland could be seriously threatened by climate change. However, models of the impact in Scotland suggest

Table 5.2. Effect of doubled CO_2 concentration on the cold hardiness of winter wheat. Plants were grown in a mix of soil:sand:peat (2:1:1) for 6 days at 20°C day/15°C night, then cold hardened at 2°C day/0°C night. Plants were exposed to either control (440 μmol mol^{-1}) or elevated CO_2 (895 μmol mol^{-1}) for 24 h d^{-1} throughout. During the day the air temperature amongst the leaf bases was 3.7°C under hardening conditions, and the photon flux density at plant height was 57 W m^{-2}. ANOVA indicated a significant ($P < 0.05$) effect of CO_2 on cold hardiness, mean data are presented ± SE of 10 independent measurements at each temperature (Andrews, unpublished)

Cultivars	Hardening period (weeks)	LT^*_{50}(°C)	
		Control CO_2	Elevated CO_2
Frederick	4	−15.7 ± 0.3	−13.9 ± 0.3
	6	−17.0 ± 0.1	−15.4 ± 0.2
	8	−18.2 ± 0.5	−17.5 ± 0.5
Norstar	4	−20.0 ± 0.3	−18.5 ± 0.2
	6	−20.9 ± 0.1	−19.8 ± 0.4
	8	−22.3 ± 0.3	−20.7 ± 0.5
		LSD($P < 0.05$) = 0.95	

LT_{50}s were calculated from the data for 10 plants frozen at each of five temperatures: −12, −14, −16, −18 and −20°C and then regrown in order to calculate LT_{50}.

that effects on the timing of budburst will depend on the chilling requirement of the species (Murray *et al.*, 1989).

CONCLUSIONS

1. Sulphur-containing air pollutants, particularly SO_2, can predispose plants to winter stress, and when plants are growing slowly in wintertime they are more sensitive to SO_2. The significance of such interactions may differ in different crops (see Davison & Barnes, 1986). However, effects of SO_2 during the winter months are believed to be a major factor driving the selection of SO_2-resistant genotypes in native plant communities. Trees that are exposed above the snowcover are at particular risk from the combined effects of SO_2 and winter stress. Even when there is genetic variation within a population of trees, the longevity of the individuals and rarity of establishment of seedlings, combined with their relatively slow growth rate, mean that there is no chance that the populations will be able to adapt to relatively rapid changes in their circumstances. The possible importance of SO_2–frost interactions is underlined by the fact that combinations of SO_2 and freezing are accepted to be one of the main factors contributing to what has been referred to as type 2 forest decline. This decline affects large areas of Norway spruce in the Ore mountains of Czech Republic and on the East German border.

2. Winter injury is increased in plants exposed to O_3, but at realistic concentrations the effects observed are slight. There is, however, considerable variation from plant to plant in the extent of the response (Barnes & Davison, 1988), which often limits the ability to detect an effect. It is possible that interactions between O_3 and winter stress are one factor, amongst many, contributing to the decline of high-elevation forests in central Europe and eastern North America. Future studies must establish the ecological significance of subtle interactions between O_3 and winter stress in a wider range of plant species, paying particular attention to the mechanisms which may underlie effects.

3. A critical review of the available literature suggests that blind acceptance of the concept that increasing levels of tissue N predispose plants to greater risk of frost injury is far from entirely justified. It is our overall opinion that most plants fertilized with N at levels which promote optimum growth will cold acclimate at a similar rate and to the same degree as plants grown under a low-fertility regime, and probably exceed the cold hardiness of plants grown under severe nutrient deficiencies. Consequently, theories that increasing levels of anthropogenic N deposition are contributing to the decline of conifers at high altitudes in eastern North America and central Europe through effects on frost hardiness should be viewed with extreme scepticism. The only groups of plants that consistently demonstrate an increase in frost sensitivity in response to increased levels of soil nitrogen are those herbaceous species that grow late in the autumn, i.e., temperate grasses and many native upland species (see Table 5.1), where low nutrient levels may favour increased frost hardiness. In addition, there is little evidence that NO_x directly affects winter hardiness at present atmospheric concentrations, although there is the possibility of effects where NO_x co-exists with phytotoxic levels of O_3 or SO_2. In the case of NH_3, there is strong evidence that concentrations >70 nmol mol^{-1} NH_3 can predispose plants to frost damage through effects on the rate of frost hardening in the autumn, and the promotion of early growth in the spring. Problems in extrapolating from N fertilization studies to the effects of long-term low-level inputs of N from the atmosphere mean that there is an urgent need for studies specifically designed to investigate the effects of excess N loading from the atmosphere on natural ecosystems, and particular attention should be paid in these studies to possible indirect effects of N on the sensitivity of plants to abiotic factors such as frost and drought.

4. There have been few studies that have investigated the interactive effects of elevated CO_2 and temperature on the growth of cool-climate species. As far as we know, those studies aimed at investigating the interaction between CO_2 and temperature have generally been conducted at temperatures higher than 10°C. Consequently, there is little information on which to base predictions of how plants will respond to rising CO_2 at low temperature, and vice versa. It is generally perceived that rising CO_2 will have little impact on plants at low temperature. However, this prediction is based on little experimental data and almost without exception the species for which CO_2/temperature response has been characterized are not active or adapted to growth at temperatures below 10°C. Indeed many studies have involved chilling sensitive species, so it is dangerous to extrapolate effects to a wider group of species. Many temperate species (e.g., winter wheat) are capable of relatively high rates of photosynthesis at low temperatures, and have a continued demand for carbon throughout the winter months. Species that accumulate fructans, for instance, may respond differently as the enzymes involved in fructan synthesis are known to be relatively insensitive to temperature (Pollock, 1986). This prompts speculation, supported by available experimental data, that in some species adapted to cool-season growth, elevated CO_2 may have a beneficial effect on growth and productivity at relatively low temperatures. There is also evidence that rising levels of CO_2 may ameliorate chilling injury in species of tropical or subtropical origin, whilst predisposing temperate species to frost damage. Future research must address the interactive effects of temperature and CO_2 if we are to predict changes in community structure with any degree of accuracy.

ACKNOWLEDGEMENTS

We are indebted to Dr C.J. Andrews, Dr S. Uren and Dr C.J. Renner for allowing us to use their unpublished data and acknowledge financial support provided by the Royal Society, UK Department of the Environment, Commission of the European Communities, the Natural Environment Research Council and the Agricultural and Food Research Council.

REFERENCES

Allen, S.G., Idso, S.B., Kimball, B.A. & Anderson, M.G. 1988. Interactive effects of CO_2 and environment on photosynthesis of *Azolla*. *Agric. For. Meteorol.* **42**: 209–217.

Allen, S.G., Idso, S.B. & Kimball, B.A. 1990. Interactive effects of CO_2 and environment on net photosynthesis of water-lily. *Agric. Ecosyst. Environ.* **30**: 81-88.

Alscher, R.G., Amundson, R.G., Cumming, J.R., Fellows, S., Fincher, J., Rubin, G., Van Leuken, P. & Weinstein, L.H. 1989. Seasonal changes in the pigments, carbohydrates and growth of red spruce as affected by ozone. *New Phytol.* **113**: 211-223.

Anonymous, 1991. Elevated CO_2 levels. *Plant, Cell Environ* (Special Issue) **14**: 729-880.

Antonnen, S. 1992. Changes in lipids of *Pinus sylvestris* needles exposed to industrial air pollution. *Ann. Bot. Fenn.* **29**: 89-99.

Aronsson, A. 1980. Frost hardiness in Scots pine (*Pinus sylvestris* L.) II. Hardiness during winter and spring in young trees of different mineral nutrient status. *Stud. For. Suecica* **155**: 2-27.

Arp, W.J. 1991. Effects of source-sink relations on photosynthetic acclimation to elevated CO_2. *Plant, Cell Environ.* **14**: 869-875.

Ashmore, M.R. & Bell, J.N.B. 1991. The role of ozone in global change. *Ann. Bot.* **67**: 39-48.

Atkinson, C.J., Wookey, P.A. & Mansfield, T.A. 1991. Atmospheric pollution and the sensitivity of stomata on barley leaves to abscisic acid and carbon dioxide. *New Phytol.* **117**: 535-541.

Balling, R.C. & Idso, S.B. 1990. Effects of greenhouse warming on maximum summer temperatures. *Agric. For. Meteorol.* **53**: 143-147.

Barber, J. & Andersson, B. 1992. Too much of a good thing: light can be bad for photosynthesis. *TIBS* **17**: 61-66.

Barnes, J.D. & Brown, K.A. 1990. The influence of ozone and acid mist on the amount and wettability of the surface waxes in Norway spruce (*Picea abies* [L.] Karst.). *New Phytol.* **114**: 531-535.

Barnes, J.D. & Davison, A.W. 1988. The influence of ozone on the winter hardiness of Norway spruce (*Picea abies* (L.) Karst.). *New Phytol.* **108**: 159-166.

Barnes, J.D., Davison, A.W. & Booth, T.A. 1988a. Ozone accelerates structural degradation of epicuticular wax on Norway spruce needles. *New Phytol.* **110**: 309-318.

Barnes, J.D., Reiling, K., Davison, A.W. & Renner, C.J. 1988b. Interaction between ozone and winter stress. *Environ. Pollut.* **53**: 235-254.

Barnes, J.D., Eamus, D. & Brown, K.A. 1990a. The influence of ozone, acid mist and soil nutrient status on Norway spruce II. Photosynthesis, dark respiration and soluble carbohydrate status of trees in late autumn. *New Phytol.* **115**: 149-156.

Barnes, J.D., Eamus, D. & Brown, K.A. 1990b. The influence of ozone, acid mist and soil nutrient status on Norway spruce [*Picea abies* (L.) Karst.]. I. Plant-water relations. *New Phytol.* **114**: 713-720.

Barnes, J.D., Eamus, D., Davison, A.W., Ro-Poulsen, R. & Mortensen, L. 1990c. Persistent effects of ozone on needle water loss and wettability in Norway spruce. *Environ. Pollut.* **63**: 345-363.

Barnes, J.D. & Pfirrmann, T. 1992. The influence of CO_2 and O_3, singly and in combination, on gas exchange, growth and nutrient status of radish (*Raphanus sativus*). *New Phytol.* **121**: 403-412.

Bazzaz, F.A. 1990. The response of natural ecosystems to the rising global level of CO_2. *Ann. Rev. Ecol. Syst.* **21**: 167-196.

Bell, J.N.B. 1993. Biotic and abiotic interactions with air pollutants. In Jäger, H.J., Unsworth, M., De Temmerman, L. & Mathy, P. (eds) *Effects of Air Pollution on Agricultural Crops in Europe*: 383-408. Air Pollution Research Report 46, CEC, Brussels.

Blank, L.W. 1985. A new type of forest decline in Germany. *Nature* **314**: 311-314.

Bowden, R.D., Geballe, G.T. & Bowden, W.B. 1989. Foliar uptake of ^{15}N from simulated cloud water by red spruce (*Picea rubens*) seedlings. *Can. J. For. Res.* **19**: 382-386.

Brown, K.A. & Roberts, T.M. 1988. Effects of ozone on foliar leaching in Norway spruce (*Picea abies* L. Karst): confounding factors due to NO_x production during ozone generation. *Environ. Pollut.* **55**: 55-73.

Brown, K.A., Roberts, T.M. & Blank, L.W. 1987. Interaction between ozone and cold sensitivity in Norway spruce: a factor contributing to the forest decline in Central Europe? *New Phytol.* **105**: 149-155.
Brunold, C. 1990. Reduction of sulphate to sulphide. In Rennenberg, H., Brunold, C., DeKok, L.J. & Stuhlen, I. (eds) *Sulphur Nutrition and Sulphur Assimilation*: 13-31. Springer-Verlag, Berlin.
Cape, J.N., Leith, I.D., Fowler, D., Murray, M.B., Sheppard, L.J., Eamus, D. & Wilson, R.H. 1991. Sulphate and ammonium in mist impair the frost hardening of red spruce seedlings. *New Phytol.* **118**: 119-126.
Caporn, S.J.M., Cockerill, G.J. & Lee, J.A. 1990. Effects of atmospheric nitrogen deposition on upland *Calluna vulgaris*. In Abstracts — *International Conference on Acidic Deposition: Its Nature and Impacts*: 113. The Royal Society of Edinburgh, Edinburgh.
Chappelka, A.H., Kush, J.S., Meldahl, R.S. & Lockaby, B.G. 1990. An ozone-low temperature interaction in loblolly pine (*Pinus taeda* L.). *New Phytol.* **114**: 721-726.
Conway, T.J., Tans, P., Waterman, L.S., Thoning, K.W., Masarie, K.A. & Gammon, R.A. 1988. Atmospheric carbon dioxide measurements in the remote global troposphere, 1981-1984. *Tellus* **40B**: 81-115.
Cooley, D.R. & Manning, W.J. 1987. The impact of ozone on assimilate partitioning in plants: A review. *Environ. Pollut.* **47**: 95-113.
Cooling, E.N. 1967. Frost resistance in *Eucalyptus grandis* following the application of fertilizer borate. *Rhodesia, Zambia and Malawi J. Agric. Res.* **5**: 97-100.
Cure, J.D. 1985. Carbon Dioxide doubling responses: a crop survey. In Strain, B.R. & Cure, J.D. (eds) *Direct Effects of Increasing Carbon Dioxide on Vegetation*: 99-116. US Department of Energy, Washington, DC.
Davies, T. 1980. Grasses more sensitive to SO_2 pollution in conditions of low irradiance and short days. *Nature* **284**: 483-485.
Davis, D.D. & Wood, F.A. 1973. The influence of environmental factors on the sensitivity of Virginia pine to ozone. *Phytopathology* **63**: 371-376.
Davison, A.W. & Bailey, I.F. 1982. SO_2 pollution reduces the freezing resistance of ryegrass. *Nature* **297**: 400-402.
Davison, A.W. & Barnes, J.D. 1986. Effects of winter stress on pollutant responses. In *How are the Effects of Air Pollutants on Agricultural Crops Influenced by the Interaction with Other Limiting Factors?*: 16-32. CEC, Brussels.
Davison, A.W. & Barnes, J.D. 1992. Patterns of air pollution: Critical loads and abatement strategies. In Newsom, M.D. (ed.) *Managing the Human Impact on the Natural Environment: Patterns and Processes*: 109-129. Bellhaven Press, London.
Davison, A.W. & Barnes, J.D. 1993. Pre-disposition to stress following exposure to air pollution. In Jackson, M.B. & Black, C.R. (eds) *Interacting Stresses on Plants in a Changing Climate*: 111-124, NATO-ASI Series v. 16, Springer-Verlag, Berlin.
Davison, A.W., Barnes, J.D. & Renner, C.J. 1987. Interactions between air pollutants and cold stress. In Schulte-Hostede, S., Darrall, N.M., Blank, L.W. & Wellburn, A.R. (eds) *Air Pollution and Plant Metabolism*: 307-328. Elsevier Applied Science, London.
DeHayes, D.H., Ingle, M.A. & Waite, C.E. 1989. Nitrogen fertilization enhances cold tolerance of red spruce seedlings. *Can. J. For. Res.* **19**: 1037-1043.
DeHayes, D.H., Thornton, F.C., Waite, C.E. & Ingle, M.A. 1991. Ambient cloud deposition reduces cold tolerance of red spruce seedlings. *Can. J. For. Res.* **21**: 1292-1295.
Drake, B.G. & Leadley, P.W. 1991. Canopy photosynthesis of crops and native plant communities exposed to long-term elevated CO_2. *Plant, Cell Environ.* **14**: 853-860.
Dueck, T.A. 1990. Effect of ammonia and sulphur dioxide on the survival and growth of *Calluna vulgaris* (L.) Hull seedlings. *Funct. Ecol.* **4**: 109-116.
Eamus, D. 1991. The interaction of rising CO_2, and temperature with water use efficiency. *Plant, Cell Environ.* **14**: 843-852.
Eamus, D., Leith, I. & Fowler, D. 1989. Water relations of red spruce seedlings treated with acid mist. *Tree Physiol.* **5**: 387-397.

Eamus, D., Barnes, J.D., Ro-Poulsen, H., Mortensen, L. & Davison, A.W. 1990. Persistent stimulation of CO_2 assimilation and stomatal conductance by summer ozone fumigation in Norway spruce. *Environ. Pollut.* **63**: 365-379.

Edwards, G.S., Pier, P.A. & Kelly, J.M. 1990. Influence of ozone and soil magnesium status on the cold hardiness of loblolly pine (*Pinus taeda* L.) seedlings. *New Phytol.* **115**: 157-164.

Eissenstat, D.M., Syvertsen, J.P., Dean, T.J., Yelenosky, G. & Johnson, J.D. 1991. Sensitivity of frost resistance and growth in citrus and avocado to chronic ozone exposure. *New Phytol.* **118**: 139-146.

Farquhar, G.D. & Von Caemmerer, S. 1982. Modelling of photosynthetic responses to environmental conditions. In Lange, O.L., Nobel, P.S., Osmond, C.B. & Ziegler, H. (eds) *Encyclopaedia of Plant Physiology, Vol. 12B, Physiological Plant Ecology II*: 549-587. Springer-Verlag, Berlin.

Fincher, J. & Alscher, R.G. 1992. The effect of long-term ozone exposure on injury in seedlings of red spruce (*Picea rubens* Sarg.). *New Phytol.* **120**: 49-59.

Fincher, J., Cumming, J.R., Alscher, R.G., Rubin, G. & Weinstein, L. 1989. Long-term ozone exposure effects winter hardiness of red spruce (*Picea rubens* Sarg.) seedlings. *New Phytol.* **113**: 85-96.

Fowler, D., Cape, J.N., Deans, J.D., Leith, I.D., Murray, M.B., Smith, R.I., Sheppard, L.J. & Unsworth, M.H. 1989. Effects of acid mist on the frost hardiness of red spruce seedlings. *New Phytol.* **113**: 321-335.

Freer-Smith, P.H. & Mansfield, T.A. 1987. The combined effects of low temperature and $SO_2 + NO_2$ pollution on the new season's growth and water relations of *Picea sitchensis*. *New Phytol.* **1106**: 237-250.

Friedland, A.J., Hawley, G.J. & Gregory, R.A. 1985. Investigations of nitrogen as a possible contributor to red spruce (*Picea rubens* Sarg.) decline. In *Proceedings of the Symposium on the Effects of Air Pollutants on Forest Ecosystems*: 83-107, University of Minnesota Press, Minneapolis.

Gillies, S.L. & Vidaver, W. 1990. Resistance to photodamage in evergreen conifers. *Physiol. Plant.* **80**: 148-153.

Grulke, N.E., Riechers, G.H., Oechel, W.C., Hjelm, U. & Jaeger, C. 1990. Carbon balance in tussock tundra under ambient and elevated atmospheric CO_2. *Oecologia* **83**: 485-494.

Hanninen, H. 1991. Does climatic warming increase risk of frost damage in northern trees? *Plant, Cell Environ.* **14**: 449-454.

Hausladen, A., Madamanchi, N.R., Alscher, R.G., Amundson, R.G. & Fellows, S. 1990. Seasonal changes in antioxidants in red spruce as affected by ozone. *New Phytol.* **115**: 447-458.

Heath, R.L. 1980. Initial injury to plants by air pollutants. *Annu. Rev. Plant Physiol.* **31**: 395-431.

Hogan, K.P., Smith, A.P. & Ziska, C.H. 1991. Potential effects of elevated changes in temperature on tropical plants. *Plant, Cell Environ.* **14**: 763-778.

Hough, A.M. & Derwent, R.G. 1990. Changes in the global concentration of tropospheric ozone due to human activities. *Nature* **344**: 645-648.

Hov, O. 1984. Ozone in the troposphere: High level pollution. *Ambio* **13**: 73-79.

Idso, S.B. 1990. Interactive effects of carbon dioxide and climate variables on plant growth. In Kimball, B.A., Rosenberg, N.J. & Allen, L.H. (eds) *Impact of Carbon Dioxide, Trace Gases, and Climate Change on Global Agriculture*: 61-70. American Society of Agronomy, Madison.

Idso, S.B. & Kimball, B.A. 1989. Growth response of carrot and radish to atmospheric CO_2 enrichment. *Environ. Exp. Bot.* **29**: 135-139.

Idso, S.B., Kimball, B.A., Anderson, M.G. & Mauney, J.R. 1987. Effects of atmospheric CO_2 enrichment on plant growth: The interactive role of air temperature. *Agric. Ecosyst. Environ.* **20**: 1-10.

IPO, 1990. *Effects of NH_3 and $(NH_4)_2$ SO_4 Deposition on Terrestrial Semi-natural Vegetation on Nutrient-poor Soils*. IPO Report R90/06 on Project 124/125. IPO, Wageningen, 312 pp.

Jahnke, L.S., Hull, M.R. & Long, S.P. 1991. Chilling stress and oxygen metabolizing enzymes in *Zea mays* and *Zea diploperennis*. *Plant, Cell Environ.* **14**: 97-104.
Jones, T. & Mansfield, T.A. 1982. The effect of SO_2 on growth and development of seedlings of *Phleum pratense* under different light and temperature environments. *Environ. Pollut.* **27**: 57-71.
Jordan, D.B. & Ogren, W.L. 1984. The CO_2/O_2 specificity of ribulose-1,5-bisphosphate concentration, pH and temperature. *Planta* **161**: 308-313.
Keeling, C.D. 1986. *Atmospheric CO_2 Concentrations Mauna Loa Observatory, Hawaii* 1858-1986. Report NDP-001/R1. Carbon Dioxide Information Centre, Oak Ridge National Laboratory, Oak Ridge, TN.
Keller, T. 1978. Wintertime atmospheric pollutants as they affect the performance of deciduous trees in the ensuing growing season. *Environ. Pollut.* **16**: 243-247.
Keller, T. 1981. Consequences of SO_2 fumigation of spruce in winter. *Gartenbauwissenschaft* **46**: 170-178.
Kerstiens, G. & Lendzian, K.J. 1989. Interactions between ozone and plant cuticles. II. Water permeability. *New Phytol.* **112**: 21-27.
Kettunen, L., Mukula, J., Pohjonen, V., Rantanen, O. & Varjo, U. 1987. The effects of climatic variations on agriculture in Finland. In Parry, M.L., Carter, T.R. & Konijn, N.T. (eds) *The Impact of Climatic Variations on Agriculture, Vol 1. Assessment in Cool Temperate and Cold Regions*: 513-614. Kluwer Academic Publishers, Dordrecht.
Kimball, B.A. 1986. CO_2 stimulation of growth and yield under environmental constraints. In Enoch, H.Z. & Kimball, B.A. (eds) *Carbon Dioxide Enrichment of Greenhouse Crops, Vol II. Physiology, Yield and Economics*: 53-67. CRC Press, Boca Raton, FL.
Kimball, B.A., Rosenberg, N.J. & Allen, L.H. 1990. *Impact of Carbon Dioxide, Trace Gases, and Climate Change on Global Agriculture.* ASA Publication 53, American Society of Agronomy, Madison.
Klein, R.M., Perkins, T.D. & Myers, H.L. 1989. Nutrient status and winter hardiness of red spruce foliage. *Can. J. For. Res.* **19**: 754-758.
Krupa, S.V. & Manning, W.J. 1988. Atmospheric ozone: formation and effects on vegetation. *Environ. Pollut.* **50**: 101-137.
L'Hirondelle, S.J., Jacobsen, J.S. & Lassoie, J.P. 1992. Acidic mist and nitrogen fertilization effects on growth, nitrate reductase activity, gas exchange, and frost hardiness of red spruce seedlings. *New Phytol.* **121**: 611-622.
Lemon, E.R. 1983. *CO_2 and Plants: The Response of Plants to Rising Levels of Atmospheric Carbon Dioxide.* Westview Press, Boulder, CO.
Lendzian, K.J. 1984. Permeability of plant cuticles to gaseous air pollutants. In Koziol, M.J. & Whatley, F.R. (eds) *Gaseous Air Pollutants and Plant Metabolism*: 77-81. Butterworths, London.
Levitt, J. 1980. *Responses of Plants to Environmental Stress, Vol 1. Chilling, Freezing and Temperature Stress.* Academic Press, New York.
Lindow, S.E., Arny, C. & Upper, C.D. 1978. Distribution of ice nucleation-active bacteria on plants in nature. *Appl. Environ. Microbiol.* **36**: 831-838.
Lindow, S.E., Arny, C. & Upper, C.D. 1982. Bacterial ice nucleation: A factor in frost injury to plants. *Plant Physiol.* **70**: 1084-1089.
Long, S.P. 1991. Modification of the response of photosynthetic productivity to rising temperature by atmospheric CO_2 concentrations: Has its importance been underestimated? *Plant, Cell Environ.* **14**: 729-730.
Long, S.P. 1994. The potential effects of concurrent increases in temperature, CO_2 and O_3 on net photosynthesis, as mediated by Rubisco. In Alscher, R.G. & Wellburn, A.R. (eds) *Plant Responses to the Gaseous Environment: Molecular, Metabolic and Physiological Aspects*: 21-38. Chapman & Hall, London.
Long, S.P. & Hutchin, P.R. 1991. Primary production in grasslands and coniferous forests with climate change: an overview. *Ecol. Applic.* **1**: 139-156.
Lucas, P.W. & Penuelas, J. 1990. Effects of ozone exposure and growth temperature on the development of winter injury in sitka spruce seedlings. In Payer, H.D., Pfirrmann, T.

& Mathy, P. (eds) *Environmental Research With Plants in Closed Chambers*: 381-394. CEC, Brussels.

Lucas, P.W., Cottam, D.A., Sheppard, L.J. & Francis, B.J. 1988. Growth responses and delayed winter hardening in sitka spruce following summer exposure to ozone. *New Phytol.* **108**: 495-504.

Lyons, J.M. 1973. Chilling injury in plants. *Annu. Rev. Plant Physiol.* **24**: 445-466.

Malhotra, S.S. & Khan, A.A. 1978. Effects of sulphur dioxide fumigation on lipid biosynthesis in pine needles. *Phytochem.* **17**: 241-244.

Mansfield, T.A., Davies, W.J. & Whitmore, M.E. 1986. Interactions between the responses of plants to air pollution and other environmental factors such as drought, light and temperature. In *How are the Effects of Air Pollutants on Agricultural Crops Influenced by the Interaction with Other Limiting Factors?*: 2-15. CEC, Brussels.

Mansfield, T.A., Wookey, P.A., Atkinson, C.J. & Lucas, P.W. 1990. Does air pollution increase the sensitivity of plants to water deficits? Paper presented at Spanish/British Workshop "Sobre Medio Ambiente". San Lorenzo del Escorial, Spain, August 1990.

Mansfield, T.A., Pearson, M., Atkinson, C.J. & Wookey, P.A. 1992. Ozone, sulphur dioxide and nitrogen oxides: some effects on the water relations of herbaceous plants and trees. In Jackson, M.B. & Black, C.R. (eds) *Interacting Stresses on Plants in a Changing Climate*: 77-88. NATO ASI Series. 16, Springer-Verlag, Berlin.

Margolis, H.A. & Vezina, L-P. 1990. Atmospheric CO_2 enrichment and the development of frost hardiness in containerized black spruce seedlings. *Can. J. For. Res.* **20**: 1392-1398.

Materna, J. 1974. Einfluss der SO_2-immissionen auf fichtenpflanzen in wintermonaten. *IXth Int. Tagung uber Luftverunreinigung und Forstwirtschaft.* Marianske Lazne, Czechoslovakia. 14-26.

McLaughlin, S.B. 1985. Effects of air pollution on forests: a critical review. *J. Air Pollut. Control Assoc.* **35**: 512-521.

Murray, M.B., Cannell, M.G.R. & Smith, R.I. 1989. Date of budburst of fifteen tree species in Britain following climatic warming. *J. Appl. Ecol.* **26**: 693-700.

Neighbour, E.A., Cottam, D.A. & Mansfield, T.A. 1988. Effects of sulphur dioxide and nitrogen dioxide on the control of water loss by birch (*Betula* spp.). *New Phytol.* **108**: 149-157.

Neighbour, E.A., Pearson, M. & Mehlhorn, H. 1990. Purafil-filtration prevents the development of ozone-induced frost injury: a potential role for nitric oxide. *Atmos. Environ.* **24**: 711-715.

Nihlgard, B. 1985. The ammonium hypothesis — an additional explanation to the forest dieback in Europe. *Ambio* **14**: 2-8.

Oechel, W.C. & Reichers, G.I. 1987. *Responses of a Tundra Ecosystem to Elevated Atmospheric* CO_2. US Department of Energy, Washington, DC.

Oechel, W.C. & Strain, B.R. 1985. Native species responses to increased atmospheric carbon dioxide concentration. In Strain, B.R. & Cure, J.D. (eds) *Direct Effects of Increasing Carbon Dioxide on Vegetation*: 119-145. US Department of Energy, Washington, DC.

Pellett, H.M. & Carter, J.V. 1981. Effect of nutritional factors on cold hardiness of plants. *Hort. Rev.* **3**: 144-177.

Percy, K.E., Jensen, K.F. & McQuattie, C.J. 1992. Effects of ozone and acidic fog on red spruce needle epicuticular wax production, chemical composition, cuticular membrane ultrastructure and needle wettability. *New Phytol.* **122**: 71-80.

Pfirrmann, T. 1992. Wechelwirkungen von ozon, kohlendioxid und wassermangel bei zwei klonen unterscheidlich mit kalium ernahrter fichten. PhD Thesis, University of Giessen.

Pollock, C.J. 1986. Fructans and metabolism of sucrose in vascular plants. *New Phytol.* **105**: 1-24.

Potvin, C. 1985. Amelioration of chilling effects by CO_2 enrichment. *Physiol. Vegetale* **23**: 345-352.

Potvin, C. 1994. Interactive effects of temperature and atmospheric CO_2 on physiology and growth. In Alscher, R.G. & Wellburn, A.R. (eds) *Plant Responses to the Gaseous Environment: Molecular, Metabolic and Physiological Aspects*: 39-54. Chapman & Hall, London.

Puempel, U.B., Gaebl, F. & Tranquillini, W. 1975. Growth, mychorrhizae and frost resistance of *Picea abies* following fertilization with different levels of N. *Europ. J. For.* **5**: 83-97.

Renner, C.J. 1990. Interactive effects of sulphur dioxide, nitrogen dioxide and the winter environment on *Lolium perenne*. PhD Thesis, University of Newcastle upon Tyne.

Robinson, D.C. & Wellburn, A.R. 1991. Seasonal changes in the pigments of Norway spruce, *Picea abies* (L.) Karst, and the influence of summer ozone exposures. *New Phytol.* **119**: 251-259.

Rotty, R.M. & Marland, G. 1986. Fossil fuel consumption: recent amounts, patterns and trends of CO_2. In Trabalka, J. & Reichle, D.E. (eds) *The Changing Carbon Cycle: A Global Analysis*: 55-78. Springer-Verlag, Berlin.

Rowland-Bamford, A.J., Baker, J.T., Allen, L.H. & Bowes, G. 1991. Acclimation of rice to changing atmospheric carbon dioxide concentration. *Plant, Cell Environ.* **14**: 577-583.

Sage, R.F., Sharkey, T.D. & Seemann, J.R. 1989. Acclimation of photosynthesis to elevated CO_2 in five C_3 species. *Plant Physiol.* **89**: 590-596.

Sage, R.F., Sharkey, T.D. & Seemann, J.R. 1990. Regulation of ribulose l, 5bisphosphate carboxylase activity in response to light intensity and carbon dioxide in the C3 annuals *Chenopodium album* L. and *Phaseolus vulgaris* L. *Plant Physiol.* **94**: 1735-1742.

Schneider, S.H. 1989. The greenhouse effect-science and policy. *Science* **243**: 771-781.

Sharkey, T.D. & Vandeveer, P.J. 1989. Stromal phosphate concentration is low during feedback limited photosynthesis. *Plant Physiol.* **91**: 679-684.

Sionit, N., Strain, B.R. & Beckford, H.A. 1981. Environmental controls on the growth and yield of okra. I. Effects of temperature and of CO_2 enrichment at cool temperatures. *Crop Sci.* **21**: 885-888.

Skärby, L., Troeng, E. & Bostrom, C.A. 1987. Ozone uptake and effects on transpiration, net photosynthesis, and dark respiration in Scots pine. *For. Sci.* **33**: 801-808.

Steponkus, P.L., Uemura, M. & Webb, M.S. 1992. Redesigning crops for increased tolerance to freezing stress. In Jackson, M.B. & Black, C.R. (eds) *Interacting Stresses on Plants in a Changing Climate*: 697-714. NATO ASI Series v.16, Springer-Verlag, Berlin.

Strain, B.R. & Cure, J.D. 1985. *Direct Effects of Increasing Carbon Dioxide on Vegetation.* US Department of Energy, Washington, DC.

Stuiver, C.E.E., De Kok, L.J. & Kuiper, P.J.C. 1992. Freezing tolerance and biochemical changes in wheat shoots as affected by H_2S fumigation. *Physiol. Biochem.* **30**: 47-55.

Tissue, D.T. & Oechel, W.C. 1987. Response of *Eriophorum vaginatum* to elevated CO_2 and temperature in the Alaskan tussock tundra. *Ecology* **68**: 401-410.

Tranquillini, W. 1982. Frost-drought and its ecological significance. In Lange, O.L., Nobel, P.S., Osmond, C.B. & Ziegler, H. (eds) *Encyclopaedia of Plant Physiology, New Series, Vol. 12B. Physiological Plant Ecology II*: 125-159. Springer-Verlag, Berlin.

UK-TERG, 1988. Interactions between air pollutants and other environmental factors. In *The Effects of Acid Deposition on the Terrestrial Environment in the U.K.*: 73-83. HMSO, P London.

Velissariou, D., Barnes, J.D., Davison, A.W., Pfirrmann, T. & Holevas, C.D. 1992. Effects of air pollution on *Pinus halepensis* Mill. 1. Pollution levels in Attica, Greece. *Atmos. Environ.* **26**: 373-380.

Wellburn, A.R., Barnes, J.D., Lucas, P.W., McLeod, A.R. & Mansfield, T.A. 1996. Controlled O_3 exposures and field observations of O_3 effects in the U.K. In Sandermann, H., Wellburn, A.R. & Heath, R.L. (eds) *Forest Decline and Ozone: A Comparison of Controlled Chamber and Field Experiments*. Ecological Studies Series, Springer-Verlag, Berlin (in press).

Whitmore, M.E. & Mansfield, T.A. 1983. Effects of long-term exposures to SO_2 and NO_2 on *Poa pratensis* and other grasses. *Environ. Pollut.* **31**: 217-235.

Wigley, T.M.L. 1985. Climatology: Impact of extreme events. *Nature* **316**: 106-107.

Wilson, J.M. 1976. Mechanism of chill and drought hardening of *Phaseolus vulgaris* leaves. *New Phytol.* **76**: 257-270.

Wilson, J.M. & Crawford, R.M.M. 1974. The acclimatization of plants to chilling temperatures in relation to the fatty-acid composition of leaf polar lipids. *New Phytol.* **73**: 805-820.

Wolfenden, J. & Mansfield, T.A. 1990. Physiological disturbances in plants caused by air pollutants. In Last, F.T. & Whatling, R. (eds) *Acidic Deposition: Its Nature and Impacts*: 117-138. The Royal Society of Edinburgh, Edinburgh.

Wolfenden, J. & Wellburn, A.R. 1991. Effects of summer ozone on membrane lipid composition during subsequent frost hardening in Norway Spruce [*Picea abies* (L.) Karst]. *New Phytol.* **118**: 323-329.

Wolfenden, J., Pearson, M. & Francis, B.J. 1991. Effects of over-winter fumigation with sulphur and nitrogen dioxides on biochemical parameters and spring growth in red spruce. *Plant, Cell Environ.* **14**: 35-45.

Ziska, L.H. & Bunce, J.A. 1993. Inhibition of whole plant respiration by elevated CO_2 as modified by growth temperature. *Physiol. Plant.* **87**: 459-466.

Ziska, L.H. & Bunce, J.A. 1994. Direct and indirect inhibition of single leaf respiration by elevated CO_2 concentrations: Interaction with temperature. *Physiol. Plant.* **90**: 130-138.

6 Barrier Properties of the Cuticle to Water, Solutes and Pest and Pathogen Penetration in Leaves of Plants Grown in Polluted Atmospheres

G. KERSTIENS

INTRODUCTION

There is a wealth of literature about effects on the appearance of surface waxes of leaves grown under the influence of air pollutants. Approximately 60 experimental papers have been published on that subject, most of them dealing with conifer needles. Comprehensive reviews were given fairly recently by Riederer (1989) and by Turunen & Huttunen (1990). The picture that emerges is that of an unspecific acceleration of the changes epicuticular waxes undergo naturally during ageing. A change in appearance, however, does not necessarily imply an impaired functionality of the protective layer beneath the surface waxes, the cuticle. This review summarizes what is known about the results of exposure to air pollutants in terms of the barrier properties of cuticles of newly grown leaves. It deals with effects on the loss of water and solutes through cuticles, and the impact on specific interactions between pests or pathogens and the plant surface. All reported effects were significant if not stated otherwise.

The cuticle, which covers all primary aerial plant organs, is a complex structure. It consists of a polymer matrix, the cutin, with polyester and sometimes polyethylene-like domains, in which a reticulum of polysaccharide fibres is embedded (Holloway, 1982; Nip *et al.*, 1986). Under physiological conditions (above pH 3) cutin carries fixed negative charges at carboxyl groups, making the cuticle a cation exchanger (Schönherr & Huber, 1977). The third constituent of the cuticle is a group of diverse, mostly long-chained saturated aliphatic compounds, usually called waxes (Baker, 1982). They are present both in and on the cuticular membrane, in a more or less distinct crystalline state. Whereas only the epicuticular waxes can be observed with the scanning electron microscope, the waxes embedded mainly in the outer part of the cuticular matrix largely determine the rate of diffusion processes through the cuticle (Schönherr & Riederer, 1989; Bauer & Schönherr, 1992). The transport of any substance through the cuticle takes place by activation energy-driven random diffusion of single molecules or ions on tortuous "paths" around areas of crystalline waxes (Riederer & Schneider, 1990;

Plant Response to Air Pollution. Edited by Mohammad Yunus and Muhammad Iqbal

Reynhardt & Riederer, 1994). The polysaccharide reticulum may perhaps ease the diffusion of water and water-soluble substances (Kerstiens & Lendzian, 1989; Kerstiens, 1994). There are no indications of volume flow of aqueous solutions or gases through pores in the cuticle (apart from the stomatal pores, of course).

The cuticular transport of a non-ionic substance is usually characterized by a permeance, which is the quotient of mass flow rate density and concentration difference across the cuticle. Thus permeances are defined in a way analogous to that used for stomatal conductance, but they apply specifically to non-porous barriers.

The understanding of cuticular transport processes is not sufficient for any conclusions to be drawn from observations of cuticle thickness, wax content or composition. Comparisons between species (Kamp, 1930; Becker *et al.*, 1986) and within species (unpublished results) failed to show a correlation between cuticle thickness and water permeability. Generally, water permeability is not related to the amount or composition of waxes (Riederer & Schneider, 1990). An observed decrease of the thickness of the cuticular membrane, or a reduction in the amount of waxes, if found after a treatment, would not indicate or even make probable a partial loss of the cuticular-barrier function to water or solutes. This point is illustrated by the work of Percy & Baker (1987, 1988) reported subsequently. One has to rely on measurements of the actual properties of interest to learn about any effects on cuticular function.

CUTICULAR WATER PERMEABILITY

Minimum conductances of whole leaves

It has been shown repeatedly that the rate of water loss of whole leaves or twigs under conditions where stomata were supposed to be closed (darkness, excision, etc.) was affected by pretreatment with air pollutants. Although this effect was attributed to an increase in cuticular transpiration by the respective authors, a contribution by incompletely closed stomata cannot be excluded in such systems. Thus the term minimum conductance is preferred here.

Cape & Fowler (1981) reported that the rate of water loss from detached needles of *Pinus sylvestris* L. grown in polluted air (about 40 ppb SO_2) was higher by 40 and 100%, respectively, in the two youngest age classes compared with needles of the same provenance taken from an unpolluted site, but reached identical values in the third year. The significance of these differences, however, was not stated. Minimum conductances were $2.6–8.3 \times 10^{-5}$ m s^{-1}.

Two investigations dealt with *Picea abies* (L.) Karst. needles. Barnes & Davison (1988) found that treating their plants with 120 ppb ozone for 70 days (November to January) did not affect the rate of water loss from excised current year's needles of any of the eight clones they used. Only the previous year's needles of one clone showed a significant increase by 22%, the other seven clones being unaffected. The control plants received charcoal-filtered air. Mengel *et al.* (1989) found a significant effect of a treatment with acidic fog of pH 3.0 (control: pH 5.0) applied for a period of 8 weeks (May to July) on water-loss rates of excised twigs of the current year. Minimum conductance increased by 44%. Barnes & Davison (1988) reported weight-loss rates of 0.6–1.2% h^{-1}, whereas Mengel *et al.* (1989) found rates of 0.35 and 0.50% h^{-1}, respectively. Taking into account that the water-vapour concentration gradient in the latter study was about three times as high as in the former, the means of the various groups of specimens in the two papers differ by up to an order of magnitude. Assuming a ratio of 5.1 mm^2 leaf surface per mg fresh weight for this species (R. Steinbrecher, pers. comm.), minimum conductances can be estimated as $1\text{–}9 \times 10^{-5}$ m s^{-1}.

Moldau *et al.* (1990) exposed *Phaseolus vulgaris* L. plants to up to 640 ppb ozone for 3–4 hours in darkness under saturating CO_2 concentrations. Conductances were found to be six times higher under these conditions than in the controls (no O_3). Minimum conductance of control plants was 5×10^{-5} m s^{-1}. Knittel & Pell (1991) treated another crop plant, *Zea mays* L., with acidic rain of pH 3.0 twice weekly for 2 hours during the summer (June to September). Minimum conductances ($3\text{–}20 \times 10^{-4}$ m s^{-1}) were determined for attached leaves under low-light conditions. They found no difference between the acidic-rain treatment and the control (rain of pH 5.0).

Except for the minimum conductances from the latter study, all reported values for the respective controls were well within the range of unambiguous values of cuticular permeances (reviewed by Lendzian & Kerstiens, 1991; Kerstiens, 1994). This does not exclude, however, the possibility that any effects on minimum conductance reflect partial losses of stomatal control. Loss of stomatal control itself, however, might be an effect of locally increased cuticular water loss, if that leads to a loss of counter-pressure for the stomatal apparatus. This indirect mechanism was suggested by Neighbour *et al.* (1988) after they had found that leaves of *Betula pubescens* Ehrh. growing in an atmosphere containing 40 ppb SO_2 and NO_2 lacked the ability to close their stomata upon excision. Micrographs taken 10 minutes after excision revealed wide-open stomata within fields of collapsed epidermal cells of fumigated leaves. When the abaxial surface of the hypostomatous leaves was covered with petroleum jelly, however, minimum rates of water loss were identical for treated and untreated leaves.

The ambiguity about the causes of increased minimum transpiration rates can only be resolved by excluding any stomata-bearing leaf surfaces from the gas exchange system. Studies using that approach are presented in the next section.

Adaxial cuticular permeances of hypostomatous leaves

Kerstiens & Lendzian (1989) investigated the effect of a combination of ozone (30–60 ppb with episodes of up to 130 ppb) and acid mist (episodes of pH 3.0, lasting 2–5 days), given alternatively, on water conductance of the astomatous adaxial surfaces of *Fagus sylvatica* and *Hedera helix* L. plants. The plants were grown in phytotrons during their natural vegetative period for 3 (*Hedera*) and 5 (*Fagus*) months, respectively, and only leaves which had expanded during the treatment were used for the experiment. In this experiment, as in the ozone treatment mentioned below, N_2O_5 potentially produced by the electric discharge ozone generator was removed from the air stream to avoid artifacts (Neighbour *et al.*, 1990). Water loss was determined on non-detached leaves using a well-sealed cup attached to the adaxial surface. Inside the cup, leaves were exposed to dry nitrogen (relative humidity $\ll 1\%$) which carried the transpired water to a highly sensitive moisture monitor. No indication of a difference between the pollutant treatment and the controls was observed with either species. Cuticular water permeances were 3.8×10^{-5} m s^{-1} and 0.3×10^{-5} m s^{-1} for *Fagus* and *Hedera*, respectively.

Kerstiens (1995) used a very similar experimental protocol to assess the effect of sulphur dioxide (35 ± 15 ppb), ozone (30–120 ppb) and enhanced carbon dioxide (600 ppm), respectively, on *Fagus* grown from bud break for several months in a greenhouse ("Solardomes"; Lucas *et al.*, 1987) designed to track ambient conditions very closely. None of the treatments affected adaxial cuticular water permeance ($2.4–2.9 \times 10^{-5}$ m s^{-1}). Treatment of four deciduous tree species with 300–400 ppb NO using a similar experimental protocol resulted in no effects or in a slight decrease in *Acer pseudoplatanus* L. (1.7×10^{-5} m s^{-1}), *Corylus avellana* L. (1.9×10^{-5} m s^{-1}) and *Fagus sylvatica* (1.6×10^{-5} m s^{-1}) (Kerstiens, 1995). An increase of cuticular water permeability of *Prunus avium* L. (1.0×10^{-5} m s^{-1}) leaves by 15% was significant when pairs of leaves of corresponding size and insertion level were compared, but insignificant if the full natural variability (coefficient of variation 40%) was taken into account. With the same technique, a clear and significant increase of cuticular water permeance from 1.0 to 2.7×10^{-5} m s^{-1} was observed with *Betula pubescens* plants treated with 50–60 ppb SO_2 and NO_2, but the highly artificial growing conditions applied in this case affected overall vitality of all plants and might have influenced the result. Likewise, only visibly damaged leaves showed an increase of cuticular permeability when the experiment was repeated under natural light (Kerstiens, 1995). Minimum conductance of visibly damaged as well as undamaged leaves, however, was strongly affected by the mixture of SO_2 and NO_2. Furthermore, fumigation with NO increased minimum conductance in *Corylus* and *Prunus* (Kerstiens, 1995).

Chamel *et al.* (1992) reported that water sorption of cuticles isolated from *Abies alba* Mill. needles of healthy or declining trees from a forest in the French alps was not correlated with tree vitality. Water sorption is one parameter determining water permeability of membranes, although due to the heterogeneity of cuticles the relationship between water uptake and permeability is much less straightforward than in synthetic polymer membranes (Lendzian & Kerstiens, 1991). Kerstiens & Lendzian (1989) examined both water sorption in and water permeability of isolated cuticles after treatment with 2.5% ozone in oxygen for one hour. Whereas with *Ficus elastica* Roxb. water sorption was more than doubled, permeability surprisingly decreased by 20%. In *Citrus aurantium* L., both water sorption and permeability increased by *ca.* 50%. *Hedera helix* cuticles showed no significant effect on either parameter.

Garrec & Kerfourn (1989) found no significant effect on water permeability of enzymatically isolated *Ilex aquifolium* L. leaf cuticles after a 1-week fumigation with 1–1.5 ppm ozone, a 1-week exposure to an acidic solution of pH 2, or a combination of both treatments, although the means of all treated groups were on average about twice as high as the control (0.8×10^{-5} m s^{-1}).

Apart from the fumigation of *Betula* with SO_2 and NO_2, where only those leaves which were visibly affected by either the fumigation or the general growing conditions showed an impact of the gas mixture on cuticular permeability, not a single study which excluded stomata-bearing leaf surfaces showed an effect of air pollutants on adaxial cuticular water permeability. It appears that effects on minimum conductance were most probably due to a loss of stomatal responsiveness to signals triggering maximum stomatal closure, such as darkness, leaf excision or high CO_2 concentration, rather than a direct effect on cuticular water permeability itself. This explanation fits in with the partial loss of stomatal responsiveness to less extreme stimuli after treatment with pollutants, as discussed by Mansfield & Pearson (this volume, Chapter 7).

SOLUTE TRANSPORT THROUGH THE CUTICLE (LEACHING)

There are four potential effects of air pollutants on leaching rates: (a) controlled or uncontrolled release of solutes into the apoplastic solution, e.g., due to damage to cell membranes, (b) damage to the cuticle resulting in higher permeabilities and/or ion exchange capacity, (c) increased wettability of the surface leading to an increase in contact area and/or time between cuticle and precipitation, and (d) in the case of ions, effects on cuticular-flow rates due to a change in ion concentration in the solution covering the cuticle (cf. Kerstiens *et al.*, 1992), brought about by dissolution and chemical reactions of acidic air pollutants (SO_2, NO_x) in the surface solution. The latter leads, for instance, to an increased exchange of protons from the surface solution for cations from the cuticle (e.g., Scherbatskoy & Tyree, 1990). As this review focuses on pollutant effects on the cuticle itself, it will concentrate on three types of experiments, excluding the first and fourth potential effects mentioned above: (a) transport experiments using isolated cuticles from plants exposed to air pollutants or acid precipitation, (b) ion uptake determinations following pollutant treatment, and (c) investigations about effects of short-term treatments of isolated cuticles with very high pollutant doses. Pollutant effects on leaf wettability are dealt with in the next section. Miller (1984), Klemm (1989) and Turunen & Huttunen (1990) discussed the subject in a broader sense.

A short report by Huttunen *et al.* (1989) appears to be the only one dealing with isolated cuticles which were exposed to pollutants prior to isolation. They state that potassium-ion penetration through *Picea abies* and *Pinus sylvestris* cuticles was several times higher if the seedlings had been treated with acidic precipitation (pH 3), compared with a "dry" treatment (exposure period not given). The needles were 30 and 20 days old, respectively, at the time of isolation of cuticles.

Evans *et al.* (1981, 1985) and Klemm (1989) applied solutions of various pH values (down to 2.7 and 1.1, respectively) containing radiolabelled anions and cations to the surfaces of *Phaseolus vulgaris* and *Picea abies* leaves, respectively. The rate of ion uptake did not significantly change during the duration of exposure (6 and 2 hours, respectively), indicating that even a solution of pH close to 1 (50 mmol l^{-1} H_2SO_4) did not affect the barrier characteristics of the cuticle during the exposure time. It is interesting to compare these results with the findings of

Foster (1990). When he analysed throughfall from *Lycopersicon esculentum* Mill. plants exposed to artificial mist of pH 2.5, 4.0 or 5.6 for 4 hours after pre-rinsing with deionized water, he found that leaching rates of potassium, magnesium, nitrate and phosphate, but not calcium, ions increased dramatically and more or less steadily during the exposure period when the most acidic mist was applied. This was taken as an indication of microscopic damage to both the cuticle and underlying cell membranes, but the fact that similar treatments did not lead to an increase in ion uptake (Evans *et al.*, 1981, 1985; Klemm, 1989) does not support his conclusion.

Krause & Kaiser (1977) dusted leaves of *Lactuca sativa* L., *Raphanus sativus radicula* L., *R. sativus oleifera* L., *Setaria italica* L. and *Tagetes* sp. with a mixture of CdO, PbO_2, CuO and MnO_2 over a 4-week period with or without concurrent fumigation with 80 ppb SO_2. Considerable uptake of these water-insoluble oxides through the leaf surfaces was observed, but no difference was found between the SO_2 treatment and the control plants.

Percy & Baker (1989) found that a treatment with acid rain of pH 2.6 for 10 weeks from bud break led to increased uptake of rubidium and sulphate ions into needles of some clones of *Picea sitchensis* (Bong.) Carr. when compared to an unpolluted rain treatment (pH 5.6). Percy & Baker (1988), treating crop plants with simulated rain between pH 5.6 and 2.6 from leaf emergence to full expansion, found contrasting effects of pH treatment on uptake of rubidium, nickel and sulphate ions in the different species. Uptake of Rb increased into *Brassica napus* L. leaves and decreased into *Phaseolus vulgaris* leaves exposed at pH 3.4 or below. Uptake of sulphate and nickel ions by *Brassica* leaves increased after exposure to rain at pH 2.6 but was unaffected in *Phaseolus*. It is interesting to note that *Phaseolus* leaves showed a significant reduction in cuticular thickness (by some 50%) and amount of waxes when pH was reduced, whereas *Brassica* cuticles were not affected (Percy & Baker, 1987).

Garrec & Kerfourn (1989) treated cuticles isolated from *Ilex aquifolium* leaves with 1–1.5 ppm ozone for 7 days, with acidic solutions of pH 2 or 3 for 7 days, or with ozone for 1 week and the pH 2 solution for another week. None of these treatments resulted in a significant effect on the permeability to potassium, calcium or magnesium ions. There was a tendency for all treatments to lead to lower permeabilities. The isoelectric point was lowered from pH 2.75 to around pH 2 if the cuticles were treated with acidic solutions.

WETTABILITY AND ATTACKS BY PESTS AND PATHOGENS

Laurence (1981), Huttunen (1984), Hasler *et al.* (1990) and Magan & McLeod (1991) extensively reviewed interactions between pathogens, air pollution and plants in general, but next to nothing is known about effects of air pollutants on the specific interactions between pathogens (fungi, bacteria) or pests (mainly insects), and plant cuticles. What follows is an attempt to bring together results from a range of studies, many unrelated to air pollutants, which might indicate potential forms of such interactions.

Primary interactions include recognition of the host plant and effects of repellent or inhibitory substances present in the cuticle or in the epicuticular wax layer. Morphological and chemical features of the leaf surface tend to be altered under the influence of air pollutants (see Riederer, 1989; Turunen & Huttunen, 1990), and such alterations could, therefore, lead to the inability of a pathogen or an insect to recognize its host plant or to a breakdown of defence substances. Osswald & Elstner (1987) found that waxes from *Picea abies* needles contained a compound with fungitoxic activity, *p*-hydroxy-acetophenol (*p*-HAP), and that the content of *p*-HAP was significantly reduced if the needles had been exposed to 45 ppb ozone.

Germination of spores depends on the presence of water, and thus any increase in surface wettability leading to an increase in contact time or area between a surface solution and the cuticle is likely to interact with germination. It appears that virtually every investigation of the effect of exposure to air pollutants on surface wettability (as measured by the contact angle of a water droplet on the surface) demonstrated an increase in hydrophilicity of the leaf surface (Turunen & Huttunen, 1990; Jagels, 1994). It is not clear, however, whether contact angle and water retention on the leaf are in fact correlated (cf. Haines *et al.*, 1985). A change in water retention or cuticular permeability for solutes (e.g., sugars) will also affect the amount of solutes deposited as airborne particles or leached from the leaf interior, which again might affect pathogen behaviour and success (cf. Fiala *et al.*, 1990).

Many fungi appear to use the enzyme cutinase to dissolve the cuticular matrix beneath an appressorium, but the respective importance of enzymatic and mechanic penetration of the cuticle in many cases is still unclear (e.g., Bonnen & Hammerschmidt, 1989; Stahl & Schäfer, 1992). If the cuticle is penetrated mechanically, be it by a probing insect or a fungal penetration hypha, its mechanical strength and thickness may play an important role (e.g., Jhooty & McKeen, 1965; Walker, 1988; Yang *et al.*, 1992). Cuticular thickness was found to be influenced positively or negatively, respectively, in two clones of *Picea sitchensis* by exposure to acid rain for 10 weeks from bud break (Percy & Baker, 1990). Thickness of the cuticle also increased in *Picea rubens* Sarg. seedlings exposed to 250 ppb ozone and acidic fog at pH 4.2 or 3.0 for 11 weeks from bud break, but decreased when the plants were exposed to 70 ppb ozone and fog at pH 3.0 (Percy *et al.*, 1992). *Phaseolus vulgaris*, *Pisum sativum* L. and *Vicia faba* but not *Brassica napus* plants responded with a decrease in cuticular thickness to a treatment with acid rain at pH 4.2 or lower (Percy & Baker, 1987). The mechanical strength of isolated cuticles from *Citrus aurantium*, *Ficus elastica* and *Hedera helix* leaves which had been treated for 10 months with intermittent exposures to ozone (up to 130 ppb) and acidic mist of pH 3.0 (see above) was not affected by the treatment (Kerstiens, 1988). A treatment of isolated cuticles with 2.5% ozone for 1 hour, however, resulted in a significantly lower mechanical stability in two out of four species tested (Kerstiens, 1988).

Amounts of waxes were sometimes found to be decreased by exposure to air pollutants (e.g., Günthardt-Goerg & Keller, 1987: *Picea abies*, 150 ppb ozone for 6 months; Percy & Baker, 1987: *Phaseolus vulgaris*, acid rain at pH $\leq$4.6 for

17 days), but were not affected in other studies (e.g., Barnes & Brown, 1990: *Picea abies*, 100 ppb ozone and/or mist at pH 3.6 for 2 years; Percy & Baker, 1990: *Picea sitchensis*, acid rain of pH 2.6–4.2 for 10 weeks from bud break; Turunen & Huttunen, 1991: *Pinus sylvestris*, acid rain at pH 3 for 4 years). Tsumuki *et al.* (1989) have shown that the amount of waxes on the surface of barley cultivars (*Hordeum* sp.) was negatively correlated with the degree of aphid infestation in the field. Eigenbrode & Shelton (1992) demonstrated that a substantial reduction in the amount of waxes in *Brassica oleracea* L. strongly reduced the acceptance of cabbage leaves by larvae of the diamondback moth (*Plutella xylostella* L.). The effect was caused by application of a thiocarbamate herbicide (not toxic to the larvae), and was accompanied by a change in wax composition. A similar effect was observed earlier on leaves of genetically glossy plants lacking the normal wax bloom (Eigenbrode *et al.*, 1991).

The accelerated fusion of wax crystallites, in particular on conifer needles, into a morphologically amorphous wax layer is a typical observation made in many air pollutant studies (Riederer, 1989; Turunen & Huttunen, 1990). Thus it is interesting to note that Carver *et al.* (1990), investigating the infection of three grass species by the powdery mildew fungus *Erysiphe graminis* D.C., found that leaf surfaces covered by amorphous sheet waxes were highly resistant, whereas such surfaces of the same species with crystalline plate waxes were infected by the fungus. Removal of the amorphous waxes led to the formation of normal infection structures. Desprez-Loustau & Le Menn (1989) made a somewhat similar observation when they treated *Pinus pinaster* Ait. with *Melampsora pinitorqua* Rostr., a fungus causing pine twisting rust. The formation of appressoria was rarely observed on shoots with a poorly developed or degraded wax structure.

CONCLUSIONS

Whereas it was shown that cuticular water permeability of healthy broadleaf plants was neither affected by SO_2, O_3 or acid precipitation, this was not the case with respect to ion permeability. Acid precipitation applied to growing leaves from bud break can affect the cuticular barrier properties to ions positively or negatively. The question remains, however, whether effects of acid precipitation on ion-transport properties of the cuticle are indeed significant in physiological terms, compared with effects on cell membrane permeability and composition of the leaf surface solution, respectively.

In contrast to findings with growing leaves, neither acid nor ozone caused effects when applied to isolated cuticles at very high concentrations for relatively short periods. This indicates that the interaction between air pollutants and plant cuticles is not mainly due to chemical reactions on the plant surface. It is rather mediated by plant metabolism. It is largely unknown, however, for how long the efficacy of the cuticle as a barrier is actively maintained by metabolic processes once a leaf has reached its final size. Riederer (1989) pointed out that the tubular wax crystals typically observed on conifer needles were inherently instable structures in terms of their energetic state, and thus the really intriguing fact was not the gradual change towards a morphologically amorphous appearance of epicuticular waxes, but the preservation of such physically labile structures for periods of months or even years by the plant. He concluded that conifers most probably continue to replace their waxes even in fully grown needles. Only when a needle can no longer sustain the metabolic effort necessary, ageing of epicuticular structures becomes observable.

Pollutant stress, as well as any other biotic or abiotic stress, might well interact with the amount of resources allocated to the maintenance of cuticular structure. Perhaps it is a telling fact that the only effect of air pollutants on cuticular water permeability was observed on plants which were in a rather bad shape. There is a wide open field for multiple-stress experiments.

There is also a wide spectrum of possible and probable effects of air pollutants on interactions between pests, pathogens and the plant surface on which research has hardly started yet. Potential effects might as well decrease or increase plant resistance to pathogens. It appears that research in this area is most urgently needed.

ACKNOWLEDGEMENTS

Thanks are due to the European Environmental Research Organisation and Lancaster University for their generous support during the preparation of this manuscript.

REFERENCES

Baker, E.A. 1982. Chemistry and morphology of plant epicuticular waxes. In Cutler, D.F., Alvin, K.L. & Price, C.E. (eds) *The Plant Cuticle*: 139-165. Academic Press, London.

Barnes, J.D. & Brown, K.A. 1990. The influence of ozone and acid mist on the amount and wettability of the surface waxes in Norway spruce [*Picea abies* (L.) Karst.]. *New Phytol.* **114**: 531-535.

Barnes, J.D. & Davison, A.W. 1988. The influence of ozone on the winter hardiness of Norway spruce [*Picea abies* (L.) Karst.]. *New Phytol.* **108**: 159-166.

Bauer, H. & Schönherr, J. 1992. Determination of mobilities of organic compounds in plant cuticles and correlation with molar volumes. *Pestic. Sci.* **35**: 1-11.

Becker, M., Kerstiens, G. & Schönherr, J. 1986. Water permeability of plant cuticles: Permeance, diffusion and partition coefficients. *Trees* **1**: 54-60.

Bonnen, A.M. & Hammerschmidt, R. 1989. Role of cutinolytic enzymes in infection of cucumber by *Colletotrichum lagenarium*. *Physiol. Mol. Plant Path.* **35**: 475-481.

Cape, J.N. & Fowler, D. 1981. Changes in epicuticular wax of *Pinus sylvestris* exposed to polluted air. *Silva Fenn.* **15**: 457-458.

Carver, T.L.W., Thomas, B.J., Ingerson-Morris, S.M. & Roderick, H.W. 1990. The role of the abaxial leaf surface waxes of *Lolium* spp. in resistance to *Erysiphe graminis*. *Plant Path.* **39**: 573-583.

Chamel, A., Escoubes, M., Baudrand, G. & Girard, G. 1992. Determination of water sorption by cuticles isolated from fir tree needles. *Trees* **6**: 109-114.

Desprez-Loustau, M.L. & Le Menn, R. 1989. Epicuticular waxes and *Melampsora pinitorqua* Rostr. pre-infection behaviour on maritime pine shoots. *Eur. J. For. Path.* **19**: 178-188.

Eigenbrode, S.D. & Shelton, A.M. 1992. Survival and behaviour of *Plutella xylostella* larvae on cabbages with leaf waxes altered by treatment with S-ethyl dipropylthiocarbamate. *Entomol. Exp. Appl.* **62**: 139-145.

Eigenbrode, S.D., Stoner, K.A., Shelton, A.M. & Kain, W.C. 1991. Characteristics of glossy leaf waxes associated with resistance to diamondback moth (Lepidoptera: Plutellidae) in *Brassica oleracea*. *J. Econ. Entomol.* **84**: 1609-1618.

Evans, L.S., Curry, T.M. & Lewin, K.F. 1981. Responses of leaves of *Phaseolus vulgaris* L. to simulated acidic rain. *New Phytol.* **88**: 403-420.

Evans, L.S., Santucci, K.A. & Patti, M.J. 1985. Interactions of simulated rain solutions and leaves of *Phaseolus vulgaris* L. *Environ. Exp. Bot.* **25**: 31-40.

Fiala, V., Glad, C., Martin, M., Jolivet, E. & Derridj, S. 1990. Occurrence of soluble carbohydrates on the phylloplane of maize (*Zea mays* L.): Variations in relation to leaf heterogeneity and position on the plant. *New Phytol.* **118**: 609-615.

Foster, J.R. 1990. Influence of pH and plant nutrient status on ion fluxes between tomato plants and simulated acid mists. *New Phytol.* **116**: 475-485.

Garrec, J.P. & Kerfourn, C. 1989. Effets de pluies acides et de l'ozone sur la perméabilité à l'eau et aux ions de cuticules isolées. *Environ. Exp. Bot.* **29**: 215-228.

Günthardt-Goerg, M.S. & Keller, T. 1987. Some effects of long-term ozone fumigation on Norway spruce. II. Epicuticular wax and stomata. *Trees* **1**: 145-150.

Haines, B.L., Jernstedt, J.A. & Neufeld, H.S. 1985. Direct foliar effects of simulated acid rain. II. Leaf surface characteristics. *New Phytol.* **99**: 407-416.

Hasler, T., Frey, B. & Schüepp, H. 1990. Interaktionen zwischen Pflanzen, Mikroorganismen, Wirkstoffen und Luftschadstoffen. *Bot. Helv.* **100**: 121-131.

Holloway, P.J. 1982. Structure and histochemistry of plant cuticular membranes: An overview. In Cutler, D.F., Alvin, K.L. & Price, C.E. (eds) *The Plant Cuticle*: 1-31. Academic Press, London.

Huttunen, S. 1984. Interactions of disease and other stress factors with atmospheric pollution. In Treshow, M. (ed.) *Air Pollution and Plant Life*: 321-356. Wiley, New York.

Huttunen, S., Turunen, M. & Reinikainen, J. 1989. Studies on Scots pine (*Pinus sylvestris* L.) and Norway spruce [*Picea abies* (L.) Karst.] needle cuticles. *Annu. Sci. For.* (suppl.) **46**: 553s-556s.

Jagels, R. 1994. Effects of air pollutants on leaf wettability. In Percy, K.E., Cape, J.N., Jagels, R. & Simpson, C.J. (eds) *Air Pollutants and the Leaf Cuticle (Ecological Sciences, vol. 36)*: 97-105. Springer-Verlag, Berlin.

Jhooty, J.S. & McKeen, W.E. 1965. Studies on powdery mildew of strawberry caused by *Sphaerotheca macularis*. *Phytopathology* **55**: 281-285.

Kamp, H. 1930. Untersuchungen über Kutikularbau und kutikuläre. Transpiration von Blättern. *Jb. wiss. Bot.* **72**: 403-465.

Kerstiens, G. 1988. Funktionelle Veränderungen der pflazlichen Kutikula durch Ozon. Doctoral Thesis, Technische Universität München, FRG.

Kerstiens, G. 1994. Air pollutants and plant cuticles: Mechanisms of gas and water transport, and effects on water permeability. In Percy, K.E., Cape, J.N., Jagels, R. & Simpson, C.J. (eds) *Air Pollutants and the Leaf Cuticle (Ecological Sciences, vol. 36)*: 39-53. Springer-Verlag, Berlin.

Kerstiens, G. 1995. Cuticular water permeance of European trees and shrubs grown in polluted and unpolluted atmospheres, and its relation to stomatal response to humidity in beech (*Fagus sylvatica* L.). *New Phytol.* **129**: 495-503.

Kerstiens, G. & Lendzian, K.J. 1989. Interactions between ozone and plant cuticles. II. Water permeability. *New Phytol.* **112**: 21-27.

Kerstiens, G., Federholzner, R. & Lendzian, K.J. 1992. Dry deposition and cuticular uptake of pollutant gases. *Agric. Ecosystems Environ.* **42**: 239-253.

Klemm, O. 1989 (with contributions by 19 co-authors). Leaching and uptake of ions through above-ground Norway spruce tree parts. In Schulze, E.-D., Lange, O.L. & Oren, R. (eds) *Forest Decline and Air Pollution* (Ecological Studies vol. 77): 210-237. Springer-Verlag, Berlin.

Knittel, R. & Pell, E.J. 1991. Effects of drought stress and simulated acidic rain on foliar conductance of *Zea mays* L. *Environ. Exp. Bot.* **31**: 79-90.

Krause, G.H.M. & Kaiser, H. 1977. Plant response to heavy metals and sulphur dioxide. *Environ. Pollut.* **12**: 63-71.

Laurence, J.A. 1981. Effects of air pollutants on plant-pathogen interactions. *Z. Pflanzenkrankh. Pflanzenschutz* **87**: 156-172.

Lendzian, K.J. & Kerstiens, G. 1991. Sorption and transport of gases and vapors in plant cuticles. *Rev. Environ. Contam. Toxicol.* **121**: 65-128.

Lucas, P.W., Cottam, D.A. & Mansfield, T.A. 1987. A large-scale fumigation system for investigating interactions between air pollution and cold stress on plants. *Environ. Pollut.* **43**: 15-28.

Magan, N. & McLeod, A.R. 1991. Effects of atmospheric pollutants on phyllosphere microbial communities. In Andrews, J.H. & Hirano, S.S. (eds) *Microbial Ecology of Leaves*: 379-400. Springer-Verlag, Berlin.

Mengel, K., Hogrebe, A.M.R. & Esch, A. 1989. Effect of acidic fog on needle surface and water relations of *Picea abies*. *Physiol. Plant.* **75**: 201-207.

Miller, H.G. 1984. Deposition-plant-soil interactions. *Phil. Trans. R. Soc. London* **B305**: 339-352.

Moldau, H., Sober, J. & Sober, A. 1990. Differential sensitivity of stomata and mesophyll to sudden exposure of bean shoots to ozone. *Photosyn.* **24**: 446-458.

Neighbour, E.A., Cottam, D.A. & Mansfield, T.A. 1988. Effects of sulphur dioxide and nitrogen dioxide on the control of water loss by birch (*Betula* spp.). *New Phytol.* **108**: 149-157.

Neighbour, E.A., Pearson, M. & Mehlhorn, H. 1990. Purafil-filtration prevents the development of ozone-induced frost injury: A potential role for nitric oxide. *Atmos. Environ.* **24A**: 711-715.

Nip, M., Tegelaar, E.W., de Leeuw, J.W. & Schenck, P.A. 1986. A new non-saponifiable highly aliphatic and resistant biopolymer in plant cuticles. *Naturwiss.* **73**: 579-585.

Osswald, W.F. & Elstner, E.F. 1987. Investigations on spruce decline in the Bavarian Forest. *Free Rad. Res. Comms.* **3**: 185-192.

Percy, K.E. & Baker, E.A. 1987. Effects of simulated acid rain on production, morphology and composition of epicuticular wax and on cuticular membrane development. *New Phytol.* **107**: 577-589.

Percy, K.E. & Baker, E.A. 1988. Effects of simulated acid rain on leaf wettability, rain retention and uptake of some inorganic ions. *New Phytol.* **108**: 75-82.

Percy, K.E. & Baker, E.A. 1989. Effect of simulated acid rain on foliar uptake of Rb^+ and SO_4^{2-} by two clones of Sitka spruce [*Picea sitchensis* (Bong.) Carr.]. In Bucher, J.B. & Bucher-Wallin, I. (eds) *Air Pollution and Forest Decline*: 493-495. Proc. 14th Int. Meeting for Specialists in Air Pollution Effects on Forest Ecosystems, EAFV, Birmensdorf, Switzerland.

Percy, K.E. & Baker, E.A. 1990. Effects of simulated acid rain on epicuticular wax production, morphology, chemical composition and on cuticular membrane thickness in two clones of Sitka spruce [*Picea sitchensis* (Bong.) Carr.]. *New Phytol.* **116**: 79-87.

Percy, K.E., Jensen, K.F. & McQuattie, C.J. 1992. Effects of ozone and acidic fog on red spruce needle epicuticular wax production, chemical composition, cuticular membrane ultrastructure, and needle wettability. *New Phytol.* **122**: 71-80.

Reynhardt, E.C. & Riederer, M. 1994. Structures and molecular dynamics of plant waxes. II. Cuticular waxes from leaves of *Fagus sylvatica* L. and *Hordeum vulgare* L. *Eur. Biophys. J.* **23**: 59-70.

Riederer, M. 1989. The cuticle of conifers: Structure, composition and transport properties. In Schulze, E.-D., Lange, O.L. & Oren, R. (eds) *Forest Decline and Air Pollution* (Ecological Studies vol. 77): 157-192. Springer-Verlag, Berlin.

Riederer, M. & Schneider, G. 1990. The effect of the environment on the permeability and composition of *Citrus* leaf cuticles. II. Composition of soluble cuticular lipids and correlation with transport properties. *Planta* **180**: 154-165.

Scherbatskoy, T. & Tyree, M.T. 1990. Kinetics of exchange of ions between artificial precipitation and maple leaf surfaces. *New Phytol.* **114**: 703-712.

Schönherr, J. & Huber, R. 1977. Plant cuticles are polyelectrolytes with isoelectric points around three. *Plant Physiol.* **59**: 145-150.

Schönherr, J. & Riederer, M. 1989. Foliar penetration and accumulation of organic chemicals in plant cuticles. *Rev. Environ. Contam. Toxicol.* **108**: 1-70.

Stahl, D.J. & Schäfer, W. 1992. Cutinase is not required for fungal pathogenicity on pea. *Plant Cell* **4**: 621-629.

Tsumuki, H., Kanehisa, K. & Kawada, K. 1989. Leaf surface wax as a possible resistance factor of barley to cereal aphids. *Appl. Ent. Zool.* **24**: 295-301.

Turunen, M. & Huttunen, S. 1990. A review of the response of epicuticular wax of conifer needles to air pollution. *J. Environ. Qual.* **19**: 35-45.

Turunen, M. & Huttunen, S. 1991. Effect of simulated acid rain on the epicuticular wax of Scots pine needles under northerly conditions. *Can. J. Bot.* **69**: 412-419.

Walker, G.P. 1988. The role of leaf cuticle in leaf age preference by bayberry whitefly (Homoptera: Aleyrodidae) on lemon. *Annu. Entomol. Soc. Am.* **81**: 365-369.

Yang, J., Verma, P.R. & Lees, G.L. 1992. The role of cuticle and epidermal cell wall in resistance of rapeseed and mustard to *Rhizoctonia solani*. *Plant and Soil* **142**: 315-321.

7 Disturbances in Stomatal Behaviour in Plants Exposed to Air Pollution

T.A. MANSFIELD & M. PEARSON

INTRODUCTION

Stomata play a vital role for higher land plants, enabling them to achieve the gaseous intake necessary for photosynthesis while providing a highly effective means of regulating water loss. The diurnal pattern of the movements of stomata often appears very simple to the casual observer, viz. they open during daytime and close at night. Detailed physiological analysis has shown, however, that there is great complexity in the control mechanisms underlying this pattern. It has been demonstrated experimentally that if stomata are given artificial treatments to cause them to open to their full aperture, normal mesophytic plants are unable to survive for long in ordinary conditions of transpiration (Mansfield & Davies, 1981). This is because the rate of loss of water through wide-open stomata exceeds the rate at which it can be supplied to the leaves via the xylem conduits, even when the roots are in moist soil.

Avoidance of water stress in plants does, therefore, involve regulatory mechanisms which achieve a restriction of stomatal opening to a level which the plant can tolerate in any particular combination of environmental (aerial and edaphic) conditions. Stomata respond to a wide range of environmental factors, especially light (photon fluence, changes in wavelength, duration), carbon dioxide, water vapour, mechanical stimuli such as wind, and changes in temperature. They are also greatly influenced by factors within the plant, principally endogenous circadian rhythms and phytohormones. The latter have a profound involvement which has begun to be revealed only in the last few years. In particular, abscisic acid (ABA) appears to play a central role in root-to-shoot signalling. When roots are in drying soil, the stomata on the leaves receive a signal (almost certainly ABA) which causes them to close partially, reducing the rate of water consumption (Davies & Zhang, 1991). It has been known for a long time that some external agents can affect stomatal behaviour. For example, several plant pathogens cause disturbances that can have serious physiological consequences (Ayres, 1981). Some of the major air pollutants (SO_2, NO_2, O_3) have also been found to induce changes in the diurnal pattern of stomatal movements. In this chapter we shall look at some of the effects of pollutants that have been explored in sufficient detail to allow phenomenological analyses to be made.

Plant Response to Air Pollution. Edited by Mohammad Yunus and Muhammad Iqbal

EFFECTS OF SULPHUR DIOXIDE POLLUTION

The action of sulphur dioxide (SO_2) on stomata cannot be summarized simply, because almost every conceivable type of response has been reported in the literature. Darrall (1989) has provided a useful summary of the outcome of studies on a wide range of species. The situation is too complex to allow satisfactory conclusions to be drawn, but in general it appears that short-term fumigations with SO_2, particularly if the concentration is low (<50 ppb), often cause wider stomatal opening, while longer-term fumigations with higher concentrations of SO_2 usually cause partial stomatal closure. The reasons for these differences cannot be fully explained, although it is possible to put forward some tentative explanations. First, it is important to recognize that there can be *direct* and *indirect* effects of external factors on stomatal guard cells. *Direct* action might involve an attack by pollutant molecules on sensitive sites in the guard cells, for example within the plasma membrane. *Indirect* action might occur if the pollutant were to perturb metabolism elsewhere in the leaf, causing changes that eventually have an impact on stomatal functioning. A particularly important form of indirect action on stomata is when there are changes in intercellular CO_2 concentration. Disturbances in respiration and/or photosynthesis usually cause changes in the intercellular CO_2 concentration. Guard cells are sensitive to CO_2 on their inward-facing surfaces (Mott, 1988) and so stomatal aperture is under partial control of the intercellular CO_2 concentration. Another indirect action is when a perturbation causes some of the epidermal cells, but not the stomatal guard cells, to lose turgor. This can lead to stomatal opening because the guard cells encounter less mechanical resistance to their lateral expansion. There are very few studies which provide any insight into the effects of SO_2 on stomata at the cellular level. The technical difficulties of performing meaningful experiments are considerable, and few investigators have attempted to surmount them. One of the most interesting and rewarding studies was performed by Taylor *et al.* (1981). They made use of detached epidermis from *Vicia faba* L. which they exposed to aqueous solutions of SO_2 (buffered H_2SO_3). When SO_2 enters the leaf by diffusing through the stomata it must dissolve in extracellular water before it can exert any effect on cellular function. Thus the use of buffered H_2SO_3 to expose cells to different concentrations of SO_2 is arguably realistic. The experimental data showed clearly that low SO_2 concentrations induced stomatal opening on detached epidermis, and that higher concentrations caused closure. These experimental results were important because they are in accord with the conclusions reached in studies on intact plants, namely that low SO_2 concentrations cause stomatal opening and high concentrations cause closure. They agree particularly well with other research also on *Vicia faba* (e.g. Black & Black, 1979) where there was a clear enhancement of opening by SO_2 concentrations as low as 17.5 ppb. Black & Black took epidermal strips from control and polluted plants of *V. faba* and examined the guard cells and their neighbours under the microscope. They found that exposure to low concentrations of SO_2 in the range 17.5 to 175 ppb had affected the survival of epidermal cells, particularly on the abaxial surface (Table 7.1).

They estimated the viability of cells using the vital stain neutral red. Black & Black pointed out difficulties in interpreting their findings. It was not clear whether the loss of viability of epidermal cells was indicative of the situation on the intact leaf, or whether the procedure of stripping the epidermis had caused the damage they observed. If the latter were the case, then the data in Table 7.1

Table 7.1. Mean leaf conductance and cell survival in epidermal strips from control and SO_2-polluted plants of *Vicia faba*. EC and GC refer to adjacent epidermal cells and guard cells respectively. The standard error of the mean is shown in parentheses (from Black & Black, 1979; by permission of Oxford University Press)

SO_2 concentration (ppb)	Leaf conductance (cm s^{-1})	Cell survival (%)			
		Adaxial		Abaxial	
		EC	GC	EC	GC
0	0.34	87.7 (1.01)	96.1 (1.61)	64.2 (2.54)	100.0 (0)
17.5	0.41	81.6 (2.51)	93.1 (3.13)	35.4 (1.02)	97.5 (2.5)
0	0.35	76.5 (2.25)	93.0 (1.23)	54.8 (2.50)	93.1 (3.12)
70	0.41	59.7 (2.91)	92.0 (3.74)	27.7 (2.82)	91.3 (2.95)
0	0.23	88.3 (2.77)	94.7 (1.53)	—	—
70	0.27	69.7 (4.70)	81.3 (2.41)	—	—

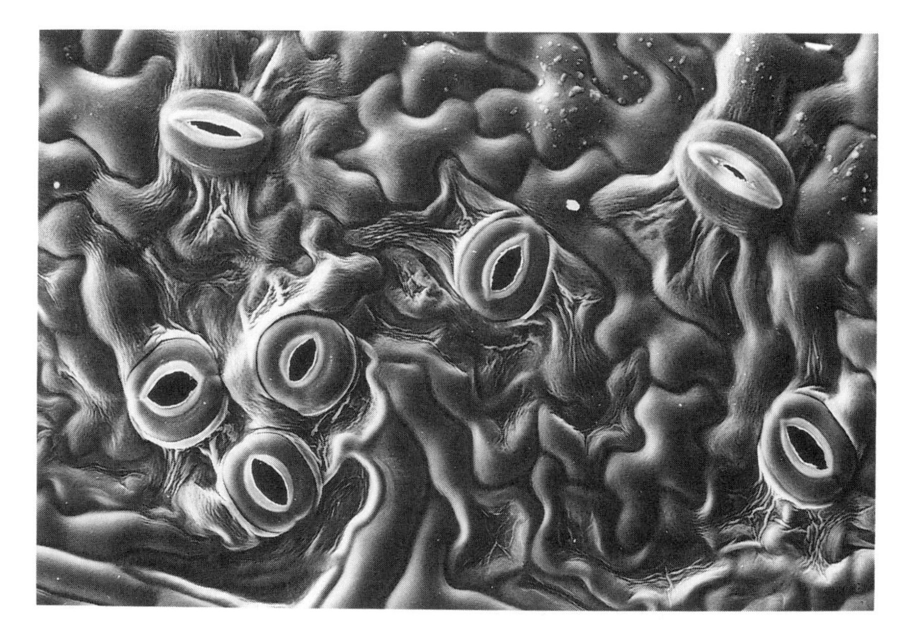

Figure 7.1. The abaxial surface of a frozen hydrated leaf of *Betula pubescens* showing the damaged epidermal cells after exposure to 40 ppb SO_2 + 40 ppb NO_2 for 40 days (courtesy of E.A. Neighbour & K. Oates, Lancaster University)

should be taken as an indication of changes in the cells that had rendered them more vulnerable to stripping. Later studies have suggested that SO_2 pollution can affect the integrity of cells of the epidermis *in situ*. Yunus *et al.* (1985) took pieces of leaf from *Calendula officinalis* L. and *Dahlia rosea* Car. and fixed them for scanning electron microscopy. SO_2-induced opening of the stomata was particularly evident in *D. rosea* and there was evidence of accompanying damage to the cells surrounding the stomata. Neighbour *et al.* (1988) made a detailed study of the water relations of young trees of *Betula pendula* Roth. and *B. pubescens* Ehr. as affected by SO_2 and NO_2 pollution. Abaxial surfaces of leaves of *B. pubescens* were examined by low-temperature scanning electron microscopy, which enabled frozen hydrated material to be studied. In leaves that had been exposed to 40 ppb $SO_2 + 40$ ppb NO_2 for 40 days, patches of stomata with abnormally wide openings were identified, and damage to nearby epidermal cells could be clearly seen (Figure 7.1). These studies by Neighbour *et al.* (1988) were coupled to an investigation of stomatal behaviour and transpiration of whole plants. It was evident that daily stomatal opening was affected by exposure to $SO_2 + NO_2$, and this had a marked influence on the rate of transpiration (Figure 7.2). When

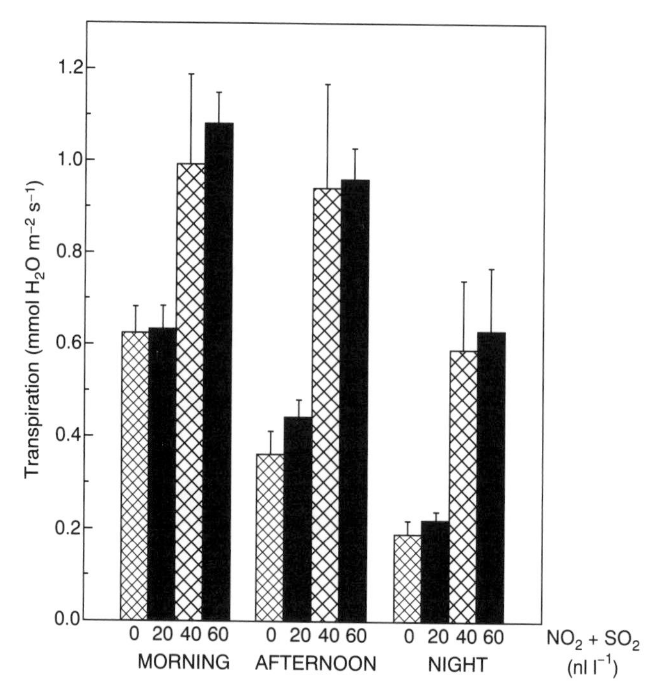

Figure 7.2. Transpiration rates of clonal *Betula pubescens* plants during the morning, afternoon and night. The plants had been fumigated in clean air, and 20, 40 or 60 ppb SO_2 and NO_2 for approximately 1 month at the time of measurement. Each column represents the mean of transpiration rates on three plants. The vertical bars represent standard errors × 2 (from Neighbour *et al.*, 1988; by permission of Cambridge University Press)

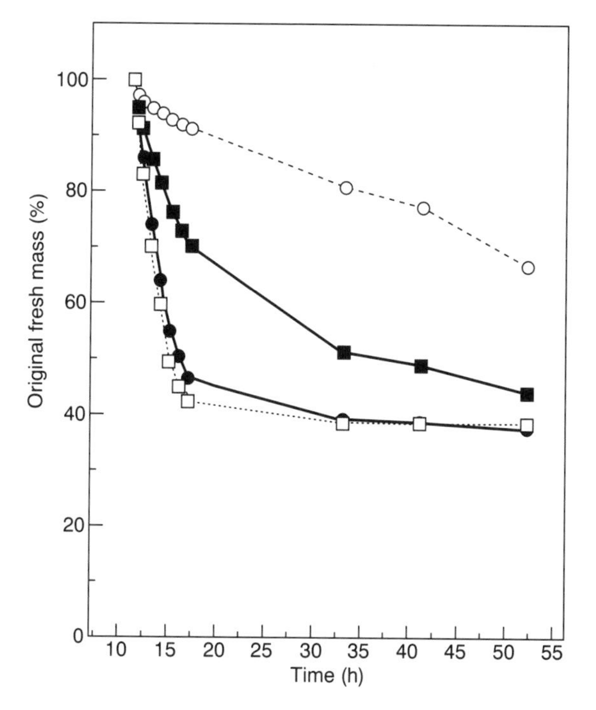

Figure 7.3. Changes in mass over time of leaves of one clonal type of *Betula pendula* after excision from plants which had been exposed for the previous 29 days to ambient air (○), or to SO_2 plus NO_2 at 20 ppb (■), 40 ppb (●) or 60 ppb (□). Values are means of nine replicates (from Neighbour *et al.*, 1988; by permission of Cambridge University Press)

leaves were detached and allowed to wilt, the rate at which the stomata closed was also greatly affected by the pollution regime under which they had grown (Figure 7.3).

Despite the clear indications of pollution-induced stomatal opening that have come from investigations such as these, there are nevertheless many reports in the literature of partial stomatal *closure* caused by SO_2 or $SO_2 + NO_2$. Atkinson *et al.* (1991) made a detailed study of spring barley (*Hordeum vulgare* L. cv. Klaxon) and found that on intact, well-watered plants there was little effect of 35 ppb $SO_2 + NO_2$, apart from a very small reduction in stomatal opening which persisted through the photoperiod (Figure 7.4). However, when a solution of abscisic acid (ABA) was fed through the transpiration stream to detached leaves from polluted and control plants, the stomata of those previously exposed to air pollution responded less rapidly to the hormone (Figure 7.5). The effects of pollution on the stomatal response to ABA raised the possibility that the mechanism of turgor reduction in the guard cells, which is an essential feature of stomatal closure, might in some way be impaired. To explore this question, the responses of stomata to CO_2 on leaves of control and polluted plants were also examined

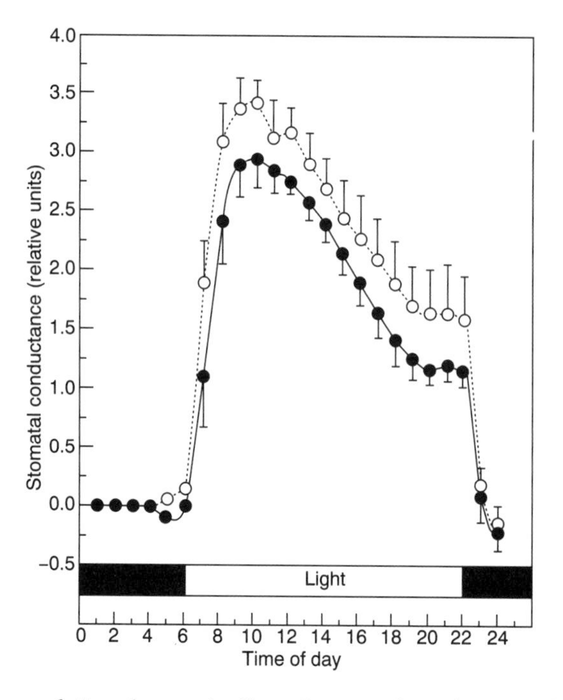

Figure 7.4. The mean relative changes in diurnal stomatal conductance for attached leaves of *Hordeum vulgare* L. cv. Klaxon, previously exposed to either ambient air (○) or to 35 ppb $SO_2 + 35$ ppb NO_2 (●). Results represent the means (±SE) of nine plants (from Atkinson *et al.*, 1991; by permission of Cambridge University Press)

by Atkinson *et al.* (1991). It was found that there was no difference in the responsiveness to CO_2 (Figure 7.6), but this experiment confirmed the conclusion from a previous one (Figure 7.4) that the treatment with $SO_2 + NO_2$ caused a slight depression of stomatal conductance in barley. Thus the effect of $SO_2 + NO_2$ on the response of stomata to ABA may be specific to this hormone. The implications of this will be discussed in detail later.

It is clear that the effects of SO_2 (alone or in combination with NO_2) on stomatal behaviour are very complex. The studies on barley perhaps do, however, provide an explanation of conflicting reports in the literature. Prior to treatment with ABA, stomatal conductance was slightly reduced in the polluted plants, but after the application of ABA the opposite was true, i.e., stomatal conductance was higher in the polluted leaves. Thus the pretreatment of the plants may be all-important in determining the nature of the stomatal response to $SO_2 + NO_2$. Well-watered plants with little endogenous ABA may show a different reaction from mildly stressed plants with an elevated ABA content.

EFFECTS OF OZONE POLLUTION

The responses of stomata to ozone (O_3) have been studied more extensively than to SO_2, and for a longer period of time. Unfortunately this has not resulted in a

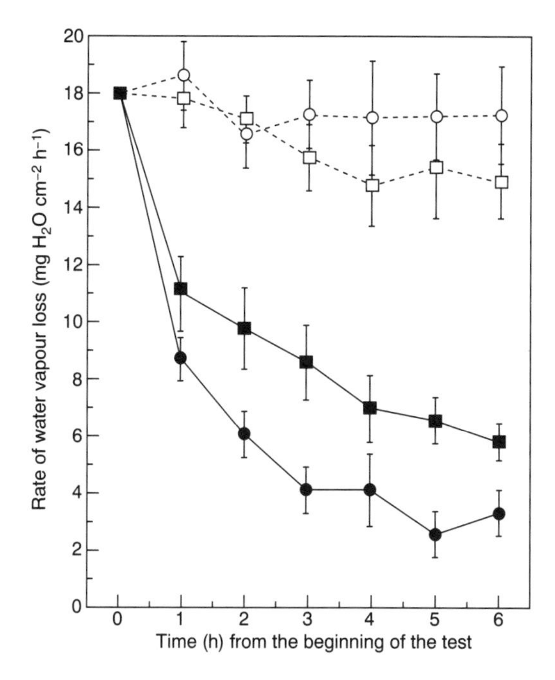

Figure 7.5. ABA response tests on *Hordeum vulgare* plants after exposure to pollution levels as in Figure 7.4 for 6–11 days. The results shown are the rates of water vapour loss per unit area (mg cm^{-2} h^{-1}) from illuminated detached leaves supplied with either distilled water or 10^{-1} mol m^{-3} ABA solution. The youngest fully expanded leaf was selected for each test. The figure shows the combined data from three separate occasions (means $\pm$ 1 SE) during which each treatment was replicated 10 times (i.e., data are shown for 30 plants per treatment). Symbols: leaves previously exposed to ambient air (controls) and supplied with distilled water (○) with ABA solution (●), or exposed to $SO_2 + NO_2$ and supplied with distilled water (□) or with ABA solution (■). Temperature was maintained at 25°C and PAR was 200 μmol quanta m^{-2} s^{-1} (from Atkinson *et al.*, 1991; by permission of Cambridge University Press)

picture that is any clearer, indeed it is remarkably similar to that for SO_2. Most studies from the 1950s onwards, conducted either with photochemical smog (real or simulated) out-of-doors, or with different ozone doses under laboratory conditions, have reported stomatal closure to varying degrees. However, many of these experiments were conducted with high concentrations of O_3 (often >500 ppb) applied for short periods of time. Reviewing the effects of more realistic doses (<200 ppb), Darrall (1989) noted that there was a diversity of response, viz. stimulation as well as suppression of stomatal conductance. A recent study illustrates this diversity. Hassan *et al.* (1994) simultaneously examined the effect of ozone on rates of gas exchange in Egyptian varieties of radish and turnip. In both species photosynthesis was found to be significantly decreased by 80 ppb O_3, but stomatal conductance was increased in radish and decreased in turnip. Scanning electron microscopy revealed that in ozone-treated radish plants there was considerable loss of turgor and collapse of epidermal cells adjacent to the stomata. This damage was not evident in the controls, or in turnip plants from either the

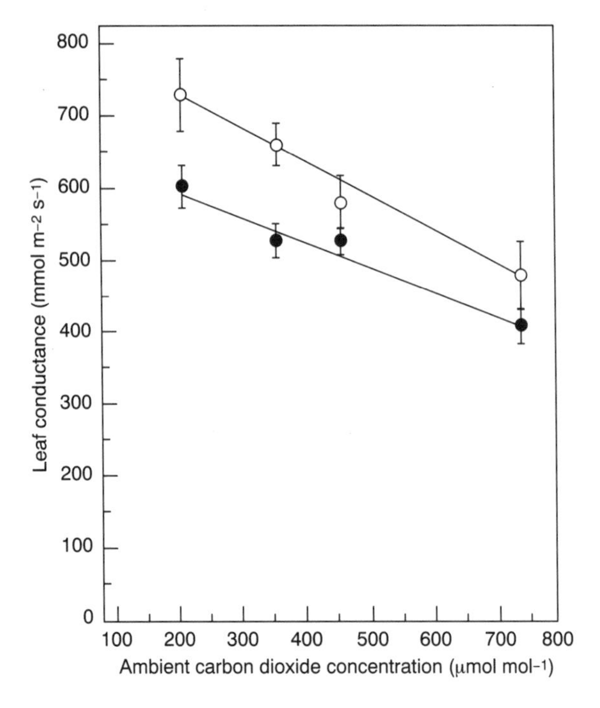

Figure 7.6. The response of leaf conductance (g) to water vapour (mmol m^{-2} s^{-1}) of *Hordeum vulgare* to different ambient CO_2 concentrations (μmol mol^{-1}). Means ±1 SE. Symbols: leaves previously exposed to ambient air (○), or to 35 ppb SO_2 + 33 ppb NO_2 (●). The slope of the lines was significantly different from zero for control and polluted plants, although there was no significant difference in the gradient between the two treatments. Regression equations are: Ambient air (control) $g = CO_2(-0.439) + 784.7$ ($R^2 = 39\%$). $SO_2 + NO_2$ (polluted) $g = CO_2(-0.319) + 629.6$ ($R^2 = 45\%$) (from Atkinson *et al.*, 1991; by permission of Cambridge University Press)

control or ozone treatment. As mentioned earlier, these epidermal cells provide mechanical resistance to the opening of stomata, and in situations where they have lost turgor there is no back pressure on the guard cells and the stomata tend to open more widely.

Ozone is a highly reactive molecule that has the potential to do great damage to many cellular activities. It can react with components of cell membranes, and in consequence produces a variety of effects such as leakage of solutes and loss of turgor control, changes in chloroplast ultrastructure and disruption of electron transport. Agents that interfere with photosynthetic CO_2 fixation are likely to cause stomatal closure simply by inducing an increase in intercellular CO_2 concentration. In view of this it is surprising how few of the researchers who have reported stomatal closure in the presence of ozone pollution have taken the trouble to perform a detailed analysis of the component processes (e.g., net photosynthesis/intercellular CO_2 concentration or A/C_i analysis). For this reason, recently completed work by Farage *et al.* (1991) is of considerable importance. Working with wheat (cv. Avalon), they observed the sequence of

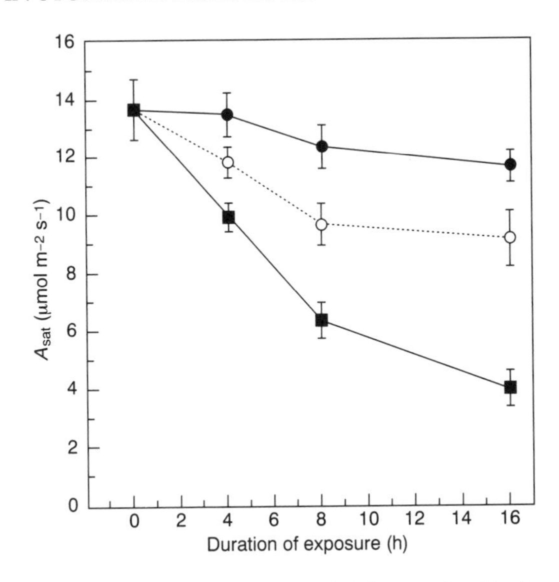

Figure 7.7. Change in the light-saturated rate of CO_2 uptake (A_{sat}) of *Triticum aestivum* leaves resulting from increasing lengths of exposure to O_3 concentrations of 200 (○) and 400 (■) ppb. Control leaves (●) were illuminated for corresponding time periods without O_3 before measurement of A_{sat}. Measurements were made at 20°C with a PPFD of 1500 µmol m^{-2} s^{-1} in an external CO_2 concentration of 340 ppm. Points are the mean of four to six replicates; vertical bars represent SE (from Farage *et al.*, 1991; by permission of American Society of Plant Physiologists)

changes in carboxylation efficiency, CO_2-saturated photosynthesis, and stomatal limitation to CO_2 uptake. They gave brief (4–16 hour) treatments with high but nevertheless fairly realistic concentrations of O_3, viz. 200 or 400 ppb. Ozone caused a substantial decrease in the light-saturated rate of photosynthesis (Figure 7.7), which occurred together with a decrease in stomatal conductance. However, stomatal limitation to CO_2 uptake could not explain the decline in the rate of photosynthesis, and it was concluded that the main causal factor was a decrease in carboxylation efficiency. The capacity for regeneration of ribulose-1,5-bisphosphate carboxylase/oxygenase (Rubisco), and the photochemical efficiency of photosystem II, were less affected by O_3 then the carboxylation efficiency. A study undertaken by Pell *et al.* (1992) revealed similar findings. They gave particular attention to the effect of ozone on the content and activity of Rubisco in hybrid poplar and radish (*Raphanus sativus* L. cv. Cherry Belle). In addition to findings concerned with effects on Rubisco, gas exchange studies revealed a decline in photosynthesis for poplar which preceded a reduction in foliar conductance. Analysis of the A/C_i response showed that there was no effect of O_3 on stomatal limitation to CO_2 assimilation except for a slight increase during the last 2 weeks of the 9-week experiment. For radish, which was grown in open-top chambers, similarly large effects on photosynthesis were observed, but in this case leaf conductance was significantly reduced. A/C_i analysis, which would have allowed determination of the stomatal limitation, was not performed on these plants. Atkinson *et al.* (1988) working on the same cultivar of radish had previously shown, however, that changes in photosynthesis caused by O_3 could not be entirely explained by reductions in stomatal conductance.

In an experiment which was specifically designed to address the question of whether stomata were directly or indirectly affected, it was found for Norway spruce that ozone caused not an increase, but a significant decrease in the stomatal limitation (Wallin *et al.*, 1992). The authors reported an additional important observation. The coupling between the stomata and photosynthesis seemed to have been affected by ozone, i.e., there was a greater spread of intercellular CO_2 concentrations. These studies by Farage *et al.* (1991), Wallin *et al.* (1992) and Pell *et al.* (1992) are of considerable importance because they suggest that stomata may be less affected directly by O_3 pollution than they are by increases in intercellular CO_2 concentration that result from inhibitory effects in the mesophyll.

Disruptions caused by ozone to the normal functioning of stomata in Norway spruce were also observed by Barnes *et al.* (1990). Ozone significantly reduced the maximum stomatal conductance in the light, and for O_3 in combination with acid mist the minimum stomatal conductance, obtained in darkness, was greater than the controls. Experiments with excised shoots showed that there was a greater rate of water loss in those trees which had previously been exposed to ozone pollution. A possible explanation for this effect was provided by Maier-Maercker (1989), who compared the degree of lignification of subsidiary and guard cell walls in plants of Norway spruce grown in control and ozone-polluted environments. Plants exposed to ozone showed reduced lignification of the external walls of the stomatal apparatus. This area of the stomatal wall is believed to be important in stomatal responses to humidity. The loss of lignification is believed to reduce the effectiveness of this sensory pathway and may thereby reduce the responsiveness of stomata. In a follow-up experiment Maier-Maercker & Koch (1991) examined the control of water balance with decreasing humidity for control and ozone-fumigated twigs of Norway spruce. The polluted plants were found to show irregular and delayed stomatal closure. Disruptions of the kind observed here may have important repercussions for plant vitality and long-term survival. Plants may become susceptible to other environmental perturbations, particularly under conditions of drought and/or decreasing humidity.

It has often been suggested that stomatal closure in the presence of a pollutant could constitute an important mechanism for avoidance of injury to internal tissues. This cannot be disputed, but if stomatal closure occurs in response to intercellular CO_2 after the mesophyll has been damaged to some degree, the protection offered may only be of limited benefit. Tingey & Hogsett (1985), in an ingenious experiment, showed that stomatal closure did confer a degree of protection from ozone. Factors within the leaf also influence the degree of sensitivity. Nearly a decade after the work of Tingey & Hogsett (1985), Guzy & Heath (1993) examined the relative sensitivity to ozone of varieties of *Phaseolus vulgaris*. They measured a set of physiological and biochemical characters which were thought to mediate the response to ozone. Their work suggested that damage appeared to result from either comparatively high stomatal conductances, or comparatively low levels of antioxidants in spite of low stomatal conductances.

Two other studies have revealed further important features of the complexity of stomatal responses to ozone. Taylor & Dobson (1989) made a detailed study of photosynthesis and gas exchange characteristics of beech (*Fagus sylvatica* L.). This is an important deciduous species in Europe which has been reported to be damaged in many locations where ozone concentrations are elevated in the summer months. Trees were grown in open-top chambers in charcoal-filtered or unfiltered air, i.e., the latter were exposed to ambient ozone pollution. Measurements of stomatal conductance were made throughout the summer from May to September, and there were differences in response to the treatments between leaves which had expanded early or late in the season. The leaves that developed during the initial shoot extension in May had lower stomatal conductances in unfiltered air, but those that developed later during the "second flush" (also known as lammas growth) had greater conductances. Determinations of net photosynthesis (A) and its relationship to intercellular CO_2 concentration (C_i) showed that there were higher rates of photosynthesis in the "lammas leaves" of plants grown in unfiltered air, which were the result of enhanced regeneration of ribulose bisphosphate (RuBP). The limitation on the regeneration of RuBP is thought to be the production of ATP and the reduction of $NADP^+$, and consequently Taylor & Dobson (1989) suggested that ozone pollution had stimulated photophosphorylation in these leaves. These observations thus revealed complexities in the responses of stomata to ozone which hitherto had not been suspected. There is no explanation at present for the differences in response between the leaves of the first and second flush, but Taylor & Dobson pointed out that the effects may be of long-term significance in a tree if they occur annually and cause temporal changes in dry matter accumulation, and in its subsequent partitioning.

Our own studies have revealed further complexities in the responses of stomata of beech to ozone (Pearson & Mansfield, 1993). Three-year-old transplants of beech were subjected to simulated summer ozone pollution in daylit fumigation chambers. Higher ozone concentrations were applied on days with greater incident sunlight (Figure 7.8), which is realistic simulation of the outside environment in those parts of northwest Europe that are subject to ozone pollution. The pollution treatments continued for 128 days beginning in mid-May, and measurements of stomatal conductance were made with a diffusion porometer on days after 1 August when the ozone concentration exceeded 100 ppb. From 38 days after 1 August, half of the trees were subjected to water stress. Leaves of the first flush on the well-watered trees displayed a decrease in stomatal conductance due to the ozone treatment, and the magnitude of the effect increased appreciably with time (Figure 7.9a). However, the situation was reversed in the case of the trees that were undergoing water stress (Figure 7.9b). Here the effect of drought on stomatal conductance was very marked, but it was less so in the polluted trees. Consequently, in trees under water stress the effect of ozone was to *increase* stomatal conductance.

The integrated effects of ozone on behaviour in beech may be of great significance as far as growth and survival are concerned. In the case of well-watered trees, ozone caused partial stomatal closure on leaves of the first flush, and therefore there may be interference with the acquisition of CO_2 for photosynthesis.

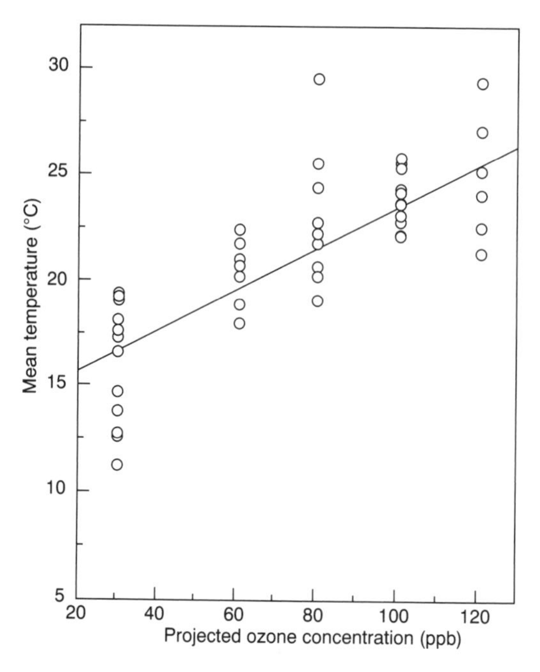

Figure 7.8. Correlation between ozone concentration and mean temperature for the experiments in Figure 7.9. Values for the regression line: $R^2 = 0.5929$, $a = 13.661$, $b = 0.0993$, $n = 58$ (from Pearson & Mansfield, 1993; by permission of Cambridge University Press)

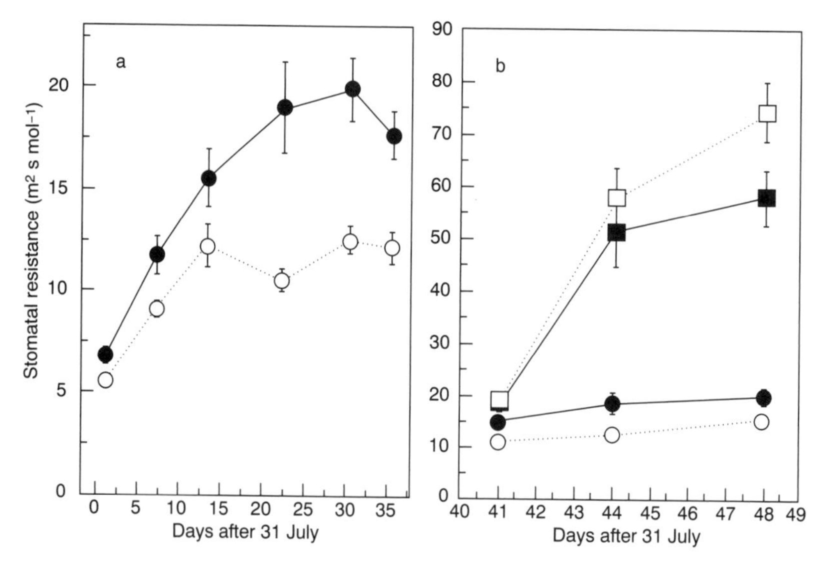

Figure 7.9. (a) The effect of episodic ozone pollution on the mean stomatal resistance in beech. Data are for well-watered trees exposed to ambient air (○) or ozone (●). (b). The interaction between ozone and water stress on stomatal resistance in beech. Data are for droughted trees exposed to ambient air (■), or ozone (□). Measurements were made after the commencement of the water stress treatment (day 39). The vertical lines represent standard errors of the means (from Pearson & Mansfield, 1993; by permission of Cambridge University Press)

But in droughted trees, the amount of stomatal closure accompanying the onset of water stress was reduced. This may represent a failure of the physiological regulation that is essential for the management of water economy at critical times.

Thus, the effects of ozone on stomatal behaviour appear to be as complex as those of SO_2. There may, however, be an important feature in common between the two pollutants: both of them are able to reduce the closing response of stomata to water stress. In both cases this effect is found with low pollutant concentrations, and it may represent a most important physiological disturbance under field conditions.

DO POLLUTANTS AFFECT STIMULUS–RESPONSE COUPLING IN GUARD CELLS?

Before attempting to answer this question it is necessary to outline current knowledge of how stomatal behaviour is regulated in plants that are under water-deficiency stress. The discovery of the physiological role of ABA in plants began to be revealed in the 1960s, and since then evaluation of its action as a "stress hormone" has been a dominant feature of the literature. Wright (1969) first reported that ABA (then known as "inhibitor β") accumulated quickly in wheat leaves when they were excised from the plant and allowed to wilt. This discovery occurred contemporaneously with the first indications that ABA is a potent inhibitor of stomatal opening (Mittelheuser & Van Steveninck, 1969). For many years thereafter it was believed that the ABA which suppresses stomatal opening is formed in the leaves when they begin to experience effects of water stress, such as a loss of turgor or an increasingly negative water potential. More recently, evidence has accumulated indicating that the ABA which initiates stomatal closure during the early stages of water stress in soil-grown plants is formed in root tips when they are in contact with drying soil (Davies & Zhang, 1991). It is transported via the xylem and arrives in the leaf epidermis where it initiates the cellular processes that lead to stomatal closure.

There have also been considerable advances recently in our understanding of events in the guard cells when ABA first arrives at the cell surface (Mansfield *et al.*, 1990). Within minutes of the application of ABA to guard cells, an increase in cytosolic free calcium concentration has been detected (McAinsh *et al.*, 1990; Schroeder & Hagiwara, 1990). It is now believed that a Ca^{2+}-dependent signal transduction system is an essential feature of the events that enable ABA to restrict the turgor of guard cells. It is also known that the initial action of ABA is at the outer surface of the plasma membrane of the guard cells (Hartung, 1983). Patch-clamp recordings of ion currents across the plasma membrane of guard cells suggested that ABA caused Ca^{2+} transients by somehow permitting passive influx of Ca^{2+} through ion channels (Schroeder & Hagiwara, 1990). Although many features of the cellular mechanisms have still to be evaluated, it is clear that the action of ABA on guard cells is intricately linked with events occurring within the plasma membrane.

McAinsh (1994) has recently obtained preliminary evidence that oxidative stress can affect Ca^{2+} homeostasis of cells, and consequently Ca^{2+}-based signal transduction may be disrupted. A dramatic effect of ozone on calcium distribution in leaf tissues was found by Fink (1991). When needles of Norway spruce [*Picea abies* (L.) Karst.] had been exposed to O_3, large deposits of calcium oxalate were located inside epidermal and subsidiary cells, rather than within the cell walls, which is their normal position within this species. For this massive relocation to occur there must have been movements of large quantities of calcium ions across cell membranes, and it is inconceivable that normal signal transduction would be unaffected.

The experimental evidence that low concentrations of SO_2 and of O_3 can reduce the responses of stomata to water stress and ABA clearly need to be examined in relation to this new information about the role of ABA transported as a signal in the plant, and its action on the guard cells themselves. Two tentative hypotheses can be put forward: (1) The pollutants cause damage to epidermal cells (cf. Figure 7.1) which form part of the pathway for the ABA as it moves towards the guard cells. In consequence the stomata respond less rapidly to endogenous ABA or to ABA applied experimentally. (2) The pollutants damage the plasma membranes of the guard cells, and the ability to respond to incoming ABA molecules is thereby reduced. Both these hypotheses seem perfectly realistic. The guard cells and the epidermal cells are directly exposed to air pollutant molecules as they enter the leaf. Future experiments may be able to put these hypotheses to critical tests.

REFERENCES

Atkinson, C.J., Robe, S.V. & Winner, W.E. 1988. The relationship between changes in photosynthesis and growth for radish plants fumigated with SO_2 and O_3. *New Phytol.* **110**: 173–184.

Atkinson, C.J., Wookey, P.A. & Mansfield, T.A. 1991. Atmospheric pollution and the sensitivity of stomata on barley leaves to abscisic acid and carbon dioxide. *New Phytol.* **117**: 535–541.

Ayres, P.G. 1981. Responses of stomata to pathogenic organisms. In Jarvis, P.G. & Mansfield, T.A. (eds) *Stomatal Physiology*: 205–221. Cambridge University Press, Cambridge.

Barnes, J.D., Eamus, D. & Brown, K.A. 1990. The influence of ozone, acid mist and soil nutrient status on Norway spruce [*Picea abies* (L.) Karst.]. I. Plant–water relations. *New Phytol.* **114**: 713–720.

Black, C.R. & Black, V.J. 1979. The effects of low concentrations of sulphur dioxide on stomatal conductance and epidermal cell survival in field beans (*Vicia faba* L.). *J. Exp. Bot.* **30**: 291–298.

Darrall, N.M. 1989. The effect of air pollutants on physiological processes in plants. *Plant, Cell Environ.* **12**: 1–30.

Davies, W.J. & Zhang, J. 1991. Root signals and the regulation of growth and development of plants in drying soil. *Annu. Rev. Plant Physiol. Plant Mol. Biol.* **42**: 55–76.

Farage, P.K., Long, S.P., Lechner, E.G. & Baker, N.R. 1991. The sequence of change within the photosynthetic apparatus of wheat following short-term exposure to ozone. *Plant Physiol.* **95**: 529–535.

Fink, S. 1991. Unusual patterns in the distribution of calcium oxalate in spruce needles and their possible relationships to the impact of pollutants. *New Phytol.* **119**: 41–51.

Guzy, M.R. & Heath, R.L. 1993. Responses to ozone of varieties of common bean (*Phaseolus vulgaris* L.). *New Phytol.* **124**: 617–626.

Hartung, W. 1983. The site of action of abscisic acid at the guard cell plasmalemma of *Valerianella locusta*. *Plant, Cell Environ.* **6**: 427-428.

Hassan, I.A., Ashmore, M.R. & Bell, J.N.B. 1994. Effects of O_3 on the stomatal behaviour of Egyptian varieties of radish (*Raphanus sativus* L. cv. Baladey) and turnip (*Brassica napa* L. cv. Sultani). *New Phytol.* **128**: 243-250.

Maier-Maercker, U. 1989. Delignification of subsidiary and guard cell walls of *Picea abies* (L.) Karst. by fumigation with ozone. *Trees* **3**: 57-64.

Maier-Maercker, U. & Koch, W. 1991. Experiments on the control capacity of stomata of *Picea abies* (L.) Karst. after fumigation with ozone and in environmentally damaged material. *Plant, Cell Environ.* **14**: 175-184.

Mansfield, T.A. & Davies, W.J. 1981. Stomata and stomatal mechanisms. In Paleg, L.G. & Aspinall, D. (eds) *The Physiology and Biochemistry of Drought Resistance in Plants*: 315-346. Academic Press, Sydney.

Mansfield, T.A., Hetherington, A.M. & Atkinson, C.J. 1990. Some current aspects of stomatal physiology. *Annu. Rev. Plant Physiol. Plant Mol. Biol.* **41**: 55-75.

McAinsh, M.R. 1994. Effects of oxidative stress on stomatal behaviour and guard cell cytosolic free Ca^{2+}. *Plant Physiol.* **S105**: 101.

McAinsh, M.R., Brownlee, C. & Hetherington, A.M. 1990. Abscisic acid-induced elevation of guard cell cytosolic Ca^{2+} precedes stomatal closure. *Nature (London)* **343**: 186-188.

Mittelheuser, C.J. & Van Steveninck, R.F.M. 1969. Stomatal closure and inhibition of transpiration induced by (RS)-abscisic acid. *Nature (London)* **221**: 281-282.

Mott, K.A. 1988. Do stomata respond to CO_2 concentration other than intercellular? *Plant Physiol.* **86**: 200-203.

Neighbour, E.A., Cottam, D.A. & Mansfield, T.A. 1988. Effects of sulphur dioxide and nitrogen dioxide on the control of water loss by birch (*Betula* spp.). *New Phytol.* **108**: 149-157.

Pearson, M. & Mansfield, T.A. 1993. Interacting effects of ozone and water stress on the stomatal resistance of beech (*Fagus sylvatica* L.) *New Phytol.* **123**: 351-358.

Pell, E.J., Eckardt, N. & Enyedi, A.J. 1992. Timing of ozone stress and resulting status of ribulose bisphosphate carboxylase/oxygenase and associated net photosynthesis. *New Phytol.* **120**: 397-406.

Schroeder, J.I. & Hagiwara, S. 1990. Repetitive increases in cytosolic Ca^{2+} of guard cells by abscisic acid activation of nonselective Ca^{2+} permeable channels. *Proc. Nat. Acad. Sci., USA* **87**: 9305-9309.

Taylor, G. & Dobson, M.C. 1989. Photosynthetic characteristics, stomatal responses and water relations of *Fagus sylvatica*: Impact of air quality at a site in southern Britain. *New Phytol.* **113**: 265-273.

Taylor, J.S., Reid, D.M. & Pharis, R.P. 1981. Mutual antagonism of sulphur dioxide and abscisic acid in the effect on stomatal aperture in broad bean (*Vicia faba* L.) epidermal strips. *Plant Physiol.* **68**: 1504-1507.

Tingey, D.T. & Hogsett, W.E. 1985. Water stress reduces ozone injury via a stomatal mechanism. *Plant Physiol.* **77**: 944-947.

Wallin, G., Ottosson, S. & Sellden, G. 1992. Long-term exposure of Norway spruce, *Picea abies* (L.) Karst., to ozone in open-top chambers. IV. Effects on the stomatal and non-stomatal limitation of photosynthesis and on the carboxylation efficiency. *New Phytol.* **121**: 395-400.

Wright, S.T.C. 1969. An increase in the "inhibitor-β" content of detached wheat leaves following a period of wilting. *Planta (Berlin)* **86**: 10-20.

Yunus, M., Singh, S.N., Srivastava, K., Kulshreshtha, K. & Ahmad, K.J. 1985. Relative sensitivity of *Calendula* and *Dahlia* to SO_2. In Rao, D.N., Ahmad, K.J., Yunus, M. & Singh, S.N. (eds) *Perspectives in Environmental Botany*: 271-284. Print House (India), Lucknow.

8 Resistance Mechanisms in Plants Against Air Pollution

M. IQBAL, M.Z. ABDIN, MAHMOODUZZAFAR, M. YUNUS & M. AGRAWAL

INTRODUCTION

The response of plants to air pollution at physiological and biochemical levels can be understood by analysing the factors that determine resistance and susceptibility. A range of strategies, including stomatal movements, enzyme actions and detoxyfying processes, as well as genetical and developmental factors, have been proposed in building resistance against air pollutants (Malhotra & Khan, 1984; Wolfenden & Mansfield, 1991). Levitt (1972) suggested *stress avoidance* and *stress tolerance* as two alternative mechanisms underlying the resistance against major types of natural stresses. Stress avoidance denotes the ability of plants to exclude the stress, whereas in stress tolerance the stress-inducing agent acts internally but without a serious bearing on the functioning of cells/tissues. The degree to which the normal functioning of cell is affected is indeed determined by *strain tolerance* or *strain avoidance*. In strain avoidance normal functioning of the cells is achieved at energy costs and in strain tolerance normal functioning is affected but not to the extent that survival is hampered. Taylor (1978) has confirmed Levitt's terminology with respect to air pollution stress.

EPIDERMAL AND STOMATAL MECHANICS

The extent of damage that air pollutant(s) can cause in a plant depends upon the pollutant influx into the leaf and the reactivity of the pollutant and/or reactivity of its reaction products with cellular constituents. Leaves are the primary route of the uptake which is controlled by the stomatal aperture and conductance to gas diffusion (Tingey & Taylor, 1982; Darrall, 1989; Mansfield & Pearson, 1993; Heath, 1994; Kangasjärvi *et al.*, 1994). The diurnal changes in stomatal aperture, controlling the movement of CO_2 inwards and O_2 outwards in the leaves, also influence the pollutant entry. The opening and closing of stomata depend on turgor changes in the guard cells and their neighbours due to massive intercellular fluxes of potassium (MacRobbie, 1988). Guard cells have to be turgid to open the stomata. This occurs during the daytime in most of the plants and, hence, pollutant gases enter mostly during the daytime. Stomata close under

Plant Response to Air Pollution. Edited by Mohammad Yunus and Muhammad Iqbal

sudden and/or heavy increase in the CO_2 concentration (Morison & Gifford, 1983; Morison, 1987; Eamus & Jarvis, 1989; Mansfield *et al.*, 1990; Norby & O'Neill, 1991).

Abscisic acid (ABA) and indol-3-acetic acid (IAA) regulate the response of the stomata to CO_2 (Snaith & Mansfield, 1982). The response may vary according to the water status of the plant (Raschke, 1975). The regulation of the CO_2-sensing ability by two hormones means that this form of transpiration control can alter with the water status of the plant (Mansfield & Freer-Smith, 1984; Davies & Mansfield, 1987). This strategy to defend the plant against excessive water loss would constitute an avoidance mechanism.

With increase in the concentration of a gas in the atmosphere, the rate of its diffusion into the leaf increases according to Fick's law (Meidner & Mansfield, 1968). The stomata should close as the leaf begins to experience stress due to the increased entry of pollutant molecules. However, since stomatal closure itself represents a stress because it reduces the availability of CO_2 for photosynthesis, the closure should be partial. It should be enough to reduce the entry of pollutant molecules to a tolerable level, but below the level sufficient to deprive the leaf of CO_2. Thus, an ideal mechanism for avoiding pollution stress would involve a balancing of the leaf's priorities between import of CO_2 for photosynthesis and the prevention of excessive water loss by transpiration (Mansfield & Freer-Smith, 1984; Darrall, 1989).

Mansfield & Freer-Smith (1984) suggested that a toxic gas may act on cells of the epidermis in two ways:

(a) Molecules diffuse in through the stomata to reach the substomatal cavity, and then enter other cells of the epidermis. The guard cells' inner surfaces are more tolerant of the strain imposed by the pollutant than are the subsidiary cells; and hence the latter become injured and lose turgor while the former retain it. Stomatal opening enlarges purely as a result of the release of surrounding turgor pressure.

(b) Molecules enter the epidermis as in (a) and cause a loss of turgor of the guard cells too; this brings about stomatal closure.

The nature of the action of the pollutant at the cellular level in both these cases could be the same, e.g., structural damage on the tonoplast membrane leading to a loss of the solutes responsible for maintaining turgor (Mansfield & Freer-Smith, 1984).

Evidence suggests that pollutants can affect both stomata and other parts of the epidermal layer (Mansfield & Freer-Smith, 1984). The cells most exposed to air pollutant action are those surrounding the substomatal cavity. It has been suggested that the effect of air pollutants on plant leaves depends on the number and size of stomata and the degree of their opening (Rich *et al.*, 1970; Yunus *et al.*, 1979). Since stomata regulate the pollutant entry into the plant body, these are believed to have an important role in determining susceptibility of plants to pollution. Kimmerer & Kozlowski (1981) suggested two types of stress avoidance response at the stomatal level: (a) *passive avoidance* through maintenance of low gas exchange capacity under favourable and unfavourable conditions and (b) *active avoidance* through reducing gas exchange during exposure period. Active avoidance is a metabolically active process or manifestation of injury to guard cells, leading to stomatal closure. Species with higher stomatal conductances are likely to be more susceptible to air pollutants (Winner & Mooney, 1980a, b; Pande & Mansfield, 1985; Reich & Amundson, 1985; Mansfield & Pearson, this volume, Chapter 7).

The foliar cuticle not only restricts water loss from leaf tissues, but also protects the plant from the effects of air pollutants. It acts as a barrier to the movement of air pollutants, albeit sorption of gases, water and dissolved solutes takes place. Thus, permeability of the cuticle is important in determining the harmful effects of pollutants on plants (Poborski, 1988; Kerstiens, this volume, Chapter 6). Permeability coefficients (P) have been described for a number of pollutants (Jain, 1972; Nobel, 1974; Price, 1982; Reed & Tukey, 1982) using $J = P(C_1 - C_2)$, where J = flux of mass, C_1 = concentration of donor solution, C_2 = concentration of acceptor solution, and P = permeability coefficient.

Lendzian (1984) published values of "P" for SO_2 which were determined using $^{35}SO_2$ in a gas/membrane/water system. Experiments indicate that SO_2 is a highly sorptive gas and a considerable amount of it remains bound on the surface of the cuticle (Poborski & Straszewski, 1983). The relative penetration of ionic substances including heavy metals through cuticle membranes has also been studied (Chamel, 1980, 1984; Chamel & Eloy 1983; Chamel *et al.*, 1984; Singh *et al.*, 1990).

Amongst the most important properties of the cuticular membranes are the thickness and tortuosity of the pathway, the wettability, the chemical composition and the arrangement of the membrane constituents (Hull, 1970; Martin & Juniper, 1970; Merida *et al.*, 1981; Price, 1982). Other factors for consideration are the molecular radius of the solute, its affinity for the cuticular membrane (the partition coefficient) and the concentrations inside and outside the tissues (Schönherr & Schmidt, 1979; Price, 1982).

Sulphur dioxide

In certain plants, such as broad bean *(Vicia faba)*, SO_2 could induce stomatal opening wider than normal (Majernik & Mansfield, 1971) without disturbing the essential diurnal functioning of the stomata. The increase in stomatal conductance was about 20% in the range of 18–350 ppb SO_2 when the plant material was raised under low light intensities (Black & Black, 1979; Black & Unsworth, 1980). At higher SO_2 concentrations of 200–700 ppb, stomatal closure has been observed (Natori & Totsuka, 1984; Taylor *et al.*, 1986). Variations in sensitivity to SO_2 have often been correlated with stomatal conductances. The enhanced stomatal opening was accompanied by extensive disorganization of epidermal cell protoplasts. Even if the epidermal cells *in situ* are not actually killed by exposure to SO_2, damage to their membranes would impair their ability to maintain full turgor. The guard cells would then be able to achieve wider stomatal apertures (Black & Black, 1979). SO_2-induced stomatal opening has been observed in several other plants also (see Yunus *et al.*, 1979; Rao & Dubey, 1988; Singh *et al.*, 1990).

Once a molecule of atmospheric SO_2 has entered through a stoma, the route to the surface of a nearby subsidiary or epidermal cell is very short and, therefore, the cells of the epidermis are in a vulnerable position. SO_2 is nearly 30 times more soluble, compared to CO_2, in aqueous medium, which suggests that SO_2 dissolves mainly on the lower inner surfaces of the guard cells where most of the H_2O is lost (Heath, 1988). These cells may also be more susceptible to SO_2 injury because they lack chloroplasts. This could deprive them of the ability to convert SO_2 to H_2S using energy from photosynthetic electron transport. Sekiya *et al.* (1981) reported that H_2S emission from cucurbit leaves is almost totally light dependent. Cells without chloroplasts may, therefore, lack an important detoxification mechanism for removing excess sulphur by the production of H_2S, and its diffusion back into the atmosphere (see Mansfield & Freer-Smith, 1984). The ability of SO_2 to damage epidermal and subsidiary cells preferentially can cause stomatal opening through reduction of the pressure normally imposed upon the guard cells by the turgor of their neighbours. This leads to wider stomatal apertures during the day, though does not necessarily impair their ability to respond to stimuli such as light and CO_2 (see Puckett *et al.*, 1977; Black & Black, 1979; Singh *et al.*, 1985).

The stomatal opening responses to SO_2 may substantially increase transpiration of a field crop (Biscoe *et al.*, 1973). Mansfield & Davies (1981) enclosed the whole plant of *Commelina communis* in a transparent container and illuminated for 2 hours. The restricted gas exchange led to conditions of high humidity and low CO_2 concentrations which stimulated wide stomatal opening. When the container was suddenly removed and the leaves were exposed to a light wind, severe wilting occurred and parts of the leaves were permanently damaged. The plant was not deprived of water and the wilting occurred because the very wide stomatal apertures led to a high rate of transpiration which exceeded the rate at which water could be replaced by the vascular system. The brown necrotic areas that developed were similar to those caused by SO_2 in many plants, thus indicating that some form of acute injury due to SO_2 could in fact be the result of excessive transpiration. Stomatal apertures in the field need to be finely regulated and any agent that interferes with this regulation by stimulating opening is likely to cause acute injury at times when transpiration is high. Stomatal opening in SO_2-polluted air obviously counters resistance mechanisms because resistance to SO_2 would need stomatal closure rather than opening (see Heath, 1988).

In this regard, water vapour content of the atmosphere also matters. Normally, an increase in the vapour pressure deficit (vpd), which is indicative of low relative humidity, brings about stomatal closure in many species. This response is thought to be due to evaporation from the guard cells themselves, probably through special unthickened areas in their walls (Appleby & Davies, 1982). The operation of a vpd-sensing mechanisms using evaporation from the guard cells implies a resistance to water movement from elsewhere in the leaf to allow the guard cells to lose turgor. The epidermis is isolated hydraulically from the rest of the leaf, and the guard cells can only acquire water via the epidermal cells. Black & Unsworth (1980) suggested that exposure to SO_2 interacts with this mechanism, and that, in vpd-sensitive species, loss of turgor by the guard cells in dry air will be greater in the presence of SO_2. In a climate where the air usually has a high vpd, the stomata may tend to have smaller apertures in the presence of SO_2, thus providing a mechanism for the avoidance of pollution stress. It is suggested that pH changes and increased levels of bisulphite and sulphite in SO_2-exposed plants disturb the fluxes of potassium, calcium, chloride, malate and protons between guard cells and their neighbours, which regulate stomatal behaviour. Calcium-regulated response of stomatal closing is thought to be interfered with by sulphite, a product of SO_2 reaction with water (Mansfield *et al.*, 1990).

The effects of SO_2 on net photosynthesis vary greatly among different species (see Black, 1982). Generally, concentrations above 0.5 ppm are strongly inhibitory. At high SO_2 concentration for longer duration, the observed stomatal closure may be both the direct effect of SO_2 on turgor of the guard cell and indirectly due to increased concentration of CO_2 in the substomatal cavity. Unsworth & Black (1981) suggested that direct stomatal responses to SO_2 depend on rates of entry into the leaf interior, sites of chemical reactions with tissue and transfer pathways of toxic products. Koch & Maier-Maercher (1986) have found three response phases of transpiration in *Picea abies* at 300 ppb SO_2 concentration — an initial reduction with a subsequent increase to high but erratic rate and a final reduction associated with visible injury. This response depicts varying mechanism of stomatal behaviour depending upon the exposure concentration. Treatment of leaves with inhibitors of photosynthetic CO_2 fixation causes stomatal closure, but this can be reversed if the leaf is flushed with CO_2-free air (Allaway & Mansfield, 1967). It is likely that the same would apply in the case of SO_2 (Mansfield & Freer-Smith, 1984). The stomatal closure that occurs because of the inhibitory action of a pollutant on photosynthesis in the mesophyll cannot be a desirable way of avoiding the pollution stress. The stomatal closure will certainly cause an increased diffusion resistance to SO_2, but a true avoidance mechanism would involve closure prior to stress in the mesophyll, rather than as an event secondary to that stress.

In some species such as silver birch (*Betula pendula*) stomatal closure in response to SO_2 possibly operates as a useful avoidance mechanism (Osonubi & Davies, 1980). The pollutant depresses net photosynthesis as well as transpiration (Figure 8.1). Where stomatal closure is the cause of changes in photosynthesis and transpiration, a smaller proportional effect would be predicted for the former than for the latter, especially in C_3 plants where 'internal resistance' to CO_2 intake is appreciable (Mansfield, 1976). Since the net sulphur flux is closely associated with the stomatal conductance to water vapour, the response of silver birch to SO_2 during the first day of exposure, as shown in Figure 8.1, seems to represent a mechanism for avoiding pollution stress without a major interference with the supply of CO_2 for photosynthesis (Mansfield & Freer-Smith, 1984).

Nitrogen oxides

NO and NO_2 are the two oxides of nitrogen that are toxic air pollutants, but NO is rapidly oxidized to NO_2 in free atmosphere. NO_2 increases internal resistance to CO_2 uptake, and the stomata close at high NO_2 concentrations (Srivastava *et al.*, 1975). The entry of NO_2 into leaves is similar to that of SO_2, however,

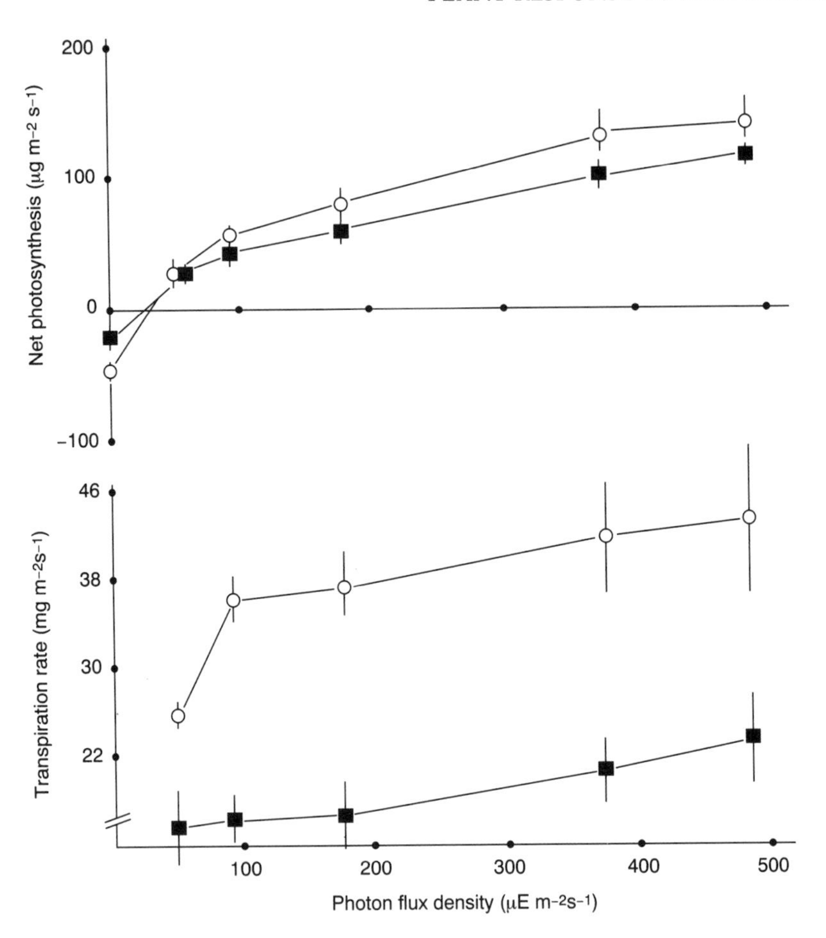

Figure 8.1. Effects of exposure to 0.07 ppm SO_2 on net photosynthesis and transpiration in silver birch. ○ —— ○, plants grown in clean air; ■ —— ■, plants placed in SO_2 after growing in clean air. Measurements were made during the first day of exposure to the pollutant. Means and standard errors are shown using vertical bars at each observation. (courtesy of Mansfield & Freer-Smith, 1984)

entry through cutice is higher as cuticular resistance against NO_2 entry is lower than against SO_2 or O_3. A single exposure of *Euonymus japonica* to 100 ppb NO_2 leads to an increase in stomatal conductance (Natori & Totsuka, 1984), but higher concentrations may reduce the stomatal conductance (Saxe, 1986a, b). A complex situation may arise when stomata are affected by two pollutants simultaneously. The stomata of pinto bean, among others, opened in response to SO_2 and closed in response to O_3. However, in a mixture of the two they closed even further than in O_3 alone (Beckerson & Hofstra, 1979a, b). Mansfield & Freer-Smith (1984) suggest that such results point to the dangers of drawing too many conclusions about effects in the field from fumigations using single pollutants. Stomatal opening caused by SO_2 may be unimportant where episodes of SO_2 pollution are accompanied by O_3.

In silver birch, $SO_2 + NO_2$ initially caused a reduction in stomatal opening comparable to that in SO_2 alone. After a 20-day exposure, however, difference between the polluted and control plants was little in the light, but stomatal closure in the dark was inhibited. This means there could be progressive injury to epidermal cells during prolonged exposures. After 20 days treatment with 0.07 ppm SO_2 + 0.07 ppm NO_2, visible foliar injury appeared. Whether this was a cause or a consequence of stomatal malfunctioning was not clear, however (Mansfield & Freer-Smith, 1984).

Ozone

Most of the reports suggest that O_3 causes at least partial closure of stomata (Gillespie & Winner, 1989; Aben *et al.*, 1990; Mansfield & Pearson, 1993) though some reports indicate increases in stomatal conductance (Barnes & Davison, 1988; Moldau *et al.*, 1990). These contradictory results may be due not only to interspecific differences but also to differences in growth and measurement conditions. Stomatal closure due to O_3 has often been correlated with reduction in photosynthesis. Farage *et al.* (1991) have shown that 38% decline in photosynthesis rate is due to stomatal closure.

Aben *et al.* (1990) have shown direct effect of O_3 on stomata as well as on photosynthesis but stomata appeared to be more sensitive. This suggests an avoidance mechanism operating in plants to check the excessive entry of O_3. The direct action of O_3 on stomatal function may be mediated through its action on cell permeability (Heath, 1980). Ozone has been found to increase membrane permeability to K^+ ions (Dominy & Heath, 1985), probably mediated through a membrane-bound K^+-ATPase activity (Tester, 1990). The fluxes of K^+ are known to regulate stomatal movements. Low O_3 concentrations have been suggested to act upon membrane permeability of subsidiary cells leading to water flow into the guard cells which open wider due to more turgidity. At higher concentrations, O_3 begins to affect guard cell permeability, and stomata close. Heath (1994) has explained the differential response of stomata to O_3 as being due to differential action of O_3 upon the membranes of each cell type.

Exogenous amendment of ABA has been shown to reduce O_3 phytotoxicity (Adedipe & Ormrod, 1972). Downton *et al.* (1988) have suggested ABA-induced modification of stomatal function being responsible for O_3 resistance in plants.

REGULATION OF POLLUTANT UPTAKE

The first barrier to gaseous pollutant is boundary layer resistance which varies with wind speed, and size, shape and orientation of leaves (Heath, 1988). At higher wind speed boundary layer resistance declines allowing more pollutant entry into the leaf. The cells most exposed to air pollutant action are epidermal cells, but waxy cuticle is a potential barrier to most pollutant gases. However, acidic gases can dissociate and react with cuticular waxes and enter leaves by penetrating through damaged cuticle. Kerstiens & Lendzian (1989) suggested that a reactive pollutant like ozone might be expected to react with the cuticular constituents, but ambient concentrations of O_3 do not seem to affect the water movement through cuticle.

The uptake of SO_2 may be a consequence of internal resistances, such as physiological processes inside the leaf limiting SO_2 influx, and/or metabolic processes counteracting SO_2 influx. At low SO_2 concentrations, enhanced S levels are frequently reported due to accumulation of sulphate (De Kok, 1990). In contrast, reductive conversion of SO_2 in the chloroplast as H_2S represents a metabolic resistance that counteracts SO_2 influx (Rennenberg, 1991).

Mesophyll resistance is an important barrier before SO_2 and its products sulphite and bisulphite reach the likely targets such as chloroplast (Heath, 1988). It has been suggested that tolerance or sensitivity to SO_2 within plants is partly due to variations in mesophyll resistance. Buffering capacities in the extracellular fluid, the cytoplasm and inside the chloroplast are critical factors for determining SO_2 sensitivity. Kropff (1991) has proposed that metabolic buffering capacity of leaf cells is related to the rates of sulphate and nitrate reduction and the import rate of organic anions in plants exposed to SO_2 for a long time.

It is unfair to make simple assumptions about the similarities of pathways for diffusion of different gases from the atmosphere to leaves. For instance, the resistance to movement through the cuticle may be less for SO_2 than for H_2O vapour

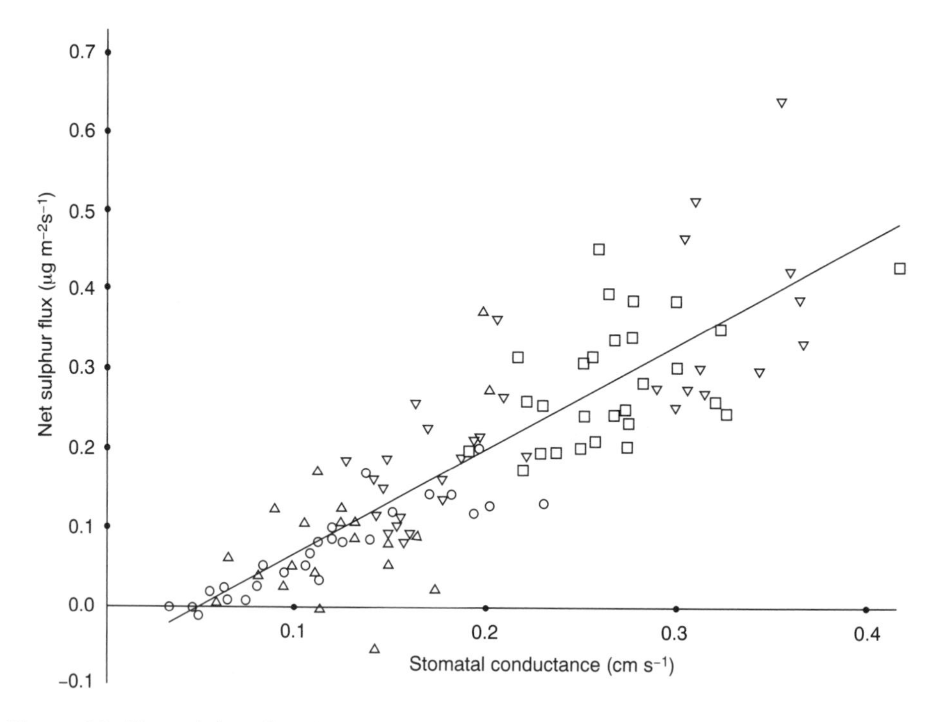

Figure 8.2. Net sulphur flux for shoots of silver birch plotted against stomatal conductance. Plants had been exposed to 0.07 ppm SO_2 + 0.07 ppm NO_2 for: ○, zero time; △, 5 days; ▽, 12 days; □, 20 days. Different photon flux densities were employed to produce changes in stomatal conductance, over the range 0–500 µE m^{-2} s^{-1} (courtesy of Mansfield & Freer-Smith, 1984)

(see Unsworth *et al.*, 1976). Uptake of CO_2 and H_2O vapour were measured simultaneously with those of gaseous sulphur compounds, taking Scots pine shoots as test material (Hällgren *et al.*, 1982). Diurnal variation in the uptake of SO_2 was prominent, with values during the day about three times those at night, but stomatal opening did not seem to be the primary controlling factor. The discrepancy could be partly due to re-emission of H_2S from the needles, which was a light-dependent process.

NO_2 and SO_2 normally occur together in urban air, and it is therefore realistic to include NO_2 in experimental fumigations with SO_2 (Mansfield & Freer-Smith, 1981). Figure 8.2 displays the findings recorded for nearly 1-month-old silver birch seedlings fumigated for different durations, with net sulphur uptake plotted against stomatal conductance to H_2O vapour. The uptake was measured for the whole shoots with solid uptake excluded. Prior to the determinations the plants were grown in 12-hour days with a photon flux density of 240 $\mu E\ m^{-2}\ s^{-1}$. The regression line is significant at $P < 0.001$ and 74% of the variance was accounted for (Mansfield & Freer-Smith, 1981). The control of net sulphur flux by factors other than stomata may also be important; the possibility of increased emissions of reduced sulphur compounds at higher light intensities (Hällgren *et al.*, 1982), and the entry of SO_2 through the cuticle, especially when the leaf is beginning to show visible injury, should not be lost sight of.

The stomatal control of SO_2 uptake is, therefore, important in silver birch, and the stomatal responses should thus be considered in determining the pollutant dose.

GENETICALLY DETERMINED RESISTANCE

The variable responses of different species and cultivars to pollutants suggest phenotypic expressions of genotypes that have evolved naturally or been introduced by selective breeding (Roose *et al.*, 1982; Pitelka, 1988). The genetic resistance enables plants to grow, survive and reproduce in polluted environments either through avoidance or tolerance mechanisms (Pitelka, 1988). Avoidance is suggested to be an effective process of adaptation to acute pollution, whereas tolerance is the mechanism of resistance with chronic pollution.

Genetic differences in susceptibility to acute injury have been noted to exist between genera in the Cucurbitaceae (Bressan *et al.*, 1978a, 1979a, 1981). Cultivars of *Cucumis sativus* (cucumber) were more sensitive than cultivars of *Cucurbita pepo* (squash and pumpkin). Differences in susceptibility occur among cultivars of the same species as well. This genetic difference in resistance to SO_2 injury behaved like a simple Mendelian trait, with resistance being dominant in cucumber (Bressan *et al.*, 1981). This genetically determined resistance could possibly be ascribed to differences in SO_2 absorption from air. The four cultivars exhibited marked resistance differences when exposed to the same external concentrations of SO_2 for the same length of time (Bressan *et al.*, 1978a). The greater the resistance of a cultivar to SO_2 injury at a given concentration, the lower would be the rate of SO_2 absorption at that concentration. Differences in resistance due to differences in uptake of a gas are usually presumed to reflect differences in stomatal numbers or apertures. When the rate of the uptake of SO_2 and the fixation of CO_2 per unit leaf area are comparable among the different cultivars, stomatal status alone may not account for uptake differences (Filner *et al.*, 1984). Some studies have suggested that the genetic differences in uptake of gaseous SO_2 reflect differences in permeability of a barrier to SO_2 entry other than the stomatal aperture. A likely candidate for this barrier is the plasma membrane (Bressan *et al.*, 1979a). The cucurbit cultivars differed in cold tolerance in the same order as they differed in SO_2 tolerance (Bressan *et al.*, 1979b). This reminds us of the idea suggested by Levitt (1972), that there is a common denominator of stress tolerance — probably some property of a membrane. The significance of the plasma membrane is paramount. If the membrane fails, nothing inside can save the cell (see Filner *et al.*, 1984). Differences in relative tolerance to SO_2 exist among *Gladiolus* cultivars also. It has been suggested that SO_2-tolerant plants develop various strategies to protect themselves against SO_2 damage and the degree of susceptibility of a plant depends greatly on its capacity to neutralize SO_2 and its toxic products in metabolism (Singh *et al.*, 1990). One such strategy involves detoxification of SO_2 by its reduction to volatile sulphur (i.e., H_2S) and its subsequent emission into the atmosphere. In duckweeds, the rate of H_2S emission is approximately four times higher in sulphite-tolerant (*Lemna voldiviana*) than in sulphite-susceptible species (*Lemna gibba, Spirodela oligorhiza*) (Takemoto *et al.*, 1986). In the field, a SO_2-tolerant conifer (e.g., *Picea pungens*) emits higher amounts of H_2S in response to atmospheric SO_2 than a SO_2-sensitive conifer (e.g., *Picea abies*) (see Rennenberg & Herschbach, this volume, Chapter 11). Ethylene production within a population in response to pollutants varies immensely (Wolfenden *et al.*, 1992).

Wounded living tissue produces ethylene by enzymatic processes (Boller & Kende, 1980), which rises and then falls with increasing injury (Elstner & Konze, 1976). Ethane emission, on the other hand, is thought to result from reactions of radicals with linolenic acid in membranes (Konze & Elstner, 1978), something unusual for a healthy intact cell. Green chloroplasts and light seem to be the requirement for ethane production (Wilson *et al.*, 1979). Ethane estimation is preferable as it has a simpler relationship with injury level. In fact, the most attractive assay is the ethane:ethylene ratio because it increases monotonically with injury and is dimensionless (Bressan *et al.*, 1979a). Therefore, it does not require measurement of the leaf area, which must be done in order to compare emission rates of either ethylene or ethane by different samples. Further, the ethane/ethylene assay requires a very small amount of tissue. Using the ethane emission assay, Filner *et al.* (1984) found that leaf discs from cucurbits with differing susceptibility to injury by bisulphite and SO_2 respond to cold, heat, salt and alcohols in the same order (Table 8.1). The discs lost their ability to fix CO_2 almost simultaneously with increase in ethane emission.

Resistant genotypes experience a metabolic cost of tolerating a pollutant (energy reduction) or a reduced rate of resource allocation (stomatal closure), suggests Pitelka (1988). Development of genetic resistance among tree population depends upon generation time, seed banks and timing of selection during the tree's life cycle. Wolfenden *et al.* (1992) have suggested that enzyme induction or

Table 8.1. Comparison of ethane emissions by cucurbit cultivars in response to various stresses (courtesy of Filner *et al.*, 1984)

Treatment[a]	Post-treatment time (h)	Ethane (nl cm^{-2} Chipper)	Ethane (% of emission by Chipper)			
			Chipper (cucumber)	National Pickling (cucumber)	Prolific Straightneck (squash)	Small Sugar (pumpkin)
Bisulphite (15 mM, 2h)	24	3.4 ± 0.8	100 ± 23.5	88.2 ± 16.7	23.5 ± 4.1	17.6 ± 8.3
Alcohol (10%, 2.5h)	23	7.2 ± 0.1	100 ± 1.4	73.6 ± 0.01	37.5 ± 0.01	30.6 ± 0.02
Heating (65°C, 3h)	24	2.4 ± 0.05	100 ± 2.1	75.0 ± 8.3	25.0 ± 8.0	8.3 ± 8.3
Chilling (2-4°C, 66h)	24	8.6 ± 2.6	100 ± 30.2	69.8 ± 23.2	6.9 ± 2.3	6.4 ± 2.9
KCl (3 mM, 22h)	0	5.7 ± 1.0	100 ± 16.9	87.7 ± 11.6	74.1 ± 7.0	70.6 ± 5.3

[a]Leaves 110-150 cm^2 were selected, usually the fourth leaf from the apex of a plant 30-35 days old. Internodes and leaves down to leaf 5 from the apex collapsed when the whole plant was heated, therefore discs were punched from leaf 7 in the case of heating. In the case of alcohol treatment, plants 50-55 days old were used and discs were punched from leaf 14 from the base. Ethane emitted after KCl treatment was calculated from triplicate samples. Variance is reported as standard error.

the regulation of the alternative biochemical pathways are under genetic control and some of these characters can be used to induce tolerance in plants.

DEVELOPMENTALLY DETERMINED RESISTANCE

Heath (1994) has reviewed the available data on effect of age/developmental stage on sensitivity of plants to O_3 and suggested that maximum visible injury occurs when the leaf is just beyond half way to its maximum size. The variations in response pattern of plants at different developmental stages is related to general metabolic demand. Higher levels of soluble sugars and amino acids were found responsible for affording protection against O_3 exposure (Dugger & Ting, 1970). Reich & Lassoie (1985) have shown that in young leaves of poplar, pigment breakdown and decline in net photosynthesis are delayed to some extent by the operation of repair mechanisms such as an increase in dark respiration, which also compensate for the shortened life span of the leaves by increasing leaf production. Pandey & Agrawal (1994) also found that leaf turnover is more rapid in the pollutant-tolerant plants.

Young leaves are injured far less than mature leaves on the same plant (Guderian, 1977; Bressan *et al.*, 1978a) although young leaves take up far more SO_2 than the mature leaves (Bressan *et al.*, 1978a). Thus the young leaves have to have a biochemically based resistance mechanism which functions after SO_2 metabolism in resistant versus sensitive, but otherwise closely similar plant material. Filner *et al.* (1984) fumigated cucurbit cultivars with $^{35}SO_2$ in a closed system and traced distribution of ^{35}S in metabolites. Only small differences in amounts of the ^{35}S metabolites were found in fractions obtained from homogenates of young and mature leaves, no matter whether from genetically resistant or sensitive plants. About 60% of the absorbed $^{35}SO_2$ was oxidized to $^{35}SO_4^{2-}$ (Sekiya *et al.*, 1978, 1982c; Filner *et al.*, 1984). Nevertheless, there was a vast difference between the emission of volatile sulphur by young, resistant leaves, and mature, sensitive leaves on the same plant. The young leaves emitted a lot of H_2S, while the mature leaves emitted very little (Sekiya *et al.*, 1982; Filner *et al.*, 1984). The H_2S emitted by young leaves was equal to more than 10% of the SO_2 absorbed. The positive correlation of H_2S emission with SO_2 resistance is a meaningful clue to the developmentally regulated, biochemically based resistance mechanism encountered in cucurbits (see Filner *et al.*, 1984).

The level of SO_2 is higher inside young leaves than inside mature leaves throughout an acute exposure to SO_2 but the mature leaves are injured far more severely (see Bressan *et al.*, 1978a; Wilson *et al.*, 1979; Filner *et al.*, 1984). This could mean that SO_2 itself is not toxic and might be transformed to something else to be able to cause injury. Conversely, it could also mean that young leaves do not have injurable sites accessible to the absorbed SO_2. Virtually all of the absorbed SO_2 is metabolized within about 3 hours (Wilson *et al.*, 1979) in light or dark (see Filner *et al.*, 1984). Bisulphite can be absorbed in the dark without causing injury; a later exposure to light leads to injury as manifested by necrosis and ethane production. Thus, injury does not seem to require the persistance of bisulphite. This could also mean that there is a primary injury event caused directly by SO_2/HSO_3^-. A (reduced?) metabolite of SO_2/HSO_3^- could form and cause injury which is latent, i.e., storable in darkness. The injury would remain latent until the light comes on and secondary consequences set in.

Oxidation of SO_2 to SO_4^{2-} is almost the same, *ca*. 60%, in resistant and sensitive cucurbit leaves (Figure 8.3; Filner *et al.*, 1984). Assuming that oxidation in both cases proceeds via the same pathway, it appears unlikely that such oxidation is the crucial detoxification process. Thus, the only striking difference between the fates of SO_2 in the leaf types is the conversion of about 10% of the absorbed SO_2 to the reduced product, H_2S, in young leaves, which does not happen in mature leaves. The emission of H_2S in young leaves of cucurbits was 100-fold higher than mature leaves. This pattern suggests that a product of reduction of SO_2 causes injury, and that either further reduction of this product, or a diversionary reduction to H_2S, of a precursor to the injurious product, protects the leaf from injury (Filner *et al.*, 1984). Another possible explanation is that SO_2/HSO_3^- causes injury in a certain compartment, in which reduction rather than oxidation occurs, and this compartment could be the chloroplast (see Rennenberg, 1991).

Furthermore, there may be more than one path for oxidation of SO_2 to SO_4^{2-} in leaves. Extracts of cucurbit leaves were found by Ream & Wilson (1982) to contain a light-dependent, partly particulate system and a light-independent soluble system. The light-dependent system is inhibited by 3-(3′,4′-dichlorophenyl)-1, 1 dimethylurea (DCMU) or superoxide dismutase whereas the light-independent, soluble enzyme system is not. This suggests that the former system involves superoxide formed by transfer of electrons from chloroplast photosystem II to the reducing site of photosystem I. However, this reaction may not necessarily occur to a significant extent in a healthy intact chloroplast, or, if it does, the superoxide

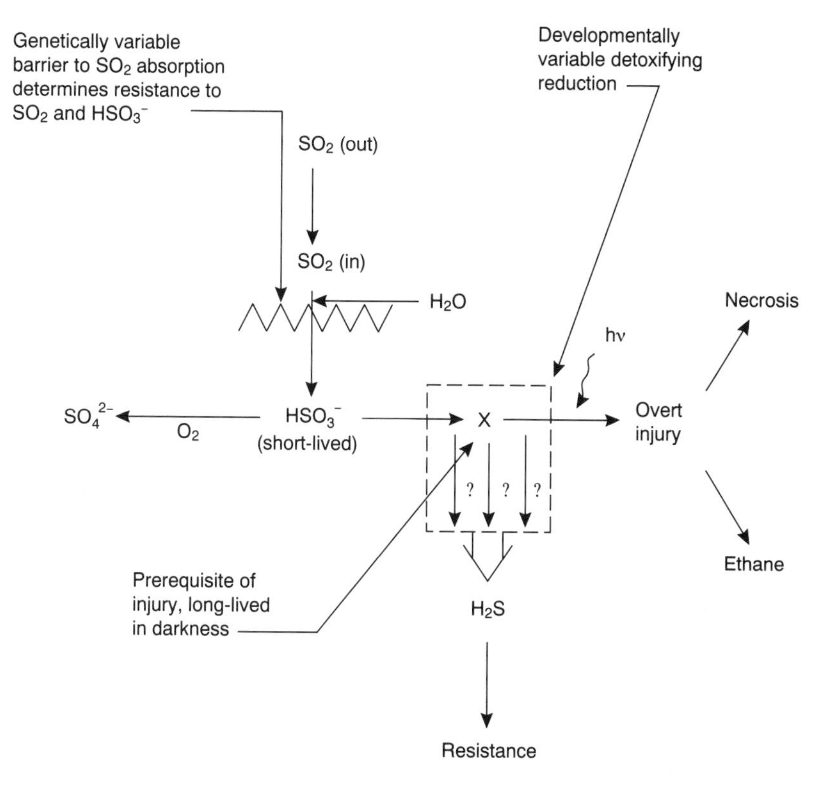

Figure 8.3. Resistance to SO_2 injury by synthesis and emission of hydrogen sulphide. X = a non-identifiable reduced product of bisulphite (HSO_3^-), which is latent in the darkness but becomes active in the light and causes injury (courtesy of Filner *et al.*, 1984)

would be left untouched by the endogenous superoxide dismutase. Therefore, the light-dependent oxidation should be viewed as most likely a test-tube artifact. It may be significant, however, that the light-independent oxidation reaction is the predominant one in extracts of young leaves, which are most resistant to acute injury by SO_2 (Ream & Wilson, 1982). Another mechanism to counteract the toxic effect of SO_2 involves oxidation of SO_3^{2-} to sulphate in the apoplastic space catalysed by apoplastic sulphite-oxidase (Pfanz *et al.*, 1990; Rennenberg & Polle, 1994). SO_4^{2-}, a non-toxic compound, is utilized in the synthesis of various organic sulphur compounds. These compounds may be used for growth and also stored in the shoots and roots.

BIOCHEMISTRY OF PLANT TOLERANCE

Gaseous pollutants after entering through the stomatal cavity dissolve in extracellular fluid and react with biological compounds to produce ionic spcies or free radicals (Peiser & Yang, 1977; Asada & Takahashi, 1987; Mehlhorn *et al.*, 1990; Heath, 1994). Free radicals interact with proteins and lipids in the cell wall and membranes leading to oxidation and lipid peroxidation and initiation of chain reactions giving rise to more free radicals followed by increases in cell permeability (Mehlhorn *et al.*, 1986; Pell & Dann, 1991). This can all lead to severe metabolic disorder and damage to genetic material (Brawn & Fridovich, 1981).

The gaseous pollutants, viz. O_3, SO_2, NO_x and HF, destroy the ultrastructural organization of leaf cells: thylakoids swell, curl and then disintegrate; the cytoplasmic and plastid matrix become granular; mitochondria injured; and ribosomes decrease in number (Blingy *et al.*, 1973; Godzik & Sassen, 1974; Malhotra, 1976; Młodzianowski & Biołobok, 1977). HF and SO_2 cause similar ultrastructural changes in the needles of pine and spruce, but injury by HF is more serious (Soikkeli, 1981). Besides ultrastructural changes, these pollutants produce free radicals (Tanaka *et al.*, 1982; Grimes *et al.*, 1983), that lead to oxidation of proteins (Mehlhorn *et al.*, 1986), peroxidation of lipids (Chia *et al.*, 1984) and DNA-strand breakage (Brawn & Fridovich, 1981). Fluoride is said to inhibit enzyme action by binding with the calcium and/or magnesium needed for enzyme activity and by the possible formation with the fluoride ion of poisonous metabolic analogues. It also reduces markedly the content of free phosphates, and causes precipitation of calcium (Borel, 1945).

The mechanism of the injurious action of SO_2 on enzymes can be explained in part by the similarity of sulphate with some other anions and in part by its competition with CO_2 in metabolic processes (Ziegler, 1973). Conversely, however, peroxidase from the needles of spruce, despite being an iron-containing (haem) enzyme is not inhibited; rather its activity increases when trees are subjected to HF, particularly when they are also fertilized with nitrogen (H. Keller, 1976). In addition new isozymes of peroxidase are induced in spruce by H^- ion treatment and the activities of esterases and β-galactosidase increase. In beech (*Fagus sylvatica*), on the other hand, the activity of peroxidase does not change in response to HF. In Scots pine, the activity of α-mannosidase increases, while those of β-galactosidase, α-galactosidase and β-glucosidase do not change significantly (Bücher-Wallin, 1976). In roots of maize (*Zea mays*), the content of ribosomal RNA decreases (Chang, 1970a) and the activity of ribonuclease increases (Chang, 1970b). In the leaves of beech and spruce, the activities of acid phosphatase and ribonuclease did not change (Bücher-Wallin, 1976), though the fluoride ion disturbed the ratios between K, Mg and Ca by precipitating them as fluoride salts, which could possibly inhibit ribonuclease which is dependent on these cations (Hanson, 1960). SO_2, both at high and low concentrations, alters the physiological and biochemical events in plants leading to visible injury symptoms (Rao & Dubey, 1988; Singh *et al.*, 1988, 1990; Dubey, 1990; Khan *et al.*, 1990; Rennenberg & Herschbach, this volume, Chapter 11).

Disintegrated organelles liberate the enzymes contained in them or on their surfaces. This may account for an increase in the activity of some enzymes and also for changes in isozyme patterns sometimes observed in extracts obtained from leaves of trees growing in regions polluted by industrial emissions (see Mejnartowicz, 1984). In some species, the subcellular distribution of certain enzymes differs in trees of varying tolerance (H. Keller, 1976; Kieliszewska-Rokicka, 1978; Młodzianowski & Młodzianowska, 1980). Horsman & Wellburn (1975) have used isoenzymes of peroxidase enzyme as useful markers of pollution sensitivity.

Ozone-generated oxyradicals

More than 90% of the damage to vegetation due to air pollution in the USA has been attributed to photochemical oxidants of which O_3 is the most important component (Heck *et al.*, 1973). Since O_3 is also a natural component of the atmosphere to which there has been evolutionary adaptation, all green plants show some tolerance to O_3 and varieties vary in their relative susceptibility or tolerance (Bennett *et al.*, 1981).

The dissolution of O_3 in water leads to decomposition products such as superoxide, peroxyl and hydroxyl radicals (Staehelin & Hoigne, 1982). At physiological pH, however, only hydroxyl radicals are detected, especially in presence of phenolics such as caffeic or ferulic acids (Grimes *et al.*, 1983). The reaction of O_3 with substrate that yields H_2O_2 can increase the O_2^- radicals and consequently lead to generation of highly reactive hydroxyl radicals. OH^- radicals may attack membranes with subsequent fixation of the damage by the formation of peroxy radicals. This involves membrane-derived organic oxygen radicals; O_2^- serves only as a chain propagator. The damage caused by O_3-induced oxyradicals may be especially prevalent in chloroplasts where their concentrations may be enhanced during illumination, resulting in light-induced injury to susceptible chloroplasts. Starch and sugars in carbohydrate-loaded chloroplasts, on the other hand, might mitigate the injury (Bennett *et al.*, 1984). Ozone and secondary products can react with a wide range of bio-organic molecules and inorganic cell constituents. High O_3 concentrations cause general destruction to the cells. Chronic low-level concentrations often present in ambient air can have more specific effects.

Plants have a range of mechanisms to detoxify the toxic molecules (Larson, 1988; Foyer *et al.*, 1994). These include enzyme systems (catalase, peroxidase, superoxide dismutase, etc.), ascorbic acid, vitamin E (α-tocopherol), peptides (glutathione), carotenoids (β-carotene), polyamines and organic buffering systems. An understanding of the biochemical markers providing protection against oxidative damage is essential for the development of plants more resistant to the growing number of environmental stresses (Monk *et al.*, 1989).

Role of enzymes

Superoxide dismutase

Activity of superoxide dismutase (SOD) in biological systems was discovered in the late 1960s during the course of research using an oxidase enzyme, viz., xanthine oxidase to generate superoxide radicals (O_2^-) in experimental fractions obtained from bovine blood (Fridovich, 1981). A haemocuprein protein, in fact a copper-containing SOD, was isolated 30 years ago and has since been under extensive study without its enzymatic role being understood. SOD metalloproteins are classified on the basis of the specific metal co-factor (iron, manganese or copper) required for activity. This family of oxyradical-scavenging enzymes is of universal occurrence in aerobic and aerotolerant cells. Zinc in the Cu–Zn-containing SOD can be replaced by a variety of other metals without the loss of enzymatic activity (Fridovich, 1981). The widespread occurrence of superoxide dismutases in aerobic organisms has brought out SODs as important cellular protective enzymes. Since the enzyme substrate, O_2^-, is less toxic than the other potential secondary oxyradicals that may be formed in cells, the protective action is believed to be largely indirect, i.e., by removing O_2^- radicals which may cause generation of more toxic chemical species. The comparatively low reactivity of the superoxide anion radical renders it capable of transmitting radical properties over wider distances in the cell. Superoxide dismutases along with catalase and the peroxidases that act on the end product (H_2O_2) of SOD activity can interact to regulate injurious oxyradical and peroxyl concentrations in cells and organelles and determine equilibrium rates (see Bennett *et al.*, 1984; Fridovich, 1986).

Tissues contain SODs in the form of water-soluble enzymes as well as metalloproteins bound to cell membranes. When associated with cell membranes, SODs appear to increase the order (decrease the entropy) of the lipid layer fluidity, causing structural alterations of membrane. This possibly decreases the susceptibility of membranes to chemical attack. Reactions of superoxide and other oxyradicals at activated membrane surfaces can be highly complex. The most damaging potential reaction products include the extremely potent OH^- free radicals and various organic radical and peroxides. Reactions, induced by hydroxyl-free radicals that mimic OH^- effects but tend to be more specific in their actions, lead to delayed toxic responses (see Bennett *et al.*, 1984). If oxyradicals lead to phytotoxicity in cells, cell defences that evolve for protection should prevent their generation and/or minimize the damage caused by those produced. Superoxide dismutases and the catalase/peroxidase enzyme systems serve as interlinked primary protection mechanisms in reducing the potential for cellular injury, which has been demonstrated by enhanced SOD and peroxidase activities related to increased tolerance in plants exposed to SO_2, NO_2 and O_3, either singly or in combinations (Asada, 1980; Decleire *et al.*, 1984; Tanaka *et al.*, 1985; Castillo *et al.*, 1987; Matters & Scandalios, 1987; Rao & Dubey, 1990; Sen Gupta *et al.*, 1991). Common metabolic reductants such as ascorbic acid, glutathione and α-tocopherol would further minimize damage (Bennett *et al.*, 1984; Asada & Takahashi, 1987; Polle & Rennenberg, 1993). This system operates in the cytoplasm and chloroplast in a series of reactions, in which the reduction of activated oxygen (O_2^- and H_2O_2) is achieved eventually at the expense of photosynthetically or enzymatically produced reductant, NAD(P)H (Polle & Rennenberg, 1993). Free radical scavengers protect against irradiation effects while radiation damage is potentiated in the presence of oxygen.

Superoxide dismutase is the most efficient known scavenger for superoxide radicals. The different isozymes of SOD occur in different subcellular compartments (Asada & Takahashi, 1987). It is estimated that chloroplasts contain SOD, ascorbate and glutathione at concentrations of 10 μM, 10–50 mM and 3 mM, respectively, and contribute in scavenging of $O_2{\cdot}^-$ in the order 10:1–5:1 (Rennenberg & Polle, 1994).

When exposed to different stresses, level of SOD isozymes increased specifically in different subcellular compartments (Perl-Treves & Galun, 1991; Tsang *et al.*, 1991). Positive effects, however, were not always observed, when SOD levels increased by constructing transgenic organisms (Malan *et al.*, 1990;

Tepperman & Dunsmuir, 1990; Bowler *et al.*, 1991; Pitcher *et al.*, 1991). In bacteria, overproduction of SOD increased oxidant toxicity (Scott *et al.*, 1987; Liochev & Fridovich, 1991). Deleterious effects also occurred in transgenic plants in which manganese SOD was slightly overproduced (Bowler *et al.*, 1991). These observations suggest that overproduction of SOD causes an increased H_2O_2 production, which overwhelmed the capacity for H_2O_2 decomposition. Consequently, $O_2{\cdot}^-$ and H_2O_2 levels are shifted towards a ratio favourable for OH production. In contrast, strong overproducers of manganese SOD are more protected from oxidative stress, than controls. It is assumed that $O_2{\cdot}^-$ concentrations are so far diminished in these plants and $O_2{\cdot}^-$ is not available to drive the Haber–Weis reaction (Bowler *et al.*, 1991). Sen Gupta *et al.* (1993) found reduction in cellular damage mediated by oxyradicals in plants with higher SOD.

Cultivars of *Conyza bonariensis, Lolium perenne* and *Nicotiana tabacum,* with concurrent increase in several components of antioxidative systems, are more resistant to environmental stresses than cultivars with low antioxidative capacity (Shaaltiel & Gressel, 1986; Shaaltiel *et al.*, 1988). Hybrids of maize cultivars with increased activity of two antioxidant enzymes, viz., SOD and glutathione reductase (GR), have improved protection from oxidative stress when compared with hybrids having either increased GR or SOD activity (Malan *et al.*, 1990). It appears that a balanced increase in antioxidants is required to obtain increased stress resistance. The increase in enzyme activities of ascorbate peroxidase and glutathione reductase are found to be correlated with increased SOD activity (Kangasjärvi *et al.* 1994).

The biochemical mechanisms of action by which leaf mesophyll tissues resist oxidant exposures deserve attention. Certain chemical compounds that protect plants against O_3 injury can sustain cellular integrity in physiologically stressed leaves and impede senescence (Pellissier *et al.*, 1972; Ormrod & Adedipe, 1974; Gilbert *et al.*, 1977; Lee *et al.*, 1981a, 1992). Chronic, low-level O_3 stress causes premature senescence in older leaves, whereas acute doses result in necrosis and chlorosis. Since comparative studies or the enzyme and metabolite status of O_3-tolerant and O_3-sensitive plants are influenced by many inherent biological and environmental factors (such as plant age and development, nutritional differences associated with mineral efficient vs. inefficient plants, and other enzyme-induction factors), leaf tissues from a normally O_3-sensitive plant variety are made highly tolerant to O_3 in a short period of time (less than a day) by the treatment of *N*-[2-(2-oxo-1-imidazolidinyl) ethyl]-*N′* phenyl urea, better known as EDU, and then compared with control plant tissue not treated with EDU. The optimal EDU dose that promoted maximum O_3 tolerance and retardation of senescence could also induce increased levels of SOD and catalase in normally sensitive (untreated) snap bean (*Phaseolus vulgaris*) leaves. How far EDU treatment affects other cellular enzymes, such as glutathione peroxidase, is not clear but it does generally enhance RNA and protein concentrations in stressed leaf discs (Lee *et al.*, 1981a). EDU treatment increased soluble carbohydrate levels in the leaves (Lee *et al.*, 1981b).

During the development of senescence, SOD and catalase activities may decrease while lipid peroxidation increases (Dhindsa *et al.*, 1982a, b). Treatment with kinetin checked the decline in enzyme activities and delayed senescence.

Treatment with ethanol or α-tocopherol (free radical scavengers) or with diphenylisobenzofuran (singlet oxygen scavenger) strongly inhibited lipid peroxidation and senescence but did not substantially affect the decline in enzyme activities. Kinetin was believed to inhibit senescence through the modulation of membrane lipid peroxidation by maintaining higher levels of cellular scavengers such as SOD, catalase and other free radical counteractants. Damaged membranes lead to leakage and disruption of cell organelles, and mixing of partitioned metabolites with degradative enzymes. This causes loss of vigour and senescence. Preservation of cell and membrane integrity is essential to the structure-function interrelations required for cellular vitality. Genetic activity, particular (constitutive or induced) enzymatic systems present, the availability of cell metabolites and turnover rates, and energy inputs and reserves constitute the control systems, components, activators and power to maintain and repair the cells (see Bennett *et al.*, 1984).

Even though oxyradicals and other active oxygens are injurious to cells, they also act as intermediates and regulators in many vital cell phenomena provided their controlled utilization and prevention of oxyradical amplification (by chain reaction mechanisms) exist in the cells. Two particularly significant cell phenomena that involve superoxide and hydroxyl radicals are the so-called pseudocyclic photophosphorylation in photosynthesizing chloroplasts, and hydroxylation reactions leading to the formation of diphenols. During pseudocyclic photophosphorylation, electrons move from ferredoxin to molecular O_2 (eventually forming H_2O) rather than to NADP (Simonis & Urback, 1973). This results in the gain of ATP.

Important diphenolic acids produced by enzymatically controlled hydroxylation of phenolics may take part in certain inducible biological reductive reactions, such as the reduction of ferric iron to active ferrous iron making it more available for metabolic reactions (Olsen *et al.*, 1981). Free radical polymerization of diphenolic acids leads to the production of lignans, lignins and other similar compounds. Lignins constitute the second largest class of naturally occurring plant structural components.

Peroxidase

Peroxidase enzyme, that belongs to the class of oxidoreductases, occurs in a wide variety of trees and shrubs and has been extensively studied in attempts to develop a method for early identification of chronic injury from air pollution. The enzyme decomposes hydrogen peroxide with a concurrent oxidation of various substrates such as aromatic acids, phenols and heterocyclic compounds (Nicholls, 1962; Saunders *et al.*, 1964; DeKock *et al.*, 1979).

Plant cells show increased peroxidase activity under a variety of stresses, such as mechanical injury, an attack by pathogen or an influence of environmental pollution (Varshney & Varshney, 1985; Beg & Farooq, 1988; Rao, 1989; Nandi *et al.*, 1990). Farkas *et al.* (1964) suggested that an increase in the activity of enzymes under stress is associated with destruction of cellular membranes and the concomitant release of previously immobilized proteins. In plant cells exposed to SO_2, peroxidase activity increases when sulphite is oxidized to sulphate in the presence of hydrogen peroxide and peroxidase (Fridovich & Handler, 1961). This increase in peroxidase activity varies with the species of tree, the season and the concentration of the gas (Nikolayevskii, 1968; Grill *et al.*, 1980). The changes in peroxidase activity are accompanied by opposite changes in catalase activity (Nikolayevskii, 1968; Judel, 1972, Nandi *et al.* 1984).

The increase in peroxidase activity was accompanied by an increased accumulation of sulphur in leaves, a decline in CO_2 assimilation and a reduced content of ascorbic acid (Keller, 1981; Pierre & Queiroz, 1981; Khan & Malhotra, 1982). Increase in peroxidase activity, preceding the appearance of symptoms of leaf injury, may occur in the leaves exposed to toxic gases such as SO_2 (T. Keller, 1976a; Pierre & Queiroz, 1981), O_3 (Curtis & Howell, 1971; Endress *et al.*, 1980) and HF (Keller & Schwager, 1971; Bücher-Wallin, 1976; H. Keller, 1976). The increase was not accompanied by the synthesis of new isozymes of peroxidase in some dicot species exposed to SO_2 and O_3 (Curtis & Howell, 1971; Kieliszewska-Rokicka, 1979), though such isozymes were formed in the needles of Norway spruce treated with HF (Bücher-Wallin, 1976; Grill *et al.*, 1980).

Leaves of the trees characterized by considerable resistance to the action of SO_2 may have a high peroxidase activity (Nikolayevskii, 1979; Khan & Malhotra, 1982), or the reverse may apply (Kieliszewska-Rokicka, 1979). A similar observation was made in a C_4 crop plant, *Zea mays*, by Varshney & Varshney (1985). Peroxidase may serve as a good indicator of the pollution caused by SO_2, HF and O_3 that usually increase peroxidase activity, prior to any visible symptoms of leaf injury (T. Keller, 1974, 1976b; Horsman & Wellburn, 1975; Bücher-Wallin, 1976; Pierre & Queiroz, 1981; Nandi *et al.*, 1990). However, peroxidase is inadequate as an indicator when dealing with air pollution with heavy metals to which it is relatively insensitive (Ernst, 1976), even though prolonged exposure to high metal concentrations may enhance peroxidase activity (Grünhagel *et al.*, 1981; Priebe *et al.*, 1981). Peroxidase activity is modified by factors such as seasonal climatic changes (Nikolayevskii, 1968; Grill, *et al.*, 1980), leaf age (Simola & Sopanen, 1970; Esterbrauer *et al.*, 1978), the state of the ontogenetic development of the tree (Conkle, 1971; Grill *et al.*, 1980), the incompatibility of the stock and scion (Copes, 1978), crown morphology and abundance of cone production and its heterozygosity (Kavac & Ronye, 1975; Grant & Mitton, 1977; Linhart *et al.*, 1979), and fertilization with nitrogen compounds (H. Keller, 1976). Also, the activity of at least some isozymes is under strict genetic control (Kawanobe & Katsuta, 1980). This multifactorial influence makes it difficult to ascertain how much of the increase in activity is due to air pollution and how much to the other exo- and endogenous factors. The contradiction in results on peroxidase activity in leaf tissues could be because of the considerable interclonal variability in trees with regard to this character (H. Keller, 1976; Niemtur, 1979). Peroxidase may be a good indicator of air pollution only when the measurement of its activity is performed on leaves of trees for which the nature of the response of the enzyme to pollutants has already been determined and whose isoenzymes, separated electrophoretically, have been genetically analysed (Mejnartowicz, 1984).

In Scots pine, young needles have an active peroxidase in the epidermis and walls of sclerenchyma cells. In mature needles, the activity occurred in the protoplast. Localization of the enzyme undergoes no change depending on whether the tree is growing in clean environment or in environment polluted with HF and SO_2 (Przymusinski, 1980). Ultrastructural studies have shown that peroxidase is localized in cell walls, endoplasmic reticulum and the Golgi apparatus (Młodzianowski & Młodzianowska, 1980). In sensitive trees, the enzyme activity is higher in mesophyll tissue, while in resistant trees the activity is higher in the epidermis, hypodermis and in the resin canals. The activity was drastically high in the cells of the hypodermis and the endodermis of trees resistant to SO_2 when they were exposed to this gas. These tissues in the needle may have significance in the process of detoxification.

Acid phosphatase

Acid phosphatase (APH) is produced in cell walls and occurs also in the cytosol of higher plants (Hasegawa *et al.*, 1976). In the leaves, it occurs primarily in the spongy parenchyma and in minor veins, but it is much less abundant in the palisade parenchyma (Besford & Syred, 1979). APH, a glycoprotein, has been separated into two components: an "acid" one and an "alkaline" one (see Watroek *et al.*, 1977; Jonsson, 1979). The two enzymes have the same molecular weight but different isoelectric points. The major metabolic function of APH is to decompose phosphate esters (Banasik *et al.*, 1980) and release inorganic phosphate in the cells (Heredia *et al.*, 1963). When the level of phosphorus declines below 0.25% in the tissue, the APH activity rises substantially (Besford, 1979).

Some ions, including those found in the ambient air polluted with industrial fumes, e.g., Al^{3+}, Fe^{3+}, fluoride, arsenate, molybdate and phosphate, have an inhibitory effect on APH activity (Lorenc-Kubis & Morawiecka, 1973; Clancy & Coffey, 1977; Watroek *et al.*, 1977). Sensitivity to inhibitors differs between the wall and the cytoplasmic enzymes (Hasegawa *et al.*, 1976). Accumulation of iron, potassium, calcium, manganese, magnesium, phosphorus, etc. occurs in the needles of several conifers growing in an environment polluted with SO_2 (see Materna, 1962; Amundson, 1978; Mahmooduzzafar, 1991). In spruce, which is very sensitive to SO_2 (Guderian, 1977), APH activity declines to 80% of the control activity in the tissue of 6-week-old needles after treatment with 0.06 ppm SO_2 48 h^{-1} (150 μg SO_2 m^{-3} 48 h^{-1}) (Rabe & Kreeb, 1975). In old needles of jack pine and Norway spruce there was a drop in APH activity following treatment with SO_2, even though APH is still much higher than in younger needles (Yee-Meiler, 1975; Malhotra & Khan, 1980). Rabe (1978) suggested, however, that acid phosphatase is not a very good indicator enzyme since changes in its activity are heavily dependent on plant species and the concentration of gases acting on the plant.

Significant differences in the frequency of allele APH-B5 were noted between two groups of trees growing in the polluted environment but differing in their degree of injury by the pollutants. The frequency was much greater in the group of more tolerant trees (Karnosky & Houston, 1978; Mejnartowicz, 1984). APH-isoenzyme and total activity assays of tissue extracts suggested that population differentiation occurs in response to pollution stress. Frequencies of isoenzymes depend on sampling time after fumigation, and their sensitivity class (Mejnartowicz, 1984). There exists a high degree of genetic similarity between the more resistant trees in the parental population and the F_1 progeny. In the F_1 progeny the most frequent genotype was APH-B5B5 and the rarest was APH-B6B6. In an environment polluted with HF and SO_2 a selection pressure appears to exist against genotype APH-B6B6 and in favour of APH-B5B5 (Mejnartowicz, 1984).

Any direct link between the presence in a tree of a particular isoenzyme of APH and the resistance of that tree to HF or SO_2 is yet to be established even though APH is clearly an important enzyme in nutrition (Hasegawa *et al.*, 1976), transport of stored metabolites (Flinn & Smith, 1967), and stomatal opening (Mishra & Panda, 1970). Disturbance of any of these processes by industrial emissions causes injuries to trees. The presence of some isoenzymes of acid phosphatase may give an early advantage at the stage of seed germination, when a higher activity of an enzyme may depend on its *de novo* synthesis (Shain & Mayer, 1968) or on the liberation and subsequent activation of partially inactive phosphatases present in various cell fractions (Meyer *et al.*, 1971). Hasegawa *et al.* (1976) observed in roots five isoenzymes of acid phosphatase differing in sensitivity to various inhibitors, including fluorides. Some other enzymes such as malate dehydrogenase and leucinaminopeptidase also react against environmental stress (see Mejnartowicz, 1984).

ROLE OF ANTIOXIDANTS AND GROWTH REGULATORS

Certain plant varieties which are naturally quite sensitive to oxidant injury can be transformed into highly tolerant ones by some chemical treatments that alter plant metabolic processes (Lee & Bennett, 1982; Lee *et al.*, 1987, 1992; Larson, 1988; Hausladen & Kunert, 1990; Rao & Dubey, 1990; Miller *et al.*, 1994). EDU could substantially enhance plant tolerance to oxidants. The enhancement in the foliar tolerance to O_3 was up to 30-fold in plants treated by EDU systemically (Carnahan *et al.*, 1978; Jenner *et al.*, 1978; Brennan *et al.*, 1987).

Measurements of stomatal resistance made on the abaxial side of the leaves showed no statistical differences between EDU treated and untreated plants (Lee & Bennett, 1979). This suggests that EDU-induced tolerance to O_3 exposure is probably a biochemical process at leaf mesophyll sites rather than physical exclusion by stomatal closure and restricted O_3 diffusion (Lee *et al.*, 1992). It was further suggested that neither membrane phospholipids nor polyamine levels within the leaf tissue are responsible for the protective nature of EDU in O_3-treated plants. Whitaker *et al.* (1990) demonstrated that EDU does not change the level of chlorophyll and carotenoid pigments in leaf, but reduces the loss of pigments due to O_3 treatment. They suggested that EDU could play a role in maintaining integrity of the foliar membranes.

Increased plant tolerance to environmental stress may also be achieved by means of plant growth regulators and antioxidant treatments (Lee *et al.*, 1987; Hausladen & Kunert, 1990). Lee *et al.* (1992) have further shown that the tolerance to air pollutant exposure induced by gibberellic acid (GA) inhibitor is biochemical at leaf mesophyll sites as stomatal closure was not involved in

GA inhibitor-induced resistance to SO_2. GA_3 inhibitors modify the biosynthesis and action of GA by affecting RNA and protein synthesis, which in turn may regulate the growth and SO_2 sensitivity.

Ascorbic acid, a natural antioxidant in plants, has been shown to play an important role in pollution tolerance (Keller & Schwager, 1977; Lee *et al.*, 1984; Varshney & Varshney, 1984; Chen *et al.*, 1990; Wolfenden *et al.*, 1992; Pandey & Agrawal, 1994). A direct relationship between endogenous levels of ascorbic acid and plant susceptibility to pollutants has been established (Lee *et al.*, 1984; Castillo *et al.*, 1987; Chen *et al.*, 1990; Polle *et al.*, 1990, 1991a, b). Varshney & Varshney (1984) reported higher ascorbic acid concentration in leaves of SO_2-resistant species compared to sensitive ones and a slower decline of ascorbic acid in the former. Ascorbic acid, being a powerful reductant, maintains the stability of cell membranes during pollution stress (Dhindsa *et al.*, 1982a) and scavenges cytotoxic free radicals (Halliwell & Gutteridge, 1989).

The apoplastic compartment contains a variety of other potential reactants for ozone, e.g., proteins such as peroxidases. Peroxidases can provide protection only if ozone yields H_2O_2 in the aqueous phase. Other ozone products such as $OH\cdot$ (Grimes *et al.*, 1983) or organic peroxides (Hewitt *et al.*, 1990) would presumably destroy peroxidase activity as observed in Norway spruce needles exposed to ozone for longer duration in open-top chambers (Ogier *et al.*, 1991) and in needles of mature forest trees exposed to elevated ozone under field conditions (Polle *et al.*, 1991b). These results suggest that peroxidase is more likely to be a target for ozone than a defence against it. In contrast, protection afforded by ascorbate would be limited by reaction products of ozone with hydrocarbons emitted by plants, viz., highly toxic peroxides, radicals, etc. (Mehlhorn & Wellburn, 1987; Hewitt *et al.*, 1990; Polle & Rennenberg, 1993). Rao & Dubey (1990) studied antioxidant system in plants exposed to relatively low concentrations of ambient air pollutants for long durations. They suggest that biochemical changes in plants can lead to air pollution tolerance possibly in two ways: (a) plant species with high antioxidant activities (e.g., *Cassia siamea*) are comparatively more tolerant even if the activity increases a little in polluted areas, and (b) some plant species (e.g., *Calotropis procera*) develop tolerance by raising the antioxidant activities substantially in polluted areas. Based on their ability to raise peroxidase and superoxide dismutase activities in response to adverse effects, the species studied are placed in order of tolerance as: *Cassia siamea* > *Calotropis procera* > *Ipomoea fistulosa* > *Dalbergia sissoo* (Rao & Dubey, 1990).

Sugars and alcohols are among cellular products in large quantities that readily scavenge the initiating $\cdot OH$ radicals. Carbohydrates in plant tissues including free sugars, starch and other cellular components may be important stabilizers of cells and organelles exposed to $\cdot OH$ radicals. Polyamines have also been shown to be antioxidants and protect membranes from peroxidation (Bors *et al.*, 1989). Exogenous application of polyamines has provided protection against O_3 pollution (Ormrod & Beckerson, 1986; Bors *et al.*, 1989). Amelioration of O_3 injury by polyamines has been linked to their ability to stabilize cell membranes (Altman *et al.*, 1977), scavenge free radicals (Drolet *et al.*, 1986) and prevent lipid peroxidation (Tadolini, 1988). Polyamine induction in response to O_3 has been suggested as an adaptive response of plants to stress (Rowland-Bamford *et al.*, 1989). A possible mechanism for preventing O_3 injury could be the reduction of ethylene formation by polyamines (Apelbaum, 1990; Langebartels *et al.*, 1991). High levels of polyamines are known to reduce 1-aminocyclopropane-1-carboxylic acid (= ACC) synthase and ACC oxidase activity (Apelbaum, 1990), which would in turn reduce the ethylene production.

Emission of hydrogen sulphide

Synthesis of H_2S in response to SO_2/HSO_3^- may follow three pathways. Much of the SO_2/HSO_3^- absorbed by leaf tissue of cucurbits is oxidized to sulphate. Light-dependent reduction of sulphate to sulphide may constitute the path of H_2S synthesis. The resultant sulphide may split off carrier-bound sulphide, and be released as H_2S. Alternatively, carrier-bound sulphide may be incorporated into cysteine, from which H_2S may be released by the action of cysteine desulphydrase. A third path of H_2S synthesis may proceed via direct reduction of SO_2/HSO_3^- (see Figure 8.3). This finds support from the developmentally determined resistance to SO_2 in cucurbits as well as from experiments using labelled $^{35}SO_4^{2-}$ (Filner *et al.*, 1984). Thus, direct reduction of SO_2/HSO_3^- to H_2S rather than reduction subsequent to an oxidation to sulphate could be the biosynthetic path leading to H_2S emission in response to SO_2/HSO_3^-.

H_2S emission from cucurbit leaves can occur not only in response to SO_2/HSO_3^- (see Wilson *et al.*, 1978; Sekiya *et al.*, 1982c), but also in response to SO_4^{2-} (Spaleny, 1977; Wilson *et al.*, 1978; Winner *et al.*, 1981; Sekiya *et al.*, 1982a) and L-cysteine (Rennenberg *et al.*, 1982; Sekiya *et al.*, 1982b). A number of other sulphur compounds (e.g., cysteamine, D-cysteine, L-cystine, L-cysteic acid, DL-cysteic acid, L-homocysteine, L-methionine, 3-mercaptopropionic acid, *O*-methyl-L-cysteine, *S*-methyl-L-cysteine, L-serine, taurine) by themselves do not cause H_2S emission. But this does not necessarily mean that volatile sulphur is not produced in response to these compounds. They may give rise to H_2S under some definite conditions (see Sekiya *et al.*, 1982a; Filner *et al.*, 1984).

The H_2S emission phenomenon in response to each precursor/effector (SO_4^{2-}, SO_2/HSO_3^-, L-cysteine) could be distinguished by inhibitor studies. The light-dependent H_2S emission in response to SO_2/HSO_3^- could be inhibited by cyanazine, a triazine herbicide, but not by amino-oxyacetic acid (Brewer *et al.*, 1979). Light-dependent H_2S emission in response to SO_4^{2-} was also inhibited by cyanazine but stimulated by amino-oxyacetic acid (Sekiya *et al.*, 1982a). Sulphate uptake could be or could not be inhibited by amino-oxyacetic acid, depending upon the developmental state of the leaf. Inhibition of sulphate uptake secondarily inhibits H_2S production. Light-independent H_2S emission from L-cysteine is not affected by cyanazine, but is strongly inhibited by amino-oxyacetic acid (Sekiya *et al.*, 1982b), while light-dependent H_2S emission from L-cysteine is inhibited by cyanazine (see Filner *et al.*, 1984).

Inhibition of H_2S emission in response to sulphate by cyanazine and other herbicides that inhibit photosynthetic electron transport (Sekiya *et al.*, 1982a) suggests that the same photosystem which functions in photosynthetic

CO_2 reduction is responsible for light-dependent reduction of sulphate and the generation of H_2S. Furthermore, a direct release of H_2S from carrier-bound sulphide or incorporation of carrier-bound sulphide into cysteine and a subsequent desulphydration are the possible paths of H_2S synthesis from sulphate. Experiments suggest that the excess of sulphate reduced in the presence of inhibitors of glutathione synthesis (a major path of sulphur assimilation in cucurbits) is released as H_2S into the atmosphere (see Rennenberg & Filner, 1982; Sekiya *et al.*, 1982a).

THE SULPHUR CYCLE IN LEAVES

Data suggest that the H_2S emitted by cucurbit cells in response to L-cysteine in the light is derived from two different processes. One part of the H_2S emitted is produced by a light-dependent process that is inhibited by cyanazine; the balance of the H_2S emitted is derived from a light-independent process that is inhibited by amino-oxyacetic acid. The light-independent process could be the desulphydration of L-cysteine, while the light-dependent process is seemingly the light-dependent reduction step in the sulphate assimilation pathway. L-cysteine is believed to degrade to sulphide in light, but only part of the sulphide thus produced is released into the atmosphere as H_2S. The balance of the H_2S produced would be oxidized to sulphate via sulphite. Subsequently, the sulphate synthesized through this pathway would be reduced so that the sulphur from L-cysteine via sulphide, sulphite and sulphate might re-enter the H_2S pool of the cells (see Filner *et al.*, 1984).

Emission of volatiles

Recognition of air pollution damage or injury is often subject to a differential diagnosis, which includes physiological and biochemical parameters (Tingey *et al.*, 1979). Since it may be more effective to avoid pollution damage than to repair it, physiological and biochemical parameters have been used for preventive measures such as evaluation of acceptable pollution levels (Bucher & Keller, 1978) and selection of resistant cultivars (Bressan *et al.*, 1978a). In this regard, metabolic products in plants could receive more attention than substances emitted by plants. The latter vary in nature. For instance, H_2S may be regarded as product of an elevated metabolic activity or a detoxifying process (De Cormis, 1968; Wilson *et al.*, 1978), ethylene may be an expression of a general stress (Abeles, 1973), while ethane (Konze & Elstner, 1976) or monoterpenes (Rasmussen & Went, 1965) may result from cellular disturbance. There are many plant volatiles which may also be affected by pollutant (e.g., acetaldehyde and ethanol; Kimmerer & Kozlowski, 1982). Of these, ethylene and monoterpenes have been studied most extensively (Abeles, 1973; Kuc & Lisker, 1978; Yang & Pratt, 1978; Lieberman, 1979; Loomis & Croteau. 1980; Hogsett *et al.*, 1981). Ethylene, monoterpenes and other volatiles released from vegetation are not only the consequence of a pollution stress, they may be pollutants themselves.

Ethylene

Elevated ethylene emissions from plants in reaction to different abiotic or biotic stresses are well known (Yang & Pratt, 1978; Lieberman, 1979; Hogsett *et al.*, 1981). The stress ethylene accelerates abscission of organs damaged by disease, insect, drought and temperature extremes (Abeles, 1973) and may contribute in growth regulation and disease-resistance mechanisms. It must, therefore, be of potential value in the assessment of plant response and the selection of resistant individuals.

Increase in ethylene production was observed in tomato (*Lycopersicon esculentum*), tobacco (*Nicotiana tabacum*) and bean (*Phaseolus vulgaris*) plants first in response to O_3 treatment (Craker, 1971). Tingey *et al.* (1976) and Hogsett *et al.* (1981) also reported ethylene release against O_3. Later, it was reported from a variety of plants, including forest trees, against SO_2 (Bressan *et al.*, 1978b, 1979a, b; Bucher & Keller, 1978; Bucher, 1979, 1981; Peiser & Yang, 1979; Kimmerer & Kozlowski, 1982), sulphur dust (Recalde-Manrique & Diaz-Miguel, 1981), chlorine (Tingey *et al.*, 1978), cadmium (Fuhrer *et al.*, 1981; Rodecap *et al.*, 1981), and car exhaust fumes (Flückiger *et al.*, 1979). Production of stress ethylene may be used as a test for assessing effects of O_3 on plants because this parameter is seemingly more sensitive, more reproducible and less subjective than assessing visible foliar injury, especially because it could be measured even before the appearance, or in the absence of leaf necrosis (Hogsett *et al.*, 1981; Stan *et al.*, 1981). The inherent capability of plants to synthesize stress ethylene due to O_3 exposure may thus be a decisive factor in plant sensitivity to ozone (Mehlhorn & Wellburn, 1987; Langebartels *et al.*, 1991; Mehlhorn *et al.*, 1991).

Ozone stress has been shown to induce phenyl propanoid and flavonoid pathways responsible for various plant defence responses (Kangasjärvi *et al.*, 1994). Increases in the activities of PAL (Rosemann *et al.*, 1991; Eckey-Kaltenbach *et al.*, 1994), CHS (Rosemann *et al.*, 1991) and CAD (Galliano *et al.*, 1993) controlling the biosynthesis pathways of phenylpropanoid, flavonoid and lignin, respectively, have been reported. Similar correlations were found to exist between SO_2 concentrations (up to 5.5 ppm for 16 h) and the amount of ethylene and ethane generated by fumigated cucumber plants (*Cucumis sativa*) (Bressan *et al.*, 1979a). In all these fumigation experiments with O_3 or SO_2, however, concentrations used were well above the ambient. Fumigation with NO_2, H_2S and chlorine also brought out a linear increase in ethylene production (Tingey *et al.*, 1978). However, experiments on forest trees fumigated in close to field conditions (up to 0.225 ppm SO_2 for several weeks), showed that stress ethylene was unsuitable as an indicator and a measure of plant response because the ethylene emissions fluctuated strongly under the persistent stress (Bucher, 1979, 1981, 1984). Vast variations in ethylene emissions were also seen in bean plants (Stan *et al.*, 1981). Since ethylene production depends on many factors, such as position of the leaf on the stem (Lavee & Martin, 1981) and leaf age (Roberts & Osborne, 1981), the pollution-induced ethylene emission also varies enormously (Tingey *et al.*, 1976). However, stress ethylene can serve as a good indicator or measure of air pollution effects under standardized conditions with high pollutant concentrations.

The fact that ethylene acts autocatalytically to induce its own production (Kende & Baumgartner, 1974) and works in close relationship with other hormones (Liebermann, 1979), makes the issue more complex (see Dörfling, 1980). Production of normal ethylene as well as stress ethylene occurs in the plant through the same biochemical pathway:

methionine → *S*-adenosylmethionine → aminocyclopropane-carboxylic acid (ACC) → ethylene

(see Adams & Yang, 1979; Boller *et al.*, 1979; Jones & Kende, 1979; Lürssen & Naumann, 1979; Yu & Yang, 1979; Hogsett *et al.*, 1981). Only living cells produce ethylene (Elstner & Konze, 1976); when most of the leaf has died (Peiser & Yang, 1979; Kobayashi *et al.*, 1981) or aged (as in late senescence), ethylene production stops (Roberts & Osborne, 1981). Disruption in the permeability of the cell membrane due to stress was thought to release cofactors activating the enzymes involved in the ethylene biosynthesis (Roberts & Osborne, 1981). However, in water-stressed plants ethylene synthesis increased prior to any change in electrolytic leakage and decreased as the leakage increased; this implies that membrane integrity must be maintained for the biosynthesis of ethylene (Kobayashi *et al.*, 1981). It is suggested that stress ethylene is likely to be produced through a *de novo* synthesis, rather than activation of ACC-synthase (Boller & Kende, 1980; Yu & Yang, 1980; Wang & Adams, 1982). The conversion of ACC to ethylene may also be a rate-limiting step under certain conditions (De Laat & van Loon, 1982; Hoffman & Yang, 1982). The sulphur dust absorbed as SO_2 may increase the ethylene precursor methionine in the cells and so induce the biosynthesis of ethylene (Recalde-Manrique & Diaz-Miguel, 1981). ACC, the direct precursor of ethylene, is affected by air pollutants and biotic stress. Sulphur dioxide fumigation at an ambient level could increase the ACC content in the leaves (De Laat & van Loon, 1982; Langebartels *et al.*, 1991).

Contrary to the observations from the tobacco mosaic virus experiment, where increase in ACC appeared prior to the visible injury with emission of the ethylene afterwards, pollutants may cause elevated ethylene emissions before or after the visible foliar symptoms have appeared (Tingey *et al.*, 1976; Bucher & Keller, 1978; Bucher, 1979, 1981; Stan *et al.*, 1981). Saltveit & Dilley (1978) found that wound ethylene was rapidly induced after the transmission of an unknown stimulus. Evidence suggests that ACC could be this signal (Bradford & Yang, 1980). In conclusion, (a) air pollutants and other stresses influence ethylene synthesis, and (b) precursor of ethylene and not ethylene itself, acts as a hormone (see Dörfling, 1980; Yang 1980; Bucher, 1984).

Ethane

Production of ethane, on the contrary, increases with increasing membrane damage and is mostly associated with the visible injury and dying cells (Konze & Elstner, 1976; Kobayashi *et al.*, 1981). For SO_2 effects (1.6 ppm for 8 h) Peiser & Yang (1979) suggested that tissue damage in alfalfa (*Medicago sativa*) plants was caused by free radicals generated through oxidation of the pollutant, which would lead to lipid peroxidation and ethane formation (Yu *et al.*, 1982). A similar mechanism may account for synthesis of the small amount of stress ethylene that may occur in dying cells (Konze, 1977).

Sensitive plants may emit more ethane than resistant ones, so that a rank order could be established (Bressan *et al.*, 1978b, 1979b). The reason for differences in resistance among cultivars could be the relative rate of absorption of the gas (Bressan *et al.*, 1978a), which might depend on plasma membrane rather than stomatal features (Bressan *et al.*, 1981). Anyway, the physiological role of ethylene or ethane with regard to resistance is unclear and the emissions of these volatiles are, therefore, taken to be the biochemical markers only (Bucher, 1984).

Monoterpenes

The significance of monoterpenes emission from plants as an indicator of resistance to air pollutants and other stresses was postulated long ago (Cvrkal, 1959). Terpenes can be used as biochemical markers in genetic studies and in the breeding of plants resistant to various pollutants (Squillace, 1977; Andersson *et al.*, 1980; Muhs, 1981). A long-term SO_2-exposure at concentrations closer to ambient (0.15 ppm for 60 days, followed by 0.5 ppm for 7 days) showed no correlation between the monoterpene pattern and the rank order of the affected individuals in a Scots pine (*Pinus sylvestris*) and western white pine (*Pinus strobus*) population (see Bucher, 1984).

Given the significant role of the volatile monoterpenes in the host–parasite relationship in tree colonization by insects (Hanover, 1975), pollution effects on the terpene metabolism might influence infestations in polluted areas (for references see Bucher, 1984). The emissions of a fertilizer plant increased the contents of camphene, β-pinene, myrcene and tricyclene in Scots pine needles depending on their degree of injury (Lehtiö, 1981). However, SO_2-fumigation experiments with spruce (Dässler, 1964) and Scots pine (Bucher, 1982) revealed a slight decrease in the oil content of the needles. Likewise, the ambient oxidant smog of the Los Angeles basin apparently hampered the production of total volatile oils in ponderosa pine though the amounts of monoterpenes were hardly affected (Miller *et al.*, 1968). Apparently pollutant stress may lead to a decreased production of monoterpenes in severely affected plants only.

Evidence suggests, however, that SO_2 can alter drastically the absolute and relative amounts of monoterpenes. Exposure of young balsam fir (*Abies balsamea*) trees to high SO_2 concentrations (2 ppm for 5 h) on three consecutive days resulted in increased production of α and β-pinene, and camphene. The SO_2-induced increase in the emissions of α-thujene, β-phellandrene (+ limonene) and bornylacetate was smaller but still significant (Renwick & Potter, 1981). Scots pine trees fumigated with low SO_2 concentrations (0.225 ppm up to 19 weeks) released significantly more α-pinene, camphene and limonene, whereas the monoterpene content in the needles tended to decrease (Bucher, 1982).

The release of volatiles is normally through the stomatal pathway, and the cuticle possibly poses a drastic barrier to emissions (Hanover, 1972). Therefore, the occurrence of necrosis, breaking open this barrier, well indicates the huge air-pollution-induced monoterpene emissions. The emissions of volatiles from a forest canopy strongly depend upon the meteorological conditions, and terpenes are not released until the leaves become older or when their cells have died (Rasmussen & Went, 1965). The difference in the relative amounts of the stress-induced monoterpenes might be because of different composition of the compounds in the secretory cells within the foliage (Hanover, 1975), and therefore depends on the part of the needle affected. The qualitative and quantitative changes of terpenes seem to be a general expression of disturbance, and related to wound-repair processes or disease-resistance mechanisms (Kuc & Lisker, 1978). The volatile monoterpenes may, apart from their fungistatic effects and the ecological function in insect attraction, have a messenger function within the plant itself; β-pinene behaves as a potent inhibitor of photosynthesis and respiration in pine (Pauly *et al.*, 1981). It may well be that plants suffering from environmental stress are able to slow down their metabolism in this way and so partly avoid the stress. Monoterpenes may also act as a sink for ozone pollution, thus having a preventive effect. Nonetheless, only terpinolene, α-phellandrene and α-terpinene consume sizeable amounts of O_3, but normally these terpenoids are minor components in the foliage of trees and can therefore be only meagre scavengers of O_3 (Tingey & Taylor, 1982).

CONCLUSIONS

1. The epidermal cells are most vulnerable and susceptible to gaseous air pollutants as they are the first to interact with the pollutants and also lack chloroplasts. The cells without chloroplasts cannot remove excess sulphur through production and release of H_2S for want of energy that comes from photosynthetic electron transport.
2. The characteristics that plants adapt to avoid or tolerate the deleterious effect of air pollution include changes at biochemical, physiological and morphological levels. Plants regulate the entry of gaseous molecules through stomatal movement, thus restricting the entry of gaseous air pollutants and protecting the vital physiological processes such as photosynthesis and transpiration. The stomatal function is normally regulated by changes in the endogenous level of growth regulators.
3. Susceptibility to acute injury by gaseous air pollutants may differ among different genera/species or among cultivars of the same species. This could be due to differences in pollutant absorption from the air which in turn may depend on stomatal number per unit area, aperture size and/or permeability of plasma membrane. Sulphur-dioxide-tolerant plants develop various strategies in order to protect themselves from the pollutant toxicity; this may involve detoxification through SO_2 reduction and expulsion of the byproduct into the atmosphere.
4. The pathways for infusion of different ambient gases to the leaf tissues are not similar. The low values of SO_2 intake during day could be partly due to the production and release of H_2S — a light-dependent process. The rate of emissions of reduced sulphur compounds at higher light intensities and

the SO_2 infusion through cuticle, especially when the leaf begins to manifest injury, are the important factors in the control of net sulphur influx.

5. Low SO_2 concentrations normally widen the stomatal opening without affecting the diurnal functioning of the stomata. The higher concentrations of SO_2 induce stomatal closure in two ways: (a) directly by causing injury to the guard cells and (b) indirectly by modulating the internal CO_2 concentration in the leaf possibly due to its inhibitory effect on photosynthetic CO_2 fixation. In a climate with high vapour pressure deficit, the stomata may, however, tend to have smaller apertures, thus providing a mechanism for the avoidance of pollutants.
6. Young leaves are usually less injured than mature leaves on the same plant, albeit the former take up more SO_2. Young leaves emit a lot more H_2S on SO_2 exposure than mature leaves of the same cultivar. This indicates that a reduction product of SO_2 causes injury, and either further reduction of this product or a diversionary reduction of a precursor to H_2S protects the leaf from injury. Alternatively, SO_2/HSO_3^- may cause injury in a certain compartment, e.g. the chloroplast, where reduction rather than oxidation occurs. Another mechanism to counter the toxic effect of SO_2 may involve oxidation of SO_3^{2-} to SO_4^{2-} in the apoplastic space catalysed by an enzyme sulphite oxidase; SO_4^{2-}, an essential and 30 times less toxic product than SO_3^{2-}, is utilized in the synthesis of various organic sulphur compounds.
7. The gaseous pollutants destroy the ultrastructural organization of leaf cells and produce oxygen-free radicals that lead to the oxidation of biomolecules such as proteins, lipids and DNA. Plants, however, possess defence mechanisms consisting of oxyradical scavenging enzymes (e.g., SOD, catalase and peroxidase) and antioxidants. SOD acts upon O_2 and converts it into H_2O_2, which is then converted into H_2O and O_2 by the enzymes catalase and peroxidase. The reduction of activated oxygen ($\cdot O_2^-$ and H_2O_2) is achieved through a series of reactions eventually at the expense of photosynthetically or enzymatically produced reductant, NAD(P)H.
8. Fluoride inhibits enzyme action by binding with the calcium and/or magnesium needed for enzyme activity, and by the possible formation with the fluoride ion of poisonous metabolic analogues. It also reduces markedly the content of free phosphates, and causes precipitation of calcium. The injurious action of SO_2 on enzymes may be due partly to the similarity of its oxidation product, SO_4^{2-}, with some other anions and partly to its competition with CO_2 in metabolic processes. Under the influence of HF the activity of some enzymes may increase and the H^- ion may induce new isozymes of peroxidase and increase the activities of esterases and β-galactosidase.
9. Peroxidase activity increases invariably against stresses but remains insensitive to gaseous pollutants with heavy metals. The enzyme activity is usually higher in mesophyll tissue, while in resistant plants the activity is higher in the epidermis, hypodermis, endodermis and in the resin canals. These tissues, therefore, must be involved in the process of detoxification. Several

other enzymes, including acid phosphatase, also react against environmental stress. Acid phosphatase, however, may not be a good indicator as its activity changes heavily with plant species and with the concentration of gases acting on the plant.

10. Increased levels of SOD in response to different stresses do not always have positive effects. The overproduction of SOD in bacteria and transgenic plants may cause an increased H_2O_2 production, which surpasses the capacity of these organisms to decompose or scavenge H_2O_2. Consequently, $O_2{\cdot}^-$ and H_2O_2 levels may be shifted towards a ratio favourable for $OH{\cdot}$ production. On the other hand, strong overproducers of manganese SOD are more protected from oxidative stress than controls. Similarly, plant species with capability to increase their antioxidant capacity on being exposed to environmental stress, are more resistant than cultivars with low antioxidative capacity.
11. Certain chemicals are known to enhance plant tolerance either by elevating the level of SOD and catalase or by stabilizing their activities. These include EDU and kinetin. EDU induces SOD and catalase, and retards senescence, while kinetin checks decline in enzyme activities and delays senescence. On the other hand, ethanol, α-tocopherol and diphenylisobenzofuran strongly inhibit lipid peroxidation and senescence, but do not affect the decline in these enzyme activities. Kinetin modulates membrane-lipid peroxidation by maintaining higher levels of cellular scavengers such as SOD, catalase and other free radical counteractants.
12. EDU stimulants and GA-inhibitors induce tolerance at biochemical level at leaf mesophyll sites. The protective nature of EDU on O_3 stress is not related to alteration of membrane phospholipids or polyamine levels within the leaf tissue, but to traits such as free radical scavenging enzymes or endogenous antioxidant compounds.
13. Oxyradicals and other active oxygen species although injurious to cells, also act as intermediates and regulators in many vital cell phenomena including the pseudocyclic photophosphorylation, and the synthesis of diphenolic acids from enzymatically controlled hydroxylation of phenolics. The diphenolic acids may take part in certain inducible biological reductive reactions such as the reduction of ferric ion to active ferrous ion, thus enhancing its availability for metabolic reactions. Free radical polymerization of diphenolic acids leads to the production of lignans, and other similar compounds.
14. Some common metabolic reductants, such as ascorbic acid, glutathione, α-tocopherol, starch and sugars, etc., also protect plants from the pollution injury. Norway spruce is remarkably resistant to ozone-mediated damage due to higher levels of ascorbate in the cell walls. These antioxidants serve as a free oxyradical scavenging system in plants exposed to gaseous air pollutants and protect membranes from peroxidation. The protection afforded by ascorbate, however, is limited by reaction products of ozone with hydrocarbons emitted by plants, which include highly toxic peroxides, radicals, etc.

15. Leaves may emit H_2S not only in response to SO_2/HSO_3, but also in response to SO_4^{2-} and L-cysteine. A variety of other sulphur compounds may also give rise to H_2S under some definite conditions. Plants also emit other volatile compounds such as ethylene, ethane, monoterpenes, acetaldehyde and ethanol in response to cellular disturbance and injury caused by air pollutants. These are not only the consequence of a pollution stress, but may be pollutants themselves. The ethylene produced accelerates abscission of organs damaged by disease, insect, drought and temperature extremes and may contribute in growth regulation and disease resistance mechanisms. Stress ethylene may be a good indicator of pollution stress under standardized conditions with high pollutant concentrations but this is not so in the field because ethylene emissions fluctuate strongly under persistent stress. Ethylene is produced by living cells only and both normal and stress ethylene are produced in the plant through the same biochemical pathway. Air pollutants including SO_2 and biotic stress increase ACC content of leaves, the direct precursor of ethylene.
16. Ethane production increases with increasing membrane damage and is mostly associated with the dying cells, a visible injury. Sensitive plants emit more ethane than the resistant ones. Ethane/ethylene ratio is the most attractive assay for assessing the injury level of a tissue exposed to air pollutants as it increases monotonically with injury and is dimensionless.
17. Terpenes are released into the atmosphere normally through the stomatal pathway, while the cuticle acts as a barrier. However, the occurrence of necrosis indicates the breakopen of this barrier due to high emissions of air-pollution-induced monoterpene. Monoterpenes also act as a sink for ozone and, thus, provide a preventive effect.

ACKNOWLEDGEMENTS

We thank Dr Hisamuddin, the Research Associate in Botany, Mr Sabah Ashraf and Mr Mohammad Ziauddin, the computer operators, Mr Sarfaraz Alam, the typist and Ms Indu Mehra, the artist, at Jamia Hamdard, New Delhi, for their valuable assistance in developing this chapter.

REFERENCES

Abeles, F.B. 1973. *Ethylene in Plant Biology*. Academic Press, New York.

Aben, J.M.M., Janssen-Jurkovicova, M. & Adema, E.H. 1990. Effects of low level ozone exposure under ambient conditions on photosynthesis and stomatal control of *Vicia faba* L. *Plant, Cell Environ.* **13**: 463–469.

Adams, D.O. & Yang, S.F. 1979. Ethylene biosynthesis: identification of 1-aminocyclopropane-1-carboxylic acid as an intermediate in the conversion of methionine to ethylene. *Proc. Natl. Acad. Sci. USA* **76**: 170–174.

Adedipe, N.O. & Ormrod, D.P. 1972. Hormonal control of ozone phytotoxicity in *Raphanus sativus*. *Z. Pflanzenphysiol.* **68**: 254–258.

Allaway, W.G. & Mansfield, T.A. 1967. Stomatal responses to changes in carbon dioxide concentration in leaves treated with 3-(4-chlorophenyl)-1,1-dimethyl urea. *New Phytol.* **66**: 57–63.

Altman, A., Kaur-Sawhney, R. & Galston, A.W. 1977. Stabilization of oat leaf protoplasts through polyamine mediated inhibition of senescence. *Plant Physiol.* **60**: 570-574.

Amundson, R.G. 1978. The effects of SO_2 on mineral nutrition and gas exchange of pine (*Pinus contorta* × *P. banksiana*). PhD Dissertation, University of Washington, Seattle.

Andersson, B.A., Holman, R.T., Lundgren, N. & Stenhagen, G. 1980. Capillary gas chromatograms of leaf volatiles. A possible aid to breeders for pest and disease resistance. *J. Agric. Food Chem.* **29**: 985-989.

Apelbaum, A. 1990. Interrelationship between polyamines and ethylene and its implication for plant growth and fruit ripening. In Flores, H.E., Arteca, R.N. & Shannon, J.C. (eds) *Polyamines and Ethylene: Biochemistry, Physiology and Interactions*: 278-294. American Society of Plant Physiologists, Rockville, MD.

Appleby, R.F. & Davies, W.J. 1982. A possible evaporation site in the guard cell wall and the influence of leaf structure on the humidity response by stomata of woody plants. *Oecologia* **56**: 30-40.

Asada, K. 1980. Formation and scavenging of superoxide in chloroplasts with relation to injury by sulphur dioxide. *Res. Rep. Natl. Inst. Environ. Stud., Japan* **11**: 165-179.

Asada, K. & Takahashi, M. 1987. Production and scavenging of active oxygen in photosynthesis. In Keyle, D.J., Osmond, C.B. & Arntzen, C.J. (eds) *Photoinhibition*: 227-287. Elsevier, Amsterdam.

Banasik, L., Rudnicki, R.M. & Saniewski, M. 1980. Studies on the physiology of hyacinth bulbs (*Hyacinthus orientalis* L.). XIII. The distribution of amylase and acid phosphatase activities and starch grains in hyacinth bulbs. *Acta Physiol. Plant.* **2**: 145-156.

Barnes, J.D. & Davison, A.W. 1988. The influence of ozone on the winter hardiness of Norway spruce [*Picea abies* (L.) Karst]. *New Phytol.* **108**: 159-166.

Beckerson, D.W. & Hofstra, G. 1979a. Stomatal responses of white bean to O_3 and SO_2 singly and in combination. *Atmos. Environ.* **13**: 533-535.

Beckerson, D.W. & Hofstra, G. 1979b. Response of leaf diffusive resistance of radish, cucumber and soybean to O_3 and SO_2 singly or in combination. *Atmos. Environ.* **13**: 1263-1268.

Beg, M.U. & Farooq, M.H. 1988. Sulphur dioxide resistance of Indian trees II. Experimental evaluation of metabolic profile. *Water, Air Soil Pollut.* **40**: 317-328.

Bennett, J.H., Lee, E.H. & Heggestad, H.E. 1981. Apparent photosynthesis and leaf stomatal diffusion in EDU treated ozone-sensitive bean plants. In Abdel-Rahman M. (ed.) *Proceedings Fifth Annual Meeting of the Plant Growth Regulator Working Group*: 242-246. Agway, Inc., Syracuse, NY.

Bennett, J.H., Lee, E.H. & Heggestad, H.E. 1984. Biochemical aspects of plant tolerance to ozone and oxyradicals: Superixode dismutase. In Koziol, M.J. & Whatley, F.R. (eds) *Gaseous Air Pollutants and Plant Metabolism*: 413-424. Butterworths, London.

Besford, R.T. 1979. Quantitative aspects of leaf acid phosphatase activity and phosphorus status of tomato plants. *Ann. Bot.* **44**: 153-161.

Besford, R.T. & Syred, A.D. 1979. Effect of phosphorus nutrition on the cellular distribution of acid phosphatase in the leaves of *Lycopersicon* L. *Ann. Bot.* **43**: 431-453.

Biscoe, P.V., Unsworth, M.H. & Pinckney, H.R. 1973. The effects of low concentrations of sulphur dioxide on stomatal behaviour in *Vicia faba*. *New Phytol.* **72**: 1299-1306.

Black, C.R. & Black, V.J. 1979. The effects of low concentrations of sulphur dioxide on stomatal conductance and epidermal cell survival in field bean (*Vicia faba* L.). *J. Exp. Bot.* **30**: 291-298.

Black, V.J. 1982. Effects of sulphur dioxide on physiological processes in plants. In Unsworth, M.H. & Ormrod, D.P. (eds) *Effects of Gaseous Air Pollution in Agriculture and Horticulture*: 67-91. Butterworths, London.

Black, V.J. & Unsworth, M.H. 1980. Stomatal responses to sulphur dioxide and vapour pressure deficit. *J. Exp. Bot.* **31**: 667-677.

Bligny, R., Bisch, A.M., Garrec, J.P. & Fourcy, A. 1973. Observations morphologigues et structurales des effects du fluor sur les cires epicuticularies et sur les chloroplastes des aiguilles de sapin (*Abies alba* Mill.). *J. Micros.* **17**: 207-214.

Boller, T. & Kende, H. 1980. Regulation of wound ethylene synthesis in plants. *Nature* **286**: 259-260.

Boller, T., Herner, R.C. & Kende, H. 1979. Assay for and enzymatic formation of an ethylene precursor, 1-aminocyclopropane-1-carboxylic acid. *Planta* **145**: 293-303.

Borel, H. 1945. Inhibition of cellular oxidation by fluoride. *Arkiv for Kemi, Mineralog och Geologi. K. Svenska Vetenskaps-Akademien. Stockholm, Uppsala Almqvist et Wiksells Boktrycher* **20A**: 1-215.

Bors, W., Langebartels, C., Michel, C. & Sandermann, H. Jr 1989. Polyamines as radical scavengers and protectants against ozone damage. *Phytochem.* **28**: 1589-1595.

Bowler, C., Slooten, L., Vandenbranden, S., De Rycke, R., Botterman, J., Sybesma, C., Van Montague, M. & Inze, D. 1991. Manganese superoxide dismutase can reduce cellular damage mediated by oxygen radicals in transgenic plants. *EMBO* **10**: 1723-1732.

Bradford, K.J. & Yang, S.F. 1980. Xylem transport of 1-aminocyclopropane-1-carboxylic acid, an ethylene precursor, in waterlogged tomato plants. *Plant Physiol.* **65**: 322-326.

Brawn, K. & Fridovich, I. 1981. DNA strand scission by enzymatically generated oxygen radicals. *Arch. Biochem. Biophys.* **206**: 414-419.

Brennan, E., Leone, I. & Clarke, B. 1987. EDU a chemical for evaluating ozone foliar injury and yield reduction in field-grown crops. *Int. J. Trop. Dis.* **5**: 35-42.

Bressan, R.A., Wilson, L.G. & Filner, P. 1978a. Mechanisms of resistance to sulphur dioxide in the Cucurbitaceae. *Plant Physiol.* **61**: 761-767.

Bressan, R.A., Wilson, L.G., Lecureux, L. & Filner, P. 1978b. Use of ethylene and ethane emissions to assay injury by SO_2. *Plant Physiol.* **61** (Suppl.): Abstract No. 509.

Bressan, R.A., Lecureux, L., Wilson, L.D. & Filner, P. 1979a. Emission of ethylene and ethane by leaf tissue exposed to injurious concentrations of sulphur dioxide or bisulfite ion. *Plant Physiol.* **63**: 924-930.

Bressan, R.A., Lecureux, L., Wilson, L.G. & Filner, P. 1979b. Release of ethane by leaf tissue in response to various environmental stresses. *Plant Physiol.* **63**: S-327.

Bressan, R.A., Lecureux, L., Wilson, L.G., Filner, P. & Baker, L.R. 1981. Inheritance of resistance to sulphur dioxide in cucumber. *Hort. Sci.* **16**: 332-333.

Brewer, P.E., Arntzen, C.J. & Slite, F.W. 1979. Effects of atrazine and procyazine in the photochemical reaction of isolated chloroplasts. *Weed Sci.* **27**: 300-308.

Bucher, J.B. 1979. SO_2-induziertes Stress-Aethylen in den Assimilationsorganen von Waldbäumen. *Zbornik (Mitteilungen) Institut Forst-und Holzwirtschaft Ljubljana (YU), IUFRO.* Bericht der X. Fachtagung in Ljubljana, pp. 93-101.

Bucher, J.B. 1981. SO_2-induced ethylene evolution of forest tree foliage and its potential use as stress-indicator. *Europ. J. For. Pathol.* **11**: 369-373.

Bucher, J.B. 1982. Einfluss von SO_2 auf Terpen-Emissionen von Kiefern (*P. sylvestris* L.). *XII Internat. Meeting on Air Pollution Damage in Forests*, IUFRO, S2.09, Oulu, Finland.

Bucher, J.B. 1984. Emissions of volatiles from plants under air pollution stress. In Koziol, M.J. & Whatley, F.R. (eds) *Gaseous Air Pollutants and Plant Metabolism*: 399-412. Butterworths, London.

Bucher, J.B. & Keller, Th. 1978. Einwirkungen niedriger SO_2-Konzentrationen im mehrwöchigen Begasungsversuch auf Waldbäume. *VDI-Bericht* **341**: 237-242.

Bücher-Wallin, I. 1976. Zur Beeinflussung des physiologischen Blattalters von Waldbäumen durch Fluor-Immissionen, Mitteilungen Eidgenossiche Anstalt für das Forstliche Versuchswesen. *Schweiz* **52**: 101-158.

Carnahan, J.E., Jenner, E.L. & Wat, E.K.W. 1978. Prevention of ozone injury to plants by a new protectant chemical. *Phytopath.* **68**: 1225-1229.

Castillo, F.J., Miller P.R. & Greppin, H. 1987. Extracellular biochemical markers of photochemical air pollution damage to Norway spruce. *Experientia* **43**: 111-220.

Chamel, A.R. 1980. Penetration du cuivre a travers des cuticules isoles de feuilles de Poirier. *Physiol Veg.* **18**: 313-323.

Chamel, A.R. 1984. Recherches sur l'absroption foliare effectue microanalyse. *VI Internat. Colloquium for the Optimization of Plant Nutrition*, 2-8 Sept., Montpellier, Actes Proceedings, 11.

Chamel, A.R. & Eloy, J.E. 1983. Some applications of the laser probe mass spectrograph in plant biology. *Scan. Elec. Micros.* **2**: 841-845.

Chamel, A.R., Gambonnet, B., Genova, C. & Jourdain, A. 1984. Cuticular behaviour of cadmium studied using isolated plant cuticles. *J. Environ. Qual.* **13**: 483-487.

Chang, C.W. 1970a. Effect of fluoride on ribosomes from corn roots. Changes with growth retardation. *Physiol. Plant.* **23**: 536-543.

Chang, C.W. 1970b. Effect of fluoride on ribosomes and ribonuclease from corn roots. *Can. J. Biochem.* **48**: 450-454.

Chen, Y.M., Lucas, P.W. & Wellburn, A.R. 1990. Relationships between foliar injury and changes in antioxidant levels in red and Norway spruce exposed to acidic mists. *Environ. Pollut.* **68**: 1-15.

Chia, L.S., Mayfield, C.I. & Thompson, J.E. 1984. Simulated acid rain induces lipid peroxidation and membrane damage in foliage. *Plant, Cell Environ.*, **7**: 333-338.

Clancy, F.G. & Coffey, M.D. 1977. Acid phosphatase and protease release by the insectivorous plant *Drosera rotundifolia*. *Can. J. Bot.* **55**: 480-488.

Conkle, M.Th. 1971. Isozyme specificity during germination and early growth of knobcone pine. *For. Sci.* **17**: 494-498.

Copes, D.L. 1978. Isoenzyme activities differ in compatible and incompatible Douglas-fir graft unions. *For. Sci.* **24**: 297-303.

Craker, L.E. 1971. Ethylene production from ozone injured plants. *Environ. Pollut.* **1**: 299-304.

Curtis, C.R. & Howell, R.K. 1971. Increase in peroxidase isoenzyme activity in bean leaves exposed to low doses of ozone. *Phytopathology*. **61**: 1306-1307.

Cvrkal, H. 1959. Biochemische Diagnose an Fichten in Rauchgebieten, Berichte Tschechoslovakischen Akademie Landwirtschaftswissenschaft. *Sektor For.* **5**: 1033-1048.

Darrall, N.M. 1989. The effect of air pollutants on physiological processes in plants. *Plant, Cell Environ.* **12**: 1-30.

Dässler, H.-G. 1964. Einfluss des Schwefeldioxides auf den Terpengehalt von Fichtennadeln. *Flora* **154**: 376-382.

Davies, W.J. & Mansfield, T.A. 1987. Auxins and stomata. In Zeigar, E., Farguhar G.D. & Cowan, I.R. (eds) *Stomatal Function*: 293-309. Stanford University Press, Stanford, CA.

Decleire, M., Decat, W., Detemmerman, L. & Baten, H. 1984. Changes of peroxidase, catalase and superoxide dismutase activities in ozone fumigated spinach leaves. *J. Plant Physiol.* **116**: 147-152.

De Cormis, L. 1968. Dégagement d'hydrogene sulphuré par des plantes soumises à une atmosphére contenant de l'anhydride sulphureux. *Com. Rendus Acad. Sci.* **266**: 682-685.

DeKock, P.C., Hall, A. & Inkson, H.E. 1979. A study of peroxidase and catalase distribution in the potato tuber. *Ann. Bot.* **43**: 295-298.

DeKok, L.J. 1990. Sulphur metabolism in plants exposed to atmospheric sulphur. In Rennenberg, H. *et al.* (eds) *Sulphur Nutrition and Sulphur Assimilation in Higher Plants*: 111-130. SPB Acad. Publ., The Hague.

De Laat, A.M.M. & van Loon, L.C. 1982. Regulation of ethylene biosynthesis in virus-infected tobacco leaves. *Plant Physiol.* **69**: 240-245.

Dhindsa, R.S., Plumb-Dhindsa, P.L. & Reid, D.M. 1982a. Leaf senescence and lipid peroxidation: Effects of some phytohormones and free radical scavengers. *Plant Physiol.* **69** (Suppl.) 10: Abstract No. 48.

Dhindsa, R.S., Plumb-Dhindsa, P.L. & Thorpe, T.A. 1982b. Leaf senescence correlated with increased levels of membrane permeability and lipid peroxidation and decreased levels of superoxide dismutase and catalase. *J. Exp. Bot.* **32**: 93-101.

Dominy, P.J. & Heath, R.L. 1985. Inhibition of the K^+ stimulated ATPase of the plasmalemma of pinto bean leaves by ozone. *Plant Physiol.* **77**: 43-45.

Dörfling, K. IV 1980. Growth. *Progress in Botany* **42**: 111-125.

Downton, W.J., Loveys, B.R. & Grant, W.J.R. 1988. Stomatal closure fully accounts for the inhibition of photosynthesis by abscisic acid. *New Phytol.* **108**: 263-266.

Drolet, G., Dumbroff, E.B., Legge, R.L. & Thompson, J.E. 1986. Radical scavenging properties of polyamines. *Phytochem.* **25**: 367-371.

Dubey, P.S. 1990. Crop response against sulphur dioxide — Indian contribution. *Indian Rev. Life Sci.* **10**: 99-120.

Dugger, W.M. & Ting, I.V. 1970. Air pollution oxidants — Their effects on metabolic processes in plants. *Annu. Rev. Plant. Physiol.* **21**: 215-234.

Eamus, D. & Jarvis, P.G. 1989. The direct effects of increase in the global atmospheric CO_2 concentration on natural and commercial temperate trees and forests. *Adv. Ecol. Res.* **16**: 1-55.

Eckey-Kaltenbach, H., Ernst, D., Heller, W. & Sandermann, H. Jr 1994. Biochemical plant responses to ozone IV. Cross-induction of defensive pathways in parsley plants. *Plant Physiol.* **104**: 67-74.

Elstner, E.F. & Konze, J.R. 1976. Effect of point freezing on ethylene and ethane production by sugar beet leaf disks. *Nature* **263**: 351-352.

Endress, A.G., Suarez, S.J. & Taylor, O.C. 1980. Peroxidase activity in plant leaves exposed to gaseous HCl or ozone. *Environ. Pollut. (Ser A)* **22**: 47-58.

Ernst, W. 1976. Physiological and biochemical aspects of metal tolerance. In Mansfield, T.A. (ed.) *Effect of Air Pollutants on Plants*: 115-133. Cambridge University Press, Cambridge.

Esterbrauer, H., Grill, D. & Zotter, M. 1978. Peroxidase in Nadeln von *Picea abies* (L.) Karst. *Biochem. Physiol. Pflanzen* **172**: 155-159.

Farage, P.K., Long, S.P., Lechner, E.G. & Baker, N.G. 1991. The sequence of changes within the photosynthetic apparatus of wheat following short term exposure to ozone. *Plant Physiol.* **95**: 529-535.

Farkas, G.L., Dezsi, L., Horvath, M., Kisban, K. & Udvardy, J. 1964. Common pattern of enzymatic changes in detached leaves and tissues attacked by parasites. *Phytopathol. Z.* **49**: 343-354.

Filner, P., Rennenberg, H., Sekiya, J., Bressan, R.A., Wilson, L.G., Le Cureux, L. & Shimei, T. 1984. Biosynthesis and emission of hydrogen sulfide by higher plants. In Koziol, M.J. & Whatley, F.R. (eds) *Gaseous Air Pollutants and Plant Metabolism*: 291-312. Butterworths, London.

Flinn, A.M. & Smith, D.L. 1967. The localization of enzymes in the cotyledons of *Pisum arvense* L. during germination. *Planta* **75**: 10-22.

Flückiger, W., Oertli, J.J., Flückiger-Keller, H. & Braun, S. 1979. Premature senescence in plants along a motorway. *Environ. Pollut.* **13**: 171-174.

Foyer, C.H., Descourvieres, P. & Kunert, K. 1994. Protection against oxygen radicals: an important defense mechanism studied in transgenic plants. *Plant, Cell Environ.* **17**: 507-523.

Fridovich, I. 1981. Role and toxicity of superoxide in cellular systems. In Rodgers M.A.J. & Powers E.L. (eds) *Oxygen and Oxy-Radicals in Chemistry and Biology*: 197-239. Academic Press, New York.

Fridovich, I. 1986. Superoxide dismutases. *Adv. Enzymol.* **58**: 62-97.

Fridovich, I. & Handler, P. 1961. Detection of free radicals generated during enzymatic oxidations by the initiation of sulphite oxidation. *J. Biol. Chem.* **236**: 1834-1840.

Fuhrer, J., Geballe, G.T. & Fries, C. 1981. Cadmium-induced change in water economy of beans: involvement of ethylene formation. *Plant Physiol.* **67** (Suppl.): Abstract No. 307.

Galliano, H., Heller, W. & Sanderman, H.I. 1993. Ozone induction and purification of spruce cinnamyl alcohol dehydrogenase. *Phytochem.* **32**: 557-563.

Gilbert, M.D., Elfving, D.C. & Lisk, D.J. 1977. Protection of plants against ozone injury using the antiozonant, *N*-(1,3-di-methylbutyl)-*N'*-phenyl-p-phenylenediamine. *Bull. Environ. Contamin. Toxicol.* **18**: 783-786.

Gillespie, C.T. & Winner, W.E. 1989. Development of lines of radish differing in resistance to ozone and sulphur dioxide. *New Phytol.* **112**: 353-361.

Godzik, S. & Sassen, M.M.A. 1974. Einwirkung von SO_2 auf die Feinstruktur der Chloroplasten von *Phaseolus vulgaris*. *Phytopathol. Z.* **79**: 155-159.

Grant, M.C. & Mitton, J.B. 1977. Genetic differentiation among growth forms of engelmann spruce and subalpine fir at tree line. *Arctic & Alpine Res.* **9**: 259-263.

Grill, D., Esterbauer, H. & Birkner, M. 1980. Untersuchungen über die Peroxidaseaktivät in Lärchennadel. *Beitr. Biol. Pflanzen.* **55**: 67-76.

Grimes, H.D., Perkins, K.K. & Boss, W.F. 1983. Ozone degrades into hydroxyl radical under physiological conditions. A spin trapping study. *Plant Physiol.* **72**: 1016-1020.

Grünhagel, L., Klein, H. & Jäger, H.J. 1981. Langzeiteinwirkungen von Frieilandrelevanten SO_2-und Cadmiumkonzentrationen an Pflanzen. Nachweis und Wirkung forstshädlicher Luftverunreinigungen, *XI Internationalen Arbeitstagung forstlicher Rauchsschadenssachverstädiger*, 1-6 September 1980, Graz, Austria, pp. 309-317.

Guderian, R. 1977. *Air Pollution. Phytotoxicity of Acidic Gases and its Significance in Air Pollution Control.* Springer-Verlag, Berlin.

Hällgren, J.E., Linder, S., Richter, A., Troeng, E. & Granat, L. 1982. Uptake of SO_2 in shoots of Scots pine: field measurements of new flux of sulphur in relation to stomatal conductance. *Plant, Cell Environ.* **5**: 75-83.

Halliwell, B. & Gutteridge, J.M.C. 1989. *Free Radicals in Medicine and Biology*, 2nd edition. Clarendon Press, Oxford.

Hanover, J.W. 1972. Factors affecting the release of volatile chemicals by forest trees. *Mitteilungen aus der forstlichen Bundssversuchsanstalt, Wien* **97**: 625-644.

Hanover, J.W. 1975. Physiology of tree resistance to insects. *Annu. Rev. Entomol.* **20**: 75-95.

Hanson, J.B. 1960. Impairment of respiration, ion accumulation and ion retention in root tissue treated with ribonuclease and ethylenediamine tetra-acetic acid. *Plant Physiol.* **35**: 372-379.

Hasegawa, Y., Lynn, K.R. & Brockbank, W.J. 1976. Isolation and partial characterization of cytoplasmic and wall-bound acid phosphatases from wheat roots. *Can. J. Bot.* **54**: 1163-1169.

Hausladen, A. & Kunert, K.J. 1990. Effects of artificially enhanced levels of ascorbate and glutathione on the enzymes monohydroascorbate reductase, hydroascorbate reductase, and glutathione reductase in spinach. *Physiol. Plant.* **79**: 384-388.

Heath, R.L. 1980. Initial events in injury to plant by air pollutants. *Annu. Rev. Plant Physiol.* **31**: 395-431.

Heath, R.L. 1988. Biochemical mechanisms of pollutant stress. In Heck, W.W., Taylor, O.C. & Tingey, D.T. (eds) *Assessment of Crop Loss from Air Pollutants*: 259-286. Elsevier Applied, New York.

Heath, R.L. 1994. Alterations of plant metabolism by ozone exposure. In Alscher, R.G. & Wellburn, A.R. (eds) *Plant Responses to the Gaseous Environment*: 121-145. Chapman & Hall, London.

Heck, W.W., Taylor, O.C. & Heggestad, H.E. 1973. Air pollution research needs: Herbaceous and ornamental plants and agriculturally generated pollutants. *J. Air Pollut. Control Assoc.* **23**: 257-266.

Heredia, C.F., Yen, F. & Sols, A. 1963. Role and formation of the acid phosphatase in yeast. *Biochem. Biophys. Res. Commun.* **10**: 14-18.

Hewitt, N., Kok, G. & Fall, R. 1990. Hydroperoxide in plants exposed to ozone mediates air pollution damage to alkene emitters. *Nature* **344**: 56-58.

Hoffman, N.E. & Yang, S.F. 1982. Enhancement of wound-induced ethylene synthesis by ethylene in preclimacteric cantaloupe. *Plant Physiol.* **69**: 317-322.

Hogsett, W.E., Raba, R.M. & Tingey, D.T. 1981. Biosynthesis of stress ethylene in soybean seedlings: Similarities to endogenous ethylene biosynthesis. *Physiol. Plant.* **53**: 307-314.

Horsman, D.C. & Wellburn, A.R. 1975. Synergistic effects of SO_2 and NO_2 polluted air upon enzyme activity in pea seedlings. *Environ. Pollut.* **8**: 123-133.

Hull, H.M. 1970. Leaf structure as related to absorption of pesticides and other compounds. *Residue Reviews* **31**: 1-55.

Jain, M.M. 1972. *The Biomolecular Lipid Membrane: A System.* Norstrand Reinhold Company, New York.

Jenner, E.L., Carnahan, J.E. & Wat, E.K.W. 1978. A new antiozone-plant protectant. In Abdel-Rahman, M. (ed.) *Proceedings of the Fifth Annual Meeting of the Plant Growth Regulator Working Group*: 238-241. Agway, Inc. Syracuse, NY.

Jones, J.F. & Kende, H. 1979. Auxin-induced ethylene biosynthesis in subapical stem sections of etiolated seedlings of *Pisum sativum* L. *Planta* **146**: 649-656.

Jonsson, J. 1979. Isolation and partial characterization isoenzymes of acid phosphatase from needles of *Pinus sylvestris* L. In Rudin, D. (ed.) *Proceedings of the Conference on Biochemical Genetics of Forest Trees*: 5-14. Report 1. Swedish University of Agricultural Sciences, Umea, Sweden.

Judel, G.K. 1972. Anderung in der Aktivität der Peroxidase und der Katalase und im Gehalt an Gesamtphenolen in der Blättern der Sonnenblume unter dem Einfluss von Kupfer- und Stickstoffmangel. *Z. Pflanzenernährung & Bodenkunde* **133**: 81-92.

Kangasjärvi, J., Talvinen, J., Utriainen, M. & Karjalainen, R. 1994. Plant defense systems induced by ozone. *Plant, Cell Environ.* **17**: 783-794.

Karnosky, D.F. & Houston, D.S. 1978. Genetics of air pollution tolerance of trees in the northeastern United States. *Proceedings of the 25th Northeastern Forest Tree Improvement Conference*: 161-178. Pennsylvania State University, USA.

Kavac, Y.E. & Ronye, V.M. 1975. Isoenzimy pyeroksidazy khvoi v populyaciyakh yeli obyknovyennoi. [Isoenzymes of peroxidases of needles in population of Norway spruce.] In *Gyenyetichyeskiye isslyedovaniya drevesnykh v Latviiskoi SSR.* [*Genetic investigation of trees in Latvian SSR.*]: 58-62. Zinatnye, Riga, USSR.

Kawanobe, K. & Katsuta, M. 1980. Genetic analysis of peroxidase isozymes in *Pinus thunbergii* Parl. self progeny. *J. Jap. For. Soc.* **62**: 357-360.

Keller, H. 1976. Histologische und physiologische Untersuchungen an Forstpflanzen in einem Fluorschadensgebiet. *Ber. Eidgen. Anstalt Forstl. Versuchswesen* **154**: 1-82.

Keller, T. 1974. The use of peroxidase activity for monitoring and mapping air pollution areas. *Europ. J. For. Pathol.* **4**: 11-19.

Keller, T. 1976a. Auswirkungen niedriger SO_2-konzentrationen auf junge Fichten. *Schweiz. Z. Forstowesen* **127**: 237-251.

Keller, T. 1976b. Der Einfluss von Schwefeldioxid als Luftverunreinigung auf die Assimilation der Fichte. *Ber. Eidgen. Anstalt Forstl. Versuchswesen* **57**: 48-53.

Keller, T. 1981. Die beeinflussung physiologischer Prozesse der Fichte durch eine Winterbegasung mit SO_2. *Mitteilungen der forstlichen Bundesrersuchstanstalt Wien* **137**: 115-120.

Keller, T. & Schwager, H. 1971. Der Nachweis unsichtbarer ("physiologischer") Fluorimmissionsschädigungen an Waldbäumen durch eine einfache kolorimetrische Bestimmung der Peroxidase-Aktivität. *Europ. J. For. Pathol.* **1**: 6-18.

Keller, T. & Schwager, H. 1977. Air pollution and ascorbic acid. *Europ. J. For. Pathol.* **7**: 338-350.

Kende, H. & Baumgartner, B. 1974. Regulation of aging flowers of *Ipomoea tricolor* by ethylene. *Planta* **116**: 279-289.

Kerstiens, G. & Lendzian, K.J. 1989. Interaction between ozone and plant cuticle. I. Ozone deposition and permeability. *New Phytol.* **112**: 13-19.

Khan, A.A. & Malhotra, S.S. 1982. Peroxidase activity as an indicator of SO_2 injury in jack pine and white birch. *Biochem. Physiol. Pflanzen* **177**: 643-650.

Khan, A.M., Pandey, V., Shukla, J., Singh, N., Yunus, M., Singh, S.N. & Ahmad, K.J. 1990. Effect of thermal power plant emissions on *Catharanthus roseus*, L. *Bull. Environ. Contam. Toxicol.* **44**: 865-870.

Kieliszewska-Rokicka, B. 1978. Relation between the activity of peroxidase and the resistance of clones of Scots pine to the action of SO_2. In Bialobok, S. (ed.) *Studies on the Effect of Sulphur Dioxide and Ozone on the Respiration and Assimilation of Trees and Shrubs in Order to Select Individuals Resistant to Action of these Gases*: 82-84. Polish Academy of Sciences, Institute of Dendrology, Kornik, Poland.

Kieliszewska-Rokicka, B. 1979. Peroxidase activity in varieties of *Weigela* and *Pinus sylvestris* resistant and susceptible to SO_2. *Arbor. Kornickie* **24**: 313-320.

Kimmerer, T.W. & Kozlowski, T.T. 1981. Stomatal conductance and sulphur uptake of five clones of *Populus tremuloides* exposed to sulphur dioxide. *Plant Physiol.* **67**: 990-995.

Kimmerer, T.W. & Kozlowski, T.T. 1982. Ethylene, ethane, acetaldehyde, and ethanol production by plants under stress. *Plant Physiol.* **69**: 840-847.

Kobayashi, K., Fuchigami, L.H. & Brainerd, K.E. 1981. Ethylene and ethane production and electrolyte leakage of water-stress in 'Pixy' plum leaves. *Hort. Sci.* **16**: 57-59.

Koch, W. & Maier-Maercher, U. 1986. Changes in the humidity response characteristics of *Picea abies* (L.) Karst produced by fumigation with SO_2. *Europ. J. For. Pathol.* **16**: 329-341.

Konze, J.R. 1977. Bildung von Aethylen und Aethan in Geweben höherer Pflanzen und in Modellreakfionen. Dissertation Universität Bochum, Germany, 69 pp.

Konze, J.R. & Elstner, E.F. 1976. Ethylene- and ethane formation in leaf disks, plastids and mitochondria. *Ber. Deuts. Bot. Gesell.* **89**: 547-553.

Konze, J.R. & Elstner, E.F. 1978. Ethane and ethylene formation by mitochondria as indication of aerobic lipid degradation in response to wounding of plant tissue. *Biochem. Biophys. Acta* **528**: 213-221.

Kropff, M.J. 1991. Long term effects of SO_2 on plants: SO_2 metabolism and regulation of intracellular pH. *Plant & Soil* **131**: 235-245.

Kuc, J. & Lisker, N. 1978. Terpenoids and their role in wounded and infected plant storage tissue. In Kahl, G. (ed.) *Biochemistry of Wounded Plant Tissues*: 203-242. de Gruyter, Berlin.

Langebartels, C., Kerner, K., Leonardi, S., Schraudner, M., Trost, M., Heller, W. & Sandermann, H. Jr 1991. Biochemical plant responses to ozone I. Differential induction of polyamine and ethylene biosynthesis in tobacco. *Plant Physiol.* **95**: 882-889.

Larson, R.A. 1988. The antioxidants of higher plants. *Phytochem.* **27**: 969-978.

Lavee, S. & Martin, G.C. 1981. Ethylene evolution from various developing organs of olive (*Olea europea*) after excision. *Physiol. Plant.* **51**: 31-38.

Lee, E.H. & Bennett, J.H. 1979. Comparative studies of foliar protection from ozone induced by EDU in greenhouse and growth chamber preconditioned plants. In Abdel-Rahman, M. (ed.) *Proceedings of the Sixth Annual Meeting of the Plant Growth Regulator Working Group*: 218. Agway, Inc., Syracuse, NY.

Lee, E.H. & Bennett, J.H. 1982. Superoxide dismutase: A possible protective enzyme against ozone injury in snap beans (*Phaseolus vulgaris* L.). *Plant Physiol.* **69**: 1444-1449.

Lee, E.H., Bennett, J.H. & Heggestad, H.E. 1981a. Retardation of senescence in red clover leaf discs by a new antiozonant *N*-[2-(2-oxo-1-imidazolidinyl)ethyl]-*N'*-phenylurea. *Plant Physiol.* **67**: 347-350.

Lee, E.H., Wang, C.Y. & Bennett, J.H. 1981b. Soluble carbohydrates in bean leaves transformed into oxidant-tolerant tissues by EDU treatment. *Chemosphere* **10**: 889-896.

Lee, E.H., Jersey, J.A., Gifford, C. & Bennett, J.H. 1984. Differential ozone tolerance in soybeans and snapbeans: Analysis of ascorbic acid in O_3-susceptible and O_3-resistant cultivars by high performance liquid chromatography. *Environ. Exp. Bot.*, **24**: 331-341.

Lee E.H., Saftner, R.A., Wilding, S.J., Clark, H.D. & Rowland, R.A. 1987. Effects of paclobutrazol on GA biosynthesis and fatty acid composition. *Proc. Plant Growth Reg. Soc. Am.* **14**: 295-302.

Lee, E.H., Kramer, G.F., Rowland, R.A. & Agrawal, M. 1992. Antioxidants and growth regulators counter the effects of O_3 and SO_2 in crop plants. *Agric. Ecosyst. Environ.* **30**: 99-106.

Lehtiö, H. 1981. Effect of air pollution on the volatile oil in needles of Scots pine (*Pinus sylvestris* L.). *Silva Fenn.* **15**: 122-129.

Lendzian, K.J. 1984. Permeability of plant cuticles to gaseous air pollutants. In Koziol, M.J. & Whatley, F.R. (eds) *Gaseous Air Pollutants and Plant Metabolism*: 77-82. Butterworths, London.

Levitt, J. 1972. *Response of Plants to Environmental Stresses.* Academic Press, New York.

Liebermann, M. 1979. Biosynthesis and action of ethylene. *Annu. Rev. Plant Physiol.* **30**: 533-591.

Linhart, B., Mitton, J.B., Bowman, D.M., Sturgeon, K.B. & Hamrick, J.L. 1979. Genetic aspects of fertility differentials in ponderosa pine. *Genet. Res. Camb.* **33**: 237-242.

Liochev, S. & Fridovich, I. 1991. Effects of overproduction of superoxide dismutase on the toxicity of paraquat towards *Escherichia coli*. *J. Biol. Chem.* **266**: 8747-8750.

Loomis, W.D. & Croteau, R. 1980. Biochemistry of terpenoids. In Stump, P.K. & Conn, E.E. (eds) *The Biochemistry of Plants*, Vol. 4: 363-415. Academic Press, New York.

Lorenc-Kubis, J. & Morawiecka, B. 1973. Phosphatase activity of *Poa pratensis* seeds. I. Preliminary studies on acid phosphatase II. *Acta Soc. Bot. Pol.* **42**: 369-377.

Lürssen, K. & Naumann, K. 1979. 1-Aminocyclopropane-1-carboxylic acid-a new intermediate of ethylene biosynthesis. *Naturwissenschaften* **66**: 264-265.

MacRobbie, E.A.C. 1988. Control of ion fluxes in stomatal guard cells. *Bot. Acta.* **101**: 140-148.

Mahmooduzzafar, 1991. Air pollution influence on morphology and growth of *Cassia sophera* L. and *Sida rhombifolia* L. PhD Thesis, Aligarh Muslim University, Aligarh, India.

Majernik, O. & Mansfield, T.A. 1971. Effects of SO_2 pollution on stomatal movements in *Vicia faba*. *Phytopathol. Z.* **71**: 123-128.

Malan, C., Greyling, M. & Gressel, J. 1990. Correlation between CuZn superoxide dismutase and glutathione reductase and environmental and xenobiotic stress tolerance in maize inbreds. *Plant Sci.* **69**: 157-166.

Malhotra, S.S. 1976. Effects of sulphur of biochemical activity and ultrastructural organization of pine needle chloroplasts. *New Phytol.* **76**: 239-245.

Malhotra, S.S. & Khan, A.A. 1980. Effects of sulphur dioxide and other air pollutants on acid phosphatase activity in pine seedlings. *Biochem. Physiol. Pflanzen* **175**: 228-236.

Malhotra, S.S. & Khan, A.A. 1984. Biochemical and physiological impact of major pollutants. In Treshow, M. (ed.) *Air Pollution and Plant Life*: 113-157. Wiley, Chichester.

Mansfield, T.A. 1976. Chemical control of stomatal movements. *Phil. Trans. Royal Soc. London, Series B* **273**: 541-550.

Mansfield, T.A. & Davies, W.J. 1981. Stomata and stomatal mechanisms. In Pale, L.G. & Aspinall, D. (eds) *The Physiology and Biochemistry of Drought Resistance in Plants*: 315-346. Academic Press, Sydney.

Mansfield, T.A. & Freer-Smith, P.H. 1981. Effects of urban air pollution on plant growth. *Biol. Rev.* **56**: 343-368.

Mansfield, T.A. & Freer-Smith, R.H. 1984. The role of stomata in resistance mechanisms. In Koziol, M.J. & Whatley, F.R. (eds) *Gaseous Air Pollutants and Plant Metabolism*: 131-146. Butterworths, London.

Mansfield, T.A. & Pearson, M. 1993. The physiological basis of stress imposed by ozone pollution. In Fowden, L., Mansfield, T. & Stoddart, J. (eds) *Plant Adaptation to Environmental Stress*: 155-170. Chapman & Hall, London.

Mansfield, T.A., Hetherington, A.M. & Atkinson, C.J. 1990. Some current aspects of stomatal physiology. *Annu. Rev. Plant Physiol. Plant Mol. Biol.* **41**: 55-75.

Martin, J.T. & Juniper, B.E. 1970. *The Cuticle of Plants*. Edward Arnold Ltd, Edinburgh.

Materna, J. 1962. Vliv kyslicniku siriciteho na mineralni slizeni smrkoveho jehlici. [Effect of sulphur dioxide on the mineral composition of spruce needles.] *Prace vyzkumnych ustavu lesnickych, CSSR* **24**: 7-36.

Matters, G.L. & Scandalios, J.G. 1987. Synthesis of isoenzymes of superoxide dismutase in maize leaves in response to ozone, sulphur dioxide and elevated oxygen. *J. Exp. Bot.* **38**: 842-852.

Mehlhorn, H. & Wellburn, A.R. 1987. Stress ethylene formation determines plant sensitivity to ozone. *Nature* **327**: 417-418.

Mehlhorn, H., Gunther, S., Schmidt, A. & Kunert, K.J. 1986. Effect of SO_2 and O_3 on production of antioxidants in conifers. *Plant Physiol.* **82**: 336-338.

Mehlhorn, H., Tabner, B. & Wellburn, A.R. 1990. Electron spin resonance evidence for the formation of free radicals in plants exposed to ozone. *Physiol. Plant.* **79**: 377-383.

Mehlhorn, H., O'Shea, J.M. & Wellburn, A.R. 1991. Atmospheric ozone interacts with stress ethylene formation by plants to cause visible plant injury. *J. Exp. Bot.* **42**: 17-24.

Meidner, H. & Mansfield, T.A. 1968. *Physiology of Stomata*. McGraw-Hill, London.
Mejnartowicz, L.E. 1984. Enzymatic investigations on tolerance in forest trees. In Koziol, M.J. & Whatley, F.R. (eds) *Gaseous Air Pollutants and Plant Metabolism*: 381-398. Butterworths, London.
Merida, T., Schönherr, J. & Schmidt, H.W. 1981. Fine structure of plant cuticles in relation to water permeability: The fine structure of the cuticle of *Clivia miniata* Reg. leaves. *Planta* **152**: 259-267.
Meyer, H., Mayer, A.M. & Harek, E. 1971. Acid phosphatases in germinating lettuce — evidence for partial activation. *Physiol. Plant.* **24**: 95-101.
Miller, J.E., Pursley, W.A. & Heagle, A.S. 1994. Effects of ethylenediurea on snap bean at a range of ozone concentrations. *J. Environ. Qual.* **23**: 1082-1089.
Miller, P.R., Cobb, F.W. Jr & Zavarin, E. III 1968. Effect of injury upon oleoresin composition, phloem carbohydrates and phloem pH. *Hilgardia* **39**: 135-140.
Mishra, D. & Panda, K.C. 1970. Acid phosphatase of rice leaves showing diurnal variation and its relation to stomatal opening. *Biochem. Physiol. Pflanzen* **161**: 532-536.
Młodzianowski, F. & Białobok, S. 1977. The effect of sulphur dioxide on ultrastructural organization of larch needles. *Acta Soc. Bot. Pol.* **46**: 629-634.
Młodzianowski, F. & Młodzianowska, L. 1980. Cytochemical localization of enzyme activity in Scots pine needles treated with SO_2. In Bialobok, S. (ed.) *Studies on the Effect of Sulphur Dioxide and Ozone on the Respiration of Trees and Shrubs in Order to Select Individuals Resistant to the Action of these Gases*: 79-81. Polish Academy of Sciences, Institute of Dendrology, Kornik, Poland.
Moldau, H., Sober, J. & Sober, A. 1990. Differential sensitivity of stomata and mesophyll to sudden exposure of bean shoots to ozone. *Photosynth.* **24**: 446-458.
Monk, L.S., Fagerstedt, K.V. & Crawford, R.M.M. 1989. Oxygen toxicity and superoxide dismutase as an antioxidant in physiological stress. *Physiol. Plant.* **76**: 456-459.
Morison, J.I.L. 1987. Intercellular CO_2 concentration and stomatal response to CO_2. In Zeiger, E., Farqukar, G.D. & Cowan, I.R. (eds) *Stomatal Function*: 229-251. Stanford University Press, Stanford, CA.
Morison, J.I.L. & Gifford, R.M. 1983. Stomatal sensitivity to CO_2 and humidity. *Plant Physiol.* **71**: 789-796.
Muhs, H.-J. 1981. Terpenes and some other substances as biochemical markers in forestry. A review of European studies since 1976. *IUFRO Congress, Kyoto, Japan, S.2-04-5. Biochem. Gen. Cytol.* (voluntary paper).
Nandi, P.K., Agrawal, M. & Rao, D.N. 1984. SO_2-induced enzymatic changes and ascorbic acid oxidation in *Oryza sativa*. *Water Air Soil Pollut.* **21**: 25-32.
Nandi, P.K., Agrawal, M., Agrawal, S.B. & Rao, D.N. 1990. Physiological responses of *Vicia faba* plants to sulphur dioxide. *Ecotoxic. Environ. Safety* **19**: 64-71.
Natori, T. & Totsuka, T. 1984. *Effects of Mixed Gas on Transpiration Rate of Several Woody Plants. I. Interspecific Differences in the Effects of Mixed Gas on Transpiration Rate.* Research Report No. 15, National Institute for Environmental Studies, Ibaraki, Japan.
Nicholls, P. 1962. Peroxidase as an oxygenase. In Hayaishi (ed.) *Oxygenase*: 274-303. Academic Press, New York.
Niemtur, S. 1979. Influence of zinc smelter emissions on peroxidase activity in Scots pine needles of various families. *Europ. J. For. Pathol.* **2**: 142-147.
Nikolayevskii, V.S. 1968. Aktivnost nekotorikh fyermyentovi gazoustoichivosti rastyenii. [Plant resistance to gaseous emissions and activity of some enzymes.] *Trudy Inst. Ekologii Rastyenii i Zhyvotnykh. Uralskii Filial AN SSR* **62**: 208-211.
Nikolayevskii, V.S. 1979. *Biologischeskiye Osnovy Gazoustoichivosti Rastyenii.* [A biological basis for plant resistance to gaseous emission.] Akadyemia Nauk SSR, Nauka Sibirskoye Otdyelyeniye, 278 pp.
Nobel, P.S. 1974. *Introduction to Biophysical Plant Physiology*. W.H. Freeman & Co., New York.

Norby, R.J. & O'Neill, E.G. 1991. Leaf area compensation and nutrient interactions in CO_2-enriched seedlings of yellow-poplar (*Liriodendron tulipifera* L.). *New Phytol.* **117**: 515-528.
Ogier, G., Greppin, H. & Gastillo, F. 1991. Ascorbate and guaiacol peroxidase capacities from apoplastic and cell material extracts of Norway spruce needles after long-term ozone exposure. In Lobarzewski, J., Greppin, H., Penel, C. & Gaspar, T. (eds) *Molecular, Biochemical and Physiological Aspects of Plant Peroxidases*: 391-400. University Press, Geneva.
Olsen, R.A., Clark, R.B. & Bennett, J.H. 1981. The enhancement of soil fertility by plant roots. *Amer. Sci.* **69**: 378-384.
Ormrod, D.P. & Adedipe, N.O. 1974. Protecting horticultural plants from atmospheric pollutants. *Hort. Sci.* **9**: 108-111.
Ormrod, D.P. & Beckerson, D.W. 1986. Polyamines as antiozonants for tomato. *Hort. Sci.* **21**: 1070-1071.
Osonubi, O. & Davies, W.J. 1980. The influence of plant water stress on stomatal control of gas exchange at different levels of atmospheric humidity. *Oecologia* **46**: 1-6.
Pande, P.C. & Mansfield, T.A. 1985. Responses of spring barley to SO_2 and NO_2 pollution. *Environ. Pollut. (Ser. A)* **38**: 87-97.
Pandey, J. & Agrawal, M. 1994. Evaluation of air pollution phytotoxicity in a seasonally dry tropical urban environment using three woody perennials. *New Phytol.* **126**: 53-61.
Pauly, G., Douce, R. & Carde, J.-P. 1981. Effects of β-pinene on spinach chloroplast photosynthesis. *Z. Pflanzenphysiol.* **104**: 199-206.
Peiser, G.D. & Yang, S.F. 1977. Chlorophyll destruction by the bisulfite oxygen system. *Plant Physiol.* **60**: 277-281.
Peiser, G.D. & Yang, S.F. 1979. Ethylene and ethane production from sulphur dioxide-injured plants. *Plant Physiol.* **63**: 142-145.
Pell, E.J. & Dann, M.S. 1991. Multiple stress induced foliar senescence and implications for whole plant longevity. In Monney, H.A., Winner, W.E. & Pell, E.J. (eds) *Response of Plants to Multiple Stresses*: 189-204. Academic Press, San Diego.
Pellissier, M., Lacasse, N.L. & Cole, H. Jr 1972. Effectiveness of benomyl and benomyl folicote treatments in reducing ozone injury to pinto beans. *J. Air Pollut. Control Assoc.* **22**: 722-725.
Perl-Treves, R. & Galun, E. 1991. The tomato Cu, Zn superoxide dismutase genes are developmentally regulated and respond to light and stress. *Plant Mol. Biol.* **17**: 745-760.
Pfanz, H., Dietz, K.J., Weinerth, J. & Oppmann, B. 1990. Detoxification of sulphur dioxide by apoplastic peroxidases. In Rennenberg *et al.* (eds) *Sulphur Nutrition and Sulphur Assimilation in Higher Plants*: 229-233. SPB Acad. Publ., The Hague.
Pierre, M. & Queiroz, O. 1981. Enzymic and metabolic changes in bean leaves during continuous pollution by subnecrotic levels of SO_2. *Environ. Pollut. (Ser. A)* **25**: 41-53.
Pitcher, L., Brennan, E., Hurley, A., Dunsmuir, P., Tepperman, J. & Zilinskas, B. 1991. Overproduction of petunia chloroplastic copper/zinc superoxide dismutase does not confer ozone tolerance in transgenic tobacco. *Plant Physiol.* **97**: 452-455.
Pitelka, L.F. 1988. Evolutionary responses of plants to anthropogenic pollutants. *Trends Ecol.* **3**: 233-236.
Poborski, P.S. 1988. Pollutant penetration through the cuticle. In Schulte-Hostede, S., Darrall, N.M., Blank, L.W. & Wellburn, A.R. (eds) *Air Pollution and Plant Metabolism*: 19-35. Elsevier Applied Science, London.
Poborski, P. & Straszewski, T. 1983. Issliedowanije nickotorych mie chanizmow adosorbeji SO_2 chwojnymi ljesami. *Ekologiczeskoja Koopleracja, Informacjonnyj Biulletien Poproblemie III. "Ochranaekosistiem/Biogeoccnochow/i landszafta"* **4**: 18-22.
Polle, A. & Rennenberg, H. 1993. The significance of antioxidants in plant adaptation to environmental stress. In Fowden, L., Mansfield, T.A. & Stoddart J. (eds) *Plant Adaptation to Environmental Stress*: 263-273. Champman & Hall, London.

Polle, A., Chakrabarti, K., Schürmann, W. & Rennenberg, H. 1990. Composition and properties of hydrogen decomposing systems in extracellular and total extracts from needles of Norway spruce (*Picea abies* L.). *Plant Physiol.* **94**: 312-319.

Polle, A., Chakrabarti, K. & Rennenberg, H. 1991a. Entgiftung von Peroxiden in Fichtennadeln (*Picea abies*, L.) am Schwerpunktstandort Kalkalpen (Wank). *GSF Bericht* **26/91**: 151-161.

Polle, A., Chakrabarti, K. & Rennenberg, H. 1991b. Extracellular and intracellular peroxidase activities in needles of Norway spruce (*Picea abies*, L.) under high elevation stress. In Lobarzewski, J., Greppin, H., Penel, C. & Gaspar, T. (eds) *Molecular, Biochemical and Physiological Aspects of Plant Peroxidases*: 477-453. University Press, Geneva.

Price, C.E. 1982. A review of the factors influencing the penetration of pesticides through plant leaves. In Cutler, D.F., Alvin, K.L. & Price, C.E. (eds). *The Plant Cuticle*: 237-252. Academic Press, London.

Priebe, A., Klein, H. & Jäger, H.J. 1981. Cadmiumanreicherung Immissionsbelasteter Fichten. Utteilungen der Forstlichen Bundes-Versuchsanstalt Wien. Nachweis and Wirkung forstschädlicher Luftferunreinigungen. IUFRO Tagungsbeiträge zur *XI Internationalen Arbeitstagung forstlicher Rauchschadenssachverständiger* 1-6 September 1980, Graz, Austria, pp. 319-326.

Przymusinski, R. 1980. Cytochemical localization of peroxidase activity in needles of Scots pine resistant and susceptible to industrial pollution. In Mejnartowicz, L. (ed.) *A Genetic Basis for the Resistance of Forest Trees to Anthropopressure, with Special Study of the Effect of Some Toxic Gases*: 38-49. Polish Academy of Sciences, Institute of Dendrology, Kornik, Poland.

Puckett, K.J., Tomassini, F.D., Nieboer, E. & Richardson, D.H.S. 1977. Potassium efflux by lichen thalli following exposure to aqueous sulphur dioxide. *New Phytol.* **179**: 135-145.

Rabe, R. 1978. Bioindication von Luftverunreinigungen auf Grund der Änderung von Enzymaktivität und Chlorophyllgehalt von Testpflanzen. In Cramer, J. (ed.) *Disertationes Botanicae*: 1-219. A.R. Gantner Verlag, Germany.

Rabe, R. & Kreeb, L. 1975. Eine Methode zur Laborbegasung von Testpflanzen mit Schwefeldioxide und ihre Anwendung being Untersuchungen zur Enzymakitivität. *Angew. Bot.* **50**: 71-78.

Rao, M.V. 1989. Air pollution stress and correlated plant response in Dewas Area. PhD Thesis, Vikram University, Ujjain, India.

Rao, M.V. & Dubey, P.S. 1988. Plant response against SO_2 in field conditions. *Asian Environ.* **3**: 1-9.

Rao, M.V. & Dubey, P.S. 1990. Biochemical aspects (antioxidants) for development of tolerance in plants growing at different low levels of ambient air pollutants. *Environ. Pollut.* **64**: 55-66.

Raschke, K. 1975. Simultaneous requirement for carbon dioxide and abscisic acid for stomatal closing in *Xanthium strumarium* L. *Planta* **125**: 243-259.

Rasmussen, R.A. & Went, F.W. 1965. Volatile organic material of plant origin in the atmosphere. *Proc. Natl. Acad. Sci. USA.* **53**: 215-220.

Ream, J.E. & Wilson, L.G. 1982. Possible significance to SO_2 injury of dark and light-dependent bisulfite oxidizing activities in *Cucumis sativa* leaves. *Plant Physiol.* **69**: S-16.

Recalde-Manrique, L. & Diaz-Miguel, M. 1981. Evolution of ethylene by sulphur dust addition. *Physiol. Plant.* **53**: 462-467.

Reed, D. & Tukey, J.R.H.B. 1982. Permeability of Brussels sprouts and carnation cuticles from leaves developed in different temperatures and light intensities. In Cutler, D.F., Alvin, K.L. & Price, C.E. (eds) *The Plant Cuticle*: 267-278. Academic Press, London.

Reich, P.B. & Amundson, R.G. 1985. Ambient levels of ozone reduce net photosynthesis in tree and crop species. *Science* **230**: 566-570.

Reich, P.B. & Lassoie, J.P. 1985. Influence of low concentrations of ozone on growth, biomass partitioning and leaf senescence in young hybrid poplar plants. *Environ. Pollut.* **39**: 39-51.

Rennenberg, H. 1991. The significance of higher plants in the emission of sulphur compounds from terrestrial ecosystems. In Sharkey, T.D. *et al.* (eds) *Trace Gas Emissions by Plants*: 217-260. Academic Press, San Diego.

Rennenberg, H. & Filner, P. 1982. Stimulation of H_2S emission from pumpkin leaves by inhibition of glutathione synthesis. *Plant Physiol.* **69**: 766-770.

Rennenberg, H. & Polle, A. 1994. Metabolic consequences of atmospheric sulphur influx into plants. In Alscher, R.G. & Wellbern, A.R. (eds) *Plant Response to the Gaseous Environment*: 165-180. Chapman & Hall, London.

Rennenberg, H., Sekiya, J., Wilson, L.G. & Filner, P. 1982. Evidence for an intracellular sulphur cycle in cucumber leaves. *Planta* **154**: 516-524.

Renwick, J.A.A. & Potter, J. 1981. Effects of sulphur dioxide on volatile terpene emission from balsam fir. *J. Air Poll. Contr. Assoc.* **31**: 65-66.

Roberts, J.A. & Osborne, D.J. 1981. Auxin and the control of ethylene production during the development and senescene of leaves and fruits. *J. Exp. Bot.* **32**: 875-887.

Rich, S., Waggoner, P.E. & Tomlinson, H. 1970. Ozone uptake by bean leaves. *Science* **169**: 79-80.

Rodecap, K.D., Tingey, D.T. & Tibbs, J.H. 1981. Cadmium-induced ethylene production in bean plants, *Z. Pflanzenphysiol.* **105**: 65-74.

Roose, M.L., Bradshaw, A.D. & Roberts, T.M. 1982. Evolution of resistance to gaseous air pollutants. In Unsworth M.H. & Ormrod, D.P. (eds). *Effects of Gaseous Air Pollution in Agriculture and Horticulture*: 379-409. Butterworth Scientific, London.

Rosemann, D., Heller, W. & Sandermann, H. Jr 1991. Biochemical responses to ozone. II Induction of stilbene biosynthesis in scots pine (*Pinus sylvestris* L.) seedlings. *Plant Physiol.* **97**: 1280-1286.

Rowland-Bamford, A.J., Borland, A.M., Lea, P.J. & Mansfield, T.A. 1989. The role of arginine decarboxylase in modulating the sensitivity of barley to ozone. *Environ. Pollut.* **61**: 95-106.

Salveit, M.E. & Dilley, D.R. 1978. Rapidly induced wound ethylene from excised segments of etiolated *Pisum sativum* L., cv. Alaska. *Plant. Physiol.* **62**: 710-712.

Saunders, B.C., Holmes-Siedle, A.G. & Stark, B.P. 1964. *Peroxidase: The Properties and Use of a Versatile Enzyme and of Some Related Catalysts*. Butterworths, London.

Saxe, H. 1986a. Stomatal dependent and stomatal independent uptake of NO_x. *New Phytol.* **103**: 199-205.

Saxe, H. 1986b. Effects of NO, NO_2 and CO_2 on net photosynthesis, dark respiration and transpiration of pot plants. *New Phytol.* **103**: 185-197.

Schönherr, J. & Schmidt, H.W. 1979. Water permeability of plant cuticles. *Planta* **144**: 391-400.

Scott, M., Meshnick, S. & Eaton, J. 1987. Superoxide dismutase rich bacteria: paradoxical increase in oxidant toxicity. *J. Biol. Chem.* **262**: 3640-3645.

Sekiya, J., Wilson, L.G. & Filner, P. 1978. Positive correlation between H_2S emission and SO_2 resistance in cucumber. *Plant Physiol.* **62**: Abstr. 74.

Sekiya, J., Schmidt, A., Rennenberg, H., Wilson, L. & Filner, P. 1981. Role of light in H_2S emission in response to sulfate. *16th Annual Report of the MSU-DOE Plant Research Laboratory*, Michigan State University, USA, 57 pp.

Sekiya, J., Schmidt, A., Rennenberg, H., Wilson, L.G. & Filner, P. 1982a. Hydrogen sulfide emission by cucumber leaves in response to sulfate in light and dark. *Phytochem.* **21**: 2173-2178.

Sekiya, J., Schmidt, A., Wilson, L.G. & Filner, P. 1982b. Emission of hydrogen sulfide by leaf tissue in response to L-cysteine. *Plant Physiol.* **70**: 430-436.

Sekiya, J., Wilson, L.G. & Filner, P. 1982c. Resistance to injury by sulphur dioxide: Correlation with its reduction to, and emission of hydrogen sulfide in Cucurbitaceae. *Plant. Physiol.* **70**: 437-441.

Sen Gupta, A., Alsher, R.G. & McCune, D. 1991. Response of photosynthesis and cellular antioxidants to ozone in populus leaves. *Plant Physiol.* **96**: 650-655.

Sen Gupta, A., Heinen, J.L., Holaday, A.S., Burke, J.J. & Allen, R.D. 1993. Increased resistance to oxidative stress in transgenic plants that overexpress chloroplastic Cu/Zn superoxide dismutase. *Proc. Nat. Acad. Sci. USA* **90**: 1629-1633.

Shaaltiel, Y. & Gressel, J. 1986. Multienzyme oxygen radical detoxifying system correlated with paraquat resistance in *Conyza bonariensis*. *Pest. Biochem. Physiol.* **26**: 22-28.

Shaaltiel, Y., Glazer, A., Bocion, P. & Gressel, J. 1988. Cross tolerance to herbicidal and environmental oxidants of plant biotypes tolerant to paraquat, sulphur dioxide and ozone. *Pest. Biochem. Physiol.* **31**: 13-23.

Shain, A. & Mayer, A.M. 1968. Activation of enzymes during germination — trypsinlike enzyme in lettuce. *Phytochem.* **7**: 1491-1498.

Simola, L.K. & Sopanen, T. 1970. Changes in the activity of certain enzymes of *Acer pseudoplatanus* L. cells of four stages of growth in suspension cultures. *Physiol. Plant.* **23**: 1212-1222.

Simonis, W. & Urback, W. 1973. Photophosphorylation in vivo. *Annu. Rev. Plant Physiol.* **24**: 89-114.

Singh, N., Singh, S.N., Srivastava, K., Yunus, M., Ahmad, K.J., Sharma, S.C. & Sharga, A.N. 1990. Relative sensitivity and tolerance of some *Gladiolus* cultivars to sulphur dioxide. *Ann. Bot.* **65**: 41-44.

Singh, S.N., Yunus, M., Srivastava, K., Kulshreshtha, K. & Ahmad, K.J. 1985. Response of *Calendula officinalis* L. to long-term fumigation with SO_2. *Environ. Pollut. (Ser. A)* **39**: 17-25.

Singh, S.N., Yunus, M., Kulshreshtha, K., Srivastava, K. & Ahmad, K.J. 1988. Effect of SO_2 on growth and development of *Dahlia rosea*. *Can. Bull. Environ. Contam. Toxicol.* **40**: 743-751.

Snaith, P.J. & Mansfield, T.A. 1982. Control of the CO_2 responses of stomata by indol-3yl-acetic acid and abscisic acid. *J. Exp. Bot.* **33**: 360-365.

Soikkeli, S. 1981. The types of ultrastructural injuries in conifer needles of northern industrial environments. *Silva Fenn.* **15**: 339-404.

Spaleny, J. 1977. Sulphate transformation to hydrogen sulphide in spruce seedling. *Plant & Soil* **48**: 557-563.

Squillace, A.E. 1977. Use of monoterpene composition in forest genetics research with slash pine. *Proceedings Fourteenth Southern Forest Tree Improvement Conference*: 227-238. Gainesville, FL.

Srivastava, H.S., Jolliffe, P.A. & Runeckles, V.C. 1975. The effects of environmental conditions on the inhibition of leaf gas exchange by NO_2. *Can. J. Bot.* **53**: 475-482.

Staehelin, J. & Hoigne, J. 1982. Decomposition of ozone in water. Rate of initiation by hydroxide ion and hydrogen peroxide. *Environ. Sci. Tech.* **16**: 676-681.

Stan, H.-J., Schicker, S. & Kassner, H. 1981. Stress ethylene evolution of bean plants — a parameter indicating ozone pollution. *Atmos. Environ.* **13**: 391-395.

Tadolini, B. 1988. Polyamine inhibition of lipoperoxidation. *Biochem. J.* **249**: 33-36.

Takemoto, B.K., Noble, R.D. & Harrington, H.M. 1986. Differential sensitivity of duckweeds (Lemnaceae) to sulphite. II. Thiol production and hydrogen sulphide emission as factors influencing phytotoxicity under low and high irradiance. *New Phytol.* **103**: 541-548.

Tanaka, K., Kondo, N. & Sugahara, K. 1982. Accumulation of hydrogen peroxide in chloroplasts of SO_2 fumigated spinach leaves. *Plant Cell Physiol.* **23**: 999-1007.

Tanaka, K., Suda, Y., Kondo, N. & Sugahara, K. 1985. Ozone tolerance and the ascorbate dependent H_2O_2 decomposing system in chloroplasts. *Plant Cell Physiol.* **36**: 1425-1431.

Taylor, G.E. 1978. Plant and leaf resistance to gaseous air pollution stress. *New Phytol.* **80**: 523-534.

Taylor, G.E., Tingey, D.T. & Gunderson, C.A. 1986. Photosynthesis, carbon allocation and growth of sulphur dioxide on ecotypes of *Geranium carolinianum* L. *Oecologia (Berlin)* **68**: 350-357.

Tepperman, J. & Dunsmuir, P. 1990. Transformed plants with elevated levels of chloroplastic SOD are not resistant to superoxide toxicity. *Plant Mol. Biol.* **14**: 501-511.

Tester, M. 1990. Plant ion channels: Whole cell and single channel studies. *New Phytol.* **114**: 305-340.

Tingey, D.T. & Taylor, G.E. Jr 1982. Variation in plant response to ozone: A conceptual model of physiological events. In Unsworth, M.H. & Ormrod, D.P. (eds) *Effects of Gaseous Air Pollution in Agriculture and Horticulture*: 113-138. Butterworths, London.

Tingey, D.T., Standley, C. & Field, R.W. 1976. Stress ethylene evolution: A measure of ozone effects on plant. *Atmos. Environ.* **10**: 969-974.

Tingey, D.T., Pettit, N. & Bard, L. 1978. Effect of chlorine on stress ethylene production. *Environ. Exp. Bot.* **18**: 61-66.

Tingey, D.T., Wilhour, R.G. & Taylor, O.C. 1979. The measurement of plant responses. In Krupa, S.V. & Lonzon, S.N. (eds) *Handbook of Methodology for the Assessment of Air Pollution Effects on Vegetation*: 7.1-7.35. Upper Midwest Section, Air Pollution Control Association, USA.

Tsang, E., Bowler, C., Herouart, D., Van Camp, W., Villaroel, R., Genetello, C., Van Montague, M. & Inze, D. 1991. Differential regulation of superoxide dismutases in plants exposed to environmental stress. *Plant Cell.* **3**: 783-792.

Unsworth, M.H. & Black, V.J. 1981. Stomatal responses to pollutants. In Jarvis, P.G. & Mansfield, T.A. (eds) *Stomatal Physiology*: 187-263. Cambridge University Press, London.

Unsworth, M.H., Biscoe, P.V. & Black, V.J. 1976. Analysis of gas exchange between plants and polluted atmospheres. In Mansfield, T.A. (ed.) *Effects of the Pollutants on Plants*: 5-16. Cambridge University Press, Cambridge.

Varshney, S.R.K. & Varshney, C.K. 1984. Effect of SO_2 on ascorbic acid in crop plants. *Environ. Pollut. (Ser. A)* **35**: 285-290.

Varshney, S.R.K. & Varshney, C.K. 1985. Response of peroxidase to low levels of SO_2. *Environ. Exp. Bot.* **25**: 107-114.

Wang, C.Y. & Adams, D.O. 1982. Chilling-induced ethylene production in cucumbers (*Cucumis sativus* L.) *Plant Physiol.* **69**: 424-427.

Watroek, W., Morawiecka, B. & Korczak, B. 1977. Acid phosphatase of the yeast *Rhodotorula rubra*: Purification and properties of the enzyme. *Acta Biochem. Pol.* **24**: 153-162.

Whitaker, B.D., Lee, E.H. & Rowland, R.A. 1990. EDU and ozone protection: Foliar glycerolipids and steryl lipids in snap bean exposed to O_3. *Physiol. Plant.* **80**: 286-293.

Wilson, L.G., Bressan, R.A. & Filner, P. 1978. Light-dependent emission of hydrogen sulfide from plants. *Plant Physiol.* **61**: 184-189.

Wilson, L.G., Bressan, R., Lecureux, L., Ream, J.E. & Filner, P. 1979. Destruction of bisulfite by young and mature leaves. *Plant Physiol.* **63**: Abstr. 59.

Winner, W.E. & Mooney, H.A. 1980a. Ecology of SO_2 resistance: II Photosynthetic changes of shrubs in relation to SO_2 absorption and stomatal behaviour. *Oecologia (Berlin)* **44**: 296-302.

Winner, W.E. & Mooney, H.A. 1980b. Ecology of SO_2 resistance. I. Effects of fumigations on gas exchange of deciduous and evergreen shrubs. *Oecologia (Berlin)* **44**: 290-295.

Winner, W.E., Smith, C.L., Koch, G.W., Mooney, H.A., Bewley, J.E. & Krouse, H.R. 1981. H_2S emission rates from plants and patterns of stable sulphur. *Nature* **289**: 672-674.

Wolfenden, L. & Mansfield, T.A. 1991. Physiological disturbances in plants caused by air pollutants. *Proc. Royal Soc. Edinb.* **97B**: 117-138.

Wolfenden, J. Wookey, P.A., Lucas, P.W. & Mansfield, T.A. 1992. Action of pollutants individually and in combination. In Barker, J.R. & Tingey, D.T. (eds) *Air Pollution Effects on Biodiversity*: 72-92. Van Nostrand Reinhold, New York.

Yang, S.F. 1980. Regulation of ethylene biosynthesis. *Hort. Sci.* **15**: 238-243.

Yang, S.F. & Pratt, H.K. 1978. The physiology of ethylene in wounded plant tissues. In Kahl, G. (ed.) *Biochemistry of Wounded Plant Tissues*: 595-622. de Gruyter, Berlin.

Yee-Meller, D. 1975. Über die Eignung von Phosphatase und Esteraseaktivitatbestimmungen an fichtennadeln und Birkenblattern zur Nachweis, unsichtbarer physiological Fluorimmissionschadigungen. *Europ. J. Forest Pathol.* **5**: 329-338.

Yu, S.W., Liu, Y., Li, Z.G., Tan, C. & Yu, Z.W. 1982. Studies on the mechanism of SO_2 injury in plants. In Unsworth, M.H. & Ormrod, D.P. (eds) *Effects of Gaseous Air Pollution in Agriculture and Horticulture*: 507-508. Butterworths, London.

Yu, Y.B. & Yang, S.F. 1979. Auxin-induced ethylene production and its inhibition by aminoethoxyivinylglycine and cobalt ion. *Plant Physiol.* **64**: 1074-1077.

Yu, Y.B. & Yang, S.F. 1980. Biosynthesis of wound ethylene. *Plant Physiol.* **66**: 281-285.

Yunus, M., Ahmad, K.J. & Gale, R. 1979. Air pollutants and epidermal traits in *Ricinus communis* L. *Environ. Pollut.* **20**: 189-198.

Ziegler, I. 1973. The effect of air-polluting gases on plant metabolism. In Coulston, F. & Korte, F. (eds) *Environmental Quality and Safety: Global Aspects of Chemistry Toxicology and Technology as Applied to the Environment*: 181-208. Academic Press Inc., New York.

9 Phenolic Compounds in Defence Against Air Pollution

A.M. ZOBEL

INTRODUCTION

As we near the end of the 20th century, air pollution has become recognized as one of the foremost ecological challenges facing mankind, and growing public concern about the problem has been mirrored by a great increase in research aimed at monitoring and measuring this phenomenon (Posthumus 1983, 1985; Adams *et al.*, 1984; Hutchinson, 1984; Bialobok & Fabijanowski, 1984; Bleweiss, *et al.*, 1985; Saleh *et al.*, 1985). Crop plants (Lee *et al.*, 1981), forests (Huttunen *et al.*, 1985; Innes, 1987), prairies (Clapperton & Parkinson, 1990), and even arctic tundra (Freedman *et al.*, 1990) have been increasingly falling victim to higher levels of atmospheric contamination. Umbach & Davis (1984), who monitored 57 tree species, found, for example, that over 4% of the leaf surface became necrotic from the effects of sulphur dioxide.

We are still, however, at an early stage of collecting data on how different ions enter and kill the plant (Weigel *et al.*, 1989). Studies are being made of the action of specific ions and the combinations of different factors, such as SO_2 and O_3 (Ormrod *et al.*, 1984; Ernst *et al.*, 1985; Marie & Ormrod, 1986), SO_2, O_3 and NO_2 (Elkiey & Ormrod, 1981; Yang *et al.*, 1983), SO_2 and the temperature regime (Shanklin & Kozlowski, 1985), SO_2 and humidity (Olszyk & Tibbitts, 1981, Norby & Kozlowski, 1983), SO_2 and drought conditions (W.S. Lee *et al.*, 1990) and SO_2 and heavy metals (Balsberg-Paahlsson, 1989).

The mechanisms involved in the interaction of different factors can be very complex (Olszyk & Tingey, 1984b; Reinert, 1984; Matsubara *et al.*, 1986). One must separate, if possible, the effects of each single factor and then investigate whether the overall mechanism triggering plant defence represents only the simple sum of the individual mechanisms or whether the factors can react synergistically, as has been observed in the reaction of some pollutants (Hutchinson, 1984). Some phenolic compounds, for instance, are known to react synergistically (Städler & Buser, 1984; Berenbaum & Neal, 1985).

PHENOLIC COMPOUNDS AS A DEFENCE SYSTEM IN PLANTS

One group of defence compounds synthesized by plants comprises phenolic compounds (Chew & Rodeman, 1979; Beier & Oertli, 1983; Harborne, 1985).

Plant Response to Air Pollution. Edited by Mohammad Yunus and Muhammad Iqbal

Their concentration increases with invasion by pathogens, a phenomenon linked to the production of elicitors (Matern *et al.*, 1988). In celery, leaves, for example, the biosynthesis of furanocoumarins increases and their concentrations rise up to 2500% after infection by fungi (Ashwood-Smith *et al.*, 1985; Surico *et al.*, 1987).

These protective compounds must be produced very rapidly if they are to be an effective response of a plant to any markedly changed conditions. There must be a mechanism of receiving information about changing conditions and a mechanism of quick response, sometimes within minutes (Chappell & Hahlbrock, 1984), hence changes in the enzyme–gene system due to pollution are not surprising (Scholz & Bergmann, 1984). In order to survive, a plant must have the genetic capacity to compensate for drastic changes in the environment. Such changes have been termed "stress", but this is an inexact term, as a plant is normally exposed to a wide gradient of changes, to all of which, except for those so extreme as to be lethal, it must adjust (W. Steck, pers. comm.). Secondary products must be the compounds which are involved in such adjustments. Their enormous number and diversity in structure (McKey, 1979) in every plant can be explained not only by coevolution (Berenbaum, 1981, 1983; Feeny, 1987; Ivie, 1987) of different species parallelling their response to a changing environment (Rhodes, 1979; Zobel *et al.*, 1994b).

Phenolic compounds comprise only one group of secondary products (Mann, 1980), but this group alone accounts for many thousands of compounds. As they can be characteristic of the species, they are valuable in chemotaxonomy (Hegnauer, 1962; Harborne & Turner, 1984). They can be tissue specific, like anthocyanins (Hrazdina *et al.*, 1982; Zobel & Hrazdina, 1995), or located in ideoblastic cells (Thielke, 1956; Zobel, 1985a, b). Their concentrations can change depending on altered natural environmental conditions (Zobel, 1991; Zobel & Brown 1993; Zobel *et al.*, 1994a, b), or artificially enhanced UV (Berenbaum *et al.*, 1986; Zangerl & Berenbaum, 1987; Zobel & Brown, 1993; Zobel *et al.*, 1994a).

It has been known for well over a century (Moeller, 1882) that tannins, now called proanthocyanidins (Stafford, 1987; Stafford *et al.*, 1989), have been "wound compounds", whose concentrations increase in the tissue surrounding a wound. This is the only long-established instance of enhanced biosynthesis due to wounding (Haslam, 1979). Such compounds normally exist in great variety in the plant, albeit sometimes in low concentrations. They can be located in every cell or only in the idioblastic cells (Zobel, 1975) where they can be present in high concentrations in the central vacuole of the pith parenchyma. By covering the wound the efflux from such cells can prevent the growth of bacterial and fungal spores.

These compounds, however, must not bring about the death of the cells which produce them, or inhibit their growth and multiplication (Camm *et al.*, 1976; Podbielkowska *et al.*, 1994), by binding to and precipitating proteins (Swain, 1979, 1985; Haslam, 1985b). This effect, familiar as the characteristic astringency produced in the mouth by catechins of tea or aged red wine, has had practical value since prehistoric times in the tanning of leather (Haslam, 1979, 1985a), and in folk medicine throughout human history for killing bacteria (Wagner, 1985). However, by this binding reaction, phenolic compounds alter the structure and hence the activity of enzymes and other proteins and, in order not to be toxic to the cell producing them, they must be compartmentalized, and thus separated from the cytoplasm. Speculation about this phenomenon was ingeniously advanced last century by Hartwich (1885), who logically deduced that, if not isolated by a membrane, phenolics would deactivate the cell's own cytoplasm, killing the cell. Such compartmentation of phenolics in the plant cell has been demonstrated experimentally (Wagner, 1982; Hrazdina & Wagner, 1985; Zobel, 1986; Zobel & March, 1993).

SITES OF LOCALIZATION OF PHENOLIC COMPOUNDS IN PLANTS

Two compartments for storage of phenolic compounds are possible: the vacuole (Zobel, 1986, 1993), which is separated from the ground cytoplasm by the membrane tonoplast, and outside the cell in the intercellular spaces (Zobel & Brown, 1988, 1989; Zobel *et al.*, 1989; Iwanowska *et al.*, 1994), where they are separated by another membrane plasmalemma. As long as these two membranes, both separating the outside compartments from the ground cytoplasm, are intact, phenolics are not toxic to their own cells. Vacuoles of different sizes containing phenolics can thus be compared to bombs which, when detonated by mechanical or enzymatic damage from wind, animal herbivores or growing hyphae, can destroy both the plant's cells and those of the invader. Although a few cells do, in the process, commit suicide, immediate invasion by microorganisms is prevented. This is the immediate reaction long before a cell receives any information about damage and a possible reaction involving genetic material. Then the cells below the dead ones, within minutes or in an hour's time, increase the production of phenolics (Chappell & Hahlbrock, 1984). The changed conditions trigger biosynthesis of protective compounds (Zobel, 1991, 1993), characteristic of the particular tissue and species.

The mechanism that increases the concentration of phenolics inside the needle cells of *Pinus* (Zobel & Nighswander, 1990, 1991) possibly accounts also for the reaction of plant cells exposed to air pollution.

The second compartment containing phenolics is outside the plasmalemma, a rediscovered compartment that is currently under investigation. Only flavonoids (Piccard, 1873; Nilsson, 1961; Wollenweber & Dietz, 1981; Wollenweber, 1984), and, recently, coumarins (Zobel & Brown, 1990a, 1991a, b; Zobel *et al.*, 1991b) have been proved to exist on the plant surface. Lignins (Swain, 1979), suberin, and cutin and waxes (Kolottukudy, 1984), as well as exin and intin in pollen, are known to be located outside the cell, and play a successful role in isolating the cytoplasm from harsh external conditions.

Surface compounds, especially volatile ones such as monoterpenes (Kelsey *et al.*, 1984), are responsible for communication between the plant and its environment, and thus must have been the first barrier of the plant's defence system. Sesquiterpenes extruded to the surface by trichomes of plants growing in semi-arid conditions are, for example, the first barrier against insects and grazing herbivores. A large group of about 450 flavonoids was found on the plant surface of many different species by Wollenweber's group (Wollenweber & Dietz, 1981;

Wollenweber, 1984, 1985). Coumarins comprise another group, whose numbers now approach 1900 identified compounds (Murray *et al.*, 1982; Murray, 1991). Only recently have a number of these coumarins been shown to be present on the plant surface of many species of the Rutaceae, Umbelliferae and Leguminosae (Zobel *et al.*, 1991b); in the case of *Ruta graveolens* leaves the concentration reached 2000 $\mu g\ g^{-1}$ fresh weight (Zobel & Brown, 1989). They are embedded in the epicuticular layer, which is comparable to the thickness of the epidermis (Zobel & Brown, 1989) and can be likened to a coat protecting the whole plant body.

The cuticle and epicuticular waxes cover the whole plant body (Wilkinson, 1979) and even the very apex of the shoot, partially preventing diffusion of water (Schonherr, 1976) or penetration of solutions into the epidermal cells (Hunt & Baker, 1982). Permeability depends on temperature and light intensity (Reed & Tukey, 1982). Already very early in the plant's ontogenesis waxes exhibit a complex structure and, as they are species-specific (Tulloch, 1987), they are a useful taxonomic feature. Although phenolic compounds are known to coexist with waxes (Baker, 1982), their identification has not been attempted until recently. But, using a new technique involving brief dipping in almost boiling water, which melts waxes and liberates co-occurring compounds, we have found furanocoumarins, some of which have been identified, on the surface of many medicinal plants and crop plants (Zobel & Brown, 1991a, b). Attempts at further identification of the numerous surface compounds are in progress. Quantitative data on the occurrence of psoralen, xanthotoxin and bergapten have been employed to investigate changes influenced by air pollution, such as by simulated acid and salt spray (Zobel *et al.*, 1991a, b).

SPRAY OF ACID AND SALT ON PLANTS

Both compartments, the vacuole in the case of pine, and the plant surface in the case of *Ruta graveolens*, have been studied with respect to their response to spraying with saturated solutions of sodium chloride (Zobel & Nighswander, 1990), aluminium salt (Zobel, Kuras and Nighwander, unpublished), and sulphuric acid solutions of different pH (Zobel & Nighswander, 1991). The European *Pinus nigra* and the American *P. resinosa* were used because they are known to differ in their resistance to salt spray (Davis & Wood, 1972), *P. nigra* being the more resistant (Kramer & Kozlowski, 1979). *Ruta graveolens*, a plant containing large concentrations of furanocoumarins on its leaf surface, was used as a model plant to investigate changes in the concentration of psoralens extruded to the surface, as well as those within the leaf after acid or salt spray (Zobel *et al.*, 1991a).

Spray on *Pinus*

A 3-hour application time of the simulated pollutant allowed the beginning of the cell's response to be detected and, after 24 hours, monitoring of the eventual accumulation of compounds which had been produced but were still not detectable macroscopically. Week-old samples permitted comparison of the tissue under and within the spot of yellow discoloration. After a month of treatment, the observed macroscopically distinguishable necroses were investigated. *Pinus nigra* and *P. resinosa* were affected by salt and acid spray. Different kinds of necroses were visible: basal, tip, bands and spotting. Similar necroses have been

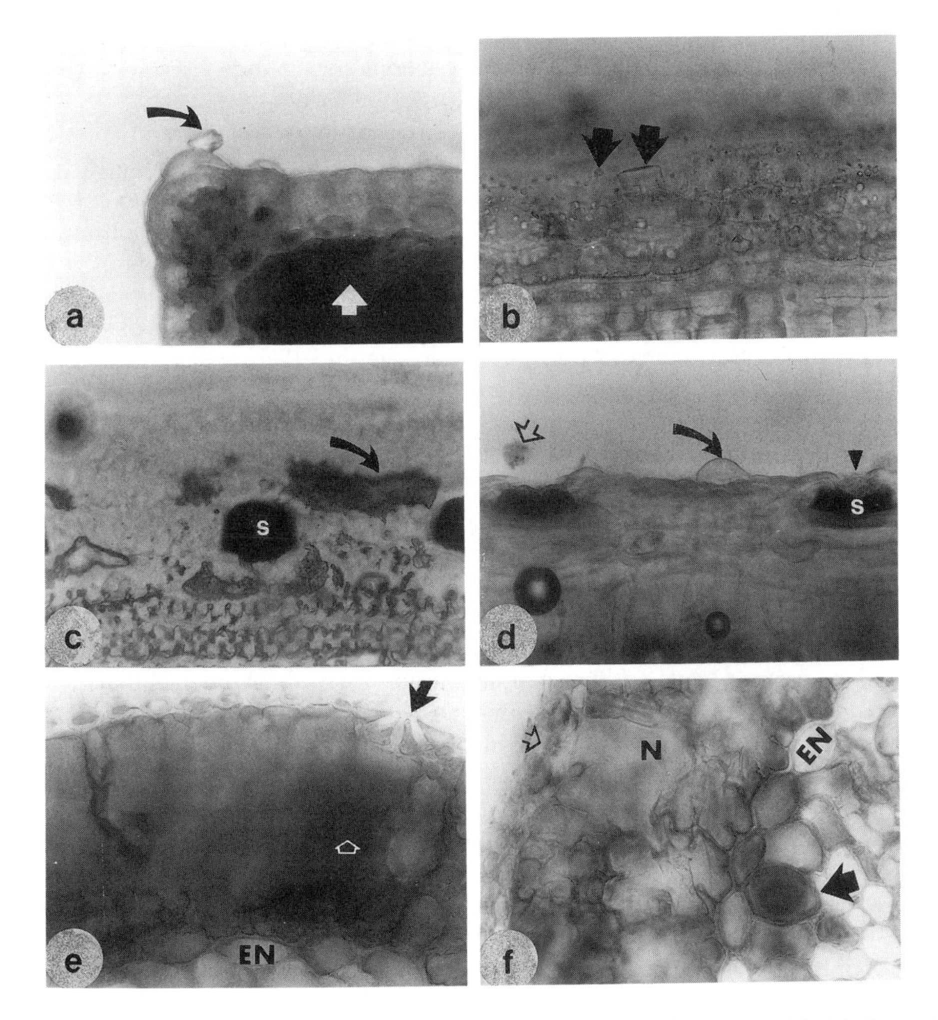

Figure 9.1. Hand-cut sections from the middle area of *Pinus* needles (*ca*. × 800). (a) Crystal of NaCl (black arrow). White arrow shows the dark reaction of phenolic compounds with 1% vanillin + HCl. Some cells of the hypodermis also contain phenolics. (b) Deposits on the surface of large (right arrow) and small crystals of aluminium sulphate. (c) Increased deposits of sulphuric acid (arrow) after 3 days of spraying. Stomata (S) contain phenolics. (d) Deposit of droplet of sulphuric acid (black arrow) with larger bottom area of attachment to cuticle. Stomata (S) contain phenolics and a lipid-like plug (arrowhead), which can be easily removed (hollow arrow). (e) Reaction for phenolic compounds (Reeve's reaction) showing stronger reaction (white arrow) in the vicinity of the stomata (black arrow), bordered by endodermis (EN). (f) Reaction of phenolics in the parenchyma cells (black arrow). Next to it lies one cell of the endodermis (EN) which contains phenolics. These cells are in the vicinity of necrosis (N) under the stomatal aperture (hollow arrow)

reported in various plants; for instance, basal blisters and basal necrosis in *Pinus* (Rice *et al.*, 1986), chlorotic mottling, flecking, and different widths of bands, as well as tip necrosis in white pine (Usher & Williams, 1982).

The reaction of *P. resinosa* seems more severe than that of *P. nigra,* but the mechanism is the same for both salt spray and sulphuric acid. Aluminium sulphate causes slightly different structural changes after a short period of exposure, but later the precipitation with phenolic compounds, lysis and necrosis of the cells are similar in all cases.

On the surface the salt is deposited as small or large crystals visible both in the light microscope (Figure 9.1a) and the SEM (Figure 9.2a, c). The acid spray disperses evenly as small drops after 24 hours (Figure 9.1b) but after a longer period larger areas are covered (Figure 9.1c), showing an increased bottom area of the drop attached to the cuticle (Figure 9.1d) arrow), and suggesting increased wettability of the leaf surface. The surface shows irregular deposits increasing with time (Figure 9.3a, d). The cuticle can crack (Figure 9.3c, arrowhead), and under a salt crystal holes can be seen after a month (Figure 9.3f). The needle-tip surface appears normal after a week, but spots are visible below the tip (Figure 9.3b, arrow). Longer exposure causes larger, rough deposits to appear (Figure 9.3b), and acid spray results in the appearance of a less smooth surface (Figure 9.3c). Small lesions are visible (Figure 9.3g, arrow), together with an increased number of cracks in necrotic spots after a month (Figure 9.3h, arrow). These cracks can be observed on cross sections as not reaching the cell interior, but they can nevertheless be deep (Figure 9.2b, arrow), and from some areas even peeling of the epicuticular layer is observed (Figure 9.2d, arrowhead).

The outer chamber of the stomata is filled with material (Figure 9.2d, arrow), and the aperture leads to a sometimes large labyrinth of intercellular spaces located below (Figure 9.2d), in contrast to a small inner cavity in the control needles (Figure 9.5b). The plugs evolve between a week and a month after spraying (Figure 9.3e, arrow), and their content must include lipids and some phenolics (Figure 9.1c, d arrowhead). The reaction for phenolic compounds is stronger in the mesophyll in the region under the stomata bordered by the endodermis (Figure 9.1e, white arrow). The reaction can be strong in some cells of the endodermis and in the parenchyma cells under it (Figure 9.1f, arrow), the cells surrounding the veins. In most cases such phenolic-containing cells lie in the vicinity of the intercellular spaces under the stomata.

The reaction of phenolic compounds with caffeine, compared to the control (Figure 9.4c), shows that there are areas of higher concentration of these compounds (Figure 9.4a) in the vicinity of the lesion (N). These lesions can reach the vein after the endodermis is broken (Figure 9.4a, hollow arrow, and Figure 9.4d). In the other type of needle injury, tip injury, the lesions can occur in the central area of the needle before the mesophyll is lysed (Figure 9.4e). Still another kind, the basal injury (Figure 9.4b), is visible in the area about 2 mm from the attachment of the needle to the stem. Beneath the epidermis are isodiametric cells containing phenolic compounds in their central vacuoles, and among them are other cells which appear empty, their thin walls broken in some places (Figure 9.4b, arrow).

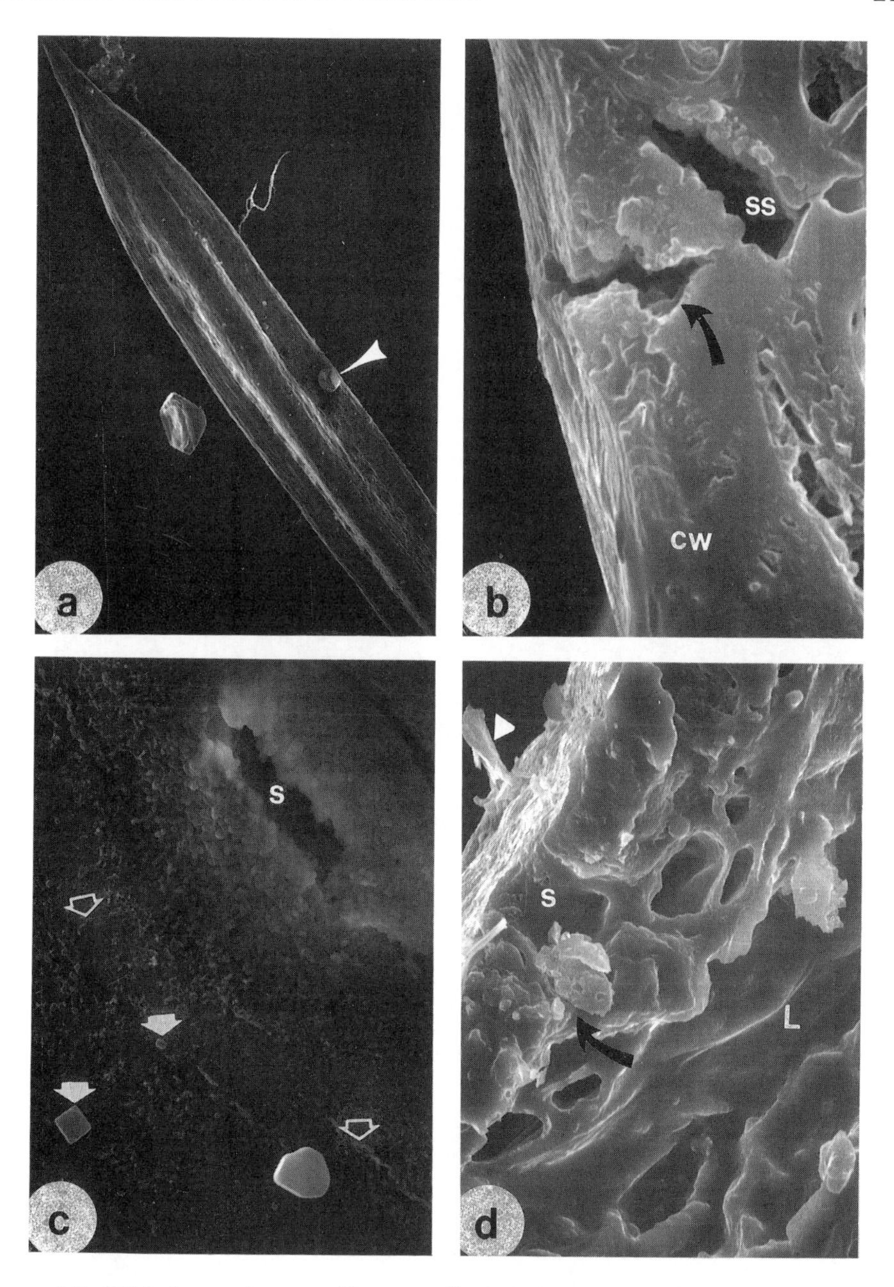

Figure 9.2. SEM observations on *Pinus* needle (×2000). (a) Tip of the needle with salt crystals (arrowhead). (b) Deep crack (arrow) in the cuticle and the outside part of the cell wall (CW) close to the stomata (SS) after 7 days of sulphuric acid spray. (c) Crystals of different size (white arrows) and cracks in cuticle (hollow arrow) in vicinity of stomata (S) after a 28-day salt spray. (d) Peeling of surface waxes of epicuticular layer (white arrowhead). Outer cavity of stomatal apparatus (S) filled with deposits (black arrow) leading to a large labyrinth (L) below, after 28 days of aluminium spray

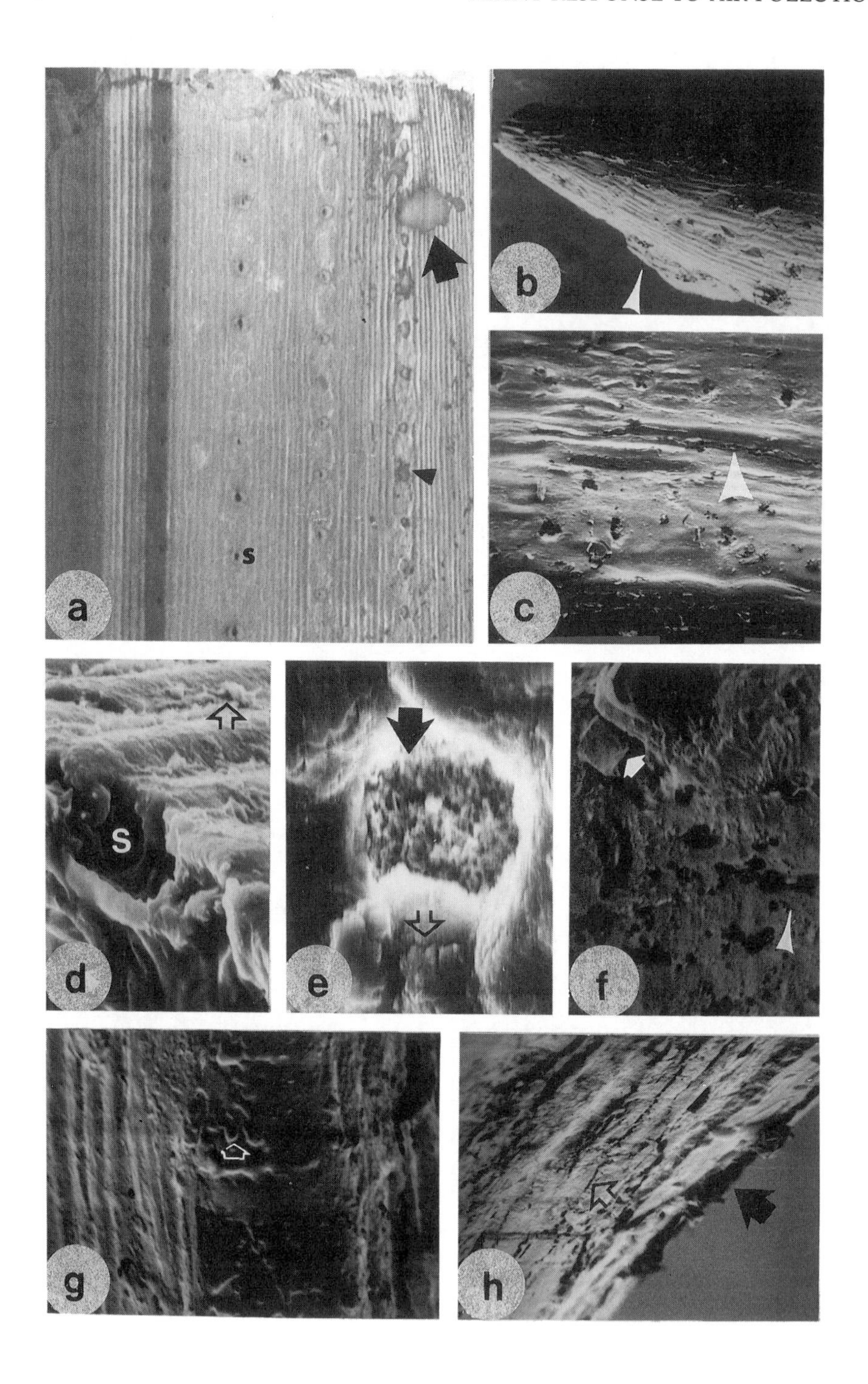
a
s
b
c
d
S
e
f
g
h

Destruction of the structure of the mesophyll cells due to air pollutants has been observed in various plants (Kaerenlampi, 1986; Psaras & Christodoulakis, 1987). The response of *Pinus* needles to pollution resembles the escalation process of senescence. The shoot apical internodes contain idioblastic cells filled with phenolic compounds (Figure 9.5a, arrow). In ontogenesis the plant produces these compounds (Hejnowicz, 1979). The control needles contain small droplets of phenolics (Figure 9.5b, arrow) but the treated needles contain large areas of tissue with many cells containing phenolics (Figure 9.5c–h). These cells can have the whole vacuole filled with electron-dense material (Figure 9.5c–f, arrows). Lesions appear first in the areas where more phenolics are stored (Figure 9.5d–g) and can occur simultaneously in many places (Figure 9.5c). The parenchyma cells around the veins, which are close to necrosis, show higher concentrations of phenolics (Figure 9.5d, arrow).

The reactions of different parts of the needle vary. In the middle part the mesophyll cells respond first, leaving the more central part still intact, protected by the endodermis (Figure 9.5f). The cells respond centripetally. The endodermis consists of one layer of cells with thin outside walls, closely packed with dense cytoplasm (Figure 9.5g, EN). These make a good temporary barrier, but then the ones in contact with affected mesophyll can begin to store phenolics, as do the parenchyma cells in the vicinity (Figure 9.5d, arrow). The basal part of the needle (Figure 9.5h) reacts to air pollution by blocking differentiation of some of the cells. Isodiametric cells contain phenolic compounds which form larger drops in the central vacuole (Figure 9.5h, arrow) than those of control cells (Figure 9.5i). Many cells have thin cell walls, some of them broken (Figure 9.5h, hollow arrow). As necrosis in this area can quickly affect large areas, basal injury progresses rapidly. The tip of the needle reacts differently (Figure 9.5c, e). There is only a thin layer of mesophyll, composed of two rows of cells between the hypodermis and endodermis. Mesophyll cells can be filled with phenolics, and the damage rapidly reaches the endodermis; then damage to the central parenchyma also proceeds quickly.

The stomatal apparatus is the gateway to the leaf (Franke, 1967; Grieve & Pitman, 1978) and response of the tissue always starts from its vicinity. We can see long necrotic tunnels up to the endodermis (Figure 9.6a). In the control plants the outside stomatal cavity can exclude microorganisms (Figure 9.6b, arrow). The interior of the stomatal cavity is bordered by a few long cells (Figure 9.6d, hollow arrow). The mesophyll cells do not contain many phenolic compounds. Between 3 and 7 days following the spray the external and internal cavities contain some

Figure 9.3. SEM observations on *Pinus* needles (×2000). (a) Surface of *P. nigra* needle after 28 days of sulphuric acid spray. Large deposits (arrow) and some stomata (S) covered with deposits (arrowhead). (b) Tip of needle with thick deposits (arrowhead). (c) Cracks visible (arrowhead) in *P. resinosa* after 7 days of sulphuric acid spray. (d) Rough surface (hollow arrow) and deposits after aluminium spray, cut through stomata (S). (e) Plug after 1 month of acid spray (arrow) and cracks in epidermis close to stomata (hollow arrow). (f) Salt crystals (white arrow) and lesions visible as black holes (arrowhead) after 1 month of aluminium spray. (g) Deposits on the surface of *P. nigra* after 7 days of acid spray. (h) Peeling off (black arrow) and cracks in cuticle (hollow arrow) after 1 month of aluminium spray

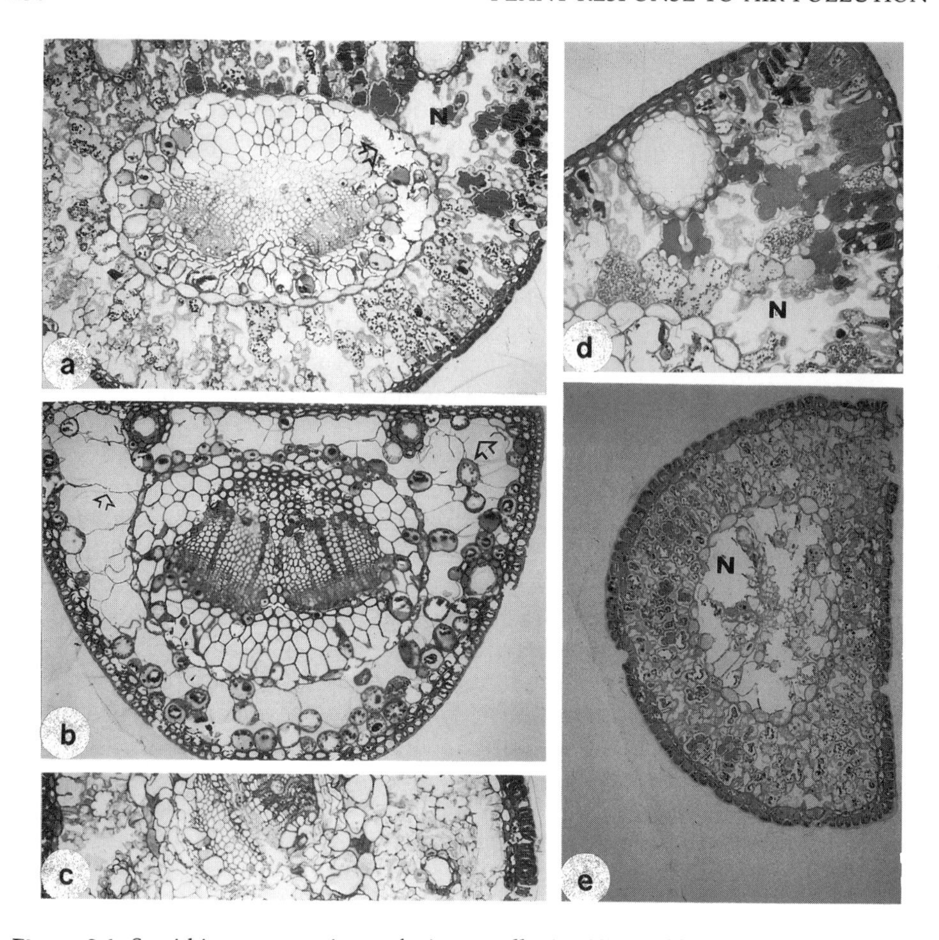

Figure 9.4. Semithin cross sections of pine needle (×400). (a) Necrosis (N) surrounded by cells filled with electron-dense phenolic compounds. Necrosis in parenchyma under endodermis (hollow arrow). (b) Basal injury. Isodiametric cells contain phenolics, empty cells with broken walls (hollow arrow). (c) Control with no phenolic compounds. (d) Middle part of the needle showing deep necrosis (N) under stomata reaching endodermis. (e) *P. nigra* closer to the tip. Necrosis (N) in the central parenchyma

debris (Figure 9.6c, arrow) and the vacuoles of the surrounding cells contain droplets of phenolics. The cell walls become broken (Figure 9.6d, arrow) and typical lysis proceeds, while some organelles can still be recognized (Figure 9.6d, hollow arrow). The last stage, after a month, is when only the cell remains are visible (Figure 9.6e, arrow), in between the large empty spaces devoid even of the debris from organelles.

In the case of aluminium-sulphate spray, the initial reaction is different. Under the stomata and under other epidermal cells, the response of the mesophyll cells is to produce small droplets of phenolic compounds dispersed throughout the vacuole (Figure 9.6f, arrow). But between a week and a month many heterogeneous vacuoles appear, containing hydrophilic droplets surrounded by phenolic compounds (Figure 9.6g, arrows). Later, necrotic areas can fuse, leaving

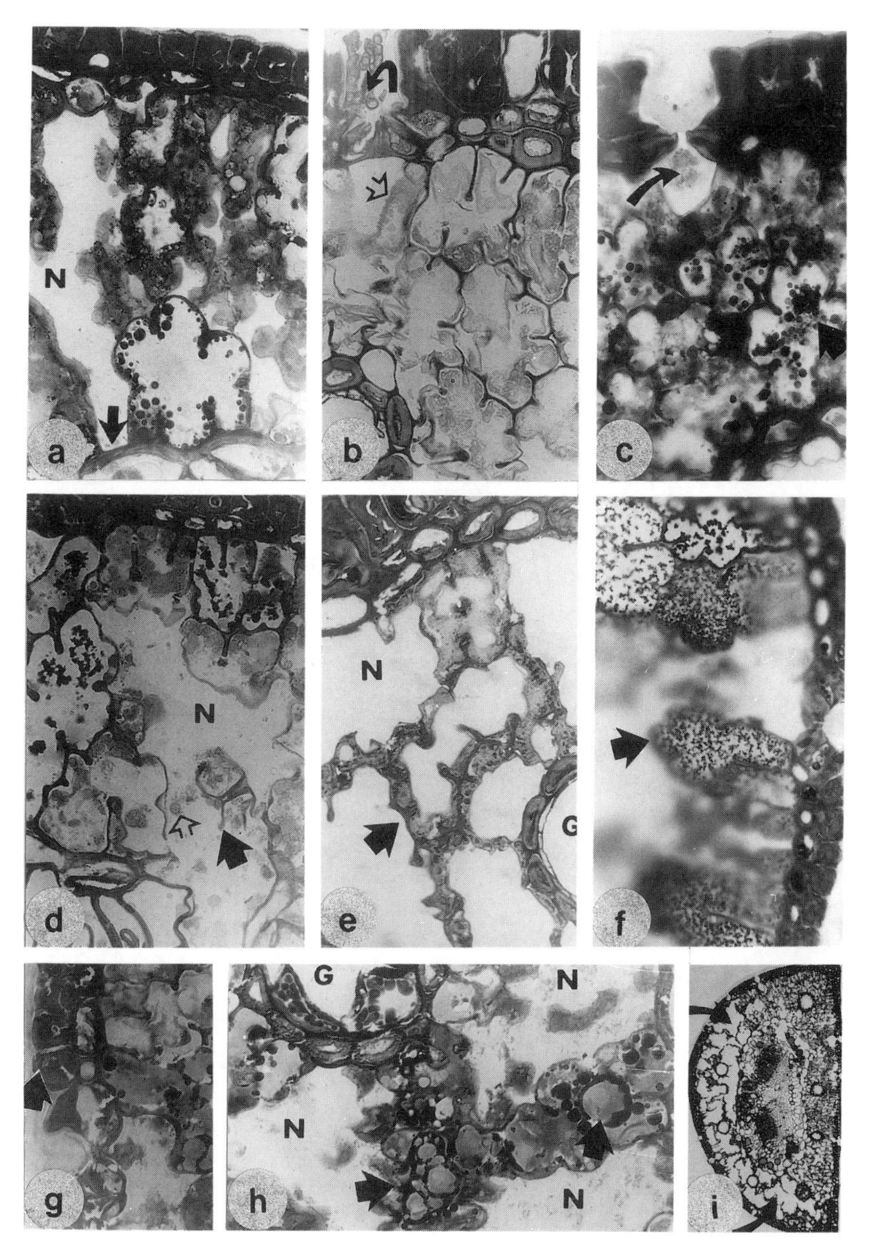

Figure 9.5. Semithin cross sections of pine needle (×1500). (a) Necrosis (N) reaching endodermis which has thick outside cell walls (arrow) after 7 days of acid spray. (b) Control with outer cavity containing spores (black arrow), and inner cavity bordered by long cells (hollow arrow). (c) Outer and inner cavity (arrow) filled with deposits after 3 days of sulphuric acid spray. (d) Very deep necrosis (N) reaching leaf interior after 7 days of acid spray. Broken cells (black arrow) and debris of organelles (hollow arrow). (e) Necrosis (N) of many cells, and only leftover parts of cell walls and cytoplasm debris (arrow) with surrounding empty spaces. Resin ducts (G) do not degenerate rapidly. (f) Necrosis (N) after 1 month of aluminium spray, showing heterogeneous vacuoles (arrows). Resin ducts are still well organized (G). (g) Dispersed droplets (arrow) of phenolic compounds in all cells under hypodermis, which also contains phenolics after 7 days of aluminium spray. (h) Many fused necroses (arrow) after 1 month of salt spray. (i) Cross section of the control needle

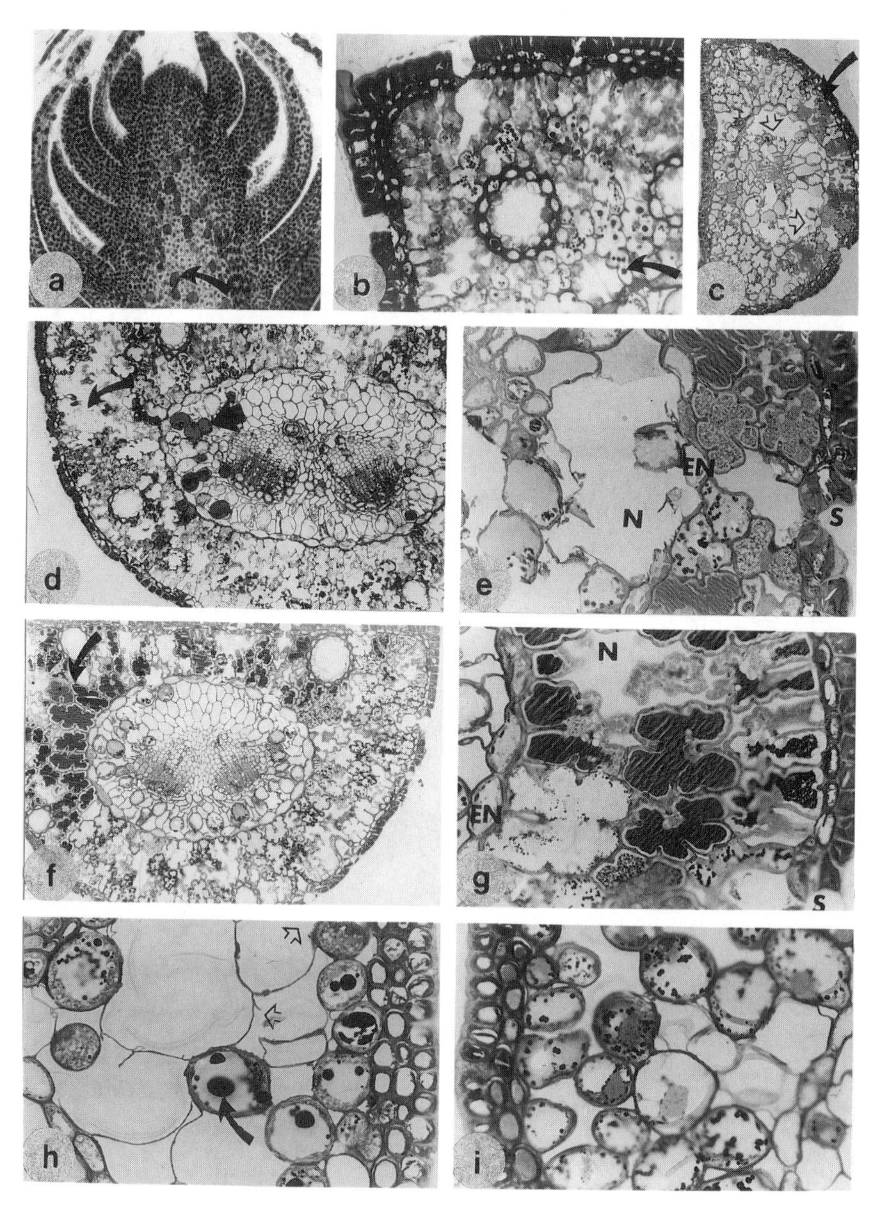

Figure 9.6. Shoot and leaves of *Pinus* (×100–800). (a) Longitudinal section of paraffin-embedded shoot apical internodes of control shoot of *P. nigra*. Arrow points to idioblastic cells filled with phenolics. (b) A few droplets (arrow) in control needle. (c) Many cells filled with phenolics (black arrow). Necrosis (hollow arrows) in parenchyma surrounding the veins after 1 month of salt spray. (d) Necrosis under stomata reaching endodermis (thin arrow). Thick black arrow points to parenchyma cells filled with phenolic compounds. (e) Tip necrosis with broken endodermis (EN) and necrosis (N) in parenchyma under affected mesophyll beneath stomata (S). (f) Large area of cells filled with dense phenolics (arrow) after 3 days of acid spray. (g) Cells filled with phenolic compounds after 7 days of salt spray. (h) Basal necrosis. Phenolics in the central vacuole (black arrow), broken cell walls (hollow arrows) in thin-walled cells. (i) Control, basal part of the needle. Small droplets of phenolics in many isodiametric cells

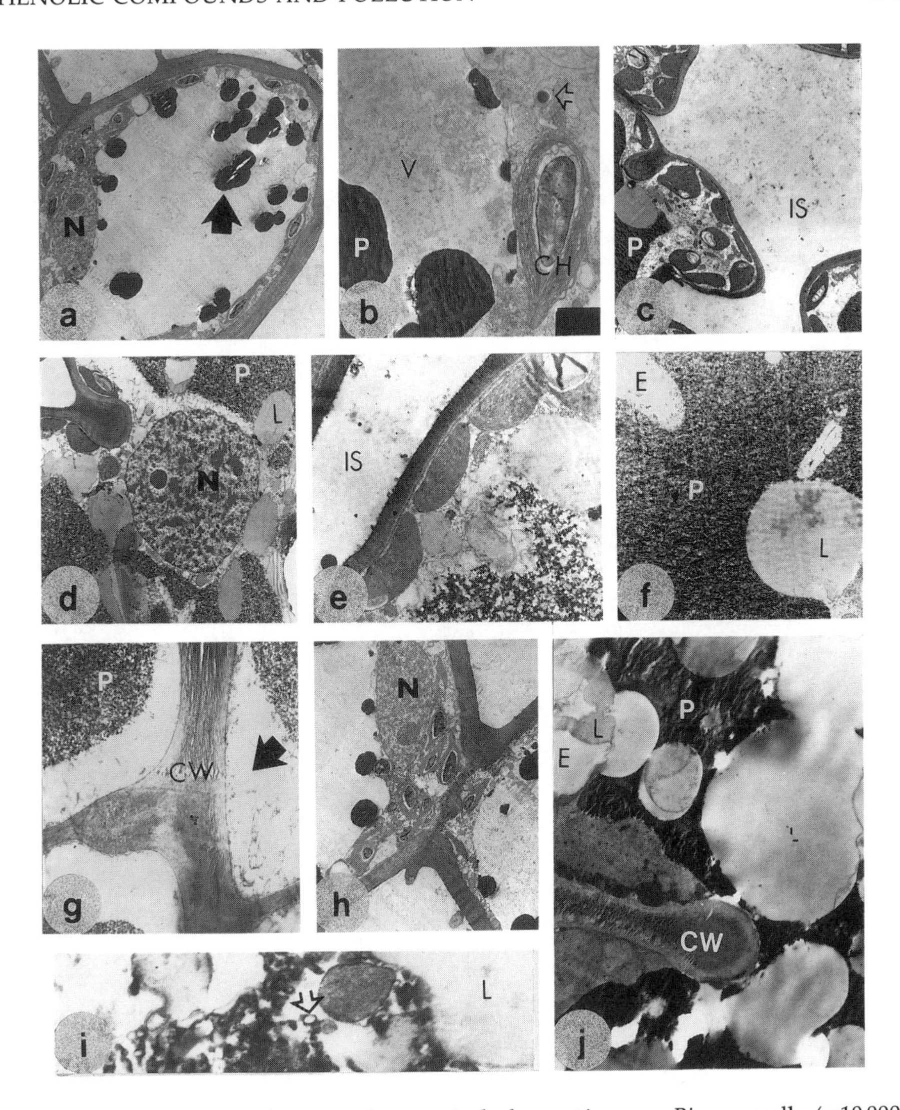

Figure 9.7. Transmission electron microscopical observations on *Pinus* needle (×10 000). (a) Isodiametric cell containing many droplets of phenolics after 1 week of salt spray. (b) In mesophyll cell vacuoles originating from endoplasmic reticulum (hollow arrow) near the chloroplast (CH). Phenolics (P) in the vacuole (V). (c) Some deposits in the intercellular spaces (IS) after 3 days of salt spray. (d) Lipophilic droplets (L) deposited into vacuole filled with phenolics (P) after a week of acid spray; N = nucleus. (e) Intercellular spaces (IS) with some debris after 3 days of acid spray. (f) Heterogeneous vacuole after 1 month of acid spray. Lipophilic areas (L) among phenolics (P), which also possess electron-empty areas (E). (g) Disappearing cytoplasm (arrow) after 1 week of salt spray. Cell walls (CW) and vacuoles with phenolics (P) are visible. (h) Phenolic droplets originating from ER, deposits in the vacuole attached to tonoplast after aluminium spray. (i) Debris in necrotic area after a month of acid spray. Lipids (L), small membrane deposits (hollow arrow) with attached phenolics. (j) Heterogeneous content of vacuole after 1 month of aluminium spray. Lipid-like droplets (L) with electron-empty areas (E) among phenolics (P)

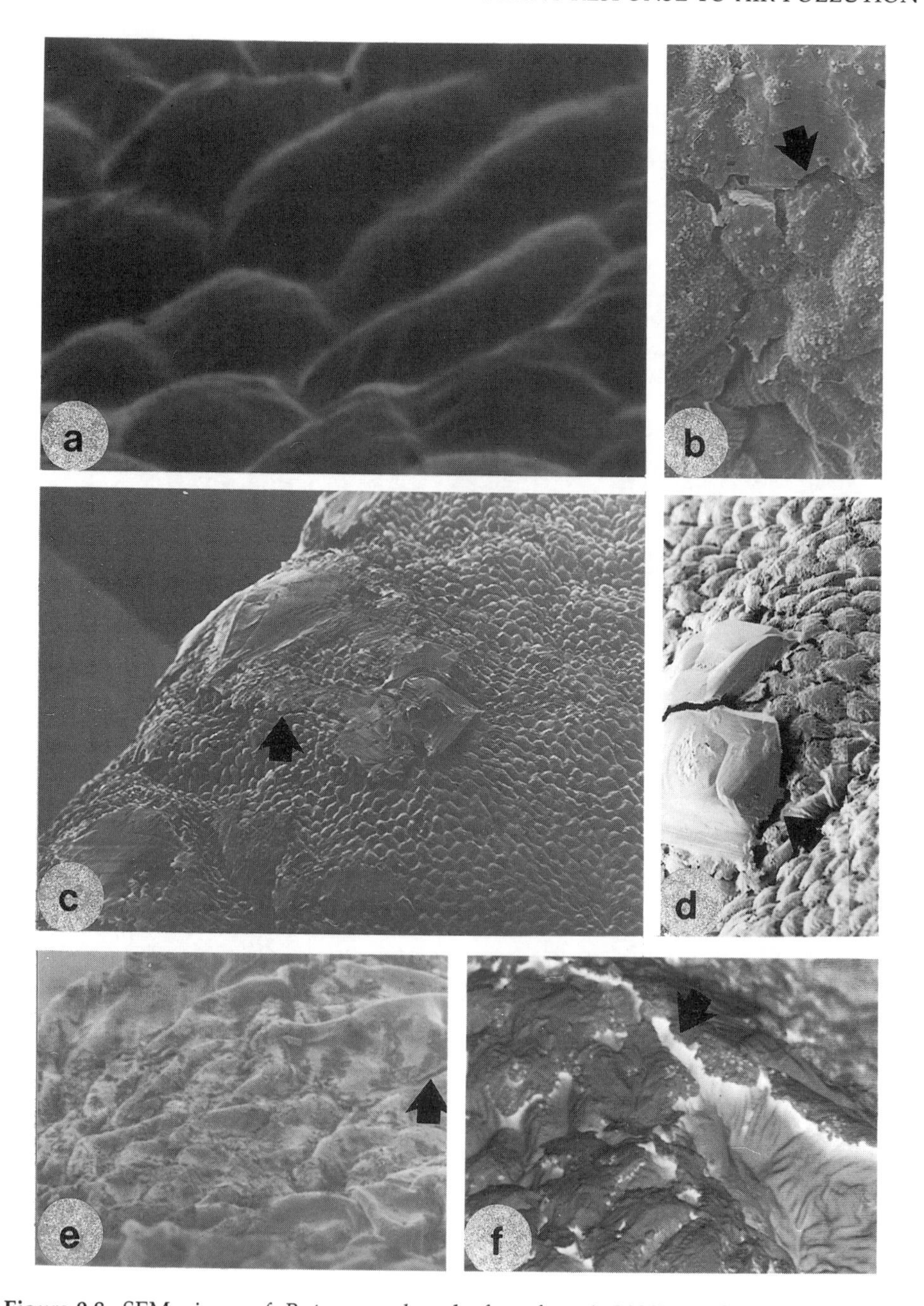

Figure 9.8. SEM views of *Ruta graveolens* leaf surface (×2000). (a) Smooth surface of control plant. (b) Crack (arrow) and deposits visible after 1 month of aluminium spray. (c) Growing crystals of salt with crack in vicinity of crystals (arrow) after 1 week of salt spray. (d) Rough surface of cuticle and cracks (arrow) under large crystals after a month of salt spray. (e) Peelings (arrow) and deposits after 1 week of acid spray. (f) Deep cracks (arrow) revealing unprotected cuticle beneath after 1 month of acid spray

substantial amounts of mesophyll tissue dead while the epidermis remains intact (Figure 9.6h).

Observations in the electron microscope confirm the above observations and reveal new facts, such as increased production of small vacuoles and vesicles filled with phenolics from the endoplasmic reticulum, and their fusion with the central vacuole, both in the isodiametric cells of the needle base (Figure 9.7a, arrow), and in mesophyll cells (Figure 9.7c, hollow arrow). After 3–7 days hydrophobic droplets became visible (Figure 9.7d, f), having been introduced into the vacuoles already containing phenolics. Intercellular spaces (Figure 9.7c, e) show some deposits at a very early stage after spraying. The cytoplasm disappears over a prolonged period (Figure 9.7g, arrow) leaving the cell walls and central vacuole with phenolic compounds. Spraying with aluminium salt causes deposits of phenolics, at first close to the tonoplast during the initial 24 hours (Figure 9.7h); in some cells lipid droplets are produced. Many central vacuoles are heterogeneous (Figure 9.7j), containing phenolic compounds (P), which are electron-dense, lipid-like droplets (L), and also electron-empty areas (E). In all treatments the last stage, lysis, leads to conglomeration of the cytoplasm and vacuole (Figure 9.7i). Different sized droplets are visible, as well as small vesicles and membranes with phenolics bound to them.

Spray on *Ruta*

On a plant surface, which appears smooth (Figure 9.8a), furanocoumarins are embedded in the wax. Salt spray (Figure 9.8c, d) leads to deposits of growing crystals. The greater part of the leaf area appears unaffected, but in the close vicinity of the crystals the epicuticular layer is broken (Figure 9.8d, arrow). Acid spray causes lesions on the surface (Figure 9.8e). After a prolonged period, cracks become visible, revealing unprotected cuticle below (Figure 9.8f, arrow). The aluminium-salt spray results in deposits of small crystal-like structures on the surface (Figure 9.8b, arrow).

Deposition of furanocoumarins on the surface is diminished by salt and sulphuric acid spray (Zobel *et al.*, 1991a), but total production after 2 weeks is drastically lowered; thus the percentage on the surface increases up to 60%. The reaction of the cells after 24 hours leads to dramatic extrusion of these compounds, which are said to have a defence role. This would be in agreement with higher production of other phenolics (epicatechins), which are extruded into the pine vacuole.

GENERALIZATIONS

A few hours after simulated air pollution the production of secondary products in the sprayed cells increases. These phenolic compounds are either stored in the vacuole (*Pinus*) or extruded to the surface (*Ruta*). The prolonged spraying time results in production of an excess of phenolic compounds by the cells, with consequent precipitation of their own cytoplasm, followed by lysis and death of the cells, visible macroscopically as necrosis. In *Ruta* the cells lower their production and extrusion of furanocoumarins within 2 weeks of spray. This suggests that these applied ions influence processes other than the biosynthesis of phenolics, such as energy production, whose lowered level could no longer support such extensive secondary product synthesis. Even if the biosynthesis is lowered extrusion still proceeds, and this is possibly because the signals for defence are still switched on. The fact that enzymes responsible for production of phenolics (Mireku & Wilkes, 1988; Karolewski & Daszkiewicz, 1990) and peroxidases (Karolewski, 1983) are stimulated by SO_2 (Karolewski, 1985), at least in some plants, supports the hypothesis that enzymes for secondary metabolism have been switched on. Increased concentrations of phenolics in leaves of poplars were influenced by SO_2 (Karolewski & Daszkiewicz, 1988). Much increased activity of glutamate dehydrogenase was observed by Jensch & Jaeger (1983) and substantial increases in peroxidase and acid phosphatase in *Lolium* and *Pisum*. The proline content increased under the influence of SO_2 in *Populus* leaves (Karolewski, 1985). In some cases decreased concentrations of protein and modification of enzyme activity, particularly isocitrate glucose-6-phosphate dehydrogenase have been noted (Pierre & Queiroz, 1988).

Experiments in our laboratory with extreme conditions such as UV radiation, and high and low temperatures, increased extrusion of furanocoumarins to the surface. Our unpublished data on chalcone synthase (a key enzyme in the biosynthesis of some phenylpropanoids) after freezing shock to *R. graveolens* leaves are also in accord with the idea that enzymes of secondary metabolism can be triggered by extreme conditions. This enzyme was activated by illumination, which is a shock to etiolated plants (Hrazdina *et al.*, 1987). Production of anthocyanins was similarly increased after the shock of illumination administered to etiolated *Fagopyrum esculentum* seedlings (Hrazdina & Zobel, 1991). The possibility of production of phenolics is visible in the tunica of promeristem cells at the very apex of the shoot, not only in mature leaves (Zobel, 1989), and chalcone synthase is located in the shoot apical dome (Zobel & Hrazdina, 1992). These facts give strong support to the suggestion that phenolics serve as the defence compounds, located in the covering tissue and on the plant surface, and thus that their synthesis is a bioindicator of drastic changes in the environment. We need more information about what defence compounds other plants produce, as well as about compounds responsible for communication between a plant and the biosystem in which it lives (Rosenthal, 1986; Zobel & Brown, 1995).

More investigation is necessary to decide whether phenolics production always increases through the influence of yet other ions and their mixtures, because there has been an instance reported when, after a forest fire, an example of extreme stress, plant shoots resprouted without secondary products (Christodoulakis *et al.*, 1986). Any such studies including attention to cell physiology and structure are extremely useful. A good start has been made on lichens, symbiotic organisms which have long been used as bioindicators of pollution because they grow under extreme conditions (Eversman & Sigal, 1987; Alebic-Juretic & Arko-Pijevac, 1989; Rope & Pearson, 1990). Investigations on the biochemical and cytological effects of sulphur dioxide are in progress (Malhotra & Hocking,

1976; Malhotra & Khan, 1984; Fangmeier, 1989), but ultrastructural changes have not yet been sufficiently investigated (Golberg & Adams, 1956; Thomson, 1975; Mlodzianowski & Bialobok, 1977; Kaerenlampi, 1986; Psaras & Christodoulakis, 1987; Dixit, 1988). And only a few data are available on isolated protoplasts of *Pinus*, which have enabled investigation of the direct influence of air pollutants on membranes and cytoplasm (Sigal *et al.*, 1988). We need more extracellular biochemical markers for damage caused by photochemical oxidant air pollution, such as those developed for Norway spruce by Castillo *et al.* (1987).

Sulphur dioxide lowers net photosynthesis, leaf respiration and transpiration of *Pinus banksiana* and *P. cartorta* (Amundson *et al.*, 1986), and assimilation rates are lowered by SO_2 fumigation in *Pinus* species and other trees (Addison *et al.*, 1984). Thus both primary metabolism (Koziol & Whatley, 1984; Darrall, 1986) and secondary metabolism (Juttner, 1988; Katoh *et al.*, 1989) are altered. Direct damage done to plant tissue can destroy the plant's ability to defend its cells against pollution stress (Lechowicz, 1987).

Far better described and understood are changes on the surface of the plant. Erosion of the leaf waxes by simulated acid rain has been described (Black & Black, 1978; Godzik, 1982; Riding & Percy, 1985; Baker & Hunt, 1986; Karku & Huttunen, 1986; Adams *et al.*, 1990). Cuticle degradation of *P. sylvestris* caused by polluted atmosphere was analysed (Fowler *et al.*, 1980). Disorganization of cuticular striation patterns and dissolution of cell walls was observed on *Ipomoea* leaves (Yunus *et al.*, 1982). Airborne pollutants change the plant surface and accelerate ageing of waxes (Huttunen & Laine, 1983). Epistomatal wax has a changed structure in *Picea* due to pollution, as observed by Sauter & Voss (1986). Wax plugs are formed in fir and spruce after acid rain spraying (Schmitt *et al.*, 1987). In view of the fact that the cuticle is responsible for allowing cation penetration (McFarlane & Berry, 1974), the condition of the cuticle will always be an indicator of environmental pollution (Sharma, 1977).

Pollutant particles landing on the surface are stored there (Mankovska *et al.*, 1988). The surface of the leaf and the air in its vicinity constitute the microenvironment (Oertli *et al.*, 1977) and its chemistry is changed after a raindrop lands on the surface (Gaber & Hutchinson, 1988). Increased wettability after acid rain spray (Adams & Hutchinson, 1987; Percy & Baker, 1988) must not only increase the area of contact with the drop, but as well must change the properties of the cuticle, allowing easier penetration by that and other pollutants. The plant's first barrier is altered, and even though the plant surface has the ability to neutralize the acid spray (Adams & Hutchinson, 1987), since cells have buffering capacity (Pfanz & Heber, 1986; Pfanz *et al.*, 1987), neutralization cannot continue indefinitely. The role of the surface is important in establishing susceptibility of plants to air pollutant injury (Swiecki *et al.*, 1982).

Decreased photosynthesis due to air pollution (Amundson *et al.*, 1986) and inhibition of total protein production (Khan & Malhotra, 1983), as well as lowered activity of particular enzymes, shown in jack pine in the cases of ribulose bisphosphate carboxylase and glycollate oxidase (Khan & Malhotra, 1982), must lower the cells' energy supply, and any protective mechanisms become insufficient to withstand continuing pollution.

There is, however, hope for the survival of some plants at least, because plants have varying tolerances to SO_2 (Tsukahara *et al.*, 1985), and some plants can adjust to air pollution (Wilson & Bell, 1985). Certain plants can even show stimulated growth, as, for example, in the response of *Sphagnum* to SO_2 and NO_2 (J.A. Lee *et al.*, 1990), and epiphytic lichens become more abundant in an increasing pollution gradient (Oksanen *et al.*, 1990). Light can reduce damage caused by SO_2 (Olszyk & Tingey, 1984a). Some pathogens are less active after spraying with SO_2 and pathogenicity is lowered (McLeod, 1988; Singh & Rai, 1989); but nevertheless, in many cases of pollution there is an increased possibility of additional infection of plants or infestation by insects (Dohmen *et al.*, 1984; Fehrmann *et al.*, 1986; Warrington, 1987; Rao *et al.*, 1990). Madden & Campbell (1987) demonstrated potential effects of air pollution on epidemics of plant diseases.

There would appear to be several possibilities of combating the effects of air pollution: to reduce its level, to develop more resistant plants, or to apply coatings of natural products able to play a protective role. This last possibility might be realistic if flavonoids were used, as they are relatively abundant in nature and exist in high concentrations on the plant surface (Wollenweber, 1985). They are also sometimes used as detoxicants or protectants; for example, they have been used to reduce the effect of mustard oil rubbed into mouse skin (Vijayaraghavan *et al.*, 1991), and butterflies use them as pigmentation on their wings to protect against UV radiation (Wilson, 1986).

We have to remember that some plants, conifers especially, are always the chief casualties of air pollution, because they are extremely well adapted to collect water from clouds and fogs, from which they can absorb ions (Board, 1988). Thus, not only acid rain, but fog, dew and frost are potentially dangerous (Wisniewski, 1982); the pH of smog over California, for instance, was as low as 1.69 (Steinfeld, 1989).

ACKNOWLEDGEMENT

I am very grateful to Professor Stewart Brown of the Department of Chemistry, Trent University, Peterborough, Ontario, for his assistance in the writing and editing of the manuscript for this chapter.

REFERENCES

Adams, C.M. & Hutchinson, T.C. 1987. Comparative abilities of leaf surface to neutralize acid raindrops II. The influence of leaf wettability, leaf age and rain duration on changes in droplet pH and chemistry of the leaf surface. *New Phytol.* **106**: 437–456.

Adams, C.M., Dengler, N.G. & Hutchinson, T.C. 1984. Acid rain effects on foliar histology of *Artemisia tilesii*. *Can. J. Bot.* **62**: 463–474.

Adams, C.M., Caporn, S.J. & Hutchinson, T.C. 1990. Crystal occurrence and wax distribution on leaf surfaces of cabbage treated with simulated acid rain. *New Phytol.* **111**: 147–158.

Addison, P.A., Malhotra, S.S. & Khan, A.A. 1984. Effect of sulphur dioxide on woody boreal forest species grown on native soils and tailings. *J. Environ. Qual.* **13**: 333–336.

Alebic-Juretic, A.N. & Arko-Pijevac, M. 1989. Air pollution damage to cell membranes in lichens — results of simple biological tests applied in Rijeka, Yugoslavia. *Water Air Soil Pollut.* **47**: 25-33.

Amundson, R.G., Walker, R.B. & Legge, A.H. 1986. Sulphur gas emissions in the boreal forest: the West Whitecourt case study. VII. Pine tree physiology. *Water Air Soil Pollut.* **29**: 129-147.

Ashwood-Smith, M.J., Ceska, O. & Chaudhary, S.K. 1985. Mechanism of photosensitivity reactions to diseased celery. *Br. Med. J.* **290**: 1249.

Baker, E.A. 1982. Chemistry and morphology of plant epicuticular waxes. In Cutler, D.F., Alvin, K.L. & Price, C.E. (eds) *The Plant Cuticle*: 139-165. Academic Press, London.

Baker, E.A. & Hunt, G.M. 1986. Erosion of leaf waxes from leaf surfaces by simulated rain. *New Phytol.* **102**: 161-173.

Balsberg-Paahlsson, A. 1989. Effects of heavy-metal and SO_2 pollution on the concentrations of carbohydrates and nitrogen in tree leaves. *Can. J. Bot.* **67**: 2106-2113.

Beier, R.C. & Oertli, E.H. 1983. Psoralen and other linear furanocoumarins as phytoalexins in celery (*Apium graveolens*). *Phytochemistry* **22**: 2595-2597.

Berenbaum, M. 1981. Patterns of furanocoumarin distribution and herbivory in the Umbelliferae: Plant chemistry and community structure. *Ecology* **62**: 1264-1266.

Berenbaum, M. 1983. Coumarins and caterpillars, a case for coevolution. *Evolution* **37**: 163-179.

Berenbaum, M. & Neal, J.J. 1985. Synergism between myristicin and xanthotoxin, a naturally occurring plant toxicant. *J. Chem. Ecol.* **11**: 1349-1358.

Berenbaum, M., Zangerl, A.R. & Nitao, J.K. 1986. Constraints on chemical coevolution: Wild parsnips (*Pastinaca sativa*) and the parsnip webworm (*Depressaria pastinacella*). *Evolution* **40**: 1215-1228.

Bialobok, S. & Fabijanowski, J. 1984. Concepts of forest management in industrial regions. *Ecol. Stud.* **49**: 231-245.

Black, C.R. & Black, V.J. 1978. Light and scanning electron microscopy of SO_2-induced injury of leaf surface of Field Bean (*Vicia faba* L.). *Plant, Cell Environ.* **2**: 329-333.

Bleweiss, G., Werker, E. & Peleg, M. 1985. Air-pollution effects on *Cedrus libani* trees — a case study. *Environ. Conserv.* **12**: 70-73.

Board, P.A. 1988. Why conifers are the chief plant casualties of atmospheric pollution. *Arboric. J.* **12**: 251-255.

Camm, E.L., Wat, C.-K. & Towers, G.H.N. 1976. An assessment of the roles of furanocoumarins in *Heracleum lanatum*. *Can. J. Bot.* **54**: 2562-2566.

Castillo, F.J., Miller, P.R. & Greppin, H. 1987. Extracellular biochemical markers of photochemical oxidant air pollution damage to Norway spruce. *Experientia* **43**: 111-115.

Chappell, J. & Hahlbrock, K. 1984. Transcription of plant defense genes in response to UV light or fungal elicitors. *Nature* **311**: 76-78.

Chew, F.S. & Rodeman, J.E. 1979. Plant responses for chemical defense. In Rosenthal, G.A. & Janzen, D.H. (eds) *Herbivores: Their Interaction with Secondary Plant Metabolites*: 271-309. Academic Press, New York.

Christodoulakis, N.S., Arianoutsou-Farggitaki, M. & Psaras, G.K. 1986. Post-fire leaf structure of two seasonally dimorphic resprouters. *Acta Oecol.* **7**: 97-102.

Clapperton, M.J. & Parkinson, D. 1990. The effect of SO_2 on the vesicular-arbuscular mycorrhizae associated with a submontane mixed grass prairie in Alberta, Canada. *Can. J. Bot.* **68**: 1646-1650.

Darrall, N.M. 1986. The sensitivity of natural photosynthesis in several plant species to short term fumigation with sulphur dioxide. *J. Exp. Bot.* **37**: 1313-1322.

Davis, D.D. & Wood, F.A. 1972. The relative susceptibility of eighteen coniferous species to ozone. *Phytopathology* **62**: 14-19.

Dixit, A.B. 1988. Effect of particulate pollutants on plants at ultrastructural and cellular levels. *Ann. Bot.* **62**: 643-651.

Dohmen, G.P., McNeill, S. & Bell, J.N.B. 1984. Air pollution increases *Aphis fabae* pest potential. *Nature* **307**: 52-53.

Elkiey, T., Ormrod, D.P. 1981. Sulphur and nitrogen nutrition and misting effects on the response of bluegrass to ozone, sulphur dioxide, nitrogen dioxide or their mixture. *Water Air Soil Pollut.* **16**: 177-186.

Ernst, W.H.O., Tonneijck, A.E.C. & Pasman, F.J.M. 1985. Ecotypic responses of *Silene cucubalus* to air pollutants (SO_2 and O_3). *J. Plant Physiol.* **118**: 439-450.

Eversman, S. & Sigal, L.L. 1987. Effect of SO_2, O_3, and SO_2 and O_3 in combination on photosynthesis and ultrastructure of two lichen species. *Can. J. Bot.* **65**: 1806-1818.

Fangmeier, A. 1989. Effect of open-top fumigations with SO_2, NO_2 and ozone on the native herb layer of a beech forest. *Environ. Exp. Bot.* **29**: 199-213.

Feeny, P.P. 1987. The roles of plant chemistry in associations between swallowtail butterflies and their host plants. In Labeyries, V., Fabes, G. & Lachaise, D. (eds) *Insects-Plants*: 353-359.

Fehrmann, H., von Tiedemann, A. & Fabian, P. 1986. Predisposition of wheat and barley to fungal leaf attack by preinoculative treatment with ozone and sulphur dioxide. *Z. Planzenkr. Pflanzenschutz.* **93**: 313-318.

Fowler, D., Cape, J.N., Nicholson, I.A., Kinnaird, J.W. & Patterson, I.S. 1980. The influence of a polluted atmosphere on cuticle degradation in Scots pine (*Pinus sylvestris*). In Drablos, D. & Tallan, A. (eds) *Proceedings of the International Conference on Ecological Impact of Acid Precipitation*: 146-156. SNSF Project, Norway.

Franke, W. 1967. Mechanisms of foliar penetration of solutions. *Annu. Rev. Plant Physiol.* **18**: 281-300.

Freedman, B., Zobens, V., Hutchinson, T.C. & Gizyn, W.I. 1990. Intense natural pollution affects arctic tundra vegetation at the Smoking Hills, Canada. *Ecology* **71**: 492-503.

Gaber, B.A. & Hutchinson, T.C. 1988. Chemical changes in simulated raindrops following contact with leaves of four boreal forest species. *Can. J. Bot.* **66**: 2445-2451.

Godzik, S. 1982. The scanning and transmission electron microscopes in use of plants as bioindicators. In Steubing, L. & Jörger, M.J. (eds) *Monitoring of Air Pollutants*: 157-180. Junk, The Hague.

Golberg, R.A. & Adams, D.F. 1956. Histological responses of some plant leaves to hydrogen fluoride and sulphur dioxide. *Am. J. Bot.* **43**: 755-760.

Grieve, A.M. & Pitman, M.G. 1978. Salinity damage to Norfolk-island pine caused by surfactants. III. Evidence for stomatal penetration as the pathway of salt spray entry to leaves. *Aust. J. Plant Physiol.* **5**: 398-413.

Harborne, J.B. 1985. Phenolics and plant defence. *Ann. Proc. Phytochem. Soc. Eur.* **25**: 393-408.

Harborne, J.B. & Turner, B.L. 1984. *Plant Chemosystematics.* Academic Press, London.

Hartwich, C. 1885. Über Gerbstoffkugeln und Ligninkörper in der Naturgeschicht der Infectoria Gallen. *Ber. Deutsch. bot. Ges.* **3**: 146-150.

Haslam, E. 1979. Vegetable tannins. *Recent Adv. Phytochem.* **12**: 475-523.

Haslam, E. 1985a. New polyphenols for old tannins. *Ann. Proc. Phytochem. Soc. Eur.* **25**: 237-257.

Haslam, E. 1985b. *Metabolites and Metabolism.* Clarendon Press, Oxford.

Hegnauer, R. (1962 *et seq.*) *Chemotaxonomie der Pflanzen* (6 vols). Birkhäuser, Basle.

Hejnowicz, A. 1979. Tannin vacuoles and starch in the development of Scots pine (*Pinus sylvestris*) vegetative buds. *Acta Soc. Bot. Pol.* **48**: 195-203.

Hrazdina, G. & Wagner, G.J. 1985. Compartmentation of plant phenolic compounds: sites of synthesis and accumulation. *Ann. Proc. Phytochem. Soc. Eur.* **25**: 119-135.

Hrazdina, G. & Zobel, A.M. 1991. Cytochemical localization of enzymes in plant cells. *Bot. Rev.* **129**: 269-322.

Hrazdina, G., Marx, G.A. & Hoch, H.C. 1982. distribution of secondary plant metabolites and their biosynthetic enzymes in pea (*Pisum sativum* L.) leaves. *Plant Physiol.* **70**: 743-745.

Hrazdina, G., Zobel, A.M. & Hoch, H.C., 1987. Biochemical, immunological, and immunochemical evidence for the association of chalcone synthase with endoplasmic reticulum membranes. *Proc. Nat. Acad. Sci. USA* **84**: 8966-8970.

Hunt, G.M. & Baker, E.A. 1982. Developmental and environmental variations in plant epicuticular waxes: some effects on penetration of naphthylacetic acid. In Cutler, D.F., Alvin, K.L.& Price, C.E. (eds) *The Plant Cuticle*: 279-292. Academic Press, New York.

Hutchinson, T.C. 1984. Adaptation of plants to atmospheric pollutants. *CIBA Found. Symp.* **102**: 52-72.

Huttunen, S. & Laine, K. 1983. Effects of air-borne pollutants on the surface wax structure of *Pinus sylvestris* needles. *Ann. Bot. Fenn.* **20**: 79-86.

Huttunen, S., Laine, K. & Torvela, H. 1985. Sulphur content of pine needles as indices of air pollution. *Ann. Bot. Fenn.* **22**: 343-359.

Innes, J.L. 1987. *Air Pollution and Forestry*. Bulletin 70, Forestry Commission, HMSO, London.

Ivie, G.W. 1987. The chemistry of plant furanocoumarins and their medical, toxicological, environmental and coevolutionary significance. *Rev. Latinoamer. Quim.* **18**: 1-6.

Iwanowska, A., Tykarska, T., Kuras, M. & Zobel, A.M. 1994. Localization of phenolic compounds in the covering tissues of the embryo of *Brassica napus* during different stages of embryogenesis and seed maturation. *Ann. Bot.* **74**: 313-320.

Jensch, U.-E. & Jaeger, H.-J. 1983. Physiological and biochemical responses of plants as indicators of an effect of air pollutants on vegetation. *Angew. Bot.* **57**: 157-171.

Juttner, F. 1988. Changes in monoterpene concentration in needles of polluted-injured *Picea abies* exhibiting mottled yellowing. *Physiol. Plant.* **72**: 48-56.

Kaerenlampi, L. 1986. Relationship between macroscopic symptoms of injury and cell structural changes in needles of ponderosa pine exposed to air pollution in California. *Ann. Bot. Fenn.* **23**: 255-264.

Karku, M. & Huttunen, S. 1986. Erosion effects of air pollution on needle surface. *Water, Air Soil Pollut.* **31**: 417-424.

Karolewski, P. 1983. Effect of sulphur dioxide on peroxidase activity in leaves of *Weigela* rooted cuttings. *Arbor. Kornickie* **28**: 113-127.

Karolewski, P. 1985. The role of free proline in the sensitivity of poplar (*Populus robusta*). *Eur. J. For. Pathol.* **15**: 199-206.

Karolewski, P. & Daszkiewicz, P. 1988. Influence of sulphur dioxide on the level of phenols in leaves of poplars differing in sensitivity to the action of this gas. *Arbor. Kornickie* **33**: 231-238.

Karolewski, P. & Daszkiewicz, P. 1990. Visible and invisible injury to Scots pine (*Pinus sylvestris*) needles caused by sulphur dioxide. *Arbor. Kornickie* **35**: 127-136.

Katoh, T., Kasuya, M., Kagamimork, S., Kazuka, H. & Kawano, S. 1989. Effects of air pollution on tannin biosynthesis and predation damage in *Cryptomeria japonica*. *Phytochem.* **28**: 439-445.

Kelsey, R.J., Reynolds, G.W. & Rodriguez, E. 1984. The chemistry of biologically active constituents secreted and stored in plant glandular trichomes. In Rodriguez, E., Healey, P.L. & Mehta, I. (eds) *Biology and Chemistry of Plant Trichomes*: 187-241. Plenum, New York.

Khan, A.A. & Malhotra, S.S. 1982. Ribulose bisphosphate carboxylase and glycolate oxidase from jack pine: effects of sulphur dioxide fumigation. *Phytochem.* **21**: 2607-2612.

Khan, A.A. & Malhotra, S.S. 1983. Protein biosynthesis in jack pine and inhibition by sulphur dioxide. *Phytochem.* **22**: 1325-1328.

Kolottukudy, P.E. 1984. Biochemistry and function of cutin and suberin. *Can. J. Bot.* **62**: 2918-2933.

Koziol, M.J. & Whatley, F.R. 1984. *Gaseous Air Pollutants and Plant Metabolism*. Butterworth, London, 480pp.

Kozlowski, T.T. 1984. Plant responses to flooding of soil. *Bioscience* **34**: 162-167.

Kramer, P.I. & Kozlowski, T.T. 1979. *Physiology of Woody Plants*. Academic Press, New York.

Lechowicz, M.J. 1987. Resource allocation by plants under air pollution stress: Implications for plant-pest-pathogen interactions. *Bot. Rev.* **53**: 281-300.

Lee, J.A., Baxter, R. & Emes, M.J. 1990. Responses of Sphagnum species to atmospheric nitrogen and sulphur deposition. *Bot. J. Linn. Soc.* **104**: 255-265.

Lee, J.J., Neely, G.E., Perrigan, S.C. & Grothaus, L.C. 1981. Effects of simulated sulphuric acid rain on yield, growth and foliar injury of several crops. *Environ. Exp. Bot.* **25**: 171-185.

Lee, W.S., Chevone, B.I. & Seiler, J.R. 1990. Growth and gas exchange of loblolly pine seedlings as influenced by drought and air pollutants. *Water, Air Soil Pollut.* **51**: 105-116.

Madden, L.V. & Campbell, C.L. 1987. Potential effects of air pollutants on epidemics of plant diseases. *Agric. Ecosyst. Environ.* **18**: 251-262.

Malhotra, S.S. & Hocking, D. 1976. Biochemical and cytological effects of sulphur dioxide on plant metabolism. *New Phytol.* **76**: 227-237.

Malhotra, S.S. & Khan, A.A. 1984. Biochemical and physiological impact of major pollutants. In Treshow, M. (ed.) *Air Pollution and Plant Life*: 113-157. Wiley, Chichester.

Mankovska, B., Peura, R. & Huttunen, F. 1988. Deposition of air-borne pollutants on the surface of spruce needles around the aluminum smeltery. *Ekologia* **7**: 291-298.

Mann, J. 1980. *Secondary Metabolism*. Clarendon Press, Oxford.

Marie, B.A. & Ormrod, D.P. 1986. Dose response relationships of the growth and injury effects of ozone and sulphur dioxide on Brassicaceae seedlings. *Can. J. Plant Sci.* **66**: 659-667.

Matern, U., Strasser, H., Wendorff, H. & Hamerski, D. 1988. Coumarins and furanocoumarins. In Vasil, I. & Constabel F. (eds) *Cell Culture and Somatic Cell Genetics of Plants*, Vol. 5: 3-21. Academic Press, New York.

Matsubara, J., Ifhioka, K., Shibata, Y. & Katoh, K. 1986. Risk analysis of multiple environmental factors: Radiation, zinc, cadmium, and calcium. *Environ. Res.* **40**: 525-530.

McFarlane, J.C. & Berry, W.L. 1974. Cation penetration through isolated cuticles. *Plant Physiol.* **53**: 723-727.

McKey, D. 1979. The distribution of secondary compounds within plants. In Rosenthal, G.A. & Janzen, D.H. (eds) *Herbivores: Their Interaction with Secondary Plant Metabolites*: 56-135. Academic Press, New York.

McLeod, A.R. 1988. Effects of open-air fumigation with sulphur dioxide on the occurrence of fungal pathogens in winter cereals. *Phytopathology* **78**: 88-94.

Mireku, E. & Wilkes, J. 1988. Production of phenols in the sapwood of *Eucalyptus maculata* after wounding and infection. *Eur. J. For. Path.* **18**: 121-127.

Mlodzianowski, F. & Bialobok, S. 1977. The effect of sulphur dioxide on ultrastructural organization of larch needles. *Acta Soc. Bot. Polon.* **46**: 629-634.

Moeller, H. 1882. Anatomische Untersuchungen über das Vorkommen der Gerbsäure. *Ber. Deutsch. Bot. Ges.* **6**: 32-42.

Murray, R.D.H. 1991. Naturally occurring plant coumarins. *Prog. Chem. Org. Nat. Prod.* **58**: 83-343.

Murray, R.D.H., Mèndez, J. & Brown, S.A. 1982. *The Natural Coumarins: Occurrence, Chemistry and Biochemistry*. Wiley, Chichester.

Nilsson, M. 1961. Chalcones from fronds of *Pityrogramma chrysophylla* var. heyderi. *Acta Chem. Scand.* **15**: 211-215.

Norby, R.J. & Kozlowski, T.T. 1983. Flooding and SO_2 stress interaction in *Betula papyrifera* and *B. nigra* seedlings. *For. Sci.* **29**: 739-750.

Oertli, J.J., Harr, J. & Guggenhein, R. 1977. The pH values as an indicator of the leaf surface microenvironment. *Z. Pflanzenkr. Pflanzenschutz.* **84**: 729-737.

Oksanen, J., Tynnyrinen, S. & Kaerenlampi, L. 1990. Testing for increased abundance of epiphytic lichen on the local pollution gradient. *Ann. Bot. Fenn.* **27**: 301-307.

Olszyk, D.M. & Tibbitts, T.W. 1981. Stomatal response and leaf injury of *Pisum sativum* L. with SO_2 and O_3 exposures. I. Influence of moisture stress and time of exposure. *Plant Physiol.* **67**: 545-549.

Olszyk, D.M. & Tingey, D.T. 1984a. Phytotoxicity of air pollutants. Evidence for photodetoxification of SO_2 but not O_3. *Plant Physiol.* **74**: 999-1005.

Olszyk, D.M. & Tingey, D.T. 1984b. Fusicoccin and air pollutant injury to plants. Evidence for enhancement of SO_2 but not O_3 injury. *Plant Physiol.* **76**: 400-402.

Ormrod, D.P., Tingey, D.T., Gumpertz, M.L. & Olszyk, D.M. 1984. Utilization of a response-surface technique in the study of plant responses to ozone and sulphur dioxide mixtures. *Plant Physiol.* **75**: 43-48.

Percy, K.E. & Baker, E.A. 1988. Effects of simulated acid rain on leaf wettability, rain retention and uptake of some inorganic ions. *New Phytol.* **108**: 75-82.

Pfanz, H. & Heber, U. 1986. Buffer capacities of leaves, leaf cells and leaf cell organelles in reaction to fluxes of potentially acidic gases. *Plant Physiol.* **81**: 597-602.

Pfanz, H., Martinoia, E., Lange, O.L. & Heber, U. 1987. Flux of SO_2 into leaf cells and cellular acidification by SO_2. *Plant Physiol.* **85**: 928-933.

Piccard, J. 1873. Über einige Bestandteile der Pappelknospen. *Chem. Ber.* **6**: 890-893.

Pierre, M. and Queiroz, O. 1988. Air pollution by SO_2 amplifies the effect of water stress on enzymes and total soluble proteins of spruce needles. *Physiol. Plant.* **73**: 412-417.

Podbielkowska, M., Kupidlowska, E., Waleza, M., Dobrzynska, K., Louis, S.A., Keightly, A. & Zobel, A.M. 1994. Coumarins as antimitotics. *Int. J. Pharmacog.* **32**: 262-273.

Posthumus, A.C. 1983. Higher plants as indicators and accumulators of gaseous air pollution. *Environ. Monit. Assess.* **3**: 263-272.

Posthumus, A.C. 1985. Plants as bioindicators of atmospheric pollution. In Nuernberg, H.W. (ed.) *Pollutants and their Ecotoxicological Significance*: 55-65. Wiley, New York.

Psaras, G.K. & Christodoulakis, N.S. 1987. Air pollution effects on the ultrastructure of *Phlomis fruticosa* mesophyll cells. *Bull. Environ. Contam. Toxicol.* **38**: 610-617.

Rao, M.V., Gupta, C.K. & Dubay, P.S. 1990. Effect of relatively low ambient air pollutants on total sucking insect populations on *Sorghum vulgare. Trop. Ecol.* **31**: 66-72.

Reed, D.W. & Tukey, H.B. Jr 1982. Permeability of Brussels sprouts and carnation cuticles from leaves developed in different temperatures and light intensities. In Cutler, D.F., Alvin, K.L. & Price, C.E. (eds) *The Plant Cuticle*: 267-278. Academic Press, New York.

Reinert, R.A. 1984. Plant response to air pollutant mixtures. *Annu. Rev. Phytopathol.* **22**: 421-442.

Rhodes, D.F. 1979. Ecological and evolutionary processes. In Rosenthal, G.A. & Janzen, D.H. (eds) *Herbivores: Their Interaction with Secondary Plant Metabolites*: 1-56. Academic Press, New York.

Rice, P.M., Carlson, C.E., Bromenshenk, J.J., Gordon, C.C. & Turangeau, P.C. 1986. Basal injury syndrome of *Pinus* needles. *Can. J. Bot.* **64**: 632-642.

Riding, R.T. & Percy, K.E. 1985. Effect of sulphur dioxide and other pollutants on the morphology of epicuticular waxes on needles of *Pinus strobus* and *Pinus banksiana. New Phytol.* **99**: 555-564.

Rope, S.K. & Pearson, L.C. 1990. Lichens as air pollution biomonitors in a semi-arid environment in Idaho. *Bryologist* **93**: 50-61.

Rosenthal, G.A. 1986. The chemical defenses of higher plants. *Sci. Amer.* **102**: 94-99.

Saleh, N.S., Hallak, A.B. & Othman, A.M. 1985. The use of palm leaves as a bioindicator of atmospheric pollution. *Int. J. Appl. Radiat. Isotopes* **36**: 321-322.

Sauter, J.J. & Voss, J.U. 1986. SEM observations on the structural degradation of epistomal waxes in *Picea abies* (L.) Karst. and its possible role in the "Fichtensterben". *Eur. J. For. Path.* **16**: 408-423.

Schmitt, V., Reutze, M. & Liese, W. 1987. Scanning electron microscopical investigations on stomatal wax plugs of fir and spruce needles after fumigation and acid rain treatment. *Eur. J. For. Path.* **17**: 1118-1124.

Scholz, F. & Bergmann, F. 1984. Selection pressure by air pollution as studied by isozyme-gene-systems in Norway spruce exposed to sulphur dioxide. *Silvae Genet.* **33**: 238-241.

Schonherr, J. 1976. Water permeability of isolated cuticular membranes: the effect of cuticular waxes on diffusion of water. *Planta* **131**: 159-164.

Shanklin, J. & Kozlowski, T.T. 1985. Effect of temperature regime on growth and subsequent responses of *Sophora japonica* seedlings to SO_2. *Plant and Soil* **88**: 399-405.

Sharma, G.K. 1977. Cuticular features as indicators of environmental pollution. *Water Air Soil Pollut.* **8**: 15-19.

Sigal, L.L., Eversman, S. & Berglund, D.L. 1988. Isolation of protoplasts from loblolly pine needles and their flow-cytometric analysis for air pollution effects. *Environ. Exp. Bot.* **28**: 151-161.

Singh, A.K. & Rai, B. 1989. In vitro growth of some fungi isolated from wheat phylloplane in relation to SO_2 treatment. *Curr. Sci.* **58**: 924-925.

Städler, E. & Buser, H.-R. 1984. Defense chemicals in the leaf surface wax synergistically accumulate oviposition by a phytophagous insect. *Experientia* **40**: 1157-1159.

Stafford, H.A. 1987. Enzymology of proanthocyanidin biosynthesis. In Hemingway, R.W. & Karchesy, J.J. (eds) *Chemistry and Significance of Condensed Tannins*: 47-70. Plenum, New York.

Stafford, H.A., Smith, E.C. & Weider, R.M. 1989. The development of proanthocyanidins (condensed tannins) and other phenolics in bark of *Pseudotsuga menziesii*. *Can. J. Bot.* **67**: 1111-1118.

Steinfeld, J.H. 1989. Urban air pollution: state of the science. *Science* **243**: 745-753.

Surico, G., Varvano, L. & Solfrizzo, M. 1987. Linear furanocoumarins accumulation in celery plants infected with *Erwinia carotovera* pathovar *carotovera*. *J. Agric. Food Chem.* **35**: 406-409.

Swain, T. 1979. Tannins and lignins. In Rosenthal, G.A. & Janzen, D.H. (eds) *Herbivores: Their Interaction with Secondary Plant Metabolites*: 657-690. Academic Press, New York.

Swain, T. 1985. Plant phenolics: Past and future. *Ann. Proc. Phytochem. Soc. Eur.* **25**: 465-468.

Swiecki, T.J., Endress, A.J. & Taylor, O.C. 1982. The role of surface wax in susceptibility of plants to air pollutant injury. *Can. J. Bot.* **60**: 316-319.

Thielke, C. 1956. Gerbstoffidioblasten in der Scheide von *Carex*. *Protoplasma* **47**: 145-155.

Thomson, W.W. 1975. Effects of air pollution on plant ultrastructure. In Mudd, J.B. & Kozlowski, T.T. (eds) *Response of Plants to Air Pollution*: 179-194. Academic Press, New York.

Tsukahara, H., Kozlowski, T.T. & Shanklin, J. 1985. Tolerance of *Pinus densiflora*, *Pinus thunbergii*, and *Larix leptolepis* seedlings to SO_2. *Plant and Soil* **88**: 385-398.

Tulloch, A.P. 1987. Epicuticular waxes of *Abies balsamea* and *Picea glauca*: occurrence of long-chain methyl esters. *Phytochem.* **26**: 1041-1043.

Umbach, D.M. & Davis, D.D. 1984. Severity and frequency of SO_2-induced leaf necrosis on seedlings of 57 tree species. *For. Sci.* **30**: 587-596.

Usher, R.W. & Williams, W.T. 1982. Acid pollution toxicity to eastern white pine in Indiana and Wisconsin. *Plant Dis.* **66**: 199-204.

Vijayaraghavan, R., Sugendran, K., Pant, S.C., Husain, K. & Malhotra, R.C. 1991. Dermal intoxication of mice with bis(2-chloroethyl)sulphide and the protective effect of flavonoids. *Toxicology* **69**: 35-42.

Wagner, G.J. 1982. Compartmentation in plant cells: the role of the vacuole. *Recent Adv. Phytochem.* **16**: 1-45.

Wagner, H. 1985. New plant phenolics of pharmaceutical interest. *Ann. Proc. Phytochem. Soc. Eur.* **25**: 409-427.

Warrington, S. 1987. Relationship between SO_2 dose and growth of the pea aphid (*Acyrthosiphon pisum*), on peas. *Environ. Pollut.* **43**: 155-162.

Weigel, H.J., Halbwachs, G. & Jaeger, H.J. 1989. The effects of air pollutants on forest trees from a plant physiological view. *Z. Pflanzenkr. Pflanzenschutz.* **96**: 203-217.

Wilkinson, H.P. 1979. The plant surface (mainly leaf). In Metcalfe, C.R. & Chalk, L. (eds) *Plant Anatomy*, 2nd edition, Vol 1. Clarendon, Oxford.

Wilson, A. 1986. Flavonoid pigments and wing color in *Melanargia galathea*. *J. Chem. Ecol.* **12**: 49-68.

Wilson, G.B. & Bell, J.N.B. 1985. Studies on the tolerance to SO_2 of grass populations in polluted areas. III. Investigations on the rate of development of tolerance. *New Phytol.* **100**: 63-77.

Wisniewski, J. 1982. The potential acidity associated with dews, frosts and fogs. *Water Air Soil Pollut.* **17**: 361-377.

Wollenweber, E. 1984. The systematic implication of flavonoids secreted by plants. In Rodriguez, E., Healey, P.L. & Mehta, I. (eds) *Biology and Chemistry of Plant Trichomes*: 53-69. Plenum, New York.

Wollenweber, E. 1985. Exkret-Flavonoide bei höheren Pflanzen arider Gebiete. *Plant Syst. Evol.* **150**: 83-88.

Wollenweber, E. & Dietz, U.H. 1981. Occurrence and distribution of free flavonoid aglycones in plants. *Phytochemistry* **20**: 869-932.

Yang, Y.S., Skelly, J.M. & Chevone, B.I. 1983. Sensitivity of eastern white pine clones to acute doses of ozone, sulphur dioxide, or nitrogen dioxide. *Phytopathology* **73**: 1234-1237.

Yunus, M., Kulshreshtha, K., Dwivedi, A.K. & Ahmad K.J. 1982. Leaf surface traits of *Ipomoea fistulosa* Mart. ex Choisy as indicators of air pollution. *New Bot.* **9**: 39-45.

Zangerl, A.R. & Berenbaum, M. 1987. Furanocoumarins in wild parsnip. Effects of photosynthetic radiation, ultraviolet light and nutrients. *Ecology* **68**: 516-520.

Zobel, A.M., 1975. Mixoploidy of tannin coenocytes in *Sambucus racemosa* L. *Acta Soc. Bot. Pol.* **44**: 491-500.

Zobel, A.M. 1985a. Ontogenesis of tannin coenocytes in *Sambucus racemosa* L. I. Development of the coenocyte from mononucleate tannin cells. *Ann. Bot.* **55**: 765-773.

Zobel, A.M. 1985b. Ontogenesis of tannin coenocytes in *Sambucus racemosa* L. II. Mother tannin cells. *Ann. Bot.* **56**: 91-104.

Zobel, A.M. 1986. Sites of localization of phenolics in tannin coenocytes in *Sambucus racemosa* L. *Ann Bot.* **57**: 801-810.

Zobel, A.M. 1989. Origin of nodes and internodes in plant shoots. II. Models of origin of nodes and internodes. *Ann. Bot.* **63**: 208-220.

Zobel, A.M. 1991. Effect of the change from field to greenhouse environment on the linear furanocoumarin levels of *Ruta chalepensis*. *J. Chem. Ecol.* **17**: 21-27.

Zobel, A.M. 1993. Phenolic compounds: An answer by the plant to air pollution. *Proceedings of Second Princess Chulabhorn Meeting, Thailand.*

Zobel, A.M. & Brown, S.A. 1988. Determination of furanocoumarins on the leaf surface of *Ruta graveolens* with an improved extraction technique. *J. Nat. Prod.* **51**: 941-946.

Zobel, A.M. & Brown, S.A. 1989. Histological localization of furanocoumarins in *Ruta graveolens* shoots. *Can. J. Bot.* **67**: 915-921.

Zobel, A.M. & Brown, S.A. 1990a. Dermatitis-inducing furanocoumarins on the leaf surfaces of eight species of rutaceous and umbelliferous plants. *J. Chem. Ecol.* **16**: 693-700.

Zobel, A.M. & Brown, S.A. 1990b. Seasonal changes of furanocoumarin concentrations in leaves of *Heracleum lanatum*. *J. Chem. Ecol.* **16**: 1623-1634.

Zobel, A.M. & Brown, S.A. 1991a. Psoralens on the surface of seeds of Rutaceae and fruits of Umbelliferae and Leguminosae. *Can. J. Bot.* **69**: 485-488.

Zobel, A.M. & Brown, S.A. 1991b. Dermatitis-inducing psoralens on the surface of seven medicinal plant species. *J. Toxicol., Cutaneous and Ocular Toxicol.* **10**: 223-231.

Zobel, A.M. & Brown, S.A. 1993. Influence of low-intensity ultraviolet radiation on extrusion of furanocoumarins to the leaf surface. *J. Chem. Ecol.* **19**: 939-952.

Zobel, A.M. & Brown, S.A. 1995. Coumarins in the interactions between the plant and its environment. *Allelopathy J.* **2**: 9-20.

Zobel, A.M. & Hrazdina, G. 1992. Chalcone synthase localization in early stages of plant development. II. The enzyme localization in shoot apices of three species. *Ann. Bot.* **70**: 423-427.

Zobel, A.M. & Hrazdina, G. 1995. Chalcone synthase localization in early stages of plant development. I. Immunohistochemical method employing plasmolysis for localization of the enzyme in cytoplasm of epidermal cells of illuminated buckwheat hypocotyls. *Biotech. Histochem.* **70**: 1-6.

Zobel, A.M. & March, R.E. 1993. Autofluorescence reveals different histological localizations of furanocoumarins in fruits of some Umbelliferae and Leguminosae. *Ann. Bot.* **71**: 251-255.

Zobel, A.M. & Nighswander, J.E. 1990. Accumulation of phenolic compounds in the necrotic areas of Austrian and red pine needles due to salt spray. *Ann. Bot.* **66**: 629-640.

Zobel, A.M. & Nighswander, J.E. 1991. Accumulation of phenolic compounds in the necrotic areas of Austrian and red pine needles due to sulphuric acid spray as a bioindicator of air pollution. *New Phytol.* **117**: 565-576.

Zobel, A.M., Kuras, M. & Tykarska, T. 1989. Cytoplastic and apoplastic location of phenolic compounds in the covering tissue of the *Brassica napus* radicle between embryogenesis and germination. *Ann. Bot.* **64**: 149-157.

Zobel, A.M., Brown, S.A. & Nighswander, J.E. 1991a. Effect of simulated acid spray on furanocoumarin concentrations on the surface of *Ruta graveolens* leaves. *Ann. Bot.* **67**: 213-218.

Zobel, A.M., Wang, J., March, R.E. & Brown, S.A. 1991b. Occurrence of other coumarins with psoralen, xanthotoxin and bergapten on leaf surfaces of seven plant species. *J. Chem. Ecol.* **17**: 1859-1870.

Zobel, A.M., Sandstrom, T., Nighswander, J.E. & Dudka, S. 1993. Uptake of metals by aquatic plants and changes in phenolic compounds. *Heavy Metals Environ.* **1**: 210-213.

Zobel, A.M., Chen, Y. & Brown, S.A. 1994a. Influence of UV on furanocoumarins in *Ruta graveolens* leaves. *Acta Hort.* 355-360.

Zobel, A.M., Crellin, J., Borwn, S.A. & Glowniak, K. 1994b. Concentrations of furanocoumarins under stress conditions and their histological localization. *Acta Hort.* 510-515.

10 Biochemical Basis for the Toxicity of Ozone

J.B. MUDD

INTRODUCTION

The effects of ozone on biochemical molecules have been studied for at least 85 years (Harries & Langheld, 1907). The effects on biological organisms have been studied for more than 40 years (Giese *et al.*, 1952), but there is no agreement on the biochemical basis of ozone toxicity. Many mechanisms have been suggested but all these have been either disputed or unconfirmed by other reports.

The reactions of ozone with biological materials or organisms can be classified in three ways:

(a) Reaction in the solid phase, e.g., with cuticular components of plant leaves, or the exposed skin in mammals.
(b) Reactions in the gas phase, e.g., reactions with the hydrocarbons emitted by plants, and by animals.
(c) Reactions in the liquid phase. These require the dissolution of ozone in aqueous media, followed by reaction with lipids, proteins, or other cellular components.

Reactions in the solid phase and in the gas phase may be specially relevant in plant damage. The mechanisms of toxicity in the liquid phase apply equally well to plants and to animals. The large surface area of the mammalian lung is analogous to the large internal area of a plant leaf. Both have the physiological purpose of rapid and efficient gas exchange. Both the cells of the alveolar epithelium and the cells in the substomatal cavity of plant leaves are covered with a thin film of aqueous solution. Harvesting of the intercellular washing fluid of barley leaves allows the demonstration of at least 50 proteins on two-dimensional electrophoresis (Pfanz *et al.*, 1990). Compounds in the thin film are susceptible to oxidation by ozone. Cross *et al.* (1992) have examined the reaction of ozone with blood serum (used as a surrogate for the lung-lining fluid) and found that while both ascorbic acid and uric acid are oxidized by ozone, the oxidation of uric acid is more important quantitatively. These results suggest two possibilities: (1) toxic reaction products generated in the fluid layer may migrate into the cells overlain by the fluid, and (2) the ozone dose reaching the cells will be attenuated by the reaction of ozone in the fluid layer.

Plant Response to Air Pollution. Edited by Mohammad Yunus and Muhammad Iqbal

REACTION IN THE SOLID PHASE

It is generally considered that the point of entry of air pollutants into the plant leaf is through the stomata. However, reactions also take place at the cuticular surface. The cuticle is composed of two basic components. Cutin overlies the membrane of the cells and is composed mainly of polymers of hydroxy-fatty acids in ester linkages. The second component is the lipid of the cuticle, which both impregnates the cutin and is overlaid as epicuticular lipids. These lipids contain a number of different structures: alkanes, branched alkanes, alkenes, esters of long-chain alcohols and fatty acids, in some cases glycolipids, and some terpenoid compounds. The synthesis of these compounds is presumably a property of the epithelial cells.

Kerstiens & Lendzian (1989) have measured the deposition rates of ozone in the leaves of several plants. There is considerable variation from plant to plant. For example, the deposition rate from an ozone exposure at 400 nl l^{-1} is 6.5 and 27 ml $g^{-1}s^{-1}$ in *Ficus elastica* and *Lycopersicon esculentum*, respectively. It was calculated that flux of ozone through the cuticle is less than 1/10 000 of that through the stomata.

Even though this rate of flux appears to be negligible, there are effects of ozone on the appearance of epicuticular lipids on leaf surfaces after exposure to ozone. Barnes *et al.* (1988) have described the effects of ozone on the epicuticular lipids of Norway spruce (*Picea abies*). In the unexposed plants the epicuticular lipid has a crystalline array of fine tubules, but the epicuticular lipid in the exposed plants has a melted amorphous structure, which tends to block stomata and hence photosynthesis. Miller *et al.* (1969) have reported that photosynthesis in Ponderosa pine (*Pinus ponderosa*) is inhibited with a concomitant decrease in carbohydrate content of the leaves. They conclude that premature leaf senescence may be related to the decreased carbohydrate content.

REACTIONS IN THE GAS PHASE

Elstner & Osswald (1984) reported that the damage of spruce leaves might be associated with the generation of hydrogen peroxide and aldehydes as a result of ozone reaction with ethylene. Gäb *et al.* (1985) discovered the formation of hydroxy-methyl hydroperoxide and bis-(hydroxymethyl)-peroxide. They concluded that these compounds were formed by the ozonolysis of alkenes emitted by the plant material. However, other alkenes of biogenic and anthropogenic origin, e.g., isoprene, would be equally susceptible to oxidation by ozone with the production of organic peroxides. In the case of the specific products mentioned above the alkene would be ethylene. The ozonolysis gives rise to an aldehyde fragment and a biradical which reacts with water to form the hydroxymethylhydroperoxide (HMHP). In classical ozonolysis schemes, this hydroxyhydroperoxide breaks down to give aldehyde and hydrogen peroxide; however, there is sufficient evidence to show that under certain conditions the hydroxyhydroperoxide is quite stable (Hellpointner & Gäb, 1989).

It should be noted that these reactions take place in the gas phase in a humid atmosphere. Mehlhorn & Wellburn (1987) reported that the synthesis of ethylene was intimately related to injury caused by ozone. They showed that the exposure of pea seedlings to increasing concentrations of ozone resulted in the production of increasing amounts of ethylene and increased foliar symptoms. Inhibition of ethylene production by aminoethoxyvinylglycine (AVG) caused a decrease in ethylene production and a decrease in foliar injury. In contrast, Mehlhorn (1990) has reported that the promotion of ascorbate peroxidase by ethylene prevents injury by hydrogen peroxide, ozone and paraquat. It appears that ethylene can have opposite effects on ozone sensitivity depending on temporal separation of the ethylene availability. Taylor *et al.* (1988) measured stomatal conductance, and carbon dioxide assimilation in the presence of AVG and ozone stress. In the presence of AVG the production of ethylene was inhibited compared to an ozone treatment without AVG. Stomatal conductance and carbon dioxide assimilation were less inhibited by ozone in the presence of AVG than in its absence. Taylor *et al.* concluded that the production of ethylene, when plants are stressed by ozone, serves as a messenger for the effects on stomatal physiology and other effects dependent on stomatal aperture. The enhancement of ozone damage in the presence of ethylene could be because of direct ozonolysis of the ethylene or as a consequence of the physiological changes induced by ethylene.

Pell and co-workers (1992) have detected a sensitivity of ribulose-1, 5-bisphosphate carboxylase (RuBP carboxylase) to ozone. There appears to be an aggregation of the enzyme in the presence of ozone (see also Nie *et al.*, 1993). Both groups of researchers believe that something other than ozone is causing the inhibition. Hewitt *et al.* (1990a) propose that the inhibitor is a product of ethylene oxidation.

Mehlhorn *et al.* (1990), using spin trapping techniques, detected radicals during the reaction of ozone with ethylene. Radical production in this reaction, and others involving unsaturated compounds, have been quantified by Atkinson *et al.* (1992). The method was to include a large amount of cyclohexane in the reaction mixture. Cyclohexane does not react with ozone but does react with OH radicals, producing cyclohexanol and cyclohexanone. The total amount of these products give the amount of OH radicals produced. In the case of ethylene, the yield of radicals was 0.12, for isoprene 0.27, and for myrecene it was 1.15. The OH radicals are derived from the biradical formed by the degradation of the initial ozonide. These reactions were carried out in the gas phase, but the yields were not changed in the presence of water vapour. Similarly, Paulson & Seinfeld (1992) found that the reaction of ozone with 1-octene gave a yield of 0.45 hydroxyl radicals over ozone reacted. However, the octene reacted could be accounted for by 80% heptanal and 10% stabilized Criegee biradical.

Hellpointner & Gäb (1989) have measured the amounts of organic peroxides in rain-water. They found that hydrogen peroxide was the predominant peroxide, but at acid pHs (6.1 and below) the organic peroxides made a significant contribution. Hewitt *et al.* (1990b) proposed that the toxicity of ozone can be attributed to the peroxides produced when alkenes, particularly isoprene, which is emitted in profuse amounts in some plants (Monson & Fall, 1989; Silver & Fall, 1991), react with ozone. They further pointed out that the hydroperoxides are quite toxic to some plant enzymes and inhibition or inactivation of these enzymes, e.g., peroxidase, may be the ultimate cause of ozone toxicity (Marklund, 1973). This hypothesis implies an important role for the activity of peroxidase in the maintenance of health of the plant. If the hydroxyhydroperoxide can diffuse to the chloroplast, it may be predicted that ascorbic acid peroxidase would be inhibited resulting in the build-up of toxic levels of hydrogen peroxide. However, Tanaka *et al.* (1982) have demonstrated that the toxic effects of hydrogen peroxide simulate those of sulphur dioxide, and it does not seem likely that the mechanism of action of ozone and sulphur dioxide can be the same, especially since neither sulphur dioxide nor hydrogen peroxide are toxic to RuBP carboxylase.

It should be pointed out that the production of hydrocarbons by plants can have two different effects. When released to the atmosphere the biogenic compounds can be significant contributors to the formation of ozone by the classical photochemical reactions involving oxides of nitrogen and hydrocarbons (Seinfeld *et al.*, 1991). In the gas spaces of plants, however, the emphasis has been laid on the consumption of ozone to produce the organic peroxides. These apparently contradictory positions may be reconciled by the decreased illumination within the gas spaces of the leaf, especially in the depths of plant canopies, which then would decrease the photochemical reactions giving rise to ozone.

Hewitt & Kok (1991) have reported the formation and presence of peroxides in the atmosphere and in laboratory experiments. In the experimental work nine hydrocarbons were used: ethene, propene, 1-butene, isoprene, *cis*-2-butene, α-pinene, β-pinene, 2-carene and limonene. The experiments were done in a 12-litre flask with a large excess of water. Nine peroxides were found as reaction products. No hydrogen peroxide was found in the reactions of ethene, propene, 1-butene and isoprene. In these cases the major product was HMHP. For 2-*cis*-butene the major product was hydrogen peroxide. Results of Simonaitis *et al.* (1991) differ from those of Hewitt & Kok (1991). All alkenes tested gave rise to hydrogen peroxide and organic peroxides. Simonaitis *et al.* also found that the terpenoid compounds give rise to organic peroxides which were not seen in the study of Hewitt & Kok. The reasons for these differences are not obvious except that the experimental conditions were markedly different. Hewitt & Kok used high concentrations of alkenes and a large amount of water in a glass reaction vessel. Simonaitis used a Teflon vessel with water vapour at 1×10^4 ppm and ozone at 58–406 ppb, and initial alkene concentrations at 40–900 ppb.

Reactions of ozone with alkenes in the gas phase may be an important factor in the toxicity of ozone, particularly in plants. The products of these ozonolyses may be the substances which eventually disrupt plant metabolism. This mechanism of toxicity should be focused not only on the gases escaping from the plants. There is a significant amount of interstitial wall fluid (IWF) in plants. This liquid contains both proteins and compounds of low molecular weight. Some of these compounds are susceptible to oxidation by ozone, and the oxidation products may be responsible for intracellular responses.

REACTIONS IN THE LIQUID PHASE

Reactions of ozone in aqueous media immediately raise the question of the active species in oxidation. Is it ozone itself, or is it oxidants derived from ozone?

In discussions of the biological effects of ozone, it is frequently assumed that radicals generated from ozone are the active agents. This assumption needs careful evaluation.

Weiss (1935) discussed his own and previous work on the decomposition of ozone in aqueous solutions. He concluded that the decomposition rate of ozone was a function of hydroxyl ion concentration. The reaction:

$$O_3 + OH^- \rightarrow O_2^- + HOO \quad (1)$$

was postulated on the basis of spectroscopic evidence: "absorption in the blue region of the visible, due presumably to KOO and O_2^- ion." Gorbenkov-Germanov & Kozlova (1973, 1974) re-examined this reaction 37 years later using modern spectroscopic instrumentation. Ozone was added to 8 M solutions of KOH at −50°C. The first products were $\dot{O}H$ radical and O_3^-. At −50°C there was degradation to O_2^-.

$$O_3 + OH^- \rightarrow O_3^- + OH\cdot \quad (2)$$

$$3O_3^- + H_2O \rightarrow O_2^- + 3O_2 + OH^- \quad (3)$$

It is important to note that the reaction of ozone with water is much more rapid at alkaline pH than at acid pH. The half-life at pH 4 is 30 minutes, whereas at pH 10 it is 0.33 minutes (Hoigné & Bader, 1976). The capability of ozone to oxidize various substrates also varies with pH. At high pH the oxidation is non-selective, very complete and independent of temperature, typical of radical-induced oxidations. At pH ≤ 7 the oxidation is selective and dependent on temperature, indicating a reaction mechanism not involving radicals (Gorbenkov-Germanov *et al.*, 1973).

There has been a great deal of research on the use of ozone in water purification, much of it being relevant to biological effects of air pollutants. Hoigné & Bader (1975), coming to a similar conclusion to that of Gorbenkov-Germanov *et al.* (1973), indicate two reaction pathways for ozone:

$$O_3 + S \rightarrow \text{direct oxidation of S, highly selective} \quad (4)$$

$$O_3 + OH^- \rightarrow O_3^- + OH\cdot \quad (5a)$$

$$OH\cdot + S \rightarrow \text{fast oxidation of S, low selectivity} \quad (5b)$$

$$O_2^- \overset{H^+}{\rightleftarrows} HOO\cdot \quad (5c)$$

$$HOO\cdot + S \rightarrow \text{oxidation/reduction of S, highly selective} \quad (5d)$$

Reaction (4) is typical of ozone oxidation at neutral pH, whereas the radical reactions of (5) are typical of those at high pH.

Reactions of ozone in the liquid phase require initial dissolution of ozone in an aqueous medium; however, the action of ozone may be attributed to diffusion of ozone into lipid, e.g., the lipid bilayer of the plasma membrane. Hill (1971) has determined the uptake of several pollutant gases by a canopy of vegetation. The amount of pollutant taken up is clearly related to the water solubility of the gas. However, ozone is anomalous in that much more ozone is taken up than would be predicted from its water solubility. In the normal case, the gas in the gaseous phase is in equilibrium with the gas in the dissolved phase. The latter phase can be depleted by reaction with chemicals in the liquid phase.

$$CO_{2(gas)} \overset{K}{\rightleftarrows} CO_{2(dissolved)} + \text{reactants} \overset{k_2}{\rightarrow} \text{products}$$

In the case of ozone the reaction k_2 is so rapid that the first equilibrium is never established. This conception is consistent with the observation of Laisk *et al.* (1989) that the concentration of ozone in leaf air spaces is close to zero.

Reaction of ozone with lipids

Two mechanisms have been suggested for the action of ozone and lipids in these liquid phases: peroxidation and ozonolysis.

The important highlights about peroxidation of unsaturated fatty acids are:

(a) The reaction starts in the presence of radicals.
(b) Molecular oxygen is taken up from the medium and incorporated into the products.
(c) The radical attack is at the methylene carbon adjacent to a double bond.

The radical can migrate causing the movement of the double bond position. In the case of polyunsaturated fatty acids, this generates a conjugated double bond system which has a characteristic absorbance at 233 nm. In fatty acids with three or more double bonds, cyclic peroxides are formed which degrade with the release of malonaldehyde.

The important points to realize about the ozonolysis reaction are:

(a) The reaction is not radical-initiated.
(b) The oxygen in the end products is derived from ozone.
(c) The basic stoichiometry is the production of two aldehydes and one hydrogen peroxide for each double bond broken, although this stoichiometry can be modified by the further oxidation of the aldehydes to acids by ozone.

It should also be noted that in polyunsaturated fatty acids, the cleavage of methylene-interrupted double bond systems gives rise to malonaldehyde. Thus the detection of malonaldehyde cannot be used to distinguish peroxidation and ozonolysis.

Although radicals have been detected either directly or by spin trapping methods, in ozonolysis reactions (Goldstein *et al.*, 1968; Grimes *et al.*, 1983;

Mehlhorn *et al.*, 1990), there is usually no quantification of the number of radicals produced per molecule of ozone. The intermediation of hydroxide radicals, hydroperoxy radicals and superoxide has been invoked, but not proven. The appealing nature of the hypothesis that the biochemical basis for ozone toxicity lies in the ability to initiate lipid peroxidation has had many followers (Pryor *et al.*, 1976), but the most common method used to detect lipid peroxidation is the measure of malonaldehyde, which we have seen to be not exclusively a characteristic of lipid peroxidation. When both hydrogen peroxide and organic peroxide are assayed after ozonolysis of lipid systems, there is none of the organic peroxide which would be predicted by a peroxidation mechanism (Heath & Tappel, 1976). Barber & Thomas (1978) measured directly the reaction of hydroxyl radicals with phosphatidylcholine bilayers. They concluded that the bilayer structure prevented the entry and movement of radicals in the bilayer. On prolonged exposure the head groups of the lipid reacted with the hydroxyl radicals, a reaction not observed when biological systems are exposed to ozone.

On the other hand, there is now sufficient evidence that ozonolysis is the pathway of reaction with biological materials. First indications of the ozonolysis were the detection of hydrogen peroxide in biological materials exposed to ozone (Mudd *et al.*, 1969). Later, Teige *et al.* (1974) demonstrated that the reaction of ozone with unsaturated lipids produced stoichiometric amounts of hydrogen peroxide and the aldehydes (and carboxylic acids) predicted for ozonolysis. Pryor *et al.* (1991b) have also reported that aldehydes and hydrogen peroxide are products of ozonolysis and these may be responsible for the toxicity of ozone (Pryor & Church, 1991).

A definitive distinction between ozonolysis and peroxidation was made by Santrock *et al.* (1992). They synthesized ozone labelled with ^{18}O and treated fatty acids and phospholipids with this labelled ozone. Treatment of 1-palmitoyl-2-oleoyl-*sn*-glycero-3-phosphocholine (POPC) gave rise to hydrogen peroxide in which both oxygen atoms were derived from ozone, aldehydes in which 50% of the oxygen was derived from ozone, and carboxylic acids in which one of the oxygen atoms was derived from aldehyde and the other from ozone. These results clearly demonstrate that ozonolysis and not peroxidation is the mechanism of action of ozone with lipids. Another very useful result obtained in these experiments was the detection of the products of the ozonolysis on an HPLC column for the separation of molecular species of PC. This showed the presence at the *sn*-2 position of nine carbon fragments: the aldehyde, the carboxylic acid and the hydroxyhydroperoxide. These products can now be used to determine quantitatively the oxidation of fatty acids when biological systems are treated with ozone. Previous attempts to measure this decrease in fatty acids after ozone treatment have been suspect because of the difficulty in measuring a small putative decrease in a large number. The separation of reaction products now makes it possible to measure the oxidation of fatty acids as an increase in a number which was previously zero.

The effect of ozone on the survival rate of *Escherichia coli* has been reported. Ohlrogge & Kernan (1983) made use of an *E. coli* mutant which was unable to synthesize unsaturated fatty acids. The supplementation of the medium with fatty acids determined the fatty-acid composition of the membrane. Thus the naturally occurring oleic acid (18:1) was replaced with either a cyclopropane (saturated) fatty acid or a more unsaturated fatty acid, linoleic acid (18:2). Analysis of membrane fatty acids confirmed that the composition had been altered by the supplements. The various forms of *E. coli* were exposed to ozone. The growth rate was inhibited by the exposure to ozone, but this was greatest in the cells containing the saturated fatty acid. This result makes it unlikely that the susceptibility of biological organisms is related to the degree of fatty acid unsaturation in the membranes.

It seems most likely that if the biochemical basis for the toxicity of ozone were reaction with lipids, then the chemical mechanism would be ozonolysis. But reaction with lipids is not the only target for the action of ozone. In the biological membrane the double bonds of the fatty acids are spatially protected from ozone.

Reaction of ozone with amino acids and proteins

Giese *et al.* (1952) reported that the absorption spectrum of bovine serum albumin, serum globulin, egg albumin and gelatin was changed after treatment by ozone. The principal changes were a decrease in absorbance at 280 nm and an increase in absorbance at longer wavelengths. These authors then exposed tryptophan to ozone, finding the decrease in absorbance at 280 nm and an increase in the absorbance in the range 300–350 nm. Similar changes were reported for indole. Oxidations of tyrosine and phenylalanine were also examined, and an increase in absorbance was observed. Previero and co-workers (1963a, b) have exposed amino acids dissolved in formic acid to ozone. They found that cysteine and cystine were converted to cysteic acid, methionine was converted to methionine sulphone, tryptophan to kynurenine, and tyrosine to aspartic and oxalic acids. All other amino acids were not reactive with ozone. Previero *et al.* (1963a, b) proceeded to show that tryptophan in peptide linkage was equally well oxidized by ozone with *N*-formyl-kynurenine being the first product. Previero & Scoffone (1963) then showed that cysteine and cystine in peptide linkage was converted to cysteic acid. By exposing human globulin, horse globulin, trypsin, lysozyme and gramicidin to ozone, Previero & Bordignon (1964) concluded that tryptophan and methionine were equally reactive and more so than cysteine and tyrosine.

Mudd *et al.* (1969) examined the oxidation of amino acids and proteins by ozone. These experiments were done in buffered solutions close to neutrality. Some differences with Previero's work were found. The order of reactivity was found to be methionine = tryptophan > tyrosine > histidine > phenylalanine > cystine. The oxidation of tyrosine, histidine and cystine were influenced by pH, being greatest at the highest pH tested, 8.7. The reduction in biotin binding by avidin after exposure of the avidin to ozone was attributed to the oxidation of tryptophan. Reaction rates of ozone and amino acids were determined by Pryor *et al.* (1984). They found that the reactivities were in the order cysteine > tryptophan > methionine > tyrosine > histidine. Methionine was equally reactive as 3-hexenoic acid, which was used as a model lipid substrate.

Meiners *et al.* (1977) measured the oxidation of tryptophan, 5-hydroxy-tryptophan, 5-hydroxy-tryptamine and 5-hydroxy-hydroxy-indoleacetic acid by observing the elimination of fluorescence. All these indole compounds were oxidized by ozone. In the case of tryptophan, it was predicted that the ozonolysis of the double bond in the nitrogen-containing ring would give rise to aldehydic and keto compounds and hydrogen peroxide. The production of *N*-formyl-kynurenine is consistent with this idea; however, the yield of hydrogen peroxide was of the order of 20% of the tryptophan oxidized, less than would be predicted by the ozonolysis reaction. The products from these various oxidations did not themselves show any new fluorescence peaks.

The reaction of ozone with proteins has had considerable study. Todd (1958) exposed urease, peroxidase, catalase and papain to ozone. Papain was 100 times more sensitive than the other enzymes, consistent with the requirement of a sulphhydryl group for the catalytic activity of papain. The other enzymes required approximately equal amounts of ozone for 50% inhibition. The *in vivo* inactivation of lysozyme in pulmonary lavage of animals exposed to ozone was reported by Holzman *et al.* (1968). This work was followed by a study of the inactivation of acid phosphatase, lysozyme and β-glucuronidase in macrophage isolated from animal lungs after ozone exposure (Hurst *et al.*, 1970). Acid phosphatase and lysozyme were both inhibited by the ozone exposure. There was recovery of these enzymes 24 hours after the exposure.

The reaction of ozone with lysozyme has also been studied in great detail. Previero *et al.* (1967) treated lysozyme with ozone in anhydrous formic acid. Residues 108 and 111 were oxidized without any loss of enzymic activity. The inactivation of lysozyme by ozone at physiological pHs is reportedly related to only one of the six tryptophan residues (see Kuroda *et al.*, 1975). In this work it could not be certain whether the residue oxidized was 62 or 63. However, Sakiyama (1977) developed a method for the cleavage of the polypeptide at the *N*-formyl-kynurenine moiety and was, therefore, able to demonstrate that the susceptible tryptophan is 62. Sakiyama *et al.* (1978) demonstrated that with ozonolysis of tryptophan in methanol, which behaves as a participating solvent, the methylhydroxy analogue of the hydroxy hydroperoxide is formed at the *N* position. This compound can be reduced by dimethylsulphide to *N*-formyl-kynurenine, but under acid conditions, kynurenine is formed.

Tamaoki *et al.* (1978) exploited the capability to convert *N*-formyl-kynurenine to kynurenine in ribonuclease T_1. The single tryptophan at position 59 was oxidized by ozone to NFK-RNase which was deformylated under acid conditions to Kyn-RNase. The Kyn-RNase was four times as active as NFK-RNase at pH 4.75, and the CD spectrum indicated conformational change in NFK-RNase but not in Kyn-RNase. It was concluded that oxidation of tryptophan did not affect the active site directly but inactivated the enzyme by a conformational change.

Dooley & Mudd (1982) confirmed the findings of Kuroda *et al.* (1975) that lysozyme is inactivated by ozone but found in addition that the specificity for tryptophan 62 is true only at low ozone concentrations. At higher concentrations of ozone, the oxidation of as many as four tryptophan residues is necessary to inactivate the enzyme completely.

The oxidation of tryptophan in the proteins of membranes has been shown by Goldstein & McDonagh (1975). Human red cells were exposed to ozone and tryptophan fluorescence measured subsequently. Levels of ozone that produced

50–70% decrease in protein fluorescence oxidized only 10% of the total tryptophan in the protein. Energy from the fluorescence of tryptophan was not transferred to the hydrophobic fluorescent probe, 8-anilo-1-naphthalene sulphonic acid. This indicates that the fluorescent tryptophan residues which are oxidized by ozone are in the hydrophilic rather than the hydrophobic regions of the protein.

NADH is readily oxidized by ozone in a classical ozonolysis reaction (Mudd *et al.*, 1974). The reaction of ozone with NADH in the presence of tryptophan, methionine and glutathione was measured. In mixtures of NADH and tryptophan, there was no oxidation of tryptophan until NADH was almost completely oxidized. Methionine does not affect the oxidation of NADH quantitatively or qualitatively. The oxidation of methionine in the mixture with NADH was inhibited by more than 50%. In low concentrations of NADH and glutathione both compounds are oxidized with a slight preference for NADH. NADH is an intracellular component, and these competition experiments become redundant if the amount of ozone that passes through the biological membrane is insignificant.

Since the reactivity of ozone makes it unlikely that a large fraction of administered ozone passes through the plasma membrane (Pryor, 1992), there should be some focus on reaction of ozone with the components of the plasma membrane. If ozone reacted with the lipids of the plasma membrane, the nine-carbon fragments retained at the *sn*-2 position would destabilize the membrane and cause lysis of the cells. However, the treatment of human red cells with ozone did not result in lysis (Teige *et al.*, 1974). In contrast, if PC was treated with ozone, the reaction products did lyse red blood cells. The conclusion is that ozone does not have ready access to the lipids of these plasma membranes. One could go further and postulate that the reaction of ozone with plasma membranes is mostly, if not exclusively, with the proteins, particularly the extra-membranal parts, of the plasma membrane. In the exposure of plant cell cultures to ozone, no effect could be detected in the capabilities of metabolism or regeneration of cell wall (Grimes *et al.*, 1983). It is conceivable, however, that there were effects on the plasma membrane which were not assayed.

Even though the fraction of ozone crossing the biological membrane may be small, it could have dramatic effects. In human red blood cells, the most dramatic enzyme inhibition is of glyceraldehyde-3-phosphate dehydrogenase (GPDH), which is generally considered to be a marker for the inside of the membrane (Steck & Kant, 1974). It is also notable that the most obvious effect of the exposure of alveolar type 2 cells to ozone is an inhibition of lipid synthesis, which is clearly a property of enzymes inside the cell (Haagsman *et al.*, 1985). In contrast, some proteins which clearly are associated with the outside of the membrane, e.g., the glucose transporter (Peters *et al.*, 1993), are not inhibited by ozone. It may be concluded that the effects of ozone on proteins are determined by three factors: (a) accessibility, (b) reactivity and (c) susceptibility. It is predicted that acetylcholine esterase has reactive amino acid residues, but these reactivities do not affect the enzymic properties of the enzyme which is, therefore, not susceptible. On the other hand, GPDH is not very accessible, but one of the oxidizable amino acid residues (cysteine) is at the active site of the enzyme and, therefore, is extremely susceptible (Knight & Mudd, 1984).

In the context of toxic effects of the products of ozone oxidation, Pryor and colleagues (1991a) have suggested that the aldehydes and hydrogen peroxide produced by ozonolysis of lipids may be the mediators of ozone toxicity. They have also demonstrated the lysis of red blood cells by addition of the aldehydes and hydrogen peroxide (Pryor *et al.*, 1991b). Kindya & Chan (1976) reported that the Na/K ATPase of red blood cells was inhibited by ozone. The same authors (Chan *et al.*, 1977), using red cell membrane fragments, also reported that the ATPase was completely inhibited by low doses of ozone, but the activity could be fully restored by the addition of phosphatidylserine (PS). With higher doses of ozone, only 40% of the ATPase was reactivable by phosphatidylserine, but further protection could be obtained by the prior incubation with semicarbazide, which was proposed to be active because of the trapping of aldehydic products of ozone oxidation. These studies did not resolve the difference between effects of ozone on lipids or on proteins. A subsequent work emphasized the reaction of ozone with phospholipid to yield products which were inhibitory to the Na/K ATPase (Kesner *et al.*, 1979). But this method of ozone exposure determines the result. The authors concluded that aldehydes and ketones produced by the ozonolysis were the inhibitory compounds. But it is not proven that these compounds are produced when ozone inhibits the ATPase of the intact cells.

EFFECTS ON PLANT PHYSIOLOGY AND BIOCHEMISTRY

The effects of ozone on plant physiology have been thoroughly reviewed (Heath, 1987; Pell *et al.*, 1992).

There are two subjects to be discussed in this review:

(a) Inhibition of RuBP carboxylase by ozone.
(b) Genetic basis for the resistance/susceptibility of plants to ozone.

Inhibition of RuBP carboxylase

The inhibition of RuBP carboxylase by ozone (Farage *et al.*, 1991; Pell *et al.*, 1992) is fundamental to the prevention of carbon dioxide fixation, and removes a sink for the assimilatory capacity (ATP and NADPH) produced by the light reactions of photosynthesis. The elevation of carbon dioxide concentration in the internal gas spaces of the leaf is proposed to be the reason for closure of stomata (Farage *et al.*, 1991). The lack of a pathway from the light reactions to the fixation of carbon dioxide may also lead to photoinhibition of the photosystems and hence to the premature senescence of the leaf. In field experiments, the inhibition of single-leaf photosynthesis and premature senescence of leaves are well-documented phenomena (Reich, 1983; Roper & Williams, 1989).

Genetic basis for plant resistance/susceptibility to ozone

It has frequently been reported that different cultivars of agricultural and horticultural plants vary greatly in their susceptibility to ozone (Engle &

Gabelman, 1966; Cameron & Taylor, 1973; Gesalman & Davis, 1978; Lee *et al.*, 1984; Heggestad, 1991). In the case of sweet corn, the heritability of resistance was determined (Cameron, 1975). Crosses of resistant and susceptible varieties gave susceptible F_1s. The distribution in F_2s was skewed to lower injury. The relationship of ozone resistance to the genetic composition of the cultivars makes it possible to correlate restriction fragment length polymorphism (RFLP) with the resistance trait. Progress in this analysis over the next few years will make it possible for plant breeders to screen new hybrids for the ozone-resistance traits. Since agricultural yields are reduced by air pollution by as much as 20% (Heck *et al.*, 1987), there is a prospect that ozone-resistant varieties of agricultural crops may be available in the future.

Genetic engineering is beginning to play a role in the study of the biochemical basis for ozone toxicity. Since there have been suggestions that superoxide, hydrogen peroxide and hydroxyl radicals are involved in ozone toxicity, it has been suggested that enzymes which detoxify superoxide or hydrogen peroxide may be responsible for resistance to ozone (Lee & Bennett, 1982; Matters & Scandalios, 1987).

$$2O_2^- + 2H^+ \rightarrow O_2 + H_2O_2 \quad \text{superoxide dismutase} \quad (6)$$

$$H_2O_2 + AA \rightarrow 2H_2O + DHA \quad \text{ascorbate peroxidase} \quad (7)$$

$$DHA + 2GSH \rightarrow AA + GSSG \quad \text{glutathione/ascorbate reductase} \quad (8)$$

$$GSSG + NADPH + H^+ \rightarrow 2GSH + NADP^+ \quad \text{glutathione reductase} \quad (9)$$

However, transformation of tobacco plants with the gene for the copper–zinc form of superoxide dismutase (chloroplast form) did not confer resistance to ozone (Pitcher *et al.*, 1991). Other researchers have transformed tobacco with glutathione reductase and found that the transformed plants are more resistant to paraquat, but not to ozone (Aono *et al.*, 1991). Since paraquat is known to stimulate the formation of superoxide, the result of protection with glutathione reductase is reasonable. However, the implication of the ozone result is that superoxide is not involved in the toxicity.

CONCLUSIONS

The biochemical mechanism of ozone toxicity to plants is still not understood.

There is an increasing belief that the first products of ozonolysis migrate to sensitive sites in the cell where inhibition occurs. If this hypothesis proves to be correct, we still need to know the compound(s) which are first oxidized by ozone. Candidates could be ethylene or other alkenes. But there are many other compounds in the interstitial wall fluid which need to be considered. These include the phenyl propenoic acids used as precursors of lignin biosynthesis.

It seems likely that some fraction of the ozone to which plants are exposed reaches the plasma membrane. Further studies are needed to determine whether reactions with the lipid or protein component are predominant.

There is some doubt that ozone can pass through the plasma membrane and have direct effects on the cellular contents. This point needs clarification. The inhibition by ozone of sensitive metabolic functions such as lipid synthesis and GPDH suggests that some small fraction of ozone does penetrate as far as the cytoplasmic contents of the cell.

Molecular genetic investigations of cultivar differences in resistance to ozone are ripe for investigation. These investigations could be either the RFLP mapping or by direct transformation of plants with genes encoding enzymes of putative protection. The former method is laborious, but of great use to seed breeders. The second method depends on a calculated or intuitive conclusion about protective enzymes. In the two cases mentioned in the text, the guesses about protective enzymes have not been verified. There will undoubtedly be more research of this type, hopefully with a positive result.

REFERENCES

Aono, M., Kubo, A., Saji, H., Natori, T., Tanaka, K. & Kondo, N. 1991. Resistance to active oxygen toxicity of transgenic *Nicotiana tabacum* that expresses the gene for glutathione reductase from *Escherichia coli*. *Plant Cell Physiol.* **32**: 691-697.

Atkinson, R., Aschmann, S.M., Arey, J. & Shorees, B. 1992. Formation of OH radicals in the gas phase reactions of O_3 with a series of terpenes. *J. Geophys. Res.* **97**: 6065-6073.

Barber, D.J.W. & Thomas, J.K. 1978. Reactions of radicals with lecithin bilayers. *Radiation Res.* **74**: 51-65.

Barnes, J.D., Davison, A.W. & Booth, T.A. 1988. Ozone accelerates structural degradation of epicuticular wax on Norway spruce needles. *New Phytol.* **110**: 309-318.

Cameron, J.W. 1975. Inheritance of sweet corn for resistance to acute ozone injury. *J. Amer. Soc. Hort. Sci.* **100**: 577-579.

Cameron, J.W. & Taylor, O.C. 1973. Injury to sweet corn inbreds and hybrids by air pollutants in the field and by ozone treatments in the greenhouse. *J. Environ. Qual.* **2**: 387-389.

Chan, P.C., Kindya, R.J. & Kesner, L. 1977. Studies on the mechanism of ozone inactivation of erythrocyte membrane ($Na^+ + K^+$)-activated ATPase. *J. Bio. Chem.* **252**: 8537-8541.

Cross, C.E., Motchnik, P.A., Bruener, B.A., Jones, D.A., Kaur, H., Ames, B.N. & Halliwell, B. 1992. Oxidative damage to plasma constituents by ozone. *Federation European Biochem. Soc.* **298**: 269-272.

Dooley, M.M. & Mudd, J.B. 1982. Reaction of ozone with lysozyme under different exposure conditions. *Arch. Biochem. Biophys.* **218**: 459-471.

Elstner, E.F. & Osswald, W. 1984. Fichtsterben in "Reinluftgebieten" struktur resistenzverlust. *Naturw. Rundschau* **37**: 52-61.

Engle, R.L. & Gabelman, W.H. 1966. Inheritance and mechanism for resistance to ozone damage in onion, *Allium cepa* L. *Amer. Soc. Hort. Sci.* **89**: 423-430.

Farage, P.K., Long, S.P., Lechner, E.G. & Baker, N.R. 1991. The sequence of change within the photosynthetic apparatus of wheat following short-term exposure to ozone. *Plant Physiol.* **95**: 529-535.

Gäb, S., Hellpointner, E., Turner, W.V. & Koŕte, F. 1985. Hydroxymethyl hydroperoxide and bis(hydroxymethyl) peroxide from gas-phase ozonolysis of naturally occurring alkenes. *Letters to Nature* **316**: 535-536.

Giese, A.C., Leighton, H.L. & Bailey, R. 1952. Changes in the absorption spectra of proteins and representative amino acids induced by ultraviolet radiations and ozone. *Arch. Biochem. Biophys.* **40**: 71-84.

Goldstein, B.D. & McDonagh, E.M. 1975. Effect of ozone on cell membrane protein fluorescence. I. *In vitro* studies utilizing the red cell membrane. *Environ. Res.* **9**: 179-186.

Goldstein, B.D., Balchum, O.J., Demopoulos, H.B. & Duke, P.S. 1968. Electron paramagnetic resonance spectroscopy. *Arch. Environ. Health* **17**: 46-49.

Gorbenkov-Germanov, D.S. & Kozlova, I.V. 1973. Mechanism of the decomposition of ozone in basic aqueous media. *Doklady Acad. Nauk USSR* **210**: 851-854. Plenum Publishing Corp., New York, pp. 456-458.

Gorbenkov-Germanov, D.S. & Kozlova, I.V. 1974. Intermediate decomposition products of ozone in alkaline aqueous media investigated by electron spin resonance. *Russian J. of Phys. Chem.* **48**: 93-95.

Gorbenkov-Germanov, D.S., Vodop'yanova, N.M., Kharina, N.M. & Gorodnov, M.M. 1973. Oxidation of some saturated organic compounds initiated by the catalytic decomposition of ozone in alkaline solution. *Doklady Acad. Nauk USSR* **210**: 1121-1123. Plenum Publishing Corp., New York, pp. 472-474.

Grimes, H.D., Perkins, K.K. & Boss, W.F. 1983. Ozone degrades into hydroxyl radical under physiological conditions. *Plant Physiol.* **72**: 1016-1020.

Haagsman, H.P., Schummans, E.A.J.M., Alink, G.M., Batenburg, J.J. & van Golde, L.M.G. 1985. Effects of ozone on phospholipid synthesis by alveolar type II cells. *Exp. Lung Res.* **9**: 67-84.

Harries, C. & Langheld, K. 1907. Über das Verhalten des Caseins gegen Ozon. *Hoppe-Seyler's Zeitschrift f. physiol. Chemie* **51**: 342-372.

Heath, R.L. 1987. Biochemical mechanisms of pollutant stress. In Heck, W.W. Taylor, O.C. & Tingey, D.T. (eds) *Assessment of Crop Loss from Air Pollutants*: 259-286. Elsevier Applied Science Publishers, London.

Heath, R.L. & Tappel, A.L. 1976. A new sensitive assay for the measurement of hydroperoxides. *Anal. Biochem.* **76**: 184-191.

Heck, W.W., Taylor, O.C. & Tingey, D.T. (eds) 1987. *Assessment of Crop Loss from Air Pollutants*. Elsevier Applied Science Publishers, London.

Heggestad, H.E. 1991. Origin of Bel-W3, Bel-C, and Bel-B tobacco varieties and their use as indicators of ozone. *Environ. Pollut.* **74**: 264-291.

Hellpointner, E. & Gäb, S. 1989. Detection of methyl, hydroxymethyl and hydroxyethyl hydroperoxides in air and precipitation. *Nature* **337**: 631-634.

Hewitt, C.N. & Kok, G.L. 1991. Formation and occurrence of organic hydroperoxides in the troposphere: Laboratory and field observations. *J. Atmos. Chem.* **12**: 181-194.

Hewitt, C.N., Kok, G.L. & Fall, R. 1990a. Hydroperoxides in plants exposed to ozone mediate air pollution damage to alkene emitters. *Nature* **344**: 56-58.

Hewitt, C.N., Lucas, P., Wellburn, A.R. & Fall, R. 1990b. Chemistry of ozone damage to plants. *Chemistry & Industry*: 478-481.

Hill, A.C. 1971. Vegetation: A sink for atmospheric pollutants. *J. Air Pollut. Control Assoc.* **21**: 341-346.

Hoigné, J. & Bader, H. 1975. Ozonation of water: Role of hydroxyl radicals as oxidizing intermediates. *Science* **190**: 782-783.

Hoigné, J. & Bader, H. 1976. The role of hydroxyl radical reactions in ozonation processes in aqueous solutions. *Water Res.* **10**: 377-386.

Holzman, R.S., Gardner, D.E. & Coffin, D.L. 1968. In vivo inactivation of lysozyme by ozone. *J. Bacteriol.* **96**: 1562-1566.

Hurst, D.J., Gardner, D.E. & Coffin, D.L. 1970. Effect of ozone on acid hydrolases of the pulmonary alveolar macrophage. *J. Reticuloendothelial Soc.* **8**: 288-300.

Kerstiens, G. & Lendzian, K.J. 1989. Interactions between ozone and plant cuticles. I. Ozone deposition and permeability. *New Phytol.* **112**: 13-19.

Kesner, L., Kindya, R.J. & Chan, P.C. 1979. Inhibition of erythrocyte membrane (Na^+ + K^+)-activated ATPase by ozone-treated phospholipids. *J. Biol. Chem.* **254**: 2705-2709.

Kindya, R.J. & Chan P.C. 1976. Effect of ozone on erythrocyte membrane adenosine triphosphatase. *Biochim. Biophys. Acta.* **429**: 608-615.

Knight, K.L. & Mudd, J.B. 1984. The reaction of ozone with glyceraldehyde-3-phosphate dehydrogenase. *Arch. Biochem. Biophys.* **229**: 259-269.

Kuroda, M., Sakiyama, F. & Narita, K. 1975. Oxidation of tryptophan in lysozyme by ozone in aqueous solution. *J. Biochem.* **78**: 641-651.

Laisk, A., Kull, O. & Moldau, H. 1989. Ozone concentration in leaf intercellular air spaces is close to zero. *Plant Physiol.* **90**: 1163-1167.

Lee, E.H. & Bennett, J.H. 1982. Superoxide dismutase. A possible protective enzyme against ozone injury in snap beans (*Phaseolus vulgaris* L.). *Plant Physiol.* **69**: 1444-1449.

Lee, E.H., Jersey, J.A., Gifford, C. & Bennett, J.H. 1984. Differential ozone tolerance in soybean and snap beans: analysis of ascorbic acid in O_3-susceptible and O_3-resistant cultivars by high-performance liquid chromatography. *Environ. Exp. Bot.* **24**: 331-341.

Marklund, S. 1973. Mechanisms of the irreversible inactivation of horseradish peroxidase caused by hydroxymethylhydroperoxide. *Arch. Biochem. Biophys.* **154**: 614-622.

Matters, G.L. & Scandalios, J.G. 1987. Synthesis of isozymes of superoxide dismutase in maize leaves in response to O_3, SO_2 and elevated O_2. *J. Exper. Bot.* **38**: 842-852.

Mehlhorn, H. 1990. Ethylene-promoted ascorbate peroxidase activity protects plants against hydrogen peroxide, ozone and paraquat. *Plant, Cell Environ.* **13**: 971-976.

Mehlhorn, H. & Wellburn, A.R. 1987. Stress ethylene formation determines plant sensitivity to ozone. *Nature* **327**: 417-418.

Mehlhorn, H., Tabner, B.J. & Wellburn, A.R. 1990. Electron spin resonance evidence for the formation of free radicals in plants exposed to ozone. *Physiol. Plant.* **79**: 377-383.

Meiners, B.A., Peters, R.E. & Mudd, J.B. 1977. Effects of ozone on indole compounds and rat lung monoamine oxidase. *Environ. Res.* **14**: 99-112.

Miller, P.R., Parmeter, J.R. Jr, Flich, B.H. & Martinez, C.W. 1969. Ozone dosage response of ponderosa pine seedlings. *J. Air Pollut. Control Assoc.* **19**: 435-438.

Monson, R.K. & Fall, R. 1989. Isoprene emission from aspen leaves. Influence of environment and relation to photosynthesis and photorespiration. *Plant Physiol.* **90**: 267-274.

Mudd, J.B., Leavitt, R., Ongun, A. & McManus, T.T. 1969. Reaction of ozone with amino acids and proteins. *Atmos. Environ.* **3**: 669-682.

Mudd, J.B., Leh, F. & McManus, T.T. 1974. Reaction of ozone with nicotinamide and its derivatives. *Arch. Biochem. Biophys.* **161**: 408-419.

Nie, G.-Y., Tomasevic, M. & Baker, N.R. 1993. Effects of ozone on the photosynthetic apparatus and leaf proteins during leaf development in wheat. *Plant, Cell Environ.* **16**: 643-651.

Ohlrogge, J.B. & Kernan, T.P. 1983. Toxicity of activated oxygen: lack of dependence on membrane unsaturated fatty acid oxidation. *BBRC* **113**: 301-308.

Paulson, S.E. & Seinfeld, J.H. 1992. Atmospheric photochemical oxidation of 1-octene: OH, O_3, and $O(^3P)$ reactions. *Environ. Sci. Technol.* **26**: 1165-1173.

Pell, E.J., Eckardt, N. & Enyedi, A.J. 1992. Timing of ozone stress and resulting status of ribulose biophosphate carboxylase/oxygenase and associated net photosynthesis. *New Phytol.* **120**: 397-405.

Peters, R.E., Inman, C., Oberg, L. & Mudd, J.B. 1993. Effect of ozone on metabolic activities of rat hepatocytes and mouse peritoneal macrophage. *Tox. Lett.* **69**: 53-61.

Pfanz, H., Dietz, K.-J., Weinerth, I. & Oppmann, B. 1990. Detoxification of sulphur dioxide by apoplastic peroxidases. In Rennenberg, H. Brunold, C. De Kok, L.J. & Stulen, I. (eds) *Sulphur Nutrition and Sulphur Assimilation in Higher Plants*: 229-233. SPB Academic Publishing, The Hague.

Pitcher, L.H., Brennan, E., Hurley, A., Dunsmuir, P., Tepperman, J.M. & Zilinskas, B.A. 1991. Overproduction of petunia chloroplastic copper/zinc superoxide dismutase does not confer ozone tolerance in transgenic tobacco. *Plant Physiol.* **97**: 452-455.

Previero, A. & Bordignon, E. 1964. Modifica controllata di triptofano, metionina, cistina e tirosina in peptidi naturali e proteine. *Gazzetta* **94**: 630-638.

Previero, A. & Scoffone, E. 1963. Indagini sulla struttura delle proteine. Nota XII. Ossidazione dei legami disolfuro con ozone in peptidi della cistina. *Gazzetta Chimica Italiana* **93**: 859-866.

Previero, A., Scoffone, E., Pajetta, P. & Benassi, C.A. 1963a. Indagini sulla struttura delle proteine. Nota X. Comportamento degli amminoacidi di fronte all'ozono. *Gazzetta Chimica Italiana* **93**: 841-848.

Previero, A., Scoffone, E., Benassi, C.A. & Pajetta, P. 1963b. Indagini sulla struttura delle proteine. Nota XI. Modificazioni del residuo del triptofano in catena peptidica. *Gazzetta Chimica Italiana* **93**: 849-858.

Previero, A., Coletti-Previero, M.-A. & Jollès, P. 1967. Localization of non-essential tryptophan residues for the biological activity of lysozyme. *J. Mol. Biol.* **24**: 261-268.

Pryor, W.A. 1992. How far does ozone penetrate into the pulmonary air/tissue boundary before it reacts? *Free Radical Biol. Med.* **12**: 83-88.

Pryor, W.A. & Church, D.F. 1991. Aldehydes, hydrogen peroxide, and organic radicals as mediators of ozone toxicity. *Free Radical Biol. Med.* **11**: 41-46.

Pryor, W.A., Stanley, J.P., Blair, E. & Cullen, G.B. 1976. Autoxidation of polyunsaturated fatty acids. Part I. Effect of ozone on the autoxidation of neat methyl linoleate and methyl linolenate. *Arch. Environ. Health* **31**: 201-210.

Pryor, W.A., Giamalva, D.H. & Church, D.F. 1984. Kinetics of ozonation. 2. Amino acids and model compounds in water and comparisons to rates in nonpolar solvents. *J. Am. Chem. Soc.* **106**: 7094-7100.

Pryor, W.A., Miki, M., Das, B. & Church, D.F. 1991a. The mixture of aldehydes and hydrogen peroxide produced in the ozonation of dioleoyl phosphatidylcholine causes hemolysis of human red blood cells. *Chem. — Biol. Interactions* **79**: 41-52.

Pryor, W.A., Das, B. & Church, D.R. 1991b. The ozonation of unsaturated fatty acids: aldehydes and hydrogen peroxide as products and possible mediators of ozone toxicity. *Chem. Res. Toxicol.* **4**: 341-348.

Reich, P.B. 1983. Effects of low concentrations of O_3 on net photosynthesis, dark respiration, and chlorophyll contents in aging hybrid poplar leaves. *Plant Physiol.* **73**: 291-296.

Roper, T.R. & Williams, L.E. 1989. Effects of ambient and acute partial pressures of ozone on leaf net CO_2 assimilation of field-grown *Vitis vinifera* L. *Plant Physiol.* **91**: 1501-1506.

Sakiyama, F. 1977. Selective chemical cleavage of the peptide bond at *N'*-formylkynurenine in ozone-oxidized hen egg-white lysozyme. *J. Biochem.* **82**: 365-375.

Sakiyama, F., Masuda, N., Nakazawa, T. & Katsuragi, Y. 1978. Quantitative ozone-oxidation of tryptophan to *N'*-formylkynurenine and kynurenine. *Chem. Letts.*: 893-896.

Santrock, J., Gorski, R.A. & O'Gara, J.F. 1992. Products and mechanism of the reaction of ozone with phospholipids in unilamellar phospholipid vesicles. *Chem. Res. Toxicol.* **5**: 134-141.

Seinfeld, J.H. (Chairman) 1991. *Rethinking the Ozone Problem in Urban and Regional Air Pollution.* Committee on Tropospheric Ozone Formation and Measurement, National Research Council, National Academy Press, Washington, DC.

Silver, G.M. & Fall, R. 1991. Enzymatic synthesis of isoprene from dimethylallyl diphosphate in aspen leaf extracts. *Plant Physiol.* **97**: 1588-1591.

Simonaitis, R., Olszyna, K.J. & Meagher, J.F. 1991. Production of hydrogen peroxide and organic peroxides in the gas phase reactions of ozone with natural alkenes. *Geophys. Res. Letts.* **18**: 9-12.

Steck, T.L. & Kant, J.A. 1974. Preparation of impermeable ghosts and inside-out vesicles from human erythrocyte membranes. *Methods Enzymol.* **31**: 172-180.

Tamaoki, H., Sakiyama, F. & Narita, K. 1978. Chemical modification of ribonuclease T_1 with ozone. *J. Biochem.* **83**: 771-781.

Tanaka, K., Otsubo, T. & Kondo, N. 1982. Participation of hydrogen peroxide in the inactivation of calvin-cycle SH enzymes in SO_2-fumigated spinach leaves. *Plant Cell Physiol.* **23**: 1009-1018.

Taylor, G.E. Jr, Ross-Todd, B.M. & Gunderson, C.A. 1988. Action of ozone on foliar gas exchange in *Glycine max* L. Merr: a potential role for endogenous stress ethylene. *New Phytol.* **11D**: 301-317.

Teige, B., McManus, T.T. & Mudd, J.B. 1974. Reaction of ozone with phosphatidylcholine liposomes and the lytic effect of products on red blood cells. *Chem. Phys. Lipids* **12**: 153–171.
Todd, G.W. 1958. Effect of low concentrations of ozone on the enzymes catalase, peroxidase, papain and urease. *Physiol. Plant.* **11**: 457–463.
Weiss, J. 1935. Investigations on the radical HO_2 in solution. *Trans. Faraday Soc.* **31**: 668–681.

11 Responses of Plants to Atmospheric Sulphur

H. RENNENBERG & C. HERSCHBACH

INTRODUCTION

Sulphur is available to plants as sulphate in the soil to the roots, and/or as gaseous sulphur compounds in the atmosphere to the shoot. Despite significant reduction of emission from anthropogenic sources due to legislation (Whelphale, 1992), sulphur dioxide (SO_2) is still the most important atmospheric sulphur compound taken up by vegetation (Fowler, 1985). Whereas sulphate influx into the roots is mainly an active, carrier-mediated process (Rennenberg, 1984; Cram, 1990; Clarkson *et al.*, 1993) controlled by sulphur nutrition of the plant (Herschbach & Rennenberg, 1991, 1994), influx of SO_2 is mainly a diffusive flux through the stomata dependent on stomatal aperture (Fowler, 1985; Rennenberg & Polle, 1994). Thus, avoidance of SO_2 influx into plants can only be achieved at the cost of reduced photosynthesis and water-vapour exchange, and hence, reduced growth. It is, therefore, not surprising that avoidance by stomatal closure is not the only strategy of plants to cope with SO_2 in the atmosphere. Depending on the species and the environmental conditions, exposure to SO_2 may result in stomatal closure, stomatal opening, or no reaction of the stomata at all (Black, 1985).

Sulphur dioxide that has entered the gas phase of a leaf rapidly dissolves in the aqueous phase of the cell wall (apoplastic space) and reacts with water to form bisulphite. The equilibrium of this reaction is far on the side of bisulphite at the pH of the apoplastic space (Rennenberg & Polle, 1994). The fate of bisulphite in this compartment of the leaf has been a matter of debate in recent literature. There are only few studies in which bisulphite or sulphite has been found in the apoplastic space. This may either be explained by rapid metabolic conversion in the apoplastic space or by rapid transport out of the apoplastic space through the plasmalemma into the cytoplasm. From the few data available, the plasmalemma appears to be impermeable for bisulphite or sulphite; but SO_2 (aq.) may pass the plasmalemma by diffusion and, therefore, by a relatively slow process (Spedding *et al.*, 1980). Rapid metabolic conversion of sulphite to sulphate may be achieved by apoplastic sulphite oxidase activity (Pfanz *et al.*, 1990). Because this conversion requires molecular oxygen, hydrogen peroxide and monophenols, the reaction is thought to be catalysed by apoplastic peroxidases. Recently, the significance of this reaction in the apoplastic space has been doubted, because (1) apoplastic

Plant Response to Air Pollution. Edited by Mohammad Yunus and Muhammad Iqbal

sulphite oxidation is competitive with the oxidation of phenolics in lignin formation, and (2) peroxidase activity is inhibited by sulphite during sulphite oxidation with mMolar k_i-values (Takahama *et al.*, 1992). However, Rennenberg & Polle (1994) calculated that the capacity of apoplastic fluids for enzymatic conversion of sulphite to sulphate is three orders of magnitude higher than the influx of atmospheric SO_2 into the leaf via the stomata at 30 ppb SO_2. Therefore, apoplastic sulphite oxidase activity may still be able to convert the bulk of the sulphite produced from SO_2 influx into the leaves at the atmospheric SO_2 concentrations plants experience even in polluted environments. The finding that sulphite accumulates in the xylem sap of red spruce trees fumigated in the winter with low SO_2 concentrations is consistent with this assumption, as low metabolic activities may have prevented rapid conversion of sulphite in the apoplastic space during this time of the year (Wolfenden *et al.*, 1991).

Sulphate produced from sulphite in the apoplastic space will enter a large apoplastic sulphate pool (0.5–2mM; Wolfenden *et al.*, 1991) transported into the leaf with the transpiration stream. From this pool sulphate derived from SO_2 will enter the symplasm by active carrier-mediated transport through the plasmalemma (Rennenberg, 1984; Cram, 1990; Clarkson *et al.*, 1993). Thus, SO_2 exposure may result in enhanced apoplastic sulphate availability and elevated sulphate influx into leaf cells. It may be considered as uncontrolled sulphur fertilization at the nutritional level and will, therefore, interact with processes that control the plant's sulphur nutrition. The present review aims to summarize present knowledge on the interaction of SO_2 influx with sulphur nutrition and its regulation.

SULPHUR EMISSIONS BY PLANTS IN RESPONSE TO SULPHUR DIOXIDE EXPOSURE

Almost three decades ago it was first observed that plants exposed to SO_2 in the atmosphere re-emit part of the sulphur taken up (Materna, 1966). Numerous studies have shown that hydrogen sulphide (H_2S) is the predominant sulphur compound released by plants in response to SO_2 and that this release of H_2S is not restricted to a particular species, but is a general phenomenon in higher plants (Rennenberg, 1991). Investigations on the metabolic paths of H_2S formation in plants have shown that besides cysteine, both sulphate and sulphite may be substrates of H_2S synthesis. Conversion of sulphate and sulphite to H_2S was found to be light dependent and was, therefore, assumed to take place in the chloroplast (Rennenberg, 1991). Thus, irrespective of whether SO_2 entering the apoplastic space directly permeates the plasmalemma, or is first metabolized to sulphate in the apoplastic space and then actively transported into the cytoplasm, it may partially be converted in the chloroplast to H_2S that can be emitted into the atmosphere.

Several pieces of evidence suggest that emission of H_2S in general, and in response to SO_2 in particular, is a means by which plants get rid of excess sulphur and, thus, control sulphur nutrition (Rennenberg, 1991). Apparently, H_2S formation is part of an intracellular sulphur cycle (Figure 11.1) operating to control homeostatically the cysteine concentration of plant cells (Rennenberg

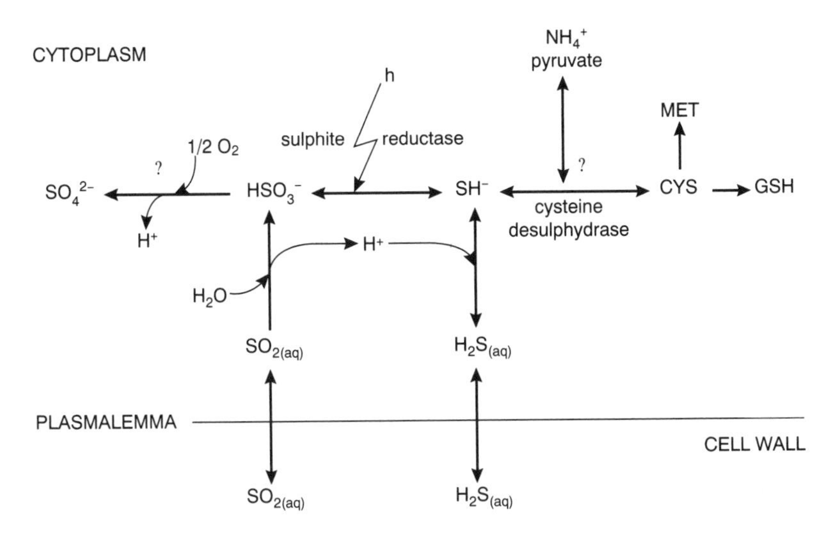

Figure 11.1. The fate of SO_2 in the cytoplasm. CYS = cysteine; MET = methionine; GSH = reduced glutathione (from Rennenberg & Polle, 1994; by permission of Chapman & Hall)

et al., 1982). This cycle will enable plant cells to use as much of the SO_2 taken up as required for the synthesis of sulphur compounds for growth and development, when other sources of sulphur are scarce. At the same time it would allow the cysteine pool to be maintained at an appropriate concentration by emitting excess sulphur as H_2S into the atmosphere. Obviously, this strategy of plants to control sulphur nutrition is dependent on a sufficient concentrations gradient for H_2S between the plant and the atmosphere. It will, therefore, not operate when a significant H_2S concentration is present in the atmosphere (cf. De Kok, 1990). The presence of an intracellular sulphur cycle may be the metabolic basis of numerous observations showing that SO_2 can be beneficial to plant when other sulphur sources are scarce (Rennenberg, 1984; De Kok, 1990).

At the metabolic level, an important consequence of reductive conversion of sulphite to sulphide rather than oxidative conversion to sulphate is a prevention of SO_2-derived acidification. Obviously, the overall acidification of the cell resulting from SO_2 influx cannot be prevented by emission of H_2S: compared to SO_2 influx, H_2S emission is of minor significance from a quantitative point of view. It accounts for only 7–15% of the SO_2 absorbed depending on the species or cultivar analysed (Rennenberg, 1991). Still, the capacity for H_2S emission was found to be connected with the sensitivity of plants or tissues to SO_2 (Rennenberg, 1991). In cucurbit plants, young leaves absorbing high amounts of SO_2 are more resistant to SO_2 than mature leaves absorbing smaller amounts of this air pollutant. In both young and mature leaves *ca.* 60% of the SO_2 absorbed was oxidized to sulphate, but young leaves emitted H_2S at a rate up to 100-fold higher than mature leaves (Filner *et al.*, 1984). In duckweeds, H_2S was emitted at a rate *ca.* four times higher in sulphite-tolerant (*Lemna valdiviana*) than in sulphite-susceptible species (*Lemna gibba, Spirodela oligorhiza*) (Takemoto *et al.*, 1986). In the field, SO_2-tolerant conifers (e.g., *Picea pungens*) emitted higher amounts of H_2S in response to atmospheric SO_2 than SO_2-sensitive conifers (e.g., *Picea abies*) (Kindermann *et al.*, 1995).

From these findings the question has to be addressed, why reduction of sulphite to sulphide and subsequent emission of H_2S can be a mechanism of SO_2 tolerance, when only a small part of the SO_2 absorbed undergoes this conversion. Light-dependent reduction of SO_2-derived sulphite to sulphide was found to proceed in the chloroplast. This compartment is also thought to be an important site of SO_2-mediated injury (Rennenberg, 1991). It may, therefore, be assumed that H_2S emission reflects sulphite reduction required to prevent injury in this cellular compartment, e.g., by acidification due to sulphite oxidation. Obviously, other strategies to achieve SO_2 resistance may operate in other cellular compartments.

ACCUMULATION AND LONG-DISTANCE TRANSPORT OF SULPHUR COMPOUNDS DERIVED FROM ATMOSPHERIC SULPHUR DIOXIDE

Numerous investigations have shown that plants are able to use atmospheric sulphur including SO_2 for growth and development (Rennenberg, 1984; De Kok, 1990). Spinach plants containing *ca.* 10 μmol sulphur per gram fresh weight, may require an atmospheric concentration of 400 ppb SO_2 to meet their needs for sulphur at a growth rate of 15% (De Kok, 1990). For slow-growing plants like spruce trees this requirement may be significantly lower (Slovik *et al.*, 1992a, b). However, inorganic and organic sulphur already accumulate in plant shoots, exposed to much lower than the calculated atmospheric SO_2 concentrations. Most of the sulphur accumulated in shoots exposed to SO_2 could be attributed to sulphate (Rennenberg, 1984; De Kok, 1990). This was not only found in fumigation experiments, but also in field studies with trees growing in different pollution climates (e.g., Kaiser *et al.*, 1993a). Still, part of the SO_2 absorbed by foliage was found to be reduced and assimilated rapidly into various organic sulphur fractions of the shoot. Elevated levels of water-soluble, non-protein thiol have commonly been observed upon SO_2 exposure (Rennenberg, 1984; De Kok, 1990; Herschbach *et al.*, 1995a). In field studies, enhanced thiol levels were also found in needles of spruce trees collected in strongly SO_2-polluted regions (Grill *et al.*, 1982). In many plants elevated thiol levels, as a consequence of SO_2 exposure, could be attributed to glutathione; in spinach, enhanced cysteine and γ-glutamylcysteine levels have also been observed (De Kok, 1990). Whereas short-term SO_2 exposure of spruce trees did not result in elevated organic sulphur content, protein-sulphhydryl increased in trees growing in SO_2-polluted areas (Grill *et al.*, 1980; Kaiser *et al.*, 1993a). This may be an indication that SO_2 absorbed by plants in the field can partially be diluted by growth, provided other nutrients are available in sufficient amounts.

Sulphur originating from SO_2 and absorbed by the leaves was found to be translocated to the roots and appeared in various sulphur fractions of the root including sulphate. Also part of the SO_2 assimilated into organic sulphur compounds in the shoot was transported as glutathione and cysteine to the roots (De Kok, 1990). Since both sulphate and glutathione are mobile in the phloem (Rennenberg, 1984; Rennenberg & Lamoureux, 1990), and are readily exchanged between phloem and xylem (Biddulph, 1956; Schupp *et al.*, 1992; Schneider *et al.*, 1994), circulation of oxidized and reduced sulphur can be assumed in the vascular system of the stem. Upon SO_2 exposure, leaves may feed sulphate and glutathione into this circular pathway; the growing parts of the stem and the roots may provide a sink for glutathione. Removal of sulphate from the vascular system by the roots is apparently too low to result in measurable accumulation of sulphate in this organ (De Kok, 1990). Thiols like glutathione, however, can accumulate in measurable amounts in the roots upon exposure of the shoot to SO_2 (Herschbach *et al.*, 1995a).

INTERACTION OF ATMOSPHERIC SULPHUR DIOXIDE WITH PROCESSES IN THE ROOT

Recent studies indicate that SO_2 exposure of the shoot significantly affects processes in the root. Glutathione was found to inhibit sulphate influx and the percentage of the sulphate taken up that was loaded into the xylem of detached roots (Herschbach & Rennenberg, 1991). Also in intact tobacco plants sulphate uptake and the sulphate transport to the shoot was reduced in the presence of glutathione (Herschbach & Rennenberg, 1994). Accumulation of SO_2 and H_2S-derived glutathione in the roots may, therefore, result in reduced sulphate influx into the roots (Figure 11.2). This assumption has recently been confirmed in fumigation experiments with tobacco and spinach plants (Herschbach *et al.*, 1995a, b). Both sulphur gases, SO_2 and H_2S, led to a reduced uptake and xylem loading of sulphate. This effect was dependent on sulphur nutrition and the plant species analysed. Whereas in spinach plants sulphate uptake was diminished if plants were grown under sulphur deficiency, it was reduced in tobacco plants only if supplied with normal amounts of sulphur. In both species the amount of sulphate transported to the shoot and the relative proportion of sulphate loaded into the xylem were reduced on fumigation with SO_2 and H_2S. However, reduction of these two transport processes was more pronounced upon H_2S then upon SO_2 fumigation. Apparently, both sulphur gases can be used as sulphur sources for sulphur nutrition and can counteract sulphate uptake and sulphate transport to the shoot (Herschbach *et al.*, 1995a, b). In this way, the negative effects of SO_2 absorption by the shoot, i.e., acidification due to sulphite oxidation and accumulation of excess sulphur, can be reduced.

Apparently, acidification due to the oxidation of SO_2-derived sulphite to sulphate in the leaves can be relieved by proton excretion out of the roots (Thomas & Runge, 1992; Kaiser *et al.*, 1993b). This excretion may be as much as 50% of the total amount of acid formed in the reaction of SO_2 absorbed by the shoot with water (Kaiser *et al.*, 1993b). As reductive conversion of sulphite may counteract acidification by *ca.* 40% of the SO_2 absorbed, pH-stat mechanisms in the leaves appear to contribute little to the prevention of acidification. The high excretion of acid by the roots of plants exposed to SO_2 via the shoot is

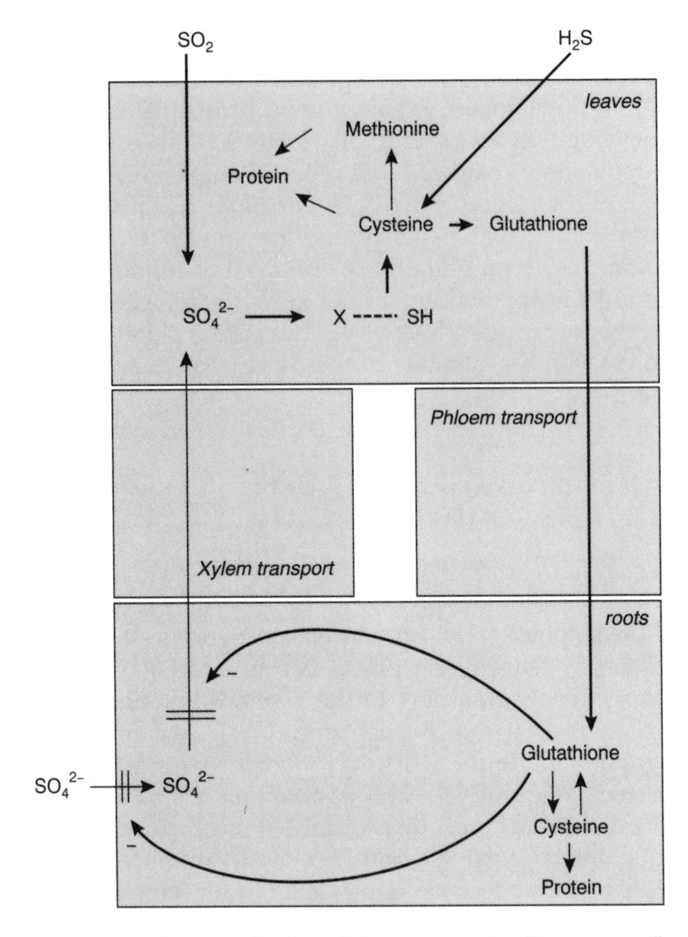

Figure 11.2. Interaction of atmospheric sulphur gases: the "inter-organ" regulation of sulphur nutrition. X - - - - SH = carrier-bound thiol

consistent with the finding that in spruce needles from heavily SO_2-polluted areas appreciable acidification of cellular pH values was not determined (Kaiser *et al.*, 1991). In contrast to previous observations (Garsed & Read, 1977a, b), excretion of acid was not accompanied by excretion of sulphate from the roots of pea and barley plants (Kaiser *et al.*, 1993b). This finding may be the consequence of sulphate lacking in the nutrient solution. Under these conditions, sulphate-transport entities may operate at maximum activity (Rennenberg, 1984; Cram, 1990), thereby preventing any net efflux of sulphate from the roots.

CONCLUSIONS

Plants contain several mechanisms to compensate negative effects of SO_2 exposure. These mechanisms include oxidation of sulphite to sulphate in the apoplastic space, as well as reductive conversion of sulphite to sulphide in the chloroplasts combined with the emission of H_2S and the synthesis of organic sulphur compounds. These compounds may not only be used for growth, but may also be stored in the shoot and the roots. In the roots, SO_2-derived organic sulphur in the form of glutathione may reduce sulphate uptake by the roots and transport into the xylem and may, therefore, facilitate enhanced reductive conversion of sulphite in the shoot. Excretion of protons derived from the reaction of SO_2 with water in the shoot is observed in the roots. This excretion seems to be the major path to prevent acidification upon SO_2 exposure. Excretion of sulphate by the roots has also been observed, but quantitative measures for the excretion of SO_2-derived sulphate from the roots have so far not been reported. At present the capacity of individual processes for the compensation of stress resulting from SO_2 exposure in a particular species is unknown. Therefore, it cannot be decided which of these processes may become limiting under conditions of prolonged SO_2 exposure. In addition, the capacities of SO_2-specific compensation mechanisms have to be evaluated by comparison with the capacity of general compensation mechanisms for oxidative stress (Polle & Rennenberg, 1993). This information is urgently needed to identify the metabolic factor(s) determining the sensitivity of plants for SO_2.

REFERENCES

Biddulph, S.F. 1956. Visual indications of ^{35}S and ^{32}P translocation in the phloem. *Am. J. Bot.* **43**: 143-148.

Black, V.J. 1985, SO_2 effects on stomatal behavior. In Winner, W.E., Mooney, H.A. & Goldstein, R.A. (eds) *Sulphur Dioxide and Vegetation*: 96-117. Stanford University Press, Stanford, CA.

Clarkson, D.T., Hawkesford, M.J. & Davidian, J.C. 1993. Membrane and long-distance transport of sulfate. In De Kok, L.J., Stulen, I., Rennenberg, H., Brunold, C. and Rauser, W.E. (eds) *Sulfur Nutrition and Assimilation in Higher Plants*. SPB Acad. Publ., The Hague.

Cram, W.J. 1990. Uptake and transport of sulfate. In Rennenberg, H., Brunold, C., De Kok, L.J. & Stulen, I. (eds) *Sulphur Nutrition and Sulphur Assimilation in Higher Plants*: 3-11. SPB Acad. Publ., The Hague.

De Kok, L.J. 1990. Sulphur metabolism in plants exposed to atmospheric sulphur. In Rennenberg, H., Brunold, C., De Kok, L.J. & Stulen, I. (eds) *Sulphur Nutrition and Sulphur Assimilation in Higher Plants*: 111-130. SPB Acad. Publ., The Hague.

Filner, P., Rennenberg, H., Sekiya, J., Bressan, R.A., Wilson, L.G., LeCureux, L. & Shimei, T. 1984. Biosynthesis and emission of hydrogen sulfide by higher plants. In Koziol, M.J. & Whatley, F.R. (eds) *Gaseous Air Pollutants and Plant Metabolism*: 291-312. Butterworth, London.

Fowler, D. 1985. Deposition of SO_2 onto plant canopies. In Winner, W.E., Mooney, H.A. & Goldstein, R.A. (eds) *Sulphur Dioxide and Vegetation*: 389-402. Stanford University Press, Stanford, CA.

Garsed, S.G. & Read, D.J. 1977a. Sulphur dioxide metabolism in soybean, *Glycine max* var. *biloxi*. I. The effects of light and dark on uptake and translocation of $^{35}SO_2$. *New Phytol.* **78**: 111-119.

Garsed, S.G. & Read, D.J. 1977b. The uptake and metabolism of $^{35}SO_2$ in plants of different sensitivity to sulphur dioxide. *Environ. Pollut.* **13**: 173-186.

Grill, D., Esterbauer, H., Scharner, M. & Felgitsch, C. 1980. Effect of sulphur-dioxide on protein-SH in needles of *Picea abies*. *Eur. J. For. Pathol.* **10**: 263-267.

Grill, D., Esterbauer, H. & Hellig, K. 1982. Further studies on the effect of SO_2-pollution on the sulphhydryl-system of plants. *Phytopathol. Z.* **104**: 264-271.

Herschbach, C. & Rennenberg, H. 1991. Influence of glutathione on sulphate influx, xylem loading and exudation in excised tobacco roots. *J. Exp. Bot.* **42**: 1021-1029.

Herschbach, C. & Rennenberg, H. 1994. Influence of glutathione (GSH) on net uptake of sulphate and sulphate transport in tobacco plants. *J. Exp. Bot.* **45**: 1069-1076.

Herschbach, C., De Kok, L.J. & Rennenberg, H. 1995a. Net uptake of sulphate and its transport to the shoot in spinach plants fumigated with H_2S or SO_2: Does atmospheric sulphur affect the "inter-organ" regulation of sulphur nutrition? *Botanica Acta* **108**: 41-46.

Herschbach, C., De Kok, L.J & Rennenberg, H. 1995b. Net uptake of sulphate and its transport to the shoot in tobacco plants fumigated with H_2S or SO_2. *Plant & Soil* **175**: 75-84.

Kaiser, W.M., Dittrich, A.P.M. & Heber, U. 1991. Sulfatakkumulation in Fichtennadeln als Folge von SO_2-Belastung. In 2. *Statusseminar der PBWU zum Forschungsschwerpunkt "Waldschäden"*. GSF-Bericht 26/91: 425-437. GSF, Neuherberg.

Kaiser, W., Dittrich, A. & Heber, U. 1993a. Sulfate concentrations in Norway spruce needles in relation to atmospheric SO_2: A comparison of trees from various forests in Germany with trees fumigated with SO_2 in growth chambers. *Tree Physiol.* **12**: 1-13.

Kaiser, W.M., Höfler, M. & Heber, U. 1993b. Can plants exposed to SO_2 excrete sulphuric acid through the roots? *Physiol. Plant.* **87**: 61-67.

Kindermann, G., Hüve, K., Slovik, S., Lux, H. & Rennenberg, H. 1995. Emission of hydrogen sulfide by twigs of conifers — a comparison of Norway spruce [*Picea abies* (L.) Karst.], Scotch pine (*Pinus sylvestris* L.) and blue spruce (*Picea pungens* Engelm.). *Plant & Soil* **168/169**: 421-423.

Materna, M. 1966. Die Ausscheidung des durch Fichtennadeln absorbierten Schwefeldioxids. *Arch. Forstwesen* **15**: 691-692.

Pfanz, H., Dietz, K.-J., Weinerth, J. & Oppmann, B. 1990. Detoxification of sulphur dioxide by apoplastic peroxidases. In Rennenberg, H., Brunold, C., De Kok, L.J. & Stulen, I. (eds) *Sulphur Nutrition and Sulphur Assimilation in Higher Plants*: 229-233. SPB Acad. Publ., The Hague.

Polle, A. & Rennenberg, H. 1993. Significance of antioxidants in plant adaptation to environmental stress. In Fowden, L., Mansfield, T. & Stoddart, J. (eds) *Plant Adaptation to Environmental Stress*: 263-273. Chapman & Hall, London.

Rennenberg, H. 1984. The fate of excess sulphur in higher plants. *Annu. Rev. Plant Physiol.* **35**: 121-153.

Rennenberg, H. 1991. The significance of higher plants in the emission of sulphur compounds from terrestrial ecosystems. In Sharkey, T.D., Holland, E.A. & Mooney, H.A. (eds) *Trace Gas Emissions by Plants*: 217-260. Academic Press, San Diego.

Rennenberg, H. & Lamoureux, G.L. 1990. Physiological processes that modulate the concentration of glutathione in plant cells. In Rennenberg, H., Brunold, C., De Kok, L.J. & Stulen, I. (eds) *Sulphur Nutrition and Sulphur Assimilation in Higher Plants*: 53-65. SPB Acad. Publ., The Hague.

Rennenberg, H. & Polle, A. 1994. Metabolic consequences of atmospheric sulphur influx into plants. In Alscher, R. & Wellburn, A. (eds) *Plant Response to the Gaseous Environment*: 165-180. Chapman & Hall, London.

Rennenberg, H., Sekiya, J., Wilson, L.G. & Filner, P. 1982. Evidence for an intracellular sulphur cycle in cucumber leaves. *Planta* **154**: 516-524.

Schneider, A., Schatten, T. & Rennenberg, H. 1994. Exchange between phloem and xylem during long distance transport of glutathione in spruce trees [*Picea abies* (Karst.) L.]. *J. Exp. Bot.* **45**: 457-462.

Schupp, R., Schatten, T., Willenbrink, J. & Rennenberg, H. 1992. Long-distance transport of reduced sulphur in spruce (*Picea abies* L.). *J. Exp. Bot.* **43**: 1243-1250.

Slovik, S., Kaiser, W.M., Körner, C., Kindermann, G. & Heber, U. 1992a. Quantifizierung der physiologischen Kausalkette von SO_2-Immissionsschäden. (I) Ableitung von SO_2-Immissionsgrenzwerten für akute Schäden an Fichte. *AFZ* **15**: 800-805.
Slovik, S., Heber, U., Kaiser, C., Kindermann, G. & Körner, C. 1992b. Quantifizierung der physiologischen Kausalkette von SO_2-Immissionsschäden. (II) Ableitung von SO_2-Immissionsgrenzwerten für chronische Schäden an Fichte. *AFZ* **15**: 913-920.
Spedding, D.J., Ziegler, I., Hampp, R. & Ziegler, H. 1980. Effect of pH on the uptake of ^{35}S-sulphur from sulfate, sulfite, and sulfide by *Chlorella vulgaris*. *Z. Pflanzenphysiol.* **97**: 205-214.
Takahama, U., Veljovic-Iovanovic, S. & Heber, U. 1992. Effect of the air pollutant SO_2 on leaves. Inhibition of sulfite oxidation in the apoplast by ascorbate and of apoplastic peroxidase by sulfite. *Plant Physiol.* **100**: 261-266.
Takemoto, B.K., Noble, R.D. & Harrington, H.M. 1986. Differential sensitivity of duckweeds (Lemnaceae) to sulphite. II. Thiol production and hydrogen sulphide emission as factors influencing phytotoxicity under low and high irradiance. *New Phytol.* **103**: 541-548.
Thomas, F.M. & Runge, M. 1992. Proton neutralization in the leaves of English oak (*Quercus robur* L.) exposed to sulphur dioxide. *J. Exp. Bot.* **43**: 803-809.
Whelphale, D.M. 1992. An overview of the atmospheric sulphur cycle. In Howarth, R.W., Stewart, J.W.B. & Ivanov, M.V. (eds) *Sulphur Cycling on the Continents*: 5-26. Scientific Committee on Problems of the Environment 48, Chichester, UK.
Wolfenden, J., Pearson, M. & Francis, J. 1991. Effects of overwinter fumigation with sulphur and nitrogen oxides on biochemical parameters and spring growth in red spruce (*Picea rubens* Sarg.). *Plant, Cell Environ.* **14**: 35-45.

12 Impact of Ozone on Carbon Metabolism in Plants

P.V. SANE, M. YUNUS & R.D. TRIPATHI

INTRODUCTION

Ozone (O_3) — a three-atom allotrope of oxygen is produced (a) within the planetary boundary layer (PBL), extending from the earth's surface to about 2 km; (b) in the free troposphere, the region between the PBL and the stratosphere; and (c) in the stratosphere.

In the stratosphere O_3 is produced through photolysis of molecular oxygen by ultraviolet ($\lambda \leq 242$ nm) solar radiations (Chapman, 1930). This process of "photodissociation" initiates a series of reactions in which O_3 is both formed and destroyed leading to a steady-state O_3 concentration (Finlayson-Pitts & Pitts, 1986; Chameides & Lodge, 1992). The stratospheric ozone serves as a shield against biologically harmful solar UV radiations because of its strong absorption in the 280–320 nm region. The downward intrusions of stratospheric ozone are responsible for the supply of O_3 to the troposphere to initiate the photochemical process (see Chapter 2 this volume).

In the troposphere (free troposphere together with PBL), the O_3 is produced through the photodissociation of NO_2 (source of atomic oxygen) at $\lambda \leq 420$ nm.

$$NO_2 + h\nu \rightarrow NO + O$$

$$O_2 + O \rightarrow O_3$$

Ozone is also produced in the troposphere, through the photo-oxidation of carbon monoxide, methane, formaldehyde and the more persistent non-methane organic compounds (Fishman & Carney, 1984; Fishman *et al.*, 1985):

- ozone from carbon monoxide (CO)

$$CO + 2O_2 + h\nu \rightarrow CO_2 + O_3$$

- from methane (CH_4)

$$CH_4 + 4O_2 + 2h\nu \rightarrow HCHO + H_2O + 2O_3$$

$$HCHO + h\nu \rightarrow H + HCO(\lambda < 330 \text{ nm})$$

$$HCO + h\nu \rightarrow H + CO(\lambda < 360 \text{ nm})$$

$$CO + 2O_2 + h\nu \rightarrow CO_2 + O_3$$

Plant Response to Air Pollution. Edited by Mohammad Yunus and Muhammad Iqbal

- from non-methane hydrocarbons

$$RH + 4O_2 + 2h\nu \rightarrow RCHO + H_2O + 2O_3$$

The increase in the ambient O_3 level in urban and even in rural areas has been primarily due to the presence of "photochemical smog" produced through increased industrial activity.

The effects of O_3 on living organisms have been studied for more than 40 years (Giese *et al.*, 1952), but the biochemical basis of ozone toxicity is still debated.

Ozone is known to react with the cuticular compounds of plant leaves and this reaction is in the solid state. Secondly ozone reacts with the hydrocarbons emitted by plants and this reaction is in the gas phase. The third type of reaction relates to the interaction of dissolved ozone with the cellular components (see Mudd, this volume, Chapter 10).

OZONE UPTAKE AND PENETRATION

The entry of O_3 and its penetration to the deeper tissues has been reviewed by Mudd (see Chapter 10, this volume). The reactivity and capacity of O_3 to form oxyradicals inevitably leads to questions as to (a) whether its effects can be attributed to molecular O_3 *per se* and/or (b) what reactions are likely to occur and where in the pathway from O_3 in air to the interior of leaf tissue.

Apoplastic space and plasmalemma were major sinks of 1 ppm O_3 in sunflower at least during the first few minutes (Laisk *et al.*, 1989). It was suggested that reactions taking place in apoplastic space and plasmalemma check penetration of O_3 *per se* into the deeper layers of cells. This suggestion appears relevant to link *in vivo* conditions if one considers many known reactions of O_3 with biological molecules.

Although many effects of O_3 on intracellular enzymes and other components have been demonstrated in both *in vivo* and *in vitro* systems, there is no direct evidence to show that *in vivo* reactions are caused by O_3 itself. The indirect evidence in favour of the reaction of ozone includes ultrastructural changes of inner membranes, e.g., indentation of chloroplast envelope, swelling of thylakoids and Golgi body cisternae, shrinkage of mitochondrial cristae, appearance of crystalline bodies inside chloroplast (Thomson *et al.*, 1974; Miyake *et al.*, 1984; Runeckles & Chevone, 1992). Changes in plasmalemma disturb the osmotic relationship causing visible injury manifestations in response to O_3 exposures (0.4–0.5 ppmv). One of the rapid responses, swelling of thylakoids, is detectable much earlier than any visible toxicity appearance, indicating the possibility that molecular O_3 penetrates beyond plasmalemma (Miyake *et al.*, 1984; Guderian *et al.*, 1985). Interaction of O_3 with chloroplast may generate a number of reactive and toxic species of oxygen (free radicals–superoxide and hydroxyl radicals and H_2O_2) deleterious to plastid function. However, the defence system of chloroplast involving reducing compounds such as ascorbate and glutathione (GSH) prevents accumulation of free radicals (Halliwell, 1982; Robinson, 1988; Salin, 1988).

EFFECT ON CHLOROPLAST, CHLOROPHYLL AND PHOTOSYSTEMS

The light reaction of photosynthesis including photochemistry is very fast and completes in about 10^{-4} s while the enzymatic dark reactions are slow and require 0.02 s to complete, Glazer, 1989. Net photosynthesis rate (P_N) is the product of these two reactions running in tandem. The most widely observed chemical change induced by ozone is the destruction of chlorophyll (Guderian *et al.*, 1985). Chlorophyll *b* is more sensitive than chlorophyll *a* and carotenoids are less sensitive to O_3 than chlorophyll (Runeckles & Chevone, 1992). The chlorophyll *a*/*b* ratio decreased under ozone treatment in *Phaseolus vulgaris* (Agrawal *et al.*, 1991). The light-saturated rate of CO_2 uptake decreases significantly when plants are exposed to elevated levels of O_3 (Atkinson *et al.*, 1988; Farage *et al.*, 1991; Baker *et al.*, 1994). Whether this decline could be a result of reductions in pigment concentration or decrease in the efficiency of energy transduction of the photosynthetic membranes was investigated by Farage *et al.* (1991); the possibility; such effects could lead to loss of active photosystem II. The long-term exposure of plants to ozone resulted in the loss of chlorophyll content which appeared very late and did not provide a satisfactory interpretation for the initial reduction in the rate of photosynthesis (Farage *et al.*, 1991).

In the chloroplast the electron transfer from ferredoxin to oxygen and the subsequent formation of H_2O_2 via superoxide dismutase necessitates a pathway to prevent buildup of H_2O_2.

$$Fd_{red} + O_2 \rightarrow Fd_{ox} + O_2^- \text{ and the subsequent reaction is:}$$

$$2O_2^- + 2H^+ \rightarrow H_2O_2 + O_2$$

Since chloroplast is deficient in catalase, alternative enzymes and metabolites that can reduce H_2O_2 perform this function (Alscher & Amthor, 1988; Runeckles & Chevone, 1992).

The detoxification of H_2O_2, as depicted in Figure 12.1, involves participation of reduced ascorbate and the glutathione system. The equations given below summarize the events:

$$H_2O_2 + AA \rightarrow dHAA + 2H_2O$$

$$dHAA + 2GSH \rightarrow AA + GSSG$$

$$GSSG + NADPH + H^+ \rightarrow 2GSH + NADP^+$$

The different energy levels involved in the intermediate reactions of photosystems I and II conveniently make it possible to supply exogenous electron acceptors or donors that can act as sinks or sources of electrons at different locations within the overall scheme. Sites of electron paramagnetic resonance (EPR) signals affected by ozone (Vaartnou, 1988) are depicted in Figure 12.2.

The effect of O_3 on isolated chloroplasts has been studied through Hill reaction by providing an appropriate exogenous electron acceptor as shown in the following equation:

$$2H_2O + 4A \rightarrow O_2 + 4AH$$

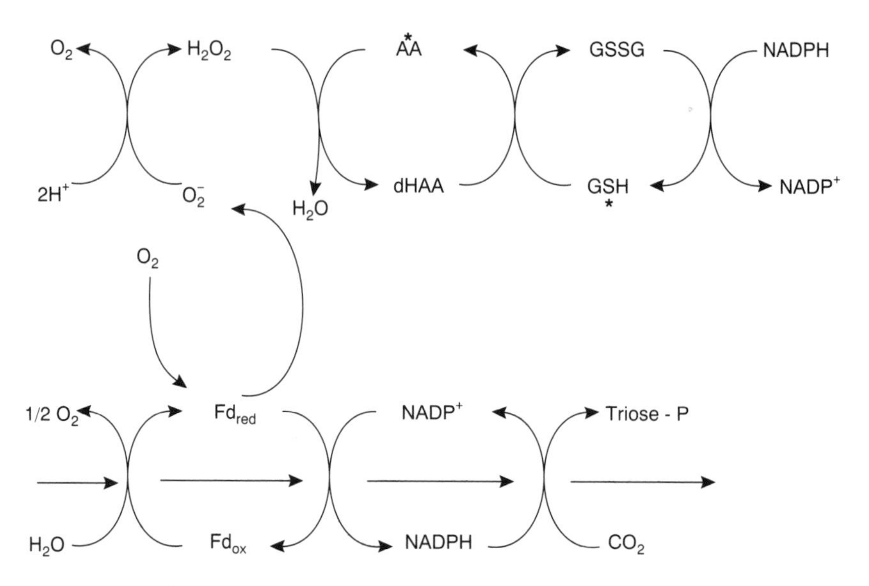

Figure 12.1. Normal and alternate electron flow in the chloroplast. Straight arrows indicate normal electron flow from water through ferredoxin (Fd) to triose phosphate of the Calvin cycle. Pseudocyclic electron flow is initiated by the single electron transfer from reduced ferredoxin (Fd_{red}) to oxygen. The enzymes involved are superoxide dismutase, ascorbate peroxidase, dehydroascorbate reductase, and glutathione reductase. Asterisks indicate potential direct interactions with ozone (after Lefohn, 1992; courtesy, Lewis Publishers, Chelsea, USA)

Decline in Hill activity of isolated chloroplasts from several species including spinach (Koiwoi & Kisaki, 1976; Sugahara *et al.*, 1984) has been reported. The photoreduction of dichlorophenol-indophenol (DCIP) by water in illuminated chloroplasts together with release of O_2 indicates activity of PS II, while photoreduction of NADP by DCIPH2 indicates the functioning of PS I. Inhibition of both photosystems was observed when chloroplasts were exposed to 0.5 ppmv O_3 for 4 h. However, there was no adverse effect on either of the photosystems at 0.1 ppmv O_3 exposure.

The *in vivo* approach involves the measurement of chlorophyll fluorescence transients (CFTs). The reaction centre chlorophylls associated with PS I and PS II (P700 and P680, respectively: Figure 12.2) are activated by photons absorbed by their respective antenna pigment systems and the light energy is utilized to drive electron transport in and between both PS I and PS II. A fraction of the absorbed energy is lost as heat and some is re-emitted as fluorescence. Changes in the predominant fluorescence arising from PS II, and the measurements associated with PS I are known (Kyle *et al.*, 1983). The O-I rise occurs as the plastoquinone pool becomes reduced and subsequent decline through S and M result from the photochemical fluorescence quenching in PS II reaction centres and non-photochemical quenching due to other pathways of de-excitation including transfer to PS I and heat loss. Schreiber *et al.* (1978) noted changes in CFTs in bean leaves and attributed them to a decrease in water-splitting capacity, leading to

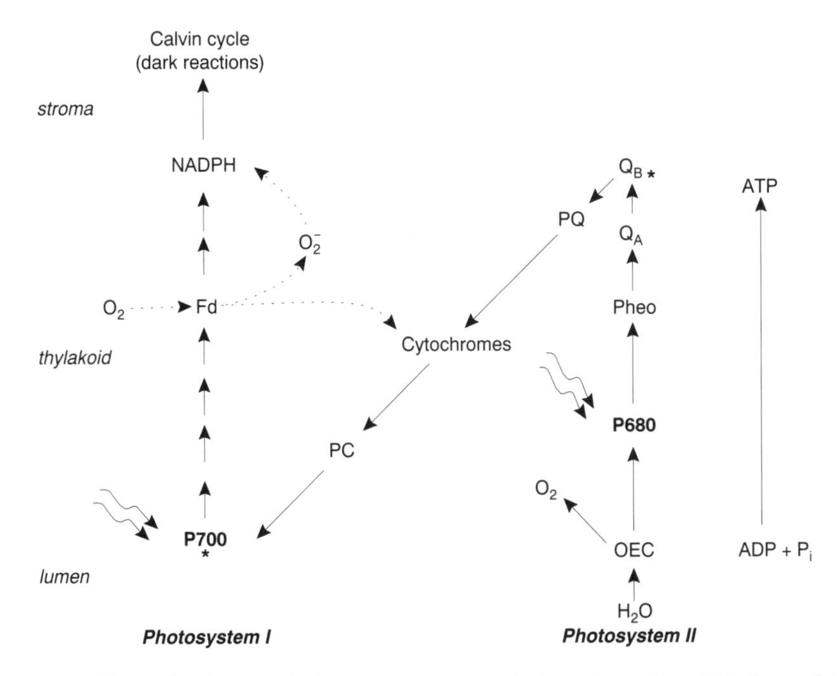

Figure 12.2. General scheme of electron transport photosystems I and II. Several intermediates are omitted for clarity. Pseudocyclic electron flow is shown by broken lines. OEC = oxygen evolving complex; P680, P700 = active chlorophyll a; PC = plastocyanin; Pheo = pheophytin; PQ = plastoquinone; Q_A, Q_B = iron-protein-bound plastoquinones. Asterisks indicate sites of electron paramagnetic resonance signals affected by ozone (after Lefohn, 1992; courtesy, Lewis Publishers, Chelsea, USA)

increased oxidation of the plastoquinone pool rather than to a direct effect on the reaction centre of PS II; these changes were also confirmed with spinach leaves (Shimazaki, 1988).

Several studies (Cape *et al.*, 1988; Davison *et al.*, 1988; Rowland-Bamford *et al.*, 1989) have tended to use CFTs measurement as general "early warning" indicators of ozone effects/responses rather than for the study of specific effects on the photosynthetic processes. Furthermore, in many studies where fluorescence ratio Fv/Fm was measured in ozone-treated foliage, either no impact or less sensitive ratios were recorded (Farage *et al.*, 1991; Godde & Buchhold, 1992). Under ozone stress (0.25 ppmv), no change in variable fluorescence in snap bean, *Phaseolus vulgaris* cv. Bush Blue Lake 290C, was observed though half rise time decreased (Agrawal *et al.*, 1991). Lack of responsiveness of photosystem II as O_3 did not alter the ability of thylakoids of wheat chloroplasts to bind atrazine was further explained by Farage *et al.* (1991).

Degradation and/or synthesis of 32 KDa D1 protein in the reaction centre of photosystem II was investigated in some plants. Neither acute exposure of wheat leaves (Farage *et al.*, 1991) nor the chronic exposures of spruce trees (Godde & Buchhold, 1992) to ozone caused any change in D1 protein content. However, the turnover rate of this protein was significantly higher suggesting that O_3 enhanced both synthesis and degradation of the protein. It is one of the few crucial reports which supports relative susceptibility of photosystem II to O_3. From these studies it is also inferred that the repair mechanisms are operative to neutralize any adverse effects on photosynthesis. However, the energy necessary for this repair should be taken into account in any cost analysis for maintenance of the plants receiving ozone insult.

Robinson & Wellburn (1983) using γ-aminoacridine as a probe, found that pulses of ozone to oat leaf thylakoid membrane caused the progressive decline in pH gradient, suggesting a decrease in ATP synthesis. Studies on the leaf tissue from plants exposed to ozone have shown both increases and decreases in the total ATP (Wellburn, 1984), although the analysis failed to differentiate among the different pools of ATP within the cell. Because of the dependence of the chloroplast ATP generation on the pH gradient, the latter should result in reduced ATP formation. Reduced levels of chloroplast ATP caused by O_3 will eventually impair the reaction of the Calvin cycle.

EFFECT ON CARBON DIOXIDE FIXATION

Many observations support the notion that reductions in net photosynthesis P_N cannot result solely from stomatal closure and its consequences as shown in Figure 12.3 (Reich & Amundson, 1985; Reich *et al.*, 1985; Lehnherr *et al.*, 1988; Runeckles & Chevone, 1992). Nevertheless, the concomitant development of a

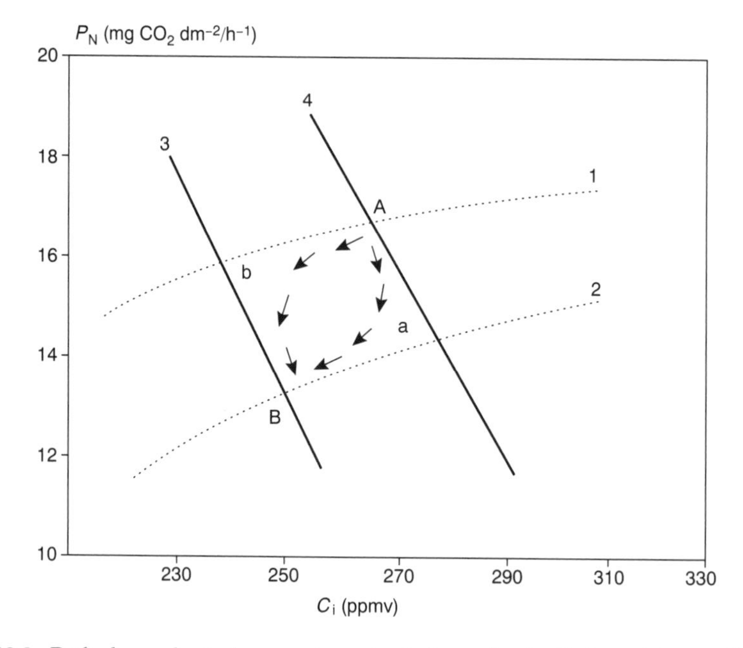

Figure 12.3. Path-dependent changes in net photosynthesis (P_N) and intercellular CO_2 concentration (C_i) when initial P_N limitation results from (a) increased mesophyll resistance, or (b) increased stomatal resistance (after Lefohn, 1992; courtesy, Lewis Publishers, Chelsea, USA)

reduced steady-state photosynthetic rate recorded during short O_3 exposures in crop and forest seedlings (Yang *et al.*, 1983a; Le Sueur-Brymer & Ormrod, 1984; Rowland-Bamford *et al.*, 1989; Runeckles & Chevone, 1992) suggests the attainment of a steady-state situation between ozone influx and P_N inhibition. Figure 12.4 represents a schematic model that summarizes the metabolic processes that seem to be important in development of such an equilibrium state. The initial entry of ozone and the concomitant interference with carbon assimilation are expressed by two series of solid arrows. In the upper sequence of events, ozone influx to the intercellular spaces [$O_{3\,int(o)}$] is controlled by the initial stomatal conductance [$g_{s(o)}$] and the external O_3 concentration [$O_{3\,ext}$]. The flux of O_3 and toxic metabolites [$Ox_{(o)}$] mesophyll is, therefore, governed by a number of reactions that lead to either an accumulation or removal of chemical species that can interfere with photosynthesis. The net rate of accumulation of $Ox_{(o)}$[$k_{n(o)}$] is the resultant of processes generating toxic substances and those involved in detoxification. As $Ox_{(o)}$ accumulates, the photosynthetic rate in salubrious air [$P_{N\,amb(o)}$] declines along the curve d. The properties of $P_{N\,amb}$ reduction depend upon the mechanism(s) involved, e.g. Rubisco inactivation, H_2O_2 inhibition of the Calvin cycle, sulphhydryl enzymes, thylakoid membrane disruption and impaired electron flow and the net rate of inhibition process(es) [$K_{r(o)}$]. The rate of inhibition also depends on the ability of cells or organelles to repair biochemical and cytological lesions.

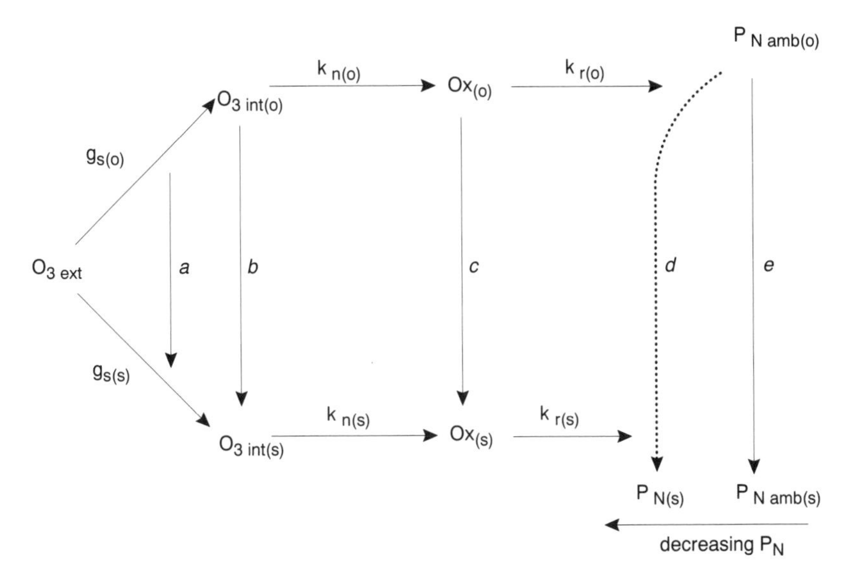

Figure 12.4. Schematic model representing interactive processes involved in an ozone-induced steady-state inhibition of net photosynthesis. $O_{3\,ext}$ = ambient ozone concentration external to the leaf; $O_{3\,int}$ = intercellular leaf ozone concentration; O_x = ozone or metabolites that inhibit P_N; $P_{N\,amb}$ = net photosynthetic rate in ambient air containing no ozone; g_s = stomatal conductance to ozone; k_n = net accumulation rate of ozone or toxic metabolites; k_r = inhibition rate of O_x or P_N; T = time; (o) subscripts represent initial events in ozone fumigation; (s) subscripts represent steady-state events (after Lefohn, 1992; courtesy, Lewis Publishers, Chelsea, USA)

The reduction in CO_2 fixation leads to an increase in C_i (intercellular CO_2) and a consequent change in stomatal opening through feedback (Runeckles & Chevone, 1992; refer to Figure 12.3). Stomatal conductance will, therefore, gradually decrease (arrow a) to a new steady-state level $g_{s(s)}$. The intercellular O_3 flux will also decline (arrow b) from $O_{3\,int(o)}$ to the equilibrium rate, $O_{3\,int(s)}$. Considering that $k_{n(o)}$ remains unchanged, the mesophyll flux $Ox_{(s)}$ of toxic compounds will then decrease (arrow c) from $Ox_{(o)}$ to $Ox_{(s)}$. The decreased flux will establish a new net inhibition rate $[K_{r(s)}]$ and net photosynthesis will attain an equilibrium rate at $P_{N\,amb(s)}$, determined by both $g_{s(s)}$ and $K_{r(s)}$ (Runeckles & Chevone, 1992).

The development of a reduced steady-state $P_{N\,amb(s)}$ rate caused by ozone exposure can, therefore, be viewed as a series of interactive processes whereby O_3 influx is regulated by inhibition of net photosynthesis, the increase in C_i and the subsequent decrease in g_s.

The variation in tolerance of net photosynthesis to ozone can be viewed as a change in component functions depicted in Figure 12.4. A tolerant genotype may possess a relatively low $g_{s(s)}$ resulting in low intercellular flux of $O_{3\,int(s)}$. Besides, high endogenous levels of ascorbate and catalase (Lee & Bennett, 1982) and glutathione (Guri, 1983), scavengers of ozone or toxic oxygen species (Mudd, 1982; Salin, 1988) may favour a lower accumulation rate (K_n) of intermediate toxic metabolites (Ox). The level of Ox may never rise to the extent so as to surpass the maintenance/repair ability within the chloroplast, and net photosynthesis will remain largely unaffected by ozone exposure.

The model presented in Figure 12.4 does not differentiate between effects of ozone or its products on the individual processes of gross photosynthesis or photorespiration, each of which is a contributor to the net photosynthetic CO_2 fixation.

An alternative approach to differentiate between direct effects on P_N and those resulting from stomatal changes utilizes measurements of discrimination against uptake of $^{13}CO_2$ relative to $^{12}CO_2$. The limit/capacity of Rubisco to fix $^{12}CO_2$ preferentially results in negative $\delta^{13}C$ values (typically −2.2 to −4.0%) in the photosynthetic products of C_3 plants (Troughton, 1979). The discrimination decreases with decreasing C_i, such as would occur with g_s in the absence of reduced P_N. Greitner & Winner (1988) found small but significant increases in $\delta^{13}C$ (i.e., less negative values) in radish (cv. Cherry belle) and soybean (cv. Williams) plants exposed to 0.12 ppmv O_3. They suggested that ozone treatment resulted in decreased intercellular CO_2 build-up due to stomatal closure. Similar findings were reported for a range of species exposed to mixtures of O_3 and SO_2 and O_3, SO_2 and NO_2 (Martin *et al.*, 1988).

Assumed relationships of P_N and the stomatal conductance formed the basis of the model of leaf response established by Schut (1985). These models of short-term leaf exposures to ozone were based on the reports of Hill & Littlefield (1969), Larsen & Heck (1976), Elkiey & Ormrod (1979) and Black *et al.* (1982). Schut (1985) hypothesized that stomatal regulation maintained intercellular CO_2 constant (C_i constant) and ozone uptake, CO_2 uptake and stomatal conductance

Table 12.1. Impact of ozone on net photosynthesis (P_N) in selected plants*

Species	P_N response	Reference
Acer saccharinum	Inhibition	Reich *et al.* (1986)
Liriodendron tulipifera	No effect	Robert (1990)
Picea abies	Inhibition	Keller & Hasler (1987)
Picea rubra	No effect	Taylor *et al.* (1986)
Pinus elliottii	Inhibition	Barnes (1972b)
Pinus ponderosa	Inhibition	Miller *et al.* (1969), Weber *et al.* (1993)
Pinus rigida	Inhibition	Barnes (1972b)
Pinus strobus	Inhibition	Yang *et al.* (1983a, b), Reich *et al.* (1987), Botkin *et al.* (1972)
Pinus sylvestris	No effect	Barnes (1972b), Skärby, *et al.* (1987)
Pinus taeda	Inhibition	Barnes (1972b), Hanson *et al.* (1988), Sasek & Richardson (1989)
Populus deltoides × *Populus cv. Caudina*	Inhibition	Sengupta *et al.* (1991)
Populus maximowizzi × *trichocarpa*	Inhibition	Pell *et al.* (1992)
Quercus alba	No effect	Foster *et al.* (1990)
Quercus rubra	No effect	Reich *et al.* (1986)
Raphanus sativus	Inhibition	Dann & Pell (1989)
Triticom sp.	Inhibition	Lehnherr *et al.* (1987)

*Partially from Chappelka & Chevone (1992).

were proportional, so that the inhibition of P_N was proportional to ozone uptake. The impact of O_3 on net photosynthesis of selected plants is shown in Table 12.1.

EFFECT ON RUBISCO PROTEIN

Rubisco is the most abundant enzyme in plants; its concentration can go up to 10 mg protein per gram fresh weight of leaf. The dominant level of Rubisco and its high affinity for phosphodiester, cause it to serve as a metabolic buffer capable of sequestering metabolites culminating in far-reaching consequences in terms of activity regulation and maintenance of given levels of metabolites in chloroplasts (Kumar & Sane, 1990). Rubisco accounts for as much as 50–70% of the total soluble leaf protein. However, the percentage pool of enzyme which is active varies with environment and genotype.

Loss in activity of Rubisco is associated with a reduction in the quantity of the protein (Dann & Pell, 1989; Pell *et al.*, 1992, 1994a, b; Baker *et al.*, 1994).

Ozone-induced reductions in Rubisco levels were noticed for alfalfa and rice plants exposed to short-term but relatively high concentrations of ozone. On the contrary, prolonged low doses of ozone decrease the level of Rubisco in potato, radish, aspen, wheat, hybrid poplar, black cherry, northern red oak and sugar maple (Dann & Pell, 1989; Pell *et al.*, 1992, 1994b, c; Baker *et al.*, 1994). Thus, it is hypothesized that Rubisco modification through oxidation is necessary for proteolysis. This is consistent with Stadman's (1988) view that specific proteases are designed to recognize oxidized enzymes. Landry & Pell (1993) exposed hybrid poplar plants to O_3 at levels which reduced quantity and the total activity of Rubisco.

Furthermore, *in vitro* treatment of leaf extracts with O_3 decreased total Rubisco and binding of enzymes transition state analog, 2-carboxyarabinital bisphosphate (CABP). O_3 increased the loss of Rubisco large subunit (rbcL) when extracts were incubated at 37°C. Treatment of isolated intact chloroplast with O_3 accelerated both loss of 55 kD a Rubisco rbcL and accumulation of Rubisco rbcL aggregates as visualized by immunoblotting (Landry & Pell, 1993). The time-dependent modification in Rubisco structure was the primary response of the isolated organelles to O_3 treatment, with little degradation of rbcL (Landry & Pell, 1993).

EFFECT ON RUBISCO SYNTHESIS AND ACTIVITY

Plants of potato subjected to O_3 stress showed rapid loss of the mRNA for both the large (rbcL) and small subunits (rbcS) of Rubisco, respectively (Reddy *et al.*, 1993; Pell *et al.*, 1994b). Recently, Glick & Pell demonstrated that chronic exposures of potato plants to O_3 (0.08 $\mu l\ l^{-1}$ for 4 h day^{-1}) rapidly decreased levels of rbcS mRNA with only minimal loss in rbcL mRNA (see Pell *et al.*, 1994a). A decline in steady-state Rubisco synthesis decreased in plants exposed to O_3 (Pell *et al.*, 1994b). These authors investigated whether loss in rbcS mRNA detected by blot analysis could be explained by reduced levels of transcription. There is a decrease in the levels of transcription of rbcS in nuclei isolated from O_3-treated plants, though the O_3-induced decrease in the level of steady-state rbcS mRNA detected by blot analysis is greater than the loss in transcription (see Pell *et al.*, 1994a).

Whether the loss in transcription is sufficient to account for the loss in rbcS mRNA remains to be established. When young (not fully expanded) and mature (fully expanded) foliage were fumigated with O_3, both showed losses in rbcS and rbcL mRNAs for Rubisco transcript messages (Reddy *et al.*, 1993).

Potato plants treated with 0.30 $\mu l\ l^{-1}$ O_3 showed a decrease in Rubisco level both in immature (unexpanded) and mature (fully expanded) leaves. The level of rbcS mRNA was higher in immature leaves. The reduction due to O_3 was significant in such leaves; and there was no significant effect on the level of rbcL mRNA (Eckardt, 1993, Pell *et al.*, 1994b). O_3 stress reduces rbcS mRNA, thus influencing the potential of the tissue to synthesize the required levels of Rubisco. Potato plants given a dark phase immediately after O_3 exposure showed decline in rbcS mRNA in all the plants thus interrupting any further synthesis of protein. After a dark phase of 48 h, a decline of 12% in the level of Rubisco was recorded in immature leaves of both O_3-treated and control plants. However, O_3 treatment resulted in 53% decrease in Rubisco level as compared to control, showing a decline of 13% in mature leaves (Eckardt, 1993; Pell *et al.*, 1994). It may be concluded that O_3 can affect both synthesis and proteolysis of Rubisco and the response of these processes may be age dependent.

When activity is assayed *in vitro*, the enzyme is extracted and activated with CO_2 and Mg prior to measurement. The resulting measure of carboxylation is a reflection of the existing pool to fully functional enzyme (henceforth referred to as total activity). When the enzyme is extracted in a CO_2-free environment, the subsequent measure of activity presumably reflects the endogenous (initial) Rubisco activity. Ozone-induced decline in total Rubisco activity has been recorded in hybrid poplar, potato, radish, rice and wheat (Nakamura & Zaka, 1978; Lehnherr *et al.*, 1987; Dann & Pell, 1989; Enyedi *et al.*, 1992; Pell *et al.*, 1992, Landry & Pell, 1993).

Reductions in both the initial (endogenous) and total Rubisco activity due to O_3 were observed in potato (Dann & Pell, 1989) albeit no effects of O_3 on the percentage activity of the enzyme were detected. Reports show that Rubisco extracted from O_3-exposed wheat plants does reflect an increase in percentage activation (Lehnherr *et al.*, 1987, 1988).

Flag leaves of wheat fumigated for a single time duration show incomplete enzyme activation rendering the total activity lower than the initial activity. It may be that at a certain stage during leaf development O_3 treatment could reflect enhanced percentage activation of Rubisco providing a way for a prolonged optimum CO_2 fixation potential of plants. However, carboxylation potential of leaves declined under O_3 stress.

Ozone-induced reduced Rubisco synthesis may affect plants through (a) reduced CO_2 fixation, and (b) premature senescence, due to reduction in Rubisco (Pell *et al.*, 1992).

Ozone exposure often causes a surge in the production of ethylene (a well-established hormone promoting fruit ripening and senescence) — *the stress ethylene* (Tingey *et al.*, 1976a; Mehlhorn & Wellburn, 1987; Mehlhorn *et al.*, 1991; Kangasjarvi *et al.*, 1994). Although the mode by which this emission result is not understood, it has been demonstrated that in the presence of inhibitors of ethylene biosynthesis such as aminooxyacetic acid or amino ethoxyvinyl glycine, toxicity of O_3 on net photosynthesis or rbcS mRNA is greatly reduced (Taylor *et al.*, 1988; Pell *et al.*, 1994a). These findings suggest that O_3 responses may be mediated via ethylene production.

EFFECT OF OZONE ON PHOTORESPIRATION AND DARK RESPIRATION

Photorespiration comprises reactions involving cooperation of three organelles (chloroplast, peroxisomes and mitochondria) and is a consequence of the oxygenase function of Rubisco. The oxygenase property of Rubisco also requires activated enzyme just as the carboxylase property requires activated enzyme. It has been established that CO_2 and O_2 compete for the same site on the enzyme and the rate of the two reactions is determined by the prevailing concentrations of the two gases. In C_3 plants, under congenial growth conditions, oxygenase activity is 22–33% of carboxylase function, but this may increase to 50–70% under stress conditions. That means nine molecules of CO_2 will be fixed instead of six in the absence of oxygenase function.

Phosphoglycolate produced during oxygenation is metabolized through a series of reactions known as the photorespiratory cycle (as it operates in the presence of light) leading to loss of fixed CO_2. The following equation summarizes the events during oxidation of 2-phosphoglycolate due to oxygenase function.

$$\begin{aligned} &2-\text{phosphoglycolate} + O_2 + \text{glutamate} + \text{ATP} \\ &\quad \rightarrow \text{PGA} + \text{ADP} + CO_2 + NH_3 + 2-\text{oxoglutarate} + H_2O \end{aligned}$$

Although this results in an overall reduction in the net efficiency of CO_2 utilization, photorespiration may offer a protective device to permit dissipation of light energy when amounts of CO_2 are limiting (Wellburn, 1988). The oxygenase function has been suggested to involve formation of the superoxide anion which is retained within the "cage" of the catalytic site on the enzyme (Lorimer, 1981). Although effects of other pollutants on photorespiration are well documented (Hällgren, 1984), studies of the effects of ozone are few probably because of difficulties in making quantitative measurements (Miller, 1987).

The effect of ozone alone or in combination on the fate of $^{13}CO_2$ assimilate extracted from leaves was examined by ^{13}C NMR spectra. Sixteen NMR signals obtained were used to estimate the pool size of ^{13}C incorporation. O_3 exposure resulted in increased pool sizes of glycine and serine due to increased photorespiration (Ito *et al.*, 1985a).

The reactions of the Krebs cycle or the citric acid cycle which include two decarboxylation reactions take place in the intra-mitochondrial matrix and require supply of acetyl CoA through the glycolytic reaction sequences (EMP pathway) taking place in chloroplast and cytoplasm. The oxidation of the Krebs cycle intermediates, involving reduction of NAD and FAD, leads to a flow of electrons to the final acceptor, O_2. This electron transport takes place in the mitochondrial cristae.

Lee (1967) found a rapid decline in oxidative phosphorylation in tobacco mitochondria from leaves subjected to high (1 ppmv) ozone stress. MacDowall (1965) also found a decrease in cytochrome oxidase activity, oxygen uptake and ATP production in tobacco, suggesting that either changes of cristal membrane led to the uncoupling of electron and proton flows or there was a shift to a widely distributed non-phosphorylating cyanide-resistant respiratory pathway (Dizengremel & Citernae, 1988). This is further strengthened from an earlier observation (Tingey *et al.*, 1975) supporting the latter view that 0.5 ppm O_3 increased glucose-6-phosphate dehydrogenase and decreased glyceraldehyde 3-phosphate dehydrogenase in soybean.

On exposure of 10-day-old kidney bean plants to 0.1, 0.2, 0.4 ppmv O_3 alone or in combination with various levels of NOx, root respiration remained relatively unchanged up to 5 days then decreased significantly on the 7th day. The decrease in respiratory CO_2 production by roots of O_3-treated bean plants (Ito *et al.*, 1985b) is considered by Miller (1987) as a consequence of reduced photosynthesis and translocation from the aerial parts of the plant. Indirect inhibitory effects of ozone on root respiration (Hofstra *et al.*, 1981), probably result from reduced supply of photoassimilates. Anderson & Taylor (1973) recorded increased respiration of non-photosynthesizing tobacco callus tissue. It is agreed that ozone concentration sufficient to cause visible injury will ultimately result in enhanced respiration. Increased dark respiration due to ozone exposure has been reported in citrus (Dugger & Ting, 1970), pine (Barnes, 1972b), bean (Pell & Brennan, 1973) and poplar (Reich, 1983). In soybean prolonged exposure to O_3 levels (0.05–0.13 ppmv) had no effect on dark respiration (Reich *et al.*, 1986). Lehnherr *et al.*, (1987) found that dark respiration of the flag leaf of wheat exposed to O_3 for 18 or more days in open-top field chambers decreased significantly with increasing O_3 concentration.

Like O_3-induced changes in photosynthesis, the short and long-term responses of respiration also vary. Low dose (0.14 ppmv O_3 for 2 h) effects on respiration rates were reversible in oats (*Avena sativa* L. cv. Titles). The CO_2 release measured immediately at zero hours by placing plants in darkness was doubled. However, the stimulation steadily decreased over the ensuing 6 h. Effect on dark respiration may be the consequence of the direct and/or indirect influence on the mitochondrion, and the enzymes of glycolysis. Indirect effects also include translocation of photoassimilates.

EFFECT ON ASSIMILATE PARTITIONING (CARBOHYDRATE/ SUGAR POOLS) AND TRANSLOCATION

Vegetative yield can be thought of in terms of dry matter production which largely comprises carbon skeleton. Plant biomass and yield are inherently linked to carbon fixation and the subsequent partitioning of the photoassimilate. The ratio of the harvested plant product to total plant weight, called the "harvest index", has increased in a few crop plants with no concurrent increases in relative growth rates and plant biomass. Such plant species are considered superior from an agricultural viewpoint with larger fruits, enhanced seed mass and increased amounts of photoassimilate sinks at the expense of vegetative growth. As exposure to O_3 results in decreased photosynthesis, a reduction in growth is also expected. The relative effects on the growth of the different parts of the plant are the result of varied O_3 impacts on the translocation of photoassimilates from the leaves.

The data generated on the effects of short-term exposures on assimilation cannot be modelled to predict effects on productivity, though some examples may be exceptions (Reich & Amundson, 1985). As the overall growth is a long-term process, multiple impacts of various types may be observed. Changes in the partitioning pattern between leaves and the rest of the plant body frequently reduce both size and longevity of individual leaves and thereby affect the plant's long-term capabilities for photosynthetic assimilate gain (Mooney & Winner, 1988).

Among the reported effects of O_3 on carbohydrate levels, is a decline in free sugars and storage polysaccharides in the roots (Jensen, 1981; McCool & Menge, 1983; Ito *et al.*, 1985b), although the level of individual sugars such as fructose and sucrose may increase (Ito *et al.*, 1985b). Responses are, however, more varied; an increase in pine needles (Barnes, 1972a; Tingey *et al.*, 1976b), an increase and decrease in bean (Ito *et al.*, 1985b) and a decrease in soybean (Tingey *et al.*, 1973). The varied responses possibly reflect the interaction (combination) of effects on the anabolic and catabolic pathways influenced by source–sink relationships (translocation and partitioning).

Besides the differential growth responses of individual plant parts (Table 12.2), few studies focus on effects of O_3 on translocation *per se*. Sucrose is the key metabolite translocated via phloem. Sucrose is loaded through a process, "phloem loading", that requires ATP and involves the co-transport of sucrose and K^+ ions (Giaquinta, 1983). It also requires functional integrity of the cell membrane of the sieve tubes and companion cells and involves carrier proteins containing essential –SH groups (potential target of O_3) on the plasmalemma surface. Even if phloem loading is not impaired, translocation of sugars and other metabolites is dependent upon their availability as they could be within the various cellular compartments; there is evidence for presence of transportable and non-transportable pools in the cell (Koziol *et al.*, 1988).

To study the O_3 effects on the distribution pattern of assimilates, labelled CO_2 was employed. Retention of photoassimilates was enhanced by 24–57% in treated bean leaves (McLaughlin & McConathy, 1983). Okano *et al.* (1984) recorded highest retention in the primary leaves which may be at the expense of downward translocation to the roots. In contrast, export from trifoliate leaves (which supply towards the upward translocation stream) was enhanced. This information provides some explanation for the observed increase of S/R ratios by O_3 (Table 12.2). However, a decreased S/R ratio was also observed in pepper (Bennett *et al.*, 1979), millet (Agrawal *et al.*, 1983) and peanut (Heagle *et al.*, 1983).

Photoassimilate partitioning varies throughout growth. The onset of reproductive growth marks the start of a new phase in partitioning. Cooley & Manning (1987) have summarized several studies with soybean, which show that while O_3 reduces overall growth, it causes enhanced partitioning to leaves during early growth; although seed set is reduced the seeds that developed were the preferred sink. In tomato, vegetative growth was more severely affected by O_3 than the fruit development (Oshima *et al.*, 1975). In many cases, for example in peppers (Bennett *et al.*, 1979) and cotton (Oshima *et al.*, 1979), O_3 severely inhibits the reproductive growth.

As in crop plants, in addition to reducing the amount of dry matter produced, ozone can affect partitioning of carbon in tree seedlings (Tingey *et al.*, 1976b; Jensen 1981; Scherzer & McClenahen, 1989; Adams *et al.*, 1990, Spence *et al.*, 1990) and in mature trees (Wilkinson & Barnes, 1973; McLaughlin *et al.*, 1982). Root carbohydrate content and the above-ground biomass of 1-year-old green ash

Table 12.2. Summary of the observed O_3 effects on root and shoot biomass in crop plants and tree seedlings*

Species	Root biomass (%)	Shoot biomass (%)	Total biomass (%)	Reference
Acer saccharinum	—	—	−41	Kress & Skelly (1982a)
	—	—	−64	Jensen (1983)
	—	—	−(2–8)	Reich *et al.* (1986)
Arachis hypogaea	−35	−49	—	Heagle *et al.* (1983)
Beta vulgaris	−67	−50	—	Ogata & Mass (1973)
Capsicum annuum	0	+10	—	Bennett *et al.* (1979)
Daucus carota	−32	+13	—	Bennett & Oshima (1976)
Festuca arundinacea	−44	+19	—	Flagler & Younger (1982)
Fraxinus americana	—	−17	—	Kress & Skelly (1982a)
Fraxinus pennsylvanica	—	—	−(14–33)	Kress & Skelly (1982a)
Glycine max	−21	−9	—	Tingey *et al.* (1973)
G. max 'Hodgson'	Decrease	Decrease	—	Reich *et al.* (1986), Amundson *et al.* (1986)
Gossypium hirsutum 'Acala SJ-2'	Decrease	Decrease	—	Oshima *et al.* (1979)
Helianthus annuus 'Russian Mammoth'	Decrease	Decrease	—	Shimizu *et al.* (1980)
Liquidambar styraciflua	—	—	−(8–24)	Kress & Skelly (1982a)
Liriodendron tulipifera	—	—	18–41	Kress & Skelly (1982a)
	—	—	14	Mahoney *et al.* (1984)
	—	—	−9	Chappelka *et al.* (1985)
Lolium multiflorum	−32	−14	—	Bennett & Runeckles (1977)
Medicago sativa	−22	−12	—	Tingey & Reinert (1975)
Nicotiana tabacum	−62	−47	—	Faensen-Thiebes (1983)
Panicum miliaceum	−23	−43	—	Agrawal *et al.* (1983)
Petroselinum crispum	−40	−1	—	Oshima *et al.* (1978)
Phaseolus vulgaris	−15	−9	—	Mass *et al.* (1973)
Picea sitchensis	—	—	−14	Wihour & Neely (1977)
Pinus echinata	—	—	−15	Chevone *et al.* (1984)
Pinus elliottii var. *densa*	—	—	−(19–48)	Hogsett *et al.* (1985)
Pinus taeda	—	—	−(10–21)	Chevone *et al.* (1984)
Pinus taeda (wild type)	—	—	−(14–28)	Kress & Skelly (1982a)
Pinus virginiana	—	—	−3	Kress & Skelly (1982a)
Populus deltoides × *trichocarpa*	—	—	−(10–13)	Reich *et al.* (1984)

continued overleaf

Table 12.2 (*continued*)

Species	Root biomass (%)	Shoot biomass (%)	Total biomass (%)	Reference
Populus tremuloides				
Clone 1	—	—	−17	Wang *et al.* (1986)
Clone 2	—	—	−14	
Pseudotsuga menziesii	—	—	−15	Wilhour & Neely (1977)
Quercus rubra	—	—	3-5	Reich *et al.* (1986)
Raphanus sativus	−50	−10	—	Tingey *et al.* (1973)
Solanum tuberosum	−60	−53	—	Foster *et al.* (1983)
Tagetes erecta	−26	−19	—	Reinert & Sanders (1982)
Trifolium incarnatum	−27	−7	—	Bennett & Runeckles (1977)
Trifolium repens	−34	−7	—	Letchworth & Blum (1977)

*Partially from Cooley & Manning (1989), Miller *et al.* (1989) and Chappelka & Chevone (1992).

exposed to O_3 (0.5 ppm O_3, 5 d week^{-1}, up to 6 weeks) was studied by Jensen (1981). Treated seedlings had less stem and leaf dry weight, less root starch and reduced sugar content. However, root reserves were assumed to contribute less to the above-ground biomass. Similar results were obtained with pitch pine seedlings (Scherzer & McClenahen, 1989). Root starch content decreased with increasing O_3 concentrations.

As the carbohydrates are key metabolites of photosynthesis and respiration, it is expected that ozone effect on these processes will affect other aspects of carbohydrate metabolism and pool sizes of various carbohydrates independent of any direct effects. Ozone may have an effect on enzymes responsible for regulating individual pools in the tissue.

Some of the enzymes within chloroplast are light modulated, e.g., glucose-6-phosphate dehydrogenase; light inactivation of this enzyme prevents carbohydrate catabolism. Ozone, therefore, impairs carbohydrate translocation through influence on light modulated enzymes.

The reduced dry matter production due to ozone alters partitioning of carbohydrate in tree seedlings. In pitch pine seedlings root starch was reduced with increased O_3 concentration (0.15, 0.2, 0.3 ppmv) (Scherzer & McClenahen, 1989). Ozone-treated plants exhibit high level of N and free amino acids in the roots. Levels of soluble sugars, starch and phenolics increased in shoots and decreased in roots in 1-week-old ponderosa pine exposed to O_3 (Tingey, 1976). The "metabolic retention in the foliage and interruption in transport to roots" was held to be the cause for this. Kress & Skelly (1982b), Hogsett *et al.* (1985) and Chappelka & Chevone (1986) found that the root grew more than the shoot of some tree species, supporting the observations of Tingey *et al.* (1976b) on metabolite levels in shoots and roots in ozone-treated ponderosa pine.

Loblolly pine seedlings exposed to two times ambient O_3 over one growing season showed lower rates of CO_2 assimilation, increased respiration and less photosynthate allocated to fine roots (Adams *et al.*, 1990). Spence *et al.* (1990) exposed loblolly pine seedlings to 0.12 ppm O_3 (7 h d^{-1}, 5 d $week^{-1}$ for 12 weeks), labelled with ^{14}C to identify the allocation pattern. O_3-treated seedlings exhibited 16% decline in CO_2 assimilation, 11% decline in phloem transport rate and a 45% decline in photosynthate allocation to roots.

Besides affecting translocation of assimilate in tree seedlings, O_3 causes alteration in allocation pattern in mature trees. When eastern white and loblolly pine branches were exposed to 0.0, 0.1 and 0.2 ppmv O_3 for 21 days to study ^{14}C fixation patterns, soluble sugar decreased, and sugar phosphates and free amino acids increased (Wilkinson & Barnes, 1973). McLaughlin *et al.* (1982) using ^{14}C noticed that leaves and branches of O_3-sensitive eastern white pine retained more photosynthate than the intermediate or tolerant ones. Dark respiration was stimulated by 40–58% when needles of spruce (*Picea abies* (L.) Karst) were exposed to 0.1 ppmv O_3; this was associated with a decrease in ethanol-soluble carbohydrate content of the needles (Barnes *et al.*, 1990).

It can be inferred from the above studies that O_3, besides affecting overall growth, can cause shifts in the assimilate-translocation pattern. Alterations triggered by O_3 exposures reduce carbohydrate levels, especially in roots. Depletion in root reserves reduces the vigour of the root system and thus accelerates the sensitivity of trees to abiotic (freezing, drought, etc.) and biotic (root diseases) stresses.

CONCLUSIONS AND FUTURE PROSPECTS

Albeit, ozone-induced injury has been studied extensively during the last two decades, the mechanism of toxicity is still not very clear. Possible modes of injury caused by O_3 directly or/and with its derivatives could be:

- induction of active defence mechanism,
- effect on normal metabolism of the cells,
- mechanical damage to the cellular structures, and
- inducement of premature senescence.

Carbon dioxide fixation is more sensitive to O_3 than any other process associated with light harvesting. The loss in quantity and associated activity of Rubisco is possibly an indirect effect of enhanced oxidative stress. Research should address the question whether the loss in Rubisco is due to onset of senescence or whether O_3 should be held responsible for promoting early and speedy senescence which eventually results in loss in Rubisco? Ethylene, an established hormone of senescence, is induced in a variety of plants against O_3 stress (see Iqbal *et al.* this volume, Chapter 8). Accumulations of pathogenesis-related (PR) protein are induced by O_3-dependent ethylene emission (Raz & Fluhr, 1993). When inhibitors of ethylene biosynthesis are provided to leaf tissue, toxic effects of O_3 on net photosynthesis or rbcS mRNA are minimized (Taylor *et al.*, 1988; Pell *et al.*, 1994b). An important area of future investigation would be whether putative phosphorylated intermediates are generated when there is efflux of O_3-induced ethylene by leaves and whether these intermediates would influence loss in rbcS mRNA.

Besides understanding the effect of O_3 on photosynthesis and the associated carbon metabolism, it would be worthwhile to explore how mitochondrial activity is affected by O_3. This aspect may be important for the repair and recovery of the damaged tissues. Furthermore, the interaction of O_3 with other factors such as increased temperature also needs to be investigated. The molecular mechanisms by which O_3 affects different processes also deserve investigation.

REFERENCES

Adams, M.B., Edwards, N.T., Taylor, G.E. & Skaggs, B.L. 1990. Whole plant ^{14}C photosynthetic allocation in *Pinus taeda*: Seasonal patterns at ambient and elevated ozone levels. *Can. J. For. Res.* **20**: 152–158.

Agrawal, M., Nandi, P.K. & Rao, D.N. 1983. Ozone and sulphur dioxide effects on *Panicum miliaceum* plants. *Bull. Torrey. bot. Club* **110**: 435–441.

Agrawal, M., Agrawal, S.B., Krizek, D.T., Kramar, G.F., Lee, E.H., Mirecki, R.M. & Rowland, R.A. 1991. Physiological and morphological responses of snap bean plants to ozone stress as influenced by pretreatment with UV-B radiation. Impact of global climatic changes on photosynthesis and plant productivity. *Proc. Indo-US Workshop* Jan. 8–12, 1991, New Delhi, pp. 133–146. Oxford and IBH Publishing Company Pvt. Ltd, New Delhi.

Alscher, R.G. & Amthor, J.S. 1988. The physiology of free radical scavenging-maintenance and repair processes. In Schulte-Hostede, S., Darall, N.M., Blank, L.W. & Wellburn, A.R. (eds) *Air Pollution and Plant Metabolism*: 94–115. Elsevier, London.

Amundson, R.G., Raba, R.M., Schoettle, A.W. & Reich, P.B. 1986. Response of soybeans to low concentrations of ozone. II. Effects on growth, biomass allocation and flowering. *J. Environ. Qual.* **15**: 161–167.

Anderson, W.C. & Taylor, O.C. 1973. Ozone-induced carbon dioxide evolution in tobacco callus cultures. *Physiol. Plant Mol. Biol.* **39**: 379–411.

Atkinson, C.J., Robe, S.V. & Winner, W.E. 1988. The relationship between changes in photosynthesis and growth for radish plants fumigated with SO_2 and O_3. *New Phytol.* **110**: 173–184.

Baker, N.R., Nie, G. & Tomasevic, M. 1994. Responses of photosynthetic light-use efficiency and chloroplast development on exposure of leaves to ozone. In Alscher, R.G. & Wellburn, A.R. (eds) *Plant Responses to the Gaseous Environment. Molecular, Metabolic and Physiological Aspects*: 219–238. Chapman & Hall, London.

Barnes, J.D., Eamus, D. & Brown, K.A. 1990. The influence of ozone, acid mist, soil nutrient status on Norway spruce *Picea abies* (L.) Karst. II. Photosynthesis, dark respiration and soluble carbohydrates of trees during late autumn. *New Phytol.* **115**: 149–156.

Barnes, R.L. 1972a. Effects of chronic exposure to ozone on soluble sugar and ascorbic acid contents of pine seedlings. *Can. J. Bot.* **50**: 215–219.

Barnes, R.L. 1972b. Effects of chronic exposure to ozone on photosynthesis and respiration of pines. *Environ. Pollut.* **3**: 133–138.

Bennett, J.P. & Oshima, R.J. 1976. Carrot injury and yield response to ozone. *J. Amer. Soc. Hort. Sci.* **101**: 638–639.

Bennett, J.P. & Runeckles, V.C. 1977. Effects of low levels of ozone on growth of crimson clover and annual ryegrass. *Crop. Sci.* **17**: 443–445.

Bennett, J.P., Oshima, R.J. & Lippert, L.F. 1979. Effects of ozone on injury and dry matter partitioning in pepper plants. *Environ. Expt. Bot.* **19**: 33–39.

Black, V.J., Ormrod, D.P. & Unsworth, M.H. 1982. Effects of low concentration of ozone, singly and in combination with sulphur dioxide on net photosynthesis rates of *Vicia faba* L. *J. Exp. Bot.* **33**: 1302–1311.

Botkin, D.B., Smith, W.H., Carlson, R.W. & Smith, T.L. 1972. Effects of ozone on white pine saplings: Variation in inhibition and recovery of net photosynthesis. *Environ. Pollut.* **3**: 273–289.

Cape, J.N., Paterson, I.S., Wellburn, A.R., Wolfenden, J., Mehlhorn, H., Freer-smith, P.H. & Fink, S. 1988. *Early Diagnosis of Forest Decline*. Institute of Terrestrial Ecology, Morlewood Research Laboratory, England.

Chameides, W.L. & Lodge, J.P. 1992. Tropospheric ozone: Formation and fate. In Lefohn, A.S. (ed.) *Surface Level Ozone Exposures and Their Effect on Vegetation*: 5-30. Lewis Publishers, Chelsea, USA.

Chapman, S. 1930. A theory of upper atmospheric ozone. *Mem. Roy. Meterol. Soc.* **3**: 103-125.

Chappelka, A.H. & Chevone, B.I. 1986. White ash seedling growth response to ozone and simulated acid rain. *Can. J. For. Res.* **16**: 31-45.

Chappelka, A.H. & Chevone, B.I. 1992. Tree responses to ozone. In Lefohn, A.S. (ed.) *Surface Level Ozone Exposures and Their Effect on Vegetation*: 5-30. Lewis Publishers, Chelsea, USA.

Chappelka, A.H., Chevone, B.I. & Burk, T.E. 1985. Growth response of yellow poplar (*Liriodendron tulipifera* L.) seedlings to ozone, sulphur dioxide and simulated acidic precipitation, alone and in combination. *Environ. Exp. Bot.* **25**: 233-234.

Chevone, B.I., Yang, Y.S. & Reddick, G.S. 1984. Acidic precipitation and ozone effects on growth of loblolly and shortleaf pine seedlings. *Phytopathology* **74**: 756 (abstract).

Cooley, D.R. & Manning, W.J. 1987. The impact of ozone on assimilate partitioning in plants: A review. *Environ. Pollut.* **47**: 95-113.

Dalling, M.J. 1987. Proteolytic enzymes and leaf senescence. In Thomson, W.W., Nothnagel, E.A. & Huffaker, R.C. (eds) *Plant Senescence: Its Biochemistry and Physiology*: 54-70. American Society of Plant Physiologists, Rockville.

Dann, M.S. & Pell, E.J. 1989. Decline of activity and quantity of ribulose bisphosphate carboxylase/oxygenase and net photosynthesis in ozone-treated potato foliage. *Plant Physiol.* **91**: 427-432.

Davison, A.W., Barnes, J.D. & Renner, C.R. 1988. Interaction between air pollutants and cold stress. In Schulte-Hostede, S., Darrall, N.M., Blank, L.W. & Wellburn, A.R. (eds) *Air Pollution and Plant Metabolism*: 307-328. Elsevier, London.

Dizengremel, P. & Citernae, A. 1988. Air pollutant effects on mitochondria and respiration. In Schulte-Hostede, S., Darrall, N.M., Blank, L.W. & Wellburn, A.R. (eds) *Air Pollution and Plant Metabolism*: 169-188. Elsevier, London.

Dugger, W.M. Jr, & Ting, I.P. 1970. Physiological and biochemical effects of air pollution oxidants on plants. *Rec. Adv. Phytochem.* **3**: 31-58.

Eckardt, N.E. 1993. Ozone-induced structural modification and enhanced degradation of Rubisco in potato foliage. PhD Thesis, The Pennsylvania State University, Pennsylvania.

Edinger, J.G. 1973. Vertical distribution of photochemical smog in Los Angeles basin. *Environ. Sci. Tech.* **7**: 247-252.

Elkiey, T. & Ormrod, D.P. 1979. Leaf diffusion resistance responses of three *Petunia* cultivars to ozone and/or sulphur dioxide. *JAPCA* **29**: 622-625.

Enyedi, A.J., Eckardt, N.A. & Pell, E.J. 1992. Activity of ribulose bisphosphate carboxylase/oxygenase from potato cultivars with differential response to ozone stress. *New Phytol.* **122**: 493-500.

Faensen-Thiebes, A. 1983. Veranderungen in Gaswechsel, Chlorophyllgehalt un Zuwachs von *Nicotiana tabacum* L. un *Phaseolus vulgaris* L. durch Ozone und deren beziehung zur ausbildung. *Blattnekrosen Angew. Botanik* **57**: 181-191.

Farage, P.K., Long, S.P., Lechner, E.G. & Baker, N.R. 1991. The sequence of the change within the photosynthetic apparatus of wheat following short-term exposure to ozone. *Plant Physiol.* **95**: 529-535.

Finlayson-Pitts, B.J. & Pitts, J.N. Jr 1986. *Atmospheric Chemistry: Fundamentals and Experimental Techniques*. John Wiley & Sons, New York, 1098 pp.

Fishman, J. & Carney, T.A. 1984. A one-dimentional photochemical model of the troposphere with planetary boundary layer parameterization. *J. Atmos. Chem.* **1**: 351-376.

Fishman, J., Vukovich, F.M. & Browell, E.V. 1985. The photochemistry of synoptic-scale ozone synthesis: Implications for the global ozone budget. *J. Atmos. Chem.* **3**: 299-320.

Flagler, R.B. & Younger, V.B. 1982. Ozone and sulphur dioxide effects on tall fescue. I. Growth and yield responses. *J. Environ. Qual.* **11**: 386-389.
Foster, J.R., Loats, K.V. & Jensen, K.F. 1990. Influence of two growing seasons of experimental ozone fumigation on photosynthetic characteristics of white oak seedlings. *Environ. Pollut.* **65**: 371-380.
Foster, K.W., Timm, H., Labanaukas, C.K. & Oshima, R.J. 1983. Effects of ozone and sulphur dioxide on tuber yield and quality of potatoes. *J. Environ. Qual.* **12**: 75-80.
Giaquinta, R.T. 1983. Phloem loading of sucrose. *Annu. Rev. Plant Physiol.* **34**: 347-387.
Giese, A.A., Leighton, H.L. & Bailey, R. 1952. Changes in the absorption spectra of proteins and representative amino acids induced by ultraviolet radiations and ozone. *Arch. Biochem. Biophys.* **40**: 71-84.
Glazer, A.N. 1989. Light guides-directional energy transfer in a photosynthetic antenna. *J. Biol. Chem.* **264**: 1-4.
Godde, D. & Buchhold, J. 1992. Effect of long term fumigation with ozone on the turnover of the D-1 reaction center polypeptide of photosystem II in spruce (*Picea abies*). *Physiol. Plant.* **86**: 568-574.
Greitner, C.S. & Winner, W.E. 1988. Increase in the $\delta^{13}C$ values of radish and soybean plants caused by ozone. *New Phytol.* **108**: 489-494.
Guderian, R., Tingey, D.T. & Rabe, R. 1985. Effects of photochemical oxidants on plants. In Guderian, R. (ed.) *Air Pollution by Photochemical Oxidants*: 111-128. Springer-Verlag, Berlin.
Guri, A. 1983. Variation in glutathione and ascorbic acid content among selected cultivars of *Phaseolus vulgaris* prior to and after exposure to ozone. *Can. J. Plant Sci.* **63**: 733-737.
Hällgren, J.E. 1984. Photosynthetic gas exchange in leaves affected by air pollutants. In Koziol, M.J. & Whatley, F.R. (eds) *Gaseous Air Pollutants and Plant Metabolism*: 147-159. Butterworths, London.
Halliwell, B. 1974. Superoxide dismutase, catalase and glutathione peroxidase: Solutions to the problems of living with oxygen. *New Phytol.* **73**: 1075-1086.
Halliwell, B. 1982. Ascorbic acid and illuminated chloroplast. *Ann. Chem. Soc. Adv. Chem. Ser.* **22**: 263-274.
Hanson, P.J., McLaughlin, S.B. & Edwards, N.T. 1988. Net CO_2 exchange of *Pinus taeda* shoots exposed to variable ozone levels and rain chemistries in field and laboratory settings. *Physiol. Plant.* **74**: 635-642.
Heagle, A.S., Letchworth, M.B. & Mitchell, C.A. 1983. Injury and yield responses of peanuts to chronic doses of ozone in open-top field chambers. *Phytopathology* **73**: 551-555.
Hill, A.C. & Littlefield, N. 1969. Ozone: Effect on apparent photosynthesis, rate of transpiration, and stomatal closure in plants. *Environ. Sci. Technol.* **3**: 52-56.
Hofstra, G., Ali, A., Wukasch, R.T. & Fletcher, R.A. 1981. The rapid inhibition of root respiration after exposure of bean (*Phaseolus vulgaris* L.) plant to ozone. *Atmos. Environ.* **15**: 483-487.
Hogsett, W.E., Plocher, M., Wildman, V., Tingey, D.T. & Bennett, J.P. 1985. Growth response of two varieties of slash pine seedlings to chronic ozone exposures. *Can. J. Bot.* **63**: 2369-2376.
Ito, O., Mitsumori, F. & Totsuka, T. 1985a. Effects of NO_2 and O_3 alone or in combination on kidney bean plants (*Phaseolus vulgaris* L.) products of $^{13}CO_2$ assimilation detected by ^{13}C nuclear magnetic resonance. *J. Exp. Bot.* **36**: 281-289.
Ito, O., Okano, K., Kuroiwa, M. & Totsuka, T. 1985b. Effects of NO_2 and O_3 alone or in combination on kidney bean plants (*Phaseolus vulgaris* L.) growth, partitioning of assimilates and root activities. *J. Exp. Bot.* **36**: 652-662.
Jensen, K.F. 1981. Air Pollutants Affect the Relative Growth Rate of Hard Wood Seedlings. USDA, Forest Service Research Paper NE-470 NE Forest Experiment Station, Broomal.
Jensen, K.F. 1983. Growth relationships in silver maple seedlings fumigated with O_3 and SO_2. *Can. J. For. Res.* **13**: 298-302.
Kangasjarvi, J., Talvinen, J., Utriainen, M. & Karjalainen, R. 1994. Plant defense system induced by ozone. *Plant Cell Environ.* **17**: 783-794.

Keller, T. & Hasler, R. 1987. Some effects of long-term ozone fumigations on Norway spruce. I. Gas exchange and stomatal response. *Trees* **1**: 129-133.

Koiwoi, A. & Kisaki, T. 1976. Effect of O_3 on photosystem II in tobacco chloroplasts in the presence of piperobutoxide. *Plant Cell Environ.* **17**: 1199-1207.

Koziol, M.J., Whatley, F.R. & Shelvey, J.D. 1988. An integrated view of the effects of gaseous air pollutants on plant carbohydrate metabolism. In Schulte-Hostede, S., Darrall, N.M., Blank, L.W. & Wellburn, A.R. (eds) *Air Pollution and Plant Metabolism*: 148-168. Elsevier, London.

Kress, L.W. & Skelly, J.M. 1982a. Response of several eastern forest tree species to chronic doses of ozone and nitrogen dioxide. *Plant Disease* **66**: 1149-1152.

Kress, L.W. & Skelly, J.M. 1982b. Relative sensitivity of 18 full-sub families of *Pinus taeda* to O_3. *Can. J. For. Res.* **12**: 203-209.

Kumar, N. & Sane, P.V. 1990. Carboxylation enzymes in photosynthesis. *Biol. Edu.* **7**: 26-34.

Kyle, D.J., Baker, N.R. & Arntzen, C.J. 1983. Spectral characterization of photosystem I fluorescence at room temperature using thylakoid protein phosphorylation. *Photobiochem. Photobiophys.* **5**: 79-86.

Laisk, A., Kull, O. & Moldau, H. 1989. Ozone concentration in leaf intercellular air spaces is close to zero. *Plant Physiol.* **90**: 1163-1167.

Landry, L.G. & Pell, E.J. 1993. Modification of Rubisco and altered proteolytic activity in O_3-stressed hybrid poplar (*Populus maximowizii* × *trichocarpa*). *Plant Physiol.* **101**: 1355-1362.

Larsen, R.I. & Heck, W.W. 1976. Air quality data analysis system for interrelating effects, standards and needed source reductions: Part III. Vegetation injury. *JAPCA* **26**: 325-333.

Lee, E.H. & Bennett, J.H. 1982. Superoxide dismutase. A possible protective enzyme against ozone injury in snap beans (*Phaseolus vulgaris* L.). *Plant Physiol.* **69**: 1444-1449.

Lee, T.T. 1967. Inhibition of oxidative phosphorylation and respiration by ozone in tobacco mitochondria. *Plant Physiol.* **42**: 691-696.

Lefohn, A.S. 1992. *Surface Level Ozone Exposures and Their Effect on Vegetation.* Lewis Publishers, Chelsea, USA.

Lehnherr, B., Grandjean, A., Machler, F. & Führer, J. 1987. The effect of ozone in ambient air on ribulose bisphosphate carboxylase/oxygenase activity decreases photosynthesis and grain yield in wheat. *J. Pl. Physiol.* **130**: 189-200.

Lehnherr, B., Machler, F., Grandjean, A. & Führer, J. 1988. The regulation of photosynthesis in leaves of field-grown spring wheat (*Triticum aestivum* L. cv. Albis) at different levels of ozone in ambient air. *Plant Physiol.* **88**: 1115-1119.

Le Sueur-Brymer, N.M. & Ormrod, D.P. 1984. CO_2 exchange rates of fruiting soybean plants exposed to O_3 and SO_2 singly or in combination. *Can. J. Plant Sci.* **64**: 69-75.

Letchworth, M.B. & Blum, U. 1977. Effects of acute ozone exposure on growth, nodulation and nitrogen content of ladino clover. *Environ. Pollut.* **14**: 303-312.

Levy, H. II. 1971. Normal atmosphere: Large radical and formaldehyde concentrations predicted. *Science* **173**: 141-143.

Lorimer, G.H. 1981. The carboxylation and oxygenation of ribulose-1,5-biphosphate: The primary events in photosynthesis and photorespiration. *Annu. Rev. Plant Physiol.* **32**: 349-383.

MacDowall, F.D.H. 1965. Stages of ozone damage to respiration of tobacco leaves. *Can. J. Bot.* **43**: 419-427.

Mahoney, M.J., Skelly, J.M., Chevone, B.I. & Moore, L.D. 1984. Responses of yellow-poplar (*Liriodendron tulipifera*) seedlings shoot growth to low concentrations of O_3, SO_2 and NO_2. *Can. J. For. Res.* **14**: 150-153.

Martin, B., Bytnerowicz, A. & Thortenson, Y.R. 1988. Effect of air pollution on composition of stable isotopes $\delta^{13}C$, of leaves and wood and on leaf injury. *Plant Physiol.* **88**: 218-223.

Mass, E.V., Hoffman, G.J., Rawlins, S.L. & Ogata, G. 1973. Salinity-ozone interactions in Pinto bean: Integrated response to ozone concentration and duration. *J. Environ. Qual.* **2**: 400-404.

McCool, P.M. & Menge, J.A. 1983. Effect of O_3 on carbon partitioning in tomato: Potential role of carbon flow in regulation of mycorrhizal symbiosis under condition of stress. *New Phytol.* **94**: 241-247.

McLaughlin, S.B. & McConathy, R.K. 1983. Effects of SO_2 and O_3 on allocation of ^{14}C-labelled photosynthate in *Phaseolus vulgaris*. *Plant Physiol.* **73**: 630-635.

McLaughlin, S.B., McConathy, R.K., Duvick, D. & Mann, L.K. 1982. Effect of chronic air pollution on photosynthesis, carbon allocation and growth of white pine trees. *For. Sci.* **28**: 60-70.

Mehlhorn, H. & Wellburn, A.R. 1987. Stress ethylene formation determines plant sensitivity to ozone. *Nature* **327**: 417-418.

Mehlhorn, H., O'Shea, J.M. & Wellburn, A.R. 1991. Atmospheric ozone interacts with stress ethylene formation by plants to cause visible plant injury. *J. Exp. Bot.* **42**: 17-24.

Miller, J.E. 1987. Effects of ozone and sulphur dioxide stress on growth and carbon allocation in plants. In Saunders, J.A., Kosak-Channing, L. & Conn, E.E. (eds) *Phytochemical Effects of Environmental Compounds*: 55-100. Plenum Press, New York.

Miller, P.R., Parmeter, J.R., Flick, B.H. & Martinez, C.W. 1969. Ozone dosage response of ponderosa pine seedlings. *JAPCA* **19**: 435-538.

Miyake, H., Furukawa, A., Totsuka, T. & Maeda, E. 1984. Differential effects of ozone and sulphur dioxide on the fine structure of spinach leaf cells. *New Phytol.* **96**: 215-228.

Mooney, H.A. & Winner, W.E. 1988. Carbon gain allocation and growth as affected by atmospheric pollutants. In Schulte-Hostede, S., Darrall, N.M., Blank, L.W. & Wellburn, A.R. (eds) *Air Pollution and Plant Metabolism*: 272-287. Elsevier, London.

Mudd, J.B. 1982. Effects of oxidants on metabolic function. In Unsworth, M.H. & Ormrod, D.P. (eds) *Effects of Gaseous Air Pollutants on Agriculture and Horticulture*: 189-201. Butterworth Scientific, London.

Nakamura, H. & Zaka, H. 1978. Photochemical oxidants injury in rice plants. *Japan. J. Crop Sci.* **47**: 704-714.

Ogata, G. & Maas, E.V. 1973. Interactive effects of salinity and ozone on growth and yield of garden beet. *J. Environ. Qual.* **2**: 518-520.

Okano, K., Ito, O., Takeba, G., Shimizu, A. & Totsuka, T. 1984. Alteration of ^{13}C-assimilate partitioning in plants of *Phaseolus vulgaris* exposed to ozone. *New Phytol.* **97**: 155-163.

Oshima, R.J., Taylor, O.C., Braegelmann, P.K. & Baldwin, D.W. 1975. Effect of ozone on the yield and plant biomass of a commercial variety of tomato. *J. Environ. Qual.* **4**: 463-464.

Oshima, R.J., Bennett, J.P. & Braegelmann, P.K. 1978. Effect of ozone on growth and assimilate partitioning in parsley. *J. Am. Soc. Hort. Sci.* **103**: 348-350.

Oshima, R.J., Braegelmann, P.K., Flagler, R.B. & Teso, R.R. 1979. The effects of ozone on the growth, yield and partitioning of dry matter in cotton. *J. Environ. Qual.* **8**: 575-579.

Pell, E.J. & Brennan, E. 1973. Changes in respiration, photosynthesis, adenosine 5'-triphosphate, and total adenylate content of ozonated pinto bean foliage as they relate to symptom expression. *Plant Physiol.* **51**: 378-381.

Pell, E.J., Eckardt, N.A. & Enyedi, A.J. 1992. Timing of ozone stress and resulting status of ribulose bisphosphate carboxylase/oxygenase and associated net photosynthesis. *New Phytol.* **120**: 397-405.

Pell, E.J., Eckardt, N.A. & Glick, R.E. 1994a. Biochemical and molecular basis for impairment of photosynthetic potential. *Photosynthesis Res.* **39**: 453-462.

Pell, E.J., Landry, L.G., Eckardt, N.A. & Glick, R.E. 1994b. Effects of gaseous air pollutants on ribulose bisphosphate carboxylase/oxygenase: Effects and implications. In Alscher, R.G. & Wellburn, A.R. (eds) *Plant Responses to the Gaseous Environment: Molecular, Metabolic and Physiological Aspects*: 239-254. Chapman & Hall, London.

Pell, E.J., Temple, P.J., Friend, A.L., Mooney, H.A. & Winner, W.E. 1994c. Compensation as a plant response to ozone and associated stresses: An analysis of ROPIS experiments. *J. Environ. Qual.* **23**: (in press)

Raz, V. & Fluhr, R. 1993. Ethylene signal is transduced via protein phosphorylation events in plants. *Plant Cell* **5**: 523-530.

Reddy, G., Arteca, R.N., Dai, Y.R., Flores, H.E., Negm, F.B. & Pell, E.J. 1993. Changes in ethylene and polyamines in relation to mRNA levels of the large and small subunits of ribulose biphosphate carboxylase/oxygenase in ozone-stressed potato foliage. *Plant Cell Environ.* **16**: 559-561.

Reich, P.B. 1983. Effects of low concentrations of O_3 on net photosynthesis, dark respiration and chlorophyll contents in ageing hybrid poplar leaves. *Plant Physiol.* **73**: 291-296.

Reich, P.B. & Amundson, R.G. 1985. Ambient levels of ozone reduce net photosynthesis in tree and crop species. *Science* **230**: 566-570.

Reich, P.B., Lassoie, J.P. & Amundson, R.G. 1984. Reduction in growth of hybrid poplar following field exposures of O_3 and/or SO_2. *Can. J. Bot.* **62**: 2835-2841.

Reich, P.B., Schoettle, A.W. & Amundson, R.G. 1985. Effects of low concentration of O_3, leaf age and water stress on leaf diffusive conductance and water use efficiency in soybean. *Physiol. Plant.* **63**: 58-64.

Reich, P.B., Schoettle, A.W., Raba, R.M. & Amundson, R.G. 1986. Response of soybean to low concentrations of ozone. I. Reductions in leaf and whole plant net photosynthesis and leaf chlorophyll content. *J. Environ. Qual.* **15**: 31-36.

Reich, P.B., Schoettle, A.W., Stroo, H.F., Troiano, J. & Amundson, R.G. 1987. Influence of O_3 and acid rain on white pine (*Pinus strobus*) seedlings grown in five soils. I. Net photosynthesis and growth. *Can. J. Bot.* **65**: 977-987.

Reinert, R.A. & Sanders, J.S. 1982. Growth of radish and marigold following repeated exposure to nitrogen dioxide and ozone. *Plant Disease* **66**: 122-124.

Robert, B.R. 1990. Physiological response of yellow-poplar seedlings to simulated acid rain, ozone fumigation and drought. *For. Ecol. Manag.* **31**: 215-224.

Robinson, D.C. & Wellburn, A.R. 1983. Light-induced changes in the quenching of 9-amino acridine fluorescence by photosynthetic membranes due to atmospheric pollutants and their products. *Environ. Pollut.* **32**: 109-120.

Robinson, J.M. 1988. Does O_2 photoreduction occur within chloroplasts? *Physiol. Plant.* **72**: 666-680.

Rowland-Bamford, A.J., Coghland, S. & Lea, P.J. 1989. Ozone-induced changes in CO_2 assimilation, O_2 evolution and chlorophyll *a* fluorescence transients in barley. *Environ. Pollut.* **59**: 129-140.

Runeckles, V.C. & Chevone, B.I. 1992. Crop responses to ozone. In Lefohn, A.S. (ed.) *Surface Level Ozone Exposures and Their Effect on Vegetation*: 189-269. Lewis Publishers, Chelsea, USA.

Salin, M.L. 1988. Toxic oxygen species and protective systems of the chloroplast. *Physiol. Plant* **72**: 681-689.

Sasek, T.W. & Richardson, C.J. 1989. Effects of chronic doses of ozone on loblolly pine: Photosynthetic characteristics in the third growing season. *For. Sci.* **35**: 745-755.

Scherzer, A.J. & McClenahen, J.R. 1989. Effects of ozone or sulphur dioxide on pitch pine seedlings. *J. Environ. Qual.* **18**: 57-61.

Schreiber, U., Vidaver, W., Runeckles, V.C. & Rosen, P. 1978. Chlorophyll assay for ozone injury in intact plants. *Plant Physiol.* **61**: 80-84.

Schut, H.E. 1985. Models for the physiological effects of short O_3 exposures on plants. *Ecol. Model* **30**: 175-207.

Sengupta, A., Alscher, R.G. & McCune, D. 1991. Response of photosynthesis and cellular antioxidants to ozone in Populus leaves. *Plant Physiol.* **96**: 650-655.

Shimazaki, K. 1988. Thylakoid membrane reactions to air pollutants. In Schulte-Hostede, S., Darrall, N.M., Blank, L.W. & Wellburn, A.R. (eds) *Air Pollution and Plant Metabolism*: 116-133. Elsevier, London.

Shimizu, H., Motohashi, S., Iwaki, H., Furukawa, A. & Totsuka, T. 1980. Effects of chronic exposures to ozone on the growth of sunflower plants. *Environ. Control Biol.* **19**: 137-147.

Skärby, L., Troeng, E. & Botstrom, C.A. 1987. Ozone uptake and effects on transpiration, net photosynthesis and dark respiration in scots pine. *For. Sci.* **33**: 810-818.

Spence, R.D., Rykiel, E.J. & Sharpe, P.J.H. 1990. Ozone alters carbon allocation in loblolly pine: assessment with carbon-11 labelling. *Environ. Pollut.* **64**: 93-106.

Stadman, E.R. 1988. Biochemical markers of ageing. *Exp. Geront.* **23**: 327-347.

Sugahara, K., Ogura, K., Takimoto, M. & Kondo, N. 1984. Effects of air pollutant mixtures on photosynthetic electron transport systems. *Res. Rep. Natl. Ins. Environ. Stud. (Japan)* **65**: 155-164.

Taylor, G.E. Jr, Norby, R.J., McLaughlin, S.B., Johnson, A.H. & Turner, R.S. 1986. Carbon dioxide assimilation and growth of red spruce (*Picea rubens* Sarg.) seedlings in response to ozone, precipitation chemistry, and soil type. *Oecologia* **70**: 163-171.

Taylor, G.E. Jr, Ross-Todd, B.M. & Gunderson, C.A. 1988. Action of ozone on foliar gas exchange in *Glycine max*. L. Merr.: a potential role for endogenous stress ethylene. *New Phytol.* **110**: 301-307.

Thomson, W.W., Nagahashi, J. & Platt, K. 1974. Further observations on the effects of ozone on the ultrastructure of leaf tissue. In Dugger, M. (ed.) *Air Pollution Effects on Plant Growth*: 83-93. *ACS Symposium Series 3*. American Chemical Society, Washington, DC.

Tingey, D.T. & Reinert, R.A. 1975. The effect of ozone and sulphur dioxide singly and in combination on plant growth. *Environ. Pollut.* **9**: 117-125.

Tingey, D.T., Reinert, R.A., Wickliff, C. & Heck, W.W. 1973. Chronic ozone or sulphur dioxide exposures, or both, affect the early vegetative growth of soybean. *Can. J. Plant Sci.* **53**: 875-879.

Tingey, D.T., Standley, C. & Field, R.W. 1976a. Stress ethylene evolution: A measure of ozone effects on plants. *Atmos. Environ.* **10**: 969-974.

Tingey, D.T., Wilhour, R.G. & Standley, C. 1976b. The effect of chronic ozone exposures on the metabolite content of ponderosa pine seedlings. *For. Sci.* **22**: 234-241.

Troughton, J.H. 1979. $\delta^{13}C$ as an indicator of carboxylation reactions. In Gibbs, M. & Latzko, E. (eds) *Photosynthesis Encyclopedia. Plant Physiol.* New Series vol. 6, 140-149. Springer-Verlag, Berlin.

Vaartnou, M. 1988. EPR Investigation of Free Radicals in Excised and Attached Leaves Subjected to Ozone and Sulphur Dioxide. PhD Thesis, University of British Columbia, Vancouver.

Wang, D., Karnosky, D.G. & Bormann, H.F. 1986. Effects of ambient ozone on the productivity of *Populus tremuloides* Mich. grown under field conditions. *Can. J. For. Res.* **16**: 47-55.

Weber, J.A., Clark, S.C. & Hogsett, W.E. 1993. Analysis of the relationships among O_3 uptake, conductance and photosynthesis in the needles of *Pinus ponderosa*. *Tree Physiol.* **13**: 157-172.

Wellburn, A.R. 1984. The influence of atmospheric pollutants and their cellular products upon photophosphorylation and related events. In Koziol, M.J. & Whatley, F.R. (eds) *Gaseous Air Pollutants and Plant Metabolism*: 203-223. Butterworths, London.

Wellburn, A.R. 1988. *Air Pollution and Acid Rain: the Biological Impact*. Longmans Scientific and Technical, Harlow, England, 274 pp.

Wilhour, R.G. & Neely, G.E. 1977. Growth response of conifer seedlings to low ozone concentrations. In *Proc. Int. Conf. Photo. Oxidant Pollution and its Control*: 635-645. USEPA Rep. No. 300/377-0106.

Wilkinson, T.G. & Barnes, R.L. 1973. Effects of ozone on $^{14}CO_2$ fixation patterns in pines. *Can. J. Bot.* **51**: 1573-1578.

Yang, Y.S., Skelly, J.M., Chevone, B.I. & Birch, J.B. 1983a. Effects of short-term ozone exposure on net photosynthesis, dark respiration and transpiration of three eastern white pine clones. *Environ. Int.* **9**: 265-269.

Yang, Y.S., Skelly, J.M., Chevone, B.I. & Birch, J.B. 1983b. Effects of long-term ozone exposure on photosynthesis, dark respiration and transpiration of eastern white pine. *Environ. Sci. Tech.* **17**: 371-373.

13 The Effect of Air Pollutants and Elevated Carbon Dioxide on Nitrogen Metabolism

P.J. LEA, A.J. ROWLAND-BAMFORD & J. WOLFENDEN

INTRODUCTION

Nitrogen is a constituent of a large number of important compounds found in plant cells, e.g., amino acids, polyamines, chlorophyll, proteins and nucleic acids. Although nitrogen is readily available in the air, only certain bacteria have the ability to carry out nitrogen fixation and synthesize ammonia. In the root nodules of legumes this is a particularly important process. However, in the majority of agricultural crop plants, nitrate is the sole source of nitrogen and is taken up from the soil by the roots. Ammonia, in the form of NH_4^+ may also be taken up by roots, and this will be discussed in a later section. Nitrogen is transported to the growing parts of the plant and is ultimately stored in the seed, where it may be of commercial value, e.g., in cereals and legumes.

Nitrate is taken up in roots by two systems. The first is a high-affinity system, which has a low K_m and is induced by nitrate. The second is a low-affinity system that operates at high external nitrate concentrations (King *et al.*, 1992; Oaks, 1992). Nitrate is reduced to ammonia by a two-step process catalysed by the enzymes nitrate reductase and nitrite reductase (Kleinhofs & Warner, 1990).

$$NO_3^- + 2H^+ + 2e^- \rightarrow NO_2^- + H_2O$$

$$NO_2^- + 8H^+ + 6e^- \rightarrow NH_4^+ + 2H_2O$$

Nitrate reduction can then take place in the root or the shoot depending upon the plant species, developmental age or nitrate supply. In general, as the external nitrate increases, the proportion that is transported to the shoot for reduction increases.

Nitrate reductase

Nitrate reductase is located in the cytoplasm of plant tissues and uses NADH as the preferred coenzyme. However, a NADPH-dependent enzyme has been isolated from a number of sources (Wray & Kinghorn, 1989; Kleinhofs & Warner, 1990). The nitrate reductase protein comprises two identical subunits

Plant Response to Air Pollution. Edited by Mohammad Yunus and Muhammad Iqbal

of 100–115 kDa, each containing the prosthetic groups FAD, haem and a molybdenum cofactor.

Although at first examination nitrate reductase appears relatively simple to assay, the enzyme is notoriously unstable (Hageman & Reed, 1980). A range of protectants have to be added to the enzyme-extraction medium, if accurate measurements of *in vitro* reductase activity are to be determined (Wray & Fido, 1990). Speed is also important to prevent loss of enzyme activity (Riens & Heldt, 1992). As will be seen in a later section, the enzyme is subject to complex regulation which may well be affected by pollutant gases. It is now also possible to determine the level of nitrate reductase protein (Remmler & Campbell, 1986; Oaks *et al.*, 1988) and mRNA (Deng *et al.*, 1990; Bowsher *et al.*, 1991).

It has been known for some time that the addition of nitrate to nitrogen-starved plants induces nitrate reductase activity (Wray & Kinghorn, 1989; Kleinhofs & Warner, 1990; Redinbaugh & Campbell, 1991), and that these findings have now been confirmed at the gene level (Cheng *et al.*, 1986; Melzer *et al.*, 1988; Crawford *et al.*, 1989; Vaucheret *et al.*, 1992). In leaves, light is also required for the induction of nitrate reductase synthesis (Deng *et al.*, 1990; Bowsher *et al.*, 1991), although in some circumstances sucrose can substitute for light (Cheng *et al.*, 1992). During the normal light/dark cycles of growth, considerable diurnal variations in the levels of nitrate reductase mRNA, protein and activity have been detected (Deng *et al.*, 1990; Bowsher *et al.*, 1991; Lillo, 1991). Considerable variations in the levels of nitrate reductase activity have also been detected in both broadleaved and coniferous trees during the growing season (Clough *et al.*, 1989; Pearson *et al.*, 1989).

As well as these advances in understanding the regulation of nitrate reductase activity at the transcriptional level, research has also shown that there may be rapid modulation of enzyme activity *in vivo*. When spinach leaves are exposed to low CO_2 in the light, nitrate reductase decreases to 10% of the original activity within 1 hour. This inhibition is reversed when the leaves are returned to normal air (Kaiser & Brendle-Behnisch, 1991). In a related series of experiments Riens & Heldt (1992) showed that nitrate reductase activity decreased dramatically following the darkening of spinach leaves for 10 minutes. They argued that such a decrease in activity prevents the build-up of toxic nitrite. It has been suggested that nitrate reductase may be reversibly deactivated following phosphorylation by ATP (Kaiser & Spill, 1991; Huber *et al.*, 1992).

Nitrite reductase

Nitrite reductase is located both in the leaf chloroplasts and root plastids. Most studies have shown that the enzyme is a monomeric protein of 60–64 kDa (Kleinhofs & Warner, 1990), which uses ferredoxin as the source of reductant to convert nitrite to ammonia. The gene for the enzyme has been cloned from a variety of plant sources (Kramer *et al.*, 1989; Vaucheret *et al.*, 1992) and the transit peptide required for transport into the chloroplast determined.

Nitrite reductase is regulated by nitrate and light in a manner similar to that of nitrate reductase (Kleinhofs & Warner, 1990). In maize, nitrite reductase mRNA is rapidly induced by nitrate, increasing first in the roots and then in the leaves.

The mRNA reaches a peak within 5 hours and then declines to a lower level (Kramer *et al.*, 1989). In maize leaves, levels of nitrite reductase mRNA, but not those of enzyme activity, follow similar diurnal oscillations to those of nitrate reductase (Bowsher *et al.*, 1991). In tobacco, Faure *et al.* (1991) have suggested that there is coregulated expression of nitrate and nitrite reductase.

The assimilation of ammonia

The action of two enzymes are required for the assimilation of ammonia into amino acids: (a) glutamine synthetase and (b) glutamate synthase (Lea *et al.*, 1990, 1992). The operation of the two enzymes in the glutamate synthase cycle is shown in Figure 13.1.

Glutamine synthetase is an octameric protein with a native molecular weight of 350–400 kDa. In leaves, two major isoenzymes have been isolated and shown to be localized in the cytoplasm (GS_1) and chloroplast (GS_2) (Lea *et al.*, 1990). The proportion of the two isoenzymic forms present varies with the plant studied (McNally *et al.*, 1983; Stewart *et al.*, 1988, 1989). The appearance of chloroplastic glutamine synthetase in leaves is apparently regulated by light. In wheat leaves, the enzyme activity increases with leaf age and there is a clear correlation with photosynthetic and photorespiratory capacity (Tobin *et al.*, 1988). The chloroplast enzyme comprises one subunit of 43–45 kDa termed δ. The polypeptide is assembled in the chloroplast following the removal of a 4–5 kDa transit peptide (Forde & Cullimore, 1989; Coruzzi, 1991a, b).

The activity of plant glutamine synthetase increases dramatically during the nodulation of many legume species. Three distinct subunits of the enzyme (designated α, β, γ) have been isolated from root nodules with a molecular weight of 43 kDa. The genes coding for the individual subunits have now been isolated from *Phaseolus vulgaris* and *Pisum sativum* and their regulation studied in detail (Cullimore & Bennett, 1992).

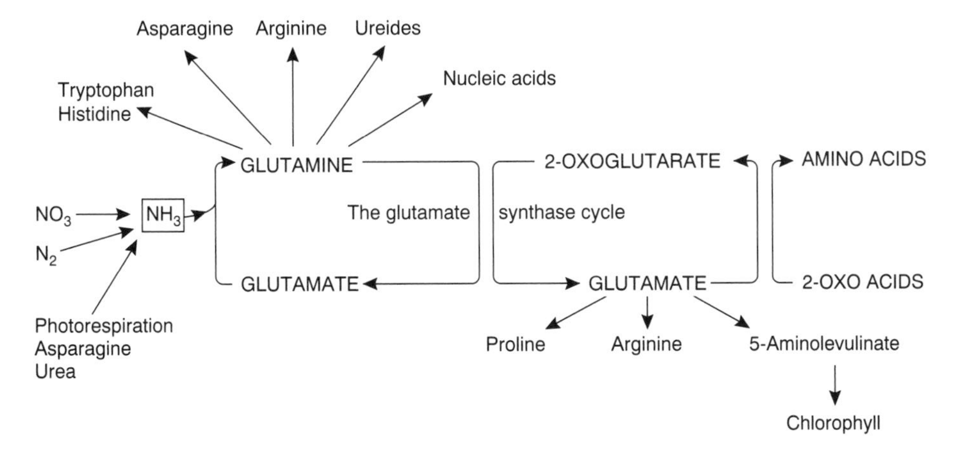

Figure 13.1. The assimilation of ammonia in higher plants via the glutamate synthase cycle

Two different forms of glutamate synthase are present in higher plants, one utilizes reduced ferredoxin as a source of reductant and the other utilizes NADH (Lea *et al.*, 1990). Ferredoxin-dependent glutamate synthase, which is present in high levels in leaves, is located solely in the chloroplast. The enzyme is an iron sulphur flavoprotein with a single polypeptide of molecular weight 140–160 kDa. NADH-dependent glutamate synthase is present in low levels in leaves but appears to play a major role in nitrogen-fixing root nodules. The enzyme from root nodules is a monomer of 200 kDa and two plastid-localized isoenzymes have been isolated in *Phaseolus vulgaris*.

The gene for ferredoxin-dependent glutamate synthase has been cloned from maize (Sakakibara *et al.*, 1991). The nucleotide sequence shows a 40% similarity to the sequence of the *Escherichia coli* NADPH-dependent enzyme. The amount of the mRNA coding for the enzyme was shown to increase 6 hours after the illumination of etiolated leaves, confirming previous data showing a light-dependent increase in enzyme activity. As yet the gene coding for the NADH-dependent enzyme has not been isolated. However, immunological studies suggest that in higher plants the ferredoxin and NADH-dependent glutamate synthase are different proteins.

Glutamate dehydrogenase catalyses the reversible amination of 2-oxoglutarate to glutamate and activity is induced in the presence of ammonia (Lea *et al.*, 1990). Until recently the enzyme had been considered to operate in the direction of glutamate synthesis. There is now good evidence that the mitochondrial enzyme operates in the direction of glutamate breakdown and is used in the synthesis of 2-oxoglutarate required for the operation of the Krebs citric acid cycle (Robinson *et al.*, 1991, 1992). The enzyme has also been used as a marker of the action of pollutant gases (Wellburn *et al.*, 1976; Borland & Lea, 1991).

The story as described above is not quite as simple as it first appears. Although the direct pathway of nitrate assimilation is nitrate → nitrite → ammonia → glutamine → glutamate, ammonia may also be generated by a number of other processes. These include nitrogen fixation, photorespiration and the metabolism of phenylpropanoids and nitrogen-transport compounds. Photorespiration occurs in the leaves of all C_3 plants and ammonia is released at a rate at least 10 times that generated by nitrate reduction (Lea *et al.*, 1990, 1992). This has been confirmed by the isolation of mutants of barley containing only 10% of the normal levels of glutamine synthetase and glutamate synthase. Such mutants can grow normally at elevated levels of CO_2 that suppresses photorespiration, but die in air (Joy *et al.*, 1992). Preliminary studies suggest that the minimum level of enzyme activity required for normal growth is approximately 50%, suggesting that there is an excess of ammonia assimilation capacity available to cope with extreme conditions (Häusler *et al.*, 1994). Raven *et al.* (1992) have recently discussed in detail, the fluxes of ammonia in both land and aquatic plants.

Amino acid synthesis

Under conditions of active growth and rapid protein synthesis, the 2-amino group of glutamate is transaminated to pyruvate and oxaloacetate to yield alanine and aspartate, respectively. However, when nitrogen has to be stored or transported, the amide nitrogen of glutamine is utilized to synthesize asparagine (Lea *et al.*, 1990; Coruzzi, 1991b) and in tropical legumes, ureides (Schubert & Boland, 1990).

A complete discussion of the biosynthesis of the remainder of the amino acids is beyond the scope of this chapter. However, it is possible to divide amino acids into families, each with a "head" or precursor, based on their biosynthetic pathway (Bryan, 1990). These divisions are artificial and there are obviously occasions when an arbitrary decision has to be taken, as to the derivation of an individual amino acid.

Aspartate:	Asparagine, lysine, threonine, methionine, isoleucine. Although leucine and valine do not derive carbon from aspartate, they share a common enzyme pathway with isoleucine.
Glutamate:	Glutamine, arginine and proline.
Pyruvate:	The group of amino acids derived from 3C precursors is the most heterogeneous and includes alanine, serine, cysteine and glycine. Pyruvate donates carbon to lysine, isoleucine and valine, and phosphoenolpyruvate is involved in the synthesis of aromatic amino acids.
Erythrose-4-phosphate:	The aromatic amino acids, phenylalanine, tyrosine and tryptophan.
Ribose-5-phosphate:	Histidine.

Although it is not intended to include sulphur metabolism in this review, some mention must be made of glutathione. This tripeptide comprises glutamate-cysteine-glycine (Rennenberg & Lamoureux, 1990), and is involved in the detoxification of harmful oxygen products via the glutathione-ascorbate cycle (Foyer & Halliwell, 1976; Asada, 1992; Polle & Rennenberg, 1992). Glutathione is maintained in the reduced state by the NADPH-dependent enzyme, glutathione reductase. A number of publications have suggested that the levels of glutathione and the activity of glutathione reductase increase following oxidative stress induced by gaseous pollutants (Mehlhorn *et al.*, 1986; Alscher *et al.*, 1987; Rowland *et al.*, 1988; Hausladen *et al.*, 1990; De Kok, 1990; Dohmen *et al.*, 1990; Price *et al.*, 1990; Rennenberg & Lamoureux, 1990; Smith *et al.*, 1990; Asada, 1992; Madamanchi *et al.*, 1992; Polle & Rennenberg, 1992).

EFFECT OF NITROGEN OXIDES, SULPHUR DIOXIDE AND OZONE ON NITROGEN METABOLISM

Nitrate reductase

Numerous factors influence the level of extractable nitrate reductase. Light intensity, time of day, leaf age, carbohydrate and nitrate supply may all have a greater effect on the levels of enzyme activity than individual pollutants.

Consequently, all these factors should be taken into account when interpreting previously published data and in the planning of fumigation experiments. The subject area has previously been reviewed in some depth by Wellburn (1990).

Induction of nitrate reductase activity by atmosphere NO_2 was first demonstrated by Zeevaart (1974) in peas grown on an ammonium-based medium. When exposed to very high levels of NO_2 (12 μl l^{-1}), rapid induction by enzyme activity was observed after only 10 minutes of fumigation. In studies on glasshouse-grown crops in CO_2-enriched atmospheres, when the level of NO can be high, Murray & Wellburn (1985) could only find significant increases in nitrate reductase activity in one cultivar (Ailsa Craig) of tomato, when exposed to 1.5 μl l^{-1} NO_2 for 18 hours. In another variety of tomato (Eurocross BB) and in two pepper varieties (Bell Boy and Rhumba), no change in root or shoot nitrate reductase activity was detected with either NO_2 or NO.

Srivastava & Ormrod (1984) showed that the large increases in shoot nitrate reductase activities were associated with increases in root nitrate supply. These were accentuated by NO_2 fumigation (0.5 μl l^{-1}), but only when the nitrogen applied to the roots was low (<1 mM). Rowland *et al.* (1987) showed that in barley, fumigation with 0.3 μl l^{-1} NO_2 for 9 days induced a significant increase in shoot nitrate reductase activity in plants grown hydroponically at low (0.01 mM) and adequate (0.1 mM) levels of nitrate. In a similar experiment, Beckett (unpublished results) examined the effect of NO_2 on barley plants grown in the presence and absence of added nitrate. The dramatic increase in leaf nitrate reductase caused by added nitrate is readily obvious in Table 13.1, along with a significant effect of NO_2. Taking into account the circadian rhythms described previously, Hufton, Wellburn & Besford (unpublished results) examined the effects of NO on nitrate reductase activity and protein levels in two varieties of lettuce grown in CO_2-enriched glasshouses. Enzyme activity increased during the light period in both varieties and decreased again at night. However, in the NO-tolerant variety of lettuce (cv. Talent), NO fumigation caused a significant increase in nitrate reductase activity, and this increase in activity was mirrored by enhanced levels of nitrate reductase protein determined by an ELISA assay. The less tolerant variety (cv. Ambassador) was less able to form fresh nitrate reductase protein.

A number of workers have examined the effect of NO_2 exposure on nitrate reductase activity in conifers because the deposition of excess nitrogen has been suggested as a possible cause of forest decline (see Wellburn, 1988). Wingsle *et al.* (1987), for example, fumigated Scots pine seedlings with 85 nl l^{-1} NO + NO_2 and,

Table 13.1. Levels of nitrate reductase (NaR) and nitrite reductase (NiR) activity (μmol g^{-1} FW h^{-1}) in the leaves of barley plants treated with NO_2 for 7 days

Treatment	NaR*	NiR*
No added nitrate	0.5	22.2
No nitrate + NO_2 (500 nl l^{-1})	3.2	36.4
Added nitrate (15 mM twice a day)	11.1	78.8
Added nitrate plus NO_2	12.2	91.1

* *In vitro* assays.

using an *in vivo* assay, detected a dramatic increase in nitrate reductase activity (from 10–20 to 400 nmol g^{-1} FW h^{-1}), 2–4 days after the onset of the exposure. After 15 days, however, the activity then decreased to 200 nmol g^{-1} FW h^{-1}. In Norway spruce seedlings exposed to 500 nl l^{-1} NO_2 for 11 weeks, shoot nitrate reductase activity trebled but the root activity fell by 60% (Tischner *et al.*, 1988). Similarly, Norby *et al.* (1989) detected a 37-fold increase in shoot nitrate reductase activity in red spruce following exposure to either NO_2 (75 nl l^{-1}) or HNO_3 vapour (75 nl l^{-1}) for just 1 day. These elevated levels of nitrate reductase activity persisted for a long time after the HNO_3 vapour treatment. More recently, Thoene *et al.* (1991) exposed branches of Norway spruce to 60 nl l^{-1} NO_2. *In vitro* nitrate reductase activity was found to increase from 5 to 27 nmol mg^{-1} protein h^{-1} in 2 days but returned to the initial rate 2 days later. They concluded that the fall in activity was caused by either a transport of the nitrate into a storage pool, or long-distance transport from the needles.

Using a range of bryophytes, Morgan *et al.* (1992) compared the action of 35 nl l^{-1} NO_2 and 35 nl l^{-1} NO. In all four species, nitrate reductase activities in NO_2-treated plants were the highest 1 day after the start of fumigation, although the magnitude of the induction varied between species. By the second day, nitrate reductase activities in the NO_2-treated plants fell, but remained higher than the control plants for 5 days. Following 3 weeks exposure to NO_2, fumigated and control plants responded differently to a single application of nitrate. Nitrate reductase activity was rapidly induced in control plants but not in the NO_2-treated plants. Furthermore, instead of the usual induction of nitrate reductase caused by NO_2, exposure to NO caused a decline in activity in *Ctenidium molluscum* to 15% of the control activity within 24 hours. In addition, exposure of all four bryophyte species to NO for 21 days led to a loss of nitrate induction of nitrate reductase activity (Morgan *et al.*, 1992).

Some of the responses of vegetation to O_3 may be attributed to detrimental effects on nitrate reduction (Tingey *et al.*, 1973; Leffler & Cherry, 1974). In a study on bush beans using combinations of O_3 (50–60 nl l^{-1}) and NO_2 (30–40 nl l^{-1}), Bender *et al.* (1991) showed that the individual effects of these gases were dependent upon the age of the plants. During vegetative growth, when nitrate reductase was already high, it was further stimulated by NO_2 but not by O_3. Following anthesis, O_3 reduced the level of extractable nitrate reductase activity. In Scots pine, Lau (unpublished results) showed that episodic treatment with O_3 (up to 100 nl l^{-1} for 6 days) also caused a sequential and rapid fall in nitrate reductase activity. However, once O_3 levels declined, recovery was equally rapid.

The effect of combinations of $SO_2 + NO_2$, $O_3 + NO_2$ and $O_3 + SO_2 + NO_2$ on nitrate reductase activity has been studied in Norway spruce (Klumpp *et al.*, 1989). The data obtained depended upon the age of the needles and the availability of Ca^{2+} and Mg^{2+}. In current-year needles, $NO_2 + O_3$ stimulated nitrate reductase activity but in 1-year-old needles there was a loss of enzyme activity particularly in the trees deficient in Ca^{2+} and Mg^{2+}. Similar data were obtained when net photosynthesis rates were measured (Kuppers & Klumpp, 1988). These studies confirm the close interaction between photosynthetic CO_2 assimilation and nitrate reduction discussed previously (Kaiser & Brendle-Behnisch, 1991).

Nitrite reductase

Yoneyama *et al.* (1979) exposed kidney beans, sunflower and maize to 4 $\mu l\ l^{-1}$ NO_2 either during the day or at night for up to 6 hours. Levels of nitrite reductase increased in each species but by different amounts, with maize showing the smallest response. During CO_2 enrichment in glasshouses (see also later section), fumigations with NO of different cultivars of tomato or lettuce induced significant levels of nitrite reductase activity (Wellburn *et al.*, 1980; Besford & Hand, 1989), although varietal differences within tomato were detected (Murray & Wellburn, 1985). Levels of activity of nitrite reductase in both Bell Boy and Rhumba cultivars of sweet pepper were severely decreased by exposure to either NO or NO_2, but unlike some cultivars of tomato, levels of nitrate reductase were unaffected (Murray & Wellburn, 1985). Anderson & Mansfield (1979) demonstrated that NO can affect the growth of different cultivars of tomato to various extents. The tomato cultivar most affected by NO (Ailsa Craig) was also the one in which the respective activities of nitrate and nitrite reductases were readily enhanced by fumigation with either NO_2 or NO (Wellburn *et al.*, 1980).

Increases in nitrite reductase activity are also accompanied by increases in the steady-state levels of enzyme protein, which indicates that the increases are due to net enzyme synthesis rather than activation of pre-existing enzyme (Besford & Hand, 1989). The increases in nitrite reductase activity in barley shown in Table 13.1 are accompanied by a concomitant increase in enzyme protein as determined by Western blot analysis (Beckett, unpublished results).

The data of Faure *et al.* (1991) propose that the genes coding for both nitrate and nitrite reductase are coordinately regulated by nitrate. It would appear that the majority of the data described above indicate that NO and NO_2 are converted to nitrate (see Wellburn, 1990), which is then able to induce enzyme synthesis via mRNA transcription.

A study of the effect of SO_2, NO_2 or $SO_2 + NO_2$ on the nitrite reductase levels of different strains of the grass *Lolium perenne* has produced some important results (Wellburn *et al.*, 1981). As demonstrated previously with other species, nitrite reductase activity increased in the presence of NO_2, while there was no effect of SO_2 alone. However, in the presence of SO_2 and NO_2 together, no increase in nitrite reductase activity was detected. Indeed with the exception of the SO_2-resistant S23 Bell *Lolium* clone, all levels of nitrite reductase were significantly below clean-air controls. It would appear that SO_2 has the capacity to interfere with the transcription/translation mechanism involved in the *de novo* synthesis of the nitrite reductase protein.

Ammonia assimilation

The precise contribution made by NO_2 to the total nitrogen intake of a plant is difficult to calculate, but it is considered by some to be small (Okano *et al.*, 1986, 1988; Nasholm *et al.*, 1991). Certainly, the amount of ammonia generated from internal metabolism greatly exceeds primary nitrate reduction (Lea *et al.*, 1990, 1992; Raven *et al.*, 1992).

No changes in glutamine synthetase activity have been observed following fumigation with NO_2 by a range of workers (Wellburn *et al.*, 1980; Srivastava & Ormrod, 1984; Thoene *et al.*, 1991). Tischner *et al.* (1988) did determine a doubling of glutamine synthetase activity in the shoots of spruce seedlings after fumigation with 500 nl l^{-1} NO_2, but no increase in the roots. SO_2 had little effect on glutamine synthetase activity in barley (Borland & Lea, 1991).

As part of a series of experiments examining the action of NO_2 and O_3 on the metabolism of *Phaseolus vulgaris*, Bender *et al.* (1991) detected some small effects on glutamine synthetase activity. During vegetative growth, activity of this enzyme was decreased 10% by O_3 and increased 20% by $O_3 + NO_2$. At anthesis all combinations of the gases decreased glutamine synthetase activity by 5–10%.

When crude extracts from spinach were treated with nitrite (5 mM), either in the light or dark, levels of glutamine synthetase and glutamate synthase activity were reduced by 26 and 55%, respectively (Yu *et al.*, 1988). At levels of 25 mM nitrite, both these enzymes were inhibited by 87% in the light and 57% in the dark. Yu *et al.* (1988) concluded that part of the toxicity ascribed to nitrite in these circumstances could be due to a failure to remove ammonia. However, it is difficult to extrapolate from the high concentrations of nitrite used to the physiological process taking place within the intact leaf.

Phaseolus vulgaris exposed to 0.02–0.5 μl l^{-1} NO_2 for 5 days showed increased glutamate synthase activity, while levels of aminotransferase activities were increased in a sensitive tomato cultivar (Ailsa Craig) when exposed to 0.2–0.5 μl l^{-1} NO for 14 days. Glutamate dehydrogenase activity was unaffected in peas when exposed to 0.2–0.5 μl l^{-1} NO_2 for 6 days, but the enzyme activity was considerably increased by exposure to SO_2, NH_3, $SO_2 + NH_3$ and $SO_2 + NO_2$ fumigations (Wellburn *et al.*, 1976). Similar increases in glutamate dehydrogenase activity following exposure to pollutants have been observed by other workers (Jager, 1975; Srivastava & Ormrod, 1984; Borland & Lea, 1991).

An increase in the ratio of glutamate dehydrogenase to glutamine synthetase activity was detected in *Lolium perenne* exposed to 0.25 μl l^{-1} NO_2 for 63 days, even though the glutamine synthetase activity was 50 times that of glutamate dehydrogenase. It is likely that pollutant gases exert a similar effect to ammonia, in perturbing the regulation of the Krebs citric acid cycle and creating a demand for 2-oxoglutarate which can be supplied by the deamination of glutamate (Robinson *et al.*, 1991, 1992).

Amino acid metabolism

Reports on the effect of pollutant gases on amino acid levels in plants are many and varied and it has been difficult to establish a consistent pattern. A large amount of the previous work has been reviewed by Rowland *et al.* (1988) and Wellburn (1990).

A frequent response of plants to NO_2 is an increase in the total amino acid content of the leaf (Prasad & Rao, 1980; Takeuchi *et al.*, 1985; Ito *et al.*, 1986; Rowland *et al.*, 1988). However, the nitrogen status of the plant must be taken into consideration. Barley exposed to NO_2 had a higher nitrogen and amino acid content than controls when under nitrogen stress, but a lower content when nitrate supply to the roots was adequate (Rowland *et al.*, 1987). Srivastava & Ormrod (1984) reported similar results with bean plants exposed to NO_2, and demonstrated that the effect of NO_2 was strongly influenced by nutrient nitrogen levels.

A number of workers have reported an increase in glutamine and asparagine following NO_2 fumigation (Prasad & Rao, 1980; Murray & Wellburn, 1985; Ito *et al.*, 1986). Such an accumulation of the amino acid amides suggests that the plants have an excess of soluble nitrogen (Lea *et al.*, 1990). However, in Scots pine, levels of arginine and glutamine were reduced by fumigation with 85 nl l^{-1} NO_2 for 10 days (Wingsle *et al.*, 1987). In similar experiments, Nasholm *et al.* (1991) exposed mycorrhizal and non-mycorrhizal Scots pine seedlings to 300 nl l^{-1} NO plus NO_2 for 39 days and found lower concentrations of γ-aminobutyrate and proline, but no significant effect on the levels of glutamine or arginine. Higher concentrations of glycine were found in exposed plants, but this represented only a small proportion of the amino acids in the shoots. Both increases and decreases in the level of glutamate have been detected following fumigation with NO_2 (Takeuchi *et al.*, 1985; Ito *et al.*, 1986).

The accumulation of proline in plant tissues is usually an indication of severe drought or salinity stress (Rhodes, 1988; Naidu *et al.*, 1992). Sulphur dioxide-treated (0.7 μl l^{-1} for up to 16 hours) poplar leaves also accumulated proline (Karolewski, 1985). In the genus *Weigela*, the accumulation of proline in plants treated with 2 μl l^{-1} SO_2 for 12 hours was higher in the more sensitive species and varieties (Karolewski, 1984).

The polyamines — putrescine, spermidine and spermine — are derived from arginine and ornithine in a pathway involving *S*-adenosylmethionine. Levels of polyamines have been shown to increase in plants subject to a variety of environmental stresses: e.g., drought, acidity, K^+ deficiency, low temperature and SO_2 pollution (Tiburcio *et al.*, 1990). Polyamines have also been shown to increase in O_3-treated barley (Rowland-Bamford *et al.*, 1988), wheat (Raab & Weinstein, 1990), beans (Manderscheid *et al.*, 1991), soybean (Kramer *et al.*, 1991) and Norway spruce (Sandermann *et al.*, 1989; Dohmen *et al.*, 1990).

In barley, the activity of arginine decarboxylase increased significantly in O_3-treated leaves, when visible injury was hardly apparent. It was proposed that the increase in the decarboxylase activity may be a mechanism to increase the polyamine levels of the leaves and so minimize the damaging effect of O_3. Supporting this, foliar application of α-difluoromethylarginine, a specific inhibitor of arginine decarboxylase, prevented the rise in enzyme activity, and visible injury was considerable on exposure to O_3 (Rowland-Bamford *et al.*, 1989a). Similarly, feeding of exogenous polyamines to tomato (Ormrod & Beckerson, 1986) and tobacco (Bors *et al.*, 1989) resulted in a significant suppression of O_3-induced leaf injury.

Polyamine metabolism has been examined in tobacco plants exposed to a single O_3 treatment (0.15 μl l^{-1} for 5–7 hours). The levels of free and conjugated putrescine increased rapidly in the O_3-tolerant cultivar Bell B and remained high for 3 days. This accumulation was preceded by a transient rise in arginine decarboxylase activity. Increases in putrescine and arginine decarboxylase activity were much lower in the O_3-sensitive cultivar, Bel W3, which developed necrotic lesions following the O_3 treatment. Monocaffeoyl-putrescine, an effective scavenger of oxyradicals was detected in the apoplastic fluid of the leaves of cultivar Bel B (Langebartels *et al.*, 1991).

Uptake and toxicity mechanisms of NO_2

With growing problems of atmospheric NO_x pollution in highly populated areas, interest has been focused on plants which have a large capacity for NO_2 uptake and tolerance, because of their ability to purify the atmosphere (Hill, 1971). This has prompted comparisons of uptake rates between many different types of plant (Okano *et al.*, 1988). Morikawa *et al.* (1993) used ^{15}N-labelled NO_2 to fumigate 142 species of both wild and cultivated herbaceous and woody plants, and found a difference of more than 560-fold in the efficiency of different species to assimilate NO_2.

Comparisons of many different species and cultivars have indicated a wide range of sensitivity to NO_2, but the factors determining the response to NO_x have not been easy to identify. Uptake of pollutant gases occurs primarily through the stomata and is therefore related to stomatal conductance. However, while some studies have suggested a link between stomatal conductance, uptake of NO_2 and species sensitivity (Okano *et al.*, 1988, 1989), there is evidence to indicate the existence of other internal regulating factors. Saxe (1986a) compared eight species and found that while the rate of uptake of NO_2 correlated with stomatal conductance, NO could be taken up equally in the dark, pointing to separate mechanisms for dealing with each of these pollutants. The rate of uptake is not correlated with sensitivity to NO_2, and internal mechanisms such as changes in leaf proteins and enzymes of nitrate assimilation are likely to be more important, especially in determining response to high concentrations of NO_x (Srivastava & Ormrod, 1984, 1989; Saxe, 1986a). In the case of NO_2 there is some evidence that these mechanisms are light-driven. Murray (1984) used a tomato mutant, 'flacca', the stomata of which remain permanently open, to measure the light response of NO_2 uptake. He found that uptake was independent of stomatal conductance, but was increased by higher light intensity, and suggested that a light-driven mechanism such as photosynthetically active ion uptake might be responsible.

Nitrate and nitrite reductase activities, as well as those of other enzymes in the nitrogen assimilation pathways, must be important in determining NO_x toxicity. Mutants of nitrate metabolism have proved useful tools in elucidating the mechanisms of action of NO_x. A large number of mutants have now been characterized which are deficient to different extents, or totally lacking, in nitrate reductase activity (Warner *et al.*, 1982; Cherel *et al.*, 1990). Using such mutants obtained from barley, Rowland-Bamford *et al.* (1989b) measured the flux of NO_2 into leaves with relation to nitrate reductase activity. NO_2 lowered nitrate and nitrite reductase activity in all the mutants, regardless of whether they were provided with NO_3^- or NH_4^+ as a nitrogen source. However, the flux of NO_2 into the leaves of mutants was the same as that in wild-type plants which had

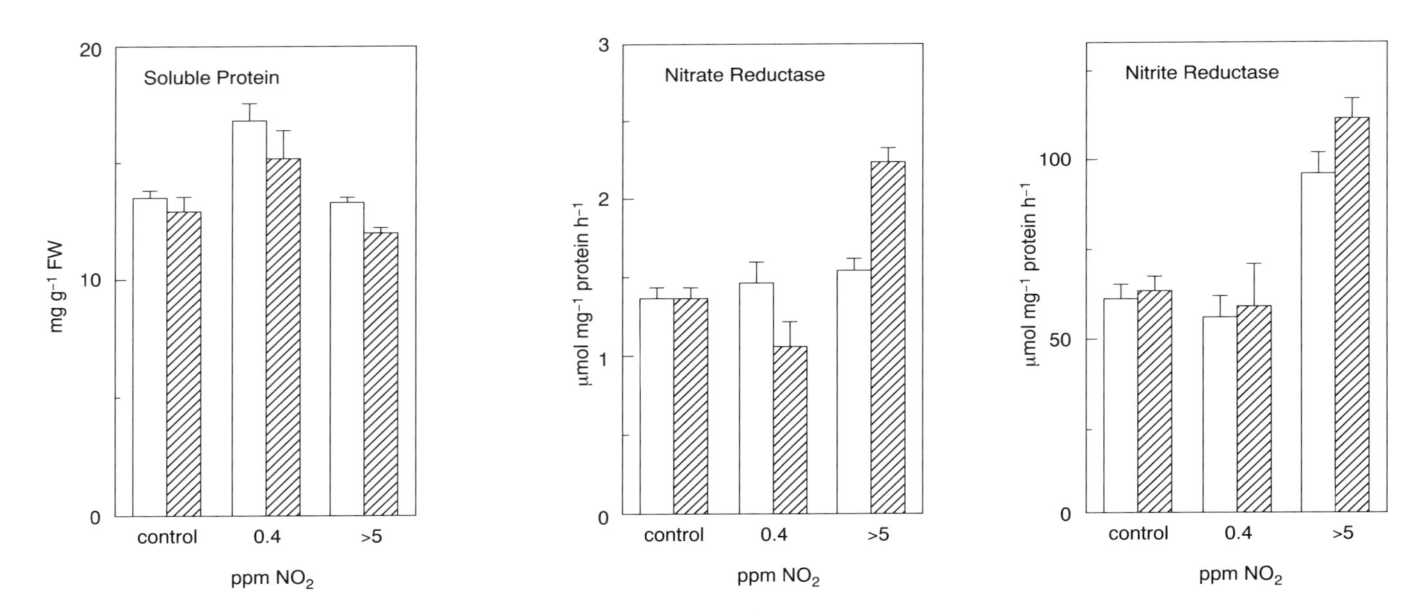

Figure 13.2. Soluble protein content and enzyme activities of barley leaf extracts from Maris Mink wild type (unshaded) and NO_2-resistant mutant W3 (shaded) after exposure to NO_2 for 18 hours

normal nitrate reductase activity, and was not dependent on stomatal conductance. Surprisingly, the mutant totally lacking in nitrate reductase and unable to grow on NO_3^- showed no injury after 9 days in 300 nl l^{-1} NO_2. Internal mechanisms other than nitrate and nitrite reductase must therefore exist, at least in the short term, to prevent toxic effect of NO_2 products.

Tolerance to NO_2 has been investigated at Lancaster University using barley mutants resistant to very high concentrations of this pollutant. Plants grown from mutated seed were screened for resistance by growing them in >5 μl l^{-1} NO_2. Such exposure usually produced two types of visible injury in normal plants: (1) whitening of the leaf from the tip downwards, usually after a few hours, and (2) dark brown lesions developing over the whole leaf several days afterwards. Mutant plants that showed neither of these symptoms after 7 days of exposure were selected. These were then grown on to produce resistant seed. In contrast to mutants deficient in nitrate reductase, when NO_2-resistant mutants growing in soil were exposed to 1 μl l^{-1} NO_2, increases in nitrate and nitrite reductase and glutamine synthetase activities and protein content were seen, which were similar to the responses observed in wild-type plants. In higher concentrations of NO_2, however, these mutants showed a significantly larger increase in specific activity of nitrate reductase than the wild type (Figure 13.2). A more detailed analysis of the mutants has appeared recently (Srivastava *et al.*, 1994).

An enhanced capacity for nitrate reduction can only partially explain the high degree of tolerance found in these mutants, and other internal factors remain to be identified. No clear differences in polypeptide content were found using SDS gel electrophoresis, nor were differences in resistance related to stomatal conductance, which was similar in all mutants and wild-type plants.

The development of plants with mutations affecting other stages of nitrogen metabolism, such as nitrite reductase, will provide improved opportunities in the search for internal mechanisms controlling response to NO_2. However, apart from the identification of barley mutants based on an accumulation of nitrite (Wray, 1989), nitrite reductase mutants have not been identified, and have not appeared during screening for nitrate reductase mutations (Pelsey *et al.*, 1991). Transgenic plants may, however, provide a useful alternative. In a recent study to determine the factors regulating expression of nitrate reductase, for example, Vaucheret *et al.* (1992) have used an antisense nitrite reductase gene in transgenic tobacco to inhibit nitrate assimilation without modifying the nitrate reductase gene itself. As yet there are no known publications describing the use of such plants in air pollution studies, apart from a preliminary report by Kamada *et al.* (1993). This group is currently investigating the NO_2 assimilation of transformed plants of *Arabidopsis thaliana* which contain an enrichment of the nitrite reductase gene.

EFFECT OF GASEOUS AMMONIA ON NITROGEN METABOLISM

Ammonia (this term will be used to denote either NH_3 or NH_4^+) is formed in the soil following the breakdown of organic material, and is normally converted to nitrate by nitrifying bacteria. In acidic and poorly aerated soils the nitrification process may be very slow. In certain areas of the Netherlands where livestock are kept in large concentrations, high levels of ammonia are generated through the decomposition of animal excreta and released into the atmosphere. Ammonia deposition in the Netherlands varies between 40 and 80 kg ha^{-1} yr^{-1}, but localized values of 500 kg ha^{-1} yr^{-1} have been recorded (Van Dijk & Roelofs, 1988; Draaijers *et al.*, 1989). Ammonia is known to be toxic to plants (Givan, 1979; Van der Eerden, 1982). Considerable damage to conifer and broadleaf trees has been detected in both upland and lowland regions of the Netherlands (Van Breeman & Van Dijk, 1988).

Ammonia uptake by the leaves

Ammonia is taken up by the leaves through the stomata and not through the cuticle. Environmental conditions, in particular humidity and light, increase the rate of uptake and, for ammonia concentrations above 10 nl l^{-1}, this rate increased linearly with the ammonia concentration (Van Hove *et al.*, 1987, 1989, 1990). Short-term experiments with *Phaseolus vulgaris* and longer-term exposure of leaves of *Populus euramericana* indicated a passive diffusion of ammonia with no effect on stomatal behaviour. In a detailed study of the long-term effects of 92 nl l^{-1} ammonia on poplar, Van Hove *et al.* (1991) reported a positive effect on the maximum photosynthetic rate, that was reversed by the addition of 122 nl l^{-1} SO_2. It is possible that cation exchange may also be involved in ammonia uptake (Roelofs *et al.*, 1985), but uptake processes vary between species, e.g., between Scots pine and Norway spruce (Wilson, 1992).

The contribution to the total nitrogen of the plant made by gaseous ammonia assimilation has been shown to vary considerably. In an excellent series of experiments, Lockyer & Whitehead (1986) grew Italian ryegrass (*Lolium multiflorum*) in adequate levels of ^{15}N-nitrate and showed that following exposure for 40 days to 749 nl l^{-1} ammonia, 35–45% of the total nitrogen was derived from the gaseous ammonia. In a follow-up series of experiments, Whitehead & Lockyer (1987) grew Italian ryegrass in an adequate and deficient nitrate supply and exposed the plants to nine concentrations of ammonia ranging from 20 to 1021 nl l^{-1}. The proportion of the total plant nitrogen that was derived from ammonia ranged from 4% at the concentration of 20 nl l^{-1} ammonia with the higher rate of nitrate addition, to 77.5% at a concentration of 1021 nl l^{-1} ammonia with the lower rate of nitrate addition. No evidence of toxicity, even after exposure of the plants to 1021 nl ammonia l^{-1}, was observed. Other experiments with *Picea* spp. have suggested that the contribution by atmospheric ammonia to the total plant nitrogen varies between 1 and 20% (Bowden *et al.*, 1989; Schulze, 1989; Garten & Hanson, 1990).

As has been discussed in the previous sections, ammonia is assimilated initially via glutamine synthetase; the enzyme is present in high activity in all tissues, particularly in leaves. Studies with glutamine synthetase-deficient mutants have shown that, in barley leaves, the enzyme is normally present at least two-fold in excess of that required for the accumulation of ammonia derived from metabolism (Lea *et al.*, 1990, 1992; Lea & Forde, 1994). At times of nitrogen excess, the amide group of glutamine is transferred into a compound that has a high N : C ratio and can act as a nitrogen store. This compound is frequently asparagine (2N : 4C), but in certain plants, arginine (4N : 6C) and ureides (4N : 4C) are also synthesized. The pioneering work of Durzan has shown that in white spruce (*Picea glauca*) arginine plays an important role in nitrogen storage (Durzan & Steward, 1967; Durzan, 1968). The amino acid accumulates at times of nitrogen excess, often brought about by the deficiencies of another nutrient, e.g., phosphate (Krupa & Branstrom, 1974; Rabe & Lovatt, 1986).

Van Dijk & Roelofs (1988) determined the amino acid levels in the needles of Scots pine in the south east of the Netherlands that showed severe yellowing (chlorotic) symptoms, presumably due to ammonia pollution. Samples were also taken from trees that contained predominantly green needles. In all the samples analysed, arginine was the predominant soluble amino acid expressed on a $\mu mol\ g^{-1}$ DW basis, suggesting that on a % N basis it would be even higher. The arginine concentration in young (1985) and old (1984) needles in Table 13.2, show clear evidence of a massive accumulation of nitrogen in arginine in the yellow chlorotic needles of both years. Small increases in histidine and lysine were also observed in the yellow needles, but there was no evidence of increase in glutamine and asparagine. Van Dijk & Roelofs (1988) proposed that arginine may function as a bioindicator of incipient ammonia stress. However, given the known accumulation of arginine as a result of other stresses (Rabe & Lovatt, 1986; Nasholm & McDonald, 1990), carefully controlled experiments are required to confirm this proposal.

In a different series of experiments, Schmeink & Wild (1990) determined the amino acid levels in needles from damaged and undamaged spruce trees (*Picea abies*) growing in the Kaufunger Forest in Germany. Pollution in this region included SO_2, NO_2 and O_3. Arginine was a minor component of the total soluble amino acid pool. Large differences in arginine content were detected in the needles of the trees within both the damaged and undamaged groups, to such an extent that no correlation between arginine concentration and damage was observed. Small but significant increases in the levels of glutamine, asparagine and proline in the damaged trees were determined.

In a similar study of healthy and damaged needles of Norway spruce and Scots pine in Sweden, arginine, glutamate, glutamine, γ-aminobutyric acid and aspartic acid were higher in the polluted sites. The highest concentration of arginine was found in needles exposed to high rates of nitrogen deposition, but again there were large variations between individuals at any one site (Edfast *et al.*, 1990).

Table 13.2. Free amino acids ($\mu mol\ g^{-1}$ DW) in the needles of Scots pine collected in April 1986

Amino acid	1984 Green	1984 Yellow	1985 Green	1985 Yellow
Arginine	31.2	146.0	7.2	177.0
Glutamine	9.6	10.5	8.3	9.5
Glutamate/Asparagine	6.0	4.8	7.0	5.5
Total*	57.88	173.45	33.81	207.68

* A full analysis of the total amino acids is given in Van Dijk & Roelofs (1988).

Ammonia uptake by the roots

It is normally stated that nitrate is the predominant form of nitrogen in soils and is, therefore, utilized most frequently by plants. This is certainly the case for most fast-growing annual crops that grow in well-aerated soils (e.g., cereals) and also for fast-growing trees (e.g., poplar and birch). Ammonia uptake from the soil tends to be carried out by perennial or slow-growing species. Such plants include the conifers, *Quercus* spp. and *Fagus sylvatica*. Stewart and his colleagues have examined the differences between plants in some detail (Stewart *et al.*, 1989, 1992).

The uptake of an NH_4^+ ion from the soil is accompanied either by the uptake of a monovalent anion or the release of a H^+ equivalent. The uptake of an NO_3^- ion is accompanied by the uptake of a monovalent cation or the release of an OH^- equivalent to the growth medium. Since nitrogen-containing ions are required in large amounts, the difference in uptake of the two ions is responsible for the alkalinization of the rhizosphere during nitrate reduction and acidification during ammonium nutrition. The complexities of ionic balance carried out by plants assimilating NO_3^- and/or NH_4^+ have been discussed in detail (Raven, 1988; Pilbeam & Kirby, 1992; Ullrich, 1992), along with the potential role of the enzyme PEP carboxylase as a pH-stat (Raven & Farquhar, 1990).

As discussed previously, nitrate may either be reduced in the root or transported to the leaf. Ammonia, on the other hand, must be assimilated immediately by the root, a process which depends on a supply of 2-oxoglutarate and other carbon skeletons. This process requires an input of one molecule of ATP and two electrons in the form of reduced ferredoxin or NADH. If nitrate is reduced in the root there is an additional energy input of eight electrons derived from NADH and reduced ferredoxin. The energetics of the different processes has been discussed in detail by Pate & Layzell (1990). In an elegant series of experiments, Bloom *et al.* (1992) monitored simultaneously the fluxes of CO_2, O_2, NH_4^+, NO_3^- and H^+ in barley roots. They calculated that under ammonia nutrition, 14% of the root-carbon catabolism was coupled to NH_4^+ absorption and assimilation, while under nitrate nutrition 5% of the root-carbon catabolism was coupled to NO_3^- absorption, 15% to NO_3^- assimilation and 3% to NH_4^+ assimilation.

As discussed above, ammonia causes an acidification of the soil, which may cause a greater solubility of aluminium and a decrease in base cations (Van Breeman & Van Dijk, 1988; Schulze, 1989). Roelofs *et al.* (1985) found that, in a field study, increased ratios of NH_4^+ to K^+, Mg^{2+} and Ca^{2+} in soil solution correlate with damage to pine forests. Boxman & Roelofs (1988) using ^{86}Rb (an analogue of potassium) showed that high ammonia concentrations caused an increase in efflux and a decrease in influx. It is likely that the decrease in uptake of K^+ and Ca^{2+} may be due to the low pH of the root environment and may be a cause of the low yield of plants grown with ammonia as a nitrogen source (Van Beusichem *et al.*, 1988; Findenegg *et al.*, 1989). Tomato plants supplied with ammonia show a greater tendency to suffer from the physiological disorder "blossom end rot" than when nitrate is supplied. This disorder is caused by a shortage of calcium in the distal end of the fruit (Mengel & Kirkby, 1987).

Ammonia and a reduced soil pH may also have an effect on root structure and the mycorrhizal associations. Boxman *et al.* (1991) have shown that in *Pinus nigra*, excess ammonia caused a decrease in the root : shoot ratio and a loss of fine roots. A reduction of mycorrhizal infection was also noted. In a range of conifers, ammonia solutions were shown to decrease the number of fine roots and mycorrhizae (Van Dijk *et al.*, 1990). Similar experiments have been carried out by Leisen *et al.* (1990) and Van der Eerden *et al.* (1992). Interestingly, Chalot *et al.* (1991) have recently shown that ectomycorrhiza isolated from Norway spruce have the capacity to assimilate externally applied $^{15}NH_4^+$ via both glutamine synthetase and glutamate dehydrogenase. The authors concluded that the fungus was able to supply nitrogen in the form of glutamine and alanine to the plant roots.

In a detailed series of experiments, Flaig & Mohr (1991, 1992) have shown that with Scots pine over a 21-day period, three times more ammonium N is taken up than nitrate N. If nitrate and ammonia were added simultaneously, they were taken up additively. In the case of high ammonia supply, nitrogen accumulated in the storage compounds glutamine, asparagine and arginine. These results conflict with the results of Boxman & Roelofs (1988), Marschner *et al.* (1991) and Peuke & Tischner (1991), who found an inhibition of NO_3^- uptake by ammonia in Scots pine and Norway spruce.

It is important to conclude this section by discussing the detailed studies on ammonia and nitrate nutrition of mustard (*Sinapis alba* L.) seedlings carried out by H. Mohr and his colleagues (see Hecht *et al.*, 1988; Mehrer & Mohr, 1989; Schmidt & Mohr, 1989; Hecht & Mohr, 1990). Ammonia exerts a toxic effect on the growth of seedlings; an action that can be reversed by nitrate, which causes a corresponding decrease in the accumulation of NH_4^+. Mehrer & Mohr (1989) examined a range of mechanisms that could cause the toxic effect of ammonia. They ruled out obvious factors such as excess nitrogen, pH changes or non-specific osmotic effects. Activities of Rubisco, plastidic NADP-dependent glyceraldehyde-3-phosphate, dehydrogenase and chalcone synthase were strongly reduced by ammonia. Cytosolic NAD-dependent glyceraldehyde-3-phosphate dehydrogenase was unaffected whereas isocitrate lyase increased following ammonia treatment. Although additional symptoms, e.g., the conversion of lipid to carbohydrate and starch accumulation were also strongly inhibited, the authors were unable to determine the key mechanism by which ammonia exerted its inhibitory action. Taking into account that the *Sinapis alba* experiments were carried out under carefully controlled conditions, it makes the task of identifying the toxic action of ammonia on the species in the Netherlands even more difficult.

EFFECT OF ELEVATED CARBON DIOXIDE LEVELS

The carbon dioxide (CO_2) concentration of the atmosphere has been rising at a increasing rate since continuous monitoring began in 1958. At that time the CO_2 concentration was about 315 $\mu l\ l^{-1}$, rising to about 350 $\mu l\ l^{-1}$ by 1988. Most of this rise is due to burning of fossil fuels. It has been predicted that the atmospheric

CO_2 could double over the next 100 years. Such an increase could enhance the productivity of C_3 plants by approximately 30% (Idso, 1988). Concurrently, with rising CO_2, the burning of fossil fuels introduces other trace gases into the atmosphere. About 1% of the gaseous component of fossil-fuel combustion is composed of oxides of sulphur (S) and nitrogen (N). Unfortunately, little research has been carried out on the potential interactions of CO_2 and other air pollutants on plants, although both originate from fossil-fuel burning. The literature is even sparser if N metabolism is also considered.

One of the few reviews on CO_2 and air pollution is that by Allen (1990). It has been generally found that tolerance to atmospheric pollutants increases with CO_2 enrichment (Heck & Dunning, 1967; Carlson, 1983; Acock & Allen, 1985). In a study by Hou *et al.* (1977) a two-fold CO_2 enrichment reduced the visible injury in alfalfa caused by SO_2 and NO_2. However, even though CO_2 may help to mitigate the effects of air pollution, it does not eliminate them. Possibly, reduced stomatal conductance in response to elevated CO_2 (Morrison, 1987) plays a role in reducing injury because most gaseous pollutants enter the leaf via the stomata (Mansfield & Freer-Smith, 1984). Allen (1990) calculated that the yield increase due to pollutant-entry reduction as a result of CO_2 enrichment would be nearly 15%.

There is less effect of CO_2 enrichment on C_4 than on C_3 species (Garbutt *et al.*, 1990), and response of C_3 and C_4 species to CO_2 enrichment in combination with air pollution are different. The growth of several C_3 species was reduced in the presence of 0.25 $\mu l\ l^{-1}$ SO_2 at CO_2 concentrations of 300 $\mu l\ l^{-1}$ but not 600 or 1200 $\mu l\ l^{-1}$ CO_2 (Carlson & Bazzaz, 1982). However, in C_4 species, SO_2 reduced growth in the two higher CO_2 concentrations. These different growth responses are thought to be due to changes in photosynthetic rate brought about by SO_2 damage to CO_2-fixing enzymes, occurring to different extents in C_3 and C_4 species. C_4-plant photosynthesis appears to be less sensitive to SO_2 than C_3-plant photosynthesis (Winner & Mooney, 1980; Carlson & Bazzaz, 1982).

The effects of horticultural greenhouse conditions, with high CO_2 enrichment and NO_x pollution from burners, have received much attention in the past. However, few of these experiments were designed to measure or predict responses of crops to rising global CO_2. In reproducing the conditions in greenhouses, often very high CO_2 concentrations were needed; about 1000 $\mu l\ l^{-1}$ (Anderson & Mansfield, 1979; Saxe, 1986b; Caporn, 1989), as recommended for greenhouse CO_2 enrichment, or 2000–4000 $\mu l\ l^{-1}$, which occur when flueless burners are a sole source of heat (Law & Mansfield, 1982). This is an important consideration since the CO_2-response curve of species can vary; with continual increase in photosynthesis, growth and yield as CO_2 increases in some species, e.g., soybean (Valle *et al.*, 1985; Allen *et al.*, 1988; Campbell *et al.*, 1990), and a dropping off in response in other species, e.g., rice (Baker *et al.*, 1990a, b; Rowland-Bamford *et al.*, 1991).

Many studies have demonstrated increases in vegetative growth and yield in response to elevated CO_2 (Sionit *et al.*, 1981; Allen *et al.*, 1988; Baker *et al.*, 1990b). This would impose an additional demand for N and indeed there are many reports in the literature on decreases in N concentration in plants exposed to CO_2

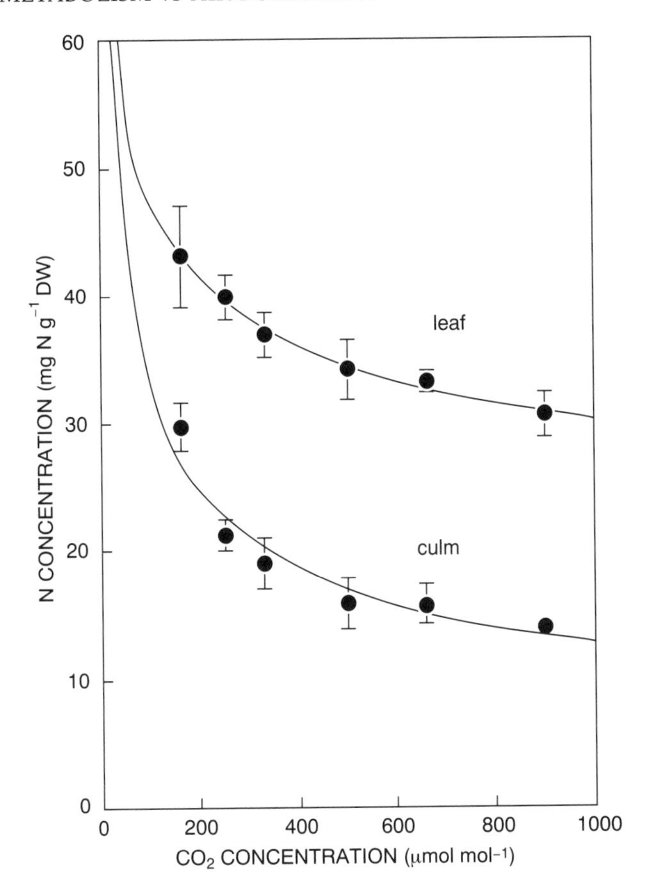

Figure 13.3. Nitrogen (N) concentration (on a dry weight basis) of 44-day-old rice plants grown in paddy culture in controlled-environment chambers under a range of atmospheric CO_2 concentrations for 33 days. Values are means ± SE (A.J. Rowland-Bamford, J.T. Baker and L.H. Allen Jr, unpublished)

enrichment (see Table 12.3). In rice, an increase in CO_2 concentration resulted in decreases in both leaf blade and culm (including leaf sheath) N concentration (Figure 13.3) and a similar response was noted over the whole season (Baker *et al.*, 1992). Tremblay *et al.* (1988) found that CO_2 enrichment up to 1000 μl l^{-1} resulted in a 10% decrease in N concentration in shoots and roots of celery (*Apium graveolens*). Garbutt *et al.* (1990) using 350, 500 and 700 μl l^{-1} CO_2 also found that there was a significant CO_2 and species interaction caused by the different rates of decline in N among five annuals tested, three C_3 and two C_4 species. Shoot % N was also reduced under elevated CO_2 in *Scirpus olneyi* (C_3) when brackish marsh communities were treated. However, there was no effect of CO_2 on two C_4 species in the community (Curtis *et al.*, 1989).

The mechanisms that produce this decline in N are not clear. Possibilities include a reduced N uptake as a result of stomatal closure at high CO_2 (Madsen, 1975), the reallocation of N to other plant parts (Cure *et al.*, 1988), or possibly a

Table 13.3. The effect of nitrogen on the response of plants to elevated CO_2

Species	CO_2 (µmol mol^{-1})	Duration (days/weeks)	[N] response (% of ambient)	N source	Reference[a]
Abutilon theophrasti *Ambrosia artemisiifolia* *Chenopodium album* *Amaranthus retroflexus* *Setaria faberii*	350, 500, 700	about 12 weeks	500 700[b] 85% 63% C_3 plant 104% 92% C_3 plant 96% 92% C_3 plant 100% 91% C_4 plant 103% 94% C_4 plant	Soil and fertilizer	A
Alnus rubra	350, 650	47 days	94% nodulated, low N 89% nodulated, high N 113% non-nodulated	Porous medium and Hoaglands nutrient solution	B
Apium graveolens	330, 1000	77 days	90% shoot 88% root	Soil watered with nutrient solution	C
Bromus mollis	350, 650	129 days	55% low N at 49 days 71% high N at 49 days 100% both N at 129 days	Porous medium and Hoaglands nutrient solution	D
Glycine max	330, 450, 600, 800	134 days	450 600 800[b] 80% 74% 90%	Soil and fertilizer	E
Glycine max	350, 700	98 days	67% 0.05 mM N 70% 1.0 mM N 69% 2.5 mM N 93% 5.0 mM N 86% 10.0 mM N	Perlite and nutrient solution	F
Glycine max	400, 650, 900	34 days	650 900[b] 91% 87% 16 days 102% 107% 34 days	Flowering nutrient solution	G
Lactuca sativa	360, 800, 1300, 1800	30 days	800 1300 1800[b] 85% 76% 70% (low nitrate) 96% 95% 92% (high nitrate)	Nutrient solution	H
Oryza sativa	330, 660	103 days	114% shoot	Soil and fertilizer	I
Plantago lanceolata	350, 700	8 weeks	76%	Soil and fertilizer	J
Quercus alba	362, 690	40 weeks	81% attached leaves 66% roots	Soil	K
Scirpus olneyi, *Spartina patens*	Ambient + 336	28 weeks	82% C_3 plant 100% C_4 plant	Natural community	L

[a] A = Garbutt *et al.* (1990); B = Arnone & Gordon (1990); C = Tremblay *et al.* (1988); D = Larigauderie *et al.* (1988); E = Allen *et al.* (1988); F = Cure *et al.* (1988); G = Vessey *et al.* (1990); H = Ikeda & Osawa (1988); I = Baker *et al.* (1990c); J = Fajer (1989); K = Norby *et al.* (1986b); L = Curtis *et al.* (1989).
[b] % of ambient for each CO_2 concentration.

lower stimulation of N assimilation than that of CO_2 assimilation (Wong, 1990). Such changes would result in a change in the C/N ratio in plants in response to CO_2 enrichment.

Elevated C/N ratios frequently occur in plants exposed to high CO_2 (Table 13.3). CO_2 enrichment acts to increase the balance of leaf storage carbohydrate versus nitrogen. Accumulation of carbohydrates, especially starch, has been demonstrated in many species in response to CO_2 enrichment (Spencer & Bowes, 1986; Vu *et al.*, 1989; Yelle *et al.*, 1989a; Johnson & Lincoln, 1991). In rice, CO_2 enrichment up to 500 $\mu l\ l^{-1}$ increased the total non-structural carbohydrate, starch and sucrose concentrations, but at higher CO_2 concentrations no further increase in carbohydrates was measured (Rowland-Bamford *et al.*, 1990). Allen *et al.* (1988) showed that the dry weight increase in soybean under CO_2 enrichment was mainly due to starch accumulation accompanied by a decrease in leaf N concentration. This indicated that CO_2 enrichment promoted higher structural matter-to-protein ratios. The accumulation of carbohydrates is thought to dilute the concentration of foliar proteins, i.e., foliar N (Sionit, 1983). Carbohydrate accumulation may be due to an excess availability of carbon (C) needed for growth or that can be used for growth. Possibly the different sinks for C and accumulation of carbohydrate in different species play a regulatory role in the photosynthetic adjustments to an altered CO_2 environment. It should be noted that such accumulation of leaf carbohydrates can result in misinterpretation of root/shoot ratios. Changes in C/N ratios can have substantial effects on the feeding of herbivores (Johnson & Lincoln, 1991), decomposers and nutrient cycling (Oechel & Strain, 1985; Taylor *et al.*, 1989). Fajer (1989) showed that larvae of *Junonia* fed on tissue grown at high CO_2 (which had higher C/N ratio) grew more slowly and experienced higher mortality than larvae fed on tissue grown at ambient CO_2. Therefore, dramatic changes in the natural ecosystem may result if future plant communities have a lower nutritional quality in enriched CO_2 environments and maintain a reduced insect herbivore population.

The response to CO_2 concentration depends on the level of other environmental factors, and the plant may be prevented from benefiting fully from high CO_2 concentration if sink strength is low because some other factor such as N supply is low. Ikeda & Osawa (1988) found that the N concentration in the nutrient solution needs to be kept relatively high to produce the yield increase by CO_2 enrichment in a number of vegetable crops. The effect of CO_2 enrichment on cotton-assimilation rate was also greater when N fertilization was high (Wong, 1979). However, over the world as a whole, much of the soil in agricultural use has a N content that is well below optimum. Nitrogen deficiency would be expected to reduce the response of the plants to CO_2 enrichment unless that enrichment made more N available via increased root growth or uptake. An increase in N acquisition can be an important determinant of the growth enhancement by enriched CO_2. Nevertheless, plants respond to CO_2 even under severe nutrient limits (Sionit *et al.*, 1981; Luxmoore *et al.*, 1986; Norby *et al.*, 1986a). Luxmoore (1981) mentioned that elevated CO_2 could increase nutrient uptake by plants in forest soil–plant systems. However, Norby *et al.* (1986a) found that total uptake of N in *Quercus alba* was not affected by CO_2 and, since growth was enhanced by 85%, tissue N concentration declined in elevated CO_2 (690 μl l^{-1}) in nutrient-deficient soil. Furthermore, very few studies have been carried out on the effect of the different forms of N source. Ikeda & Osawa (1988) suggested the possibility that NH_4^+ could be a more effective N source than NO_3^- under higher atmospheric CO_2 concentration. Bazzaz *et al.* (1989) showed that *Amaranthus retroflexus*, a C_4 annual, had a much higher rate of N uptake per unit root than did *Abutilon theophrasti*, a C_3 annual. This was due in part to a decreased biomass allocation to roots of *Amaranthus* with CO_2 enrichment. Decreased leaf N concentration was accompanied by increased root/shoot ratio in response to elevated CO_2 in *Bromus mollis* (Larigauderie *et al.*, 1988). Other stimulations of root growth at high CO_2 concentration have been observed (Luxmoore *et al.*, 1986; Norby *et al.*, 1986a; Cure *et al.*, 1988). Curtis *et al.* (1990) demonstrated an increase in root dry mass accompanied by a lower %N in roots of *Scirpus* grown under elevated CO_2 compared with roots from ambient-grown plants. Improvements in N-use efficiency have also been recorded in response to CO_2 enrichment. An increase in N-use efficiency was apparent in *Quercus alba* and hence an increased growth in response to increased CO_2 availability was possible in a N-limited system (Norby *et al.*, 1986a). Soybean (*Glycine max*) grown under 350 and 700 μl l^{-1} CO_2 and on a range of nitrate nutrient solutions had increased N-utilization efficiency under elevated CO_2 at all N levels except the lowest (Cure *et al.*, 1988). Others have demonstrated greater dry matter accumulation per unit assimilated N; i.e., apparent higher N-utilization efficiency, in CO_2-enriched plants (Mousseau & Enoch, 1989). A long-term result of the increase in N-use efficiency in CO_2-enriched trees may be that the amount of N sequestered in perennial tissues increases, with a concomitant decrease in the amount of N available for internal or external cycling (Norby *et al.*, 1986b).

Very few studies have measured the response of enzymes involved in N metabolism to CO_2 enrichment. Besford & Hand (1989) found that lettuce grown at 1000 μl l^{-1} CO_2 had reduced nitrite reductase activity which resulted from increased net enzyme synthesis rather than activation of pre-existing enzymes. Reduced levels of other Calvin cycle enzymes, 3PGA kinase and NADP-G3P dehydrogenase, were also observed in leaves grown in the high CO_2. However, when NO_x gases were also included, these enzyme activities were restored to levels similar to those in plants grown at a normal level of CO_2. This may be due to NO_x-enhanced N-metabolizing pathways (Wellburn *et al.*, 1980; Rowland *et al.*, 1987), thereby raising the levels of proteins in general. Nitrogen metabolism and various processes of carbon metabolism are closely linked (Stulen, 1986). Nitrate reduction is dependent on carbon metabolism for (a) the supply of carbon skeletons and energy for enzyme synthesis, (b) reducing equivalents and (c) for the provision of carbon skeletons to accept the reduced N (Beevers & Hageman, 1969). Warner & Kleinhofs (1974), using chlorophyll-deficient mutants of barley, found significant correlations between *in vitro* activities of nitrate and nitrite reductases and ribulose-1,5-bisphosphate carboxylase oxygenase (Rubisco), suggesting a close functional relationship between photosynthesis or chloroplast development and the enzymes for nitrate reduction. Furthermore, *de novo* synthesis of Rubisco and glutamine synthetase are closely linked (Hirel *et al.*, 1984). This suggests that a common mechanism may regulate the synthesis of both enzymes (see also Tobin *et al.*, 1988; Coruzzi, 1991a, b).

The decreases in leaf N observed in many species under CO_2 enrichment, as discussed above, could be due to redistribution of N due to the reduced requirement for Rubisco protein in an elevated CO_2 atmosphere. Rubisco comprises up to 65% of the soluble leaf protein in healthy plants (Ellis, 1979) and is, therefore, a major sink for N. Furthermore, it is the major enzyme responsible for carbon fixation in plants with the C_3 photosynthetic pathway. CO_2 enrichment can result in more efficient C fixation by Rubisco, so that less enzyme, and, therefore, less protein N is required to produce a given amount of dry matter. There is evidence that in a range of species the amount and activity of Rubisco decreases when plants are grown in high CO_2 (Spencer & Bowes, 1986; Yelle *et al.*, 1989b; Bowes, 1991; Rowland-Bamford *et al.*, 1991), although this is absent in other species (Campbell *et al.*, 1988). Besford *et al.* (1984) found that the fall in the assimilation rate in CO_2-enriched plants was correlated with the decrease in Rubisco activity. This reflected a drop in the amount of Rubisco protein. Yelle *et al.* (1989a, b) found that tomato exposed to either 330 or 990 $\mu l\ l^{-1}$ CO_2 for 10 weeks showed a decline in total Rubisco activity and an increase in starch accumulation under CO_2 enrichment. Cotton and rice also showed an accumulation of starch and a reduction in Rubisco activity and protein (Wong, 1979, 1990; Rowland-Bamford *et al.*, 1990, 1991). A detailed analysis of the effects of elevated CO_2 on photosynthetic carbon metabolism has recently been presented by Stitt (1991). Tissue N may be reallocated to new growth as less is invested in Rubisco under elevated CO_2. Consequently, CO_2 enrichment can improve the N-use efficiency of plants, enabling growth under marginally N-deficient conditions to be comparable to that in soils of adequate N status.

Nitrogen and carbon metabolism are linked in several ways. Interactions between nitrate assimilation and carbon metabolism are found at various levels and great progress has been made in elucidating the interactions involved (Stulen, 1986). However, little of this information and expertise has been utilized in studying the interactions of CO_2 enrichment and N metabolism. The powerful tools of molecular biology have yet to be exploited in this area. Antibody and gene probes could be used to determine specific changes in the levels of enzyme proteins and mRNA when plants are exposed to high CO_2 and/or NO_x.

In general, there is little information about the interactions between CO_2 concentration and other environmental factors that affect plant growth. Further investigation of these interactions is needed using both field enclosures and controlled-environment chambers to obtain more data about the interactions and to elucidate the mechanisms. There is an extreme dearth of information on the mechanisms underlying the responses to elevated CO_2 and interactions with other environmental factors such as N nutrition.

ACKNOWLEDGEMENTS

We are thankful to Dr P.W. Lucas, Professor T.A. Mansfield, Professor A.R. Wellburn and several other colleagues at the University of Lancaster for considerable advice and stimulating discussion, to Dr J. Pearson of University College, London, for allowing us access to an unpublished review, and to Annie Brotheridge for her tireless patience in typing this manuscript. We acknowledge the receipt of research grants from the AFRC, USDE, USDA, Wolfson Foundation and NERC.

REFERENCES

Acock, B. & Allen, L.H. Jr. 1985. Crop responses to elevated carbon dioxide concentrations. In Strain, B.R. & Cure, J.D. (eds) *Direct Effects of Increasing Carbon Dioxide on Vegetation*: 53-97. DOE/ER-0238, US Dept. of Energy, Carbon Dioxide Research Division, Washington, DC.

Allen, L.H. Jr 1990. Plant responses to rising carbon dioxide and potential interactions with air pollutants. *J. Environ. Qual.* **19**: 15-34.

Allen, L.H. Jr, Vu, J.V.C., Valle, R.R., Boote, K.J. & Jones, P.H. 1988. Nonstructural carbohydrates and nitrogen of soybean grown under carbon dioxide enrichment. *Crop. Sci.* **28**: 84-94.

Alscher, R., Bower, J.L. & Zipfel, W. 1987. The basis for different sensitivities of photosynthesis to SO_2 in two cultivars of pea. *J. Exp. Bot.* **38**: 99-108.

Anderson, L.S. & Mansfield, T.A. 1979. The effects of nitric oxide pollution on the growth of tomato. *Environ. Pollut.* **20**: 113-121.

Arnone, J.A. III & Gordon, J.C. 1990. Effect of nodulation, nitrogen fixation and CO_2 enrichment on the physiology, growth and dry mass allocation of seedlings of *Alnus rubra* Bong. *New Phytol.* **116**: 55-66.

Asada, K. 1992. Ascorbate peroxidase-A hydrogen peroxide scavenging enzyme in plants. *Physiol. Plant.* **85**: 235-241.

Baker, J.T., Allen, L.H. Jr & Boote, K.J. 1990a. Growth and yield responses of rice to carbon dioxide concentration. *J. Agri. Sci.* **115**: 313-320.

Baker, J.T., Allen, L.H. Jr, Boote, K.J., Jones, J.W. & Jones, P. 1990b. Rice photosynthesis and evapotranspiration in ambient and superambient carbon dioxide concentrations. *Agron. J.* **82**: 834-840.

Baker, J.T., Allen, L.H. Jr, Broote, K.J., Jones, J.W., Jones, P., Waschmann, R.S., Albrecht, S.L. & Kamuru, F. 1990c. Response of vegetation to carbon dioxide, series O62. Carbon dioxide effects on growth, photosynthesis and evapotranspiration of rice at three nitrogen fertilizer levels. Joint programme of USDE and USDA on Plant Stress and Protection. University of Florida, Gainsville, FL.

Baker, J.T., Laugel, F., Boote, K.J. & Allen, L.H. Jr 1992. Effects of daytime carbon dioxide concentration on dark respiration in rice. *Plant, Cell Environ.* **15**: 231-239.

Bazzaz, F.A., Garbutt, K., Reekie, E.G. & Williams, W.E. 1989. Using growth analysis to interpret competition between a C_3 and a C_4 annual under ambient and elevated CO_2. *Oecologia* **79**: 223-235.

Beevers, L. & Hageman, R.H. 1969. Nitrate reductase in higher plants. *Annu. Rev. Plant. Physiol.* **20**: 495-522.

Bender, H., Weigel, H.J. & Jager, H.J. 1991. Response to nitrogen metabolism in leaves (*Phaseolus vulgaris* L.) after exposure to ozone and nitrogen dioxide alone and in sequence. *New Phytol.* **119**: 261-267.

Besford, R.T. & Hand, D.W. 1989. The effects of CO_2 enrichment and nitrogen oxides on some calvin cycle enzymes and nitrite reductase in glass house lettuce. *J. Expt. Bot.* **40**: 329-336.

Besford, R.T., Withers, A.C. & Ludwig, L.J. 1984. Ribulose bisphosphate carboxylase and net CO_2 fixations in tomato leaves. In Sybesma, C. (ed.) *Advances in Photosynthetic Research, III*: 779-782. Martinus Nijhoff/Dr. W. Junk, The Hague.

Bloom, A.J., Sukrapanna, S.S. & Warner, R.L. 1992. Root respiration associated with ammonium and nitrate absorption and assimilation by barley. *Plant Physiol.* **99**: 1294-1301.

Borland, A.M. & Lea, P.J. 1991. The response of enzymes of nitrogen and sulphur metabolism in barley to low doses of sulphur dioxide. *Agric. Ecosyst. Environ.* **33**: 281-292.

Bors, W., Langebartels, C., Michel, C. & Sandermann, H. 1989. Polyamines as radical scavengers and protectants against ozone damage. *Phytochem.* **28**: 1589-1595.

Bowden, R.D., Geballe, G.T. & Bowden, W.B. 1989. Foliar uptake of ^{15}N from simulated cloud water by red spruce (*Picea rubens*) seedlings. *Can. J. For. Res.* **19**: 382-386.

Bowes, G. 1991. Growth at elevated CO_2: Photosynthetic responses mediated through Rubisco. *Plant, Cell Environ.* **14**: 795-806.

Bowsher, C.G., Long, D.M., Oaks, A. & Rothstein, S.J. 1991. Effects of light/dark cycles on expression of nitrate assimilatory genes in maize shoots and roots. *Plant Physiol.* **95**: 281-285.

Boxman, A.W. & Roelofs, J.G.M. 1988. Some effects of nitrate versus ammonium nutrition on the nutrient fluxes in *Pinus sylvestris* seedlings. Effects of mycorrhizal infection. *Can. J. Bot.* **66**: 1091-1097.

Boxman, A.W., Krabbendam, H., Bellenmakers, M.J.S. & Roelofs, J.G.M. 1991. Effects of ammonium and aluminium on the development and nutrition of *Pinus nigra* in hydroculture. *Environ. Pollut.* **73**: 119-133.

Bryan, J.K. 1990. Advances in the biochemistry of amino acid biosynthesis. In Miflin, B.J. & Lea, P.J. (eds) *The Biochemistry of Plants*, Vol. 16: 161-195. Academic Press, San Diego.

Campbell, W.J., Allen, L.H. Jr & Bowes, G. 1988. Effects of CO_2 concentration on Rubisco activity, amount and photosynthesis in soybean leaves. *Plant Physiol.* **88**: 1310-1316.

Campbell, W.J., Allen, L.H. Jr & Bowes, G. 1990. Response of soybean canopy photosynthesis to CO_2 concentration, light and temperature. *J. Expt. Bot.* **41**: 427-433.

Caporn, S.J.M. 1989. The effects of oxides of nitrogen and carbon dioxide enrichment on photosynthesis and growth of lettuce (*Lactuca sativa* L.). *New Phytol.* **111**: 473-483.

Carlson, R.W. 1983. The effects of SO_2 on photosynthesis and leaf resistance at varying concentrations of CO_2. *Environ. Pollut. (A)* **30**: 309-321.

Carlson, R.W. & Bazzaz, F.A. 1982. Photosynthetic and growth response to fumigation with SO_2 at elevated CO_2 for C_3 and C_4 plants. *Oecologia* **54**: 50-54.

Chalot, M., Stewart, G.R., Brun, A., Martin, F. & Botton, B. 1991. Ammonium assimilation by spruce-*Hebeloma* sp. ectomycorrhizas. *New Phytol.* **119**: 541-550.

Cheng, C-L., Dewdrey, J., Kleinhofs, A. & Goodman, H.M. 1986. Cloning and nitrate induction of nitrate reductase in mRNA. *Proc. Nat. Acad. Sci.* **83**: 6825-6828.

Cheng, C.-L., Acedo, G.W., Cristinsin, M. & Conkling, M.A. 1992. Sucrose mimics the light induction of *Arabidopsis* nitrate reductase gene transcription. *Proc. Nat. Acad. Sci.* **89**: 1861-1864.

Cherel, I., Gonneau, M., Meyer, C., Pelsy, F. & Caboche, M. 1990. Biochemical and immunological characterisation of nitrate reductase deficient mutants of *Nicotiana plumbaginifolia*. *Plant Physiol.* **92**: 659-665.

Clough, E.C.M., Pearson, J. & Stewart, G.R. 1989. Nitrate utilization and nitrogen status in English Woodland communities. *Annu. Sci. For.* **46S**: 669-672S.

Coruzzi, G.M. 1991a. A gene network controlling glutamine and asparagine biosynthesis in plants. *Plant. J.* **1**: 275-280.

Coruzzi, G.M. 1991b. Molecular approaches to the study of amino acid biosynthesis in plants. *Plant. Sci.* **74**: 144-155.

Crawford, N.G., Smith, M., Bellisimo, D. & Davis, R.W. 1989. Sequence and nitrate regulation of the *Arabidopsis thaliana* mRNA encoding nitrate reductase: A metalloflavoprotein with three domains. *Proc. Nat. Acad. Sci.* **85**: 5006-5010.

Cullimore, J.V. & Bennett, M.J. 1992. Nitrogen assimilation in the legume root nodule-current status of the molecular biology of the plant enzymes. *Can. J. Microbiol.* **38**: 461-466.

Cure, J.D., Israel, D.W. & Rufty, T.W. Jr 1988. Nitrogen stress effects on growth and seed yield of non-nodulated soybean exposed to elevated carbon dioxide. *Crop. Sci.* **28**: 671-677.

Curtis, P.S., Drake, B.G. & Whigham, D.F. 1989. Nitrogen and carbon dynamics in C_3 and C_4 estuarine marsh plants grown under elevated CO_2 *in situ*. *Oecologia* **78**: 297-301.

Curtis, P.S., Balduman, L.M., Drake, B.G. & Whigham, D.F. 1990. Elevated atmospheric CO_2 effects on belowground processes in C_3 and C_4 estuarine marsh communities. *Ecology* **71**: 2001-2006.

De Kok, L.J. 1990. Sulphur metabolism in plants exposed to atmospheric sulphur. In Rennenberg, H., Brunold, C., De Kok, L.J. & Stulen, I. (eds) *Sulphur Nutrition and Sulphur Assimilation in Higher Plants*: 111-130. SBP Academic, The Hague.

Deng, M-D., Moureaux, T., Leydecker, M.T. & Caboche, M. 1990. Nitrate reductase expression is under the control of a circadian rhythm and is light inducible in *Nicotiana tabaccum* leaves. *Planta* **180**: 257-261.

Dohmen, G.P., Koppers, A. & Langebartels, C. 1990. Biochemical response of Norway spruce towards 14-month exposure to ozone and acid mist: effects on amino acid, glutathione and polyamine titers. *Environ. Pollut.* **64**: 375-383.

Draaijers, G.P.J., Ivens, W.P.M.F., Bos, M.M. & Bleuten, W. 1989. The contribution of ammonia emissions from agriculture to the deposition of acidifying and eutrophying compounds into one forest. *Environ. Pollut.* **60**: 55-66.

Durzan, D.J. 1968. Nitrogen metabolism of *Picea glauca*. I. Seasonal changes of free amino acids in buds, shoot apices and leaves. The metabolism of uniformly labelled ^{14}C-L-arginine by buds during the onset of dormancy. *Can. J. Bot.* **46**: 909-919.

Durzan, D.J. & Steward, F.C. 1967. The nitrogen metabolism of *Picea glauca* and *Pinus banksiana*, as influenced by mineral nutrition. *Can. J. Bot.* **45**: 695-710.

Edfast, A.B., Nasholm, T. & Ericsson, A. 1990. Free amino acid concentrations in needles of Norway spruce and Scots pine trees on different sites in areas with two levels of nitrogen deposition. *Can. J. For. Res.* **20**: 1132-1136.

Ellis, R.J. 1979. The most abundant protein in the world. *TIBS* **4**: 241-244.

Fajer, E.D. 1989. The effects of enriched CO_2 atmospheres on plant-insect herbivore interactions: growth responses of larvae of the specialist butterfly, *Junonia coenia* (Lepidoptera: Nymphalodae). *Oecologia* **81**: 514-520.

Faure, J.D., Vincentz, M., Kronenberger, J. & Caboche, M. 1991. Co-regulated expression of nitrate and nitrite reductase. *Plant J.* **1**: 107-114.

Findenegg, G.R., Nelemans, J.A. & Arnozis, P.A. 1989. Effects of external pH and Cl^- on the accumulation of NH_4^+ ions in the leaves of sugar beet. *J. Plant Nutrition* **12**: 593-602.

Flaig, H. & Mohr, H. 1991. Effects of high ammonium supply in Scots pine seedling (*Pinus sylvestris*). *Allgemeine Forst und Jagdzeitung* **162**: 35-42.

Flaig, H. & Mohr, H. 1992. Assimilation of nitrate and ammonium by the Scots pine (*Pinus sylvestris*). *Physiol. Plant.* **84**: 568-576.

Forde, B.J. & Cullimore, J.V. 1989. The molecular biology of glutamine synthetase in higher plants. In Miflin, B.J. (ed.) *Oxford Surveys of Plant Molecular and Cell Biology*, Vol. 5: 246-294. Oxford University Press, Oxford.

Foyer, C. & Halliwell, B. 1976. The presence of glutathione and glutathione reductase in chloroplasts: A proposed role in ascorbic acid metabolism. *Planta* **133**: 21-25.

Garbutt, K., Williams, W.E. & Bazzaz, F.A. 1990. Analysis of the differential response of five annuals to elevated CO_2 during growth. *Ecology* **71**: 1185-1194.

Garten, C.T. & Hanson, P.J. 1990. Foliar retention of ^{15}N-nitrate and ^{15}N-ammonium by red maple (*Acer rubrum*) and white oak (*Quercus alba*) leaves from simulated rain. *Environ. Exp. Bot.* **30**: 333-342.

Givan, C.V. 1979. Metabolic detoxification of ammonia in tissues of higher plants. *Phytochemistry* **18**: 375-382.

Hageman, R.H. & Reed, A.J. 1980. Nitrate reductase from higher plants. *Methods in Enzymology* **69**: 270-280.

Hausladen, A., Madamanchi, N., Fellows, S., Alscher, R. & Amundson, R. 1990. Seasonal changes in antioxidants in red spruce as affected by ozone. *New Phytol.* **115**: 447-456.

Häusler, R.E., Blackwell, R.O., Lea, P.J. & Leegood, R.C. 1994. Control of photosynthesis in barley leaves with reduced activity in glutamine synthetase on glutamate synthase. *Planta* **194**: 406-417.

Hecht, U. & Mohr, H. 1990. Factors controlling nitrate and ammonium accumulation in mustard (*Sinapis alba*) seedlings. *Physiol. Plant.* **78**: 379-387.

Hecht, U., Oelmuller, R., Schmidt, S. & Mohr, H. 1988. Action of light, nitrate and ammonium on the levels of NADH and ferredoxin-dependent glutamate synthetase in the cotyledons of mustard seedlings. *Planta* **175**: 130-138.

Heck, W.W. & Dunning, J.A. 1967. The effects of ozone on tobacco and pinto bean as conditioned by several ecological factors. *J. Air Pollut. Control Assoc.* **16**: 112-114.

Hill, A.C. 1971. Vegetation: A sink for atmospheric pollutants. *J. Air Pollut. Control Assoc.* **21**: 341-346.

Hirel, B., Nato, A. & Martin, F. 1984. Glutamine synthetase in ribulose-1,5-bisphosphate carboxylase/oxygenase deficient tobacco mutants in cell suspension culture. *Plant Cell Rep.* **3**: 106-108.

Hou, L., Hill, A.C. & Soleimani, A. 1977. Influence of CO_2 on the effects of SO_2 and NO_2 on alfalfa. *Environ. Pollut.* **12**: 7-16.

Huber, J.L., Huber, S.C., Campbell, W. & Redinbaugh, M.G. 1992. Reversible light/dark modulation of spinach leaf nitrate reductase activity involves protein phosphorylation. *Arch. Biochem. Biophys.* **296**: 58-65.

Idso, S.B. 1988. Three phases of plant response to atmospheric CO_2 enrichment. *Plant Physiol.* **87**: 5-7.

Ikeda, H. & Osawa, T. 1988. Effects of CO_2 concentration in the air, and shading, on the utilization of NO_3 and NH_4^+ by vegetable crops. *Japan Soc. Hort. Sci.* **57**: 52-61.

Ito, O., Okano, K. & Totsuka, T. 1986. Effects of NO_2 and O_3 alone or in combination on kidney bean plants. II. Amino acid pool size and composition. *Soil Sci. Plant Nutrition* **32**: 351-363.

Jager, H.J. 1975. Effects of SO_2-fumigation on the activities of enzymes of amino acid metabolism and the free amino acid contents in plants of different resistance. *Z. Pfl. Krankh, PFL Schutz* **82**: 139-148.

Johnson, R.H. & Lincoln, D.R. 1991. Sagebrush carbon allocation patterns and grasshopper nutrition: the influence of CO_2 enrichment and soil mineral limitation. *Oecologia* **87**: 127-134.

Joy, K.W., Blackwell, R.D. & Lea, P.J. 1992. Assimilation of nitrogen in mutants lacking enzymes of the glutamate synthase cycle. *J. Exp. Bot.* **43**: 139-146.

Kaiser, W.M. & Brendle-Behnisch, E. 1991. Rapid modulation of spinach leaf nitrate reductase activity by photosynthesis I. Modulation *in vivo* by CO_2 availability. *Plant Physiol.* **96**: 363-367.

Kaiser, W.M. & Spill, D. 1991. Rapid modulation of spinach leaf nitrate reductase by photosynthesis II. *In vitro* modulation by ATP and AMP. *Plant Physiol.* **96**: 368-375.

Kamada, M., Jin, Y., Ayabe, M., Seki, M., Sawasaki, T., Ida, S. & Morikawa, H. 1993. Transgenic "air pollutant-philic plants" produced by particle bombardment. In Murata, N. (ed.) *Proceedings of the IXth International Congress on Photosynthesis*. Vol. III: 262-265. Kluwer Academic, Dordrecht.

Karolewski, P. 1984. Effect of SO_2 on changes in proline and hydroxyproline content in leaves of eight species and varieties from the genus *Weigela*. *Arbosetum Kornickie* **29**: 119-129.

Karolewski, P. 1985. The role of free proline in the sensitivity of poplar plants to the action of SO_2. *Eur. J. For. Path.* **15**: 199-206.

King, B.J., Siddiqi, M.Y. & Glass, D.M. 1992. Studies of the uptake of nitrate in barley. *Plant Physiol.* **99**: 1582-1589.

Kleinhofs, A. & Warner, R.L. 1990. Advances in nitrate assimilation. In Miflin, B.J. & Lea, P.J. (eds) *The Biochemistry of Plants*, Vol. 16: 89-120. Academic Press, San Diego.

Klumpp, G., Kuppers, K. & Guderian, R. 1989. Nitrate reductase activity of needles of Norway spruce fumigated with different mixtures of ozone, sulphur dioxide and nitrogen dioxide. *Environ. Pollut.* **58**: 261-271.

Kramer, G.F., Lee, E.H., Rowland, R.A. & Mulchi, C.L. 1991. Effects of elevated CO_2 concentration on the polyamine levels of field grown soybean at three O_3 regimes. *Environ. Pollut.* **73**: 137-152.

Kramer, V., Lahners, K., Back, E., Privalle, L.S. & Rothstein, S. 1989. Transient accumulation of nitrite reductase in mRNA in maize following the addition of nitrate. *Plant Physiol.* **90**: 1214-1220.

Krupa, S. & Branstrom, G. 1974. Studies on the nitrogen metabolism in Ectomycorrhizae. *Physiol. Plant.* **31**: 279-283.

Kuppers, K. & Klumpp, G. 1988. Effects of ozone, sulphur dioxide and nitrogen dioxide on gas exchange and starch economy in Norway spruce (*Picea abies*). *Geo J.* **17**: 271-275.

Langebartels, C., Kerner, K., Leonardi, S., Schraudner, M., Trost, M., Heller, W. & Sandermann, H. 1991. Biochemical plant responses to ozone. I. Differential induction of polyamine and ethylene biosynthesis in tobacco. *Plant Physiol.* **95**: 882-889.

Larigauderie, A., Hilbert, D.W. & Oechel, W.C. 1988. Effect of CO_2 enrichment and nitrogen availability on resource allocation in a grass, *Bromus mollis*. *Oecologia* **77**: 544-549.

Law, R.W. & Mansfield, T.A. 1982. Oxides of nitrogen and the greenhouse atmosphere. In Unsworth, M.H. & Ormrod, D.P. (eds) *Effects of Gaseous Air Pollutions in Agriculture and Horticulture*: 93-112. Butterworth, London.

Lea, P.J., & Forde, B.G. 1994. The use of mutants and transgenic plants to study amino acid metabolism. *Plant, Cell Environ.* **17**: 541-556.

Lea, P.J., Robinson, S.A. & Stewart, G.R. 1990. The enzymology and metabolism of glutamine, glutamate and asparagine. In Miflin, B.J. & Lea, P.J. (eds) *The Biochemistry of Plants*, Vol. 16: 121-159. Academic Press, San Diego.

Lea, P.J., Blackwell, R.D. & Joy, K.W. 1992. Ammonia assimilation in higher plants. In Mengal, K. & Pilbeam, D. (eds) *Nitrogen Metabolism of Plants*: 153-186. Oxford University Press, Oxford.

Leffler, H.R. & Cherry, J.H. 1974. Destruction of enzymatic activity of corn and soybean leaves exposed to ozone. *Can. J. Bot.* **52**: 1233-1238.

Leisen, E., Haussling, M. & Marchaer, H. 1990. Effects of form and concentration of nitrogen and acidic mist treatment on rhizosphere pH and Norway spruce (*Picea abies* L. Karst.). *Forstwiss. Centr.* **109**: 275-286.

Lillo, C. 1991. Diurnal variation of corn leaf nitrate reductase: an experimental distinction between transcriptional and post-transcriptional control. *Plant Sci.* **73**: 149-154.

Lockyer, D.R. & Whitehead, D.C. 1986. The uptake of gaseous ammonia by the leaves of Italian ryegrass. *J. Exp. Bot.* **37**: 919-927.

Luxmoore, R.L. 1981. CO_2 and phytomass. *Bioscience* **31**: 626.

Luxmoore, R.L., O'Neill, E.G., Ellis, J.M. & Rogers, H.H. 1986. Nutrient uptake and growth responses of Virginia pine to elevated atmospheric carbon dioxide. *J. Environ. Qual.* **15**: 244-251.

Madamanchi, N.R., Anderson, J.V., Alscher, R.G., Cramer, C.L. & Hess, J.L. 1992. Purification of multiple forms of glutathione reductase from pea seedlings and enzyme levels in ozone fumigated pea leaves. *Plant Physiol.* **100**: 138-145.

Madsen, E. 1975. Effect of CO_2 enrichment on growth, development, fruit production and fruit quality of tomato from a physiological viewpoint. *Phytotronics* **3**: 318-330.

Manderscheid, R., Bender, J., Weigel, H.-J. & Jager, H.-J. 1991. Low doses of ozone affect nitrogen metabolism in bean (*Phaseolus vulgaris* L.) *Biochem. Physiol. Pflanzen.* **187**: 283-291.

Mansfield, T.A. & Freer-Smith, P.H. 1984. The role of stomata in resistance mechanisms. In Koziol, M.J. & Whatley, F.R. (eds) *Gaseous Pollutants and Plant Metabolism*: 131-146. Butterworths Scientific, London.

Marschner, H., Haussling, M. & George, E. 1991. Ammonium and nitrate uptake rates and rhizosphere pH in non-mycorrhizal roots of Norway spruce (*Picea abies* L. Karst.). *Trees* **5**: 14-21.

McNally, S.F., Hirel, B., Gadal, P., Mann, A.F. & Stewart, G.R. 1983. Glutamine synthetase of higher plants. Evidence for a scientific isoform content related to their possible physiological role and their compartmentation within the leaf. *Plant Physiol.* **72**: 22-25.

Mehlhorn, H., Seufert, G., Schmidt, A. & Kunert, K. 1986. Effect of SO_2 and O_3 on production of antioxidants in conifers. *Plant Physiol.* **82**: 336-338.

Mehrer, I. & Mohr, H. 1989. Ammonium toxicity — description of the syndrome in *Sinapis alba* and the search for its causation. *Physiol. Plant.* **77**: 545-554.

Melzer, J.M., Kleinhofs, A., Kudrna, D.A., Warner, R.L. & Blake, T.K. 1988. Genetic mapping of the barley nitrate reductase deficient *nar* 1 and *nar* 2 loci. *Theor. Applied Genetics* **75**: 767-771.

Mengel, K. & Kirkby, E.A. 1987. *Principles of Plant Nutrition*, 4th edition. International Potash Institute, Berne.

Morgan, S.M., Lee, J.A. & Ashenden, T.W. 1992. Effect of nitrogen oxides on nitrate assimilation in bryophytes. *New Phytol.* **120**: 89-97.

Morikawa, H., Higaki, A., Nohno, M., Kamada, M., Nakata, M., Toyohara, G. & Fujita, K. 1993. Air pollutant-philic plants from nature. In Murata, N. (ed.) *Proceedings of the IXth International Congress on Photosynthesis*, Vol. III (18): 897-900. Kluwer Academic, Dordrecht.

Morrison, J.I.L. 1987. Intercellular CO_2 concentration and stomatal response to CO_2. In Zeiger, E., Farquhar, G.D. & Cowen, I.R. (eds) *Stomatal Function*: 229-251. Stanford University Press, Stanford, CA.

Mousseau, M. & Enoch, H.Z. 1989. Carbon dioxide enrichment reduces shoot growth in sweet chestnut seedlings (*Castanea sativa* Mill.). *Plant, Cell Environ.* **12**: 928-934.

Murray, A.J.S. 1984. Light affects the deposition of NO_2 to the flacca mutants of tomato without affecting the rate of transpiration. *New Phytol.* **98**: 447-450.

Murray, A.J.S. & Wellburn, A.R. 1985. Differences in nitrogen metabolism between cultivars of tomato and pepper during exposure to glasshouse atmospheres containing oxides of nitrogen. *Environ. Pollut.* **39**: 303-316.

Naidu, B.P., Aspinall, D. & Paleg, L.G. 1992. Variability in proline accumulating ability of barley cultivars induced by vapour pressure deficit. *Plant Physiol.* **98**: 716-722.

Nasholm, T. & McDonald, A.J.S. 1990. Dependence of amino acid composition upon nitrogen availability in birch (*Betula pendula*). *Physiol. Plant.* **80**: 507-514.

Nasholm, T., Hogberg, P.& Edfast, A.-B. 1991. Uptake of NO_x by mycorrhizal and non-mycorrhizal Scots pine seedlings: quantities and effect on amino acid and protein concentrations. *New Phytol.* **119**: 83-92.

Norby, R.J., O'Neill, E.G. & Luxmoore, R.J. 1986a. Effects of atmospheric CO_2 enrichment on the growth and mineral nutrition of *Quercus alba* seedlings in nutrient-poor soil. *Plant Physiol.* **82**: 83-89.

Norby, R.J., Pastor, J. & Melilio, J.M. 1986b. Carbon-nitrogen interactions in CO_2-enriched white oak: physiological and long-term perspectives. *Tree Physiol.* **2**: 233-241.

Norby, R.J., Weerasurija, Y. & Hanson, P.J. 1989. Induction of nitrate reductase activity in red spruce needles by NO_2 and HNO_3 vapour. *Can. J. For. Res.* **19**: 889-896.

Oaks, A. 1992. A re-evaluation of nitrogen assimilation in roots. *Bioscience* **42**: 103-111.

Oaks, A., Poulle, M., Goodfellow, V.J., Cass, L.A. & Deising, H. 1988. The role of nitrate and ammonium ions and light on the induction of nitrate reductase in maize leaves. *Plant Physiol.* **88**: 1067-1072.

Oechel, W. & Strain, B.R. 1985. Native species responses to increasing carbon dioxide concentrations. In Strain, B.R. & Cure, J.D. (eds) *Direct Effects of Increasing Carbon Dioxide on Vegetation*: 117-154. DOE/ER-0238 US Dept. of Energy, Carbon Dioxide Research Division, Washington, DC.

Okano, K. & Totsuka, T. 1986. Absorption of nitrogen dioxide by sunflower plants grown at various levels of nitrate. *New Phytol.* **102**: 551-562.

Okano, K., Fukuzawa, R., Tazaki, T. & Totsuka, T. 1986. ^{15}N dilution method of estimating the absorption of atmospheric NO_2 by plants. *New Phytol.* **102**: 73-84.
Okano, K., Machida, T. & Totsuka, T. 1988. Absorption of atmospheric NO_2 by several species: estimation by the ^{15}N dilution method. *New Phytol.* **109**: 203-210.
Okano, K., Machida, T. & Totsuka, T. 1989. Differences in ability of NO_2 absorption in various broad-leaved tree species. *Environ. Pollut.* **58**: 1-17.
Ormrod, D.P. & Beckerson, D.W. 1986. Polyamines as antiozonants for tomato. *Hort. Sci.* **21**: 1070-1071.
Pate, J.S. & Layzell, D.B. 1990. Energetics and biological costs of nitrogen assimilation. In Miflin, B.J. & Lea, P.J. (eds) *The Biochemistry of Plants*, Vol. 16: 1-42. Academic Press, San Diego.
Pearson, J., Clough, E.C.M. & Kershaw, J.L. 1989. Comparative study of nitrogen assimilation in woodland species. *Annu. Sci. For.* **46S**: 663-665.
Pelsey, F., Kronenberger, J., Pollien, J.M. & Caboche, M. 1991. M_2 seed screening for nitrate reductase deficiency in *Nicotiana plumbaginifolia*. *Plant Sci.* **76**: 109-114.
Peuke, A.D. & Tischner, R. 1991. Nitrate uptake and reduction of aseptically cultivated spruce seedlings *Picea abies* (L.) Karst. *J. Exp. Bot.* **42**: 723-728.
Pilbeam, D.J. & Kirkby, E.A. 1992. Some aspects of the utilisation of nitrate and ammonium by plants. In Mengel, K. & Pibeam, D.J. (eds) *Nitrogen Metabolism of Plants*: 55-70. Clarendon Press, Oxford.
Polle, A. & Rennenberg, H. 1992. Field studies on Norway spruce trees at high altitudes: II Defence systems against oxidative stress in needles. *New Phytol.* **121**: 635-642.
Prasad, B.J. & Rao, D.N. 1980. Alteration in metabolic pools of nitrogen dioxide exposed wheat plants. *Ind. J. Exp. Biol.* **18**: 879-882.
Price, A., Lucas, P.W. & Lea, P.J. 1990. Age dependent damage and glutathione metabolism in ozone fumigated barley: a leaf section approach. *J. Exp. Bot.* **41**: 1309-1317.
Raab, M.M. & Weinstein, L.H. 1990. Polyamine and ethylene metabolism in *Triticum aestivum*. In Flores, H.E., Arteca, R.N. & Shannon, J.C. (eds) *Polyamines and Ethylene: Biochemistry, Physiology and Interactions*: 408-410. American Society of Plant Physiologists, Rockville, USA.
Rabe, E. & Lovatt, C.J. 1986. Increased arginine biosynthesis during phosphorus deficiency. *Plant Physiol.* **81**: 774-779.
Raven, J.A. 1988. Acquisition of nitrogen by the shoots of land plants: Its occurrence and implications for acid — base regulation. *New Phytol.* **109**: 1-20.
Raven, J.A. & Farquhar, G.D. 1990. The influence of N metabolism and organic acid synthesis on the natural abundance of isotopes of carbon in plants. *New Phytol.* **116**: 505-529.
Raven, J.A., Wollenweber, B. & Handley, L.L. 1992. Ammonia and ammonium fluxes between photolithotrophs and the environment in relation to the global nitrogen cycle. *New Phytol.* **121**: 5-18.
Redinbaugh, M.G. & Campbell, W.J. 1991. Higher plants response to environmental nitrate. *Physiol. Plant.* **82**: 640-650.
Remmler, J.L. & Campbell, W.H. 1986. Regulation of corn leaf nitrate reductase. II Synthesis and turnover of the enzyme activity and protein. *Plant Physiol.* **80**: 442-447.
Rennenberg, H. & Lamoureux, G.L. 1990. Physiological process that modulate the concentration of glutathione in plant cells. In Rennenberg, H., Brunold, C., DeKok, L.J. & Stulen, I. (eds) *Sulphur Nutrition and Sulphur Assimilation in Higher Plants*: 53-66. SBP Academic, The Hague.
Rhodes, D. 1988. Metabolic response to stress. In Davies, D.D. (ed.) *The Biochemistry of Plants*, Vol. 12: 201-241. Academic Press, New York.
Riens, B. & Heldt, H.W. 1992. Decrease of nitrate reductase activity in spinach leaves during a light/dark transition. *Plant Physiol.* **98**: 573-577.
Robinson, S.A., Slade, A.P., Fox, G.G., Phillips, R., Ratcliffe, R.G. & Stewart, G.R. 1991. The role of glutamate dehydrogenase in plant metabolism. *Plant Physiol.* **95**: 509-516.

Robinson, S.A., Stewart, G.R. & Philips, R. 1992. Regulation of glutamate dehydrogenase activity in relation to carbon limited and protein catabolism in carrot cell suspension culture. *Plant Physiol.* **98**: 1190-1195.

Roelofs, J.G.M., Kempers, A.J., Houdijk, A.L.F.M. & Jansen, J. 1985. The effect of airborne ammonium sulphate of *Pinus nigra* in the Netherlands. *Plant Soil* **84**: 45-56.

Rowland, A.J., Drew, M.C. & Wellburn, A.R. 1987. Foliar entry and incorporation of atmospheric nitrogen dioxide into barley plants of different nitrogen status. *New Phytol.* **107**: 357-371.

Rowland, A.J., Borland, A.M. & Lea, P.J. 1988. Changes in amino acids, amines and proteins in response to air pollutants. In Schulte-Hostede, S., Darrall, N.M., Blank, L.W. & Wellburn, A.R. (eds) *Air, Pollution and Plant Metabolism*: 189-221. Elsevier, London.

Rowland-Bamford, A.J., Borland, A.M., & Lea, P.J. 1988. The effect of air pollution on polyamine metabolism. *Environ. Pollut.* **53**: 410-412.

Rowland-Bamford, A.J., Borland, A.M., Lea, P.J. & Mansfield, T.A. 1989a. The role of arginine decarboxylase in modulating the sensitivity of barley to ozone. *Environ. Pollut.* **61**: 95-106.

Rowland-Bamford, A.J., Lea, P.J. & Wellburn, A.R. 1989b. NO_2 flux into leaves of nitrate reductase-deficient barley mutants and corresponding changes in nitrate reductase activity. *Environ. Exp. Bot.* **29**: 439-444.

Rowland-Bamford, A.J., Allen, L.H. Jr & Baker, J.T. 1990. Carbon dioxide effects on carbohydrate status and partitioning in rice. *J. Exp. Bot.* **41**: 1601-1608.

Rowland-Bamford, A.J., Baker, J.T. & Allen, L.H. Jr 1991. Acclimation of rice to changing atmospheric carbon dioxide concentration. *Plant, Cell Environ.* **14**: 577-583.

Sakakibara, H., Watanabe, M., Hase, T. & Sugiyama, T. 1991. Molecular cloning and characterisation of complementary DNA encoding for ferredoxin-dependant glutamate synthase in maize. *J. Biol. Chem.* **226**: 2028-2034.

Sandermann, H., Schmitt, R., Heller, W., Rosemann, D. & Langebartels, C. 1989. Ozone-induced early biochemical reactions in conifers. In Longhurst, J.W.S. (ed.) *Acid Deposition, Sources, Effects and Controls*: 243-254. British Library, London.

Saxe, H. 1986a. Stomatal-dependent and stomatal-independent uptake of NO_x. *New Phytol.* **103**: 199-205.

Saxe, H. 1986b. Effects of NO, NO_2 and CO_2 on net photosynthesis, dark respiration and transpiration of pot plants. *New Phytol.* **103**: 185-197.

Schmeink, B. & Wild, A. 1990. Studies on the content of free amino acid in needles of undamaged and damaged trees at a natural habitat. *J. Plant Physiol.* **136**: 66-71.

Schmidt, S. & Mohr, H. 1989. Regulation of the appearance of glutamine synthetase in mustard (*Sinapis alba* L.) cotyledons by light, nitrate and ammonium. *Planta* **177**: 526-534.

Schubert, K.R. & Boland, M.J. 1990. The Ureides. In Miflin, B.J. & Lea, P.J. (eds) *The Biochemistry of Plants*, vol. 16: 197-282. Academic Press, San Diego.

Schulze, E.D. 1989. Air pollution and forest decline in a spruce (*Picea abies*) forest. *Science* **244**: 776-783.

Sionit, N. 1983. Response of soybean to two levels of mineral nutrition in CO_2-enriched atmospheres. *Crop Sci.* **23**: 321-333.

Sionit, N., Mortensen, D.A., Strain, B.R. & Hellmers, H. 1981. Growth response of wheat to CO_2 enrichment and different levels of mineral nutrition. *Agron J.* **73**: 1023-1027.

Smith, I.F., Polle, A. & Rennenberg, H. 1990. Glutathione. In Aslcher, R. & Cumming, J. (eds) *Stress Responses in Plants: Adaption and Acclimation Mechanisms*: 201-217. Alan Liss Inc., New York.

Spencer, W. & Bowes, G. 1986. Photosynthesis and growth of water hyacinth under CO_2 enrichment. *Plant Physiol.* **82**: 528-533.

Srivastava, H.S. & Ormrod, D.P. 1984. Effects of nitrogen dioxide and nitrate nutrition on growth and nitrate assimilation in bean leaves. *Plant Physiol.* **76**: 418-423.

Srivastava, H.S. & Ormrod, D.P. 1989. Nitrogen dioxide and nitrate reductase activity and nitrate content of bean leaves. *Environ. Exp. Bot.* **29**: 433-438.

Srivastava, H.S., Wolfenden, J., Lea, P.J. & Wellburn, A.R. 1994. Differential responses of growth and nitrate reductase activity in wild type and NO_2-tolerant barley mutants to atmospheric NO_2 and nutrient nitrate. *J. Plant Physiol.* **143**: 738-743.

Stewart, G.R., Hagarty, E.E. & Sprecht, R.L. 1988. Inorganic nitrogen assimilation in plants of Australian rain forest communities. *Physiol. Plant.* **74**: 26-33.

Stewart, G.R., Pearson, J., Kershaw, J.L. & Clough, E.C.M. 1989. Biochemical aspects of inorganic nitrogen assimilation by woody plants. In Deyer, E. (ed.) *Forest Tree Physiology, Annals der Sciences Forestieres*, Vol. 46: 631-647. Elsevier, INRA, Paris.

Stewart, G.R., Joly, C.A. & Smirnoff, N. 1992. Partitioning of inorganic nitrogen assimilation between roots and shoots of cerrado and forest trees of contrasting plant-communities of South East Brazil. *Oecologia* **91**: 511-517.

Stitt, M. 1991. Rising CO_2 levels and their potential significance for carbon flow in photosynthesis. *Plant, Cell Environ.* **14**: 741-762.

Stulen, I. 1986. Interactions between nitrogen and carbon metabolism in a whole plant content. In Lambers, H., Neeteson, J.J. & Stulen, I. (eds) *Fundamental, Ecological and Agricultural Aspects of Nitrogen Metabolism in Higher Plants*: 261-278. Martinus Nijhoff, The Hague.

Takeuchi, Y., Nihira, J., Kondo, N. & Tesuka, T. 1985. Changes in nitrate-reducing activity in squash cotyledons with NO_2 fumigation. *Plant Cell Physiol.* **26**: 1027-1035.

Taylor, B.R., Parkinson, D. & Parsons, W.F.J. 1989. Nitrogen and lignin content as predictors of litter decay rates: a microcosm test. *Ecology* **70**: 97-104.

Thoene, B., Schroder, P., Papen, H., Egger, A. & Rennenberg, H. 1991. Absorption of atmospheric NO_2 by spruce (*Picea abies* L. Karst) trees. I. NO_2 influx and its correlation with nitrate reduction. *New Phytol.* **117**: 575-585.

Tiburcio, A.F., Kaur-Sawhney, R. & Galston, A.W. 1990. Polyamine metabolism. In Miflin, B.J. & Lea, P.J. (eds) *The Biochemistry of Plants*, Vol. 16: 283-325. Academic Press, San Diego.

Tingey, D.T., Fites, R.C. & Wickliff, C. 1973. Ozone alteration of nitrate reduction in soybean. *Physiol. Plant.* **29**: 33-38.

Tischner, R., Peuke, A., Godbold, D.L., Feig, R., Merg, G. & Huttermann, A. 1988. The effect of NO_2 fumigation on aseptically grown spruce seedlings. *J. Plant Physiol.* **133**: 243-246.

Tobin, A.K., Sumar, N., Patel, M., Moore, A.L. & Stewart, G.R. 1988. Development of photorespiration during chloroplast biogenesis in wheat leaves. *J. Exp. Bot.* **39**: 833-843.

Tremblay, N., Yelle, S. & Gosselin, A. 1988. Effects of CO_2 enrichment, nitrogen and phosphorus fertilization during the nursery period on mineral composition of celery. *J. Plant Nutrit.* **11**: 37-49.

Ullrich, W.R. 1992. Transport of nitrate and ammonium through plant membranes. In Mengel, K. & Pibeam, D.J. (eds) *Nitrogen Metabolism of Plants*: 121-137. Clarendon Press, Oxford.

Valle, R., Mishoe, J.W., Campbell, W.J., Jones, J.W. & Allen, L.H. Jr 1985. Photosynthetic responses of "Bragg" soybean leaves adapted to different CO_2 environments. *Crop Sci.* **25**: 333.

Van Beusichem, M.L., Kirkby, E.A. & Baas, R. 1988. Influence of nitrate and ammonium nutrition on the uptake, assimilation and distribution of nutrients in *Ricinus communis*. *Plant Physiol.* **86**: 914-921.

Van Breeman, N. & Van Dijk, J.F.G. 1988. Ecosystem effects of atmospheric deposition of nitrogen in the Netherlands. *Environ. Pollut.* **54**: 249-274.

Van der Eerden, L.J. 1982. Toxicity of ammonia to plants. *Agric. Environ.* **7**: 223-235.

Van der Eerden, L.J., Lekkerkerk, L.J.A., Smeulders, S.M. & Jansen, A.E. 1992. Effects of atmospheric ammonia and ammonium sulphate on Douglas fir (*Pseudotsuga menziesi*). *Environ. Pollut.* **76**: 1-9.

Van Dijk, H.F.G. & Roelofs, J.G.M. 1988. Effect of excessive ammonium deposition on the nutritional status and condition of pine needles. *Physiol. Plant.* **73**: 494-501.

Van Dijk, H.F.G. de Louw, M.H.J., Roelofs, J.G.M. & Verburgh, J.J. 1990. Impact of artificial, ammonium enriched rainwater on soils and young coniferous trees in a greenhouse. Part II-Effects on the trees. *Environ. Pollut.* **63**: 41-59.

Van Hove, L.W.A., Koops, A.J., Adema, E.H., Vredenberg, W.J. & Pieters, G.A. 1987. Analysis of the uptake of atmosphere ammonia by leaves of *Phaseolus vulgaris* L. *Atmos. Environ.* **21**: 1759-1763.

Van Hove, L.W.A., Van Kooten, O., Adema, E.H., Vredenberg, W.J. & Pieters, G.A. 1989. Physiological effects of long term exposure to low and moderate concentrations of atmospheric NH_3 on poplar leaves. *Plant, Cell Environ.* **12**: 899-908.

Van Hove, L.W.A., Vredenberg, W.J. & Adema, E.H. 1990. The effect of wind velocity, air temperature and humidity on NH_3 and SO_2 transfer into bean leaves (*Phaseolus vulgaris* L.). *Atmos. Environ.* **24A**: 1263-1270.

Van Hove, L.W.A., Van Kooten, O., Van Wijk, K.J., Vredenberg, W.J., Adema, E.M. & Pieters, G.A. 1991. Physiological effects of long term exposure to low concentrations of SO_2 and NH_3 on poplar leaves. *Physiol. Plant.* **82**: 32-40.

Vaucheret, H., Kronenberger, J., Lepingle, A., Vilaine, F., Boutin, J.-P. & Caboche, M. 1992. Inhibition of tobacco nitrite reductase activity by expression of antisense RNA. *The Plant J.* **2**: 425-434.

Vessey, J.K., Henry, L.T. & Raper, C.D. 1990. Nitrogen nutrition and temporal effects of enhanced carbon dioxide on soybean growth. *Crop. Sci.* **30**: 287-294.

Vu, J.V.C., Allen, L.H. Jr. & Bowes, G. 1989. Leaf ultrastructure, carbohydrates and protein of soybeans grown under CO_2 enrichment. *Environ. Exp. Bot.* **29**: 141-147.

Warner, R.L. & Kleinhofs, A. 1974. Relationships between nitrate reductase, nitrite reductase and ribulose diphosphate carboxylase activities in chlorophyll-deficient mutants of barley. *Crop. Sci.* **14**: 654-658.

Warner, R.L., Kleinhofs, A. & Muehllbauer, F.J. 1982. Characterisation of nitrate reductase-deficient mutants in pea. *Crop Sci.* **22**: 389-392.

Wellburn, A.R. 1988. Why are atmospheric oxides of nitrogen usually phytotoxic and not alternative fertilizers. *New Phytol.* **115**: 395-429.

Wellburn, A.R. 1990. *Air Pollution and Acid Rain: The Biological Impact.* Longman Scientific and Technical, London.

Wellburn, A.R., Capron, T.M., Chan, H.-S. & Horsman, D.C. 1976. Biochemical effects of atmospheric pollutants on plants. In Mansfield, T.A. (ed.) *Effects of Air Pollutants on Plants*: 105-114. Cambridge University Press, Cambridge.

Wellburn, A.R., Wilson, J. & Aldridge, P.H. 1980. Biochemical responses to nitric oxide polluted atmospheres. *Environ. Pollut.* **22**: 219-228.

Wellburn, A.R., Higginson, C., Robinson, D. & Walmsley, C. 1981. Biochemical explanation of more than addative inhibitory lower atmospheric levels of SO_2 + NO_2 upon plants. *New Phytol.* **88**: 223-237.

Whitehead, D.C. & Lockyer, D.R. 1987. The influence of the concentration of gaseous ammonia on its uptake by the leaves of Italian ryegrass with and without an adequate supply of nitrogen to the roots. *J. Exp. Bot.* **38**: 818-827.

Wilson, E.J. 1992. Foliar uptake and release of inorganic nitrogen compounds in *Pinus sylvestris* L. and *Picea abies* Karst. *New Phytol.* **120**: 407-416.

Wingsle, G., Nasholm, T., Lundmark, T. & Ericson, A. 1987. Induction of nitrate reductase in needles of Scots pine seedlings by NO_x and NO_3^-. *Physiol. Plant.* **70**: 399-403.

Winner, W.E. & Mooney, H.A. 1980. Ecology of SO_2 resistance. III. Metabolic changes of C_3 and C_4 *Atriplex* species due to SO_2 fumigations. *Oecologia* **46**: 49-54.

Wong, S.C. 1979. Elevated atmospheric partial pressure of CO_2 and plant growth. I. Interaction of nitrogen nutrition and photosynthetic capacity in C_3 and C_4 plants. *Oecologia* **44**: 68-74.

Wong, S.C. 1990. Elevated atmospheric partial pressure of CO_2 and plant growth. II. Non-structural carbohydrate content in cotton plants and its effect on growth parameters. *Photosynth. Res.* **23**: 171-180.

Wray, J. 1989. Molecular and genetic aspects of nitrite reduction in higher plants. In Wray, J. & Kinghorn, J. (eds) *Molecular and Genetic Aspects of Nitrate Assimilation*: 244-262. Oxford Science Publications, Oxford.

Wray, J.L. & Fido, R. 1990. Nitrate reductase and nitrite reductase. In Lea, P.J. (ed.) *Methods in Plant Biochemistry*, Vol. 3: 241-256. Academic Press, London.

Wray, J.L. & Kinghorn, J.R. 1989. *Molecular and Genetic Aspects of Nitrate Assimilation*. Oxford Science Publications, Oxford.

Yelle, S., Beeson, R.C., Trudel, M.J. & Gosselin, A. 1989a. Acclimation of two tomato species to high atmospheric CO_2. I. Sugar and starch concentration. *Plant Physiol.* **90**: 1465-1472.

Yelle, S., Beeson, R.C., Trudel, M.J. & Gosselin, A. 1989b. Acclimation of two tomato species to high atmospheric CO_2. II. Ribulose-1,5-bisphosphate carboxylase/oxygenase and phosphoenolpyruvate carboxylase. *Plant Physiol.* **90**: 1473-1477.

Yoneyama, T., Saskawa, H., Ishizuka, S. & Totsuka, T. 1979. Nitrate accumulation, nitrite reductase activity and diurnal change of NO_2 absorption in leaves. *Soil Sci. Plant Nutrition* **25**: 267-276.

Yu, S.-W., Li, L. & Schimazaki, K.-I. 1988. Responses of spinach and kidney beans to nitrogen dioxide. *Environ. Pollut.* **55**: 1-13.

Zeevaart, A.J. 1974. Induction of nitrate reductase by NO_2. *Acta Bot. Neerl.* **23**: 345-346.

14 Lipid Metabolism and Oxidant Air Pollutants

R.L. HEATH

INTRODUCTION

The history of how varied pollutants alter lipid metabolism shows swings in emphasis. Early in oxidant pollutant investigations, most researchers felt that the lipids were primary targets due to their unsaturated bonds. In that era one spoke of lipid peroxidation disrupting the membranes. Unfortunately, both the term "lipid peroxidation" and the concept were wrong. As often stated since, ozonolysis of unsaturated fatty acids has only a few compounds in common with lipid peroxidation. One of the simplest common molecules (malondialdehyde, MDA) derived from lipid peroxidation and ozonolysis is also one of the simplest molecules to measure; hence, the linkage which does not exist was established by accident in the literature.

The earlier investigators were not fully aware of the structure of the membrane and so believed that ozone in water behaved as ozone in non-polar solvents, including its solubility. Again as often stated, ozone solubility varies in different phases since ozone exists as a zwitterion, and because of its character ozone chemistry is quite distinct in different solvents. The reason for these misunderstandings is much more clear and is important for understanding of future work. In essence, the questions of pollutant pathology were formulated well before techniques that would allow their answers were "discovered". Research questions can only be answered with the proper technology and, more important, the proper physiological and biochemical concepts (or paradigms). Membranes could only be successfully investigated when we had the Singer–Nichelson Model. And the alterations of lipid metabolism by toxic compounds can be understood only when we have a good concept of how the lipids are formed and degraded.

This is not to say that such investigations in the absence of a proper paradigm are wrong or worthless. They help formulate a paradigm, but they can be misleading after their publication time. In fact, much of the earlier work is interesting in that it makes more sense in light of the newer concepts. Early work should not be discarded but rather re-evaluated in light of today's knowledge. The current knowledge of biochemistry makes it possible to understand past and current investigations. This chapter will first elucidate a few definitions of air pollutants and their observed pathological effects, and then attempt to (a) present a brief outline of what is known about plant lipid metabolism, and

Plant Response to Air Pollution. Edited by Mohammad Yunus and Muhammad Iqbal

(b) integrate the current and past research of pollutant effects with (a), in order to try to make sense of the diverse findings and advise on what future research should be.

Some abbreviations that will be used frequently in this chapter are given below: DG = 1,2-diacyl glycerol; DGDG = digalactosidyl diglyceride; FA = fatty acids; MDA = malonyldialdehyde; MGDG = monogalactosidyl diglyceride; NL = neutral lipids; PA = phosphatidic acid; PC = phosphatidyl choline; PE = phosphatidyl ethanolamine; PG = phosphatidyl glycerol; PL = polar lipid; SL = sulpho-lipid; TG = triacyl glycerol; TP = triose-3-phosphate.

TERMS AND CONCEPTS

There are numerous air pollutants but many heavy metals, sulphur dioxide and fluoride, have been or can be controlled at their source. I will not discuss these. At least in the developed nations they are in the process of being controlled. This review will concentrate on only two different types of pollutants which seem to be harder to control and are certainly widespread — ozone and the oxides of nitrogen (NO_x). They are all readily formed in urban industrial-based atmospheres, generally due to the operation of internal combustion engines. They are somewhat different in their chemistry and how they can be handled by plants, but they have been reported to generate activated oxygen species (such as superoxide and hydrogen peroxide) and radicals (such as hydroxyl).

Our "touch-stone" of pollutant effects will be the generation of visible injury, which is generally a region of the leaf with a chlorosis or necrosis. This is the common denominator between older and current literature. To be sure, there are other effects under conditions where pollutants are injurious to the economic health of agriculture, such as inhibition of photosynthesis leading to lowered productivity (Miller, 1987). Yet inhibition of photosynthesis of a leaf does not necessarily lead to lowered productivity over a growing season, but photosynthesis is potentially a cause-and-effect link. Lowered productivity has been well studied but requires a full growing season in the field and so does not lend itself to biochemical studies. Thus, most biochemical investigations are done with short-term experiments on a plant in a definite developmental state exposed to a chronic fumigation (from a single day up to a week of exposures). Longer-time investigations lead to problems of where in the normal developmental sequence the plant is and how the pollutant may have shifted that sequence.

One developing hypothesis is that conditions which cause visible injury are more extreme than those which cause a decline in productivity, although there are, of course, exposure indices which overlap. Mehlhorn & Wellburn (1987) have postulated that visible injury under ozone fumigation is brought upon by the release of wound-induced ethylene. If the ethylene release can be prevented, visible injury is not observed. It has been argued that the interaction of ethylene and ozone within the substomatal cavity causes the production of small molecular-weight aldehydes/alcohols (see Mudd, this volume, Chapter 10), which easily penetrate the cell membrane and move into the cytoplasm and even into the chloroplast where they can directly attack the enzymatic machinery.

It seems clear from chemistry and some experiments that ozone itself cannot move into the cytoplasm; although there are reports of ozone inducing alterations of proteins inside the plasma membrane in red blood cells (Freeman *et al.*, 1979). Certainly, ozone can alter plasma-membrane proteins and wall proteins *in vivo* as well as *in vitro*. Yet penetration is not deep into the cell. Under mild ozone stress, there are few data that suggest a real ozonolysis of unsaturated fatty acid double bonds (see Heath, 1984, 1987). The possibility does exist for multiple reactions between ozone and double bonds once the biochemical is released from the cell structure upon gross disruption of the structure. As shown in red blood cells, ozone reacts slowly with intact cells but rapidly with lysed cells (Heath, 1987). Furthermore, ozone cannot easily penetrate through membranes to react with components inside. Coulson & Heath (1974) showed that intact chloroplasts lost their ability to fix CO_2 when subjected to ozone bubbled into solution. Also isolated grana would lose the capacity for electron transport when subjected to a similar ozone dose. However, the grana within the intact chloroplasts when subjected to the same fumigation remained fully functional, after lysis of the intact plastids. The ozone was reacting with the envelope and with the stroma biochemicals (which include sulphhydryls and ascorbate) and seemingly could not penetrate the short distance to the grana.

Most of the postulated breakdown products (induced by hydroxyl ion or water) of ozone are highly reactive with existing biochemicals or enzymes; thus, their diffusional path length would be expected to be short (Heath, 1987; Mudd, this volume, Chapter 10). For other reactions to take place within the cell, ozone's products must move into the cell.

As well summarized by Mudd (this volume, Chapter 10), there are few data which suggest that the initial attack of ozone is on the double bonds of fatty acids. Indeed, the evidence seems to suggest that amino acids of proteins are much more likely targets due to their proximity and rate of reactivity. While there is a lingering question of ozone solubility within membranes, this may be a lesser problem. Matheson & King (1978) showed that the solubility of gases (argon, oxygen and ethane) was higher (nearly three-fold) in micelles of detergents. It did not seem to matter whether the gas was polar or not, but rather the amount of gas within the micelle was more dependent upon the internal forces within the micelle. The structure of the micelles seems to be different to that of membranes because the molecules making up a micelle are in a dynamic equilibrium with external monomers of the detergent (Peters & Kimmich, 1978). How these results are translated to membranes are at present unknown, but from Matheson & King (1978) gas solubility within the membrane would be predicted to be at least as high as it is in solution.

The problem is: how does ozone form a compound which is stable enough to move within organelles without reacting with molecules lying between, and yet be toxic to the specific organelle? It is not an easy problem to solve. Whatever compounds are produced have not yet been detected. Actually there is another possibility which has been proposed.

Primary initial effects of ozone

We now strongly suspect that ozone's first effect is to modify the cell membrane such that its permeability to ions is increased while inhibiting transport proteins via sulphhydryl oxidation (Heath, 1987, 1994). The mechanism for permeability alterations is not known, but it may be by oxidations of critical groups on or within ionic channels. The path to these critical groups for the toxicants produced by ozone is short. Thus, the toxicant's lifetime is no longer a problem. It is, however, the loss of the membrane's selectivity which causes the problem for the cell. The loss of K^+ from and the gain of Ca^{2+} to the cytoplasm has been documented and those changes could be expected to cause many alterations of metabolism by a shift in cellular ionic balance (Hepler & Wayne, 1985).

Rather than discussing the kind of ozone-derived molecule, I will argue that the ionic disruption is more detrimental to the cellular homeostasis. However, under high levels of ozone, where gross tissue disruption occurs (above 50% visible injury), attack of released biochemicals by ozone may occur directly. The question of how deep ozone can penetrate into the membrane and the subsequent cytoplasm remains still largely unanswered, but it is difficult to imagine that ozone can penetrate into the cytoplasm far enough to alter and to penetrate the envelope of the chloroplast. As we shall see, much of the lipid metabolism that has been shown to be altered resides within the chloroplast and on the envelope of the chloroplast.

LIPID SYNTHESIS

Progress in understanding lipid synthesis in plants has been slow. The process of building understanding based upon microbial metabolism has not necessarily served plant science well. Many false starts were generated. Furthermore, plants have a high concentration of galactosidyl lipids and a very high level of polyunsaturated fatty acids (associated with the chloroplast), which are not common in microbes. Over the years, progress has been made and in rough outline we are beginning to understand much of lipid metabolism (for a general review of lipids see Stumpf *et al.*, 1982; Moore, 1993).

Early work in oxidant interactions with plants suggested that those lipids associated with the chloroplast were more "at risk" than others (Tomlinson & Rich, 1971; Fong & Heath, 1981). Also the inhibition of photosynthesis was likewise an early event in ozone exposure (Miller, 1987). Photosynthesis is closely linked with the energy and reducing power needed for fatty acid synthesis. The obvious place to examine for lipid metabolic changes was the sites of lipid synthesis within the chloroplast. Furthermore, as will be discussed later, the levels of galactolipids changed more readily than other polar lipids.

As we now know, the production of monogalactosidyl diglyceride (MGDG, see Figure 14.1 for structures) occurs within the chloroplast envelope (the membrane surrounding the organelle) from a precursor of 1,2-diglyceride (DG), generally containing 18:1 at sn-1; 16:0 at sn-2 (positions are numbered from the top of the glycerol to the head group on the glycerol at 3). Synthesis of digalactosidyl

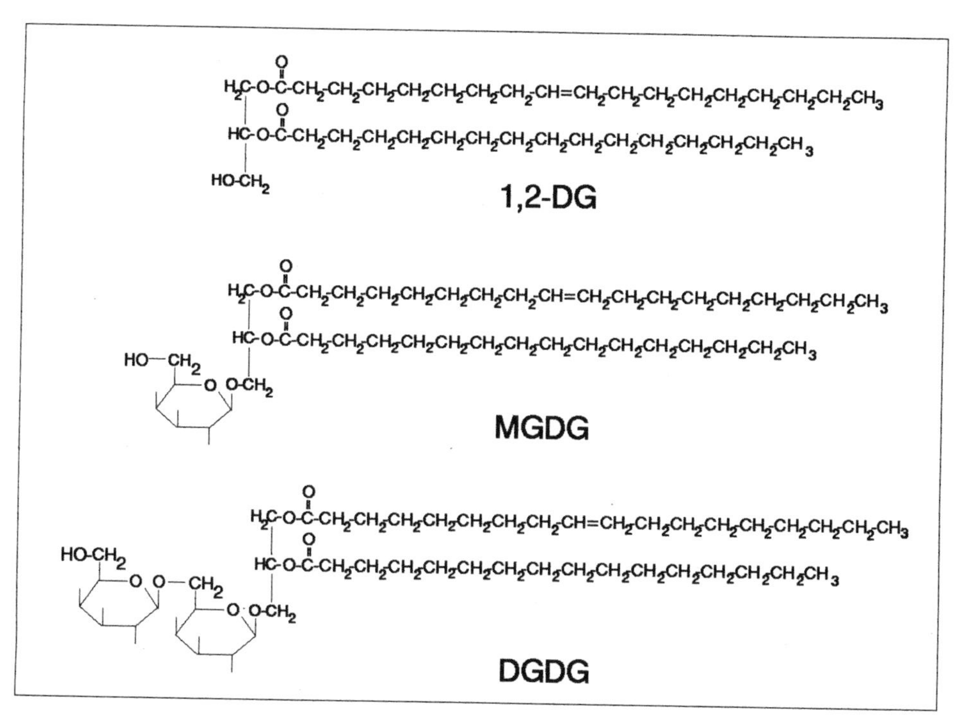

Figure 14.1. Structures of the major lipids mentioned in the chapter. Note that the fatty acid in the sn-1 position is C18:1 (linoleic acid) while the fatty acid in the sn-2 position is C16:0 (palmitic acid). The linkages of the galactoses are $1 \rightarrow 3'$ and $1 \rightarrow 6'$. For more information see Harwood (1980)

diglyceride (DGDG) occurs in the same location using the recently made MGDG unit (Figure 14.1). While there is some uncertainty about how that may occur, it is currently believed that the second galactose unit for DGDG comes from a second MGDG (with its transfer to the first MGDG unit) releasing a 1,2-DG moiety (Heemskerk *et al.*, 1986).

$$2\text{MGDG} \rightarrow \text{DGDG} + 1{,}2\text{-DG} \qquad (1)$$

Further unsaturation of the fatty acid residues occurs *in situ* generating acyl groups with up to three double bonds (e.g., C18:3; Roughan, 1986).

Figure 14.2 shows the pathway for the formation of MGDG and DGDG on the envelope of the chloroplast. DG arrives from either the endomembrane system or the stroma and the enzyme UDP-galactose:1,2-diacylglycerol galactosyltransferase (UDGT) forms MGDG with galactose from UDP-galactose. This enzyme seems to be sensitive to free fatty acids (which cause an inhibition) and Mg^{2+} (which causes a stimulation) and has a critical sulphhydryl (which can be blocked by *N*-ethyl maleimide). Two molecules of MGDG react with the enzyme, galactolipid: galactolipid galactosyltransferase GGGT, to form DGDG releasing 1,2-DG. The activity of that enzyme is stimulated by Ca^{2+} and Mg^{2+} and free fatty acids (Sakaki *et al.*, 1990b). A free sulphhydryl seems to be important for that enzyme

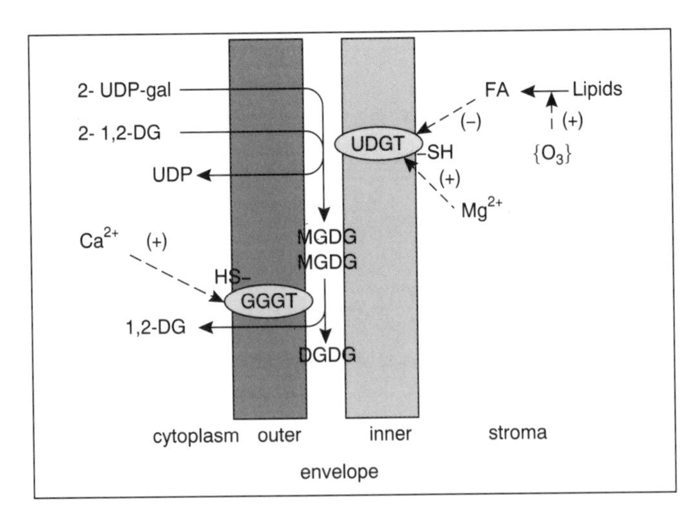

Figure 14.2. The synthetic pathway for galactolipids in plants. The two enzymes (UDGT, UDP-galactose: 1,2-diacylglycerol galactosyltransferase; GGGT, galactolipid: galactolipid galactosyltransferase) are located within the membrane of the chloroplast envelope and produce MGDG and DGDG from UDP-galactose and 1,2-DG (1,2-diacylglycerol). The sites of activators(+) and inhibitors(−) are shown as dotted lines. The effect of ozone was shown by Sakaki *et al.* (1990a, c) as {O3} since they do not believe that ozone directly affects this step

also. There are reports that UDP-gal is the sugar donor rather than MGDG, but they may be artifacts of the isolated system (see Moore, 1986; Roughan, 1986).

There are two metabolic routes for polar lipid synthesis within most plants — procaryotic and eucaryotic derived. The former occurs within the chloroplast and is most similar to procaryotic metabolism (Figure 14.3). The latter occurs outside the chloroplast within the endomembrane systems (endoplasmic reticulum and Golgi). The procaryotic system begins with acetate (or pyruvate) and leads to acyl-ACP (acyl-carrier protein). The fatty acids thus produced move out into an export product (as unesterified fatty acid) or into the acyl groups of 1,2-DG production within the chloroplast. The branch point is apparently controlled by glycerol 3-phosphate (G3P) (high levels cause an increase in the path: glycerides → phosphatidyl choline). At the same time, fatty acids are produced using acetyl-CoA as a carbon source within the cytosol. These acyl units are synthesized from cytoplasmic resources and are indicative of that source. Thus, in spinach, the fatty acids of MGDG have a variable source, from the cytoplasmic and chloroplast systems.

Figure 14.3 shows the flow of acyl units within the two regions of the cell. Modification of the acyl units, including desaturation, occurs on a glycerol backbone of varied lipids. As can be seen from the diagram, the type of fatty acid (C16 verse C18) at the sn-2 position of most polar lipids is indicative of where that acyl (and lipid) unit arose. The head group changes as the lipid "moves through" the pathway and thus the lipid can be shuttled to other final "destinations" (Moore, 1986).

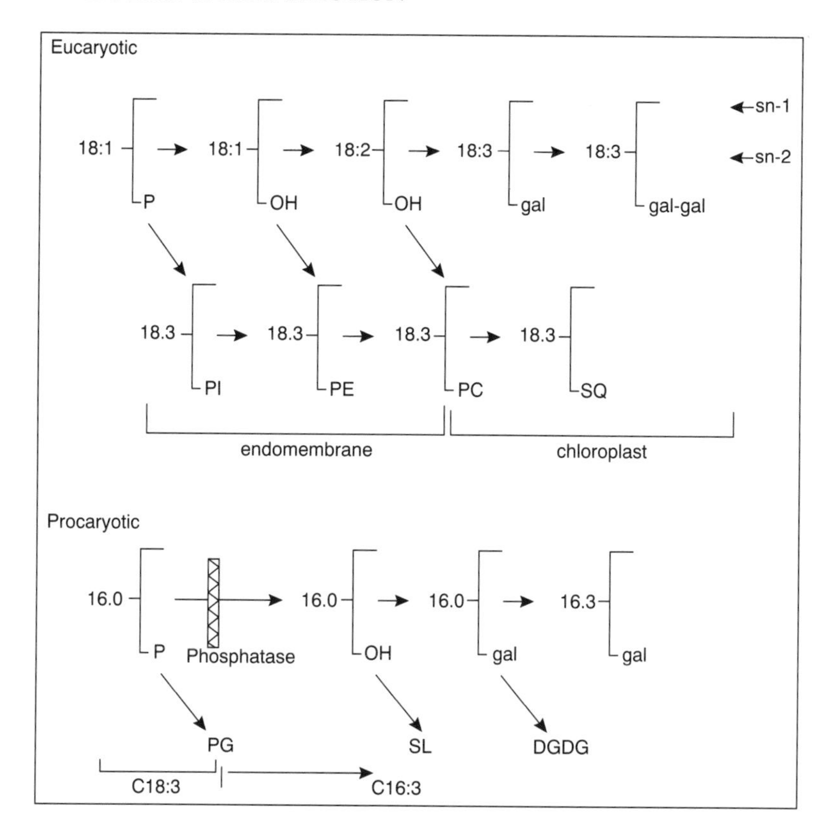

Figure 14.3. Two pathways for placement of fatty acids at the sn-2 position in galactolipids (modified from Roughan, 1986). The schematic diagram represents the glycerol backbone of polar lipids and the placement of the two different fatty acids (16:0 and 18:1). The top (sn-1) position has some specificity also, but is not important for our discussion. The eucaryotic pathway takes place up to the latter stages in the cytoplasm and its membrane system, while the production of the galactolipids takes place in the chloroplast. The difference between C16 and C18 plants is due to the lack of the phosphatase in the procaryotic system (which occurs only in the chloroplast)

There was an apparent loss of a critical phosphatide phosphatase from within the chloroplast through evolution and two types of plants now exist: C16:3 and C18:3. A C16:3 type (spinach) has the phosphatase; while a C18:3 type (maize) forms the acyl units of MGDG from only the eucaryotic (cytosolic) pathway since it is unable to transform the phosphatidic acid into 1,2-DG within the chloroplast (Roughan, 1986).

Roughan (1986) postulates four points which aid in our understanding of the alterations of lipids under oxidant stress.

(1) Fatty acids in an O-ester linkage do not move once formed. The first postulate does not state that lipids cannot be degraded but merely that fatty acid residues do not "hop" from lipid to lipid during the lipid's lifetime. This eliminates any belief that ozone exposure would cause a movement of the acyl units between polar lipids to so alter the membrane structure.

(2) C16 fatty acids are exclusively in the sn-2 position for eucaryotic synthesis while C18 fatty acids are exclusively in the sn-2 position for procaryotic synthesis. Thus, changes in the total fatty acids themselves cannot help in our understanding of alterations; any changes must be linked to its position on the polar lipids. This should allow discrimination between ozone-induced events in the chloroplast and cytoplasm.

(3) C18:1 may be desaturated in both positions while C16:0 may be desaturated in the sn-2 position of MGDG and PG. Again isolation of the fatty acids on the polar lipids can aid in discrimination of which pathways are being altered. Unfortunately, we do not currently know enough about the desaturation events but they do occur on membranes and seem to be sensitive to other metabolic events. There seems to be an indication (Frederick & Heath, 1975; Sakaki *et al.*, 1983) that the ozone fumigation alters the degree of saturation of fatty acids and so the unsaturation metabolism. In the two cases quoted above, one study found an increase in the percentage saturation (*Chlorella*; Frederick & Heath, 1975), while the other found a decrease saturation in some lipids classes (spinach; Sakaki *et al.*, 1983).

(4) MGDG, DGDG, and SL are synthesized from 1,2-DG from either pathway. Studies on lipid changes in plants have not been aware of the two distinct pathways and how different observed effects may be species-linked. Maize and peas (eucaryotic), spinach and tobacco (eucaryotic and procaryotic), parsley and *Arabidopsis* (procaryotic) cannot be compared equally with respect to induced changes in MGDG/DGDG (from Roughan, 1986).

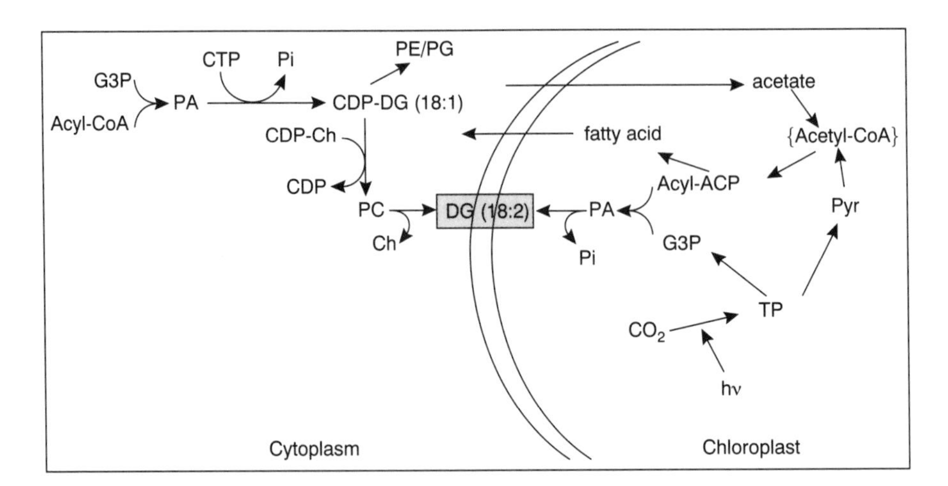

Figure 14.4. The spacial cellular distribution of the synthetic pathway for polar lipids (taken from Moore (1986), Somerville *et al.* (1986) and Harwood (1988)). The two regions for the synthesis of polar lipids with abbreviated pathways are shown, using the following abbreviations. G3P = glycerol-3-phosphate; PA = phosphatidic acid; DG = 1,2-diacyl-glycerol; PC = phosphatidyl choline; PE = phosphatidyl ethanolamine; PG = phosphatidyl glycerol; Pyr = pyruvic acid; TP = triose-3-phosphate

The fully operational systems for polar lipid synthesis are shown in Figure 14.4. Both pathways of galactolipid synthesis begin with an acylation of glycerol-phosphate (G3P) to form phosphatidic acid (PA). The difference seems to be that the eucaryotic or endomembrane system uses Co-enzyme A as the carrier of the acyl group (acyl-CoA) while the procaryotic or chloroplast system uses acyl-carrier-protein (acyl-ACP). Both acetate and glycerol are derived from triose-3-phosphate (TP), but how they are regulated seem to be quite different. The chloroplast generates TP directly from the Calvin Cycle which is regulated by light (and reactants from the light reactions, such as pH and redox level). The endomembrane system receives TP from glycolysis and from transport out of the chloroplast (in exchange for cytoplasmic Pi; Woodrow & Berry, 1988). The regulations for both systems are quite different and complex. The PA is derivatized by CTP to form CDP-DG which is the central pool for other polar lipid synthesis. Other polar lipids, especially PC and PE for the plasma membrane, are generated from this CDP-DG (Moore, 1986). The exact pathways are very dependent upon which researcher is speaking since they are not well established, but the details are not so critical for understanding some pollutant-induced shifts.

CHANGES IN LIPID SYNTHESIS BY OZONE

There are several earlier reports of lipid alterations induced by ozone exposure which have been summarized (Heath, 1984). Most of the reports have centred about the production of MDA, as a measure of lipid oxidation, and the loss of unsaturated fatty acids. With little understanding of lipid metabolism it was not surprising that the experiments were not well directed, yet the fundamental point seemed to emerge that lipid changes were really only detected under high levels of ozone exposure.

General metabolism under different exposures

A series of experiments have been run by Sakaki *et al.* since 1983. They have concentrated upon one type of fumigation system and one metabolic pathway. Thus, this literature gives the best, most complete story with regards to lipid metabolism and ozone fumigation.

They used spinach, which is a sensitive plant, but one which has not had much work with respect to ozone fumigation. The growing conditions seem to be consistent over the papers. While the level of ozone was high (0.5 ppm), enough work has been done to be able to "tease apart" what is happening. Unfortunately, no work has been done on photosynthetic rates nor on chronic, repeated exposures.

The first paper clearly shows that chlorophyll bleaching does not begin until the plants have been exposed to over 10 hours of ozone, while some MDA production begins as early as 6 hours of exposure (Sakaki *et al.*, 1983). Consistent production of MDA, indicative of gross disruptions, really only occurred after 8 hours of exposure (Sakaki *et al.*, 1985). This was within the time scale when chlorophyll and carotenoid levels were first noted to decline. Within the same time frame, the total fatty acids (FA) dropped from 481 to 358 nmol cm^{-2} as the

MDA increased from 0.6 to 2.4 nmol cm^{-2}, yielding a ratio of MDA/FA loss of 1.4%, lower than for direct ozonolysis (Heath, 1978). Fatty acid composition of total lipids did not change over 8 hours (suggesting no compensating changes). The loss of carotenoids begins with chlorophyll loss (Sakaki *et al.*, 1983).

They also watched the development of changes by cutting discs from the exposed leaves and floating them on water solutions for varied time (up to 24 hours). This technique has both pros and cons. The technique allows feeding experiments to be done easily while the cutting gives rise to an additional wound response and eliminates metabolite movement to and from other portions of the plant.

Nonetheless, after exposure the floating experiments showed that scavengers of singlet oxygen ($^1O_2$, D_2O) and of hydroxyl radical ($HO\cdot$, benzoate and formate) have no effect on the development of the MDA response (after 8 hours of *in vivo* fumigation), while scavengers of superoxide (O_2^-, tiron and ascorbate) lowered the amount of MDA formed. The experiments also showed that oxygen itself was needed for the production of MDA since a nitrogen atmosphere drastically lowered the amount of MDA. It must be remembered that this is post-fumigation incubation and would not be expected to show effects of ozone within the plant directly since all the species have extremely short half-lives. However, it is clear that post-incubation reactions most probably involve oxygen and superoxide.

By measuring metabolites immediately after cessation of fumigation they were able to show that ascorbate loss began with the onset of fumigation as did the loss of superoxide dismutase. A lag time was associated with the production of dehydroascorbate suggesting that the reaction of ascorbate with fumigation did not immediately produce the oxidation product (first 4 hours of exposure yield 30 nmol cm^{-2} of ascorbate loss with 5 nmol cm^{-2} of dehydroascorbate production, while the second 4 hours of exposure yielded 20 nmol cm^{-2} of ascorbate loss with 20 nmol cm^{-2} of dehydroascorbate production).

Changes in lipid metabolism

These studies were continued by Sakaki and his co-workers in a 1990 series of papers. They continued to use spinach in which lipid synthesis was relatively well understood. The growth and exposure conditions were fixed and carried out routinely so that comparisons could be done with the earlier work. Again lipid alterations were observed only after a somewhat severe exposure (Sakaki *et al.*, 1990a, b, c).

Sakaki *et al.* (1985) suggested a decline in polar lipids from 0 to 4 hours (unfortunately, no statistics were done and the drop was small). However, a large decline can be seen from 4 to 8 hours. Neutral lipids (NL) rose from 2 to 6 hours, then levelled off. A rise in the levels of free fatty acids (FFA) and MGDG was observed from 4 to 8 hours. But then polar lipids (PL) dropped (from *ca.* 400 to 250 nmol cm^{-2}) while the FFA and MDGD rose (from $75 + 80 = 160$ nmol cm^{-2}), suggesting a balance.

Their data do not show a linear decline with time. Thus, the changes observed from 0 to 4 hours may be of different origins to those from 6 to 8 hours. But

Table 14.1. Alteration in galactosidyl diglyceride ratio due to ozone exposure *in vivo*

Time (hours)[a]	DGDG/MGDG[b]	
0	1.82	initial
2	1.88	initial
4	1.68	initial
6	1.37	later

[a]Spinach plants were exposed to 0.5 ppm for the time indicated.
[b]The ratio was calculated from the data from Sakaki *et al.* (1985).

examining data from 0 to 6 hours, we observe that C16:3 and C18:3 have a 50% loss in level with smaller losses in levels of C16:0 and C18:2. Other changes are suspect due to statistical variation (usually only two runs were included in the data). The actual declines of C16:0, C16:3 and C18:3 in PL (in nmol cm^{-2}) are matched by FA increases in NL. From 6 to 8 hours of fumigation the decline seems to continue for C16:3 and C18:3 in PL, while the rise continues for the same FA in NL.

Initially the level of MGDG falls while the level of DGDG remains somewhat constant. It is instructive to show DGDG/MGDG ratio (Table 14.1) in order to compare with other investigations. These data were taken immediately after the fumigation and should be contrasted with the data on bean from Fong & Heath (1981) who, after a brief pulse of ozone (1 hour), monitored the ratio and found a rise in the ratio only after 6 hours post-fumigation, while after 24 hours (when visible injury began to appear) the ratio in ozone-treated plants fell somewhat. Again there may be different phases which cause different responses. Thus, it is important to note when the observations were made-immediately after fumigation or later after possible recovery. The data for 2 hours may indicate a stimulation, but there is little statistical analysis. There seems to be a difference in the two sets of data which may be due to fumigation protocol or species.

These *in vivo* data should be compared with *in vitro* experiments on bleaching of chlorophyll in isolated chloroplasts; the ratio of MGDG/DGDG does not change until after 10% of the red band of chlorophyll is lost (Dominy & Williams, 1986). This indicates that bleaching precedes MGDG/DGDG changes and *in vitro* experiments on lipid synthesis are probably not well coordinated with *in vivo* work. Lipid metabolism is linked to a coordinated effort of several organelles and is connected to intact membrane systems. Furthermore, without a connected energy source, synthesis may be inhibited and so Dominy & Williams (1986) observed a loss of chlorophyll which then led to a spread in injury to galactolipids (for reviews on photoinhibition see Heath, 1973; Kyle *et al.*, 1987), as opposed to metabolic shifts observed by others using *in vivo* systems. In the future we may possess good *in vitro* systems for such work, but those systems are not available currently.

The levels of free fatty acids did not rise (and may have actually fallen) until after 8 hours of fumigation. Thus, increases in free fatty acids cannot play much of a role initially. The levels of TG and DG rose after 2 hours with an indication of a levelling off of the rise in TG after 6 hours. Thus, it seems that the pool of MGDG (25 nmol cm^{-2}) is falling before any change in DGDG can be observed. Much of that fall is matched by a rise in DGDG (25 nmol cm^{-2}). The level of triglycerides (TG, 40 nmol cm^{-2}) also rose, but it was not clear from where those acyl groups were coming. It could be that they are newly synthesized. However, these are pool size measurements and do not necessarily represent flows of carbon.

The MGDG rapidly loses C18:3 in real and relative terms compared to C16:3. That fact indicates a metabolic shift rather than chemical oxidation since the number of double bonds play the important role in chemical oxidation compared with chain length. Again the TG seems to be gaining the C18:3 acyl units. The authors suggest that the C18:3 residue in TG is derived from the MGDG. Other data suggested that TG was gaining the C18:3 units from MGDG, but through a pathway using 1,2-DG rather than PA. The authors suggest from the literature that the increase in osmophilic granules (thought to be TG storage) upon fumigation was due to the observed increase in TG. Ozone seems to change dramatically the normal movement of compounds through the envelope since one of the first "symptoms" of ozone fumigation in carbon metabolism is an increase in starch in the chloroplast (Miller, 1987).

Nouchi & Toyama (1988) exposed two different plants to ozone (at 0.15 ppm) — morning glory and kidney bean. Under these conditions little visible injury was found initially (up to 4 hours) while the injury was great after 8 hours of exposure (*ca.* 50%). Morning glory produced more MDA (but not very high levels) than kidney bean (which produced no more than that of the zero time control). In morning glory, there was a slight drop in MDGD (5%) while the amounts of PC, PG, PE, and PI rose at 4 hours. Twenty-four hours later, the drop in MGDG was much larger and was thought to be an inhibition of UDP-galactose galactotransferase (UDGT) due to a rise in free fatty acids in the chloroplast (see above for the pathway). There was still a slight stimulation of PI (at 2 hours). The fatty acids showed a drop in only C18:3 and C16:3 at 8 hours indicating an effect on the eucaryotic pathway (since C16:0 $\rightarrow$ C16:3 was not altered). The same results were observed for kidney bean but to a lesser extent. There was a slight drop in the level of MGDG with a transient drop in PC immediately after fumigation. Again a sizeable rise in PI was noted. The same effects were noted 24 hours later, but the level of PE also rose and the change of PI was much more obvious. Here, there was no significant change in total fatty acids. It is important to note again from these experiments that there are two quite distinct time scales involved in ozone fumigation — immediately post-fumigation and a day or so later, once metabolism can alter and respond to the fumigation event. Furthermore, while different species can respond differently in terms of when events occur (time and dose of ozone), similar effects are still seen.

Enzyme changes

Sakaki and his coworkers suggest that the inhibition of UDGT was due to a release of free fatty acids (FFA) from within the chloroplast. These free fatty acids are inhibitory to UDGT, but not GGGT (Sakaki *et al.*, 1990b). GGGT is stimulated by high concentrations of Mg^{2+}, but there is a complex relationship between

FFA and Mg^{2+} (Sakaki *et al.*, 1990b). As the concentration of Mg^{2+} is raised from 1 to 10 mM, the activity of GGGT is increased two-fold at the optimum FFA concentration (300 μM), but nearly 10-fold at a lower Mg^{2+} concentration (50 μM). The inhibition of UDGT by FFA is nearly 75% at 10 mM Mg^{2+}, while it is only about 40% at 1 mM Mg^{2+} with a lower actual rate of UDGT at the lower concentration. This complex interaction demonstrates the difficulty of predicting what enzymatic steps are being changed by the measurement of a pool (e.g., MGDG). The size of a pool is dependent upon both inflow and outflow and since both are being changed by ionic and FFA concentration, what is the real cause? Convincingly, the measured activities of both enzymes isolated after fumigation are not affected by ozone fumigation *in vivo* (Sakaki *et al.*, 1990b). Yet both have sensitive sulphydryls and are located on the envelope. If ozone or its oxidizing products are not reaching these sulphhydryls, it is hard to believe that ozone or its products are inhibiting enzymes located within the envelope, such as the oxygenase/carboxylase of photosynthesis (for a complete discussion see Pell *et al.*, 1994).

The regulation of ion transport by the chloroplast is well known (Maury *et al.*, 1981) although more work has been done on pH changes. Clearly, however, most of the reactions within the chloroplast are regulated by pH, pMg, and pCa (Kaiser *et al.*, 1980). Thus, any change in ionic balance between the cytoplasm and chloroplast would be expected to alter reactions within the chloroplast and it is believed that ozone fumigation causes a loss of K^+ and Ca^{2+} from cells (Heath, 1987, 1994; Castillo & Heath, 1990). The role of Ca^{2+} is important in regulation of many cellular events, but its role within the chloroplast has not yet been fully elucidated.

Fatty acid labelling

In order to understand turnover of theses lipids Sakaki *et al.* (1990a) fed acetate directly to leaves by the application of a small drop of solution before fumigation (2 days). This allowed the acetate to enter the leaf and partially equilibriate with the acetate pools within the leaf. Fong (1975) attempted early experiments along this line but was frustrated by variability when the labelling was done just before fumigation. Evidently under those conditions, permeability and other ozone-induced metabolic changes altered the flow of acetate and so the specific activity within the leaf was uneven. In the data collected by Sakaki *et al.* (1990a), this did not seem to be a problem since the relative amounts of fatty acids within the lipids as measured by radioactivity was uniform for both fumigated and non-fumigated cases. Unfortunately, the specific activity of the fatty acids was not stated, but it would be expected to be lower than that of the fed acetate, due to metabolism and multiple pools within the leaf. The distribution throughout the leaf was likewise not noted, but would be expected to be somewhat uneven.

Ozone fumigation dropped the amount of label within MGDG within 2 hours while the amount of label in TG rose rapidly after a lag of about 2 hours. The label in 1,2-DG showed a slower rise with a longer lag. The level of FFA began to rise only after 4 hours of fumigation. Thus, as 1,2-DG is formed, ozone-fumigation must stimulate the conversion to TG in order to keep the pool of 1,2-DG low initially. Generally the flow of label through an intermediate pool:

$$\text{MGDG} \rightarrow \text{1,2-DG} \rightarrow \text{TG} \tag{2}$$

causes a transient rise in that pool. The authors argue that the flow, as suggested by equation (2), is correct since the level of unsaturation rises by ozone fumigation in both cases. Their data also suggest that on a longer time scale, PC was reduced while PA was increased (after 4+ hours). That suggests that the lipids in the plasma membrane may be involved only as some longer-term effects.

They also studied the formation of FFA within intact chloroplasts, by isolating chloroplasts after fumigation of the plants. The rate of production of FFA was twice as fast after 2 hours of fumigation (the rate of release could be roughly calculated to be 0.5 μmol mg^{-1} chloro h^{-1} for the fumigated plants). It is hard to estimate the "real rise" within a chloroplast because they also showed that the rate of production of FFA could be reduced in both cases by the addition of Co-A and ATP, suggesting an energy problem with the intact chloroplasts.

On the basis of their results, Sakaki *et al.* (1990a, b, c) postulated that ozone stimulated the following reactions:

1. GL:GL galactosyl transferase:
   $$[\text{2MGDG} \rightarrow \text{1,2-DG} + \text{DGDG}] \tag{3}$$
2. Galactolipase;
   $$[\text{MGDG} \rightarrow \text{FFA} + \text{1,2-DG}] \tag{4}$$
3. Acetyl-CoA synthetase:
   $$[\text{Acetyl-CoA} + \text{FFA} \rightarrow \text{acyl-CoA}] \tag{5}$$
4. Diacylglycerol acyltransferase:
   $$[\text{1,2-DG} + \text{acyl-CoA} \rightarrow \text{TG}] \tag{6}$$

Certainly at this time these four reactions can produce the major effects seen. It would be important to follow these observations with some good experiments in intact plastids on how these enzymes are regulated by ions. It may be that changes in the pH or Ca^{2+}/Mg^{2+} balance may shift these enzymes to produce the observed effects. Indeed the authors suggest that changes in the ionic environment may account for a majority of the effects.

Tocopherol and sterols

It has been thought for years that tocopherols functioned as antioxidants in biological systems (Tappel, 1972). Indeed they function very well to interrupt the chain reactions in lipid peroxidation reactions in solution (Foote *et al.*, 1978). The data regarding their role in membrane protection are less compelling. Certainly, most membranes contain tocopherol (to a level of about 1/1000 unsaturated fatty acid double bonds, mol mol^{-1}, Gruger & Tappel, 1971). Some animals are unable to synthesize the ring compound of tocopherol and so it is an essential vitamin for them and feeding studies can be easily done (Draper & Csallany,

1969). Even so our understanding of the role of tocopherol is not especially good. In plants tocopherol molecules are synthesized from erythrose-4-phosphate and PEP through the shikimic acid pathway (see Figure 14.5) or from tyrosine along a well-established pathway within the chloroplast (Lichtenthaler, 1977; Schultz *et al.*, 1985). Yet we know little about regulation of the pathways or the role of tocopherol. When speaking of ozone injury and the role of antioxidants, tocopherol is always mentioned with little data to suggest any effect.

Hausladen *et al.* (1990) examined the role of antioxidants in red spruce by following seasonal changes. Of particular note is one of the few reports of changes in vitamin E (tocopherol). They fit the level of tocopherol within the needles (g^{-1} FW) to the time of the year and found little changes (as level $= A + Bt + Ct^2$). The values of the fit for carbon filtered versus three times ambient were as shown in Table 14.2. From this empirical fit, the constant (A) is lower in higher atmospheric levels of ozone. The seasonal variation (B and C coefficients) are likewise lower, suggesting that there is less tocopherol all year long. Variation with the season is not particularly surprising given that phytochrome action may be linked to tocopherol biosynthesis (Lichtenthaler, 1977). The authors' conclusion was that there was a significant (5%) trend in a difference between the high and low levels of treatment; although there was no discussion of why it occurred or what it meant in relation to metabolism. Their major point was that there are

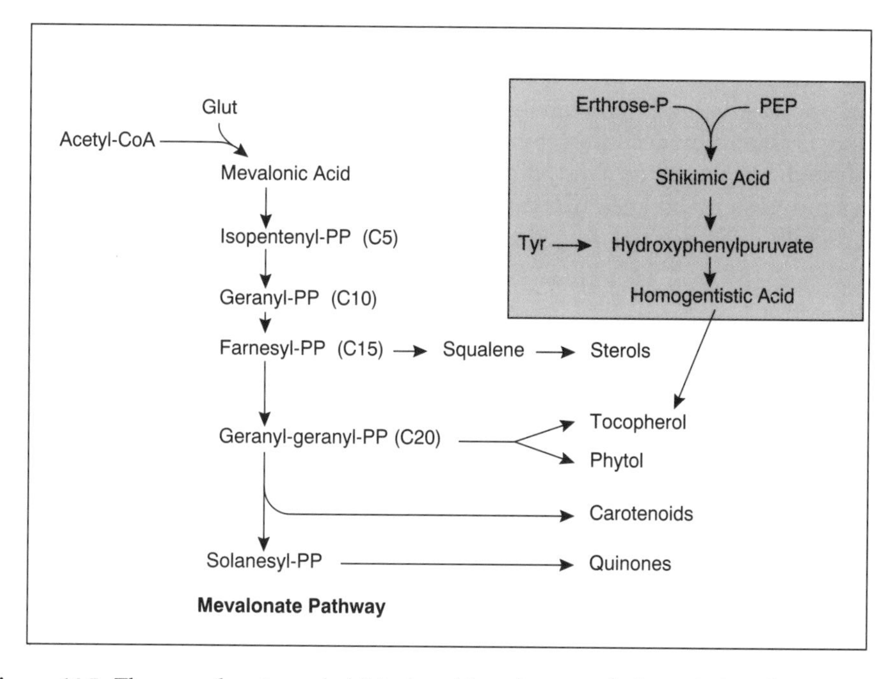

Figure 14.5. The mevalonate and shikimic acid pathways of plants (taken from Lichtenthaler, 1977 and Schultz *et al.*, 1985). The mevalonate pathway forms both sterols and the phytol chain for the tocopherols as well as carotenoids. The ring structure of tocopherol is produced from either tyrosine or erythrose-phosphate and PEP

Table 14.2. Empirical description of seasonal tocopherol changes as influenced by ozone fumigation[a]

Coefficient	CF	3 × ambient
A	-1816 ± 605	-678 ± 465
B	15.5 ± 4.8	5.7 ± 3.7
C	0.028 ± 0.009	0.009 ± 0.007

[a]The exposure to red spruce was done in open-top chambers over the course of a year. Data taken from Hausladen *et al.* (1990) and expressed as nmol g^{-1} FW versus time of year (t) (see text). Data were available for 1 × and 2 × ambient also, but the most significant (at 5% level) difference was observed above.

antioxidant changes due to ozone exposure which may cause a decline in cold hardiness (Davison *et al.*, 1988).

Tocopherol has been argued to play a role in metabolism at several levels: as a lipid antioxidant and as a stabilizer of membranes. In this latter regard, sterols have been investigated by several groups. There is a connection in metabolism between the two biosynthetic groups since the phytol chain on tocopherol follows a similar pathway to that of sterols through the mevalonate pathway (see Figure 14.5). The data on them have been somewhat mixed. Tomlinson & Rich (1971; in common bean using 0.25 ppm for 3 hours) and Grunwald & Endress (1985; in soybeans using 0.07 ppm for 6 hours, repeated over 48 days) reported an increase in free sterols with a decline in esterified sterols, while Trevanthen *et al.* (1979; in tobacco using 0.3 ppm for 6 hours) reported the opposite. None of the groups believed that ozone attacked the sterols directly but believed that these changes were involved in metabolism and membrane stability. While membrane stability is altered by declining sterol content, the exact relationships are yet to be established. On the other hand, if ozone induced a metabolic shift such that the polar lipid to sterol balance was disturbed, membrane reactions to other stresses such as cold tolerance would certainly be affected, perhaps detrimentally.

ALTERATION OF LIPID SYNTHESIS — OXIDES OF NITROGEN

The story for the interaction of NO_x with lipids has very different conclusions depending upon whether chemistry or biology is examined. It is clear that under high concentrations, NO_2 and NO will react quite well with lipids and unsaturated fatty acids (Hicks & Gebicki, 1981; Pryor *et al.*, 1982; Giamalva *et al.*, 1987). The reactions, however, are not linear with concentration. As the level of NO_x reaches the normal atmospheric levels (even in polluted urban areas) the reactions become extremely slow. Most authors (see Mudd *et al.*, 1984, for a review) feel that NO_2 especially will not react directly with lipids in biological membranes.

There is another problem, however, with NO_2. As well stated in the review by Wellburn (1990), the nitrogen from NO_2 seems to easily enter metabolic pathways of nitrogen and increase the pool sizes. That in turn, feeds back into metabolic controls and alters the homeostatic state of the plant. Thus, changes in lipid metabolism would be expected to be caused by increased nitrogen in the plant.

Space does not allow a complete discussion of how lipid pathways could be changed by nitrogen status; on the other hand, there are few investigations into this area (but see Davies, 1979; Hunt *et al.*, 1985).

NO, however, is a free radical and would be expected to interact with any free radical formed within the cell. Furthermore, it may induce other free radical reactions. However, in most investigations not enough attention has been paid to the relative levels of NO_2 and NO in NO_x atmospheres. This has arisen from the manner in which NO_x is measured. Most instruments do not discriminate and so the reported concentration is as NO_x. Also the oxides of nitrogen can relatively easily be interconverted in gaseous phase, especially with ozone present, and so much of the measured reactions (with mixtures of gases) are suspect (Anonymous, 1982).

CONCLUSIONS

While the data from the Sakaki group are very convincing with respect to the alteration expected in lipid metabolism under a severe exposure, the data are for only one species and it is not clear how much other pathways are altered under these conditions. One would expect little disturbance to lipid metabolism under a mild stress. While chronic exposures would be expected to alter many metabolisms, much may be traceable to a slight shift in carbon balance and so a shift in developmental age or state. The unwinding of those responses will require much work. It may, however, be worth it if the nutrient state of the crop (in terms of lipid/carbohydrate content) is altered.

A sequential development of how plants are altered by oxidant stress is shown in Table 14.3. The table is arranged in severity of stress (across) and biochemical developments (down). Under some stresses the trigger of metabolic change may be linked to membrane alterations while for others, gross disruption of structure may hold the key for understanding changes. In any case, this table places observed events into a measure of dose, either accumulated or transient. Clearly, however, there may be more important metabolic disturbances induced by oxidants than lipid metabolism.

A more interesting concept which is being exploited by several research groups in Europe is that ozone stress alters the membrane structure slightly (perhaps by shifting degree of unsaturation) and so makes the plant more sensitive to other stresses. The stress that they are working on is cold sensitivity (Lynch & Steponkus, 1986). There is a report that a more unsaturated fatty acid (C18:4) arises upon ozone exposure, and that this confers a freeze sensitivity upon spruce (Wolfenden & Wellburn, 1991). This is a nice linkage between lipid metabolism and ozone-induced sensitivity. Furthermore, it demonstrates an important current thrust of research-interactions between multiple stresses, such as insects and air pollutants.

While there is much that can be done on lipid metabolism *per se*, it is important to develop a coherent attack such that critical questions are asked. There are too many different research roads which can be taken and many will lead to results which are not very important to stress prevention. It seems that studies of lipid

Table 14.3. Sequence of the development of responses with an ozone exposure. For more discussion, see Heath (1987, 1994)

Phases	I	II	III
Dose	Chronic; Initial acute	High chronic; Acute	Time passage; Light exposure
Time of exposure	2 hours	4–6 hours	Over 8 hours
Gross effects	Membrane/wall reactions	Ethylene production	Light inhibition of photosystem II
Biochemical response	Ionic balance disruption	Toxics production	Photoperoxidation
	Depression of photosynthesis	Visible injury (wilting, water logging)	Chlorophyll loss; death
	Lack of ATP		
Lipid response	Slowing of FA synthesis	Inhibition of enzymes (UGGT and GGGT)	Fatty acid build-up
			Membrane structure distruption
		Stimulation of free fatty acid production	Unsaturated fatty acid loss with MDA formation

metabolism and air pollutant interaction should progress slowly and then only after more fundamental areas are understood. For example:

- How badly is photosynthesis disrupted by oxidants and how is that alteration linked to lipid metabolism?
- How is the flow of lipids from the sites of synthesis to the membrane regulated and controlled, and do ionic balances affect that flow?
- How is lipid turnover (synthesis and degradation) altered by the state of other metabolic pathways for carbon and other elements? Do the effects generated by oxidants on these pathways alter the control sites for net turnover?

All areas of metabolism (in terms of control and regulation) must be investigated sequentially with lipid metabolism in order to show that changes in lipids induced by oxidants are not only small perturbations in more gross changes in the total homeostasis of the plant.

REFERENCES

Anonymous, 1982. Air quality criteria for oxides of nitrogen. Environmental Criteria and Assessment Office, US Environmental Protection Agency, Washington, DC. EPA-600/8-82-026 (EPA-600/8-9/049b, A document in draft form, 1991; to be released later).

Castillo, F.J. & Heath, R.L. 1990. Ca^{+2} transport in membrane vesicles from pinto bean leaves and its alteration after ozone exposure. *Plant Physiol.* **94**: 788–795.

Coulson, C. & Heath, R.L. 1974. Inhibition of the photosynthetic capacity of isolated chloroplasts by ozone. *Plant Physiol.* **53**: 32-38.
Davies, D.D. 1979. Factors affecting protein turnover in plants. In Hewitt, E.J. & Cutting, C.V. (eds) *Nitrogen Assimilation of Plants*. 6th Long Ashton Symposium: 369-396. Academic Press, New York.
Davison, A.W., Barnes, J.D. & Renner, C.J. 1988. Interaction between air pollutants and cold stress. In Schulte-Hoslede, S., Darrall, N.M., Blank, L.W. & Wellburn, A.R. (eds) *Air Pollution and Plant Metabolism*: 307-328. Elsevier Applied Publ., London.
Demel, R.A. 1986. Structural and dynamic aspects of membrane lipids. In Stumpf, P.K., Mudd, J.B. & Nes, W.D. (eds) *The Metabolism, Structure, and Function of Plant Lipids*: 145-152. Plenum Press, New York.
Dominy, P.J. & Williams, W.P. 1986. Is monogalactosyl diacylglcyerol involved in the packaging of light-harvesting chlorophyll proteins in the thylakoid membrane? In Stumpf, P.K., Mudd, J.B. & Nes, W.D. (eds) *The Metabolism, Structure, and Function of Plant Lipids*: 185-188. Plenum Press, New York.
Draper, H.H. & Csallany, A.S. 1969. Metabolism and function of vitamin E. *Fed. Proc.* **28**: 1690-1701.
Fong, F. 1975. Phospholipid metabolism in bean leaves (*Phaseolus vulgaris*). PhD Thesis, University of California, Riverside, 213 pp.
Fong, F. & Heath, R.L. 1981. Lipid content in the primary leaf of bean (*Phaseolus vulgaris*) after ozone fumigation. *Z. Pflanzenphysiol.* **104**: 109-115.
Foote, C.S., Clough, R.L. & Yee, B.G. 1978. Photooxidation of tocopherols. In deDuve, C. & Hayaishi, O. (eds) *Tocopherols, Oxygen, and Biomembranes*: 13-21. Elsevier Press, London.
Frederick, P.E. & Heath, R.L. 1975. Ozone-induced fatty acid and viability changes in *Chlorella. Plant Physiol.* **55**: 15-19.
Freeman, B., McSharman, A. & Mudd, J.B. 1979. Reactions of ozone with phospholipids of vesicles and humanerythrocyteghosts. *Arch. Biochem. Biophys.* **197**: 264-272.
Giamalva, D.H., Kenion, G.B., Church, D.F. & Pryor, W.A. 1987. Rates and mechanisms of reactions of nitrogen dioxide with alkenes in solution. *J. Amer. Chem. Soc.* **109**: 7059-7061.
Gruger, E.H. Jr & Tappel, A.L. 1971. Reactions of biological antioxidants: III Composition of biological membranes. *Lipids* **6**: 147-148.
Grunwald, C. & Endress, A.G. 1985. Foliar sterols in soybeans exposed to chronic levels of ozone. *Plant Physiol.* **77**: 245-247.
Harwood, J.L. 1980. Plant acyl lipids: Structure, distribution and analysis. In Stumpf, P.K. (ed.) *The Biochemistry of Plants: Lipids: Structure and Function*, Vol. 4: 1-56. Academic Press, New York.
Harwood, J.L. 1988. Fatty acid metabolism. *Annu. Rev. Plant Physiol. Plant Mol. Biol.* **39**: 101-138.
Hausladen, A., Madamanchi, N.R., Fellows, S., Alscher, R.G. & Amundson, R.G. 1990. Season changes in antioxidants in red spruce as affected by ozone. *New Phytol.* **115**: 447-458.
Heath, R.L. 1973. The energy state and structure of isolated chloroplasts: The oxidative reactions involving the water-splitting step of photosynthesis. *Intl. Rev. Cytol.* **34**: 49-102.
Heath, R.L. 1978. The reaction stoichiometry between ozone and unsaturated fatty acids in an aqueous environment. *Chem. Phys. Lipids* **22**: 25-37.
Heath, R.L. 1984. Air pollutant effects on biochemicals derived from metabolism: organic, fatty and amino acids. In Koziol, M.J. & Whatley, F.R. (eds) *Gaseous Air Pollutants and Plant Metabolism*. Proceedings of 1st International Symposium on Air Pollutants: 275-290. Butterworths Scientific Press, London.
Heath, R.L. 1987. The biochemistry of ozone attack on the plasma membrane of plant cells. *Rec. Adv. Phytochem.* **21**: 29-54.
Heath, R.L. 1994. Alterations of plant metabolism by ozone exposure. In Alscher, R. & Wellburn, A.R. (eds) *Plant Responses to the Gaseous Air Pollutants*: 121-146. Chapman & Hall, London.

Heemskerk, J.W.M., Scheijen, M.A.M., Jacobs, F.H.H. & Wintermans, J.F.G.M. 1986. Characterization of galactosyltransferases in spinach chloroplast envelope membranes — Application of an assay for UDPGal: diacylglycerol galactosyltransferase. In Stumpf, P.K., Mudd, J.B. & Nes, W.D. (eds) *The Metabolism, Structure, and Function of Plant Lipids*: 301-304. Plenum Press, New York.

Hepler, P.K. & Wayne, R.O. 1985. Calcium and plant development. *Annu. Rev. Plant Physiol.* **36**: 397-439.

Hicks, M. & Gebicki, J.M. 1981. Inhibition of peroxidation in linoleic acid membranes by nitroxide radicals, butylated hydroxytoluene, and α-tocopherol. *Arch. Biochem. Biophys.* **210**: 56-63.

Hunt, E.R. Jr, Weber, J.A. & Gates, D.M. 1985. Effects of nitrate applications on *Amaranthus powellii* Wats. 1. Changes in photosynthesis, growth rates, and leaf area. *Plant Physiol.* **79**: 609-613.

Kaiser, W.M., Urbach, W. & Gimmler, H. 1980. The role of monovalent cations for photosynthesis of isolated intact chloroplasts. *Planta* **149**: 170-175.

Kyle, D.J., Osmond, C.B. & Arntzen, C.J. (eds) 1987. *Photoinhibition*. Elsevier Science Publishers, London, 486 pp.

Lichtenthaler, H.K. 1977. Regulation of prenylquinone synthesis in higher plants. In Tevini, M. & Lichtenthaler, H.K. (eds) *Lipids and Lipid Polymers in Higher Plants*: 231-258. Springer-Verlag, Berlin.

Lynch, D.V. & Steponkus, P.L. 1986. Plasma membrane lipid alterations following cold acclimation: possible relevance to freeze tolerance. In Stumpf, P.K., Mudd, J.B. & Nes, W.D. (eds) *The Metabolism, Structure, and Function of Plant Lipids*: 213-216. Plenum Press, New York.

Matheson, I.B.C. & King, A.D. 1978. Solubility of gases in micellar solution. *J. Colloid Interfacial Sci.* **66**: 464-469.

Maury, W.J., Huber, S.C. & Moreland, D.E. 1981. Effects of magnesium on intact chloroplasts: II. Cation specificity and involvement of the envelope ATPase in (sodium) potassium/proton exchange across the envelope. *Plant Physiol.* **68**: 1257-1263.

Mehorn, H. & Wellburn, A.R. 1987. Stress ethylene formation determines plant sensitivity to ozone. *Nature* **327**: 417-418.

Miller, J.E. 1987. Effects on photosynthesis, carbon allocation, and plant growth associated with air pollutant stress. In Heck, W.W., Taylor, O.C. & Tingey, D.T. (eds) *Assessment of Crop Loss from Air Pollutant*: 287-314. Elsevier Applied Sciences, London.

Moore, T.S. Jr 1986. Regulation of phospholipid headgroup composition in castor bean endosperm. In Stumpf, P.K., Mudd, J.B. & Nes, W.D. (eds) *The Metabolism, Structure, and Function of Plant Lipids*: 265-272. Plenum Press, New York.

Moore, T.S. Jr (ed.) 1993. *Lipid Metabolism in Plants*. CRC Press, Boca Raton, FL.

Mudd, J.B., Banerjee, S.K., Dooley, M.M. & Knight, K.L. 1984. Pollutants and plant cells. In Koziol, M.J. & Whatley, F.R. (eds) *Gaseous Air Pollutants and Plant Metabolism*. Proceedings of 1st International Symposium on Air Pollutants: 105-116. Butterworths Scientific Press, London.

Nouchi, I. & Toyama, S. 1988. Effects of ozone and peroxyacetyl nitrate on polar lipids and fatty acids in leaves of morning glory and kidney bean. *Plant Physiol.* **87**: 638-646.

Pell, E.J., Landry, L.G., Eckardt, N.A. & Glick, R.E. 1994. Air pollution and Rubisco: Effects and implications in plant responses to the gaseous air pollutants. In Alscher, R. & Wellburn, A.R. (eds) *Plant Responses to the Gaseous Air Pollutants*: 239-254. Chapman & Hall, London.

Peters, A. & Kimmich, R. 1978. The heterogeneous solubility of oxygen in aqueous lecithin dispersions and its relation to chain mobility. *Biophys. Struct. Mechanism* **4**: 67-85.

Pryor, E.A., Church, D.F., Govindan, C.K. & Crank, G. 1982. Oxidation of thiols by nitric oxide and nitrogen dioxide: synthetic utility and toxicological implications. *J. Org. Chem.* **47**: 156-159.

Roughan, G. 1986. On the control of fatty acid composition of plant glycerolipids. In Stumpf, P.K., Mudd, J.B. & Nes, W.D. (eds) *The Metabolism, Structure, and Function of Plant Lipids*: 247-254. Plenum Press, New York.

Sakaki, T., Kondo, N. & Sugahara, K. 1983. Breakdown of photosynthetic pigments and lipids in spinach leaves with ozone fumigation: role of active oxygens. *Physiol. Plant.* **59**: 28-34.

Sakaki, T., Ohnishi, J., Kondo, N. & Yamada, M. 1985. Polar and neutral lipid changes in spinach leaves with ozone fumigation: Triacylglycerol synthesis from polar lipids. *Plant Cell Physiol.* **26**: 253-262.

Sakaki, T., Kondo, N. & Yamada, M. 1990a. Pathway for the synthesis of triacylglycerols from monogalactosyldiacylglycerols in ozone fumigated spinach leaves. *Plant Physiol.* **94**: 773-780.

Sakaki, T., Kondo, N. & Yamada, M. 1990b. Free fatty acids regulate two galactosyltransferases in chloroplast envelope membranes isolated from spinach leaves. *Plant Physiol.* **94**: 781-787.

Sakaki, T., Saito, K., Kawaguchi, A., Kondo, N. & Yamada, M. 1990c. Conversion of monogalactosyldiacylglycerols to triacylglycerols in ozone fumigated spinach leaves. *Plant Physiol.* **94**: 766-772.

Schultz, G., Soll, J., Fiedler, E. & Schulze-Sibert, D. 1985. Synthesis of prenylquinones in chloroplasts. *Physiol. Plant.* **64**: 123-129.

Somerville, C.R., McCourt, P., Kunst, L. & Browse, J. 1986. Mutants of *Arabidopsis* deficient in fatty acid desaturation. In Stumpf, P.K., Mudd, J.B. & Nes, W.D. (eds) *The Metabolism, Structure, and Function of Plant Lipids*: 683-688. Plenum Press, New York.

Stumpf, P., Shimakata, T., Eastwell, K., Murphy, D.J., Liedvogel, B., Ohlrogge, J. & Kuhn, D. 1982. Biosynthesis of fatty acids in a leaf cell. In Wintermans, J.F.G.M. & Kuiper, P.J.C. (eds) *Biochemistry and Metabolism of Plant Lipids*: 3-11. Elsevier Medical Press, New York.

Tappel, A.L. 1972. Vitamin E and free radical peroxidation of lipids. *Ann. NY Acad. Sci.* **203**: 12-28.

Tomlinson, H. & Rich, S. 1971. Effect of ozone on sterols and sterol derivatives in bean leaves. *Phytopath.* **61**: 1404-1405.

Trevathan, L.E., Moore, L.D. & Orcutt, R.M. 1979. Symptom expression and free sterol and fatty acid composition of flue-cured tobacco plants exposed to ozone. *Phytopath.* **69**: 582-585.

Wellburn, A.R. 1990. Why are atmospheric oxides of nitrogen usually phytotoxic and not alterative fertilizers? *New Phytol.* **115**: 395-429.

Wolfenden, J. & Wellburn, A.R. 1991. Effects of summer ozone on membrane lipid composition during subsequent frost hardening in Norway spruce (*Picea abies* L.) (Karst.). *New Phytol.* **118**: 323-329.

Woodrow, I.E. & Berry, J.A. 1988. Enzymatic regulation of photosynthetic CO_2 fixation in C_3 plants. *Annu. Rev. Plant Physiol. Plant Mol. Biol.* **39**: 533-594.

15 Influence of Air Pollution on Root Physiology and Growth

G. TAYLOR & R. FERRIS

INTRODUCTION

Much of the early work on air pollution was directed at determining whether plants could be damaged in the absence of visible symptoms, such as the studies on grasses by Cohen & Rushton (1925) and Bleasedale (1952). This was followed by extensive research which aimed to quantify the impact of air pollution on crop growth and yield, including transect studies (Ashmore & Dalpra, 1985) and filtration experiments (Crittenden & Read, 1978; Fowler *et al.*, 1988). Since most harvestable products form some part of the plant shoot, roots were often neglected. The first reports that root growth could be particularly sensitive to atmospheric pollution were published over 20 years ago (Tingey *et al.*, 1971), but despite this there has been relatively little detailed research on root growth and physiology in polluted environments. Such work that has been done has often only considered the crudest measure of growth with little thought for changes in root morphology and physiology which could be of great importance for water and nutrient uptake and in determining the susceptibility of the plant to other forms of stress, particularly drought. Current emphasis is now focused on determining how crops, forests and natural ecosystems will respond to the environmental changes predicted in the next century. These include increased temperatures and changed patterns of rainfall (Houghton *et al.*, 1990) as well as a rise in the concentration of CO_2 and other greenhouse gases such as tropospheric ozone and methane (Fowler, 1990).

The aim of this chapter is to review the available information on the influence of air pollution, including elevated CO_2, on root growth and physiology. We do not consider the influence of air pollution on root–microbial symbiosis, a topic beyond the scope of this review. Historically, concern over the concentrations of SO_2 and later NO_x resulted in much research on these primary pollutants, with considerable emphasis on O_3 in recent years. These three pollutants will be considered here. The widespread decline of forests in areas of central Europe and the USA has yielded valuable information on the biological responses of trees from cellular to ecosystem level, and this includes some excellent data on roots (Meyer *et al.*, 1988; Schulze *et al.*, 1989). This will be considered here, as will the contrasting effects that dry and wet-deposited pollution may have on roots. These ideas are illustrated in Figure 15.1, where the potential routes for

Plant Response to Air Pollution. Edited by Mohammad Yunus and Muhammad Iqbal

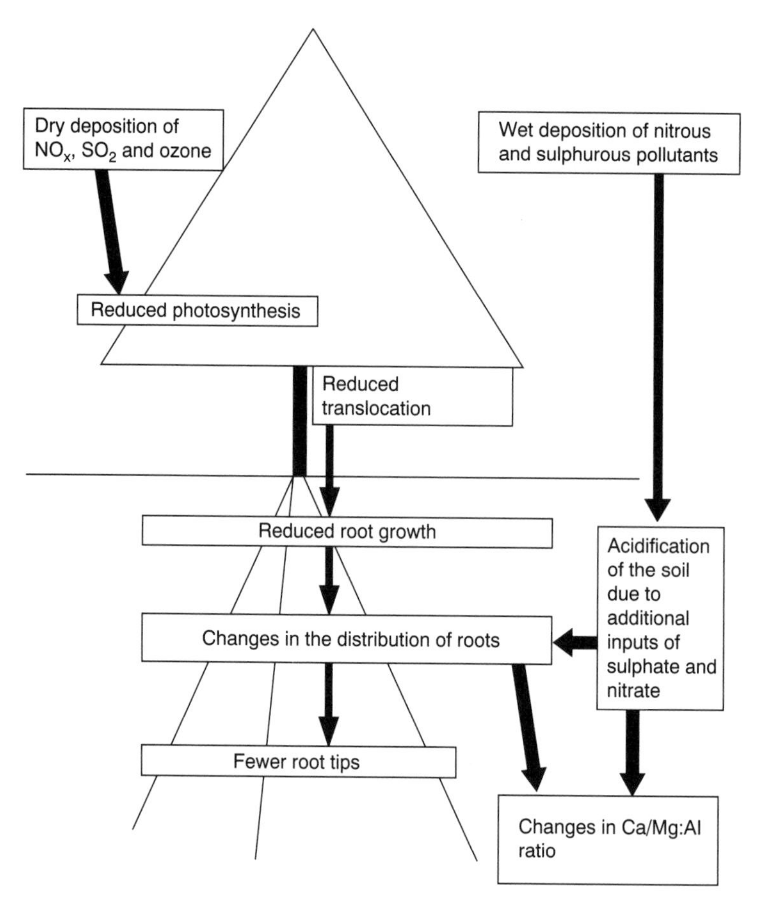

Figure 15.1. Pathways for damage to plant roots by air pollution

pollutants are shown and the mechanisms through which damage may occur are given. The relative importance of gaseous pollution deposited onto foliage and the long-term acidification of soils which may occur will vary, depending on the plant system and timescale being considered. For annuals and crops grown in base-rich soils, the influence of gaseous pollution and indirect secondary effects on roots may be of overriding importance. Conversely, in forested ecosystems which are often planted on soils already susceptible to acidification, the effects of additional inputs of nitrate and sulphate and concomitant decreases in availability of magnesium and calcium may lead to primary effects on ion uptake and root growth. The available literature on these effects is reviewed, but we begin by considering briefly some important aspects of root function and growth.

ROOT FUNCTION AND GROWTH

The root is the plant organ responsible for supplying the shoot with adequate water and nutrients. Roots also anchor plants to the soil, a function which is important, especially in large trees which may be subjected to extreme conditions

(Coutts & Philipson, 1987). Here we are primarily concerned with aspects of root growth and function which may be affected by atmospheric pollution. Root physiology and function is an extensive subject which has been reviewed on a number of occasions (Passioura, 1988) and the reader may refer to these reviews, particularly when considering water and nutrient uptake. Here we consider *root-to-shoot signalling* and *the biophysics of root cell expansion,* since both areas of research have provided information which may be relevant when considering the impact of pollution.

Root-to-shoot signals

Our understanding of the way in which plants respond to drying soil has been questioned during the last decade with the realization that plant roots may play an important role in "sensing" or "measuring" soil moisture and producing a chemical message which moves to shoots in the transpiration stream, resulting in modified shoot growth and function (Zhang & Davies, 1989, 1990). Prior to this work, the general view was that stomata responded primarily to changes in shoot water relations and concentrations of growth regulators produced within leaves. Evidence collected by a number of groups suggests that abscisic acid (ABA) is the plant-growth regulator which is produced in root tips subjected to soil drying, although the change in concentration necessary to alter stomatal conductance and leaf growth may vary in different species (Munns & King, 1988). These are important findings when considering the action of pollution on root growth, since any change in root-tip numbers would presumably alter the quantity of ABA moving in the xylem. There is no published information on abscisic acid concentrations for plants exposed to pollution, an area where future work should yield exciting information. A number of reports have, however, shown that the stomatal regulation of water loss may be impaired by exposure to pollution (Neighbour *et al.*, 1988; Lucas, 1990), although as yet no role for a root signal has been suggested.

Root growth and the biophysics of cell expansion

Root-cell expansion and division are both necessary for continued root elongation, although there is little evidence to suggest that one or other of these processes is more sensitive to atmospheric pollution. When plants are subjected to soil-moisture deficits, shoot growth is often inhibited whilst root growth may be unaltered or even stimulated (Sharp & Davies, 1979; Bradford & Hsiao, 1982). Recent evidence suggests that ABA may also play a role in mediating this growth response since at low water potentials ABA-deficient mutants were characterized by an inhibition of root growth and a promotion of shoot growth with the opposite growth response observed at the same water potentials in wild-type plants (Saab *et al.*, 1990). Work on the biophysics of root-cell expansion suggests root cells may maintain a positive turgor for cell growth, irrespective of external conditions. All plant cells expand irreversibly due to a combination of water uptake and cell-wall extension (Lockhart, 1965). Cell-wall characteristics which are important in regulating plant growth include the coefficient of extensibility,

m ($MP^{-1}h^{-1}$) and the yield turgor of the cell (Y, MPa). Cell-wall properties may be important for the regulation of growth in roots; for example, Wyn-Jones & Pritchard (1989) showed that the turgor pressure of wheat roots was maintained at approximately 0.6 MPa when roots were subjected to a variety of solutions, including ABA. Growth rate appeared to decline as a result of decreased cell-wall plasticity. In contrast to growing leaf and stem cells, root cells also have small yield turgors (Y, MPa), again supporting the idea that changes in root growth could occur without any large change in turgor pressure (Wyn-Jones & Pritchard, 1989). There is no information at present available on the regulation of root-cell expansion for plants exposed to atmospheric change. It is likely that changes in photoassimilate transport could alter solute regulation and membrane processes important for maintaining favourable water relations for continued growth. This remains an important area for future investigation.

INFLUENCE OF SO_2, NO_X, O_3 AND THEIR MIXTURES ON ROOTS

One of the first reports that roots could be more sensitive than shoots to gaseous air pollution was made by Tingey *et al.* (1971) following exposure of radish to 50 ppb ozone or 50 ppb SO_2. This work showed that root weight, length and diameter were all decreased when plants were exposed to either gas. Interestingly, effects were additive or less than additive when plants were exposed to both SO_2 and O_3 simultaneously. Since then a large number of studies have confirmed that both O_3 and SO_2 cause reductions in root growth which may be detected more quickly than effects on shoot growth (Jones & Mansfield, 1982) and which are often of a greater magnitude than effects on shoots (Bell, 1982). Some of the findings on roots are summarized in Table 15.1, although an extensive review of the literature on the effects of pollution on root:shoot ratios is given by Darrall (1989). The table illustrates that in contrast to SO_2 and O_3, NO_x has little effect on root growth, although in combination with SO_2, greater than additive effects have been observed (e.g., Pande & Mansfield, 1985). The table also illustrates that reduced root growth has been detected in a wide range of species including crops, grasses and trees, implying that there may be a common physiological basis for this growth response. Richards (1989) has even proposed that the sensitivity of root responses to pollution could be used as a marker for physiological damage to trees prior to the onset of visible decline symptoms. Using techniques such as root in-growth cores, soil and root psychrometry, minirhizotrons and 32P/33P dual-isotope labelling, changes in root activity could be detected in healthy and declining forest stands. The limitations to such a study would appear to be our knowledge of root behaviour in the field in the absence of pollution and also the variability of root systems which has been demonstrated in other studies. Nevertheless, this contribution highlights the limitations to much of the research done to date and suggests areas for future study.

Although both carbon fixation and allocation have been shown to be affected by exposure to atmospheric pollution, in reviewing the literature Darrall (1989) showed clearly that assimilate partitioning was more sensitive to pollution than rates of apparent photosynthesis. For example, carbohydrate movement

Table 15.1. Influence of gaseous air pollution on root growth

Species	Pollutant	Exposure	Effect	Reference
Glycine max	Filtered, unfiltered and O_3	Maximum 90 ppb for growing season	Decreased root length and weight in O_3. Increased root length in unfiltered air compared with filtered air	Heggestad *et al.* (1988)
Hordeum vulgare	$SO_2 + NO_2$	Maximum 120 ppb of both gases	Increased root : shoot ratio	Pande & Mansfield (1985)
Phaseolus vulgaris	SO_2, O_3	50 ppb SO_2, 400 ppb O_3, for 4 hours	Decreased transport of ^{14}C to roots. Greater effect of O_3	McLaughlin & McConathy (1983)
Pheleum pratense	SO_2	Maximum 120 ppb during growing season	Root growth decreased before any effect on shoots detected. Assimilate transport to roots reduced	Jones & Mansfield (1982)
Phleum pratense	$SO_2 + NO_2$	Maximum 240 + 172 ppb during 40-day exposure	Increased shoot : root ratio in well-watered and unwatered plants exposed to $SO_2 + NO_2$. Greater rate of soil water depletion for polluted plants	Lucas (1990)
Raphanus sativus	O_3, SO_2	Maximum 50 ppb O_3, 50 ppb SO_2	Decreased root length, width and weight by SO_2 and O_3	Tingey *et al.* (1971)
Trifolium repens	O_3	50–150 ppb O_3, 4 hours per day for 6 days	Root growth reduced by 42% shoot growth by 24%	Blum *et al.* (1983)
Various deciduous tree species	$SO_2 + NO_2$	68 ppb SO_2, 68 ppb NO_2, during 14-month fumigation	Decreased root : shoot ratio	Freer-Smith (1984)
Various grass species	SO_2	Varying concentrations	Decreased root growth to a greater extent than shoot growth, although exceptions do exist	Review by Bell (1982)

in *Phaseolus vulgaris* beans was altered by exposure to 100 ppb SO_2, whilst the rate of net photosynthesis was only reduced at 1000 ppb SO_2 (Noyes, 1980). Jones & Mansfield (1982) showed that the transport of $^{14}CO_2$ from the first fully mature leaves of *Phleum pratense* was reduced in SO_2-treated plants (120 ppb), although transport to developing parts of the shoot was increased. Blum *et al.* (1983) also found that exposure of *Trifolium repens* to varying concentrations of O_3 had an effect on the proportion of carbon allocated to roots, with the highest concentrations (100 and 150 ppb) causing decreases in the transport of $^{14}CO_2$ to roots. It has been suggested that these and similar results indicate that pollutants such as O_3 and SO_2 may have a direct effect on some aspect of translocation, possibly on phloem loading as originally proposed by Noyes (1980). In a more recent paper Gould & Mansfield (1988) also showed that transport of $^{14}CO_2$ to roots of *Triticum aestivum* was inhibited following exposure to 80–100 ppb SO_2 and NO_2. It is possible that SO_2 and O_3 could directly alter patterns of phloem loading since sucrose movement into sieve elements requires metabolic energy to maintain proton gradients for co-transport. However, responses to pollution are often complex since more photoassimilate may be transported to developing leaf tissues and under field conditions; increased photoassimilate transport to roots has even been reported following exposure of *Agropyron smithii* to 75 ppb SO_2 (Milchunas *et al.*, 1982). A similar observation was made by Blum *et al.* (1983) following exposure of *Trifolium repens* to varying concentrations of O_3. At small concentrations (below 100 ppb) photoassimilate transport to roots was greater than in the controls, whilst at larger concentrations photoassimilate transport to roots was inhibited. These results may help to explain some of the stimulations in root growth observed in ambient air which are reported below.

Ambient air pollution in southern England and root growth in tree species

Unlike many controlled fumigations, plants are usually subjected to mixtures of pollutants which vary both temporally and spatially (Last, 1991). One approach to studying the effects of such pollution climates is to subject plants to ambient mixtures of pollution and to compare responses with plants grown in filtered air (Fowler *et al.*, 1988). When beech (*Fagus sylvatica*) and Sitka spruce (*Picea sitchensis*) were exposed to ambient conditions in southern England during 1987–1989, root growth was found to be significantly affected on numerous occasions (Taylor *et al.*, 1989; Taylor & Davies, 1990). This was in contrast to effects on shoot growth which were rarely observed, although shoot physiology, including photosynthesis was often altered by ambient pollution (Taylor & Dobson, 1988).

This experimental work confirms that even when exposed to the modest concentrations of pollution in southern England, roots are generally more susceptible to pollution than shoots. Interestingly, the study also showed that changes in root length (m) and specific root length (m g^{-1}) could occur in the absence of any change in biomass. Root length and specific root length were often enhanced following exposure to ambient pollution (Table 15.2). An increase in specific root length suggests that roots were thinner or more branched; the

Table 15.2. The effects of filtration on the growth of roots of beech ($N = 8$) and Sitka spruce ($N = 4$). Taken from Taylor *et al.* (1989)

	Root weight (g)		Root length (m)		Specific root length (m g^{-1})		Root:shoot	
	F	UF	F	UF	F	UF	F	UF
Beech	4.13	3.52	97.5	129.3*	26.7	39.3*	0.571	0.498
Sitka spruce	2.12	2.96	23.6	34.8*	12.1	12.3	0.250	0.400**

F = filtered air; UF = unfiltered air.
$^{*}P < 0.05$, $^{**}P < 0.001$.

long-term consequences of such changes in root morphology will be altered root function. Landsberg & Fowkes (1978) have suggested that a longer root system with thinner roots is generally advantageous in moist soil since the amount of root surface in contact with a given soil volume is greater and water extraction might be more effective. However, as the soil dries, the high resistance to water flow, characteristic of long thin roots, might be disadvantageous. Some evidence to support these ideas was gained from the experiment with beech, since soil-moisture content was significantly less when measured in soil profiles taken from the unfiltered (ambient) treatment as compared with the filtered treatment (Figure 15.2a), an effect which was particularly noticeable in the surface layers of the soil. The influence of ambient pollution on root length persisted for the duration of the experiment with the final harvest made on beech trees 16 months after the initiation of the experiment (Figure 15.2b).

Very few measurements of root length and none of specific root length have been made following exposure to pollution. Unlike many of the fumigation studies reported in the previous section, concentrations of pollution in the unfiltered treatments were small, NO_2 and SO_2 were usually less than 10 ppb. Background concentrations of O_3 varied between 20 and 40 ppb, although peak concentrations in excess of 100 ppb were measured on a few days in each year. Daily maximum concentrations of O_3 exceeded 60 ppb on approximately 15 occasions during each year of the 2-year experiment. Given these small concentrations of pollution it seems likely that photoassimilate transport to roots was either unaffected or stimulated in ambient conditions (Milchunas *et al.*, 1982; Blum *et al.*, 1983), resulting in changed root morphology, with little change in root biomass or root : shoot ratio. An increase in root extension in unfiltered air or in air containing only small concentrations of SO_2 and NO_2 have been observed by other workers. For example, Lucas (1990) reported an increase in root weight and a decline in shoot:root ratio when *Phleum pratense* was exposed to 30 ppb NO_2 and 30 ppb SO_2. At larger concentrations of these two gases, the effect was reversed such that root weight decreased and the shoot:root ratio increased in plants exposed to pollution. Lucas also observed a larger rate of soil-moisture depletion for trees exposed to these small concentrations of pollution and suggested this was due to increased stomatal conductance, although it seems likely that there was also a root effect explaining at least some of the observed increase in water uptake by the plants. Heggestad *et al.* (1988) also reported increased root length in *Glycine max* exposed to unfiltered air when compared to

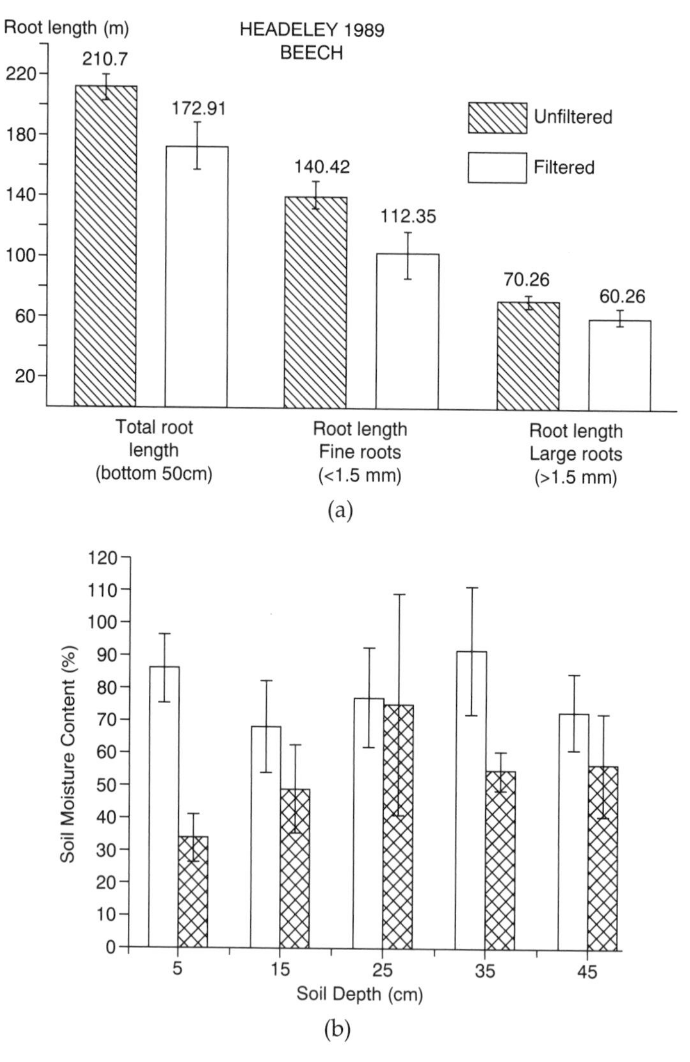

Figure 15.2. (a) Influence of filtration on root growth in *Fagus sylvatica*. Root length measured non-destructively during July 1989 following 16 months exposure to ambient pollution in Hampshire, southern Britain. Each bar represents the mean root length taken from four trees, standard errors are also shown. Hatched bars for trees grown in unfiltered air, open bars for trees grown in filtered air (b). Soil-moisture, contents for trees exposed to filtered or unfiltered air. Symbols and statistics as in (a). $P < 0.05^{*}$; $P < 0.01^{**}$; $P < 0.001^{***}$

the clean air controls, although this was not statistically significant. The results suggest that changes in root length and morphology are a sensitive response to naturally occurring episodes of NO_x, SO_2 and O_3.

INFLUENCE OF ACIDIFICATION ON ROOTS

The widespread forest declines observed during the 1970s and 1980s in continental Europe stimulated a vast amount of scientific research on the forest ecosystem (Rehfuess, 1987; Roberts *et al.*, 1989; Schulze *et al.*, 1989). There are now at least five "damage types" which are recognized (Schulze *et al.*, 1989), although the mid-altitude yellowing and loss of foliage which predominates in Norway spruce is perhaps the decline type which has received most attention (Roberts *et al.*, 1989). At least five major hypotheses have also been proposed to explain this decline (Schulze & Freer-Smith, 1991). The role of air pollution is now seen as central, although the progression, extent and outcome of the decline are linked tightly to other abiotic and biotic influences which may act as predisposing or inciting factors (McLaughlin, 1985). The importance of the below-ground response in declining trees has been the subject of speculation since the early 1980s when Ulrich first proposed that increased soil aluminium, present due to additional inputs of nitrate and sulphate, was directly toxic to roots and resulted in the death of fine roots and decreased uptake of magnesium, an ion known to be deficient in the foliage of declining trees. This hypothesis was supported by work on root proliferation at two sites which displayed contrasting amounts of damage. Root growth of Norway spruce into in-growth cores was greater for trees from the healthier site and these trees also responded more favourably to increasing the pH of the soil through liming (Matzner *et al.*, 1986). This was attributed to the higher Ca:Al ratio, known to inhibit root growth at a value below 1 (McLaughlin, 1985; Matzner, *et al.*, 1986;). Other studies on root growth contradicted the observations of Ulrich in The Solling. For example, root extension in white oak and loblolly pine were not influenced by soil pH, although ectomycorrhizal development was greater for loblolly pine in the most acidic treatment (McLaughlin, 1985). Evidence was also presented to show that the concentrations of aluminium in the soils found below Norway spruce could not, alone, account for the death of fine roots and loss of root function predicted by Ulrich.

A contrasting hypothesis to explain forest decline suggested that root function was impaired beneath damaged stands because of reduced transport of carbohydrates due to direct effects of gaseous air pollution on tree foliage. This included effects of O_3 and SO_2 on photosynthesis and photoassimilate transport. Magnesium deficiencies were considered to be due to the combined effects of reduced uptake and enhanced leaching from needles exposed to the stress of ozone pollution. The empirical information given in the preceeding sections leaves little doubt that this mechanism for damage to roots may be important and a number of reviewers have reached similar conclusions, particularly when considering the growth and productivity of relatively short-lived crop plants (Gould & Mansfield, 1988). However, the complexity of the forest ecosystem is such that the balance of direct soil-mediated effects and indirect shoot-mediated effects may be unlike that for other plant systems, especially since planting often occurs over areas where base-poor soils which are susceptible to acidification, may dominate. An extensive study of two sites, one with a healthy and the other with declining stand of Norway spruce in the Fichtelgebirge was reported by Schulze *et al.* (1989). This study provided convincing evidence that concentrations of carbohydrates were similar in both healthy and declining trees and that these were seasonal patterns of change in structural and non-structural carbohydrates (Oren, *et al.*, 1988). The distribution of fine roots, number of root tips and degree of mycorrhizal infection were closely related to the soil chemical environment (Schneider *et al.*, 1989), with the Ca:Al ratio providing the strongest indicator of unfavourable soil conditions.

ELEVATED CO_2 AND ROOT GROWTH

Exposure of plant shoots to elevated CO_2 often stimulates root growth; an observation now confirmed for a wide range of plant species, including woody perennials such as *Pinus echinata, Quercus alba* (Norby *et al.*, 1987) *Populus* hybrids (Bosac *et al.*, 1995) and sour orange, *Citrus aurantium* (Idso & Kimball, 1991), crops such as *Glycine max* (Rogers *et al.*, 1992), and herbaceous species including *Senecio vulgaris* (Berntson & Woodward, 1992). As illustrated in Table 15.3, the effect of CO_2 on root growth occurs in response to concentrations of CO_2 varying between 600 and 700 ppm; our own work reported below also illustrates increased extension of roots following exposure to 590 ppm. The influence of CO_2 is also persistent with stimulations of root growth reported following as little as 28 days of exposure to elevated CO_2 (Berntson & Woodward, 1992) but maintained for at least 1000 days in the sour orange study of Idso & Kimball (1991). Although the evidence is contradictory (see Farrar & Williams, 1991), it seems likely that root growth is stimulated in elevated CO_2 as photoassimilate transport to growing roots is increased, a response which may be modified by any factor which changes the relationship between source and sink (Arp, 1991). An important question in this context is whether plants generally grow bigger in elevated CO_2 or whether the partitioning of biomass between roots and shoots is altered (Farrar & Williams, 1991). Eamus & Jarvis (1989) suggest that this may depend on the conditions to which roots are subjected. For example when trees are exposed to elevated CO_2 in nutrient-poor soils, partitioning to roots is increased (Luxmoore *et al.*, 1986; Norby *et al.*, 1987), but partitioning may remain unaltered when nutrients are supplied adequately (Taylor *et al.*, 1994, 1995).

The work on sour oranges (Idso & Kimball, 1991) suggested that root density was increased in high CO_2 since soil cores taken from field-grown trees contained more root length. In contrast, a detailed study of root architecture of *Senecio vulgaris* revealed that there were no effects of CO_2 on root density but there were significant increases in the branching and horizontal spread of roots in the high CO_2 treatment (Berntson & Woodward, 1992). These authors suggested that this could improve the "exploitation potential" of the root system rather than altering the "exploitation efficiency". Fitter (1987) has developed the concept that changes in root topology may represent functional differences between different root systems, where less branched, more herringbone-type systems exploit large amounts of soil per unit root length and, therefore, have a high exploitation efficiency, whilst those roots which are highly branched (as in elevated CO_2) may have a greater exploitation potential.

In an interesting paper by Curtis *et al.* (1990), the differences between plants which have C_3 and C_4 mechanisms for CO_2 fixation was highlighted. When a

Table 15.3. Influence of elevated carbon dioxide on root growth and morphology

Species	Duration of exposure (days)	CO_2 concentration (ppm)	Effect	Reference
Citrus aurantium	1000	645	175% increase in fine root biomass. Very few large roots	Idso & Kimball (1991)
Glycine max	18	700	110% increase in root length, 143% increase in root weight	Rogers *et al*. (1992)
Pinus echinata	287	695	Greater root biomass. ^{14}C in fine roots and enhanced exudation	Norby *et al*. (1987)
Populus deltoides × *P. nigra* 'Primo'	60	600	Increased root length and number of root tips	Bosac *et al*. (1995)
Quercus alba	280	690	Increase in weight of fine roots and tap roots	Norby *et al*. (1986)
Scirpus oheyi (C_3) *Spartina patens* (C_4)	210	681	Increase in total biomass for C_3 species. No effect for C_4 species	Curtis *et al*. (1990)
Senecio vulgaris	28	700	Increased branching and root length. No effect on root density	Bernston & Woodward (1992)

salt-marsh community dominated by the C_3 perennial *Scirpus olneyi* was exposed to elevated CO_2 using open-top chambers, root biomass was increased, but when the vegetation was dominated by the C_4 grass *Spartina patens*, no significant effect of elevated CO_2 on root or shoot growth was observed. This study provides evidence for the idea that the nature of plant communities containing both C_3 and C_4 species could change during the coming decades (Woodward *et al.*, 1991), although it is difficult to generalize from the few studies so far conducted. One aspect of the salt-marsh study (Drake & Leadley, 1991) is that, unlike previous controlled-environment studies, field-grown plants may not exhibit a down regulation of photosynthesis at elevated CO_2. After 4 years of exposure,

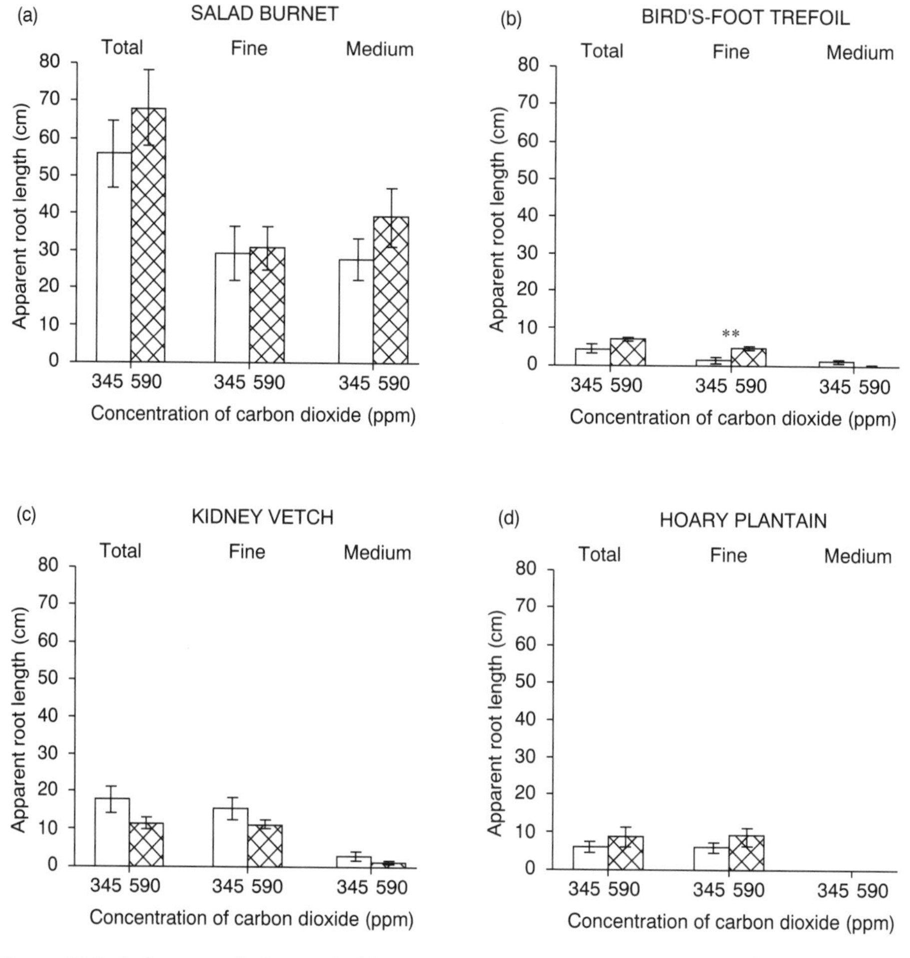

Figure 15.3. Influence of elevated CO_2 on total apparent root length of four downland plant species after 35 days exposure. (a) Salad burnet (*Sanguisorba minor*); (b) bird's-foot trefoil (*Lotus corniculatus*); (c) kidney vetch (*Anthyllis vulneraria*); (d) hoary plantain (*Plantago media*). Open bars are for plants exposed to ambient (345 ppm) CO_2; closed bars for plants exposed to elevated (590 ppm) CO_2. Mean values of 10 plants are given with standard errors. Statistics are as for Figure 15.2

photosynthetic capacity of the leaves was maintained in elevated CO_2 suggesting that there could be a long-term increase in root biomass in response to increasing CO_2. Clearly there remain important areas for further experiments aimed at elucidating the functional and ecological significance of changes in rooting which occur in response to CO_2.

Our own work has illustrated some interesting species differences which may occur in response to CO_2 (Ferris & Taylor, 1993). Chalk grassland is an important plant community in southern England where it dominates much of the landscape (Mitchley, 1983; Wells, 1989). Such species-rich grassland also occurs over areas of Europe (Bobbink, 1991), being characterized by large numbers of species which co-exist over extremely small areas of land. It is not unusual to find 20 species in an area of only 25 cm^2. In such a plant community, competition for both above- and below-ground resources is often intense, with the delicate balance of the community maintained by a combination of low nitrogen and phosphorus status of the soil which acts to check the growth of the more productive species, and regular grazing which has a similar effect on the vegetation.

Any environmental change which alters the morphology and growth of any of the component species in such a community may influence the long-term structure of the vegetation and its ability to support both invertebrate and vertebrate species. A good example of this is seen in the Dutch chalk grasslands where the increase in abundance of *Brachypodium pinnatum*, which is thought to have occurred following excessive emissions of nitrogen from agricultural practices, has resulted in decreased species diversity (Bobbink & Willems 1987; Bobbink, 1991).

When salad burnet (*Sanguisorba minor*), kidney vetch (*Anthyllis vulneraria*), hoary plantain (*Plantago media*) and bird's-foot trefoil (*Lotus corniculatus*) were grown in ambient (345 ppm) and elevated (590 ppm) CO_2, patterns of root growth were altered in all four species, an effect which varied depending on the duration of exposure. For all species, fine roots constituted the majority of root length, and CO_2 had little effect on the growth of roots other than fine roots (diameter < 1 mm). Fine root growth after 35 days of exposure to elevated CO_2 was stimulated in one of the four species, bird's-foot trefoil (Figure 15.3). After 100 days, total root length (Figure 15.4) varied considerably between species. Salad burnet had the longest root systems, approximately 20 m; hoary plantain at least 12 m; bird's-foot trefoil and kidney vetch with 5–10 m. The influence of CO_2 was more pronounced in the species which developed large root systems during a relatively short period of time, particularly salad burnet but also hoary plantain. For salad burnet there was a highly significant increase in fine root length ($P < 0.001$) at four of the five depths where root length was measured. A similar pattern of growth was observed in hoary plantain. After 100 days root growth was also promoted in the slower-growing bird's-foot trefoil and kidney vetch although this was mainly for surface roots only (Figure 15.5).

This work has illustrated that contrasting responses may be observed when different species within a plant community are exposed to elevated CO_2. Further studies are in progress to establish how changes in root growth alter root function and the ability of roots to cope with restricted soil moisture.

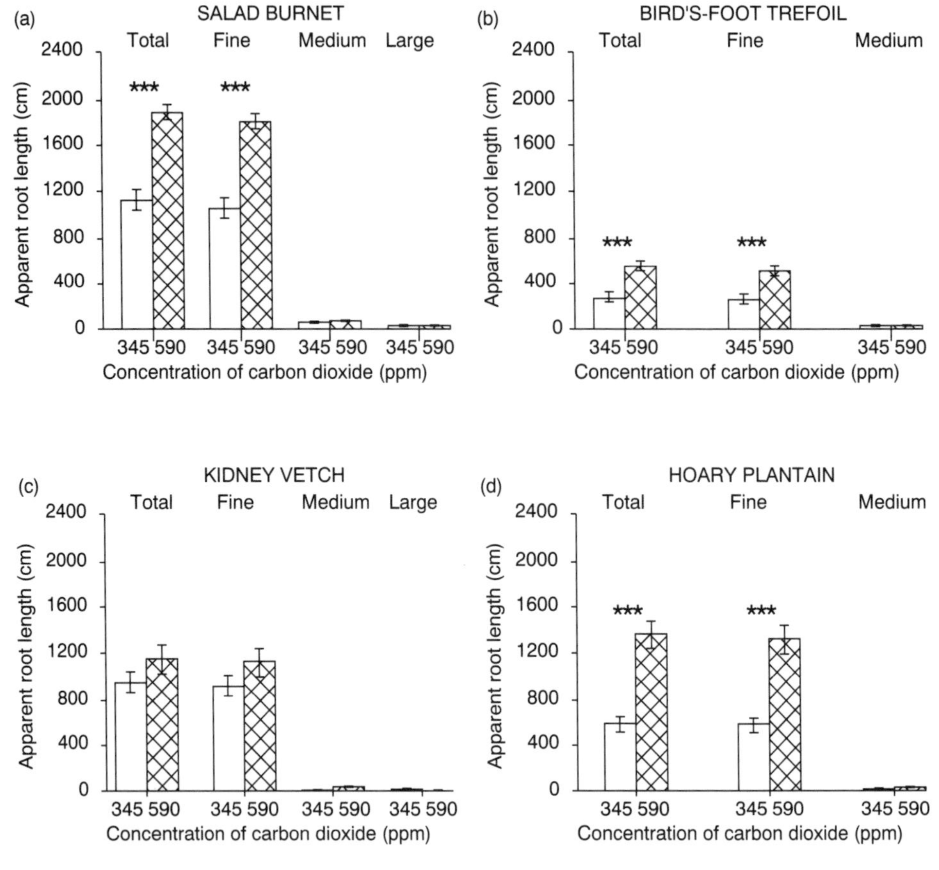

Figure 15.4. Influence of elevated CO_2 on total apparent root length after 100-day exposure to either ambient (345 ppm) or elevated (590 ppm) CO_2. (a) Salad burnet; (b) bird's-foot trefoil; (c) kidney vetch; (d) hoary plantain. Hatchings and statistics as for Figure 15.3

This review has highlighted some effects of atmospheric change on root physiology and growth. Given the sensitivity of roots to pollutants such as CO_2, NO_2 and O_3 and the large effects on root growth which are apparent following exposure to elevated CO_2, it seems likely that root responses to atmospheric change will be important in determining the effects of change on individual plants and plant communities. Future work should focus on determining the impact of atmospheric change on root function, the biophysics of root-cell expansion and on root-to-shoot signalling in plants exposed to soil-moisture deficits.

CONCLUSIONS

Roots are responsible for water and nutrient uptake and for anchoring plants to the soil. The first observations that root growth was more sensitive to air pollution than shoot growth appeared over 20 years ago and it is now known that exposure to either SO_2 or O_3 can lead to reduced assimilate transport and

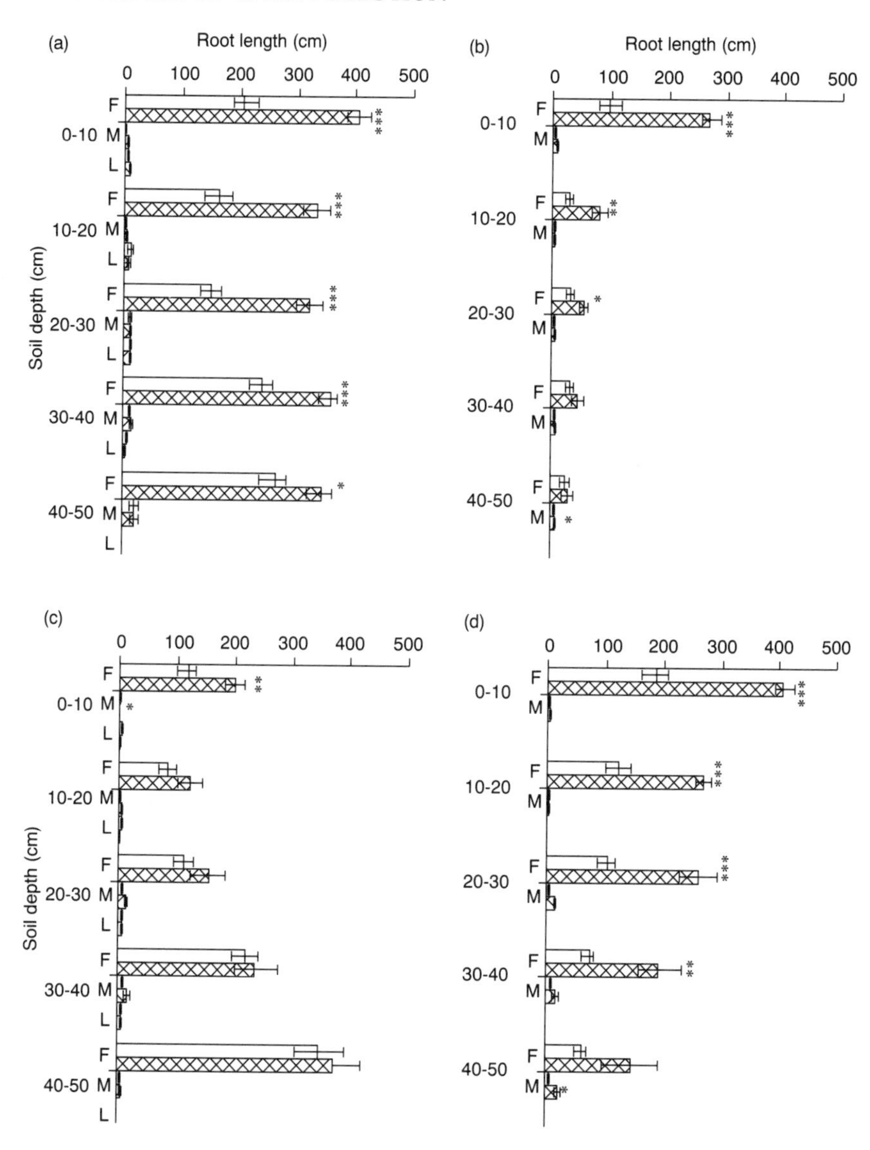

Figure 15.5. Influence of elevated CO_2 on apparent root length profiles after 100-day exposure in ambient (345 ppm) or elevated (590 ppm) CO_2. (a) Salad burnet; (b) bird's-foot trefoil; (c) kidney vetch; (d) hoary plantain. Hatchings and statistics as for Figure 15.3. Apparent fine (F), medium (M) and large (L) root lengths (in cm) are shown. Modified from Ferris & Taylor (1993)

partitioning to roots. This change is often observed prior to any alteration of the rate of net photosynthesis. Root growth may also be impaired when soil chemical processes are altered by inputs of nitrate and sulphate from wet deposition to canopies and soils.

In contrast to other forms of atmospheric pollution, exposure to elevated CO_2 at all concentrations studied results in enhanced root growth, although the magnitude of this effect may vary between species. Both concentrations of CO_2 and O_3 will probably continue to increase and be accompanied by increased temperature and altered patterns of rainfall during the coming decades. The responses of plant roots to environmental change will play an important part in determining growth, survival and abundance of individual plants and plant communities.

REFERENCES

Arp, W.J. 1991. Effects of source-sink relations on photosynthetic acclimation to CO_2. *Plant, Cell Environ.* **14**: 869-875.

Ashmore, M.R. & Dalpra, C. 1985. Effects of London's air on plant growth. *London Environ. Bull.* **3**: 4-5.

Bell, J.N.B. 1982. Sulphur dioxide and the growth of grasses. In Unsworth, M.H. & Ormrod, D.P. (eds) *Effects of Gaseous Pollution in Agriculture and Horticulture*: 225-236. Butterworths, London.

Berntson, G.M. & Woodward, F.I. 1992. The root system architecture and development of *Senecio vulgaris* in elevated CO_2 and drought. *Funct. Ecol.* **6**: 324-333.

Bleasedale, J.K.A. 1952. Atmospheric pollution and plant growth. *Nature* **169**: 376-377.

Blum, U., Mrozek, E. & Johnson, E. 1983. Investigation of ozone effects on ^{14}C distribution in landino clover. *Environ. Exp. Bot.* **23**: 369-378.

Bobbink, R. 1991. Effects of nutrient enrichment in Dutch chalk grassland. *J. Appl. Ecol.* **28**: 28-41.

Bobbink, R. & Willems, J.H. 1987. Increasing dominance of *Brachypodium pinnatum* (L.) Beauv. in chalk grasslands: a threat to a species-rich ecosystem. *Biol. Conservation* **40**: 301-314.

Bosac, C., Gardner, S.D.L., Taylor, G. & Wilkins, D. 1995. Elevated CO_2 and the growth of hybrid poplar: A detailed investigation of root and shoot growth and physiology of *P. interamericana*, 'Primo'. *Forest Ecol. Manag.* **74**: 103-116.

Bradford, K.C. & Hsiao, T.C. 1982. Physiological responses to moderate water stress. In Lange, O.L., Nobel, P.S., Osmond, C.B. & Zeigler, H. (eds) *Encyclopedia of Plant Physiology*, vol. 12B: 263-324. Springer-Verlag, Berlin.

Cohen, J.B. & Rushton, A.G. 1925. *Smoke. A Study of Town Air*. Edward Arnold, London.

Coutts, M.P. & Philipson, J.J. 1987. Structure and physiology of Sitka spruce roots. *Proc. Royal Soc. Edinb.* **93B**: 131-144.

Crittenden, P.D. & Read, D.J. 1978. The effects of air pollution on plant growth with special reference to sulphur dioxide. II. Growth studies with *Lolium perenne* L. *New Phytol.* **83**: 645-651.

Curtis, P.S., Balduman, L.M., Drake, B.G. & Wigham, D.F. 1990. Elevated atmospheric CO_2 effects on belowground processes in C_3 and C_4 estuarine marsh communities. *Ecology* **71**: 2001-2006.

Darrall, N.M. 1989. The effects of air pollutants on physiological processes in plants. *Plant, Cell Environ.* **12**: 1-30.

Drake, B.G. & Leadley, P.W. 1991. Canopy photosynthesis of crops and native plant communities exposed to long-term elevated CO_2. *Plant, Cell Environ.* **14**: 853-860.

Eamus, D. & Jarvis, P.G. 1989. The direct effects of increase in global atmospheric CO_2 concentration on the natural and commercial temperate trees and forests. *Adv. Ecol. Res.* **19**: 1-55.

Farrar, J.F. & Williams, M.L. 1991. The effects of increased atmospheric carbon dioxide and temperature on carbon partitioning, source-sink relations and respiration. *Plant, Cell Environ.* **14**: 819-830.

Ferris, R. & Taylor, G. 1993. Contrasting effects of elevated CO_2 on the root and shoot growth of four native herbs commonly found in chalk grassland. *New Phytol.* **125**: 855-866.

Fitter, A.H. 1987. Functional significance of root morphology. *New Phytol.* **106**: 87-106.
Fowler, D., Cape, J.N., Leith, I.D., Patterson, I.S., Kinnaird, J.W. & Nicolson, I.A. 1988. Effects of air filtration at small SO_2 and NO_2 concentrations on the yield of barley. *Environ. Poll. (Series A)* **53**: 135-149.
Fowler, D.F. 1990. Methane, ozone, nitrous oxide and chlorofluorocarbons. In Cannell, M.G.R. & Hooper, M.D. (eds) *The Greenhouse Effect and Terrestrial Ecosystems of the UK*: 10-13. HMSO, London.
Freer-Smith, P.H. 1984. The responses of six broadleaved trees during long-term exposure to SO_2 and NO_2. *New Phytol.* **97**: 49-61.
Gould, R.P. & Mansfield, T.A. 1988. Effects of sulphur dioxide and nitrogen dioxide on growth and translocation in winter wheat. *J. Exp. Bot.* **39**: 389-399.
Heggestad, H.E., Anderson, E.L., Gish, T.J. & Lee, E.H. 1988. Effects of ozone and soil water deficit on roots and shoots of field-grown soybeans. *Environ. Poll.* **50**: 259-278.
Houghton, J.T., Jenkins, G.J. & Ephraums, J.J. 1990. *Climate Change, The IPCC Scientific Assessment*. Cambridge University Press, Cambridge.
Idso, S.D. & Kimball, B.A. 1991. Effects of two and a half years of atmospheric CO_2 enrichment on the root density distribution of three year old sour orange trees. *Agric. Forest Meteorol.* **55**: 345-349.
Jones, T. & Mansfield, T.A. 1982. Studies on dry matter partitioning and distribution of ^{14}C-labelled assimilates in plants of *Phleum pratense* exposed to SO_2 pollution. *Environ. Poll. (Series A)* **28**: 199-207.
Landsberg, J.J. & Fowkes, N.D. 1978. Water movement through plant roots. *Ann. Bot.* **42**: 493-508.
Last, F.T. 1991. Critique. In Last, F.T. & Whatling, R. (eds) *Acid Deposition. Its Nature and Impacts*: 273-324. The Royal Society of Edinburgh, Edinburgh.
Lockhart, J.A. 1965. An analysis of irreversible plant cell elongation. *J. Theor. Biol.* **8**: 264-275.
Lucas, P.W. 1990. The effects of prior exposure to sulphur dioxide and nitrogen dioxide on the water relations of Timothy grass (*Phleum pratense*) under drought conditions. *Environ. Poll. (Series A)* **66**: 117-138.
Luxmoore, R.J., O'Neil, E.G., Ellis, J.M. & Rogers, H.H. 1986. Nutrient uptake and growth responses of Virginia pine to elevated atmospheric carbon dioxide. *J. Environ. Qual.* **15**: 244-251.
Matzner, E., Murach, D. & Fortmann, H. 1986. Soil acidity and its relationship to root growth in declining forest stands in Germany. *Water, Air Soil Poll.* **31**: 273-282.
McLaughlin, S.B. 1985. Effects of air pollution on forests. *J. Air Poll. Control Assoc.* **35**: 512-533.
McLaughlin, S.B. & McConathy, R.K. 1983. Effects of SO_2 and O_3 on allocation of ^{14}C-labeled photosynthate in *Phaseolus vulgaris*. *Plant Physiol.* **73**: 630-635.
Meyer, J., Schneider, B.U., Werk, K.S., Oren, R. & Schulze, E.-D. 1988. Performance of two *Picea abies* (L.) Karst stands at different stages of decline. V. Root tip ectomycorrhiza development and their relation to above-ground and soil nutrients. *Oecologia* **77**: 7-13.
Milchunas, D.G., Lauenroth, W.K. & Dodd, J.L. 1982. The effect of SO_2 on ^{14}C translocation in *Agropyron smithii* Rydb. *Environ. Exp. Bot.* **22**: 81-91.
Mitchley, J. 1983. The distribution and control of the relative abundance of perennials in chalk grassland. Ph.D. Thesis, University of Cambridge.
Munns, R. & King, R.W. 1988. Abscisic acid is not the only stomatal inhibitor in the transpiration stream. *Plant Physiol.* **88**: 703-708.
Neighbour, E.A., Cottam, D.A. & Mansfield, T.A. 1988. Effects of sulphur dioxide and nitrogen dioxide on the control of water loss by birch (*Betula* spp.). *New Phytol.* **108**: 149-157.
Norby, R.J., O'Neill, E.O., & Luxmore, R.J. 1986. Effects of atmospheric CO_2 enrichment on the growth and mineral nutrition of *Quercus alba* seedlings grown under elevated CO_2. *Tree Physiol.* **3**: 203-210.

Norby, R.J., O'Neill, E.O., Hood, W.G. & Luxmore, R.J. 1987. Carbon allocation, root exudation and mycorrhizal colonisation on *Pinus echinata* seedlings grown under CO_2 enrichment. *Tree Physiol.* **3**: 203-210.

Noyes, R.D. 1980. The comparative effects of sulphur dioxide on photosynthesis and translocation in bean. *Physiol. Plant Pathol.* **16**: 73-79.

Oren, R., Schulze, E.-D., Werk, K.S. & Meyer, J. 1988. Performance of two *Picea abies* (L.) Karst. stands at different stages of decline. VII Nutrient relations and growth. *Oecologia* **77**: 163-173.

Pande, P.C. & Mansfield, T.A. 1985. Responses of spring barley to SO_2 and NO_2 pollution. *Environ. Poll. (Series A)* **38**: 87-97.

Passioura, J.B. 1988. Water transport in and to roots. *Annu. Rev. Plant Physiol. Molec. Biol.* **39**: 245-65.

Rehfuess, K.E. 1987. Perception on forest declines in central Europe. *Forestry* **60**: 1-11.

Richards, J.H. 1989. Evaluation of root functioning of trees exposed to air pollutants. In *Biologic Markers of Air Pollution Stress and Damage in Forests* (ed. by Committee on Biologic Markers of Air-Pollution Damage in Trees). National Academic Press, Washington, DC.

Roberts, T.M., Skeffington, R.A. & Blank, L.W. 1989. Causes of type 1 spruce decline in Europe. *Forestry* **62**: 179-231.

Rogers, H.H., Peterson, C.M., McCrimmon, J.N. & Cure, J.D. 1992. Response of plant roots to elevated atmospheric carbon dioxide. *Plant, Cell Environ.* **15**: 749-752.

Saab, I.N., Sharp, R.E., Pritchard, J. & Voetberg, G.S. 1990. Increased endogenous ABA maintains primary root growth and inhibits shoot growth of maize seedlings at low water potentials. *Plant Physiol.* **93**: 1329-1336.

Schneider, B.U., Meyer, J., Schulze, E.-D. & Zech, W. 1989. Root and mycorrhizal development in healthy and declining Norway spruce stands. In Schulze, E.-D., Lange, O.L. & Oren, R. (eds) *Forest Decline and Air Pollution*: 370-391. Springer-Verlag, Berlin.

Schulze, E.-D. & Freer-Smith, P.H. 1991. An evaluation of forest decline based on field observations focused on Norway spruce, *Picea abies*. In Last, F.T. & Whatling, R. (eds) *Acidic Deposition, Its Nature and Impacts*: 155-168. The Royal Society of Edinburgh, Edinburgh.

Schulze, E.-D., Lange, O.L. & Oren, R. 1989. *Forest decline and Air Pollution: A Study of Spruce on Acid Soils*. Springer-Verlag, Berlin.

Sharp, R.E. & Davies, W.J. 1979. Solute regulation and growth by water stressed roots and shoots of maize plants. *Planta* **147**: 43-49.

Taylor, G. & Davies, W.J. 1990. Root growth of *Fagus sylvatica*: Impact of air quality and drought at a site in southern Britain. *New Phytol.* **116**: 457-464.

Taylor, G. & Dobson, M.C. 1989. Photosynthetic characteristics, stomatal responses and water relations of *Fagus sylvatica*: Impact of air quality at a site in southern Britain. *New Phytol.* **113**: 265-273.

Taylor, G., Dobson, M.C., Freer-Smith, P.H. & Davies W.J. 1989. Tree physiology and air pollution in southern Britain. *Res. Inform. Note* **145**, Forestry Commission, UK.

Taylor, G., Ranasinghe, S., Bosac, C., Gardner, S.D.L. & Ferris, R. 1994. Elevated CO_2 and plant growth: Cellular mechanisms and responses of whole plants. *J. Exp. Bot.* **45**: 1761-1774.

Taylor, G., Gardner, S.D.L., Bosac, C., Flowers, T.J., Crookshanks, M. & Dolan, L. 1995. Effects of elevated CO_2 on cellular mechanisms, growth and development of trees with particular reference to hybrid poplar. *Forestry* **68**: 379-390.

Tingey, D.T., Heck, W.W. & Reinhart, R.A. 1971. Effects of low concentrations of ozone and sulphur dioxide on foliage, growth and yield of radish. *J. Am. Soc. Hort. Sci.* **96**: 369-371.

Wells, T.C.E. 1989. The re-creation of grassland habitats. *The Entomol.* **108**: 125-132.

Woodward, F.I., Thompson, G.B. & McKee, I.F. 1991. The effects of elevated concentrations of carbon dioxide on individual plants, populations, communities and ecosystems. *Ann. Bot.* **67**: 23-38.

Wyn-Jones, R.G. & Pritchard, J. 1989. Stress, membranes and cell walls. In Jones, H.G., Flowers, T.J. & Jones, M.B. (eds) *Plants under Stress*: 95–114. Cambridge University Press, Cambridge.

Zhang, J. & Davies, W.J. 1989. Abscisic acid produced in dehydrating roots may enable the plant to measure the water status of the soil. *Plant, Cell Environ.* **12**: 73–81.

Zhang, J. & Davies, W.J. 1990. Changes in the concentration of ABA in the xylem sap as a function of changing soil water status can account for changes in leaf conductance and growth. *Plant, Cell Environ.* **13**: 271–285.

16 Development, Structure and Properties of Wood from Trees Affected by Air Pollution

A. POŽGAJ, M. IQBAL & L.J. KUCERA

INTRODUCTION

With increasing industrialization, forests are being affected adversely because of contamination of air, water and soil by growth-inhibiting or generally toxic substances that include gases, acids and particles, etc. According to sources from industry and environmental agencies, there are currently about 60–65 000 artificial chemical compounds incorporated in a variety of products in world-wide public use, and the number of these partly toxic substances is yearly increasing by 500. Most air pollutants decrease photosynthetic activity of woody plants, affect internal physiological changes leading to growth inhibition, and cause visible injury and death of the plants. Air pollutants rarely exist singly; the combined pollutants may have synergistic, additive or antagonistic effects.

Forests act as the source of raw roundwood, the generators of oxygen (O_2) and consumers of carbon dioxide (CO_2), the automatic controller of water regime and so on. The forests affected by air pollution are less effective; absorption of CO_2 in the diseased forest is reduced to two-thirds and the release of O_2 to half as compared to the healthy forest.

A steady diminution in the amount of sawn timber, and a gradual increase in the processing of wood for its fibre and chemical constituents during the next century is clearly foreseen. The situation warrants a search for appropriate means and measures to (a) check death of the forests, and (b) utilize the wood of diseased and dead trees.

PHYSIOLOGY OF XYLEM FORMATION

There are three main parts of the growing tree, namely, crown (branches and leaves), stem and roots (Figure 16.1), each with a different task to perform. Assimilates are produced in the crown. The crown is supported by the stem which conducts water and minerals from the roots upward, and food and hormones from places where these are manufactured down to those points where they are used in growth or stored for future use. The stems incorporate most of the wood

Plant Response to Air Pollution. Edited by Mohammad Yunus and Muhammad Iqbal

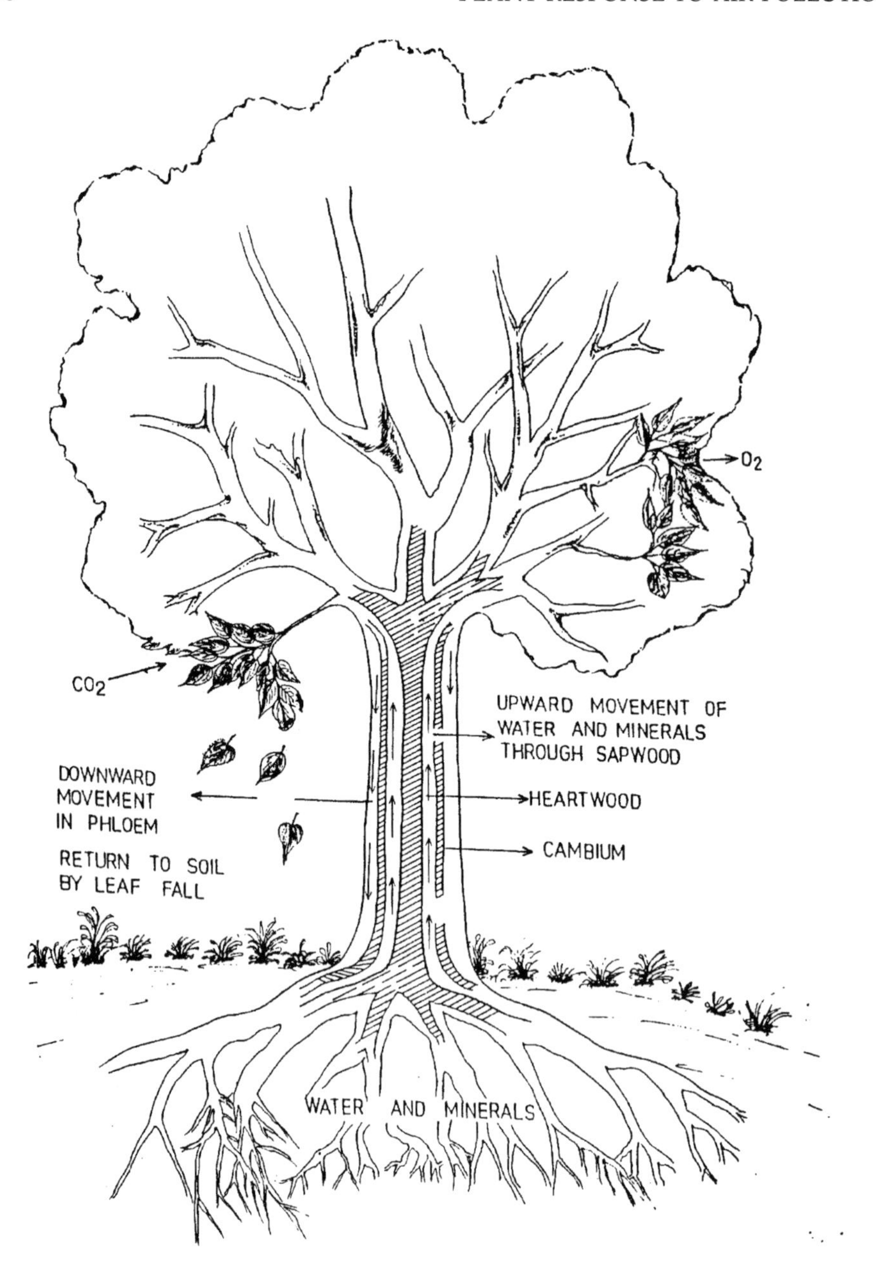

Figure 16.1. Semidiagrammatic sketch showing circulation of minerals within a tree and their return to the soil

produced in the growing trees. The roots are meant for anchorage, absorption of water and minerals from the soil, and for storage of reserve foods.

Light energy is trapped by the crown and used to synthesize reduced carbon compounds from carbon dioxide and water through photosynthesis. The process occurs only in illuminated green tissue, because chlorophyll plays an essential role in the conversion of light energy to chemical energy. Thus, photosynthesis reduces the atmospheric CO_2 to carbohydrate by using light energy, and releases oxygen from water. The following generalized equation represents this reaction:

$$CO_2 + 2H_2O \xrightarrow[\text{chlorophyll}]{\text{light}} (CH_2O) + O_2 + H_2O$$

The primary products are immediately converted into glucose and other compounds such as lipids, organic acids and amino acids, and then assimilated to form new protoplasm, cell wall and other substances required for a new tissue. The energy required for maintaining the structural integrity of the existing protoplasm is released through respiration which can be defined as oxidation of substrate in living cells, bringing about release of energy.

Water is lost from plants in the form of vapour through transpiration which is basically an evaporation process and produces energy gradients that cause movement of water through plants. Water moves from soil into the roots, to the aerial plant parts (stem and crown), and then out into the air through a series of closely interrelated processes. This is sometimes called the soil–plant–atmosphere–continuum concept (Philip, 1966). Essentially, water moves from roots to leaves through the xylem that provides the path with least resistance to flow. The movement is most rapid through the two to four outer annual rings of the young xylem.

Organic substances are translocated through the phloem down to the stem and roots and consumed in the meristems (see Iqbal, 1994, 1995). The unused materials accumulate in parenchyma cells of the xylem and phloem as the reserves. This, however, is not a one-way traffic. The organic compounds, amino acids, glutamate, aspartate, organic acids and cytokinins also move partly from roots to shoots (Kolek *et al.*, 1988). The concentration of carbohydrates is often very high in the bark but in many species the total amount in the bark is less than that in the wood. Ray parenchyma serves as the major reservoir, the amount and type of the reserves varying with distance from the cambium (see Kozlowski, 1992).

Plants grow in height (primary growth) and diameter (secondary growth) through the activity of meristems which comprise a very small fraction of the total mass. The diameter of the plant axis increases chiefly by the activity of the vascular cambium, a cylindrical lateral meristem that is located between the xylem and phloem in the stem, branches and roots. The cambium usually produces xylem on its inner side and phloem on the other side; the new annual increments of xylem and phloem are inserted between old layers of these tissues and the cambium (Iqbal & Ghouse, 1990; Romberger *et al.*, 1993; Iqbal, 1994, 1995). The xylem to phloem ratio often declines under environmental stress. The mature xylem cells are usually thick-walled and lignified. The degree of lignification varies among species, cells and different parts of the same cell (see Wardrop, 1981; Berlyn *et al.*, 1995). Most phloem cells remain soft-walled and eventually collapse or become greatly distorted (Roth, 1981; Iqbal & Ghouse, 1982; Kucera & Bergamin 1990; Iqbal & Zahur, 1995).

XYLEM FORMATION IN CONDITIONS AFFECTED BY INDUSTRIAL EMISSION

Formation of annual rings due to earlywood and latewood differentiation is common in temperate-zone trees. This phenomenon is affected by changes in physiological conditions of the plants. Air pollutants such as SO_2, HF and NO_x influence plant physiology on their own or in association with certain climatic factors (e.g., drought). SO_2 affects needles/leaves of trees and, therefore, hinders the assimilation process (see Linzon, 1978; Maas, 1987; Rennenberg & Herschbach, this volume, Chapter 11). Conifers are especially sensitive to such conditions.

Exposure to SO_2 hampers the CO_2 uptake in young spruces (Keller, 1980). The reduced photosynthetic activity in turn retards cambial activity and consequently wood production. The latter may decline markedly even in the absence of visible injury. Concentration of SO_2 is more decisive than duration of exposure in affecting photosynthesis (Guderian, 1970).

Trees are considered as active bioindicators of air pollution (Eckstein *et al.*, 1981; Braun & Lewark, 1986) as they react sensitively to all interferences. The interference may severely reduce carbohydrate contents of the roots and disturb cell differentiation of the secondary xylem. Defoliation affected the concentration of starch and soluble sugars in the xylem of *Acer saccharum* (Gregory *et al.*, 1986). Decrease in the annual ring width was apparent.

Not only needles, but even the root system, is affected by acid rains in industrial regions. Transpiration rate decreases and assimilation processes are starved of minerals (Halbwachs, 1970). These physiological disturbances possibly hamper the photosynthetic capacity of trees. The activity of cambium depends on availability of water, starch, soluble sugars, minerals and growth hormones, etc. (Riding & Little, 1984; Berlyn & Battey, 1985; Iqbal, 1995). Reduced photosynthesis results in reduced accumulation of carbohydrates which eventually retards cambial growth and wood cell lignification.

In the pollution-affected forest, four types of conifer stands may be distinguished:

(a) Healthy trees with dense foliage (needles) in the crown without any obvious symptoms.

(b) Healthy trees with needle loss not exceeding 30–40%. The wood from these trees is not altered.

(c) Dying or diseased trees with over 30–40% loss of needles. Their resistance against fungal attack seemingly decreases but moisture content of the stem does not yet create suitable conditions for fungal growth. Gross wood features are altered.

(d) Dead trees with living needles; moisture content goes below 50% (in the summer) providing optimum growth conditions for wood-destroying fungi. At this stage, acidity in the stem increases, forming favourable environments for cellulose-decomposing fungi. The drying checks in the stem also reduce wood quality and create favourable conditions for fungal and bacterial infestation.

Table 16.1. Water content of healthy, diseased and very diseased trees of *Abies alba* (from Shortle & Bauch, 1986; by permission of IAWA)

Tree	Water content (% oven-dry wood)	
	Outer sapwood 1978–1974	Heartwood 1938–1934
Healthy	170	58
Diseased	121	132
Very diseased	91	85

In a study of conifers, De Kort (1986) categorized trees as healthy, diseased, very diseased, dying and dead depending on crown appearance, which is determined by the density and condition of the needles.

Water content

The sapwood primarily conducts water and to some extent serves as a food reservoir. Decreased transpiration reduces the water and mineral uptake. Water content of active phloem and cambium in healthy as well as diseased spruce (*Picea abies* L.) trees was comparable. It declined, however, in the sapwood, particularly in the inner zone, of the diseased trees (Bauch *et al.*, 1979; Frühwald *et al.*, 1984; Sell *et al.*, 1989). The difference was most significant between sapwoods of healthy and diseased trees of fir, *Abies alba* Mill. (Brill *et al.*, 1981; Fink, 1986a; Shortle & Bauch, 1986); the moisture content was 150–170% (of dry weight) in the normal sapwood and about 50–100% in the diseased sapwood (Table 16.1) (Bauch *et al.*, 1979; Brill *et al.*, 1981; Shortle & Bauch, 1986).

The moisture content did not vary significantly between healthy and declining trees of *Abies balsamea* (L.) Mill. (Shortle & Bauch, 1986). However, it decreased significantly (up to 90%) in the sapwood of the "very diseased" fir trees (*Abies alba*) (Frühwald *et al.*, 1984; Bauch, 1986a; Böttcher, 1986; Frühwald, 1986). The difference was to the tune of 20–25% between healthy and diseased spruces (Frühwald, 1986; Rademacher *et al.*, 1986; Schulz, 1986; Sell *et al.*, 1987).

In diseased pine trees, sapwood moisture content could be reduced by 20% (Frühwald, 1986). The maximum moisture ranges from 130 to 175% in the sapwood and from 24 to 79% in the heartwood. No difference of moisture content could be found between healthy and diseased beech trees except for the outermost layer in most severely damaged beeches (Bucher & Kucera, 1991). As in the pines, the moisture content of the sapwood and the degree of crown damage were not correlated (Aszmutat *et al.*, 1986; Aufsess & Schulz, 1986; Fink, 1986a; Frühwald, 1986; Schulz, 1986).

Soluble sugars and starch

Carbohydrates, the direct products of photosynthesis, are the basic energy-storing compounds that accumulate as reserve foods (Dietrichs & Schaich, 1965; Kramer, 1969). The state of the carbohydrates could be a good pointer to the physiological processes in trees. Many carbohydrates in plants are continually undergoing conversion from one form to another. Starch–sucrose conversion occurs in both vegetative and reproductive tissues.

The amount of starch in the xylem varies during the year (Ziegler, 1964; Höll, 1985; Gregory *et al.*, 1986; Puls & Rademacher, 1988). The spruce trees have the lowest starch reserves in early spring and the highest reserves from May to late autumn. The proportion of xylem starch is lower in diseased trees than in healthy ones (Fink, 1986a; Puls & Rademacher, 1986, 1988; Rademacher *et al.*, 1986; Shortle & Bauch, 1986; Bues *et al.*, 1989).

The concentration of soluble sugars (sucrose, glucose, fructose) is indicative of the physiological activity of a tree and varies during the active season (Puls & Rademacher, 1988). In the case of spruce, the concentration seems to be highest in January–March and then in October. On the whole, the carbohydrate content is much greater in the diseased trees than in the healthy ones (Bauch, 1986a; Plus & Rademacher, 1986, 1988; Rademacher *et al.*, 1986; Shortle & Bauch, 1986). However, it was considerably reduced in the sapwood of "very diseased" spruce trees (Fink, 1986b, Rademacher *et al.*, 1986).

The enzyme activity of ATPase and phosphatases is also lowered in diseased trees (Fink, 1986b). Along the radius of the trunk, the carbohydrate content decreases with increasing distance from the cambium (Puls & Rademacher, 1986). Defoliation causes a decrease of starch and soluble sugars (Gregory *et al.*, 1986) as do the pollutants. That the area and capacity of plant for assimilation decrease under the influence of air pollution was evident from experiments with CO_2 (Keller, 1980).

Chemical constituents

The chemical properties of wood of the diseased trees can be better characterized by alkaline extracts. Easily soluble polyose, eventually short-chain cellulose and the alkaline-soluble lignin can be found in these extracts. The xylem of the diseased spruce and pine trees showed no significant difference in the contents of cellulose, hemicelluloses and lignin in comparison with healthy wood (see Fengel, 1985; Puls & Rademacher, 1986). However, the changes were obvious in dead spruce trees. The dead trees contained a low proportion of polysaccharides extracted through a benzene–ethanol compound. Water extracts showed no differences. The yield of lignin in dioxin extraction was higher in healthy wood. The lower precipitated portion of dioxin-soluble lignin in the diseased tree could be influenced through sulphur content. The content of methyl groups in dioxin-lignin is not different. The air pollutants do not affect the chemical composition of cellulose and hemicelluloses. Lignin content is normal in the middle layer of the secondary wall (S_2), primary wall and middle lamella, in both the healthy and diseased trees (see Bauch, 1986b; Shortle & Bauch, 1986). The acidity is higher in dead spruce trees, as in trees with needles. Since the higher acidity in dead trees is possibly caused through unceasingly growing content of SO_2 and NO_x, the pH of sap is positively correlated with soil pH. There was a decreasing gradient in sap pH from crown to stump, and with regard to stem diameter from sapwood to pith (Bosshard *et al.*, 1986; Cutter & Guyette, 1990).

Lignin formation in xylem cells results from the activity of the protoplasm during cell differentiation. The cells die away after lignification of the secondary wall, and the remaining cytoplasmic debris is deposited on lumen walls. If peroxidase activity is considered as the base of lignification in wood cells, there is no chemical difference between xylem of the healthy and diseased trees. Nonetheless, substantial difference figures in dead spruce trees (see Fengel, 1985).

EFFECT OF INDUSTRIAL EMISSIONS ON WOOD STRUCTURE

Gross structure

The width of the sapwood band varies greatly among species, being relatively wider in stems of young trees and narrower in old trees. Sapwood proportion also increases with the height of the stem (see Bosshard *et al.*, 1986; Nair, 1995). The change in sapwood portion may give an idea of the rate of radial growth in plants and can thus be a good indicator of the influence of air pollution on wood formation. Sapwood portion in spruce trees decreases with loss of needles; a little loss of needles may bring about evident changes in sapwood width. Thus, sapwood width decreases with increasing damage to trees. This should hamper water conduction and then transpiration in the trees. A reduction of the sapwood width and sapwood percentage in cross section is frequently combined with narrow annual rings in peripheral position (Kucera & Bosshard, 1989), thus illustrating the interaction between photosynthesis, cambial activity and sapwood physiology (Figure 16.2). No changes arise due to pollution in the heartwood of spruce trees.

The portion of active sapwood varies in healthy fir trees with penetration of "wetwood" into the sapwood. It decreases by 20–40% in the diseased fir trees (Table 16.2). The sapwood reduction in the diseased pine trees is only by 15% (Frühwald *et al.*, 1986a, b; Bues & Schulz, 1988). In Douglas fir, the decrease is significant only in very diseased trees. Thus, the spruce and fir trees react to air pollution by showing a decrease in sapwood width that leads to a reduced water transport in young xylem of the stem. This must reduce tree vitality.

The reduction of sapwood is relatively low in hardwood species. It is indeed inversely proportional to the degree of damage in beech trees (Aszmutat *et al.*, 1986; Bucher & Kucera 1991) and oak trees (Bues & Schulz, 1990). Therefore, the formation of coloured heartwood (redheart) is somewhat enhanced in diseased beeches (Bucher & Kucera, 1991).

Table 16.2. Mean sapwood portion of healthy, diseased and very diseased trees of *Abies alba* and *Abies balsamea* (from Shortle & Bauch, 1986; by permission of IAWA)

Species	Sapwood portion (% basal area)		
	Healthy	Diseased	Very diseased
Abies alba	71	55	26
Abies balsamea	33	23	—

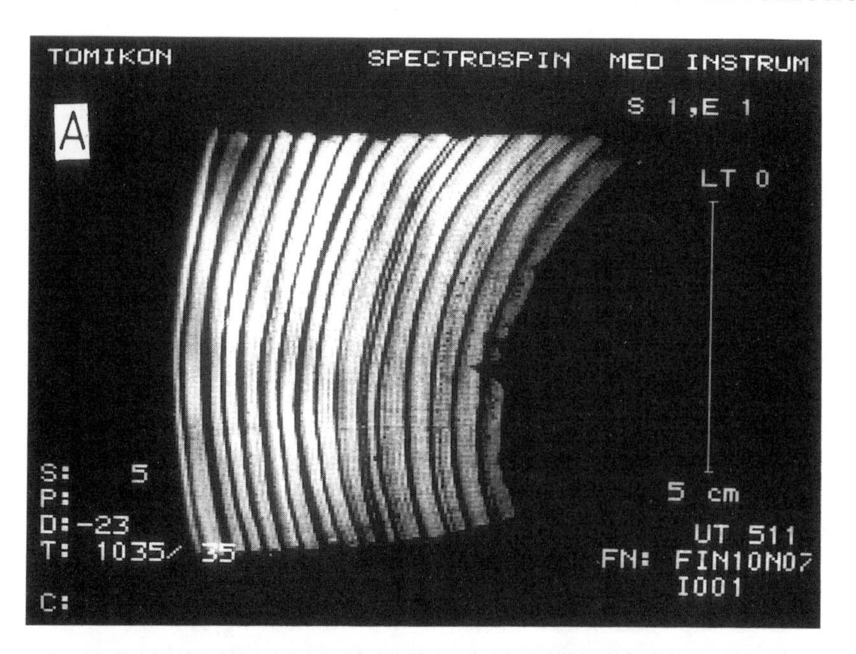

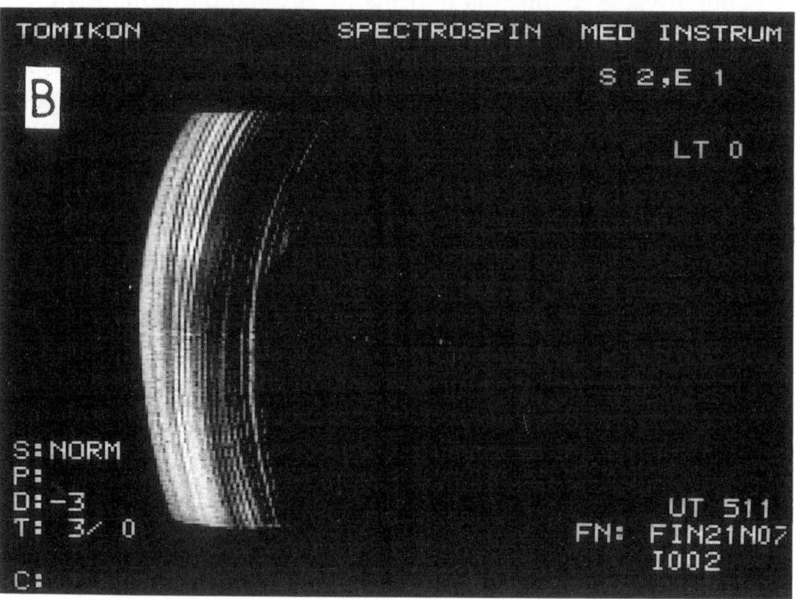

Figure 16.2. Nuclear magnetic resonance tomographs of sealed spruce wood samples. The bright zones indicate the earlywood in the sapwood area with high content of free water. (A) Spruce with 0% needle loss, wide annual rings, wide sapwood area, and a regular water distribution throughout the sapwood. (B) Spruce with 65% needle loss, narrow annual rings, narrow sapwood area, and an irregular water distribution particularly in the inner sapwood (from Kucera & Bosshard, 1989; by permission of Birkhäuser Verlag)

Annual rings, earlywood and latewood

The healthy wood of coniferous species has a limit to the minimum and maximum number of annual rings per centimetre. Therefore, there is an optimum width of the annual rings that may characterize the best quality of wood. When these limits (maximum and minimum) are exceeded, properties of the wood change.

The annual ring width increases at the beginning of tree growth, then gradually decreases, and later becomes constant (see Ghouse & Iqbal, 1981; Creber & Chaloner, 1990). The natural pattern of the annual ring growth may change with drought, side of the trunk, and habitat environment. The ring width also varies with tree height, generally increasing from base of the stem to the crown. Exceptions do exist, however. The variation in tree growth can be identified, dated, and related to crown condition, site, region, and the ecological factors (see Aufsess, 1981; Bauch, 1986b; Göttsche-Kühn *et al.*, 1988). The different species react differently to the environmental conditions.

The width of the annual rings decreases in relation to air pollution (Figure 16.3), the most in coniferous trees (Aufsess 1981; Eckstein *et al.*, 1981; Bauch & Frühwald, 1983; Bosshard *et al.*, 1986; Fink, 1986a; Frühwald, 1986; Schweingruber, 1986; Torelli *et al.*, 1986; Wagenführ, 1987; Kucera & Bosshard 1989; Sell *et al.*, 1989). The decrease is correlated to the density of needles in the crown (Bauch *et al.*, 1986, 1988; Bosshard *et al.*, 1986; Schweingruber, 1986). Spruce and fir react to air pollution by losing needles and ring width (see Shortle & Bauch, 1986). The patterns of decline are different for the two species even though both grow at the same site (Eckstein *et al.*, 1981). Spruce trees react to air

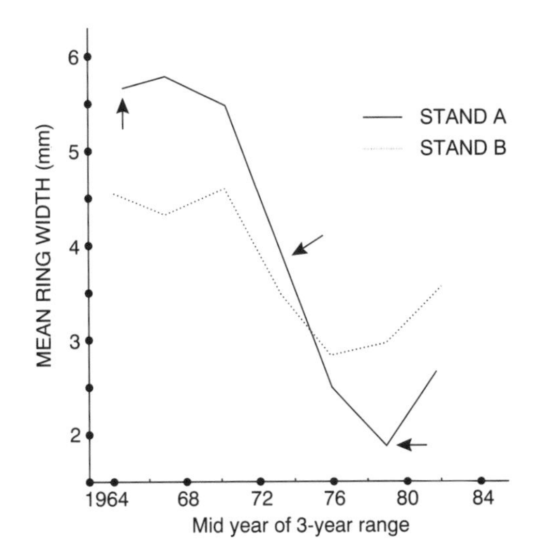

Figure 16.3. Variation in the mean ring width for three successive years in Norway spruce (*Picea abies*) trees grown in two stands located at a distance of 0.74 km (stand A) and 2.13 km (stand B) from a fertilizer factory (the pollution source). The factory (a) began operation in Ireland in 1965, (b) changed its manufacturing method in 1972 and (c) was closed down in 1981, as indicated by the arrows. Both stands were planted in 1952 (after Evertsen *et al.*, 1986; by permission of IAWA)

Table 16.3. (a) Mass increment (kg) in the stemwood of healthy and diseased spruce trees during the years 1981-1983 (based on data from Rademacher *et al.*, 1986)

	Healthy	Slightly diseased	Diseased	Very diseased
Wood mass (kg)	23.3	18.5	13.5	9
Wood mass (%)	100	79	58	39

(b) Total biomass (kg) of needles, branches, thick roots (≥2 mm) and fine roots (<2 mm) of a normal and a "very diseased" spruce

	Biomass	
	Healthy/slightly diseased	Very diseased
Needles	85	36
Branches	240	115
Thick roots	245	62
Fine roots	21	11

pollution at a much earlier stage (Bauch & Frühwald, 1983; Bauch *et al.*, 1986). The pollution affects the radial growth long before any visible symptoms appear in the needles (Poleno, 1982). The growth reduction is prominent in the very diseased spruce trees (Bauch *et al.*, 1988). The difference in wood mass between healthy and diseased spruce trees is up to 61% and the biomass reduction is greater in thick roots than in needles or branches (Table 16.3). Decrease in growth of diseased Douglas fir trees has been noticed during the last 20-30 years (De Kort, 1986). In fact, no wood has formed at the stem base of very diseased trees over the last 4-10 years. On the other hand, beech trees do not react significantly to air pollution. Both gain and loss of the ring width have been noticed (Frühwald, 1986; Bräker, 1991). Differences in annual ring width were not significant even along tree height (Wahlmann *et al.*, 1986). Changes in ring width of declining trees might correlate to the density of wood (Keller, 1980; Knigge *et al.*, 1985; Bauch *et al.*, 1986; Kucera & Bosshard, 1989).

The annual rings in temperate-zone trees consist of earlywood and latewood. The upward transport of water solution in the coniferous/dicotyledonous trees occurs through the tracheids/vessels of earlywood. The proportion of latewood is correlated to the width of the annual rings (Ohta, 1978; Grosser *et al.*, 1985; Yokobori, 1985; Shortle & Bauch, 1986; Bues *et al.*, 1989). The changes in ring width and latewood proportion also correlate with concentration of the air pollutants (Figure 16.4). Reduction in width of the annual ring is accompanied by a decline in latewood density (Table 16.4), which possibly depends on the quantity of the photosynthetic products.

Proportion of latewood is indicative of the mechanical properties of the wood. The quantity of wood substance in the total wood volume primarily depends on the mechanical elements; the larger the proportion of the elements, the greater would be the mechanical strength of the wood. In healthy trees, the latewood proportion is greater in narrower rings (Kollmann, 1951), but this probably does

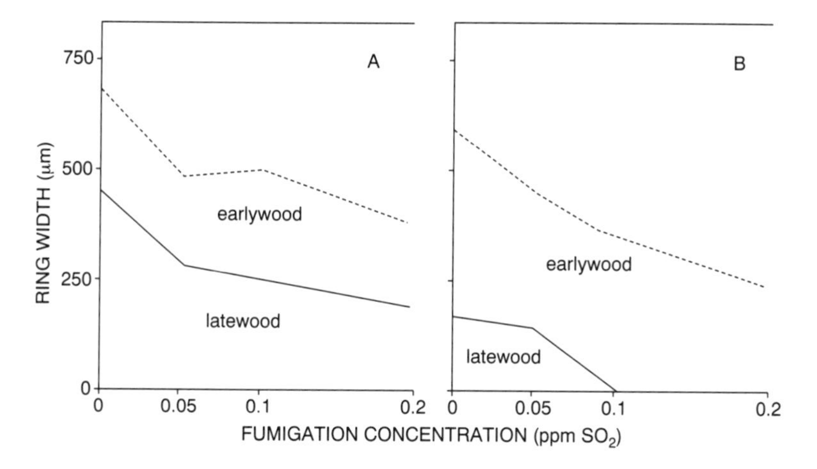

Figure 16.4. The effect of SO_2 fumigation on average ring width of earlywood and latewood in two spruce clones. Broken line indicates the total ring width (based on data from Keller, 1980)

Table 16.4. The effect of SO_2 on the latewood density in spruce trees (from Keller, 1980; by permission of National Research Council of Canada)

SO_2 concentration (ppm)	Maximum density (g cm^{-3})	
	Clone A	Clone B
0 (control)	0.74	0.67
0.05	0.71	0.61
0.1	0.68	0.59
0.2	0.69*	0.54**

*Difference from control (0 ppm) significant with $P \leq 0.05$.
**Difference from control (0 ppm) significant with $P \leq 0.01$.

not hold for very diseased trees (Bosshard *et al.*, 1986; De Kort, 1986; Wagenführ, 1987). Growth-ring reduction is significant in coniferous trees (especially spruce and fir) and the latewood may be prominently reduced in narrow rings of very diseased trees (when the loss of needles is more than 30–40%) (Kucera & Bosshard, 1989).

Microscopic structure of wood

Conifers

In softwoods, as much as 90% of the wood (xylem) is made up of vertically arranged tracheids that form almost uniform radial rows. The thin-walled earlywood tracheids form the main channel for ascent of sap. The latewood comprises thick-walled tracheids that contribute more to wood strength.

The tracheid length depends on the age of the cambium and on the growth conditions (see reviews by Philipson *et al.*, 1971; Iqbal, 1990, 1995). The length increases with the age of the cambium (Kucera & Bosshard, 1989) but decreases

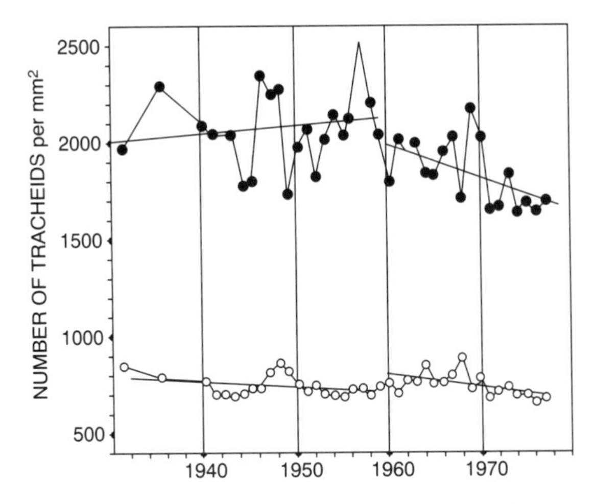

Figure 16.5. Number of tracheids per square millimetre of earlywood and latewood of the spruce tree (after Eckstein *et al.*, 1981; by permission of Springer-Verlag). ● = Latewood, ○ = Earlywood

with ring width. It may decrease (Eckstein *et al.*, 1981; Aszmutat *et al.*, 1986; Wagenführ, 1987) or increase (Bauch, 1986; Kucera & Bosshard, 1989) with the growing age of declining trees. The influence of air pollution on tracheid length varies with site, species and the concentration of pollutants.

Reduction of tracheid length due to air pollution is not significant in pine trees (Bues *et al.*, 1989; Nimmann & Knigge, 1989). The length was as usual in the diseased fir trees (Eckstein *et al.*, 1981) but unusual in very diseased trees. The length decreased in the outer zone of the sapwood in the dying trees of Douglas fir (*Pseudotsuga menziesii*) (De Kort, 1986).

The wall thickness of latewood tracheids remains nearly unchanged in spruce (Sell *et al.*, 1989; Kucera & Bosshard, 1989) though it is affected in earlywood tracheids because of heavy air pollution (Grosser *et al.*, 1985; Bauch, 1986a). Also, the air pollutants reduce the number of tracheids per square millimetre of latewood (Figure 16.5) (Eckstein *et al.*, 1981; Wagenführ, 1987). The number increases, however, in diseased fir trees. Changes due to pollution occur in the radial diameter of the earlywood tracheids (Göttsche-Kühn *et al.*, 1988; Kartusch, 1988). The fine structure of cell walls and pits of diseased conifers does not differ from that in healthy trees (Fengel, 1985).

In the wood of branches from declining trees, tracheid dimensions (width, length and wall thickness) are significantly reduced. The tracheid width in the young sapwood is variable under the influence of heavy pollution.

Hardwoods

The structure of hardwoods is more complex because the elements that compose the wood are more varied. In the hardwoods, sap conduction occurs through long tubes known as "vessels", while mechanical strength comes mainly from fibres. Compared with conifers, the hardwoods are not that sensitive to air pollution.

The average fibre length, vessel diameter and proportion of vessels per unit area in healthy and polluted trees may not evince a drastic difference (see Koltzenburg & Knigge, 1987). However, significant increase in the vessel width coupled with a reduced frequency of vessels per unit transectional area has been reported in a number of shrubs and trees (Ghouse *et al.*, 1985, 1986a, b; Mahmooduzzafar *et al.*, 1986; Gupta & Ghouse, 1987; Iqbal *et al.*, 1987a; Ansari *et al.*, 1993), whereas the reverse applies to certain other species (see Iqbal *et al.*, 1986a, b, 1987b). Likewise, the fibre length experiences a drastic increase in some species (see Ghouse *et al.*, 1986a, b, 1989; Iqbal *et al.*, 1986a, 1987b) and a marked decline in others (see Iqbal *et al.*, 1986b, 1987a; Mahmooduzzafar *et al.*, 1986; Ansari *et al.*, 1993). The air pollution caused by coal combustion influenced the size and amounts of wood components across the trunk of *Syzygium cumini*. Compared to the control, the vessel elements in different positions along the wood radius (pith to cambium) in trunks of trees growing in the polluted atmosphere were significantly shorter, wider and also fewer per unit transectional area. The fibres also were significantly shorter in all the positions analysed across the tree trunk in the polluted zone. The fibres and wood rays occupied greater transectional area in the polluted samples than in the control. The reverse was the case with vessels and the axial wood parenchyma (Mahmooduzzafar & Iqbal, 1993). The annual rings in beech trees experiencing polluted atmosphere for many years are reduced (Böttcher, 1986), and so is the cell-wall portion and the vessel diameter (Bergamin & Kucera, unpublished results). Other parameters such as vessel density and total conducting area (vessel area) may remain unchanged, except in the most severely affected trees. The annual increments of wood were reduced significantly in *Dalbergia sissoo* and *Tectona grandis* due to air pollution (Khan, 1982; Ghouse *et al.*, 1984a, b).

Trabeculae that occur as solitary rods in short to very long files that traverse two or more radially successive cells have been observed both in softwoods and hardwoods experiencing air pollution. A file may continue across many annual rings. Occurrence of trabeculae can be linked with wound tissue formation (Sachsse & Hapla, 1986). Trabeculae were more frequent in compression wood than in the normal wood, and in the annual rings of trees from the declining forests (Grosser, 1986; Chovanec & Korytárová, 1990). Air pollution stress could cause this abnormality. Such stresses reduce the radial increment (the annual ring width) and possibly disturb the normal cell division (Grosser, 1986).

EFFECT OF INDUSTRIAL EMISSION ON WOOD PROPERTIES

The uses of timber depend on its physical properties. Physical properties of spruce wood affected by air pollution (SO_2 and fluorides) are given in Table 16.5.

Density of wood may be a fundamental factor to evaluate the physical properties of wood. It is the mass of a unit volume of wood, eventually of the materials deposited in cell walls or in the lumen (e.g., water). Different species vary enormously in their wood density. The variations reflect the amount of cell-wall substance relative to the amount of air space in the wood. The density can be a good quantitative pointer of cambial activity and the influence of air pollution on it.

Table 16.5. Comparison of physical properties of wood from healthy and diseased spruce (*Picea abies* L.) (Požgaj & Kurjatko, 1986b; by permission of IAWA)

Categories	Statistical parameter*	Density (g cm^{-3})	Shrinkage (%)			Fibre saturation point (%)	Width of annual rings (mm)	Latewood portion (%)
			Radial	Tangential	Volume			
Diseased	$\bar{x}$	0.364	3.45	8.33	11.72	25.78	1.82	—
	s	0.023	0.502	0.807	1.142	1.433	0.432	—
	n	307	307	307	307	358	73	—
Dying	$\bar{x}$	0.354	3.58	7.46	10.97	23.33	1.79	19.03
	s	0.023	0.702	0.827	1.376	3.105	0.389	5.82
	n	84	74	74	74	90	84	28
Dead	$\bar{x}$	0.339	3.57	7.42	10.82	24.23	1.67	17.72
	s	0.420	0.756	0.953	1.420	3.841	0.395	3.40
	n	88	77	77	77	90	88	25
Healthy (Koželouh, 1968)	$\bar{x}$	0.428					—	
	s	0.066					—	
	n	3245					—	
Healthy (Kantor *et al.*, 1974)	$\bar{x}$	0.420					3.49	
	s	0.038					1.085	
	n	112					—	

*$\bar{x}$ = mean value of samples; s = standard deviation; n = number of samples.

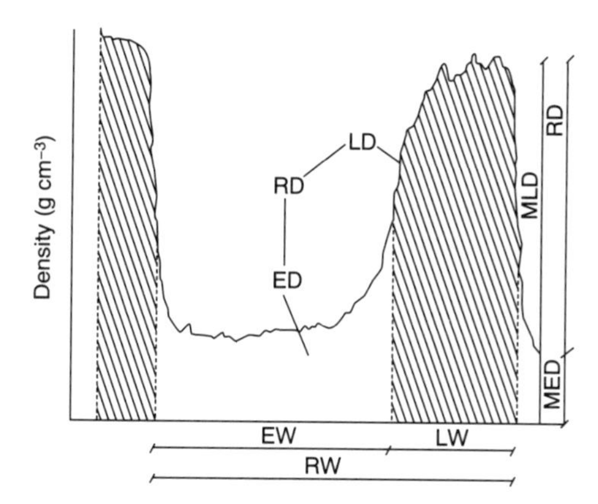

Figure 16.6. Density profile of an annual ring of the spruce wood (*Picea alba* L.). ED = density of earlywood, EW = earlywood width, LD = density of latewood, LW = latewood width, MED = maximum density of earlywood, MLD = maximum density of latewood, RD = ring density, RW = ring width (after Bodner, 1988; by permission of Oesterreichischer Agrarverlag)

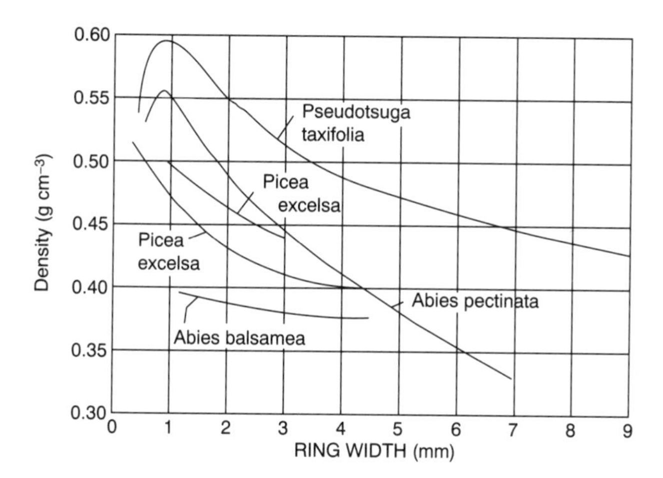

Figure 16.7. Relationship between annual ring width and density for Douglas fir (*Pseudotsuga*), spruce (*Picea*) and fir (*Abies*) (after Kollmann, 1951; by permission of Springer-Verlag)

In the annual rings of coniferous wood, the density of the latewood may be as much as three times the density of the earlywood (Bodner, 1988; Figure 16.6). Ring width and wood density in healthy coniferous trees are negatively correlated; the density decreases with increasing ring width (Burger, 1953; Bernhard, 1964; Figure 16.7). This relation is not always confirmed in very diseased spruce trees where sometimes the reverse seems to apply (Bosshard *et al.*, 1986; Evertsen *et al.*, 1986; Bodner, 1988). Moderate needle loss (up to 40%) may be correlated with increased wood density values because the annual rings produced are narrow, and the ring width is negatively correlated with density. In trees with severe needle loss (from 40% on), however, density is invariably reduced

Table 16.6. Moisture diffusion coefficients of wood from healthy and diseased spruce trees (from Požgaj & Kurjatko, 1986b; by pewrmission of IAWA)

Temperature (°C)	Statistical parameter*	Wood from diseased trees: Direction of diffusion			Wood from healthy trees: Direction of diffusion		
		Longitudinal	Radial	Tangential	Longitudinal	Radial	Tangential
20	$\bar{x}$	20.25	3.93	4.20	35.11	2.33	1.19
	s	0.490	0.228	0.150	13.932	0.837	0.704
	n	25	25	25	25	24	25
40	$\bar{x}$	54.72	8.12	7.93	112.17	5.65	4.12
	s	1.148	0.241	0.333	29.958	1.709	1.213
	n	25	25	25	26	28	26
60	$\bar{x}$	135.39	25.29	22.39	175.12	11.36	10.22
	s	1.951	0.450	0.422	58.714	2.040	1.452
	n	24	23	25	24	20	19

*$\bar{x}$ = mean value of samples; s = standard deviation; n = number of samples.

indicating an insufficient photosynthetic activity of the crown and a shortage of the photosynthetates in the cambium (Kucera & Bosshard, 1989).

The density of latewood is either decreased (Keller, 1980; Bauch, 1986a; Lewark, 1986) or unchanged (Eckstein *et al.*, 1981). There are, however, cases wherein the density of the latewood decreases and that of the earlywood increases, e.g., in the basal part of spruce stem (Bodner, 1988). In very diseased conifers rings may be extremely narrow, and wood density reduced. In some cases, however, reduction in ring width has nothing to do with density. Air pollution decreases wood density as, for instance, in certain oaks (Hughes & Woods, 1986) and spruce (Požgaj *et al.*, 1984; Požgaj & Kurjatko, 1986a, b; see also Table 16.5). Several topographical factors, in addition to the polluted condition, may be responsible for this variation.

The wood increases in volume when it absorbs water. This is "swelling". When water is expelled from the cell wall, wood volume decreases leading to "shrinkage". These changes in size are known as "movement" of timber. The shrinkage is greater in the tangential direction than in the radial direction in spruce trees. "Wood movement" was slightly increased in the narrow rings of pollution-affected trees of beech (Frühwald, 1988), spruce and fir (Sell *et al.*, 1989). This could be due to the higher density of the rings or to some other factors such as fungal attack, etc.

Wood preservation processes involve penetration of liquids or gas by a non-pressure process (diffusion, sap replacement) or by application of pressure (permeability). Table 16.6 summarizes results of the study of moisture diffusion in wood from heavily damaged and the healthy spruce. The affected spruce showed a slightly lower fibre saturation point, compared to the healthy spruce. This may be due to the activity of the cellulose-decomposing fungi which first attack hemicelluloses, the most hygroscopic wall substance.

The permeability of wood in spruce (Liese, 1986b; Kucera & Bosshard, 1989) as well as beech (Buchmüller, 1992) is not significantly dependent on the degree of tree damage by pollution. This may vary with individual tree, species, soil and microclimate, etc. The penetration of preservatives into the logs by application of pressure is deeper in firs than in spruce wood (Sell *et al.*, 1989). The wood of diseased firs has a higher penetration than the wood of diseased spruce trees. This can be explained by the occurrence of the "wet heartwood" in the former. The pits of tracheids of the fir "wetwood" are more permeable than those of spruce, as demonstrated by the vacuum-pressure method of impregnation (Klein *et al.*, 1979; Brill *et al.*, 1981; Frühwald *et al.*, 1981; Bauch, 1986a). Beech wood has a very limited natural durability, and it requires a protective treatment with creosote oil for outdoor uses such as railway sleepers. The penetration of this preservative into beech wood from healthy or diseased trees showed no significance differences (Buchmüller & Osusky, 1992).

Mechanical properties of wood

The mechanical properties of wood are influenced by pollution and these properties might in turn influence the wood processing. The final result could be a decline in the quality of wood products. It appears from Table 16.7 (Požgaj *et al.*,

Table 16.7. Comparison of some mechanical properties of the wood of healthy and diseased spruce trees (Požgaj & Kurjatko, 1986b; by permission of IAWA)

Categories	Statistical parameter*	Compression strength parallel to the grain (MPa)	Compression strength perpendicular to the grain (MPa)	Bending strength (MPa)	Modulus of elasticity in bending (MPa)	Toughness (J cm^{-2})
Diseased	$\bar{x}$	37.0	4.0	68.9	9642	4.6
	s	4.8	0.7	9.3	1461	1.9
	n	110	33	40	40	43
Dying	$\bar{x}$	40.3	4.4	67.7	9440	4.1
	s	5.4	0.9	7.8	1136	1.5
	n	34	29	15	15	41
Dead	$\bar{x}$	36.4	3.4	62.9	9774	3.5
	s	9.0	1.2	13.5	1236	2.1
	n	30	24	12	12	13
Healthy (Koželouh, 1968)	$\bar{x}$	46.4	—	82.6	10935	—
	s	6.8	—	12.4	1898	—
	n	3381	—	3524	3264	—
Healthy (Regináč & Požgaj, 1976)	$\bar{x}$	41.4	3.41	72.8	9730	5.8
	s	6.1	0.8	11.8	1739	2.54
	n	1271	292	1227	1240	306

*$\bar{x}$ = mean value of samples; s = standard deviation; n = number of samples.

1984; Požgaj & Kurjatko, 1986b) that strength values of the wood from damaged trees are lower than those from healthy trees. The differences are statistically significant. It is, however, important to note that these differences are largely due to differences in the density. If, for instance, one compares the average values for compression and bending strength, healthy spruce from Slovakia (Regináč & Požgaj, 1976) with its lower average density does not differ substantially from the affected spruce, although the difference remains statistically significant. Several authors (Kúchtík, 1982; Frühwald *et al.*, 1984; Hurda, 1985; Popper & Eberle, 1992) hold that if woods of healthy and diseased trees are of the same density, their mechanical properties do not differ substantially. Similarly, no significant differences figure in the tensile strength, elastic modulus in bending, compressive strength, toughness and shear, etc. (Požgaj *et al.*, 1984; Požgaj & Kurjatko, 1986a; Frühwald, 1988).

Most of the mechanical strength of wood is derived from the fibres — the long, narrow, thick-walled cells with pointed ends. Therefore, the density and strength of a timber depends largely on the amount of fibres it contains. The relationship between strength and density is close and almost linear. The linear relationship can be an adequate indicator of the influence of industrial emissions on mechanical properties of wood. Experiments show (Eckstein *et al.*, 1981; Schulz 1984; Knigge *et al.*, 1985; Bosshard *et al.*, 1986; Bues & Schulz, 1988, 1990) that correlations between density and strength of wood of affected spruce and fir trees are linear and the difference is not significant at varying stages of tree development. Similarly, the relationship between bending strength (also elastic modulus) and density of the affected pine trees is linear (Bues & Schulz, 1990). The correlation coefficients are high and significant.

In beech wood, a hardwood, the difference in bending strength and elastic modulus of healthy and diseased trees was not significant although some trees had apparent leaf losses (Mahler *et al.*, 1988). Other properties, namely compression strength, shear strength and toughness, were also not significantly altered (Frühwald, 1988; Popper & Eberle, 1992). However, variability of values increased. In oak wood also, the values do not differ significantly between the healthy and the damaged trees. There is a linear regression between density and bending strength in the very damaged trees (Bues & Schulz, 1990).

Some studies (Bauch & Frühwald, 1983; Knigge *et al.*, 1985; Bosshard *et al.*, 1986; Frühwald *et al.*, 1986; Sell *et al.*, 1989) have shown that strength properties are higher in wood from declining trees than in the normal wood. This is possibly in the cases where loss of needles does not exceed 30–40% of the crown density, because in this condition latewood proportion is high in the annual rings. The relationship between density and strength properties is not significantly different in healthy and diseased trees.

Toughness or shock resistance of wood from healthy or diseased coniferous trees is similar (Požgaj *et al.*, 1984; Bosshard *et al.*, 1986; Sell *et al.*, 1989 — spruce; Klein *et al.*, 1979; Frühwald *et al.*, 1981 — fir; Aufsess, 1981; Eckstein *et al.*, 1981 — spruce and fir; Frühwald, 1986 — pine; Frühwald, 1988 — beech and oak). Although values on toughness were lower in diseased spruce trees, the difference was not significant (Table 16.7) (Požgaj *et al.*, 1984).

SECONDARY CHANGES IN THE WOOD AFTER STORAGE

Wood substance with its polysaccharides and lignin content is the main source of nutrition for wood-destroying fungi. Wood-staining fungi and moulds derive their nourishment from the contents (sugar, fat, protein) of parenchyma cells. Wood extractives protect the wood by killing the decay fungi. Fungi need moisture and oxygen for their survival; 35 to 50% moisture content is most favourable for fungal growth. Most fungi require atmospheric oxygen for an optimum growth. Atmospheric temperature is another factor significant for the growth of fungi, insects or moulds. The optimum growth occurs at 25 to 30°C.

Moisture content of the sapwood is always greater than that of the heartwood (spruce, fir, pine). It is generally reduced in coniferous trees affected by air pollution. The difference of moisture content in sapwood of healthy and diseased trees is considerable in fir (*Abies alba*). In the diseased trees, the moisture content is mostly 50–100% in the sapwood (Bauch, 1986a) as against 150–175% in the healthy trees. The difference is significant in *Picea abies* and *Pinus sylvestris* also, though the water deficiency in the diseased trees is not that severe (Frühwald *et al.*, 1984, 1986a, b; Schulz, 1984, 1986). *Abies alba* is one of the most sensitive trees that react to environmental changes. Development of "wetwood" which spreads into the sapwood is a striking alteration. This considerably reduces water-conducting area in the trunk. The "wetwood" is associated with bacteria which act on the parenchyma cells. Also, the phenomenon is accompanied by degeneration of cell nuclei, occurrence of incrustations, and aspiration of pit membranes.

Roundwood in storage

The roundwood of diseased coniferous trees develops visible discoloured areas on storage (Frühwald, 1988). The affected trees already have a reduced activity of sapwood and a higher concentration of bacteria. Blue-staining (*Ceratocystis piceae*) and red sapstain (*Stereum sanquinolentum*) are common fungal infections (Aszmutat *et al.*, 1986; Aufsess, 1986; Schmidt *et al.*, 1986). However, experiments on the storage of spruce logs did not confirm correlation of the degree of tree damage with reduction in radial growth and wood quality (Schmidt *et al.*, 1986). Similarly, negative results were obtained for spruce in laboratory tests (Kucera & Bosshard, 1989) and for beech in field trials (Popper & Osusky, 1992). It appears that wood from declining trees has the same natural durability (resistance against fungal attack) as wood from healthy trees.

The moisture content of logs and roundwood from diseased trees can decline during storage. To avoid fungal and bacterial attacks, the roundwood from diseased trees is removed from the forest before the end of winter or the logs are kept permanently in water-saturated conditions (see Frühwald, 1988). The mechanical properties such as bending strength, elastic modulus and toughness (Frühwald, 1986), compression strength and density (Aufsess, 1981; Aszmutat *et al.*, 1986; Nimmann & Knigge, 1989; Weiss & Knigge, 1989) are not significantly altered. Shear strength and torsion in the radial plane, however, decreased significantly after 2 years of storage in water (Aufsess, 1986).

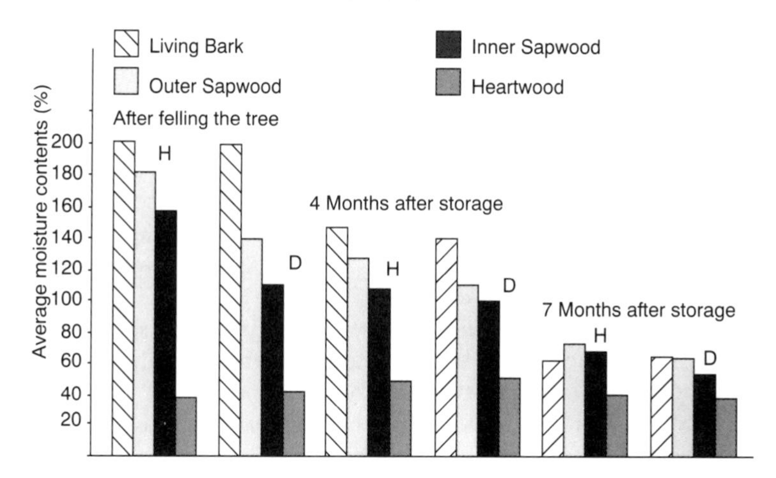

Figure 16.8. Moisture content in different zones of the cross section of spruce roundwood at different stages of maintenance. H = Healthy tree, D = Diseased tree (based on data from Frühwald, 1988)

Beech roundwood is very sensitive during storage, particularly during warm summer days. Brown-stained zones appear and spread in the outer sapwood during the storage period. Blue-staining fungi and white-rot fungi may also flourish (Aufsess, 1986; Aufsess & Schulz, 1986).

Experiments have shown that the wood of diseased trees is as resistant to fungal attack as the healthy wood (Klein *et al.*, 1979; Liese, 1986b; Schmidt *et al.*, 1986; Kucera & Bosshard, 1989; Popper & Osusky, 1992). The fungal growth is related to moisture content of the wood and then to temperature and wood exudates, etc. Oxygen content of wood depends on its water content. The larger the quantity of water in wood, the less would be the area that is filled with air, and hence the less would be the probability of fungal infection. Dead trees may frequently become invaded by wood-destroying fungi even before harvesting. But this is because of the excessive loss of water content in those trees. Wood-inhabiting microorganisms (fungi, bacteria) in samples from healthy and moderately diseased beech trees showed no significant differences (Kucera & Walter, 1992).

Since the moisture content of the sapwood of diseased coniferous trees is markedly reduced, particularly during storage (Figure 16.8), this helps fungal growth. The optimal conditions for fungal growth are created in the dead trees during summer days.

INDUSTRIAL PROCESSING OF WOOD FROM DAMAGED TREES

Since the chemical composition and fine structure of wood are not significantly altered in diseased trees, the physical and mechanical properties of wood undergo little change. However, the diseased trees may have a reduced sapwood (e.g., spruce) and an enhanced wetwood (e.g., fir).

The other changes brought about by air pollutants include a decrease in annual ring width, and ring width to density ratio. These changes may not be substantial enough to influence the processing technologies. The decisive factor in preserving quality of the roundwood from diseased trees during storage and afterwards is a permanently saturated condition. The xylem of living trees contains toxic materials that protect wood from fungi. In dead trees not only the moisture content decreases, the natural protection in the form of bark is also lost. Consequently, stems become infected with fungal spores which penetrate into sapwood, and by way of rays, into the whole stem.

Bacteria act on parenchyma cells and usually cause a decrease of pH value (Brill *et al.*, 1981). The acidity of wood creates suitable conditions for cellulose-decomposing fungi, which first attack the hemicelluloses.

The degree of tree illness is reflected in the quality of the sawn timber-boards. Grading is based on strength. The significant difference of timber quality is between healthy and very diseased or dead trees (Požgaj & Rampas, 1990). The quality decreases from bottom to top of the tree (see Palovic, 1967). This holds with respect to the yield of class I boards and sawnwaste.

An experiment with sawmilling logs showed a significantly higher production of class I boards from the affected trees (with needles) than from the dead trees (without needles), the latter with a higher portion of sawnwaste (Bučko *et al.*, 1987). The output of plywood from logs decreases with the degree of damage in spruce trees. The plywood production from logs of dead trees exhibits a low yield and this increases manufacturing costs. Manufactured boards, e.g., particle boards, fibre boards, plywood, etc. produced from wood of the affected/dead trees are similar in physical and mechanical properties to those from the healthy wood (Buchholzer & Harbs, 1986).

Since wood from diseased/dead trees has little difference with reference to length of tracheids and fibres, the quality of paper production is not significantly altered. The wood of dead trees is "cooked" quicker than healthy wood but consumption of active chemical substances during the "cooking" is higher (Bučko *et al.*, 1987). The pulp produced from the diseased/dead tree is as adequate as that from the healthy tree. However, the overall wood consumption per tonne of the bleached wood pulp increases, thus leading to increased production costs (Bučko *et al.*, 1987).

The sapwood of very diseased or dead trees is often attacked by blue-staining fungi; wood toughness decreases by 10–15%. It is possible that such wood is also attacked by fungi that destroy wood cell walls. In any case, the wood from the dying and dead trees is more "brittle" than from the healthy ones (see Schulz, 1984); this could be related to moisture content also. Conventional air-drying of beech boards from healthy and diseased trees leads to similar results with regard to drying speed and check development (Mayrhofer, 1992). The tensile shearing strength of laminated beams from spruce has shown no difference with regard to the origin of the wood from either healthy or diseased spruce trees (Kucera & Bosshard, 1989).

FINAL CONSIDERATIONS

Trees absorb water and minerals through the roots, whereas light energy and atmospheric CO_2 is absorbed by the crown during the process of photosynthesis. Roots and crown in a tree are thus mutually interdependent. The atmospheric pollutants affect needles/leaves and often appreciably inhibit photosynthesis both directly and indirectly. When needles/leaves are affected significantly, transpiration rate is reduced. Many trees, such as spruce and fir, react to the decrease in transpiration by reducing area of sapwood. This trend may not be significant in some species such as beech. The decrease in water-transport pathway is presumably related to the decreased quantity of water that is transported upward. Trees attempt to keep equilibrium between crown and roots and, therefore, modify the transport area in stems, i.e., the active sapwood.

The cambial activity depends on several parameters such as water, soluble sugar and starch contents. The atmospheric pollutants affect photosynthesis and, thus, decrease the amount of carbohydrates that are needed for a normal cambial activity. Thus, reduced photosynthetic activity leads to a reduced cambial growth and production of wood may drop significantly. Structure of xylem may also be affected in the diseased trees. The cell-wall width and the number of tracheids per square millimetre are reduced. However, changes in wood structure are largely quantitative ones that affect the amounts of earlywood and latewood.

With a 30–40% loss of needles, the tree tends to adjust to the altered condition. It reacts to the loss by decreasing the ring width; the latewood portion and the density of ring increase. Further loss of needles (more than 30–40%) brings anomaly to the cambial growth in spruce. The cambium produces fewer latewood cells, density of annual rings decreases, and eventually the annual rings are lost as, for example, in Douglas firs. Other wood properties, such as swelling, shrinkage, diffusion etc, are not significantly altered. Similarly, mechanical properties of wood from diseased trees are not hampered. The variations are within the normal range known for healthy wood.

The relationship between density and strength of wood is positively linear in both diseased and healthy trees. The strength declines at the same density in the very diseased trees. The wood of diseased or healthy trees is equally resistant against fungal attack.

The physical and mechanical properties of plywood, particle boards and fibre boards and the properties of wood pulp obtained from the wood of healthy or affected trees are comparable. Consumption of raw timber per cubic metre of boards is higher in diseased trees. The difference is especially significant between the healthy and the dead trees. Wood from the diseased and dead trees gives a relatively higher proportion of little fraction (waste) in the process of disintegration into chips, hoggs or fibres.

The dying or dead trees are prone to wood-staining and wood-destroying fungi and eventually to bacteria and wood-boring insects. Such trees, if left in the forest, would deteriorate health of the forest environment. Logs from diseased trees should be stored with particular care, because a low water content of sapwood fascilitates fungal attack. Particularly during the warm part of the year, wet storage of the logs is highly advisable.

ACKNOWLEDGEMENTS

We are thankful to Ms E. Kavciakova and Mr I. Kovacik of the University of Forestry & Wood Technology, Zvolen, Czechoslovakia and to Dr Mahmooduzzafar, Mr Sabah Ashraf and Mr Sarfaraz Alam of Hamdard University, New Delhi, India, for their valuable assistance in preparing this manuscript. Illustrations have been drawn by Ms Indu Mehra.

REFERENCES

Ansari, M.K., Mahmooduzzafar & Iqbal, M. 1993. Structural responses of a medicinally important plant, *Xanthium strumarium* L. to coal-smoke pollution. In Govil, J.N., Singh, V.K. & Hashmi, S. (eds) *Medicinal Plants: New Vistas of Research. Glimpses in Plant Research*, Vol. 11: 555-559. Today & Tomorrow, New Delhi.

Aszmutat, H., Koltzenburg, Ch. & Weiß, W.J. 1986. Untersuchung der Holzeigenschaften von Fichte und Buche aus immissionsexponierten Beständen von Hils und Solling. *Holz als Roh- und Werkstoff* **44**: 301.

Aufsess, H. von 1981. Über die Auswirkungen des Tannensterbens auf die Holzeigenschaften geschädigter Bäume. *Forst. Cbl.* **100**: 217-228.

Aufsess, H. von 1986. Lagerverhalten von Stammholz aus gesunden und erkrankten Kiefern, Fichten und Buchen. *Holz als Roh- und Werkstoff* **44**: 325.

Aufsess, H. von & Schulz, H. 1986. Investigations on the storage behaviour of roundwood from healthy and diseased beeches. In Bass, P. & Bauch, J. (eds) *The Effects of Environmental Pollution on Wood Structure and Quality*: 411-415. International Association of Wood Anatomists, Leiden, The Netherlands.

Bauch, J. 1986a. Biologische Eigenschaften des Holzes kranker Fichten aus Waldschadensgebieten zur Beurteilung der Schadensursachen und Schadwirkungen. *Holz als Roh- und Werkstoff* **44**: 302.

Bauch, J. 1986b. Characteristics and response of wood in declining trees from forests affected by pollution. In Bass, P. & Bauch, J. (eds) *The Effects of Environmental Pollution on Wood Structure and Quality*: 269-276. International Association of Wood Anatomists, Leiden, The Netherlands.

Bauch, J. & Frühwald, A. 1983. Waldschäden und Holzqualität. Experimentelle Befunde über die Eigenschaften von Nadelholz aus Waldschadensgebieten. *Holz-Zentralblatt* **109**: 2161-2162.

Bauch, J., Klein, P., Frühwald, A. & Brill, H. 1979. Alterations of wood characteristics in *Abies alba* Mill. due to "fir-dying" and considerations concerning its origin. *Eur. J. For. Pathol.* **9**: 321-331.

Bauch, J., Göttsche-Kühn, H. & Rademacher, P. 1986. Anatomische Untersuchungen am Holz von gesunden und kranken Bäumen aus Waldschadensgebieten. *Holzforschung* **40**: 281-288.

Bauch, J., Göttsche-Kühn, H. & Riehl, G. 1988. Zuwachsverlust bei Fichte auf der Waldschadensfläche "Postturm" im Forstamt Farchau/Ratzeburg. In Bauch, J. & Michaelis, W. (eds) *Das Forschungsprogramm Waldschäden am Standort "Postturm", Forstamt Farchau/Ratzeburg*: 289-304. GKSS-Forschungszentrum Geesthacht GmbH, BRD.

Berlyn, G.P. & Battey, Y.C. 1985. Metabolism and synthetic function of cambial tissue. In *Biosynthesis and Biodegradation of Wood Components*: 63-85. Academic Press, New York.

Berlyn, G.P., Beck, R.C. & Wolter, K.E. 1995. Cell wall structure, function and degradation. In: Iqbal, M. (ed.) *The Cambial Derivatives. (Encyclopedia of Plant Anatomy)*: 131-148. Gebrüder Borntraeger, Stuttgart.

Bernhard, A. 1964. Über die Rohdichte von Fichtenholz. *Holz als Roh- und Werkstoff* **22**: 215-228.
Bodner, J. 1988. Waldschäden und Holzqualität. *Holzforsch. und Holzverwert.* **40**: 97-100.
Bosshard, H.H., Kučera, L.J., Stoll, A. & Mochfegh, K. 1986. Holzkundliche und holztechnologische Untersuchungen an geschädigten Fichten und Tannen. *Schweiz. Z Forstwes.* **137**: 463-478.
Böttcher, P. 1986. Auswirkungen von Waldschäden auf einige physikalische und mechanische Holzeigenschaften von Fichte und Buche. *Holz als Roh- und Werkstoff* **44**: 302.
Bräker, O.U. 1991. Der Radialzuwachs an unterschiedlich belaubten Buchen in zwei Beständen bei Zürich und Basel. *Schweiz. Z. Forst.* **142**: 427-433.
Braun, E. & Lewark, S. 1986. Jahrringanalysen an Bohrkernen aus den Berliner Forsten. *Holz als Roh- und Werkstoff* **44**: 326-327.
Brill, H., Bock, E. & Bauch, J. 1981. Über die Bedeutung von Mikroorganismen im Holz von *Abies alba* Mill. für das Tannensterben. *Forst. Cbl.* **100**: 195-206.
Bucher, H.P. & Kučera, L.J. 1991. Vergleich der Holzeigenschaften gesunder und geschädigter Buchen (*Fagus sylvatica* L.): Feuchtegehalt und Feuchteverteilung, Vorkommen von Farbkernholz. *Schweiz. Z. Forst.* **142**: 415-426.
Buchholzer, P. & Harbs, C. 1986: Eigang von Fichtenholz aus Waldschadensgebiet zur Herstellung von Holzspanplatten. *Holz als Roh- und Werkstoff* **8**: 281-285.
Buchmüller, K.St. 1992. Vergleich der Holzeigenschaften gesunder und geschädigter Buchen (*Fagus sylvatica* L.): Durchlässigkeit. *Schweiz. Z. Forst.* **143**: 343-346.
Buchmüller, K.St. & Osusky, A. 1992. Vergleich der Holzeigenschaften gesunder und geschädigter Buchen (*Fagus sylvatica* L.): Imprägnierbarkeit. *Schweiz. Z. Forst.* **143**: 377-379.
Bučko, J., Krutel, F. & Požgaj, A. 1987. *Vlastnosti a využitie mrekového dreva napadnutého imisiami.* VŠLD Zvolen.
Bues, C.T. & Schulz, H. 1988. Festigkeit und Feuchtegehalt von Kiefernholz aus Waldschadensgebieten. *Holz als Roh- und Werkstoff* **46**: 41-45.
Bues, C.T. & Schulz, H. 1990. Festigkeit und Feuchtegehalt von Eichenholz aus Waldschadensgebieten. *Holz als Roh- und Werkstoff* **48**: 85-89.
Bues, C.T., Fengel, D., Grabner, I., Heindl, S., Schots-v.d. Zee, M. & Tang, J. 1989. Untersuchungen an Kronengeschädigten Kiefern (*Pinus sylvestris*) in Nordost Bayern. Teil II: *Holzqualität Forstarchiv* **60**: 69-73.
Burger, H. 1953. Blattmenge und Zuwachs-Fichten im gleichaltrigen Hochwald. *Eidgenössische Anstalt für forstliches Versuchswesen* **29**: 38-130.
Chovanec, D. & Korytárová, O. 1990. Morfológia trabekul v porovnaní s podobnými útvarmi v dreve pri pozorovaní v rastrovacom elecktrónovom mikroskope. *Zborník vedeckých prác VŠLD Zvolen* **11**: 27-33 (with English summary).
Creber, G.T. & Chaloner, W.G. 1990. Environmental influences on cambial activity. In Iqbal, M. (ed.) *The Vascular Cambium*: 159-199. Research Studies Press, Taunton, UK.
Cutter, B.E. & Guyette, R.P. 1990. A note on sap pH in eastern red cedar, *Juniperus virginiana* L. *Wood & Fiber Sci.* **22**: 109-112.
De Kort, I. 1986. Wood structure and growth ring width of vital and non-vital Douglas fir (*Pseudotsuga menziesii*) from a single stand in the Netherlands. In Baas, P. & Bauch, J. (eds) *The Effects of Environmental Pollution on Wood Structure and Quality*: 309-318. International Association of Wood Anatomists, Leiden, The Netherlands.
Dietrichs, H.H. & Schaich, E. 1965. Type, proportion and distribution of low molecular carbohydrates in *Fagus sylvatica*. *For. Abstr.* 1999 (26).
Eckstein, D., Greve, U. & Frühwald, A. 1981. Anatomische und mechanisch-technologische Untersuchungen am Holz einer SO_2-geschädigten Fichte und Tanne. *Holz als Roh- und Werkstoff* **39**: 477-487.
Evertsen, A.J., MacSiurtain, M.P. & Gardiner, J.J. 1986. The effect of industrial emission of wood quality in Norway spruce (*Picea abies*). In Baas, P. & Bauch, J. (eds) *The Effects of Environmental Pollution on Wood Structure and Quality*: 399-404. International Association of Wood Anatomists, Leiden, The Netherlands.

Fengel, D. 1985. Strukturuntersuchungen an den Zellwänden im Holz erkrankter Nadelbäume zur Beurteilung der Holzqualität. In Michaelis, M. Herausgeber (eds) *Projekt Europäisches Forschungszentrum für Maßnahmen zur Leftreinhaltung (REF)*: 275-284, 1. Statuskolloquium, Kernforschungszentrum Karlsruhe.

Fink, S. 1986a. Microscopical investigations on wood formation and function in diseased trees. In Baas, P. & Bauch, J. (eds) *The Effects of Environmental Pollution on Wood Structure and Quality*: 351-355. International Association of Wood Anatomists, Leiden, The Netherlands.

Fink, S. 1986b. Histochemische Untersuchungen im Bereich von Holz, Rinde und Kambium bei gesunden und geschädigten Waldbäumen. *Holz als Roh- und Werkstoff* **44**: 327.

Frühwald, A. 1986. Technological properties of wood from trees in polluted regions. In Baas, P. & Bauch, J. (eds) *The Effects of Environmental Pollution on Wood Structure and Quality*: 389-397. International Association of Wood Anatomists, Leiden, The Netherlands.

Frühwald, A. 1988. Die Qualitätsentwicklung bei Trocken- und Naßlagerung von Nadelrundholz aus Waldschadensgebieten. In Bauch, J. & Michaelis, W. (eds) *Das Forschungsprogramm Waldschäden am Standort "Postturm", Forstamt Farchau/Ratzeburg*: 349-363. GKSS-Forschungszentrum Geesthacht GmbH, BRD.

Frühwald, A., Klein, P. & Bauch, J. 1981. Über die Holzeigenschaften der vom Tannensterben betroffenen Bäume (*Abies alba* Mill.). *Forst. Cbl.* **100**: 207-217.

Frühwald, A., Bauch, J. & Göttsche-Kühn, H. 1984. Die Holzeigenschaften von Fichten aus Waldschadensgebieten. Teil 1: Untersuchungen an frisch gefälltem Holz. *Holz als Roh- und Werkstoff* **42**: 441-449.

Frühwald, A., Schwab, E., Mehringer, H. & Krause, H.-A. 1986a. Untersuchungen der physikalischen und technologischen Eigenschaften des Holzes von Kiefern unterschiedlicher Schadstufen aus Waldschadensgebieten. *Holz als Roh- und Werkstoff* **44**: 299.

Frühwald, A., Schwab, E. & Göttsche-Kühn, H. 1986b. Technologische Eigenschaften des Holzes von Fichten unterschiedlichen Erkrankungszustands. *Holz als Roh- und Werkstoff* **44**: 299-300.

Ghouse, A.K.M. & Iqbal, M. 1981. Cell length variation within bark and wood with respect to the development of trees. In Khosla, P.K. (ed.) *Advances in Forest Genetics*: 192-212. Ambika Publications, New Delhi.

Ghouse, A.K.M., Khan, F.A. & Pasha, M.J. 1984a. Effect of air pollution on wood formation in *Dalbergia sissoo*, a timber tree of Gangetic plain. *J. Tree Sci.* **3**: 140-142.

Ghouse, A.K.M., Khan, F.A., Salahuddin, M. & Rasheed, M.A. 1984b. Effect of air pollution on wood formation in *Tectona grandis*. *Indian J. Bot.* **7**: 84-86.

Ghouse, A.K.M., Khan, F.A., Khair, S., Usmani, N.R. & Sulaiman, I.M. 1985. Anatomical responses of *Chenopodium album* to air pollution caused by coal burning. *Acta Bot. Indica* **13**: 287-288.

Ghouse, A.K.M., Khan, F.A., Khair, S. & Usmani, N.R. 1986a. Micromorphological variations in *Croton bonplandianum* (L.) Baill. as induced by pollutants resulting from fossil fuel firing. *Res. J. Pl. Environ.* **3**: 93-95.

Ghouse, A.K.M., Saquib, M., Ahmad, Z. & Khan, M.S. 1986b. Air pollution and wood formation in *Mangifera indica* Linn. *Indian J. Appld Pure Biol.* **1**: 37-39.

Ghouse, A.K.M., Mahmooduzzafar, Iqbal, M. & Dastgiri, P. 1989. Effect of coal-smoke pollution on the stem anatomy of *Cajanus cajan* (L.) Mill. *Indian J. Appld Pure Biol.* **4**: 147-149.

Göttsche-Kühn, H., Bauch, J. & Feuerstack, M. 1988. Anatomische Untersuchungen am Holz gesunder und geschädigter Fichten aus einem Waldschadensgebiet im Forstamt Farchau/Ratzeburg. In Bauch, J. & Michaelis, W. (eds) *Das Forschungsprogramm Waldschäden am Standort "Postturm", Forstamt Farchau/Ratzeburg*: 311-325. GKSS-Forschungszentrum Geesthacht GmbH, BRD.

Gregory, R.A., Williams, M.W. Jr, Wong, B.L. & Hawley, G.J. 1986. Proposed scenario for dieback and decline of *Acer saccharum* in northeastern U.S.A. and southeastern Canada. In Baas, P. & Bauch, J. (eds) *The Effects of Environmental Pollution on Wood Structure and Quality*: 357-369. International Association of Wood Anatomists, Leiden, The Netherlands.

Grosser, D. 1986. On the occurrence of trabeculae with special consideration of diseased trees. In Baas, P. & Bauch, J. (eds) *The Effects of Environmental Pollution on Wood Structure and Quality*: 319-341. International Association of Wood Anatomists, Leiden, The Netherlands.

Grosser, D., Schulz, H. & Utschig, H. 1985. Mögliche anatomische Veränderungen in erkrankten Nadelbäumen. *Holz als Roh- und Werkstoff* **43**: 315-323.

Guderian, R. 1970. Untersuchungen über quantitative Beziehung zwischen dem Schwefelgehalt von Pflanzen und dem Schwefeldioxidgehalt der Luft. III. Teil. *Z. Pfl. Krankh.* **77**: 387-399.

Gupta, M.C. & Ghouse, A.K.M. 1987. Cuticular geography, pigment contents and anatomical traits of *Ficus benghalensis* L. under the influence of coal smoke pollutants. *J. Tree Sci.* **6**: 106-110.

Halbwachs, G. 1970. Vergleichende Untersuchungen über die Wasserbewegung in gesunden und fluorgeschädigten Holzgewächsen. *Cbl. ges. Forstw.* **87**: 1-2.

Höll, W. 1985. Seasonal fluctuation of reserve materials in the trunkwood of Spruce (*Picea abies* (L.) Karst.). *J. Plant Physiol.* **117**: 355-362.

Hughes, M.L. & Woods, F.W. 1986. Wood density of three oak species in Tennessee, 1940-45 and 1970-75. *Forest Ecol. Manag.* **17**: 253-259.

Hurda, B. 1985. Štruktúra a vlastnosti dřeva smrku ze stromu z porostu poškozeného imisemi. [Structure and properties of spruce wood of trees from a stand damaged by emissions]. *Rozvoj Techniky a Ekonomiky v Dřevarském Pùmyslu* **2**: 29-44.

Iqbal, M. 1990. *The Vascular Cambium*. Research Studies Press, Taunton, UK.

Iqbal, M. 1994. Structural and operational specializations of the vascular cambium of seed plants. In: Iqbal, M. (ed.) *Growth Patterns in Vascular Plants*: 211-271. Dioscorides Press, Portland, USA.

Iqbal, M. 1995. Structure and behaviour of vascular cambium and the mechanism and control of cambial growth. In: Iqbal, M. (ed.) *The Cambial Derivatives (Encyclopedia of Plant Anatomy)*: 1-67. Gebrüder Borntraeger, Stuttgart, Germany.

Iqbal, M. & Ghouse, A.K.M. 1982. Comparative features of some arid zone species of *Acacia* and *Prosopis*. *Phytomorphology* **32**: 373-380.

Iqbal, M. & Ghouse, A.K.M. 1990. Cambial concept and organization. In: Iqbal, M. (ed.) *The Vascular Cambium*: 1-36. Research Studies Press, Taunton, UK.

Iqbal, M. & Zahur, M.S. 1995. Secondary phloem: Origin, structure and specialization. In: Iqbal, M. (ed.) *The Cambial Derivatives (Encyclopedia of Plant Anatomy)*: 187-240. Gebrüder Borntraeger, Stuttgart, Germany.

Iqbal, M., Ahmad, Z., Kabeer, I., Mahmooduzzafar & Kalimullah. 1986a. Stem anatomy of *Datura inoxia* Mill. in relation to coal-smoke pollution. *J. Sci. Res.* **8**: 103-105.

Iqbal, M., Mahmooduzzafar, Kalimullah & Ahmad, Z. 1986b. A plant grows under the stress of environmental pollution. *Res. J. Pl. Environ.* **3**: 5-7.

Iqbal, M., Mahmooduzzafar & Ghouse, A.K.M. 1987a. Impact of air pollution on the anatomy of *Cassia occidentalis* L. and *Cassia tora* L. *Indian J. Appld Pure Biol.* **2**: 23-26.

Iqbal, M., Mahmooduzzafar, Kabeer, I., Kalimullah & Ahmad, Z. 1987b. The effect of air pollution on the stem anatomy of *Lantana camara* L. *J. Sci. Res.* **9**: 121-122.

Kantor, V., Duožáček, A. & Matovič, A. 1974. Some characteristics of subalpine ecotype of *Picea excelsa* L. in the Jesenik Mts., *Acta Univ. Agric. Brno. Fac. Silvicult.* **43**(2): 11-137 (with English summary).

Kartusch, B. 1988. Holzanatomische Untersuchung an Unterschiedlich mit Fluor belasteten Kiefern (*Pinus sylvestris* L.). *Angew. Botanik* **62**: 183-192.

Keller, T. 1980. The effect of continuous springtime fumigation with SO_2 on CO_2 uptake and structure of the annual ring in spruce. *Can. J. For. Res.* **10**: 1-6.

Khan, A.U. 1982. Studies on the effect of air pollution on the growth activity of *Dalbergia sissoo*. Roxb. Ph.D. Thesis, Aligarh Muslim University, Aligarh, India.

Klein, P., Bauch, J. & Frühwald, A. 1979. Naßkerneigenschaften von Tannenholz. *Holz-Zentralblatt* **105**: 1465-1466.

Knigge, W., Aszmutat, H. & Weiss, W.J. 1985. Verändern die Immissionsbelastungen die Holzeigenschaften unserer Wälder? *Forstarchiv* **56**: 65-70.

Kolek, J., Kozinka, V. *et al.* 1988. *Fyziológia koreňového systému rastlín.* 1 vyd. VSAV, Brastislava.

Kollmann, F. 1951. *Technologie des Holzes und der Holzwerkstoffe* Vol. 1, 2nd edition. Springer-Verlag, Berlin.

Koltzenburg, Ch. & Knigge, W. 1987. Holzeigenschaften von Buchen aus immissions geschädigten Beständen. *Holz als Roh- und Werkstoff* **45**: 81-87.

Koželouh, B. 1968. *Výpočet dřevěných konstrukcií podle medzních stavu.* Štátny drevársky výskumný ústav Bratislava, ČSFR.

Kozlowski, T.T. 1992. Carbohydrate sources and sinks in woody plants. *Bot. Rev.* **58**: 107-222.

Kramer, P.J. 1969. *Plant and Soil Water Relationships. A Modern Synthesis.* McGraw-Hill, New York.

Kucera, L.J. 1989. Biologische und technologische Eigenschaften des Holzes geschädigter Fichten. *Schweiz. Z. Forst.* **140**: 203-215.

Kucera, L.J. & Bergamin, L. 1990. Struktur, Funktion und physikalische Eigenschaften der Rinde. Kapitel 2 aus: Baumrinden, V.H. (ed.): 15-32. Ferdinand Enke Verlag, Stuttgart.

Kucera, L.J. & Bosshard, H.H. 1989. *Holzeigenschaften geschädigter Fichten.* Birkhäuser Verlag, Basle.

Kucera, L.J. & Walter, M. 1992. Vergleich der Holzeigenschaften gesunder und geschädigter Buchen (*Fagus sylvatica* L.): Mikroorganismen im frischen Holz und Lagerfähigkeit. *Schweiz. Z. Forst.* **143**: 347-359.

Kúchtík, J. 1982. Přímé účinky exhalátu na dřevo. *Zdroje dřevní suroviny.* [Direct effects of emissions on wood. Collections of lectures: Sources of raw timber]. *Jablonec nad Nisou*: 103-110 (without English summary).

Lewark, S. 1986. Die Methode der Röntgendensitometrie von Holz und ihre Anwendung an Holz Immissionen ausgesetzter Bäume. *Forstarchiv* **57**: 105-107.

Liese, W. 1986a. Biologische Resistenz und Tränkbarkeit von Fichtenholz aus Waldschadensgebieten. *Holz als Roh- und Werkstoff* **44**: 325-326.

Liese, W. 1986b. Zur biologischen Resistenz von Fichtenholz aus Waldschadengebieten. *Holz-Zentralblatt* **112**.

Linzon, S.N. 1978. Effects of airborne sulphur pollutants on plants. In Nriagu, J.O. (ed.) *Sulphur in the Environment: Part II. Ecological Impacts*: 109-161. Wiley, New York.

Maas, F.M. 1987. *Responses of Plants to Sulphur Containing Air Pollutants* (H_2S *and* SO_2). Drukkerij van Denderen B.V., Groningen, The Netherlands.

Mahler, G., Klebes, J. & Höwecke, B. 1988. Holztechnologische Eigenschaften von Buchen mit neuartigen Waldschäden. *Holz-Zentralblatt* **114**: 402-404.

Mahmooduzzafar & Iqbal, M. 1993. Coal-smoke pollution affects size and amount of wood elements across the trunk of *Syzygium cumini* Skeel. *15th Int. Bot. Cong.*, Yokohama, Japan, p. 335 (Abst. 3035).

Mahmooduzzafar, Iqbal, M., Ahmad, Z. & Kalimullah, 1986. Anatomical variation in *Sida spinosa* L. due to coal-smoke pollution. *Res. J. Pl. Environ.* **3**: 9-11.

Mayrhofer, A. 1992. Vergleich der Holzeigenschaften gesunder und geschädigter Buchen (*Fagus sylvatica* L.): Trocknungsverhalten bei Freiluftlagerung. *Schweiz. Z. Forst.* **143**: 361-376.

Nair, M.N.B. 1995. Sapwood and heartwood. In: Iqbal, M. (ed.) *The Cambial Derivatives (Encyclopedia of Plant Anatomy)*: 149-172. Gebrüder Borntraeger, Stuttgart, Germany.

Nimmann, B. & Knigge, W. 1989. Anatomische Holzeigenschaften und Lagerungsverhalten von Kiefern aus immissionsbelasteten Standorten der Norddeutschen Tiefebene. *Forstarchiv* **60**: 78-83.

Ohta, S. 1978. The observation of tree ring structure by soft X-ray densitometry (I). The effects of air pollution on annual ring structure. *J. Jap. Wood Res. Soc.* **24**: 429-434.

Palovic, J. 1967. *Nová piliarska technológia ihlicnatych drevín*, 1st edition SNTL Bratislava.

Philip, J.R. 1966. Plant water relations. Some physical aspects. *Annu. Rev. Plant Physiol.* **17**: 245-268.

Philipson, W.R., Ward, J.M. & Butterfield, B.G. 1971. *The Vascular Cambium: Its Development and Activity*. Chapman & Hall, London.

Poleno, Z. 1982. Technické vlastnosti poškozené dřevní hmoty ze souší v imisnych oblastech. *Zdroje dřevní* suroviny. [Technical properties of damaged wood from dry stands in emission areas. Collection of lectures. Sources of raw timber]. *Jablonec nad Nisou*. 26-37 (without English summary).

Popper, R.R. & Eberle, G. 1992. Vergleich der Holzeigenschaften gesunder und geschädigter Buchen (*Fagus sylvatica* L.): Physikalisch-mechanische Eigenschaften. *Schweiz. Z. Forst.* **143**: 333-338.

Popper, R.R. & Osusky, A. 1992. Vergleich der Holzeigenschaften gesunder und geschädigter Buchen (*Fagus sylvatica* L.): Bewitterungs- und Erd-Eingrabungsversuch. *Schweiz. Z. Forst.* **143**: 339-342.

Požgaj, A. & Kurjatko, S. 1986a. Effects of atmospheric emissions on physical and mechanical properties of spruce wood. *18th IUFRO Congress*: 222-226, Ljubljana, Yugaslavia.

Požgaj, A. & Kurjatko, S. 1986b. Wood properties of spruce from forests affected by pollution in Czechoslovakia. In Baas, P. & Bauch, J. (eds) *The Effects of Environmental Pollution on Wood Structure and Quality*: 405-410. International Association of Wood Anatomists, Leiden, The Netherlands.

Požgaj, A. & Rampas, L. 1990. Vplyv emisií na kvalitatívne zastúpenie smrekového reziva. DF VŠLD Zvolen, 50 pp. (Internal Research Report, without English summary.)

Požgaj, A. *et al.* 1984. Stanovenie fyzikálnych, elastickych a mechanickych vlastností imisiami postihnutého smrekoveho dreva. VŠLD DF Zvolen, ŠDVU Bratislava. (Internal Research Report, without English summary.)

Puls, J. & Rademacher, P. 1986. Chemische Untersuchungen an Fichten aus Waldschadensgebieten. *Holz als Roh- und Werkstoff* **44**: 307-312.

Puls, J. & Rademacher, P. 1988. Jahreszeitliche Schwankungen im Gehalt von löslichen Zuckern, Stärke und Extraktstoffen im Holz gesunder und geschädigter Fichten (*Picea abies* (L.) Karst.). In Bauch, J. & Michaelis, W. (eds) *Das Forschungsprogramm Waldschäden am Standort "Postturm", Forstamt Farchau/Ratzeburg*: 327-336. GKSS-Forchungszentrum Geesthacht GmbH, BRD.

Rademacher, P., Bauch, J. & Puls, J. 1986. Biological and chemical investigations of the wood from pollution-affected spruce (*Picea abies* (L.) Karst.). *Holzforschung* **40**: 331-338.

Regináč, L. & Požgaj, A. 1976a. *Anizotropia vlastnostívne zastúpenie smrekového reziva*. DF VŠLD (Diplomová práca), 50 pp.

Regináč, L. & Požgaj, A. 1976b. *Anizotropia vlastností smrekového dreva*. Výskumná správa, nepublikovaná. Drevárska fakulta VŠLD, Zvolen.

Riding, R.T. & Little, C.H.A. 1984. Anatomy and histochemistry of *Abies balsamea* cambial zone cells during the onset and breaking of dormancy. *Can. J. Bot.* **62**: 2570-2579.

Romberger, J.A., Hejnowicz, Z. & Hill, J.F. 1993. *Plant Structure: Function and Development*. Springer-Verlag, Heidelberg, Germany.

Roth, I. 1981. *Structural Patterns of Tropical Barks (Encyclopedia of Plant Anatomy Series)*. Gebrüder Borntraeger, Berlin.

Sachsse, H. & Hapla, F. 1986. Ist im Holz immissionsgeschädigter Fichten mit Veränderungen der Zellwandstruktur zu rechnen? *Forstarchiv* **57**: 12-14.

Schmidt, O., Bauch, J., Rademacher, P. & Göttsche-Kühn, H. 1986. Mikrobiologische Untersuchungen an frischem und gelagertem Holz von Bäumen aus Waldschadensgebieten und Prüfung der Pilzresistenz des frischen Holzes. *Holz als Roh- und Werkstoff* **44**: 319-327.

Schulz, H. 1984. Immissionen-Waldschäden und Holzqualität. *Holz-Zentralblatt* **110**: 1493-1495.

Schulz, H. 1986. Festigkeit und Wassergehalt in Fichten, Kiefern und Buchen unterschiedlicher Schadstufen. *Holz als Roh- und Werkstoff* **44**: 300-301.

Schweingruber, F.H. 1986. Abrupt growth changes in conifers. In Baas, P. & Bauch, J. (eds) *The Effects of Environmental Pollution on Wood Structure and Quality*: 277-283. International Association of Wood Anatomists, Leiden, The Netherlands.

Sell, J., Schnell, G.R. & Arnold, M. 1987. Anteil und Wassergehalt des Splints von Schweizerischen Fichten und Tannen unterschiedlicher Vitalität. *Holz als Roh- und Werkstoff* **45**: 329-334.

Sell, J., Schnell, G. & Arnold, M. 1989. Vergleichende Untersuchung der Holzqualität gesunder und kranker Fichten und Tannen von 6 schweizerischen Standorten. *Holz als Roh- und Werkstoff* **47**: 87-92.

Shortle, W.C. & Bauch, J. 1986. Wood characteristics of *Abies balsamea* in the New England States compared to *Abies alba* from sites in Europe with decline problems. In Baas, P. & Bauch, J. (eds) *The Effects of Environmental Pollution on Wood Structure and Quality*: 375-387. International Association of Wood Anatomists, Leiden, The Netherlands.

Torelli, N., Čufar, K. & Robič, D. 1986. Some wood anatomical, physiological, and silvicultural aspects of silver fir dieback in Slovenia (NW Yugoslavia). In Baas, P. & Bauch, J. (eds) *The Effects of Environmental Pollution on Wood Structure and Quality*: 343-350. International Association of Wood Anatomists, Leiden, The Netherlands.

Wagenführ, R. 1987. Struktursuchungen immisionsgeschädigten Holzes. *Holztechnologie* **1**: 12.

Wahlmann, B., Braun, E. & Lewark, S. 1986. Radial increment in different tree heights in beech stands affected by air pollution. In Baas, P. & Bauch, J. (eds) *The Effects of Environmental Pollution on Wood Structure and Quality*: 285-288. International Association of Wood Anatomists Leiden, The Netherlands.

Wardrop, A.B. 1981. Lignification and xylogenesis. In Barnett, J.R. (ed.) *Xylem Cell Development* Castle House Publ. Ltd, Tunbridge Wells, UK: 115-152.

Weiss, W.J. & Knigge, W. 1989. Auswirkung von Immissionsbelastung auf einige technologische Eigenskchaften und das Lagerungsverhalten von Stammholz der Fichte (*Picea abies* L. Karst). *Forstarchiv* **60**: 22-27.

Yokobori, M. 1985. Studies on appraising the effects of air pollution upon trees. Diss. Univ. of Osaka, Japan.

Ziegler, H. 1964. Storage, mobilization and distribution of reserve material in trees. In Zimmermann, M.H. (ed.) *The Formation of Wood in Forest Trees*: 303-320. Academic Press, New York.

17 Air Pollution and Seed Growth and Development

D.P. ORMROD

INTRODUCTION

Gaseous air pollutants decrease crop yield in many parts of the world. Air pollution resulting from industrial and urban development has resulted in both localized and regional damage in spite of government abatement policies. The main air pollutants toxic to plants are ozone, sulphur dioxide and nitrogen dioxide. In recent years, the greatest concern has been about the effects of the regional episodes of tropospheric ozone that occur when atmospheric conditions are optimal for photochemical synthesis of ozone — when gaseous hydrocarbons and nitrogen oxides accumulate under hot, humid and hazy weather conditions. Ozone is considered to cause more world-wide damage to vegetation than all other air pollutants combined (Heagle, 1989). Sulphur dioxide is also of considerable concern because of the incidence of localized industrial fumigations that damage crops, while only a few localized incidences of nitrogen dioxide injury have been reported.

Research by the National Crop Loss Assessment Network (NCLAN) indicated that tropospheric ozone pollution in the United States of America results in very substantial crop losses (Preston & Tingey, 1988). A 40% decrease in ambient ozone would have net benefits in increasing yield by almost 3 billion dollars per year (Adams *et al.*, 1989) in the USA. There is ample evidence that gaseous air pollution has adverse effects on leaves and on total yields and that these effects vary enormously among genotypes and environments. However, less research has been directed to the correlation of the leaf injury and its associated photosynthesis impairment with air pollution effects on flowering, pollination and seed set, and on carbon allocation from leaves to developing fruit.

Even though many of the yield losses have been demonstrated with seed crops such as wheat, maize and soybeans, little attention has been given to the effect of gaseous air pollutants on the seed development process in such species, in terms of either physiological mechanisms or morphological changes. Few investigators have studied either the direct or indirect effects of ozone and other air pollutants on seed growth and development. The research that has been conducted indicates that, depending on species and genotype within species, ozone reduces the number of flowers, fruits and seeds, and that the allocation of dry matter to developing seed is affected indirectly through the effects of the

Plant Response to Air Pollution. Edited by Mohammad Yunus and Muhammad Iqbal

air pollutants on relative sink strength, translocation patterns and photosynthetic activity (Miller, 1988).

While much research has been conducted to prove that ozone and other air pollutants impair photosynthesis in plant leaves (Heath, 1980; Darrall, 1989), the same attention has not been given to direct effects of air pollutants on reproductive growth from pollen germination and pollen-tube growth to the morphological, anatomical and physiological aspects of ovule and fruit growth. This is perhaps because of the assumption that the impacts of air pollution on seeds are indirect and only involve changes in translocation of photosynthates from injured sites in leaves. The fact that developing seeds are generally well protected from the external gaseous environment may have led researchers to decide that seed structures will not be directly exposed to pollutant gases and that there will not be morphological and anatomical effects on seeds. A similar rationale may explain why physiological investigations of developing seeds in polluted environments have had little study.

Many questions arise when decreased yield of seed crops due to air pollution is considered. Is the decreased yield associated more with seed number or seed weight? Is decreased yield in terms of reduced seed number due to direct effects of the air pollution on flowering? On pollination? On seed set? Is photosynthesis impairment in injured leaves the main reason for seed-weight decrease? Is carbon allocation to the seed decreased by the air pollutants and thereby detrimental to seed and growth development?

The creation of gas pollution treatments for specific well-defined experimental environments poses particular problems in that a separate enclosed environment must be created for each level of air pollution. The rapid dispersion and mixing of introduced experimental gases means that containment and separation of treatments must be an important feature of research facilities. Controlled amounts of pollutant gas are released into an enclosed environment to simulate a particular concentration level and pattern that may be found in polluted air under outdoor conditions. A thorough review of air pollution exposure facilities and experimental protocols for exposure has been developed by Hogsett *et al.* (1987). They describe and evaluate existing exposure facilities, monitoring equipment and microenvironmental sampling. Research methodology for field studies with open-top chambers has been reviewed by Heagle *et al.* (1988). They discuss how these chambers affect microclimate, plant gas exchange and crop yield. The utilization of open outdoor natural environments for research on gaseous air pollution effects on plants raises problems in interpretation because comparative treatments at the same location cannot be created. Interacting environmental factors also may complicate studies conducted entirely under outdoor conditions.

EFFECTS OF AIR POLLUTANTS ON SEED YIELD AND YIELD COMPONENTS

The initial impact of gaseous pollutants leading to yield effects may be on pollen germination and growth. The deleterious effects of ozone on pollen germination and ultrastructure were demonstrated in early research (Harrison & Feder,

1974). Facteau & Rowe (1981) observed a retardation of sweet cherry and apricot pollen-tube growth in response to SO_2 and developed a mathematical relationship between increasing dose and decreasing pollen-tube growth. Thompson *et al.* (1976) grew two sweet corn hybrid cultivars (genotypes) in ambient air containing photochemical air pollutants (mostly ozone) and compared seed set and development, growth and yield with plants grown in charcoal-filtered air. While both cultivars sustained foliar injury, only one of them had substantial reductions in seed development in ambient air (Table 17.1). This research clearly indicated that the number and fresh weight of marketable ears (inflorescences) and the number of fully developed seeds (fruits) can be severely reduced by air pollution and that large differences between genotypes can be demonstrated. While Thompson *et al.* (1976) did not investigate the mechanistic basis for the yield reductions, they did hypothesize that the yield effects were most likely related to impaired functioning of leaves. The supply of photosynthate would be decreased by leaf injury, potentially decreasing energy supply for pollination, seed growth and final ear weight, but pollination also could be impaired more directly as well, affecting seed set and ultimate seed number.

Another study of the effect of ozone on yield and yield components was conducted by Heagle *et al.* (1979) using soft red winter wheat cultivars. Plants were exposed to ozone in field chambers for an extended period of the growth cycle. Yields were reduced by ozone exposure compared with plants grown in charcoal-filtered air. Head weight per plant, seed weight per plant and seed size (100-seed weight) were all decreased similarly by increased ozone concentration (Table 17.2). In this study, the differences in yield sensitivity to ozone among the four cultivars were not marked and did not relate to the relative sensitivity of growth and leaf injury effects of ozone on young plants of the same cultivars (data not shown). This study demonstrated yet another dimension of the complexity of air pollutant-yield interactions, namely, that growth effects on young plants and visible injury to plant foliage are not necessarily related to the sensitivity of seed growth and development to air pollution. The effect of ozone

Table 17.1. Effect of ozone on seed set and development on primary ears of two sweet corn hybrids ("Monarch Advance" and "Bonanza") grown in ambient and filtered air. Data from Thompson *et al.* (1976)

	Filtered air		Ambient air	
	"Monarch Advance"	"Bonanza"	"Monarch Advance"	"Bonanza"
Total no. of ovules per ear	811	627	817	600
Number of seeds per ear				
Fully developed	757	604	440**	544**
Prematurely shrivelled	0	0	148	0
Number of ovules not set				
Basal two-thirds	11	15	9	27
Apical one-third	43	8	220	29
Fresh wt per ear (g)	344	256	248**	232**

**Significantly different from filtered air ($P = 0.01$).

on wheat kernel growth components also has been reported by Slaughter *et al.* (1993). Increasing ozone decreased seed growth rate, seed-fill duration, effective filling period and assimilate utilization in both cultivars under study (Table 17.3).

In studies of ozone effects on tomato fruit growth, Oshima *et al.* (1975) found that the leaves and stems were more sensitive than the developing fruit (Table 17.4). A much higher ozone concentration was required to decrease the

Table 17.2. Mean effects of ozone on yield parameters of wheat in open-top field chambers. Data from Heagle *et al.* (1979)

	Ozone concentration*				
	30	60	100	130	LSD (0.05)
Head wt per plant (g)	6.52	6.32	5.65	4.51	0.42
Seed wt per plant (g)	5.08	4.90	4.28	3.38	0.53
100-seed wt (g)	3.49	3.52	3.43	2.87	0.08

*7-hour mean daily ozone concentrations (nl l^{-1}).

Table 17.3. Effects of ozone and genotype on wheat kernel growth components. Data from Slaughter *et al.* (1993)

Cultivar (genotype)	Ozone concentration*				
	32.1	48.6	60.5	72.6	92.7
Kernel growth rate (mg kernel^{-1} d^{-1})					
Severn	1.06	1.04	0.99	0.73	0.72
MD5518308	1.09	0.99	1.01	0.93	0.77
Kernel-fill duration (days)					
Severn	30.4	35.4	33.4	30.5	29.4
MD5518308	26.9	28.3	30.5	30.1	28.9
Effective filling period (days)					
Severn	26.7	28.9	28.3	26.2	25.5
MD5518308	22.3	25.3	27.4	28.7	23.8
Assimilate utilization (g m^{-2} d^{-1})					
Severn	7.2	7.1	7.0	4.5	5.0
MD5518308	5.4	5.6	4.7	4.8	3.5

*4-hour mean daily ozone concentrations (nl l^{-1}).

Table 17.4. Effects of ozone on tomato fruit production. Data from Oshima *et al.* (1975)

	Ozone treatments (nl l^{-1})*		
	0	200	350
Leaves and stems (g plant^{-1})	167a**	114b	46c
Immature fruit (g plant^{-1})	17.3a	17.6a	7.3b
Mature fruit			
Fruit height (cm)	4.6a	4.9a	4.5a
Fruit diameter (cm)	5.5a	5.7a	5.3a
Fruit weight (g)	76a	86a	68a
Number of fruit (plant^{-1})	11.3a	9.9a	6.9b

*2.5 hours per day, 3 days per week, for 15 weeks.

**Mean separation in rows by Duncan's multiple range test ($P = 0.01$). Means followed by the same letter are not significantly different.

weights of immature fruit than to decrease the weight of leaves and stems. The size of mature fruit was not affected by ozone treatments but the number of fruit was decreased. Further evidence that ozone may affect fruit number but not size is found in studies with cotton (Oshima *et al.*, 1979). Ozone exposure markedly reduced the number of bolls (fruits) in cotton as well as total weight per plant of bolls, seed and ginned fibre (Table 17.5) but the weight per boll of lint and seed and proportion of boll weight in ginned fibre and seed were unaffected. In contrast, Endress & Grunwald (1985) found that all components of soybean yield were reduced by ozone exposure, including numbers of pods, filled pods and seeds and weight of seed per plant (Table 17.6). In this species the number of fruits (pods) was the least sensitive parameter, in contrast to the studies of tomato and cotton which indicated that fruit number per plant is markedly decreased by ozone while fruit size is sustained.

Environmental factors such as water stress, temperature and radiation have their own effects on seed growth and development and may be strongly

Table 17.5. Effect of ozone exposure on yield of Alcala SJ-2 cotton. Data from Oshima *et al.* (1979)

	Ozone treatments*		
	0	1	2
Number of bolls			
Open	6.6a**	3.2b	3.0b
Closed	0.2	0.0	1.0
Dry weights			
Open bolls	55.21a	22.06b	22.41b
Ginned fibre	17.09a	6.41b	7.01b
Seed	25.28a	10.20b	10.40b
Residue	12.34a	5.45b	5.00b
Weight per boll			
Lint	2.59a	2.30a	2.34a
Seed	3.91a	3.19a	3.47a
Proportion of boll weight			
Ginned fibre	30.96a	29.06a	31.28a
Seed	46.69a	46.24a	46.41a
Residue	22.35a	24.70a	22.31a

* Treatment 1: Twice weekly 6 hours per day, 250 nl l^{-1}, for 13 weeks. Treatment 2: As 1 but for 19 weeks.
**Mean separation in rows by Duncan's multiple range test ($P = 0.05$). Means followed by the same letter are not significantly different.

Table 17.6. Components of soybean yield expressed as percentage of a 20 nl l^{-1} ozone ambient air treatment. Data from Endress & Grunwald (1985)

Ozone concentration (nl l^{-1})	Pods per plant	Filled pods per plant	Seeds per plant	Seed weight per plant
46	116.5*	105.8	108.2	115.1
70	91.9	61.9*	65.0*	66.0*
97	68.2*	65.8*	64.6*	60.4*

*Significantly different from ambient air treatment ($P = 0.05$).

Table 17.7. Effects of ozone exposure and water supply on reproductive growth of soybean. Data from Amundson *et al.* (1986)

	Ozone concentration (nl l^{-1})		
	50*	90	130
Well-watered			
Pod number	29.0 ± 6.6**	18.7 ± 6.4	31.3 ± 8.6
Pod dry weight (g $plant^{-1}$)	1.20 ± 0.30	0.70 ± 0.23	0.99 ± 0.28
Dry weight per pod (g)	0.040 ± 0.002	0.040 ± 0.005	0.030 ± 0.003
Water-stressed			
Pod number	3.2 ± 1.1	3.3 ± 1.2	3.5 ± 1.3
Pod dry weight (g $plant^{-1}$)	0.038 ± 0.016	0.065 ± 0.029	0.060 ± 0.024
Dry weight per pod (g)	0.012 ± 0.002	0.016 ± 0.003	0.015 ± 0.002

*Plants were exposed to ozone for 6.8 hours daily in controlled environment chambers.
**Means ± SE.

Table 17.8. Ozone effects on reproductive and vegetative organs*. Data from Cooley & Manning (1987)

Species	Weight reduction	
	Flower, fruit or seed (%)	Vegetative (%)
Bean (*Phaseolus vulgaris*)	24	19
Corn, sweet (*Zea mays*)	30	17
Corn, field (*Zea mays*)	4	24
Cotton (*Gossypium hirsutum*)	40	30
Millet (*Panicum miliaceum*)	56	38
Peanut (*Arachis hypogea*)	0	11
Pepper (*Capsicum anuum*)	54	0
Soybean (*Glycine max*)	3	23
Tomato (*Lycopersicon esculentum*)	7	18
Wheat (*Triticum aestivum*)	16	14

*Ozone doses varied among species.

interacting elements in determining ozone effects. Water-stress effects dominated over ozone stress effects in a study of reproductive growth of soybean (Amundson *et al.*, 1986). Water-stressed plants were not adversely affected by ozone (Table 17.7), while plants growing without water stress had fewer and smaller fruits (pods) in some ozone treatments.

Cooley & Manning (1987), in a major review compared the reported effects of ozone on reproductive and vegetative growth for several species (Table 17.8). Reproductive and vegetative growth effects vary greatly among species. The review also summarized ozone effects on components of yield. For example, ozone decreases most yield components in soybean including pods per plant, seeds per plant, seed weight per plant and individual seed weight. Ozone apparently decreases the ability of soybean to set seed, but, once set, the seeds receive preferential carbon allocation compared with other plant parts. Peanut seeds also have a higher allocation priority than the vegetative plant parts. In wheat, relatively more carbon goes to seeds although there are fewer seeds to fill. In these seed crops, the greater sink strength of developing seeds than other parts of the

Table 17.9. Effects of ozone exposure on wheat yield during the grain fill period and flour quality. Data from Slaughter *et al.* (1989)

	Ozone concentration*				
	32	49	60	73	96
Grain yield (g m^{-2})	172a**	152a	165a	152a	102b
Kernel wt (g 100^{-1})	2.8a	2.7a	2.7a	2.5ab	2.2b
Kernels per spike	30.3a	29.5a	26.0a	31.3a	30.5a
Flour protein (g kg^{-1})	146a	149a	146a	153a	161b
AWRC# (%)	54.6a	54.6a	54.7a	54.7a	56.0b
Baking quality score	98.9a	98.9a	98.9a	97.9a	91.9b
PSI## (%)	43.1c	44.3bc	43.9c	46.0b	48.7a

*4-hour mean daily ozone concentrations in nl l^{-1}.
**Means followed by the same letter in a row are not significantly different ($P = 0.05$).
#Alkaline-water retention capacity.
##Particle size index.

plant plays a major role in ensuring that seeds are relatively robust in the face of widespread injury to other plant parts. The minimal yield losses due to ozone in some seed crops may be due to reductions in some yield components being compensated for by increases in others. On the basis of this rationale, Cooley & Manning (1987) suggest that if ozone affects a given stage of seed development more than other stages, later stages in the development sequence would tend to compensate to some extent, minimizing the overall yield loss.

Effects of ozone on seed quality

Seed quality parameters may also be affected by ozone exposure. Mulchi *et al.* (1986) exposed winter wheat to ozone during anthesis and, with increased ozone, found higher grain protein, particle size index, and alkaline-water retention capacity in the flour made from the seeds; that is, generally reduced flour quality for baking purposes. Ozone treatments during grain fill, after anthesis, also reduced seed quality (Slaughter *et al.*, 1989). Grain yield and seed weight were decreased by ozone but the number of seeds per spike was unaffected while flour protein, alkaline-water retention capacity and particle size index increased and baking quality score decreased (Table 17.9). This research indicates that such changes in grain quality need to be considered along with changes in yield and yield components as the seed grows and develops under ozone stress.

PHOTOSYNTHATE ALLOCATION TO SEEDS IN INJURED PLANTS

Ozone may suppress translocation of photosynthate to roots depressing root growth more than shoot growth. Carbon allocation to reproductive structures also may be reduced in some instances, although increases or no effects have also been demonstrated. Increases in allocation to shoots have been attributed to the need for photosynthate to provide for the metabolic costs of repair of the stress damage in shoots and the general precedence of shoots over roots as primary sinks during periods of stress or limited periods of photosynthate production. Also, the distribution of carbon between sucrose and starch in the leaf may influence relative root and shoot allocation. Thus any air-pollution-induced shift in sucrose–starch metabolism might affect carbohydrate allocation patterns. Regardless of other factors, the outcome appears to be preferential translocation of photosynthate to shoot tissues or retention in the shoot under pollutant stress.

In a major review of factors affecting crop productivity and photosynthate allocation, Gifford *et al.* (1984) set out the major components involved in this complex as interception of radiation by foliage, conversion of the intercepted radiation to photosynthetic assimilates, and allocation of photoassimilates to organs of economic interest. In this scheme ozone-induced visible chlorotic and necrotic injury to leaves can be expected to reduce interception of radiation and injury to leaf cells will impair photosynthetic carbon assimilation. The allocation of available photoassimilates to reproductive growth is likely to be of primary importance in the study of air pollution effects on seed yield. There are a number of aspects relating to seed development that are relatively unexplored from the standpoint of understanding air pollutant effects on carbon allocation. For example, the mechanisms controlling phloem loading in source regions and unloading in sink regions, including fruit and seed tissues, are not well understood despite their obvious importance as factors influencing allocation patterns. Unloading into the apoplast and through the symplast (via plasmodesmata) are both known to occur. No direct information is available concerning air pollutant effects on phloem unloading in fruit and seed tissues.

Not all assimilated carbon is immediately metabolized or translocated to its final sink. Temporary storage pools serve to promote biochemical stability during changing conditions and may be of considerable interest in studies of air pollution stress effects on carbon allocation. Long-term storage and remobilization may be important in crop plants during the stages of reproductive development or seed fill when the demand for assimilate is high, but the production is low due to air-pollution-induced leaf injury. McLaughlin & McConathy (1983), using radioactively labelled CO_2, demonstrated that leaves and small fruit are stronger sinks for photosynthate under ozone stress while large fruit do not draw as heavily upon photosynthate when under ozone stress (Table 17.10). The large fruit must be utilizing remobilized carbohydrate stored elsewhere in the plant.

Miller (1988), in order to demonstrate the relative magnitude of the flow of carbon in a crop plant, utilized the estimates of Gifford *et al.* (1984). Of the total

Table 17.10. Effect of ozone on photosynthate allocation in bush bean. Activity apportionment at harvest (% of total activity). Data from McLaughlin & McConathy (1983)

	Ozone concentration (nl l^{-1})*		
	0	100	150
Leaves	36.2 ± 11.7**	38.1 ± 9.0	56.7 ± 10.0
Small pods	1.1 ± 1.0	1.1 ± 1.2	3.7 ± 2.0
Large pods	62.7 ± 11.7	60.9 ± 9.0	39.6 ± 11.8

*18-hour exposure at 5 weeks.
**Means ± SD.

CO_2 that is photoassimilated during the life cycle of a cereal crop, it was estimated that 15 to 20% is lost due to photorespiration and approximately 34 to 40% due to true respiration (catabolism). Of the 40 to 51% of remaining dry matter, 10 to 23% (of the total carbon fixed) is thought to go to grain production. The compounding effects of stresses such as air pollution on these allocation patterns can be envisaged, even when effects on total carbon assimilation are ignored. For example, a 10% increase in total amount of carbon lost to respiration associated with air pollution injury and repair processes might result in a 15 to 30% reduction in grain yield if the relative partitioning of carbon among the plant tissues remains the same. Miller (1988) has developed a model for observing the effects of air pollutants on carbon allocation in plants. In this model air-pollutant-induced leaf injury causes physiological perturbations that result in decreased photosynthesis and increased maintenance and repair process energy costs in leaves. The net effect is reduced available photosynthate levels in terms of sucrose and starch in leaves. Under these stressful circumstances, the allocation priorities change and there is decreased allocation to the root and either increased or decreased allocation to the reproductive structures.

Even though Cooley & Manning (1987) noted substantial species differences, a comprehensive model of assimilate partitioning patterns emerged from their review. Their analysis of more than 10 reports on effects of ozone on soybean fruiting led them to describe carbon allocation as a two-phase process, the second phase of which is flowering and fruit development. During the second phase, carbon from the leaves is preferentially allocated to the seeds rather than to other leaves, stems and roots.

The majority of the research concerning ozone effects on allocation of carbon among plant tissues or organs has been with pot-grown plants in controlled environments. Under these conditions the results vary depending on the species, the environmental conditions of the study, and the characteristics of the pollutant exposures.

NEEDED RESEARCH

Research on seed growth and development should emphasize anatomical and morphological changes and physiological functions of individual tissues and be based upon the establishment of specific goals with statements of questions to be addressed (Ormrod & Hale Marie, 1990). Appropriate experimental materials and methods will need to be selected. Particular challenges lie in efforts to correlate seed initiation and growth with photosynthesis and to relate allocation of carbon to initiation and development of reproductive tissue. Studies of source loading of photosynthate into the translocation system and translocation control by the reproductive tissue will be of particular value.

Specific research questions that should be asked include:

1. Is the air pollutant effect due to induced changes within or close to the seed itself and not due simply to the effect of the availability of photosynthate?

An approach to answering this question would be to use sensitive and insensitive cultivars and:

(a) impose partial defoliation and/or shading treatments, along with differential ozone treatments;

(b) alter the demand for photosynthate by removal of some fruit, along with differential ozone treatment.

Do such leaf and fruit manipulations affect the initiation and development of fruit and the reduction in fruit dry weight attributable to ozone exposure?

2. Does the air pollutant somehow truncate the translocation period to the developing fruit? Experiments are needed with discontinuous O_3 treatments, which would be episodic in relation to the sequence of fruit growth. Utilization of data taken in long-term yield studies may help answer this question but is not likely to help explain the mechanism(s) involved.

3. Are there definable genetic differences in air pollutant effects on leaf injury and fruit growth rates that can be exploited to help minimize effects on fruit development? The answer to this question could lead to the application of biotechnology and utilization of gene transfers among cultivars and species.

4. Do air pollutant stresses affect the morphology and anatomy of pollen, the pollination and seed set process and developing seeds and fruits? This will require the utilization of microscopic and ultrastructural research techniques hitherto not widely employed in air pollution research.

Experimental requirements are demanding for all these studies. The separation of direct and indirect effects will be a particular challenge. Knowledge of the mechanisms involved will serve to verify seed development effects on crop yields and help facilitate future approaches to controlling the enormous crop loss due to air pollutants. Such research is imperative in an urban/industrial world in which food supply continues to be a primary requisite for human life.

REFERENCES

Adams, R.M., Glyer, J.D., Johnson, S.L. & McCarl, B.A. 1989. A reassessment of the economic effects of ozone on U.S. agriculture. *J. Air Pollut. Contr. Assoc.* **39**: 960-968.

Amundson, R.G., Raba, R.M., Schoettle, A.W. & Reich, P.B. 1986. Response of soybean to low concentrations of ozone: II. Effects on growth, biomass allocation and flowering. *J. Environ. Qual.* **15**: 161-167.

Cooley, D.R. & Manning, W.J. 1987. The impact of ozone on assimilate partitioning in plants: A review. *Environ. Pollut.* **47**: 95-113.

Darrall, N.M. 1989. The effect of air pollutants on physiological processes in plants. *Plant, Cell Environ.* **12**: 1-30.

Endress, A.G. & Grunwald, C. 1985. Impact of chronic ozone on soybean growth and biomass partitioning. *Agric. Ecosyst. Environ.* **13**: 9-23.

Facteau, T.J. & Rowe, K. 1981. Response of sweet cherry and apricot pollen tube growth to high levels of sulphur dioxide. *J. Amer. Soc. Hortic. Sci.* **106**: 77-79.

Gifford, R.M., Thorne, J.M., Hitz, W.D. & Giaquinta, R.T. 1984. Crop productivity and photoassimilate partitioning. *Science* **225**: 801-808.

Harrison, B.H. & Feder, W.A. 1974. Ultrastructural changes in pollen exposed to ozone. *Phytopathology* **64**: 257-258.

Heagle, A.S. 1989. Ozone and crop yield. *Annu. Rev. Phytopathol.* **27**: 397-423.
Heagle, A.S., Spencer, S. & Letchworth, M.B. 1979. Yield response of winter wheat to chronic doses of ozone. *Can. J. Bot.* **57**: 1999-2005.
Heagle, A.S., Kress, L.W., Temple, P.J., Kohut, R.J., Miller, J.E. & Heggestad, H.E. 1988. Factors influencing ozone dose — yield response relationships in open-top field chamber studies. In Heck, W.H. *et al.* (eds) *Assessment of Crop Loss from Air Pollutants*: 141-179. Elsevier-Applied Science Publishers, London.
Heath, R.L. 1980. Initial events in injury to plants by air pollutants. *Annu. Rev. Plant Physiol.* **31**: 391-431.
Hogsett, W.E., Olszyk, D., Ormrod, D.P., Taylor, G.E. Jr & Tingey, D.T. 1987. *Air-pollution Exposure Systems and Experimental Protocols: Vol. 1. A Review and Evaluation of Performance and Vol. 2 Description of Facilities.* US Environmental Protection Agency EPA/600/3-87/037 a/b, Corvallis, Oregon.
McLaughlin, S.B. & McConathy, R.K. 1983. Effects of SO_2 and O_3 on allocation of ^{14}C-labeled photosynthate in *Phaseolus vulgaris*. *Plant Physiol.* **73**: 630-635.
Miller, J.E. 1988. Effects on photosynthesis, carbon allocation, and plant growth associated with air pollutant stress. In Heck, W.H. *et al.* (eds) *Assessment of Crop Loss from Air Pollutants*: 287-314. Elsevier-Applied Science Publishers, London.
Mulchi, C.L., Sammons, D.J. & Baenziger, P.S. 1986. Yield and grain quality responses of soft red winter wheat exposed to ozone during anthesis. *Agron. J.* **78**: 593-600.
Ormrod, D.P. & Hale Marie, B. 1990. *Impact of Gaseous Air Pollution on Seed Development: Research Imperatives.* Proc. Crop Science Society of America 1990 Annual Meeting. Paper #57, Madison, Wisconsin.
Oshima, R.J., Taylor, O.C., Braegelmann, P.K. & Baldwin, D.W. 1975. Effect of ozone on the yield and plant biomass of a commercial variety of tomato. *J. Environ. Qual.* **4**: 463-464.
Oshima, R.J., Braegelmann, P.K., Flagler, R.B. & Teso, R.R. 1979. The effects of ozone on the growth, yield and partitioning of dry matter in cotton. *J. Environ. Qual.* **8**: 474-479.
Preston, E.M. & Tingey, D.T. 1988. The NCLAN program for crop loss assessment. In Heck, W.H. *et al.* (eds) *Assessment of Crop Loss from Air Pollutants*: 45-62. Elsevier-Applied Science Publishers, London.
Slaughter, L.H., Mulchi, C.L., Lee, E.H. & Tuthill, K. 1989. Chronic ozone stress effects on yield and grain quality of soft red winter wheat. *Crop Sci.* **29**: 1251-1255.
Slaughter, L.H., Mulchi, C.L. & Lee, E.H. 1993. Wheat-kernel growth characteristics during exposure to chronic ozone pollution. *Environ. Pollut.* **81**: 73-79.
Thompson, C.R., Kats, G. & Cameron, J.W. 1976. Effects of ambient photochemical air pollutants on growth, yield, and ear characters of two sweet corn hybrids. *J. Environ. Qual.* **5**: 410-412.

18 Forest Growth and Decline: What is the Role of Air Pollutants?

P.H. FREER-SMITH

INTRODUCTION

Forest declines have occurred for centuries and may be a normal part of ecosystem response to environmental factors. However, in the last 25 years there has been major public concern over forest declines in Europe and North America which have been thought to be pollutant related. The use of fossil fuels is increasing in the developing countries, such as those of South and East Asia. It is thus as important as ever to establish clearly how and to what extent pollutant depositions influence forest growth and their potential role in causing declines. This chapter concentrates on recent European declines and then draws out the key conclusions which have emerged from the recent and ongoing European research. This review has had to be selective, but the conclusions drawn can usefully guide evaluation of the potential threat to forest ecosystems world-wide, and should be considered in managing forest ecosystems.

The term decline is a general description for the progressive worsening of forest stands. The term implies long-term damage and a progressive deterioration in tree condition. Dieback usually refers to the death of branches from their tips back towards the main stem often in the upper crown (branch dieback). Dieback is usually applied to single trees but is sometimes used to describe a stand in which edge trees are dying and, to avoid confusion, it is better to specify stand dieback in these circumstances. Figure 18.1a shows a typical example of forest decline in a Norway spruce [*Picea abies* (L.) Karst.] stand; the condition of some trees in the stand has progressively deteriorated over a number of years until bark beetle (*Dendroctonus micans*) infestation has eventually killed selected groups of trees within the stand (Figure 18.1b). In contrast, Figure 18.2 shows an example of classic stand dieback. At this high-elevation Norway spruce forest in central Europe, the edge trees died each year and were harvested. The slopes in the background had previously been completely forested.

Rates of tree growth are modified by a number of biological and environmental factors. The main biological or biotic factors are phenotype (species and provenance), insects, fungal and other pathogens, and mammals (particularly squirrels and deer). Important abiotic factors are climate (rainfall, frost, wind, winter cold, temperature and photoenvironment), soil type (and the inferred nutritional and moisture supply), altitude and topography. In plantation forestry,

Plant Response to Air Pollution. Edited by Mohammad Yunus and Muhammad Iqbal

the intention is to select species and provenance that are suited to the particular conditions of the site; yield and productivity are thus maximized; damage, dieback and tree death are minimal or absent. Sights like that seen in Figure 18.2 will occur, but they will be infrequent, limited in extent and can be minimized by good management. In unmanaged forests, plant competition and, in the long term, natural selection will control species composition. After the climax high forest has established, there should again not be major areas of extensive decline or dieback except from catastrophe and where decline is a phase of natural succession (see below). Without thinning, harvesting at maturity and sanitation felling there will usually be scattered dead trees in stands.

Time scale, and in particular the periodicity of extreme events, is important in the consideration of forest declines. In unmanaged forests with cyclic successions major declines, dieback and catastrophic destruction, by forest fires for example, may be part of the cycle. In the Pacific region there are a number of well-documented forest declines associated with natural successions (Huttl & Mueller-Dombois, 1992). However, many temperate and boreal forest ecosystems progress towards climax forest which is unlikely to exhibit major, extensive tree decline unless environmental conditions change. Over the period for which instrumental records exist (*ca.* 300 years) and indeed within historic times, the climate record has been relatively uniform (Manley, 1974) and we are not accustomed to major shifts in vegetation cover. Forest ecosystems have evolved over millennia and are considered to be stable; they are well suited to the environmental conditions and thus unlikely to decline unless these change. Exceptions to this are species at the edge of their natural distribution, for which the occasional infrequent extreme event may kill plants which have become established in more benign intervening years. Until the recent concern over climate change, increases of pollutant depositions have been considered to be the main anthropogenic factor likely to result in change to forest ecosystems. They may still be today.

CLASSIC FOREST DECLINES

Silver fir (*Abies alba* Mill.)

In central Europe there have been local, regional declines of silver fir dating back at least to the 18th and 19th centuries; a history which has been thoroughly reviewed by Watcher (1978). As will be seen, there are a number of parallels between these declines and the contemporary declines which have occurred in Europe in the last 20 years. The symptoms were non-specific, going through a sequence of needle browning and loss; a canopy morphology known as stalk's nest occurs and this condition becomes progressively severe up until eventual tree death. There were geographical and temporal patterns of decline which can be related, with varying confidence, to the natural distribution of silver fir, provenance selection, site conditions and climatic conditions. In some regions bad declines have been correlated with summer droughts. The reduction in annual ring widths which accompanied decline usually started in years of serious drought. There were rarely single, undisputed causes, but drought, extreme winter temperatures, late frosts, insects (aphids, bark beetles and bud moths), fungal pathogens (*Armillaria* and *Lophodermium*) and pollution have all been suggested as important causes. Silvicultural practices, particularly clear-felling and inadequate thinning, and low potassium availability have also been identified as causal factors. Low foliar potassium contents have been associated with increased sensitivity to frost.

Arguments have occurred over which factors are primary or "triggering" and which are secondary and, as early as 1928, a disease "chain" was considered (Dieterich, 1928). In spite of the regional differences, most authors have put main emphasis on drought, with insects and pathogens as secondary. Partly because of these historical problems, silver fir is known as a "difficult" species, sensitive to site conditions and silvicultural practice.

Acute pollution damage

Pollutant emissions were also a known cause of damage to silver fir locally. In the erstwhile Czechoslovakia industrial emissions were considered as primary causes of decline, although the interacting effects of abiotic and biotic factors have almost always been involved. Acute pollutant damage was also well known for other species when grown in highly polluted air close to emission sources. In Europe, examples have usually resulted from emissions in or around general industrial areas such as the Rhine and Ruhr valleys in North Rhine-Westphalia or the southern Pennines of England. In the United States point emission sources (such as smelters, aluminium reduction plants or power-stations) have also produced acute symptoms to trees and resulted in zones denuded of vegetation in their vicinity. Field studies have been conducted in many of these areas (see review by Miller & McBride, 1975; Smith, 1981) and the results of these can be used to draw up tables of relative susceptibility to acute injury attributed mainly to sulphur dioxide (Last, 1982). Conifers tend to be more sensitive than broadleaves; Norway spruce (*Picea abies*), Scots pine (*Pinus sylvestris*), larch (*Larix decidua*) and Sitka spruce (*Picea sitchensis*) are classified as sensitive. Acute injury close to local sources is still seen today, although examples are infrequent and when they occur they are usually associated with failure of emission-control equipment.

RECENT EUROPEAN DECLINES

The above description places the events in central Europe during the late 1970s and 1980s in context. Public attention was drawn to the deterioration of Germany's forests by a series of articles in the magazine *Der Spiegel*, which reported that damage was on an unprecedented scale. Damage to silver fir was the first to be observed, but it quickly spread to Norway spruce, beech (*Fagus sylvatica* L.) and other species. Again terminology was important. In Germany the emotive term "*Waldsterben*", meaning "forest death" was initially used because many feared that tree condition would continue to worsen in the way illustrated in Figure 18.2. However, during the early 1980s it became clear that this would not happen on a wide scale and the term *neuartige Waldschaden*, translated to "new type forest declines" was introduced. This term is also somewhat misleading. Although the combinations of species and geographical areas affected was new, the symptoms were not.

Main symptoms are defoliation and foliar discoloration and these have since been observed and monitored for a wide range of tree species in Austria, the Benelux countries, the Czech and Slovak Republics, France, Germany, Poland, Scandinavia, Switzerland and the UK (UNECE/CEC Report, 1992). Degree of defoliation has usually been estimated by assessing the percentage needle or leaf loss (loss in crown density). Initially five categories of needle loss were used following the system introduced by the Germans in 1984. Categories were as follows: (0) 0 to 9% needle loss, classed as "healthy", (1) 10 to 25% needle loss, "slightly damaged", (2) 26 to 60% needle loss, "moderate damage", (3) 61 to 100% needle loss, "severe damage" and (4) dead trees. In 1986 the term for 10 to 25% needle loss was changed to "warning category" since it had become clear that needle losses of up to 25% can be quite natural and may not indicate damage. More recently still, the terms not defoliated, slightly, moderately, and severely defoliated have been used in recognition that defoliation even above 25% may not be synonymous with "damage". Similarly it is now recognized that branch structure, internode lengths of needles and crown form all influence crown density which thus should not be assumed to indicate needle loss. The term "needle loss" and the percentage categories are, however, still widely used.

Although forest death ((*Waldsterben*) did not occur on a wide scale, strong public concern led to intensive international research. During the early 1980s there were a large number of hypotheses proposed to explain forest decline. A common approach in reviews has been to examine each hypothesis and the related scientific literature in turn (see, for example, Rehfuess, 1985). Various hypotheses have developed and of these a large number have been eliminated so that this approach is no longer efficient in reviewing the literature on this subject. Indeed research and discussion of individual hypotheses has hindered progress in understanding forest decline. However, three general hypothesis as described by Schulze & Freer-Smith (1992) remain important as general perspectives. Firstly, it was thought that the direct effects of anthropogenic pollutants on above-ground parts of plants could be the primary cause of problems. It was suggested that gaseous air pollutants (SO_2, NO_x, O_3, NH_3), acid mist or organic pesticides could be having physiological and biochemical effects which, over the long term (i.e., successive years), were producing visible dieback and decline. Secondly, the pollutants listed above along with those in rain could be having long-term soil-mediated effects. Such effects would be operating through soil acidification, aluminium toxicity, the leaching of base cations and nutrient deficiency. Thirdly, decline could be the consequence of complex interactions involving biotic and abiotic factors including drought, frost and pollutant depositions.

The range of techniques used to address the various hypotheses has been large with valuable work being done using closed fumigation chambers, open-top chambers, leaf and branch chambers, field or forest fumigation and irrigation systems. Recently rain exclusion and roof experiments have also started to contribute to understanding. These experimental approaches have been able to identify likely mechanisms of damage (see Freer-Smith, 1992). This information has been supplemented by surveys of forest condition, diagnostic surveys (Cape *et al.*, 1990) and by detailed regional studies. The discussion that follows illustrates how this combination of approaches has proved effective. We now have a reasonable knowledge of the condition of European forests and of how this condition alters over time. The physiological processes that determine growth rate, condition and stability have been identified, and the progress that has been made in the process-based modelling of tree and stand growth is perhaps the best illustration of this. The ways in which different combinations of biotic and abiotic factors influence forests and woodland have been documented for various geographic regions (Table 18.1). Uncertainties remain, and it is still difficult to give prognoses for forest ecosystems largely because environmental conditions (pollutant depositions, carbon dioxide concentrations, pathogen occurrence and climate) continue to change.

Table 18.1. Some European examples of recent regional declines with summaries of the probable causal sequence and selected references to key publications on each

Region	Probable or possible causal sequence and comment
Harz, Solling and Hils, northern Germany	Wet and dry-deposited sulphur and nitrogen; soil acidification; leaching of base cations; aluminium mobilization; root damage and nutritional problems. (Ulrich *et al.*, 1980; Ulrich, 1989)
Erzgebirge, and parts of Czechoslovakia, Poland and the old DDR	SO_2; heavy metals; fluorine; and other gaseous air pollutants and frost.
Fichtelberg	SO_2, and wet and dry-deposited sulphur and nitrogen; Ca/Al and Mg/Al in soil, effects on mycorrhiza and tree nutrition especially Mg and N. (Schulze *et al.*, 1989)
Black Forest and the Vosges	Dry and occult acidic deposition (to a lesser extent O_3, SO_2, NO_x and NH_3); leaching of soil Mg and Ca at rates which exceed supply by weathering; Mg, K and Mn deficiency in foliage (accelerated foliar leaching?) and drought; *Armillaria* and bark beetle in the later stages of decline. (Zottle & Huttl, 1991; Landmann, 1991)
Inner Bavarian Forest	Acid mist; frost shocks; not drought but some examples of poor provenance selection; plus some of the soils/nutritional problems described above. (Rehfuess, 1985)
The Netherlands	NH_3, O_3 and drought. (Roelofs *et al.*, 1985)
Afan Forest, South Wales	"Bent top" of Sitka spruce; nutrition; waterlogging; *Elatobium* and possibly air pollutant combinations. (Coutts, 1995)
Athens Basin	O_3 and other oxidant damage to *Pinus pinaster*.

SURVEYS OF FOREST CONDITION

Because of concern over forest decline an increasing number of European countries have undertaken systematic surveys of forest condition since 1984. These surveys have been coordinated internationally by the United Nations Economic Commission for Europe (UNECE), through the activity of the Convention on Long Range Transboundary Air Pollution, and by the Commission of the European Communities (CEC) through the Air Pollution Working Group of the standing forestry committee. In 1991 the UNECE and CEC combined their reports for the first time. (UNECE/CEC, 1992).

Figure 18.3 shows British survey data for the period 1987 to 1992 (Innes & Boswell, 1989, 1990, 1991). The categories of crown density ("damage" or "defoliation") as described above have been used, with classes 0 and 1, 2, and 3 and 4 combined. There is a strong case for considering trees in class 3 and higher (>60% loss of crown density) as indicative of significant defoliation in the UK. It can be seen from this summary of a very large data set (some 2000 trees of each species are examined annually), that crown density does fluctuate significantly from year to year. For example, the increase in crown thinning of Sitka spruce in 1989 was associated with a mild winter in 1988/89 and a consequential increase in populations of green spruce aphid (*Elatobium abietinum*). The general worsening in condition of all three species in 1991 was associated with late spring frosts, heavy snow and strong winds in the winter of 1990/91. Although only 6 years of comparable data that exist for the UK; the indication from these data is that there is not a long-term trend of deterioration superimposed on these year-to-year fluctuations.

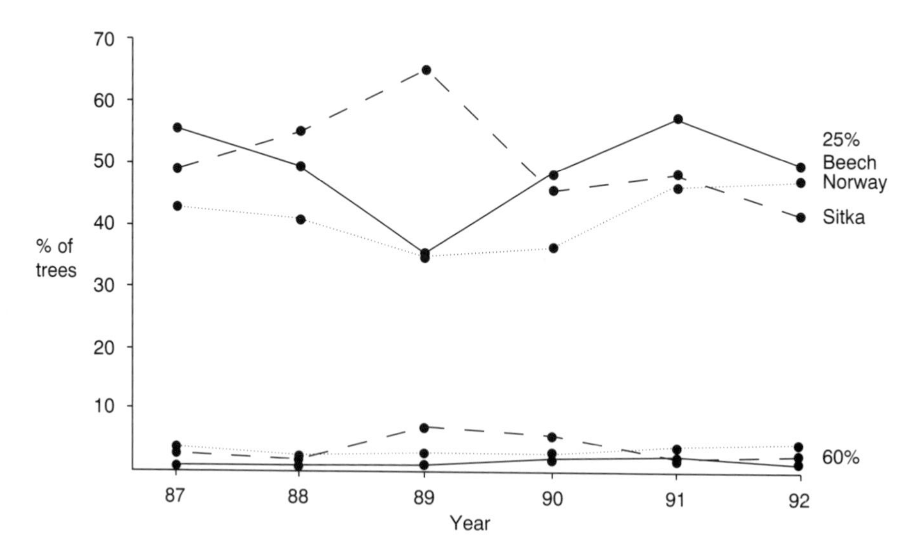

Figure 18.3. Summary of the annual survey of forest condition in the UK from 1987 to 1992. The percentage of trees out of *ca*. 2000 beech (*Fagus sylvatica*), Sitka spruce (*Picea sitchensis*) and Norway spruce (*Picea abies*) with greater than 25% loss of crown density (crown thinning) and with greater than 60% loss of crown density compared with "perfect" fully foliated trees. See text for further details

Figure 18.1. (a and b) Forest decline in Norway spruce *(Picea abies)* at low and middle elevations in central Europe, 1990

Figure 18.2. Stand dieback of Norway spruce *(Picea abies)* at high elevations in the Beskydy Mountains, 1980. At this site the main anthropogenic factor was sulphur deposition

Figure 18.4.(a) Type 1 decline of Norway spruce *(Picea abies)*. Yellowing of older needle classes associated with foliar magnesium deficiency (<ca. 0.07%). Photographed in The Fichtelberg (E. Bavaria) in 1988. (b) Type 3 decline of *Picea abies*. Reddening of older needle classes followed by needle shedding. Photographed in The Inner Bavarian Forest in 1989

Survey data have been analysed in a number of ways; for example Schulze & Freer-Smith (1992) plotted the percentage area of the forest regions in Germany in damage classes 2 to 4 (>26% needle loss) against the district mean total acid deposition (mmol sulphur plus nitrogen m^{-2} yr^{-1}) and the average summer ozone concentrations. Both regressions showed considerable scatter, but did show significant regressions with the largest percentage area of damage where pollutant inputs were greatest. In these multiple regressions acidic deposition and O_3 accounted for about 50% of the variation in percentage area damaged. These correlations are not proof of cause and effect, but they do indicate that it is sensible to examine the role of total sulphur and nitrogen depositions and of air pollutant concentrations in *neuartige Waldschaden* (forest decline).

The comparison of German, British and other European data gives a clear indication of regional and spatial differences in the condition of forests. Such analyses (UNECE/CEC, 1992) have illustrated the considerable contrasts in the relative importance of various environmental factors in the different geographic regions of Europe. In central and eastern European countries, air pollution is considered to be the most likely factor affecting forest condition. Elsewhere in Europe it is usually thought to be one of among several factors which predispose forests to decline. In several regions, including Britain, factors other than pollutant depositions are considered to be more important in determining forest condition. Such conclusions are based on both analyses of survey data of the type outlined above and on more detailed and site-specific studies. The results from specific site investigations have often been able to identify regional problems (Table 18.1). Clearly, climate, pollutant depositions, geographical, topographical and soil conditions can be sufficiently similar on a regional scale to allow such extrapolation.

SYMPTOMATOLOGY AND REGIONAL SPECIFICITY

A major step forward in understanding the processes leading to forest decline was made when symptomatology was properly classified into regional decline types (Blank *et al.*, 1989). Decline types are now recognized for a number of species including European beech, silver fir and Norway spruce. The second report of the Research Advisory Council of the German government on forest damage/air pollutants described five decline types for Norway spruce and these are summarized in Table 18.2 (Blank *et al.*, 1988). The two most widespread decline types are illustrated in Figure 18.4. Each decline type consists of an official description, main symptoms and geographical occurrence. The association between symptoms and geographic occurrence is important; it reflects the contribution of climatic factors, soil type, topography and pollutant depositions in producing a particular decline type. The concept of pollution climate, meaning the unique combination of climatic conditions and pollutant deposition associated with a particular geographical area (CEC, 1989), was developed at about the same time that symptomatologies were first described with reasonable accuracy.

Table 18.2. The descriptions of Norway spruce decline types as formulated by the FRD's Research Advisory Council in 1986 (Rehfuess, 1985)

Type	Official description	Main symptoms	Geographical occurrence
1	Needle yellowing at higher elevations	Chlorosis and subsequent loss of older needles, associated with foliar magnesium deficiency	Higher altitudes of medium-range mountains in central and southern Germany (e.g., Black Forest, Bavarian Forest)
2	Thinning of tree crowns at medium-high altitudes	Loss of needles, but little chlorosis	Medium-high altitudes (approximately 400-600 metres), particularly in central and northern Germany (e.g., Hils, Odenwald)
3	Needle reddening of older stands in southern Germany	Yellowing, then reddening of older needles, followed by needle shedding	Mainly in the foothills of the Bavarian Alps and at high elevations of the Inner Bavarian Forest
4	Chlorosis of needles in the Bavarian Alps	Chlorosis and subsequent loss of needles, starting with the youngest needles. Associated with K and/or Mn deficiency	Calcareous Alps, above approximately 1000 metres
5	Thinning of tree crowns in northern coastal areas	Reduced growth, loss of needles	Northern German Plain

The accurate description and classification of decline types also resulted in the recognition that there was no particular decline occurring in response to a single causal factor, but rather a number of regional and species-specific declines. The general symptoms were non-specific, certain factors being common to one or more decline type and regional decline. There may well also be common synchronizing factors. A description of symptomatology, an understanding of mechanisms identified from experimental studies and detailed investigation of a number of forest stands has resulted in the identification of the main causes of a number of regional declines. European examples of these are listed in Table 18.1, which is based on selected literature from the last 10 years of research in drawing up the probable causal sequences. However, the interpretations placed on the published accounts are inevitably a personal view and different opinions remain, particularly over the relative importance of the various factors involved in each decline.

In all the regional declines detailed in Table 18.1, pest and disease attack are likely to be associated with the final stages of the decline sequence (e.g. needle cast fungi — *Lophodermium*, and honey fungus — *Armillaria*), and dying or dead trees often show bark beetle infestation (*Dendroctanus* and *Ips* spp.).

The symptom or ecosystem approach

One of the most important factors in identifying regional differences of decline type has been the quantification of elemental inputs from the atmosphere. Understanding of the significance of nutrients from the atmosphere has arisen from the traditional ecological approach of quantifying elemental sinks and fluxes. The application of this approach to forest decline can be attributed primarily to Ulrich (see for example Ulrich, 1989). However, work on forest nutrient budgets and on elemental losses from forest ecosystems (thus of sustainability) (e.g., Adamson *et al.*, 1987) provided relevant data during the same period. The development of this approach also showed that research directed at specific nutritional problems (Zottle & Huttl, 1991) was complementary to the mass-balance approach (Ulrich *et al.*, 1980). These hypotheses were initially seen as competing, but foliar nutrient deficiencies can be linked with acidification and the leaching of base cations from soil by mobile anions such as sulphate.

Perhaps the clearest example of this approach is the magnesium budgets which were put together by Roberts *et al.* (1989) in their review of type 1 spruce decline. These authors examined the magnesium mass balance of six spruce ecosystems (1 — Solling, north Germany; 2 — Oberwarmensteinach, southeast Germany; 3 — Wulfersreuth, southeast Germany; 4 — Garsjon, Sweden; 5 — Beddgelert, UK; and 6 — Chesuncook, USA). Beddgelert in North Wales was the only one of these sites at which inputs of magnesium in wet and dry deposition exceeded outputs by leaching and biomass uptake. This gave a positive mass balance of 4.9 kg ha^{-1} yr^{-1}. Outputs exceeded inputs by between −0.84 and −6.95 kg ha^{-1} yr^{-1} at the other sites. This gave a strong mechanistic explanation for the previously unexplained observation that type 1 decline of spruce (needle yellowing associated with magnesium deficiency) does not occur in maritime regions.

CONCLUSIONS

This review has shown that good progress in the understanding of forest decline has been made in the last 10 years. Our knowledge of the role of elemental inputs, including anthropogenically derived sulphur and nitrogen compounds, has advanced substantially. We have much improved information on forest condition, pollutant depositions, elemental budgets, and on biochemical, physiological, soil and atmospheric processes. The ecosystem approach, based on the quantification of pools and fluxes of nutrients, has proved particularly valuable. It has brought together hypotheses that had previously focused on stand and soil processes in isolation. Similarly, the concept of a causal sequence, in which a range of biotic and abiotic factors contribute to a specific-systems effect, has allowed the principal mechanisms of damage to be identified and placed in context. These achievements have led to the identification of a number of regional declines for which causal relationships are understood to a varying degree.

Most forest scientists familiar with this area now accept that, over the long term, continuing small inputs of anthropogenically derived sulphur and nitrogen and gaseous pollutants, such as ozone, will exert an influence on forest ecosystems. However, it is also clear that the widespread and catastrophic "forest death" which was feared has not materialized. Effects of pollutant depositions are subtle and, at least in the short term, usually have little commercial impact relative to the effects of climatic extremes such as wind-throw, drought and outbreaks of fungal or insect pests. In many managed forests the nutritional problems which depositions create can be addressed by using fertilizers. However, abatement of pollutant emission would clearly be a preferred option.

As our understanding of the role of pollutant depositions in influencing forest condition has advanced, so deposition patterns have been altering. In Europe sulphur emissions have declined significantly in the last 20 years (e.g., by some 50% in the UK), while those of the oxides of nitrogen and of ammonia have been increasing. Our understanding of symptomatology, of response mechanisms, and of cause–effect relationships has allowed considerable progress to be made in identifying the threshold values at which pollutant depositions are harmful to forest ecosystems and other terrestrial ecosystems (Ashmore, 1994). The critical loads and levels approach has developed from this understanding (Freer-Smith & Benham, 1994). Critical levels for forest damage are exceeded in many areas of Europe, and it is clear that pollutant depositions must be included amongst the abiotic factors which have significant effects on forest ecosystems. In spite of this, forecast or prognosis of forest condition, stability and growth remain difficult. This is because important gaps remain in our understanding, hence the uncertainty attached to even the best described explanations of specific forest declines.

Perhaps even more importantly, it is now recognized that rising carbon dioxide concentrations have influenced trees in the past (Woodward, 1987) and are likely to affect forest ecosystems in the foreseeable future. It seems increasingly likely that, in addition to the direct effects of CO_2, this and other greenhouse gases are resulting in climate change (IPCC Working Group 1, 1992) which will also have major effects on forest ecosystems in the future. The predicted climate scenarios are not yet of sufficient spatial resolution to allow proper examination of the threats which climate change may pose. Similarly there is not enough known about the direct effects of CO_2 on trees or of the way in which these effects interact with the other environmental factors which influence forest ecosystems.

REFERENCES

Adamson, J.K., Hornung, M., Pyatt, D.G. & Anderson, A.R. 1987. Changes in solute chemistry of drainage waters following the clearfelling of a Sitka spruce forest. *Forestry* **60**: 165–177.

Ashmore, M. 1994. *Proceedings of the UNECE Workshop, Egham 1992*. HMSO, DoE Publication, London.

Blank, L.W., Roberts, T.M. & Skeffington, R.A. 1988. New perspectives on forest decline. *Nature* **336**: 27–30.

Cape, J.N., Freer-Smith, P.H., Paterson, I.S., Parkinson, J.A. & Wolfenden, J. 1990. The nutritional status of *Picea abies* (L). Karst across Europe, and impliations for "forest decline". *Trees* **4**: 211–224.

CEC, 1989. *Pollution Climates in Europe and their Perception by Terrestrial Ecosystems*. CEC Air Pollution Research Report 6.

Coutts, M.P. 1995. *The Decline in Sitka Spruce on the South Wales Coalfield*. Technical Paper 9, The Forestry Commission, Edinburgh.

Dieterich, V. 1928. Silver fir decline. *Silva* **16**: 81-86.

Freer-Smith, P.H. 1992. *Acid Deposition, its Nature and Impacts: Atmospheric Pollution and Forests*. Forest Ecology and Management, Volume 51 (Special Issue).

Freer-Smith, P.H. & Benham, S. 1994. Critical levels of air pollutants for direct effects on forests and woodlands. In Ashmore, M. (ed.) *Proceedings of the UNECE Workshop, Egham 1992*: 94-108, NMSO, DoE Publication, London.

Huttl, R. & Mueller-Dombois, D. 1992. *Forest Decline in the Atlantic and Pacific Region*. Springer-Verlag, Berlin.

Innes, J.L. & Boswell, R.C. 1989. *Monitoring of Forest Condition in the United Kingdom 1988*. Forestry Commission Bulletin 88, HMSO.

Innes, J.L. & Boswell, R.C. 1990. *Monitoring of Forest Condition in the United Kingdom 1989*. Forestry Commission Bulletin 94, HMSO.

Innes, J.L. & Boswell, R.C. 1991. *Monitoring of Forest Condition in the United Kingdom 1990*. Forestry Commission Bulletin 98, HMSO.

IPCC Working Group 1. 1992. *Scientific Assessment of Climate Change (Supplement)*. WMO and UNEP.

Landmann, G. 1991. *French Research into Forest Decline*. DEFORPA Programme 2nd Report. Pub. ENGREF, Nancy, France.

Last, F.T. 1982. Effects of atmospheric sulphur compounds on natural and man-made terrestrial and aquatic ecosystems. *Agric. Environ*. **7**: 229-387.

Manley, G. 1974. Central England temperatures: Monthly means 1659-1973. *Quart. J. Royal Meteor. Soc*. **100**: 389-405.

Miller, P.R. & McBride, J.R. 1975. Effects of air pollutants on forests. In Mudd, J.B. & Kozlowski, T.T. (eds) *Responses of Plants to Air Pollutants*: 195-235. Academic Press, New York.

Rehfuess, K.E. 1985. On the causes of decline of Norway spruce (*Picea abies* Karst) in Central Europe. *Soil Use Manag*. **1**: 30-32.

Roberts, T.M., Skeffington, R.A. & Blank, L.W. 1989. Causes of type 1 spruce decline in Europe. *Forestry* **62**: 179-222.

Roelofs, J.G.M., Kempers, A.J., Houdijk, A.L.F.M. & Jansen, J. 1985. The effect of air-borne ammonium sulphate on *Pinus nigra* var. *maritima* in The Netherlands. *Plant & Soil* **84**: 45-56.

Schulze, E.D. & Freer-Smith, P.H. 1992. An evaluation of forest decline based on field observations focussed on Norway spruce, *Picea abies*. In Last, F.T. & Watling, R. (eds) *Acidic Deposition: Its Nature and Impacts. Proc. Royal Soc. Edinb., Section B*. **97**: 155-168.

Schulze, E.D., Lange, O.L. & Oren, R. 1989. *Processes leading to Forest Decline. A Synthesis*. Ecological Studies 77. Springer-Verlag, Berlin.

Smith, W.H. 1981. *Air Pollution and Forests: Interactions between Air Contaminants and Forest Ecosystems*. Springer-Verlag, Berlin.

Ulrich, B. 1989. Effects of acidic precipitation on forest ecosystems in Europe. In Adriano, D.C. & Johnson, A.H. (eds) *Acidic Precipitation*. Vol. 2, *Biological and Ecological Effects*: 189-269. Springer-Verlag, Berlin.

Ulrich, B., Mayer, R. & Khanna, P.K. 1980. Chemical changes due to acid precipitation in a loess-derived soil in central Europe. *Soil Science* **30**: 193-199.

UNECE/CEC, 1992. *Forest Health and Air Pollution Report 1991*. United Nations Economic Commission for Europe and Commission of the European Communities.

Watcher, A. 1978. German literature on silver fir decline (1830-1978). *Z. Pflanzenkrank. Pflanzenchutz* **85**: 361-381.

Woodward, F.I. 1987. Stomatal numbers are sensitive to increases in CO_2 from pre-industrial levels. *Nature* **327**, 617.

Zottle, H.W. & Huttl, R.F. 1991. *Management of Nutrition in Forests under Stress*. Kluwer Academic Publishers, Dodrecht.

19 Physiological and Biochemical Tools in Diagnosis of Forest Decline and Air Pollution Injury to Plants

H. SAXE

INTRODUCTION

Some of the major objectives for investigating plant responses to gaseous air pollutants are: (a) to protect vegetation from damage; (b) to identify stress factors, spatial and temporal patterns, their mechanisms of injury; and (c) to select appropriate cultivars/provenances. The methods applied towards these objectives have involved the use of indicator plants, bioindication, early tests, early detection, early diagnosis, biological markers, risk index and diagnostic tools, all of which are more or less overlapping in their use of particular applications of visual, physiological, (bio)chemical, genetical and (ultra)structural studies. Most investigations include a dose-response strategy, or a comparison of plants along environmental gradients.

Diagnostic tests were reviewed by Darrall & Jäger (1984), and more recently by Garrec *et al.* (1990), Saxe (1991), and Wild & Schmitt (1995). The methods of plant bioindication of air pollutants and other environmental stresses have greatly expanded in the last few years, not the least in forest research, where sensitive cultivars or provenances cannot be substituted by resistant provenances at short notice. The potential value of biological markers in air pollution research is now widely accepted (Bolhàr-Nordenkampf, 1989; Grossblatt, 1989). This chapter, a sequel to the review by Saxe (1991), summarizes a wide range of bioindicative methods published since then.

Bioindication may be used as early warning for a potential injury, or as a diagnostic tool to determine the relative importance of different stress factors, their spatial and temporal distribution, or their mechanisms of injury to particular cultivars, appropriate countermeasures, or to select the most tolerant cultivars relative to the prevailing or predicted environment.

DEVELOPMENT OF INJURY

To understand bioindication, it is important to have an operative, multifarious concept for the characteristics and development of injury in an organism: plant, stand or ecosystem. Injury by an air pollutant or by any other type of stress

Plant Response to Air Pollution. Edited by Mohammad Yunus and Muhammad Iqbal

to plants may be caused by both *chronic* and *acute* assaults, and by natural and anthropogenic stress. The result of acute stress is relative to the resilience, resistance and protections of the organism, which are positively or negatively modified by the chronic load.

This is illustrated by Figure 19.1, where the stressed organism is represented by an elastic (plant resilience) cup of a certain size (plant resistance) with a definite

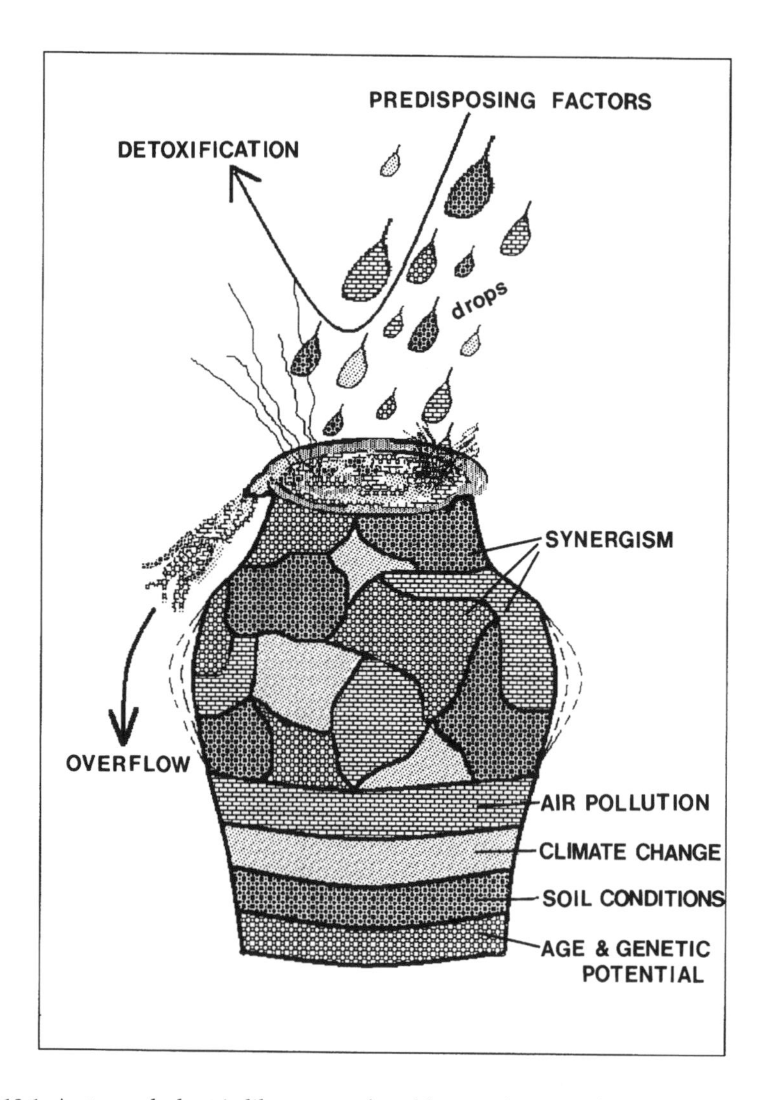

Figure 19.1. A stressed plant is like a cup of problems: when it is full, it only takes the last drop to make it flow over; but is this arbitrary drop so important? The drops symbolize predisposing, triggering and contributing stress factors which interact with positive or negative synergism inside the cup (the problems expand or dwindle), and by that the concoction may flow over (visible injury); evaporation from the cup symbolizes detoxification and metabolism of the stress factors or repair of injury, under normal circumstances preventing the cup from flowing over

opening, into which several, disparate, small drops (multiple *predisposing* factors, chronic attack) are constantly trickling, filling the "cup of problems" (chronic load). The components in the concoction interact (positive or negative synergism), making it expand or shrink. This weakened state of the cup invites new types of stress (*contributing* factors). The opening represents the potential for uptake, on a plant level controlled by the boundary layer, the stomata and cuticle (plant protections), and is moderated by the prevailing habitat, which also regulates the number, variety, size and elasticity of cups. The large drops symbolize acute attacks. When evaporation of the concoction does not keep up with the input of drops, the cup eventually flows over, and the injury becomes visible, and readily observed. The evaporation symbolizes detoxification and metabolism of the stress factors, or repair of injury.

An old English proverb says that *it is the last drop that makes the cup flow over*. This particular drop is the *triggering* or *inciting* factor in plant injury, but it may be infinitely tiny and its nature may be completely arbitrary. The drop could be something new, or any of the known ingredients. Any predisposing factor can also be a triggering drop, if the timing and its arrival is appropriate. The size and elasticity of the cup, the evaporation, and the outline of the opening, together with timing, size and velocity of the drop, decide whether a particular drop will make a particular cup overflow. The drop velocity symbolizes the shape of the stress pulse, for example a high air pollution concentration over a short period. It is very difficult to decide which of the many drops, falling in close succession into the cup, is the one to make the cup flow over.

Predisposing factors, such as genetic potential, age, climate change, soil fertility, compaction and moisture-holding capacity, and direct and indirect effects of air pollution, exert long-term and permanent stress, and predispose an organism for injury. Inciting factors of short duration produce a severe injury, especially to organisms with reduced capacity for regeneration caused by predisposing stress. Inciting factors include air pollutants, drought, frost, insects and pathogens. Contributing factors join the decline spiral (Manion, 1981; Manion & Lachance, 1992) and may finally lead to serious degeneration and death of the organism. Contributing factors in trees include bark beetles, root-rotting fungi, canker fungi, viruses, and mycoplasma- and Rickettsia-like bacteria (Neinhaus, 1985). Only when the cup runs over do we become aware that there is a problem. That, in my view, is the relation of predisposing, triggering and contributing factors in forest decline.

DIAGNOSIS BY VISIBLE EVIDENCE

In many cases plant injuries are induced by causes or complex combinations of causes, which are never identified (Innes, 1993). The categories of visible injury are relatively few, whereas their possible causes are multiple. The most common visible injuries include scorching of leaves (e.g., Hanisch & Kilz, 1990), reduction of growth, and changes in the root : shoot ratio. Visible injuries are typically *non-specific*, and may indicate damage caused by various air pollutants, pathogens, insects and abiotic stresses such as nutrient deficiency, drought, frost or shading.

Exempt from this are the so-called *indicator plants*, sensitive cultivars which respond to specific air pollutants by their occurrence and growth, or with specific visible injuries (Donagi & Goren, 1979; Nouchi & Aoki, 1979; Manning & Feder, 1980; Hirata & Kunishige, 1981; Kress *et al.*, 1982; Posthumus, 1982; Bennett & Stolte, 1985; Laurence *et al.*, 1985; Keller, 1988; Luethy-Krause *et al.*, 1989; Cronan *et al.*, 1989; Roloff, 1989a, b; Kromroy *et al.*, 1990; Weinstein *et al.*, 1990; Heggestad, 1991; Mignanego *et al.*, 1992; Heagle *et al.*, 1994). Visible scorching or loss of leaves, cultivars or species indicate the integrated load of pollutants, but the injury says little about what may happen to other cultivars or species under the same environmental conditions. The use of indicator plants is limited to air pollution doses high enough to cause visible injury. Most plants, however, even the sensitive indicator cultivars, are affected by doses lower than those which cause visible injury. The use of indicator plants, therefore, does not appear to be a sufficiently sensitive method to efficiently assist us in protecting vegetation.

Visible observations are often inaccurate, as illustrated by novel decline in European and East American conifer forests, both described by a discoloration and premature loss of needles (Skelly, 1992). But while novel decline in European forests typically involves a premature senescence of old needles often associated with nutrient deficiencies, the typical syndrome in the eastern USA typically involves injuries to the current-year needles often associated with extreme climatic conditions. European novel decline typically strikes conifers from-the-inside-out, and from-the-bottom-up, while East American novel decline often strikes conifers from-the-outside-in, and from-the-top-down. The European syndrome began in the mid-1970s, while the East American syndrome began in the mid-1960s. Clearly, these represent at least two different syndromes, and possibly many more, though originally listed under the common name of novel decline.

Visible plant injuries do not necessarily lead to a reduced productivity. A loss of old leaves may relieve the plant of a transpiratory and respiratory burden, giving new foliage better conditions.

DIAGNOSIS BY INVISIBLE EVIDENCE

Before visible injury becomes apparent, plants react to environmental stress with complex, endogenous response patterns. To protect vegetation, we must describe and predict plant injuries by precise, objective measures, and by parameters which are sensitive, react specifically, and which effectively lead to injuries. The primary targets when air pollutants injure plants are biomembranes and enzymes. Subsequently, the biochemistry, the photosynthesis, respiration and transpiration processes, and the structural and chemical components are affected.

Not all of the early, invisible responses triggered by the broad range of environmental factors are useful indicators for potential injury, but several physiological, (ultra)structural and (bio)chemical responses have been successfully used in *bioindication*, though when used separately, they were not always specific. One problem is the partitioning of anthropogenic and natural effects in the field, both inducing similar responses; another is that different species react differently. The current trend, therefore, is to include several responses into composite response models for bioindication or early tests for defined sites, stands or cultivars (Elstner *et al.*, 1985; Bender *et al.*, 1986; Bolhàr-Nordenkampf & Lechner, 1989; Jones & Coleman, 1989; Cape *et al.*, 1990; Saxe, 1990; Markan & Fischer, 1991; Riitters *et al.*, 1991; Tynnyrinen *et al.*, 1992). On an ecosystem level, the trend is to include several different organisms. An essential prerequisite for the use of diagnostic tools is that the diurnal and seasonal variations are well characterized, and that the organisms used are genetically well defined.

The bioindicative methods described below are classified by the mechanisms involved (Table 19.1).

Photosynthesis and stomatal conductance

The use of photosynthesis (PS) and stomatal conductance (SC) as early detectors of potential stress and air pollution injury to plants seem appropriate since: (a) the relationship between PS and SC is under biological control and is related to productivity; (b) PS and SC are known to change quickly to stress and air pollutants; (c) SC is a measure for uptake of gaseous air pollutants (modified by surface and transcuticular deposition, Cadle *et al.*, 1991), and thereby for potential injury; and (d) PS and SC can be measured with non-destructive techniques. Saxe (1991) reviewed a large number of chronic and acute injuries of air pollutants to PS and SC.

There are a number of reasons, however, why PS and SC measurements are difficult to use as biomarkers for air pollution in the field, particularly in forest decline. Most important, changes in PS and SC are non-specific for exposures to pollutants (Winner, 1989), secondly they are rather time consuming to measure on large trees in the field, and finally they are greatly influenced by age (Wallin *et al.*, 1990) and a broad range of environmental (Chappelka *et al.*, 1990; Dobson *et al.*, 1990; Tjoelker and Luxmoore, 1991) and endogenous (Eamus and Murray, 1991) influences.

By comparing PS and SC in paired sets of field-grown *Picea abies*, pairs of 20–25-year-old healthy and damaged trees in a damaged stand, and pairs of 18-year-old healthy and damaged trees in a healthy stand, Benner *et al.* (1988) demonstrated that the damaged trees had a significantly lower water use efficiency (WUE) and apparent photorespiration, increased light and CO_2-compensation points and dark respiration. A study over three seasons showed that the gas exchange rates of the trees were markedly influenced by the climatic conditions in their natural habitat. The prerequisite for the use of PS and SC as injury-specific bioindicators was, therefore, the inclusion of healthy trees as "internal references".

A damaged *Picea abies* tree showed a greater reduction of water potential during the course of the day (Maier-Maercker & Koch, 1991). Recovery in the evenings was slower and often incomplete. With increasing degree of damage to a tree, the maximum and minimum values of the osmotic potential in needles deviated more strongly above and below the mean. The disturbance of hydroregulation could be due to a lack of stomatal control, caused by long-term air pollution. The reduced tolerance of drought in damaged trees may be used as a relative bioindication of injury.

Table 19.1. Methods used for bioindication of air pollutants and environmental stresses and the resulting plant injuries*

Reference	Species and cultivar	Inducer used in the test	Response parameter	Use and value of bioindication
Photosynthesis and stomatal conductance				
Benner *et al.* (1988)	*Picea abies* L.	Field conditions	PS, TR (with "internal reference")	Indicates forest decline
Lange *et al.* (1986)	*Picea abies* L.	Field conditions	Photosynthetic capacity in cut twigs	Indicates forest decline
Larsen & Wellendorf (1990)	*Picea abies* L. (16 full-sub families)	None	WUE, R : PS	Early test to find trees with a high growth potential
Larsen *et al.* (1988)	*Abies alba* Mill. (12 provenances)	SO_2	PS	Selects adaptable cultivars
Larsen *et al.* (1990)	*Abies alba* Mill. (6 provenances)	O_3	PS, transpiration	Selects adaptable cultivars
Maier-Maercker & Koch (1991)	*Picea abies* L.	Field conditions	Water and osmotic potential, water content in needles	Indicates forest decline
Saxe (1990)	*Picea abies* L. (11 half-sub families)	SO_2, NO_2, $SO_2 + NO_2$, O_3	PS, SC	Selects forest-decline-resistant cultivars ($P < 0.005$–0.05)
Leaf pigments				
Heagle *et al.* (1994)	*Trifolium repens* L., sensitive + resistant clone	O_3	Relative chlorophyll content	Indicates O_3-sensitive crop loss
Köstner *et al.* (1990)	*Picea abies* L.	Field conditions	Chlorophyll content, and formation in old needles	Indicates novel decline (Fichtelgebirge)
Moroni *et al.* (1991)	*Triticum aestivum* L.	Toxic levels of Mn	Chlorophyll content	Selection for Mn tolerance
Tschaplinski & Norby (1991)	*Platanus occidentalis* L.	Urea N fertilization	Pigment content, PS	Indicates N fertilization (and N deposition?)

Chlorophyll fluorescence				
Baillon *et al.* (1988)	*Picea abies* L.	Mg and K deficiency	Fluorescence index, photochemical and non-photochemical quenching	Early detection of mineral deficiency
Clement (1996)	*Triticum aestivum* L.	Controlled freezing	Chlorophyll fluorescence	Evaluates frost hardiness
Lindgren & Hällgren (1993)	*Pinus contorta* *P. sylvestris*	Freezing	Chlorophyll fluorescence	Indicates freezing injury, cold acclimation
Ruth & Weisel (1993)	Spruce	Acute levels of O_3 vs. novel decline	Fluorescence induction kinetics	Indicates a limited role for acute O_3 levels in novel decline
Element content				
Bondietti *et al.* (1989)	*Picea rubens, Tsuga canadensis*	Field conditions (East American sites)	Al : Ca and Al : Mg in annual rings, and radial growth	Identifies acid-deposition-related forest decline
Cape *et al.* (1990)	*Picea abies* L.	Field conditions (across Europe)	Time-dependent nutrient content and ratios ("risk index")	Indicates existing and potential forest decline across Europe
Gärtner *et al.* (1990)	Spruce	Field conditions (in Hessen)	Mn content in 2nd-year needles of 7th whirl	Indicates needle loss
Gebauer & Schulze (1991)	*Picea abies* L.	Field conditions on healthy and declining sites	$\delta^{15}N$	Indicates existing forest decline ($P < 0.0001$)
Glooschenko (1989)	*Sphagnum fuscum*	Field conditions	Cd content	Maps Cd deposition across Canada
Granier & Chevreuil (1992)	*Platanus vulgaris*	Field conditions	PCBs and lindane contents	Maps deposition of organochlorines
Landolt *et al.* (1989a)	*Picea abies* L.	Field conditions	S, Cl, F contents, "heavy metals"	Maps industrial pollution

continued overleaf

Table **19.1.** *(continued)*

Reference	Species and cultivar	Inducer used in the test	Response parameter	Use and value of bioindication
Long & Davis (1989)	*Quercus alba* L.	Field conditions (with increasing distance from a power-plant)	Sr content and other "ICP-minerals"	Indicates historical fly-ash deposition
Matyssek *et al.* (1992)	*Betula pendula*	O_3	$\delta^{13}C$	Indicates O_3 injury
Mössnang (1990/91)	*Picea abies* L.	Field conditions (altitudinal gradient)	N and S contents and other "ICP-minerals"	Excludes SO_2 and NO_x as responsible for novel decline
Oren *et al.* (1993)	*Picea abies*	Acid/alkaline soil processes	Chlorophyll and nutrient analysis	Indicates spruce decline
Polamäki *et al.* (1992)	*Hypogymnia physodes* (on *P. abies* branches)	Strip-mine area	F and S contents	Maps depositions from a fertilizer plant and a strip mine in Finland
Riitters *et al.* (1991)	*Abies balsamea, Acer saccharum, Pinus banksiana, P. resinosa, Populus tremuloides*	Field conditions	12 nutrient elements and their ratios ("DRIS")	Indicates existing and potential forest decline (crowns)
Ruhling (1992)	Moss species	Field conditions in Europe	Heavy metal content	Integrated heavy metal air pollution in Europe
Saxe (1993)	*Picea abies*	Field conditions	Mn/nutrient analysis	Indicates "Red" syndrome spruce decline
Steinnes *et al.* (1992)	*Hylocomium splendens*	Field conditions, 500 sites	Contents of 26 elements	Identifies long-range transport of pollutants in Norway
Weikert *et al.* (1989)	*Picea abies*	Terminal bud removal	Mineral analysis, chlorophyll content	Diagnoses the immediate cause of forest decline

Weinstein *et al.* (1990)	Native crops and trees	Field conditions	F content	Maps F deposition in the Rhône valley in Switzerland
Metabolite content				
Balsberg-Påhlsson (1989)	*Picea abies, Pinus sylvestris, Betula pendula, B. pubescens*	Industrial pollution	Starch, sugars	Assesses tree injury
Ericsson *et al.* (1993)	*Picea abies*	Airborne N deposition	Arginine concentration	Identifies mineral nutrient imbalances
Forschner *et al.* (1989)	*Picea abies*	Field conditions	Starch and phloem cell injury	Assesses forest decline
Friend *et al.* (1992)	*Pinus taeda* L.	O_3	Starch in current-year foliage	Assesses pollutant impacts in forest ecosystems
Gezelius & Näsholm (1993)	*Pinus sylvestris*	N content in hydroculture	Arginine and glutamine concentration	Identifies nutrient availability
Näsholm *et al.* (1994)	*Vaccinium myrtillus, V. vitis-idaea, Deschampsia flexuosa, Epilobium angustifolium, Pleurozium schreberi, Polytrichum commune*	Nitrogen availability	Arginine, glutamine, asparagine concentration	Identifies nitrogen-enriched ecosystems
Jones & Coleman (1989)	*Populus deltoides*	Model tested with O_3	Mobile and polymerized CSM, mobile and total N	Early warning of stress injury, and diagnosis of stress factors
Jensen & Løkke (1990)	*Picea abies* L.	O_3, drought, acid rain and soil, atrazine (field, open-top-chamber, climate chamber)	4-hydroxyacetophenone spruce decline ($P < 0.0004$)	Early warning of O_3 injury
Landolt *et al.* (1989b)	*Pinus sylvestris*	O_3	Inositol content in needles	Early warning of O_3 injury

continued overleaf

Table 19.1. *(continued)*

Reference	Species and cultivar	Inducer used in the test	Response parameter	Use and value of bioindication
Pietilä *et al*. (1991)	*Pinus sylvestris* L.	Field conditions	Arginine, glutamine, total soluble proteins	Indicates NH_3 deposition
Ranieri *et al*. (1990)	*Zea mays*	SO_2	Free amino acids	Indicates plant injury
Sandermann *et al*. (1989, 1992)	Spruce, tobacco	O_3	Amines in apoplast	Diagnosis of O_3 injury
Santerre *et al*. (1990a)	*Picea abies*	Acid rain conditions	Polyamines	Assesses forest decline
Weidmann *et al*. (1990)	*Picea abies*	O_3, SO_2 stress, ambient conditions	ATP/ADP ratio	Identifies spruce decline
Zaerr & Schill (1991)	*Picea abies* L.	Acid rain + car exhaust	Unidentified constituents (HPLC)	Early diagnosis of damage
Enzyme activity				
Hernández *et al*. (1994)	*Vigna unguiculata*	Salt stress	SOD isozymes activity	Identifies salt stress *in vitro*
Karjalainen *et al*. (1992)	*Picea abies*	SO_2, NO_2, $SO_2 + NO_2$	PO and Rubisco activity, soluble protein	Diagnosis of SO_2, NO_2, or $SO_2 + NO_2$ stress (in the laboratory)
Luethy-Krause *et al*. (1990)	*Picea abies*, *Pinus sylvestris*	O_3	Decreased SA, QA; increased pH, MA, CA; decreased MDH activity; increased cyt. C.ox., PEPC and SHDH activity	Sensitive, non-specific, O_3 indicator
Markkola *et al*. (1990)	*Pinus sylvestris* L.	Field conditions	PO activity in fine roots and mycorrhiza	Non-specific diagnosis of air pollution stress
Norby (1989)	*Picea rubens*	Laboratory: NO_2, HNO_3 vapour	Nitrate reductase activity	Early warning of potential injury by "N-overloading"
Richardson *et al*. (1989, 1990)	Loblolly pine (3 half-sub families)	O_3, acid rain	SOD, PO and GSH activity, MDA, PS, α-toc, β-carotenoid, CAT, AscA	Early warning of oxidative stress

Sandermann *et al.* (1989, 1992)	Tobacco, pine	O_3	β-1,3-glucanase Stilbene synthase (and stilbenes)	Diagnosis of O_3 injury
	Spruce, pine	O_3	CAD	Diagnosis of O_3 injury
Schmiedel & Dässler (1989)	*Picea abies*	SO_2, natural senescence	Non-specific acid phosphatase and esterase activity, CAT and GOT: multiple forms and rhythmic activity	Early diagnosis of air-pollution-induced forest decline
Schröder & Debus (1990)	*Picea abies* L.	Halone 1211	GST activity	Biomarker of halone injury
Morphological analyses				
Cornelius (1994)	Forest trees	Field conditions	Yield, height, stem diameter, volume	Select more productive trees
Stribley (1993)	*Fagus sylvatica*	Field conditions	Crown structure	Classification and detection of forest health
Ultrastructure and histopathology				
Fink (1992, 1993)	*Picea abies*	O_3, acid rain, Mg and K deficiency	Histopathological changes, starch grains, calcium oxalate	Diagnosis of cause of decline
Holopainen *et al.* (1992)	*Pinus sylvestris*	Air pollution and mineral deficiency	Leaf ultrastructure: chloroplasts and other organelles	Indicates specific injuries by SO_2, HF, O_3, and N, P, K, and Mg deficiency (diagnose cause)
Huettl *et al.* (1990), Zoettl *et al.* (1989)	*Picea abies*, *Pinus ponderosa*	Acid mist, O_3, field conditions	Histological analysis and diagnostic fertilization	Diagnosis of cause of decline and revitalization of trees
Schmitt *et al.* (1986)	*Picea abies*	Field conditions	Transfusion parenchyma and endodermis plastids, sieve elements	Diagnosis of cause of decline
Sutinen & Koivisto (1992)	Spruce, pine	Field conditions, SO_2, O_3 acid rain	Leaf ultrastructure: chloroplasts and other organelles	Indicates environmental stress, including air pollution

continued overleaf

Table 19.1. *(continued)*

Reference	Species and cultivar	Inducer used in the test	Response parameter	Use and value of bioindication
Wulff (1992)	Spruce	O_3, acid rain	Ca in outer epidermal cell wall	Diagnose acid mist injury
Zobel & Nighswander (1991a)	*Pinus nigra*, *P. resinosa*	Acid rain: pH 2.5, 3.5, 4.5	Phenolic compounds in mesophyll	Early detection of acid stress
Zobel & Nighswander (1991b)	*Pinus nigra*, *P. resinosa*	Saturated NaCl solution	Phenolic compounds in mesophyll	Early detection of salt stress
Genetical analysis				
Geburek *et al.* (1986)	*Picea abies* L.	Al in hydroponic solutions	Isozyme gene marker: GOT	Selects Al-tolerant trees
Guangrong *et al.* (1990)	*Vicia faba*	Industrial air: xylene, HF	LTMT	Genotoxic test
Müller *et al.* (1994)	*Picea abies*	O_3, CO_2, soil type, provenance	Root meristem chromosome abnormalities	Identifies pollution stress
Schmidtling (1994)	*Picea abies*, *Pinus taeda*	Provenance/ temperature/ transplanting	Growth	Identifies effects of temperature change
Scholz & Bergman (1992)	Spruce, beech, pine	SO_2, O_3, HF and field conditions	2–3 combined gene loci (GOT, GRD, etc.)	Diagnose pollution stress
Sieffert (1988)	*Picea abies* L. 3 clones	Clonal differences	2-dimensional electrophoresis	Gene expression suggested for use as early test for sensitivity
Miscellaneous methods				
Andersson (1992)	*Pinus sylvestris*	Short-term freezing	Tissue discoloration	Forecast field mortality
Davidson (1992)	*Picea abies*	Field conditions	2-D electrophoresis	Assessment of individual tree health

Egnell & Örlander (1993)	*Picea abies*, *Pinus sylvestris*	Infrared thermography	Growth and visible damage	Assessment of seedling viability
Fry (1989)	Old trees	Field conditions	$\delta^{13}C$ in tree rings annual increments	Indicates human perturbation
Gamon *et al*. (1990)	*Helianthus annuus*	Field conditions	Xanthophyll, chlorophyll fluorescence	Evaluates photosynthetic performance
Härtel & Fuchshofer (1987)	Conifers	SO_2, HF, O_3	Turbidity in needle extract	Early detection of air pollution injury to conifers
Larsen & Wellendorf (1990)	*Picea abies* L. (16 full-sib families)	None	Frost resistance	Early test to find trees with a high growth potential
MacDonald *et al*. (1989)	*Populus deltoides*, *Prunus serotina*, *Robinia pseudoacacia*, *Celtis occidentalis*, *Acer negundo*, *Liquidambar styraciflua*	Field conditions (NO_2)	Aerobic ethanol production	Diagnostic indicator of acidic pollution stress in trees
Rock *et al*. (1988)	Forest trees	Field conditions	Spectral reflectance ("blue shift")	Indicates/maps existing forest decline
Salemaa & Jukola-Sulonen (1990)	*Picea abies* L.	Polluted air near airport	Electrical impedance in living inner bark tissue	Indicates defoliation ($P < 0.001$)
Schmuck (1990)	*Picea abies*	Field conditions	Chlorophyll fluorescence	Assesses forest decline
Skrøppa (1991)	*Picea abies* L.	Laboratory freezing test	Frost sensitivity in laboratory test	Predicts height growth potential in the field
Sulzer *et al*. (1993)	*Picea mariana*	Height, diameter, freezing test, photosynthesis, transpiration	Field growth over 3 and 10 years	Early selection of tall-growing spruce
Williams & Ashenden (1992)	*Trifolium repens* L. cv. Grasslands Huia	O_3	Spectral reflectance (multiple wavelengths)	Early detection of O_3 injury (dry weight loss)

continued overleaf

Table 19.1. *(continued)*

Reference	Species and cultivar	Inducer used in the test	Response parameter	Use and value of bioindication
Multivariate methods				
Bender *et al.* (1986)	*Picea abies* L., *Abies alba* Mill., *Fagus sylvatica* L.	SO_2 and/or O_3, + acid rain	Ca, Mg, K, Mn content, inorganic and organic S, GSH, buffering capacity, chlorophyll content	Early diagnosis of forest decline
Bolhàr-Nordenkampf (1989)	*Picea abies*	Field conditions (with increasing altitude)	K, P, Mg, Ca, S, Pb, Cd contents, PS, chlorophyll fluorescence, entomological and mycorrhiza investigations	Early detection of O_3-induced forest decline
Elstner *et al.* (1985)	*Picea abies* L.	Field conditions on 18 different locations	Chlorophyll content, ethane, malonyl-ACC, ascorbate, xanthoxin, K, Na, Ca, Mg, Mn content, SOD, CAT, wax layer	Bioindication of photooxidants ACC
Goldstein & Ferson (1994)	*Acer saccharum*, *Picea rubens*, *Pinus ponderosa*, *P. taeda*, *Populus tremuloides*, *Raphanus sativus*	Open-top chamber exposures	Plant biochemistry and physiology, biogeochemistry	Predicts plant response to interacting stresses
Stöcker & Gluch (1990)	Pine, spruce	Acid deposition	Bark: EC, S and Ca contents, pH	Bioindication of acid deposition
Tynnyrinen *et al.* (1992)	*Picea abies* and *Hypogymnia physodes*	Fertilizer plant and strip mining	Ultrastructure, element contents	Maps affected area

*α-toc = α-tocopherol; ACC = 1-aminocyclopropane-a-carboxylic acid; AscA = ascorbic acid; β-car = β-carotene; BC = buffering capacity; CA = citric acid; CAD = cinnamyl alcohol dehydrogenase; CAT = catalase; chl. = chlorophyll; CSM = carbon-based secondary metabolites; Cyt.C-ox. = cytochrome C oxidase; DRIS = Diagnosis and Recommendation Integrated System; EC = electric conductivity; GRD = glutathion reductase; GDH = glutamate dehydrogenase; GOT = glutamase oxaloacetate transaminase; GSH = glutathione; GST = glutathione-S-transferase; HPLC = high performance liquid chromatography; ICP = inductively coupled plasma atomic emission spectroscopy; LTMT = leaf tip cell micronucleus test; MA = malic acid; MDA = malondialdehyde; MDH = malate dehydrogenase; N = Nitrogen; NR = nitrate reductase; PEPC = Phosphoenol pyrovate carboxylase; PO = peroxydase; PS = photosynthesis; R = respiration; SA = shikimic acid; SC = stomatal conductance; SHDH = shikimic acid dehydrogenase; SOD = superoxide dismutase; TR = transpiration; WUE = water use efficiency; QA = quinic acid.

Lange *et al.* (1986) introduced a rapid method for field determination of photosynthetic capacity of cut spruce twigs. As this method may be used to quickly scan many samples per day, it could be useful in diagnosing decline, since it overcomes the problem of having to sample many trees, and several layers of the crown.

The investigator may him or herself introduce the necessary "specificity", by investigating correlations between observed, quantified injuries to mother trees, and responses of their half-sib or full-sib offspring. Significant correlations may be used to empirically select resistant clones or cultivars, or to identify the relative importance of likely stress factors in the observed injury.

Abies alba (silver fir) provenance trials in Italy and Denmark demonstrated that seed sources from southern Italy (Calabria) and southeastern Europe were characterized by a high vitality and growth vigour, whereas all provenances from central Europe in these comparative field trials showed pronounced signs of "silver fir dieback". Larsen (1986) proposed that the decline of silver fir in Europe is basically caused by insufficient genetical variation in central European provenances, which in turn causes lack of adaptability. Provenances may have lost adaptability (diminished genetic pool) if they are the descendants of only a few parents, when the central European forest populations were segregated and diminished during the Glacial Age. The PS response to one month of high SO_2 exposure was found to be significant in most tested *A. alba* clones, but it recovered sooner in Calabrian and southeast European provenances, than in central and northeast European provenances (Larsen *et al.*, 1988). The SO_2-stressed photosynthesis response was thus demonstrated to be a diagnostic tool in selecting potentially resistant (adaptable) provenances. Likewise, Larsen *et al.* (1990) found both PS and transpiration in central European provenances of *A. alba* to be significantly inhibited by acute levels of O_3 (10 d, 250 ppb), while south Italian provenances were not significantly affected. The O_3-stressed PS and SC response was demonstrated to be a diagnostic tool in selecting potentially resistant (adaptable) provenances.

Larsen & Wellendorf (1990) found a significant correlation in 16 *Picea abies* families between growth (height, diameter, wood density) and transpiration in 13-year-old field-grown trees, and the WUE of their 3-year-old full-sib seedlings. They suggested the use of WUE as a relative index for selecting *P. abies* in an early test for growth and efficient water conservation in the field. The respiration : photosynthesis ratio (R : PS ratio), but not the photosynthesis itself, was also found useful for early tests. Saxe (1990) found significant correlations between PS and transpiration responses in 11 provenances of 3–4-year-old *P. abies* half-sib seedlings exposed to acute levels of SO_2, NO_2, $SO_2 + NO_2$ and O_3, and the health in cloned mother trees. The stress–physiological responses were combined in a model that also included morphological qualities, which could be used to select half-sib families of east European (sensitive) spruce, for resistance to novel decline symptoms (premature senescence of older leaf classes). The strength of this model was that SO_2 and NO_2 induced the opposite response to that of O_3, giving their combined use a high degree of efficiency. The model did not assume air pollutants to be of special importance in novel decline.

On the contrary, the method permitted the selection of resistant seedlings with no knowledge of predisposing, triggering and inciting factors in the observed decline. The method is non-destructive.

Leaf pigments

Measuring the loss of chlorophyll from plants is an objective measure of most visible leaf injuries, and directly corresponds to damage. Unfortunately, like the visible injuries, chlorosis is not specific for air pollutant stress, and varies greatly with the time of year.

The photosynthetic pigment concentrations were successfully applied, however, together with the net photosynthetic rate, to indicate a nitrogen fertilization response in *Platanus occidentalis* (American sycamore, Tschaplinski & Norby, 1991). This bioindication could be used to optimize the timing of fertilization. The trees responded to applications as low as the atmospheric nitrogen deposition in polluted areas.

Chlorophyll formation ceased in the older needle classes of chlorotic trees when the new flush was sprouting, thus indicating that nutritional deficiencies affect needle yellowing more than possible direct needle damage by air pollution (Köstner *et al.*, 1990). Forest decline of this type could, therefore, be indicated by the amount and timing of production of chlorophyll in old needles. Chlorophyll content in Mn-stressed seedlings could be an effective diagnostic tool in selecting Mn-tolerant cultivars of *Triticum aestivum* (Moroni *et al.*, 1991). Chlorophyll content in an ozone-sensitive clone, compared to an ozone-resistant clone of *Trifolium repens* indicated specific growth reductions caused by atmospheric ozone during the summer (Heagle *et al.*, 1994).

Chlorophyll fluorescence

The function of the pigment system, such as chlorophyll fluorescence, is a much more sensitive bioindicator than pigment contents (Heath, 1989), and potentially more specific than the response of the photosynthetic rate. But fluorescence responses are difficult to interpret, particularly in the field (Massacci & Jones, 1990), where they are affected by several environmental factors (Renger & Schreiber, 1986). It only takes low concentrations of SO_2, however, to affect photochemical and non-photochemical quenching (Hove *et al.*, 1991).

Magnesium (Mg) deficiency affects chlorophyll fluorescence in young *Picea abies* needles (decrease in the fluorescence index and photochemical quenching, and an increase in non-photochemical quenching) with little effect on water content and the photosynthetic rate (Baillon *et al.*, 1988). While Mg deficiency affected older needles first, potassium (K) deficiency affected younger and older needles simultaneously, and while the response to Mg deficiency was not followed by a recovery, the response to K deficiency was followed by a high recovery rate, all signifying the potential specificity of chlorophyll fluorescence in bioindication. The investigations, however, were only carried out under laboratory conditions.

Damaged spruce trees at two field sites could easily be distinguished from healthy trees by several chlorophyll fluorescence characteristics (Lichtenthaler &

Rinderle, 1988). By comparing different parameters of chlorophyll fluorescence induction kinetics in spruce, Ruth & Weisel (1993) concluded that acute ozone fumigation (600–1200 $\mu g\ m^{-3}$, 3–21 d) could not simulate all effects on the photosynthetic system detected in spruce of different damage classes of novel decline. It would be more relevant in future studies to compare the same responses in trees exposed to long-term, low levels of O_3 with declining trees, in order to determine a more realistic role of O_3 in forest decline. This type of comparative bioindication may also be used to determine the qualitative role of other stress parameters in forest decline.

Chlorophyll fluorescence was indicated as a potential technique for rapid detection of freezing injury in *Pinus contorta* and *P. sylvestris*, and for ranking needle material with respect to development of cold acclimation (Lindgren & Hällgren, 1993). It was used as a reliable parameter to determine the frost hardiness of wheat leaves (Clement, 1996).

Element content

Elements from those air pollutants which accumulate in plants (S from SO_2, F from HF, Na, Cl and Mg from sea salts, heavy metals from anthropogenic emissions, and stable organic compounds from pesticides) may be used for specific bioindication for possible effects caused by these pollutants, and for mapping their spatial and temporal distribution. Measuring accumulated components in plants rather than directly in the environment, provides the dual advantages of an integrated rather than a momentary value, and of a biological content, rather than a mere concentration (Stöcker & Gluch, 1990).

Landolt *et al.* (1989a) used needle analysis in *Picea abies,* to describe large-scale air pollution, primarily with S, F and heavy metals, in Switzerland. Weinstein *et al.* (1990) mapped fluorine pollution in the Rhône valley in Switzerland using selected crops and trees. Granier & Chevreuil (1992) mapped the deposition of PCBs (polychlorinated biphenyls) and lindane which were found to accumulate in plane trees with a factor of 10^4 and 10^5, respectively. Analysing element contents in spruce needles, Mössnang (1990/91) demonstrated that SO_2 was not responsible for increasing novel forest decline along an altitudinal gradient (800–1720 m) in Bavaria, since the low S concentration in the needles excluded a burden by air pollution caused by SO_2. Long & Davis (1989) quantified the contents of minerals emitted from a power-plant in *Quercus alba* xylem tissues by ICP (inductively coupled plasma atomic emission spectroscopy) analysis, and concluded that only xylem-Sr among the 18 analysed minerals may be used as a sensitive bioindicator of historical fly-ash deposition.

Lichens (Glooschenko, 1989; Palomäki *et al.*, 1992) and bryophytes (Ruhling, 1992; Steinnes *et al.*, 1992) have an advantage in element deposition surveys, since they have no proper root system for selective uptake of available ions, but are more dependent upon direct deposition from the air. A number of studies suggested that biochemical indication with cryptogamic plants may have considerable future potential (Brown, 1992). Other authors stress the advantage of complex interactions in bioindication, since complexity leaves room for higher sensitive and specificity. Fellner (1989), for example, encouraged myco-bioindication of air pollutants using mycorrhiza-forming fungi. Though this method is presumed to be very sensitive, its specificity is not yet known.

Analysis of accumulations does not provide information on the occurrence of oxides of nitrogen and ozone, since nitrogen (unlike sulphur) is easily translocated in plants, and its content, therefore, is an unreliable indicator for NO_x, NH_3 and HNO_3 pollution, and ozone does not accumulate. Even the nutritional disturbances caused by ozone are reported with conflicting results. Tingey *et al.* (1986) found O_3-induced changes in mineral content in bean tissues qualitatively different from reported changes in other species and cultivars.

Manganese content in second-year spruce needles in the seventh whirl was highly correlated with needle loss at sites with mature trees in Hessen (Gärtner *et al.*, 1990). Elevated Mn contents were also found by Saxe in declining *Picea abies* (Saxe, 1993) and *P. omorica* (not published) relative to neighbouring healthy trees of the same species and provenances in Danish forests. The reason for this decline is still unknown. Increased content of Mn is possibly an indicator of novel forest decline. Further studies are needed.

Nutrient ratios seem to have a much higher capability than nutrient contents for bioindication of non-accumulating pollutants, but analysis becomes more complex, and frequent sampling more important. Luxmoore (1989) concluded that the nutrient use efficiency (NUE), the ratio of plant dry weight gain (the net carbon fixed) to total nutrient uptake, is a feasible bioindicator, but only with frequent tissue or solution sampling. The diurnal and annual rhythms, and the influence of all the conditions of the ambient environment, make it a rather unpractical and unspecific indicator of stress under field conditions, particularly for trees, since whole plants should be included in the test.

Bondietti *et al.* (1989) found strong circumstantial evidence that the increases in Al : Ca and Al : Mg ratios in undisturbed *Picea rubens* (red spruce) and *Tsuga canadensis* (eastern hemlock) annual rings in the Great Smoky Mountains National Park have occurred in recent decades caused by acid deposition (SO_4^{-2}, NO_3^-, H^+) and related aluminium mobilization and base-cation leaching. Since an increase in aluminium availability and changes in alkaline-earth behaviour do not necessarily lead to injury, the authors only recommend the use of the Al : Ca and Al : Mg ratio in annual rings as a promising method for identifying acid air-pollution-induced forest decline, with simultaneous measurements of radial growth. This method may be used in the field, except in areas which are intensively managed.

Cape *et al.* (1990) defined a "risk index" for trees showing no visible decline symptoms based upon nutrient content and nutrient ratios, suggested useful in identifying sites liable to experience deterioration in crown conditions (early detection). The index is empirical, and not necessarily related to potential effects of air pollution, since crown condition depends on the species and cultivar relative to all the surrounding conditions. With the exception of one German site, where only a few "poor" trees were observed, the defined index increased from the northern (Scotland–England) to the central European (the Netherlands, Germany) sites. These authors further suggested that the time-dependent content of foliar nutrients could be useful in diagnosis. At sites with trees having "poor" crown conditions, even apparently healthy trees showed a lack in increase in calcium content with needle age, and decreases in nitrogen content, and very large decreases in magnesium content with needle age, all demonstrating the importance of sampling several year classes of foliage from coniferous trees when determining the nutrient status of a tree.

Oren *et al.* (1993) suggest the use of foliar chlorophyll and nutrient analysis of *Picea abies* (Norway spruce) needles representing a range in severity of chlorosis as a quick method of identifying the soil processes which, in some areas, cause forest decline.

Another similar index called "DRIS" (Diagnosis and Recommendation Integrated System) was based on analysis of 12 elements and their ratios in increment cores of five forest tree species in Minnesota, Wisconsin and Michigan, USA (Riitters *et al.*, 1991). The multivariate index was used for bioindication of environmental stresses such as air pollution to forests. The indicative value of multivariate indices is far above that of individual parameters. Further multivariate methods are referred to in a later section of this chapter, where several types of response parameters are combined.

Gebauer & Schulze (1991) found *Picea abies* needles from trees on healthy sites to be more depleted in ^{15}N than needles from declining sites particularly with the older needle classes, while neither the nitrogen concentration nor the $\delta^{13}C$ values were affected. While the carbon metabolism of the healthy and the declining stands were similar, the nitrogen assimilation was affected at the declining site. The highly significant difference in $\delta^{15}N$ between healthy and declining sites suggested that it may be used for early detection, and the results are significant, since they indicate nitrogen assimilation problems to be part of the early events in forest decline. Other authors reported $\delta^{13}C$ in plants to be a sensitive bioindicator of low levels of O_3 (Matyssek *et al.*, 1992). But a general effect by global environmental change on $\delta^{13}C$ casts doubt on the specificity of this method relative to ozone (Beerling, 1994). The specificity of stable isotope techniques in bioindication is not yet clear, but their application on a large scale would be rather costly.

The elegant technique of terminal bud removal in *Picea abies* was used to diagnose the immediate cause of novel decline in a polluted area (Fichtelgebirge, $SO_2 + NO_x$). The authors concluded that it was not the direct impact of pollutants that caused the decline, but an acute nutrient deficiency, particularly Mg. The removal of terminal buds relieved particularly the youngest needle class of nutrient loss through export of mobile ions, which in damaged trees was the cause of injury to these needles (Weikert *et al.*, 1989).

Metabolite content

Tissue damage initiates processes of repair and defence, which may result in mobilization of carbon and nitrogen to the site of damage for repair, and polymerization of carbon-based secondary metabolites (CSM) at the site of damage. Jones & Coleman (1989) suggested a model with different perturbations as dependent variables, and mobile and polymerized CSM, total and mobile nitrogen (N), and time, as independent variables for prediction of injury. They used laboratory data from an experiment where two clones of *Populus deltoides* (cottonwood) were exposed to an acute dose of O_3 (200 ppb, 5 h), which had no significant effect on the growth of the plant, and did not cause visible injury to the leaves. Leaves from exposed plants and from controls in charcoal-filtered air were analysed for one class of mobile CSM, phenol glycosides, polymerized phenolics, total N and polar (mobile) N. The data confirmed the model, but its use remains to be proven under field conditions.

The biosynthesis and metabolism of polysaccharides may be disturbed by air pollution, resulting in higher levels of glucose and fructose in declining than in healthy *Picea abies* trees (Villanueva *et al.*, 1987). Ozone-decreased starch content of current-year foliage in *Pinus taeda* (loblolly pine) was suggested as a useful tool in assessing impacts of pollutants in forest ecosystems (Friend *et al.*, 1992). However, a 3-year study of declining Norway spruce trees indicated that starch mobilization was delayed by a reduction in assimilate transport capacity of the needle phloem, causing a build-up of starch in declining compared with healthy trees (Forschner *et al.*, 1989). Similar results were found for *Picea abies, Pinus sylvestris, Betula pendula* and *B. pubescens* where starch built up in an industrially polluted area compared to a cleaner area (Balsberg-Påhlsson, 1989). Starch increased, while glucose and fructose were lower in the polluted area. Landolt *et al.* (1989b) found inositol to be a very sensitive bioindicator of ozone effects in *Pinus sylvestris* (Scots pine). Other sugars (sucrose, glucose, fructose) were not significantly affected, demonstrating that the O_3 effects were not serious, and that the inositol response, therefore, was a more sensitive marker than other sugars, and could be used in early detection. However, since sugar levels change considerably with the time of year and with light, the use of inositol and other sugars as bioindicators is time and habitat dependent, as well as non-specific.

Pietilä *et al.* (1991) measured total soluble proteins and amino acid concentrations in *P. sylvestris* growing in the vicinity of a fur farm in Finland, and found that arginine increased 10^2–10^3-fold compared with control trees. Total soluble proteins and other amino acids (including glutamine) increased to a lesser extent. Ericsson *et al.* (1993) found a high arginine level to indicate a low P : N ratio in needles of *Picea abies*. Similarly, Gezelius & Näsholm (1993) found that *Pinus sylvestris* seedlings grown at low levels of nitrogen contain high levels of arginine, whereas those grown at high levels of nitrogen contain high levels of glutamine. In three species of boreal forest floor plants (*Vaccinium myrtillus, V. vitis-idaea* and *Epilobium angustifolium*) *both* arginine and glutamine increased in response to NH_4NO_3 fertilization (500 kg N ha^{-1}), while a fourth species (*Deschampsia flexuosa*) gained in asparagine and glutamine (Näsholm *et al.*, 1994). In long-term experiments the dominating amino acid increase (if any) varied with the plant species. The use of arginine and other free amino acids as bioindicators is non-specific, since their endogenous concentration responds to all types of massive nitrogen deposition. For a secluded or known nitrogen-pollution source, however, it may be used to identify the area which is affected by this. Other types of stress and pollutants (see Ranieri *et al.*, 1990) may also affect the free amino acid pool.

Apoplast amines (polyamine and tyramine metabolism) have been considered as markers of ozone injury in tobacco and spruce (see Sandermann *et al.*, 1989,

1992). The putrescine and spermidine response in soybean depended on CO_2 levels (Kramer *et al.*, 1991). Villanueva & Santerre (1989) suggested putrescine, polyamines, tryptophan and free sugars (glucose and fructose) to be involved in the adaptive metabolism of healthy resistant *Picea abies* trees exposed to SO_2 air pollution, and to represent early biochemical markers of stress. This work was followed up by Santerre *et al.* (1990a) using acid rain conditions as the ambient stress. This bioindication was suggested to be useful for selection of resistant trees, which are able to live in polluted areas. The method is able to select resistant and sensitive provenances, but not to give absolute estimates of survival rates and health in a polluted environment, since these are always decided by multiple factors. Increased polyamine level does not seem to be specific for a particular type of environmental stress, but is a rather general indicator of perturbations. The mineral contents were measured in the same needles, but had no avail for bioindication (Santerre *et al.*, 1990b).

Fatty acids in *Pinus taeda* responded similarly to ozone in four different half-sib families. But since the fatty acid composition varied with physiological age and environmental conditions, it does not seem to be a useful biomarker in the field (Fangmeier *et al.*, 1990).

HAP (4-hydroxyacetophenone), synthesized in stressed conifers, and its aglucone picein were studied by Jensen & Løkke (1990), and concluded to have a potential role as indicators of chemical stress in spruce. HAP can inhibit growth, and induce discoloration of needles and needle fall. The authors found significantly altered HAP and picein levels when *Picea abies* cuttings were exposed to ozone, but only minor responses to acid rain, acid soil, drought and to the herbicide atrazine. Further studies are needed to determine the specificity of HAP and picein as indicators of ozone stress in *P. abies* in the field.

The identity of an endogenous substance does not have to be known to be used for early detection of injury. Zaerr & Schill (1991) found significantly different (canonical discriminant analysis) constituents in 4-year-old *Picea abies* trees exposed to acid rain (spraying once a day to runoff for 30 d, pH 2.5) followed by one episode (1 h) of car exhaust (Audi 100 with catalyser) identified by HPLC. Since visible symptoms never became apparent, the authors are correct in concluding that the method was sufficiently sensitive for early detection purposes. On the other hand, the induced changes were not demonstrated to be of significance to or to correlate with the health and growth of the seedlings. The same method was used to discriminate different clones, emphasizing that this bioindication is at best only valid for a specified clone. The test was never tried in the field.

The ATP/ADP ratio was higher in needles from declining spruce trees and fumigated with low O_3 and SO_2 concentrations than in controls (Weidmann *et al.*, 1990). Since the nucleotide phosphates vary considerably with season, it can at best be used as a relative measure of decline and stress.

Enzyme activity

All of the most important air pollutants believed to be involved in the decline of forests (O_3, SO_2, NO_x) can, when taken up by plants, induce production of toxic oxygen radicals. These compounds, which are extremely reactive because of their unpaired electron, can cause direct injury to enzymes, membranes and DNA. Richardson *et al.* (1989, 1990) suggested that basic elements of the defence mechanisms against the free radicals, i.e., superoxide dismutase (SOD), peroxidase (PO), catalase (CAT), α-tocopherol, β-carotenes, glutathione (GSH), ascorbic acid and malondialdehyde (MDA), can be used in combination as early warning biomarkers of oxidative stress in trees. They demonstrated the method *in vitro* using three half-sib families of loblolly pine seedlings exposed for one growing season to three levels of O_3 (charcoal-filtered, 1.5×, and 3.0× ambient) combined with two acid rain treatments in open-top chambers. The antioxidant/MDA test was found to be (only) as sensitive as the photosynthesis response, and the test was not applied *in vivo*.

Ozone exposure during the summer and autumn may lead to changes in carbohydrate metabolism associated with winter hardening in conifers and to cell damage during the late autumn and early winter (Alscher *et al.*, 1989). Oxidant pollution may also overload the cell's antioxidant resistance mechanisms past some critical "threshold". But other types of stress, for example drought and deficient nutritional levels and ratios, will change the "thresholds" for both antioxidant and carbohydrate status. This type of bioindication for an ozone-induced increase of frost sensitivity of field-grown conifers, therefore, is neither specific nor absolute.

Atmospheric deposition of nitrogen to forests has been implicated as predisposing trees to environmental stress. Since the enzyme nitrate reductase (NR) is the rate-limiting step in the assimilation of nitrate into organic compounds, and it is substrate-inducible, Norby (1989) suggested it as an appropriate marker for nitrate metabolism disturbed by NO_2 pollution, provided the diurnal and seasonal variations are well characterized (Tjoelker *et al.*, 1992). In laboratory studies, NR was not induced in *Picea rubens* foliage exposed to NO_3^- in acid mist, but NR activity increased dramatically in spruce seedlings exposed to NO_2 or HNO_3 vapour.

Peroxidase (PO) activity (with 3,3′-dimethoxybenzidine substrate) in fine roots and mycorrhizae of *Pinus sylvestris* was significantly higher in air-pollution-exposed Scots pines than in controls (Markkola *et al.*, 1990). The PO activities correlated significantly and positively with total N and S in the tissues, and Cu and NH_4 in the humus layer, and negatively with the total number of root tips and mycorrhizae density. PO activity, however, is a non-specific resistance system for several types of stress, and may only be used by itself as a good indicator of air pollution when other environmental factors can be excluded, or when combined with other bioindicators, and when the peroxidase isozymes have been analysed.

Karjalainen *et al.* (1992) reported peroxidase, Rubisco and soluble protein to be combined specific indicators of SO_2 and NO_2. Acute levels of SO_2, particularly when combined with NO_2, strongly induced PO activity. Chronic levels of SO_2 alone slightly increased Rubisco activity and the soluble protein content, while the concurrent exposure to chronic levels of both gases decreased both Rubisco activity and soluble protein. In the field, however, the responses of these biochemical markers are greatly influenced by the environment, including temperature and light, as well as the genetic stature of the investigated spruce clones.

Vigna unguiculata (cowpea) responded to salt stress by increased superoxide dismutase (SOD) isozymes in leaves and protoplasts (Hernández *et al.*, 1994). The study, however, was carried out only *in vitro*. In the field the SOD response will be modified by light, temperature and nutrient availability.

Exposure to O_3 (100–200 $\mu g\ m^{-3}$) decreased contents of shikimic acid and quinic acid (SA and QA), and increased pH and contents of malic and citric acid (MA and CA) in spruce needles, while in Scots pine there were increased activities of cytochrome C oxidase (cyt. C ox.), PEP carboxylase (PEPC) and shikimic acid dehydrogenase (SHDH), and a decreased activity of malate dehydrogenase (MDH). Though the responses were sensitive, they were not specific for ozone, and, therefore, not adequate for independent bioindication in the field (Luethy-Krause *et al.*, 1990).

Halone 1211 (difluoro-chloro-bromo-methane) represents a large group of halo-organics of recent anthropogenic origin. These chemicals are a part of pesticides, solvents, heat exchangers, propellants and fire repellants. Some are toxic to plants, have lifetimes in the range of years or even decades, and have only recently been put under international control. Schröder & Debus (1990) found glutathione-*S*-transferase (GST) activity in crude enzyme extracts of *Picea abies* trees to respond significantly to Halone 1211 (10 ppb, 41 d), with much higher specificity than pigment patterns and protein contents. They suggested it as an effective biomarker for direct effects caused by halo-organics, though the GST response to these substances has not yet been verified for a large number of species.

Sandermann *et al.* (1992) reported β-1,3-glucanase in tobacco, cinnamyl alcohol dehydrogenase (CAD) in spruce and pine, and stilbene synthase and stilbenes in pine to be specific markers for O_3 injury, when combined with endogenous contents of polyamines. CAD participates in the formation of polymers such as lignin and suberin which are accumulated in disease reactions. The markers must be measured at a particular time, since some are short-lived and others long-lived.

While most bioindications involving enzymes are typically complicated by daily and seasonal rhythms, non-specificity for air pollutants and clonal specificity, Schmiedel & Dässler (1989) suggested precisely the use of multiple enzyme forms and their rhythmic activity for the characterization of the degree of damage done to emmission-affected plants. At this level of complexity, these authors believe that enzymes may be used as specific, sensitive and effective bioindicators. They proved their point by electrophoretic analysis of non-specific acid phosphatase, non-specific esterase, CAT and glutamate oxaloacetate transaminase (GOT) in *Picea abies* needles affected by SO_2 and natural senescence. If used simply for bioindication, however, this method seems rather demanding and inscrutable. A more effective use of enzymatic analysis is in the search for mechanisms of injury and specific elicitors.

Morphological analyses

For 40 years tree breeders have used methods of plus-tree selection, i.e., first-generation selection of the better phenotypes in wild populations in terms of yield, height, stem diameter, volume, etc. (Cornelius, 1994). This method may give genetic gains of up to 15% in height and diameter and 35% in volume per unit area. However, the amount of gain from any particular plus-tree selection system depends on the values of the parameters that determine the response to selection (selection intensity, genetic variance, heritability).

In unfavourable situations, the gain can be close to zero. As an example, east European provenances were imported to Denmark during the 1960s and 1970s because of their high growth rates in early tests. During the 1980s it became clear that they did not remain healthy with age. Today they are no longer planted in Denmark. Potential health aspects should be incorporated into the plus-tree concept.

The crown morphology method described by Roloff (1986) was used by Stribley (1993) to score premature ageing in 35–50-year-old beech trees. Stribley further defined four morphological criteria for future studies which correlated with the Roloff score: (1) annual primary shoot growth, (2) total secondary shoot length relative to a standard primary shoot length, (3) mean number of subsidiary shoots per year, and (4) proportion of shoots growing at 40° from the main stem or less.

Seed weight is used under some circumstances in early selection of forest trees, since heavier seeds are known to give better initial growth of the seedlings. Sorensen & Campbell (1993), however, found "environmental" rather than "genetic" contributions of seed weight to seedling height in *Pseudotsuga menziesii* (coastal Douglas fir). The use of seed weight to select fast-growing fir, therefore, is not recommended.

Ultrastructure and histopathology

Since leaf cuticles are continually exposed to the environment, and are the first part of the plant to meet air pollutants, they might serve as early indicators of plant response to these. But significant alterations of the ultrastructure of surface waxes with field or laboratory exposure to air pollutants are not found by all investigators. It is possible that most changes in the surface structure are largely a product of pollutant-induced endogenous stress, rather than a direct consequence of exposure of the cuticle itself. The healthiest trees in a polluted environment may show less alteration of cuticles than the least healthy trees of a less polluted environment. Cuticular wax in *Picea abies* shows a very large variation in total weight corresponding to sample area and needle age, while changes corresponding to decline were minor (Prügel *et al.*, 1994). Because the effects on cuticles are non-specific, mostly resembling premature senescence, it is not possible to use them as specific indicators for air pollutants, and maybe not even for general stress (Berg, 1989). The more complex analysis of tissue structures and the ultrastructure of cell contents, on the other hand, may possibly be used for bioindication.

Comparative histological analysis of needles could be used to diagnose the cause of forest decline (Huettl *et al.*, 1990). Ozone injuries in *Pinus ponderosa* look very different from nutrient imbalance-induced decline in *Picea abies* in southwest Germany. These authors, therefore, considered O_3 to be only a predisposing factor in central European forest decline. Fertilizing forest soils was used both in diagnosis of the nutrient imbalance, and for amelioration of the decline problem.

The correct fertilization could both prevent the development of injury, and revitalize moderately yellowed needles, even on a histological level (Zoettl *et al.*, 1989). Janicki & Jones (1990/91) used a similar diagnostic fertilization of *P. abies* in western Quebec (Canada), and confirmed forest revitalization by K, N and Mg application. Schmitt *et al.* (1986) confirmed that some histological forest decline symptoms are different form air pollution symptoms, finding reduced grana and increased number of plastoglobules in transfusion parenchyma, injured endodermis plastids and reduced lumina in sieve elements adjacent to the cambium in declining *P. abies*, while none of these symptoms were reported in trees (experimentally) exposed to air pollution.

Increased deposition of phenolic compounds in mesophyll cells, beginning directly under the stomata, and disorganization of membranes are an early response in young *Pinus nigra* and *P. resinosa* to acid rain (Zobel & Nighswander, 1991a). Ultrastructural investigations were performed after 1, 3 and 7 days, and histological investigations after a month. Similar results were visible after spraying with saturated NaCl (Zobel & Nighswander, 1991b) and aluminium salt solutions, but the response has not yet been tested with gaseous pollutants. Katoh *et al.* (1989) reported a decrease in foliar tannin levels correlating with increasing sulphur levels in Japanese cedars in polluted areas, which may leave the trees with less defence against insect predation and fungal degradation.

Specific histopathological changes were induced in *Picea abies* by gaseous pollutants (primarily ozone), acid precipitation, and mineral (Mg and K) deficiencies under laboratory conditions (Fink, 1992, 1993). Acid rain, which is not thought to pass through stomata, primarily injured epidermal cells. Ozone primarily injured mesophyll cells, as histochemically indicated by an accumulation of starch grains in individual cells. Mineral deficiencies primarily induced the collapse of central phloem cells and a general accumulation of starch in all mesophyll cells. Fink (1992, 1993) demonstrated the localization of calcium oxalate in spruce needles to be a specific bioindicator: acid precipitation made epidermal deposits disappear, ozone induced a breakthrough of the normal extra cellular crystal growth into the cells. Using this method in the field, he demonstrated that spruce decline was only correlated with Mg deficiency (and partially K) and biotic factors, and not with ozone or acid rain. Wulff (1992) collaborated the specificity of a Ca response, when reporting that calcium in the outer epidermal cell walls in spruce needles was decreased by acid mist (pH 2.5, 6 months), with no effects caused by ozone. Fink (1992) did not believe in chloroplast ultrastructure as a specific bioindicator for air pollutants.

This was contradicted by Sutinen & Koivisto (1992), who compared the structure (LM) and ultrastructure (TEM) in both spruce and pine needles, and noticed specific changes induced by ambient levels of SO_2, O_3 and acid rain particularly in chloroplasts. SO_2 lightened the plastoglobuli. Ozone reduced chloroplast size, particularly with simultaneous exposures to SO_2. Acid rain caused vacuolization in the cytoplasm. Extreme frost during hardening induced rupture of organelles. Sutinen & Koivisto (1992) believe ultrastructural diagnostic tests to be specific for individual air pollutants, with little influence by tree age, soil type, geographic area and site altitude. Holopainen *et al.* (1992) reviewed ultrastructural injuries in trees caused by air pollutants and mineral deficiencies, and concluded it was possible not only to distinguish between injuries caused by SO_2, O_3 and HF, but also between N, P, K and Mg deficiency. Acid rain injuries, however, were rather unspecific at the ultrastructural level, but they were thoroughly described by Bäck & Huttunen (1992). Holopainen *et al.* (1992) recommended the use of ultrastructural and histopathological methods together with chemical analysis and observations of visible injuries, when diagnostic characters for different stresses are sought, and together with physiological and biochemical measurements, when the injury mechanisms are studied. At the ecosystem level, the diagnostic value of structural studies can be improved by studying different biological units, e.g., needles/leaves together with epiphytic lichens and roots.

Huttunen *et al.* (1992), on the other hand, found interpretation of ultrastructural symptoms in the field to be difficult. Ultrastructural investigations are, at any rate, rather complicated and costly if employed on a large scale for bioindicative purposes.

Genetical analysis

Numerous studies on forest tree species employing isozyme gene markers showed that in fact most natural populations contain a high degree of genetic variation consisting of a great allelic diversity at many polymorphic loci and a high proportion of individual heterozygosity (El-Kassaby, 1991). Bergman *et al.* (1990) suggested that allelic variation at enzyme loci, which are functionally equivalent or nearly equivalent in moderate environments, may become adaptive if the environmental conditions, e.g., due to air pollutants, change drastically. This so-called latent genetic potential is assumed to be the basis for the adaptability to rapid and particularly unexpected environmental changes, which are perceived as stress conditions by long-lived organisms like the forest tree species. Scholz & Bergman (1992) reviewed genetic effects of environmental pollution on tree populations, and reported that certain heterozygous genotypes of Norway spruce, beech and Scots pine were more often found in tolerant than in sensitive subpopulations, under controlled conditions as well as under different pollution loads occurring in the field (Harz Mountains, Fichtelgebirge). Effects of SO_2, HF and O_3 could be distinguished by looking at only two or three gene loci. This method, therefore, has great potential as a specific bioindicator for stress caused by particular air pollutants in forest trees, for early diagnosis in known provenances, and for early selection of resistant clones. The bottom line, however, is that genetic multiplicity is the best basis for tolerance, and this must be kept in mind when we select trees for pollution tolerance.

Sieffert (1988) used two-dimensional electrophoretic isozyme expression as an efficient tool for studying modifications of gene expression in *Picea abies* in response to climatic and pollution stresses. Certain heterozygous *P. abies* genotypes were found much more frequently among the healthy than among air-pollution-damaged trees (Ruetz & Bergmann, 1989). Geburek *et al.* (1986) found significant differences between aluminium sensitive and tolerant *P. abies* trees at the gene loci of GOT. Saxe (not published) attempted by isozyme expression to fingerprint 62 individual *P. abies* trees according to their symptoms of forest decline (novel decline and top dying), but found no correlations. The individual sensitivity, however, was genetically determined, since the decline symptoms were not related to tree height (which relates to exposure), social status or the degree of exposure, and individual trees and provenances in the same area were affected to a different degree. Hattemer & Müller-Starck (1989) reviewed recent literature on adaptivity of individual isozyme loci in tree populations under the influence of air pollution.

Wilson & Bell (1990) demonstrated in a half-diallel crossing programme that the SO_2 tolerance in *Lolium perenne* is of genetic nature, but concluded that genetic control of acute SO_2 tolerance appeared to be highly variable among species. This could make it difficult to find a standard test on the genetic level for diagnostic purposes. Severe limitations of molecular-marker-aided selection may occur in forest tree breeding, making it difficult to select for wood quality, resistance to biotrophic pathogens, and resistance to air pollution (Strauss *et al.*, 1992).

Schmidtling (1994) calculated the effect of 4°C elevated temperature on *Picea abies* and *Pinus taeda* (loblolly pine), using data from American and European provenance transplanting experiments. The model predicts a 5–10% loss in yield with this particular climate change.

Guangrong *et al.* (1990) reported the use of the *Vicia* leaf tip cell micronucleus test (LTMT) for significant bioindication of xylene and HF effects. This is one of the few recognized genotoxicity tests for air pollutants. Analysis of chromosomal abnormalities in *Picea abies* demonstrated that O_3 and CO_2 had effects on root meristems several months after fumigation which were much larger than effects of soil and provenance (Müller *et al.*, 1994). The chromosomal abnormalities are suggested as being useful to confirm and classify environmental types of injury.

Miscellaneous methods

Salemaa & Jukola-Sulonen (1990) constructed a "vitality index" for mature *Picea abies*, in which one parameter was the electrical impedance in the inner bark tissue of living trees, indicating the physiological status of the tree. The parameter, however, was never demonstrated to be specific for air pollution, and varied with tree size. Bark pH in *Pinus densiflora* and *Quercus mongolica* was found to be a better bioindicator than the soluble sulphur content in leaves (Cha & Lee, 1991).

Skrøppa (1991) reported a significant correlation between laboratory freezing tests of full-sib seedlings from two sets of 12 *Picea abies* families, and the height growth of young trees in the field. Larsen & Wellendorf (1990) found a significant correlation in 16 *Picea abies* populations between growth (height, diameter, wood density) in 13-year-old field-grown trees, and frost resistance of their 3-year-old full-sib seedlings, suggesting frost resistance in the offspring to be a useful early test for selecting *P. abies*. This parameter may be considered specific for the present environment in the field. Andersson (1992) used freezing tests to forecast field mortality of *Pinus sylvestris*. Sulzer *et al.* (1993) found cold hardiness of seedlings to correlate with the growth of 3-year-old and 10-year-old *Picea mariana* (black spruce) trees of the same seed stock. Photosynthesis, transpiration and "instantaneous water use efficiency" correlated with the growth of 3-year-old, but not with the growth of 10-year-old trees. Height and diameter of the seedlings, however, had the best predictive value of growth in the 3 and 10-year-old trees.

Rock *et al.* (1988) described a "blue shift" in the spectral response of forest canopies, using airborne and satellite sensor data, in forest canopies in Vermont

(USA) and Germany, where damage due to air pollution was suspected. The "blue shift" was associated with a decrease in needle chlorophyll. The early detection aspect of this method is that large areas may be frequently investigated, and injuries discovered and mapped, sooner than they are reported by observers on the ground. Gamon *et al.* (1990) used remote sensing of the xanthophyll cycle and chlorophyll fluorescence to detect photosynthetic performance in sunflower canopy, while Schmuck (1990) suggested the use of laser-induced chlorophyll fluorescence in forest decline research.

Williams & Ashenden (1992) reported changes in canopy reflectance, particularly in the green and near infrared wavebands, to specifically detect O_3-related injuries to *Trifolium repens* (white clover), with no significant effects caused by SO_2 (40 ppb) and NO_2 (40 ppb) alone or combined, or by acid mist (6 mm per week, pH 2.5, 3.5, 4.5 or 5.6), after 12 weeks of exposure. The vegetation index, defined by canopy reflectance, correlated with loss in dry weight. The investigators advocated reflectance measurements for early detection of plant injuries caused by air pollution.

Egnell & Örlander (1993) demonstrated that infrared thermography of *Pinus sylvestris* and *Picea abies* seedlings before planting was a powerful early test of annual height increment in the first growing season, as well as damage caused by the ambient environment. The validity of the method to long-term growth and health, however, is not yet known.

Davidson (1992) has described a method to diagnose forest decline in *Picea abies* by two-dimensional electrophoresis. The method, however, is in an early stage of development, and has not yet been tested in the field.

Stable isotopes of C, N and S can be used in historical studies of human perturbation to the biogeochemical cycles in ecosystems. Human activity, e.g., combustion of fossil fuels, disturbs the naturally occurring ratios of stable isotopes. The lower $\delta^{13}C$ values of "fossil CO_2" may be used to determine when a tree was first affected by air pollution (Fry, 1989). Physiological variations related to CO_2-fixation processes, however, can lead to larger $\delta^{13}C$ changes than the fossil fuel signals, so the method must be used with care.

Ethanol production was reported to occur in several tree species in the Ohio River valley as a result of acidic pollution ($SO_2 + NO_x$, MacDonald *et al.*, 1989). The aerobic ethanol production, proposed to be the result of cell acidification, was neither triggered by ozone nor by natural stresses such as flooding, herbivory, fungal infection or high temperature. This ethanol production was, therefore, believed to be a specific bioindication, which, when used in conjunction with other measures of stress, can provide a fairly simple indication of stress caused by acidic, gaseous pollutants.

The well-known turbidity test for bioindication of effects of air pollutants on conifers (Härtel, 1972) was chemically explained by Fuchshofer & Härtel (1985) as either a conversion of Ca^{2+} into insoluble compounds ($CaSO_4$, insoluble Ca–F compounds) in the case of SO_2 or HF pollution, or by leaching out Ca^{2+} with O_3 pollution. Ca^{2+} ions modify the permeability of the outermost cell-wall layers of the needles, thereby controlling the efflux of the substances which determine the turbidity of the extracts obtained in the Härtel test. Härtel & Fuchshofer (1987) improved the method by specifying sampling and extraction procedures, as well as wavelength restrictions and a standard calibration for the photometer. Though this method is sensitive to effects on conifers by SO_2, HF and O_3, and has been widely applied in the field, it has never been proven to be specific for these gases.

Multivariate methods

A larger specificity and sensitivity of bioindicative methods may be obtained by combining several response parameters.

In an integrated American programme the response of plants to interacting stresses (ROPIS) is used in process-level models of material flow within a plant and between the plant and the surrounding environment to integrate and predict plant response to interacting stresses of natural and anthropogenic origin. The programme includes measurements of (1) exposure–response studies, (2) biochemical responses in leaf tissues, (3) physiological responses of photosynthesis, respiration and winter hardening, (4) whole-plant developmental processes linking soil and climate influences, (5) material cycling throughout the air–plant–soil continuum, (6) computer modelling of carbon, water and nutrient dynamics, and (7) adaptive responses of plants to assault by different stresses using an evolutionary approach (Goldstein & Ferson, 1994). The programme focuses on regionally important tree species. All effects are integrated in the TREGRO model (Weinstein & Yanai, 1994).

Elstner *et al.* (1985) sampled spruce needles from 18 different locations in Upper Bavaria for bioindication of photooxidants, analysing chlorophyll content, ethane evolution in $NaHSO_3$-incubated needles, ACC and malonyl-ACC (ethylene precursors), ascorbate, xanthoxin, K, Na, Ca, Mg and Mn contents, SOD, CAT, loss of wax layer integrity, and fungi infections. Since several of these parameters were affected independently of needle weight, needle length and water content, they could be used for early detection.

Bender *et al.* (1986) assessed the effects caused by SO_2 and O_3 singly or in combination (+ acid rain), by measuring the following parameters in needles or leaves of young fir, spruce and beech trees in open-top chambers: element concentrations (K, Ca, Mg, Mn), inorganic and organic sulphur, glutathione, buffering capacity, chlorophyll, Lowry protein, proline and peroxidase and glutamate dehydrogenase activity. Together they could be used for early detection.

The analysis of clearly defined pine bark layers, showed that accumulation is restricted to 3–4 mm on the external surface. Stöcker & Gluch (1990) sampled this volume and analysed it for pH (acid deposition), electric conductivity (accumulated electrolytes), sulphur content (SO_2 deposition) and Ca content (alkaline flue dusts). The analysis was used for biomonitoring with the purpose of computer-aided recognition (cluster, variance and discriminance analysis) of spatial and temporal patterns of pollutant depositions on forest ecosystems and individual forest stands.

Tynnyrinen *et al.* (1992) compared visible and ultrastructural injuries and element contents in spruce needles and the lichen *Hypogymnia physodes*. They used the results to map the area influenced by a local fertilizer plant and a strip mine.

The multivariate methods at their best include multiple analysis on an ecosystem level, incorporating investigations of the soil, the fauna, the flora (physiology, biochemistry, ultrastructure), along with entomological and mycorrhiza investigations, analysis of atmospheric contents and all seasonal variations (Bolhàr-Nordenkampf, 1989). Early diagnosis aspects may include physiological behaviour in dependence of ontogenesis and seasonal variations of stress factors, to determine the extent of stress with respect to time-dependent modifications (Bolhàr-Nordenkampf & Lechner, 1989).

None of the above authors, however, specified a precise index as a result of their investigations, since such an index will always depend on the purpose of the bioindication, local conditions, species, etc. Multivariate analyses serve an even wider interest than bioindication; they aim at providing an understanding of the function of complete ecosystems.

CONCLUSIONS

The development of new bioindicative methods occurs so rapidly at present, that any review of the subject will be outdated before it is published. The methods are multitudinous, and, in order to limit the volume of the present review, it was written as a sequel to the review by Saxe (1991), where 40 methods of bioindication were listed; the present review more than triples the total number of described bioindicative methods.

The general impression of most of these methods, is that they are often cross-sensitive to a broad range of environmental conditions in the field, dependent on plant age and sampling time, often unspecific when used alone, only valid for the specified cultivars/provenances, and frequently costly, demanding and inscrutable.

These shortcomings, however, may be overcome. The cross-sensitivity and non-specificity can be reduced by defining strict protocols for sampling and by combining several response parameters into multivariate indices, or by working with genetically related offspring in controlled environments. A degree of environmental control may be obtained even in the field.

Bioindicative methods are becoming exceedingly useful for several purposes: early diagnosis of injury, selection of resistant and productive cultivars/provenances for forest replantings, identification of stress mechanisms, identification of stress parameters, and mapping of the spatial and temporal distribution of stress parameters and deposition, including pollutants.

Development of bioindicative techniques has a particular relevance to forest management, since trees are only slowly replaced by the next generation, are continuously exposed to environmental stresses over decades or centuries, and yet, are rarely nurtured like agricultural crops, though often exploited as such.

Efficient use of bioindication is both needed and attainable in future management of plant growth, forest management, production and protection. I strongly recommend that bioindicative methods are included in the currently developed international definitions of critical levels and loads (Hornung *et al.*, 1995) for environmental stress on terrestrial ecosystems.

ACKNOWLEDGMENTS

This work was supported by the Danish Agricultural and Veterinary Research Council (J.no. 5.23.26.03-1), The National Forest and Nature Agency, and IE-Consult. The library at the Royal Veterinary and Agricultural University provided helpful assistance.

REFERENCES

Alscher, R.G., Cumming, J.R. & Fincher, J. 1989. Air pollutant-low temperature interaction in trees. In Grossblatt, N. (ed.) *Biological Markers of Air-Pollution Stress and Damage in Forests*: 341-345. National Academy Press, Washington, DC.

Andersson, B. 1992. Forecasting *Pinus sylvestris* field mortality by freezing tests — methods and applications. Thesis, The Swedish University of Agricultural Sciences, Dept. of Forest Genetics and Plant Physiology, Umeå, Sweden.

Bäck, J. & Huttunen, S. 1992. Structural responses of needles of conifer seedlings to acid rain treatment. *New Phytol.* **120**: 77-88.

Baillon, F., Dalschaert, X., Grassi, S. & Geiss, F. 1988. Spruce photosynthesis: Possibility of early damage diagnosis due to exposure to magnesium or potassium deficiency. *Trees* **2**: 173-179.

Balsberg-Påhlsson, A.-M. 1989. Effects of heavy-metal and SO_2 pollution on the concentrations of carbohydrates and nitrogen in tree leaves. *Can. J. Bot.* **67**: 2106-2113.

Beerling, D.J. 1994. Predicting leaf gas exchange and $\delta^{13}C$ responses to the past 30000 years of global environmental change. *New Phytol.* **128**: 425-433.

Bender, J., Jäger, H.-J., Seufert, G. & Arndt, U. 1986. Untersuchungen zur Einzel- und Kombinationswirkung von SO_2 und O_3 auf den Stoffwechsel von Waldbäumen in Open-top-Kammern. *Angew. Botanik* **60**: 461-479.

Benner, P., Sabel, P. & Wild, A. 1988. Photosynthesis and transpiration of healthy and diseased spruce trees in the course of three vegetation periods. *Trees* **2**: 223-232.

Bennett, J.P. & Stolte, K.W. 1985. *Using Vegetation Biomonitors to Assess Air Pollution Injury in National Parks, Milkweed Survey*. Natural Resources Report Series No. 85-1. National Park Service, Air Quality Division, Denver, CO.

Berg, V.S. 1989. Leaf cuticles as potential markers of air pollutant exposure in trees. In Grossblatt, N. (ed.) *Biological Markers of Air-Pollution Stress and Damage in Forests*: 333-339. National Academy Press, Washington, DC.

Bergmann, F., Gregorius, H.-R. & Larsen, J.B. 1990. Levels of genetic variation in European silver fir (*Abies alba*). Are they related to the species' decline? *Genetica* **82**: 1-10.

Bolhàr-Nordenkampf, R. 1989. Stress-physiological ecosystem research. Altitude profile Zillertal. *Phyton* (special issue) **29** (3): 11-302.

Bolhàr-Nordenkampf, R. & Lechner, E.G. 1989. Synopse stressbedingter Modifikationen der Anatomie und Physiologie von Nadeln als Frühdiagnose einer Disposition zur Schadensentwicklung bei Fichte. *Phyton* **29**: 255-301.

Bondietti, E.A., Baes III, C.F. & McLaughlin, S.B. 1989. The potential of trees to record aluminum mobilization and changes in alkaline earth availability. In Grossblatt, N. (ed.) *Biological Markers of Air-Pollution Stress and Damage in Forests*: 281-292. National Academy Press, Washington, DC.

Brown, D. 1992. Biochemical assessment of environmental stress in bryophytes and lichens. In Roy, S., Käranlampi, L. & Hänninen, O. (eds) *7th International Bioindicators Symposium and Workshop on Environmental Health*, Abstract Book: 26. Kuopio Univ. Publ. C., Natural and Environmental Science 7, Finland.

Cadle, S.H., Marshall, J.D. & Mulawa, P.A. 1991. A laboratory investigation of the routes of HNO_3 dry deposition to coniferous seedlings. *Environ. Pollut.* **72**: 287-305.

Cape, J.N., Freer-Smith, P.H., Paterson, I.S., Parkinson, J.A. & Wolfenden, J. 1990. The nutritional status of *Picea abies* (L.) Karst. across Europe, and implications for "forest decline". *Trees* **4**: 211-224.

Cha, Y.J. & Lee, K.J. 1991. Relationships between air pollution by sulphur dioxide and soluble sulphur in the leaves and bark pH in urban forest trees. *J. Korean For. Soc.* **80**: 279-286.

Chappelka, A.H., Kush, J.S., Meldahl, R.S. & Lockaby, B.G. 1990. An ozone-low temperature interaction in loblolly pine (*Pinus taeda* L.). *New Phytol.* **114**: 721-726.

Clement, H. 1996. Interaction of atmospheric ammonia pollution with frost tolerance of plants. A study of winter wheat and Scots pine. Doctoral thesis, Rijksuniversiteit, Groningen, The Netherlands.

Cornelius, J. 1994. The effectiveness of plus-tree selection for yield. *For. Ecol. Manag.* **67**: 23-34.

Cronan, C.S., April, R., Bartlett, R.J., Bloom, P.R., Driscoll, C.T., Gherini, S.A., Henderson, G.S., Joslin, J.D., Kelly, J.M., Newton, R.M., Pernell, R.A., Patterson, H.H., Raynal, D.J., Schaedle, M., Schofield, C.L., Sucoff, E.I., Tepper, H.B. & Thornton, F.C. 1989. Aluminum toxicity in forests exposed to acidic deposition: the ALBIOS results. *Water, Air, Soil Pollut.* **48**: 181-192.

Darrall, N.M. & Jäger, H.J. 1984. Biochemical diagnostic tests for the effect of air pollution on plants. In Koziol, M.J. & Whatley, F.R. (eds) *Gaseous Air Pollutants and Plant Metabolism*: 333-349.

Davidson, N. 1992. Biokjemisk diagnose av skogskader. *Aktuelt fra Skogforsk* **16**: 38-42.

Dobson, M.C., Taylor, G. & Freer-Smith, P.H. 1990. The control of ozone uptake by *Picea abies* (L.) Karst. and *P. sitchensis* (Bong.) Carr. during drought and interacting effects on shoot water relations. *New Phytol.* **116**: 465-474.

Donagi, A.E. & Goren, A.I. 1979. Use of indicator plants to evaluate atmospheric levels of nitrogen dioxide in the vicinity of a chemical plant. *Environ. Sci. Tech.* **13**: 986-989.

Eamus, D. & Murray, M. 1991. Photosynthetic and stomatal conductance responses of Norway spruce and beech to ozone, acid mist and frost — a conceptual model. *Environ. Pollut.* **72**: 23-44.

Egnell, G. & Örlander, G. 1993. Using infrared thermography to assess viability of *Pinus sylvestris* and *Picea abies* seedlings before planting. *Can. J. For. Res.* **23**: 1737-1743.

El-Kassaby, Y.A. 1991. Genetic variation within and among conifer populations: Review and evaluation of methods. In Fineschi, S., Malvolti, M.E., Cannata, F. & Hattemer, H.H. (eds) *Biochemical Markers in the Population Genetics of Forest Trees*: 61-76. SPB Academic Publ. BV, The Hague.

Elstner, E.F., Osswald, W. & Youngman, R.J. 1985. Basic mechanisms of pigment bleaching and loss of structural resistance in spruce (*Picea abies*) needles: Advances in phytomedical diagnostics. *Experientia* **41**: 591-597.

Ericsson, A., Nordén, L-G., Näsholm, T. & Walheim, M. 1993. Mineral nutrient imbalances and arginine concentrations in needles of *Picea abies* (L.) Karst. from two areas with different levels of airbone deposition. *Trees* **8**: 67-74.

Fangmeier, A., Kress, L.W., Lepper, P. & Heck, W.W. 1990. Ozone effects on the fatty acid composition of loblolly pine needles (*Pinus taeda* L.). *New Phytol.* **115**: 639-647.

Fellner, R. 1989. Mycorrhiza-forming fungi as bioindicators of air pollution. *Agric. Ecosys. Environ.* **28**: 115-120.

Fink, S. 1992. Histopathological changes in tree foliage as a diagnostic feature for differentiation among various abiotic and biotic damage factors. In Roy, S., Käranlampi, L. & Hänninen, O. (eds) *7th International Bioindicators Symposium and Workshop on Environmental Health*, Abstract Book: 24. Kuopio Univ. Publ. C., Natural and Environmental Science 7, Finland.

Fink, S. 1993. Microscopic criteria for the diagnosis of abiotic injuries to conifer needles. In Huettl R.F. & Muller-Dombois (eds) *Forest Decline in the Atlantic and Pacific Region*: 175-188. Springer-Verlag, Berlin.

Forschner, W., Schmitt, V. & Wild, A. 1989. Investigations on the starch content and ultrastructure of spruce needles relative to the occurrence of novel forest decline. *Botanica Acta* **102**: 208-221.

Friend, A.L., Tomlinson, P.T., Dickson, R.E., O'Neill, E.G., Edwards, N.T. & Taylor, G.E. Jr 1992. Biochemical composition of loblolly pine reflects pollutant exposure. *Tree Physiol.* **11**: 35-47.

Fry, B. 1989. Human perturbation of C, N, and S biogeochemical cycles: Historical studies with stable isotopes. In Grossblatt, N. (ed.) *Biological Markers of Air-Pollution Stress and Damage in Forests*: 143-156. National Academy Press, Washington, DC.

Fuchshofer, H. & Härtel, O. 1985. Zur Physiologie des Trübungstests, einer Methode zur Bioindikation von Abgaswirkungen auf Koniferen. *Phyton* **25**: 277-291.

Gamon, J.A., Field, C.B., Bilger, W., Björkman, O., Fredeen, A.L. & Penuelas, J. 1990. Remote sensing of the xanthophyll cycle and chlorophyll fluorescence in sunflower leaves and canopies. *Oecologia* **85**: 1-7.

Garrec, H.-P., Maout, Le, L. & Rose, C. 1990. Possibilité d'application des tests physiologiques pour le diagnostic précoce du dépérissement forestier. *Ann. Gembloux* **96**: 55-77.

Gärtner, E.J., Urfer, W., Eichhorn, J., Grabowski, H. & Huss, H. 1990. Mangan-ein Bioindikator für den derzeitigen Schadzustand mittelalter Fichten in Hessen. *Forstarch.* **61**: 229-233.

Gebauer, G. & Schulze, E.-D. 1991. Carbon and nitrogen isotope ratios in different compartments of a healthy and a declining *Picea abies* forest in the Fichtelgebirge, NE Bavaria. *Oecologia* **87**: 198-207.

Geburek, Th., Scholz, F. & Bergmann, F. 1986. Variation in aluminum-sensitivity among *Picea abies* (L.) Karst. seedlings and genetic differences between their mother trees as studied by isozyme-gene-markers. *Angew. Botanik* **60**: 451-460.

Gezelius, K. & Näsholm, T. 1993. Free amino acids and protein in Scots pine seedlings cultivated at different nutrient availabilities. *Tree Physiol.* **13**: 71-86.

Glooschenko, W.A. 1989. *Sphagnum fascum* moss as an indicator of atmospheric cadmium deposition across Canada. *Environ. Pollut.* **57**: 27-33.

Goldstein, R. & Ferson, S. 1994. Response of plants to interacting stresses (ROPIS): Program, rationale, design, and implications. *J. Environ. Qual.* **23**: 407-411.

Granier, L. & Chevreuil, M. 1992. On the use of tree leaves as bioindicators of the contamination of air by organochlorines in France. *Water, Air, Soil Pollut.* **64**: 575-584.

Grossblatt, N. (ed.) 1989. *Biological Markers of Air-Pollution Stress and Damage in Forests.* Committee on Biological Markers of Air-Pollution Damaging Trees, Board on Environmental Studies and Toxicology, Commission on Life Sciences, National Research Council, National Academy Press, Washington, DC.

Guangrong, C., Bo, J., Ming, L. & Xingguo, W. 1990. The *Vicia* leaf tip cell-micronucleus bioassay and air pollution monitoring. In Wang, W., Gorsuch, J.W. & Lower, W.R. (eds) *Plants for Toxicity Assessment*: 170-174. American Society for Testing and Materials, Philadelphia.

Hanisch, B. & Kilz, E. 1990. *Waldschäden erkennen, Fichte und Kiefer*. Verlag Eugen Ulmer, Stuttgart.

Härtel, O. 1972. Langjährige Messreihen mit dem Trübungstest an Abgas geschädigten Fichten. *Oecologia* **9**: 103-111.

Härtel, O. & Fuchshofer, H. 1987. Methodische Ergänzungen zum Trübungstest zur Bioindikation von Abgaswirkungen auf Koniferen. *Phyton* **26**: 235-246.

Hattemer, H.H. & Müller-Starck, G. 1989. Das Waldsterben als Anpassungsprozess. *Allg. Forst- u. J.-Ztg.* **160**: 222-229.

Heagle, A.S., Miller, J.E. & Sherrill, D.E. 1994. A white clover system to estimate effects of tropospheric ozone on plants. *J. Environ. Qual.* **23**: 613-621.

Heath, R.L. 1989. Alteration of chlorophyll upon air pollutants exposure. In Grossblatt, N. (ed.) *Biological Markers of Air-Pollution Stress and Damage in Forests*: 347-356. National Academy Press, Washington, DC.

Heggestad, H.E. 1991. Origin of Bel-W3, Bel-C and Bel-B tobacco varieties and their use as indicators of ozone. *Environ. Pollut.* **74**: 264-291.

Hernández, J.A., Del Rio, L.A. & Sevilla, F. 1994. Salt stress-induced changes in superoxide dismutase isozymes in leaves and mesophyll protoplasts from *Vigna unguiculata* (L.) Walp. *New Phytol.* **126**: 37-44.

Hirata, Y. & Kunishige, M. 1981. Studies on the survey and application of new indicate plants to sulphur dioxide. II. The responses of *Hydrangea* "Enziandom" (*Hydrangea hortensis* SM.) as an indicator plant at an air polluted area. *Bull. Veg. Orn. Crops Res. Sta.* **5C**: 51-62.

Holopainen, T., Anttonen, S., Wulff, A., Palomäki, V. & Kärenlampi, L. 1992. Comparative evaluation of the effects of gaseous pollutants, acidic deposition and mineral deficiencies: Structural changes in the cells of forest plants. *Agric. Ecosyst. Environ.* **42**: 365-398.

Hornung, H., Sutton, M. & Wilson, R. 1995. Mapping and Modelling of Critical Loads for Nitrogen: A Workshop Report. 207 pp. *Proceedings of the Grange-over-Sands Workshop, Oct. 1994*. Institute of Terrestrial Ecology, UK.

Hove, W.A. van, Kooten, O. van, Wijk, K.J. van, Vredenberg, W.J., Adema, E.H. & Pieters, G.A. 1991. Physiological effects of long term exposure to low concentrations of SO_2 and NH_3 on poplar leaves. *Physiol. Plant.* **82**: 32-40.

Huettl, R.F., Fink, S., Lutz, H.-J., Poth, M. & Wisniewski, J. 1990. Forest decline, nutrient supply and diagnostic fertilization in Southwestern Germany and in Southern California. *For. Ecol. Managem.* **30**: 341-350.

Huttunen, S., Turunen, M. & Bäck, J. 1992. Indications of acid rain induced stress on needles. In Roy, S., Käranlampi, L. & Hänninen, O. (eds) *7th International Bioindicators Symposium and Workshop on Environmental Health*, Abstract Book: 40. Kuopio Univ. Publ. C., Natural and Environmental Science 7, Finland.

Innes, J.L. 1993. *Forest Health: Its Assessment and Status.* CAB International, Wallingford, UK.

Janicki, W. & Jones, A.R.C. 1990/91. Nutrient response to diagnostic fertilization of Norway spruce *Picea abies* (L.) Karst plantations in Western Quebec, Canada. *Water, Air, Soil Pollut.* **54**: 113-118.

Jensen, J.S. & Løkke, H. 1990. 4-hydroxyacetophenone and its glucoside picein as chemical indicators for stress in *Picea abies*. *J. Plant Dis. Protect.* **97**: 328-338.

Jones, C.G. & Coleman, J.S. 1989. Biochemical indicators of air pollution effects in trees: Unambiguous signals based on secondary metabolites and nitrogen in fast-growing species? In Grossblatt, N. (ed.) *Biological Markers of Air-Pollution Stress and Damage in Forests*: 261-273. National Academy Press, Washington, DC.

Karjalainen, R., Jokinen, J., Laine, E. & Martikainen, V. 1992. Changes in enzyme activity as an indicator of air pollution stress in woody plants. In Roy, S., Käranlampi, L. & Hänninen, O. (eds) *7th International Bioindicators Symposium and Workshop on Environmental Health*, Abstract Book: 35. Kuopio Univ. Publ. C., Natural and Environmental Science 7, Finland.

Katoh, T., Kasuya, M., Kagamimori, S, Kozuka, H. & Kawano, S. 1989. Effects of air pollution on tannin biosynthesis and predation damage in *Cryptomeria japonica*. *Phytochem.* **28**: 439-445.

Keller, T. 1988. Growth and premature leaf fall in American aspen as bio-indications for ozone. *Environ. Pollut.* **52**: 183-192.

Köstner, B., Czygan, F.-C. & Lange, O.L. 1990. An analysis of needle yellowing in healthy and chlorotic Norway spruce (*Picea abies*) in a forest decline area in the Fichtelgebirge (N.E. Bavaria). *Trees* **4**: 55-67.

Kramer, G.F., Lee, E.H. & Rowland, R.A. 1991. Effects of elevated CO_2 concentration on the polyamine levels of field-grown soybean at three O_3 regimes. *Environ. Pollut.* **73**: 137-152.

Kress, L.W., Skelly, J.M. & Hinkelmann, K.H. 1982. Relative sensitivity of 18 full-sib families of *Pinus taeda* to O_3. *Can. J. For. Res.* **12**: 203-209.

Kromroy, K.W., Olson, M.F., Grigal, D.F., Teng, P.S., French, D.R. & Amundson, G.H. 1990. A bioindicator system assessing air quality within Minnesota. In Wang, W.,

Gorsuch, J.W. & Lower, W.R. (eds) *Plants for Toxicity Assessment*: 156-169. American Society for Testing and Materials, Philadelphia.

Landolt, W., Guecheva, M. & Bucher, J.B. 1989a. The spatial distribution of different elements in and on the foliage of Norway spruce growing in Switzerland. *Environ. Pollut.* **56**: 155-167.

Landolt, W., Pfenninger, I. & Luethy-Krause, B. 1989b. The effect of ozone and season on the pool sizes of cyclitols in Scots pine (*Pinus sylvestris*). *Trees* **3**: 85-88.

Lange, O.L., Führer, G. & Gebel, J. 1986. Rapid field determination of photosynthetic capacity of cut twigs (*Picea abies*) at saturating ambient CO_2. *Trees* **I**: 70-77.

Larsen, J.B. 1986. Das Tannensterben: Eine neue Hypothese zur Klärung des Hintergrundes dieser rätselhaften Komplexkrankheit der Weisstanne (*Abies alba* Mill.). *Forstwiss. Cbl.* **105**: 381-396.

Larsen, J.B. & Wellendorf, H. 1990. Early test in *Picea abies* full sibs by applying gas exchange, frost resistance and growth measurements. *Scand. J. For. Res.* **5**: 369-380.

Larsen, J.B., Qian, X.M., Scholz, F. & Wagner, I. 1988. Ecophysiological reactions of different provenances of European silver fir (*Abies alba* Mill.) to SO_2 exposure during winter. *Eur. J. For. Path.* **18**: 44-50.

Larsen, J.B., Yang, W. & Tiedemann, A.V. 1990. Effects of ozone on gas exchange, frost resistance, flushing and growth of different provenances of European silver fir (*Abies alba* Mill.). *Eur. J. For. Path.* **20**: 211-218.

Laurence, J.A., Reynolds, K.L. & Greitner, C.S. 1985. Bioindicators of SO_2 and HF. *Water, Air, Soil Pollut.* **17**: 399-407.

Lichtenthaler, H.K. & Rinderle, U. 1988. Chlorophyll fluorescence signatures as vitality indicator in forest decline research. In Lichtenthaler, H.K. (ed.) *Applications of Chlorophyll Fluorescence*: 143-149. Kluwer Academic Publishers, Dordrecht.

Lindgren, K. & Hällgren, J.-E. 1993. Cold acclimation of *Pinus contorta* and *Pinus sylvestris* assessed by chlorophyll fluorescence. *Tree Physiol.* **13**: 97-106.

Long, R.P. & Davis, D.D. 1989. Major and trace element concentrations in surface organic layers, mineral soil, and white oak xylem downwind from a coal-fired power plant. *Can. J. For. Res.* **19**: 1603-1615.

Luethy-Krause, B., Bleuler, P. & Landolt, W. 1989. Black poplar and red clover as bioindicators for ozone at a forest site. *Angew. Botanik* **63**: 111-118.

Luethy-Krause, B., Pfenninger, I. & Landolt, W. 1990. Effects of ozone on organic acids in needles of Norway spruce and Scots pine. *Trees* **4**: 198-204.

Luxmoore, R.J. 1989. Nutrient-use efficiency as an indicator of stress effects in forest trees. In Grossblatt, N. (ed.) *Biological Markers of Air-Pollution Stress and Damage in Forests*: 317-331. National Academy Press, Washington, DC.

MacDonald, R.C., Kimmerer, T.W. & Razzaghi, M. 1989. Aerobic ethanol production by leaves: Evidence for air pollution stress in trees of the Ohio River Valley, USA. *Environ. Pollut.* **62**: 337-351.

Maier-Maercker, U. & Koch, W. 1991. Experiments on the water budget of densely and sparsely needled spruces (*Picea abies* (L.) Karst.) in a declining stand. *Trees* **5**: 164-170.

Manion, P.D. 1981. Decline diseases of complex biotic and abiotic origin. In Manion, P. (ed.) *Tree Disease Concepts*: 324-339. Prentice Hall, Englewood Cliffs, NJ.

Manion, P.D. & Lachance, D. 1992. Forest decline concepts: An overview. In Manion, P.D. & Lachance, D. (eds) *Forest Decline Concepts*: 181-190. APS Press, St Paul, MN.

Manning, W.J. & Feder, W.A. 1980. *Biomonitoring Air Pollutants with Plants.* Applied Science Publishers, Ltd, London, 142 pp.

Markan, K. & Fischer, U. 1991. Untersuchungen zur Immissionsbelastung der Berliner Forsten: Deposition und Bioindikation. *Dissertationes Botanicæ*: 170-258.

Markkola, A.M., Ohtonen, R. & Tarvainen, O. 1990. Peroxidase activity as an indicator of pollution stress in the fine roots of *Pinus sylvestris. Water, Air, Soil Pollut.* **52**: 149-156.

Massacci, A. & Jones, H.G. 1990. Use of simultaneous analysis of gas-exchange and chlorophyll fluorescence quenching for analyzing the effects of water stress on photosynthesis in apple leaves. *Trees* **4**: 1-8.

Matyssek, R., Günthardt-George, M.S., Saurer, M. & Keller, T. 1992. Seasonal growth, $\delta^{13}C$ in leaves and stem, and phloem structure of birch (*Betula pendula*) under low ozone concentrations. *Trees* **6**: 69-76.

Mignanego, L., Biondi, F. & Schenone, G. 1992. Ozone biomonitoring in Northern Italy. *Environ. Monitor. Asses.* **21**: 141-151.

Moroni, J.S., Briggs, K.G. & Taylor, G.J. 1991. Chlorophyll content and leaf elongation rate in wheat seedlings a measure of manganese tolerance. *Plant and Soil* **136**: 1-9.

Mössnang, M. 1990/91. Element contents of spruce needles (*Picea abies*) along an altitudinal gradient in the Bavarian alps. *Water, Air, Soil Pollut.* **54**: 107-112.

Müller, M., Köhler, B., Grill, D., Guttenberger, H. & Lütz, C. 1994. The effects of various soils, different provenances and air pollution on root tip chromosomes in Norway spruce. *Trees* **9**: 73-79.

Näsholm, T., Edfast, A.B., Ericsson, A. & Nordén, L.-G. 1994. Accumulation of amino acids in some boreal forest plants in response to increased nitrogen availability. *New Phytol.* **126**: 137-143.

Nienhaus, F. 1985. Infectous diseases in forest trees caused by viruses, mycoplasma-like organisms and primitive bacteria. *Experientia* **41**: 597-603.

Norby, R.J. 1989. Foliar nitrate reductase: A marker for assimilation of atmospheric nitrogen oxides. In Grossblatt, N. (ed.) *Biological Markers of Air-Pollution Stress and Damage in Forests*: 245-250. National Academy Press, Washington, DC.

Nouchi, I. & Aoki, K. 1979. Morning glory as a photochemical oxidant indicator. *Environ. Pollut.* **18**: 289-303.

Oren, R., Werk, K.S., Buchmann, N. & Zimmermann, R. 1993. Chlorophyll-nutrient relationships identify nutritionally caused decline in *Picea abies* stands. *Can. J. For. Res.* **23**: 1187-1195.

Palomäki, V., Tynnyrinen, S. & Holopainen, T. 1992. Lichen transplantation in monitoring fluoride and sulphur deposition in the surroundings of a fertilizer plant and a strip mine at Siilinjärvi. *Ann. Bot. Fennici* **29**: 25-34.

Pietilä, M., Lähdesmäki, P., Pietiläinen, P., Ferm, A, Hytönen, J. & Pätilä, A. 1991. High nitrogen deposition causes changes in amino acid concentrations and protein spectra in needles of Scots pine (*Pinus sylvestris*). *Environ. Pollut.* **72**: 103-115.

Posthumus, A.C. 1982. Biological indicators of air pollution. In Unsworth, M.H. & Ormrod, D.P. (eds) *Effects of Gaseous Air Pollution in Agriculture and Horticulture*: 27-42. Butterworths, London.

Prügel, B., Loosveldt, P. & Garrec, J.-P. 1994. Changes in the content and constituents of the cuticular wax of *Picea abies* (L.) Karst. in relation to needle ageing and tree decline in five European forest areas. *Trees* **9**: 80-87.

Ranieri, A., Bernardi, R., Pisanelli, A., Lorenzini, G. & Soldatini, G.F. 1990. Changes in the free amino acid pool and in the protein pattern of maize leaves under continuous SO_2 fumigation. *Plant Physiol. Biochem.* **28**: 601-607.

Renger, G. & Schreiber, U. 1986. Practical applications of fluorometric methods to algae and higher plant research. In Govindjee, A.J. & Fork, D.C. (eds) *Light Emission by Plants and Bacteria*: 587-616. Academic Press, London.

Richardson, C.J., Di Giulio, R.T. & Tandy, N.E. 1989. Free-radical mediated processes as markers of air pollution stress in trees. In Grossblatt, N. (ed.) *Biological Markers of Air-Pollution Stress and Damage in Forests*: 245-250. National Academy Press, Washington, DC.

Richardson, C.J., Sasek, T.W. & Di Giulio, R.T. 1990. Use of physiological and biochemical markers for assessing air pollution stress in trees. In Wang, W., Gorsuch, J.W. & Lower, W.R. (eds) *Plants for Toxicity Assessment*: 143-155. American Society for Testing and Materials, Philadelphia.

Riitters, K.H., Ohmann, L.F. and Grigal, D.F. 1991. Woody tissue analysis using an element ratio technique (DRIS). *Can. J. For. Res.* **21**: 1270-1277.

Rock, B.N., Hoshizaki, T. & Miller, J.R. 1988. Comparison of *in-situ* and airborne spectral measurements of the blue shift associated with forest decline. *Remote Sens. Environ.* **24**: 109-128.

Roloff, A. 1986. Morphologies der Kronenentwicklung von *Fagus sylvatica* L. (Rotbuche) unter besonderer Berücksichtigung möglicherweise neuartiger Veränderungen. *Berichte des Forschungszentrum Waldökosysteme/Waldsterben* **18**. Universität Göttingen.

Roloff, A. 1989a. Pflanzen als Bioindikatoren für Umweltbelastungen. I. Prinzipien der Bioindikation und Beispiel Waldbodenvegetation. *Forstarch.* **60**: 184-188.

Roloff, A. 1989b. Pflanzen als Bioindikatoren für Umweltbelastungen. II. Moose und Flechten. *Forstarch.* **62**: 227-232.

Ruetz, W.F. & Bergmann, F. 1989. Möglichkeiten zum Nachweis von autochthonen Hochlagenbeständen der Fichte (*Picea abies*) in den Berchtesgadener Alpen. *Forstw. Cbl.* **108**: 164-174.

Ruhling, Å. 1992. Experiences and results of a large-scale program for monitoring atmospheric heavy metal deposition. In Roy, S., Käranlampi, L. & Hänninen, O. (eds) *7th International Bioindicators Symposium and Workshop on Environmental Health,* Abstract Book: 56. Kuopio Univ. Publ. C., Natural and Environmental Science 7, Finland.

Ruth, B. & Weisel, B. 1993. Investigations on the photosynthetic system of spruce affected by forest decline and ozone fumigation in closed chambers. *Environ. Pollut.* **79**: 31-35.

Salemaa, M. & Jukola-Sulonen, E.-L. 1990. Vitality rating of *Picea abies* by defoliation class and other vigour indicators. *Scand. J. For. Res.* **5**: 413-426.

Sandermann, H. Jr, Schmitt, R., Heller, W., Rosemann, D. & Langebartels, C. 1989. Ozone-induced early biochemical reaction in conifers. In Longhurst, J.W.S. (ed.) *Acid Deposition; Sources, Effects and Controls*: 243-254. British Library Technical Communications, Information Press Ltd, Oxford.

Sandermann, H. Jr, Ernst, D., Heller, W. & Langebartels, C. 1992. Molecular biomarkers for air pollution in plants. In Roy, S., Kärenlampi, L. & Hänninen, O. (eds) *7th International Bioindicators Symposium and Workshop on Environmental Health,* Abstract Book: 28. Kuopio Univ. Publ. C., Natural and Environmental Science 7, Finland.

Santerre, A., Markiewicz, M. & Willanueva, V.R. 1990a. Effects of acid rain on poluamines in *Picea*. *Phytochem.* **29**: 1767-1769.

Santerre, A., Mermet, J.M. & Villanueva, V.R. 1990b. Comparative time-course mineral content study between healthy and diseased *Picea* trees from polluted areas. *Water, Air, Soil Pollut.* **52**: 157-174.

Saxe, H. 1990. Extended summary in Air pollution, primary plant physiological responses, and diagnostic tools (tillaeg iii): 2-3, Dr Agro. Thesis, The Royal Veterinary & Agricultural University, Copenhagen, 234 pp.

Saxe, H. 1991. Photosynthesis and stomatal responses to polluted air, and the use of physiological and biochemical responses for early detection and diagnostic tools. *Adv. Bot. Res.* **18**: 1-128.

Saxe, H. 1993. Triggering and predisposing factors in the "Red" decline syndrome of Norway spruce (*Picea abies*). *Trees* **8**: 39-48.

Schmidtling, R.C. 1994. Use of provenance tests to predict response to climatic change: loblolly pine and Norway spruce. *Tree Physiol.* **14**: 805-817.

Schmiedel, Th. & Dässler, H.-G. 1989. Seneszenzuntersuchungen immissionsgeschädigter Fichtennadeln mit Hilfe der Enzymanalytik. *Beiträge für die Forstwirt.* **23**: 155-158.

Schmitt, U., Liese, W. & Ruetze, M. 1986. Ultrastrukturelle Veränderungen in grünen Nadeln geschädigter Fichten. *Angew. Botanik* **60**: 441-450.

Schmuck, G. 1990. Applications of *in vivo* chlorophyll fluorescence in forest decline research. *Int. J. Remote Sensing* **11**: 1165-1177.

Scholz, F. & Bergman, F. 1992. Genetic effects of environmental pollution on tree populations. In Kim, Z.S. & Hattemer, H.H. (eds) *Conservation and Manipulation of Genetic Resources in Forest Trees*: 37-49. Kwangmungak Publ. Co., Seoul.

Schröder, P. & Debus, R. 1990. Responses of spruce trees (*Picea abies*) to fumigation with Halone 1211-first results of a pilot study. In Wang, W., Gorsuch, J.W. & Lower, W.R. (eds) *Plants for Toxicity Assessment*: 258-266. American Society for Testing and Materials, Philadelphia.

Sieffert, A. 1988. Genetic polymorphism in Norway spruce, *Picea abies* (L.) Karst, assessed by two-dimensional gel electrophoresis of needle proteins. *Trees* **2**: 188-193.

Skelly, J.M. 1992. A closer look at forest decline: A need for more accurate diagnostics. In Manion, P.D. & Lachance, D. (eds) *Forest Decline Concepts*: 85-107. APS Press, St Paul, MN.

Skrøppa, T. 1991. Within-population variation in autumn frost hardiness and its relationship to bud-set and height growth in *Picea abies*. *Scand. J. For. Res.* **6**: 353-363.

Sorensen, F.C. & Campbell, R.K. 1993. Seed weight-seedling size correlation in coastal Douglas-fir: genetic and environmental components. *Can. J. For. Res.* **23**: 275-285.

Steinnes, E., Rambaek, J.P. & Hanssen, J.E. 1992. Large scale multi-element survey of atmospheric deposition using naturally growing moss as biomonitor. *Chemosphere* **25**: 735-752.

Stöcker, G. & Gluch, W. 1990. Bioindication of acid deposition on forest ecosystems-recognition of local and regional patterns. *Arch. Nat. Schutz Landsch. forsch.* **30**: 3-12.

Strauss, S.H., Lande, R. & Namkoong, G. 1992. Limitations of molecular-marker-aided selection in forest tree breeding. *Can. J. For. Res.* **22**: 1050-1061.

Stribley, G.H. 1993. Studies on the health of beech trees in Surrey, England: Relationships between winter canopy assessment by Roloff's method and twig analysis. *Forestry* **66**: 1-26.

Sulzer, A.M., Greenwood, M.S. & Livingston, W.H. 1993. Early selection of black spruce using physiological and morphological criteria. *Can. J. For. Res.* **23**: 657-664.

Sutinen, S. & Koivisto, L. (1992). Microscopic structure of conifer needles as a diagnostic tool in the field. In Roy, S., Käranlampi, L. & Hänninen, O. (eds) *7th International Bioindicators Symposium and Workshop on Environmental Health*, Abstract Book: 106. Kuopio Univ. Publ. C., Natural and Environmental Science 7, Finland.

Tingey, D.T., Rodecap, K.D., Lee, E.H., Moser, T.J. & Hogsett, W.E. 1986. Ozone alters the concentrations of nutrients in bean tissue. *Angew. Botanik* **60**: 481-493.

Tjoelker, M.G. & Luxmoore, R.J. 1991. Soil nitrogen and chronic ozone stress influence physiology, growth and nutrient status of *Pinus taeda* L. and *Liriodendron tulipifera* L. seedlings. *New Phytol.* **119**: 69-81.

Tjoelker, M.G., McLaughlin, S.B., DiCosty, R.J., Lindberg, S.E. & Norby, R.J. 1992. Seasonal variation in nitrate reductase activity in needles of high-elevation red spruce trees. *Can. J. For. Res.* **22**: 375-380.

Tschaplinski, T.J. & Norby, R.J. 1991. Physiological indicators of nitrogen response in short rotation sycamore plantation. I. CO_2 assimilation, photosynthetic pigments and soluble carbohydrates. *Physiol. Plant.* **82**: 117-126.

Tynnyrinen, S., Palomäki, V., Holopainen, T. & Kärenlampi, L. 1992. Comparison of several bioindicator methods in monitoring the effects on forest of a fertilizer plant and a strip mine. *Ann. Bot. Fennici* **29**: 11-24.

Villanueva, V.R. & Santerre, A. 1989. On the mechanism of adaptive metabolism of healthy-resistant trees from forest polluted areas. *Water, Air, Soil Pollut.* **48**: 59-75.

Villanueva, V.R., Goff, M.Th.Le, Mardon, M. & Moncelon, F. 1987. High-performance liquid chromatographic method for sugar analysis of crude deproteinized extracts of needles of air-polluted healthy and damaged *Picea* trees. *J. Chromatography* **393**: 115-121.

Wallin, G., Skärby, L. & Selldén, G. 1990. Long-term exposure of Norway spruce, *Picea abies* (L.) Karst., to ozone in open-top chambers. I. Effects on the capacity of net photosynthesis, dark respiration and leaf conductance of shoots of different ages. *New Phytol.* **115**: 335-344.

Weidmann, P., Einig, W., Egger, B. & Hampp, R. 1990. Contents of ATP and ADP in needles of Norway spruce in relation to their development, age, and to symptoms of forest decline. *Trees* **4**: 68-74.

Weikert, R.M., Wedler, M., Lippert, M., Schramel, P. & Lange, O.L. 1989. Phosynthetic performance, chloroplast pigments, and mineral content of various needle age classes of spruce (*Picea abies*) with and without the new flush: An experimental approach for analyzing forest decline phenomena. *Trees* **3**: 161-172.

Weinstein, D.A. & Yanai, R.D. 1994. Integrating the effects of simultaneous multiple stresses on plants using the simulation model TREGRO. *J. Environ. Qual.* **23**: 407-411.
Weinstein, L.H., Laurence, J.A., Mandl, R.H. & Wälti, K. 1990. Use of native and cultivated plants as bioindicators and biomonitors of pollution damage. In Wang, W., Gorsuch, J.W. & Lower, W.R. (eds) *Plants for Toxicity Assessment*: 117-126. American Society for Testing and Materials, Philadelphia.
Wild, A. & Schmidt, V. 1995. Diagnosis of damage to Norway spruce (*Picea abies*) through biochemical criteria. *Physiol. Plant.* **93**: 375-382.
Williams, J.H. & Ashenden, T.W. 1992. Differences in the spectral characteristics of white clover exposed to gaseous pollutants and acid mist. *New Phytol.* **120**: 69-75.
Wilson, G.B. & Bell, J.N.B. 1990. Studies on the tolerance to sulphur dioxide of grass populations in polluted areas. VI. The genetic nature of tolerance in *Lolium perenne* L. *New Phytol.* **116**: 313-317.
Winner, W.E. 1989. Photosynthesis and transpiration measurements as biomarkers of air pollution effects on forests. In Grossblatt, N. (ed.) *Biological Markers of Air-Pollution Stress and Damage in Forests*: 303-316. National Academy Press, Washington, DC.
Wulff, A. 1992. Evaluation of acid mist effect on the calcium distribution of spruce needles with light microscopy. In Roy, S., Kärenlampi, L. & Hänninen, O. (eds) *7th International Bioindicators Symposium and Workshop on Environmental Health*, Abstract Book: 114. Kuopio Univ. Publ. C., Natural and Environmental Science 7, Finland.
Zaerr, J.B. & Schill, H. 1991. Early detection of effects of acid deposition and automobile exhaust on young spruce (*Picea abies* (L.) Karst.) trees. *Eur. J. For. Path.* **21**: 301-307.
Zobel, A. & Nighswander, J.E. 1991a. Accumulation of phenolic compounds in the necrotic areas of Austrian and red pine needles after spraying with sulphuric acid: A possible bioindicator of air pollution. *New Phytol.* **117**: 565-574.
Zobel, A. & Nighswander, J.E. 1991b. Accumulation of phenolic compounds in the necrotic areas of Australian and Red pine needles due to salt spray. *Ann. Bot.* **66**: 29-40.
Zoettl, H.W., Huettl, R.F., Fink, S., Tomlinson, G.H. & Wisniewski, J. 1989. Nutritional disturbances and histological changes in declining forest. *Water, Air, Soil Pollut.* **48**: 87-109.

20 Potential Areas of Research

M. YUNUS & M. IQBAL

INTRODUCTION

Since the atmospheric pollution is chiefly an outcome of the processes of industrialization and urbanization of the modern world, studies of pollution impact assessment and pollution control are of a relatively recent origin. However, because of the world-wide attention towards environmental health and environmental monitoring programmes, enormous quantities of relevant data have been generated by researchers during the last three decades. Although effects of pollution on living beings, and more specifically on plants, have been and are being studied enthusiastically all over the world, and much of what was hidden and unknown about the pollution–plant relationship a few decades ago has become known and explainable today, several interactions and the underlying mechanisms and factors still need to be elucidated and understood in order to facilitate emergence of valid generalizations.

MAJOR HIGHLIGHTS

The following points indicate important trends of the present-day research and identify areas that require further attention.

- The data available on the status of air quality at various locales in the developed countries are sufficient but such information about most of the developing world is fragmentary and not enough. A comprehensive air quality monitoring programme needs to be drawn to generate data for the developing world for making the global evaluation of air pollutant levels more authentic.
- The available quantitative interpretations for the dynamics of the atmosphere–biosphere (crop canopy) interface are not very satisfactory because (a) air pollution effects on crop growth and productivity under ambient conditions are considered to be based on a single point system (season-end yield), (b) the rule that under ambient conditions, control for one independent variable generally does not serve as a control for another variable is yet to be fully accepted, and (c) retrospective numerical analyses of previously collected field data are still relied upon. Further research on "open-field pollutant exposure vs. crop responses" is imperative to assess and apportion the effect of

Plant Response to Air Pollution. Edited by Mohammad Yunus and Muhammad Iqbal

each factor (stress plus biotic and abiotic) on growth and productivity within an integrated system.

- Variations in weather conditions have been shown to cause temporary shifts in supply and demand ratios of mineral nutrition in plants which impel changes in growth. A long-term monitoring of forest conditions (stemwood growth, needle colour, canopy leaf area) and relevant weather conditions (temperature and moisture regimes) would help evaluating "nutritional disharmony" as a concept for explaining several of the forest decline phenomena. The Huang's "nutritional disharmony" model on forest decline, which was designed to evaluate the relationship between weather and soil conditions and the seasonal nutrition pattern and decline symptoms on trees of different ages, still requires validation.
- Studies to investigate the effect of excess nitrogen loading from the atmosphere on natural ecosystems are urgently needed so as to extrapolate the possible indirect effect of nitrogen on the sensitivity of plants to abiotic factors such as frost and drought.
- Many of the conclusions drawn from the proposed models for growth pattern of the terrestrial vegetation under elevated CO_2 are not in agreement with the observed plant response. The reasons include extent of climate change, its influence on weather patterns and the potential for elevation in CO_2 compensate. The knowledge of elevated CO_2 effects, acquired largely from short-term pot experiments, cannot support prediction of effects on woody perennials and forest trees as sinks for CO_2. Experiments with long-term exposure of woody perennials to elevated CO_2, in conditions where the underground biomass increment is not restricted, may show some structural changes in xylem differentiation that may lead to important functional changes as in hydraulic conductivity (which is related to whole-plant water-use efficiency), a parameter certainly missed in a short-term potted experiments. Characters like hydraulic balances, changes in anatomy and function, potential for local and regional climate changes, changes in nutrient supply, etc., are the parameters to be incorporated in designing more versatile and meaningful models.
- Relationship between CO_2 and temperature has been studied at temperatures above 10°C. Little is known about the effects of elevated CO_2 on plants under winter conditions. Information on the mechanism underlying the responses to elevated CO_2 and interactions with other environmental factors such as nitrogen nutrition are almost completely lacking. In addition to the direct effects of CO_2 on terrestrial flora/forest ecosystems, other greenhouse gases also cause climate change and are affecting or may affect forests in the future. The predicted climate-change scenarios are not yet of sufficient spatial resolution to allow proper examination of the threats which it may pose.
- Interaction between air pollutants and plant cuticles is not primarily based on chemical reactions on the plant surface but mediated by plant metabolism. Aspects of the longevity of this relationship are, however, obscure. Many of the probable effects of air pollutants on interactions between pests, pathogens

and plant surfaces remain ravelled. Pollutant stress is likely to interact with the amount of resources allocated to the maintenance of cuticular structure. Multiple stress might alter plant resistance to pathogens. Much is yet to be understood on this aspect.

- Effects of SO_2 (alone or in combination with NO_2) as well as O_3 on stomatal behaviour are complex and inconsistent. Both can reduce, in low concentrations, the closing response of stomata to water stress. This may represent an important physiological disturbance under field conditions. Pretreatment of plants and availability of water seem to play important roles in determining the nature of the stomatal response to these pollutants. Endogenous levels of hormones such as abscisic acid (ABA) also influence the stomatal behaviour. Application of ABA to guard cells increases cytosolic free calcium concentration. Action of ABA on guard cells is intricately linked with events occurring within the plasma membrane; details of the mechanism involved need to be ellocidated critically.
- Details of the mechanism of O_3 toxicity to plants are not known. It is still unclear what compound(s) is oxidized first: ethylene, alkenes or phenyl propanoic acid. There is doubt that O_3 can pass through the plasma membrane and affect the cellular contents. Further studies are needed to determine whether reactions with lipids or protein components are predominant. The cause of the differential resistance in cultivars against O_3 at molecular level is also not known. These investigations could be the restriction fragment length polymorphism (RFLP) mapping or by direct transformation of plants with gene encoding enzymes of putative protection.
- O_3 can cause shifts in the assimilate translocation patterns and reduce carbohydrate level, especially in roots. Production of ethylene under O_3 stress is well known. It, however, merits investigation whether putative phosphorylated intermediates are generated when there is efflux of O_3-induced ethylene by leaves and whether these intermediates would influence loss in mRNA of small Rubisco subunits (rbcS).
- Ozone stress possibly does alter membrane structure, making the plant more sensitive to other stresses such as cold. Evidence suggests linkage between lipid metabolism and freeze sensitivity. Various aspects of lipid metabolism, both in terms of control and regulation, need to be investigated sequentially to show that changes induced by oxidants are not small perturbations in the total homeostasis of the plant. Particular attention is required on how oxidants hamper photosynthesis and whether this has a bearing on lipid metabolism. It may also be worked out if ionic balances affect the flow of lipid from the sites of synthesis to the membrane.
- The capacity of individual processes for compensation of stress resulting from SO_2 exposure in a particular species is not known fully, and thus it is hard to pinpoint a process that would become limiting under prolonged SO_2 exposures. Moreover, SO_2-specific compensation mechanisms need to be evaluated

by comparison with the capacity of general compensation mechanisms for oxidative stress. This information can lead to identification of the metabolic factor(s) determining the sensitivity of plants for SO_2.

- The atmospheric pollutants affect photosynthesis and decrease availability of carbohydrates required for a normal cambial growth, thus leading to a slow production of wood and bark. Changes in wood structure due to pollutional stress are largely quantitative and related to the annual increment of wood and the proportions of earlywood and latewood. The available information on performance of the cambium and quality of the cambium-derived tissues under the influence of air pollution is too meagre to allow any generalizations to be made. Both *in vivo* and *in vitro* studies need to be undertaken to determine the rate of cambial cell division, properties of differentiation of the derivative cells, physical and functional properties of wood and bark, and the production of secondary metabolites in these tissues under the influence of common pollutants.
- Air pollution effects on pollen, pollination, seed set, and the development of seeds and fruits are little explored. Correlations between seed initiation/growth and photosynthesis and between allocation of carbon and the initiation and development of reproductive tissues need to be established. It may be seen whether the gaseous air pollutants cut down the period of assimilate translocation to developing fruits. Mechanisms that control phloem loading in source regions and unloading in sink regions including fruit and seed also deserve attention.
- Even fog, dew and frost can be as potentially damaging to plants as acid rain because plants can absorb ions from them. Flavonoids, which exist in nature in abundance and occur on plant surfaces in high concentration, can play a protective role. Evidence suggests that enzymes of secondary metabolism can be triggered by extreme conditions, and may cause production of phenolics which possibly serve as defence compounds against biotic stress in particular. Further research on this aspect may unravel interesting facts.
- Defence strategies in plants include activation of oxyradical-scavenging enzymes and antioxidants. Enzymes such as superoxide dismutase, catalase, peroxidase and phosphatase react against environmental stress. Their specificity for pollutants including heavy metals has room for further research. Mechanisms by which certain chemicals, such as N-[2-(2-oxo-1-imidazolidinyl)ethyl]-N′-phenylurea (= EDU), kinetin, ethanol, etc., elevate or stabilize activity of tolerance-enhancing enzymes still have steps to be genuinely identified and studied.
- Leaves may emit volatile compounds such as ethylene, ethane, monoterpenes, acetaldehyde and ethanol in response to injury by air pollutants. The relative proportions of such emissions may be considered in assessing injury levels, and a study of their correlations with pollutant dose, exposure duration and developmental stages of the plant may produce useful results.

- Efficient use of bioindication is both needed and attainable in future management of plant growth, and the development and protection of forests.

REFERENCE

Huang, C. 1995. Modeling physiological and ecological effects of nutritional disharmony on spruce-fir forests. Doctoral thesis, Duke University, Durham, NC.

Subject Index

Author Index

Page numbers in bold type refer to references.

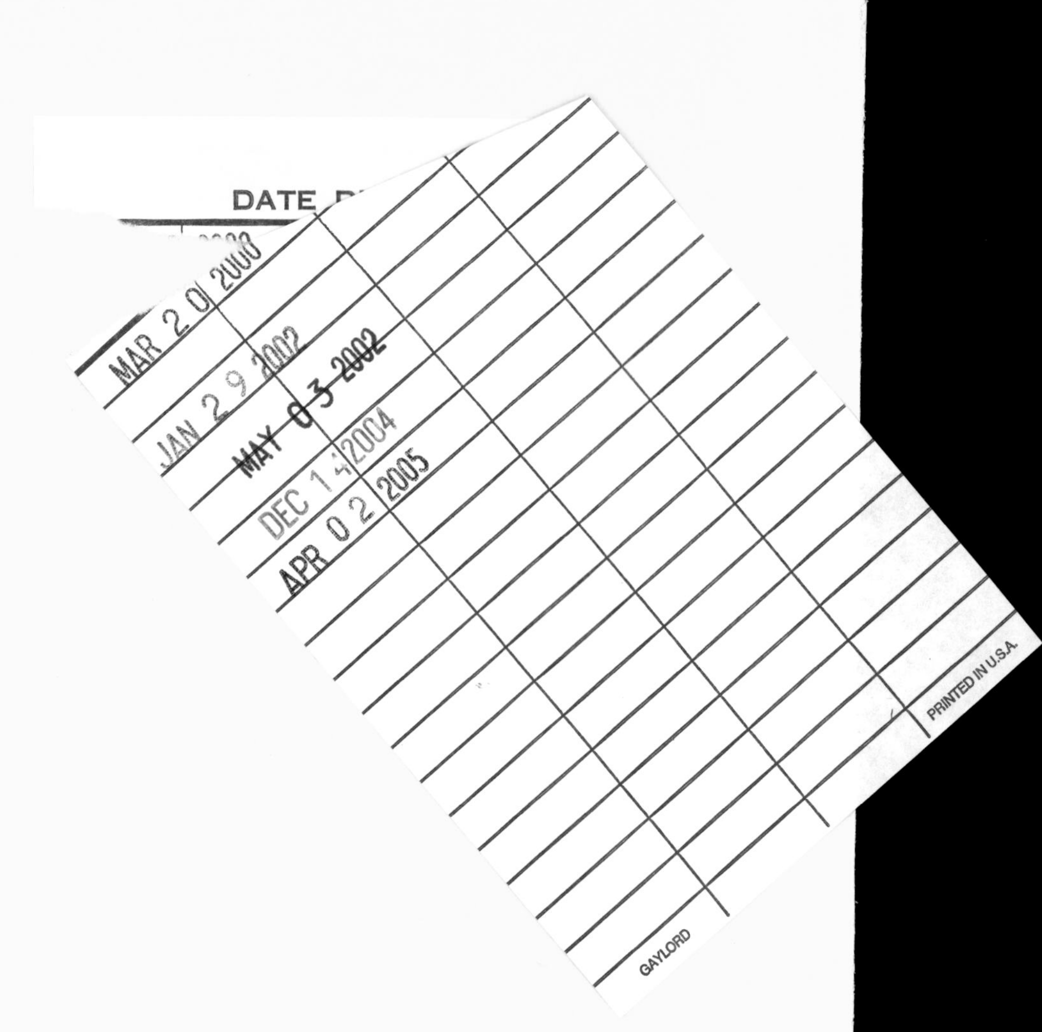

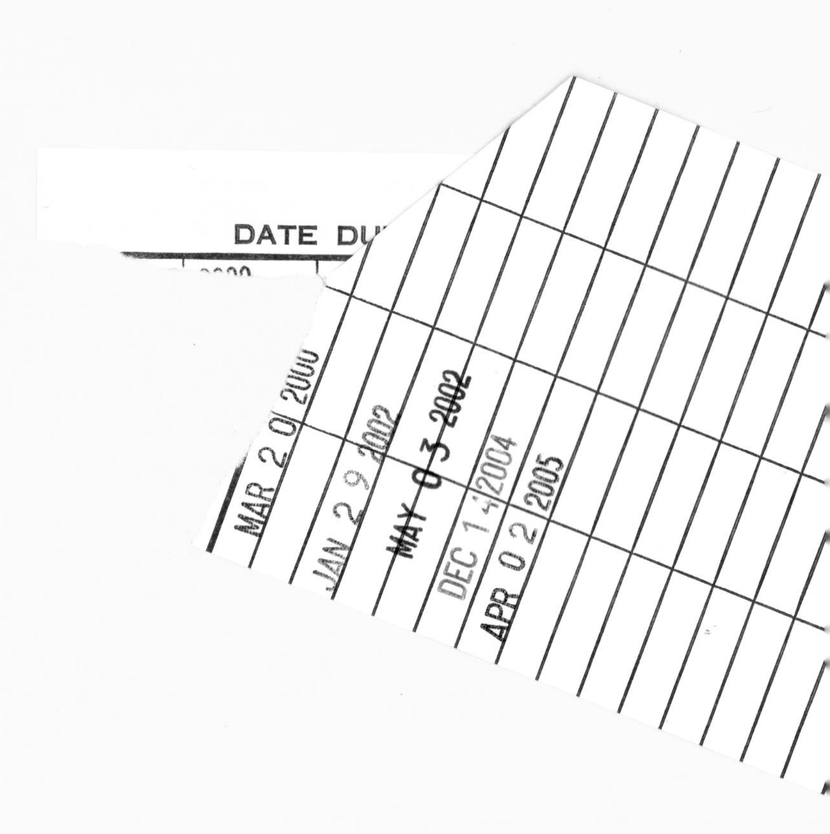